·

PROCEEDINGS OF THE
OREGON MEETING

PROCEEDINGS OF THE
OREGON MEETING

Annual Meeting of the Division of Particles and Fields of the American Physical Society

Eugene, Oregon
August 12–15, 1985

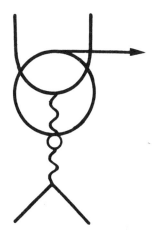

Edited by Rudolph C. Hwa

World Scientific

QC
793
.A47
1985

Published by

World Scientific Publishing Co. Pte. Ltd.
P. O. Box 128, Farrer Road, Singapore 9 128

Library of Congress Cataloging-in-Publication data is available.

PROCEEDINGS OF THE OREGON MEETING

ISBN 9971-50-041-8
    9971-50-046-9 pbk

Printed in Singapore by Kim Hup Lee Printing Co Pte Ltd.

The 1985 Annual DPF Meeting was

Hosted by the University of Oregon

Sponsored by the United States Department of Energy

Program Committee

S.L. Adler
J.W. Cronin
N.G. Deshpande
T. Ferbel
J. Gunion
I. Hinchliffe
D. Hitlin
R.C. Hwa (Chairman)
A. Kernan
T. Kirk
H. J. Lubatti
D.E. Soper

Session Organizers

M.A. Abolins
C. Baltay
L.S. Brown
J. Donoghue
J. Dorfan
F. Halzen
D. Hitlin
G. Kane
A.H. Mueller
B. Ovrut
J.L. Rosner
L. Teng

Local Organizing Committee

N.G. Deshpande
G. Eilam
R.C. Hwa (Chairman)
D.E. Soper

Meeting Staff

Sharon Finke
Katherine Wilson
Laila Zakis

PREFACE

The 1985 Annual Meeting of the Division of Particles and Fields was held during August 12-15 at the University of Oregon in Eugene, Oregon, the first time in many years that the Meeting has been held during the summer. A number of reasons can be given to suggest that the choice of time was good as well as bad. Certainly those who participated in the Meeting can recall how beautiful the weather was. Many have commented, both verbally and in writing later, on how much they enjoyed both the Meeting and the State of Oregon at its summer-time best; one participant's child was overheard to say, "Let's come back to Oregon for the Meeting next year". Some drawbacks of this timing, however, are not so obvious.

The time and place of the Meeting were actually determined by the fact that the 1985 International Symposium on Lepton and Photon Interactions at High Energies was to be held in Kyoto during August 19-25. The Executive Committee of the Division felt that Oregon, so "out of the way" under normal circumstances, would be a good venue to catch those on their way to Japan. That idea turned out to be only partially right. Many had to decline our invitation to speak in Oregon because they had agreed to participate in two pre-Kyoto satellite conferences in Japan. On the matter of conflicts, there were at least four other conferences and workshops being held in the U.S. alone during that same week: one Gordon Conference, two workshops in Santa Barbara, and a major meeting on Cosmic Rays in San Diego. Our scientific program and the extent to which the Division members could participate were definitely affected by those conflicts. It is strongly advised that the Executive Committee of the Division recommend to the members that SPIRES (CONF) be consulted for non-conflicting dates in planning future conferences, and then be informed of their choices of dates.

The previous Meeting in 1984 turned out to be an unreliable guide for us to predict the number of participants. The eleventh-hour spurt in registration at Santa Fe never occurred in Eugene. We suspect that the time of the year was the principal factor, since summer plans are rarely made at the last minute. The total number of registered participants this time is 361. Among them, 92 gave talks of one kind or another.

From the very beginning the Program Committee decided that this Meeting should be broad enough in scope to encompass all areas of high energy physics that are actively being nvestigated by members of the Division. Therefore, the parallel sessions were divided into six concurrent mini-conferences, each of which was held during four half-days. A quick glance at the table of contents will reveal the breadth of the subjects covered. It is not usual in Divisional meetings that papers ranging from superstrings to novel accelerators are simultaneously presented. As it turned out, this diversity worked out very well. In fact, there were so many speakers at some of those mini-conferences that additional sessions running in parallel had to be added.

To make the Meeting more open while guaranteeing a high standard for the presented papers, the format of mixing invited and contributed papers in the same sessions was adopted and found to be rather successful. This system can, however, be defeated in two possible ways. The organizer of a session can set a standard so high or a scope so restricted that no one apart from his invited speakers is given any time. Or an experimental group can submit a large number of contributed papers and dominate a whole session. Though the latter occurred in one session, the former was prevented from happening in this Meeting by the cooperation of the organizers, whose agreement with the concept of open participation was first sought before they accepted their responsibilities. Without their dedicated efforts to carry out the arduous task of inviting outstanding speakers and then mixing them well with those giving contributed papers, the Meeting could not have been a success. The annual DPF Meeting can become a truly significant event if someday leading physicists, shedding their self-image of importance, can participate in giving papers without specific invitation, and thereby blur the artificial line between invited and contributed papers.

Due to limitation in space, only the plenary and invited speakers have been asked to write up their talks for these Proceedings. Two of the plenary speakers (C. Baltay and L. Teng) declined because the contents of their talks have already been published elsewhere. A number of contributed papers have such high quality that their written versions are also included in these proceedings. Numerous excellent review articles, more than half of them presented at the parallel sessions, are the tangible results of this highly stimulating meeting. We feel that the size of this volume is well justified by the usefulness of the articles it contains.

The social and recreational programs elicited enthusiastic participation and appreciative comments. They were unusual in many respects, not the least of which was that they began the day before the start of the Meeting. Many participants arrived two days early and went hiking in the Cascade Mountains or white-water rafting on the McKenzie River. On Monday evening Mozart's music played over the gold fish pool by candlelight was a first even for the University of Oregon Museum of Art. The performance of the <u>Fiddler on the Roof</u> the following evening, the last of a sold-out summer season, passed the cultural and ethnic authenticity test of several discriminating experts among the physicists. Many were surprised that a city as small as Eugene (approximately 100,000) can support an outstanding musical theater group. But the bigger surprise (at least to those who are not accustomed to the weather of Northern Pacific) came the next day when the participants in the trip to the Oregon coast discovered with almost disbelief the seriousness of the repeated exhortations to wear warm clothes. The dense fog drawn in by the heat in the Willamette Valley lowered the temperature by 30°F towards the evening. Fortunately, there were bountiful scenic places to absorb the attention of all visitors and to challenge the vigor of the active hikers, some of whom puffed to the top of Cape Perpetua or walked through the entire extent of the dunes area from Honeyman Park to the beach. The freedom to choose the duration spent at each of the six planned stops was greatly appreciated. When asked where he was going, Jim Cronin responded by saying that there wasn't enough time for all six, so "I am going to do just the sand dunes, and do it well". It was gratifying that no one got so lost as to miss the last bus back to the Woahink Lake picnic ground for a salmon dinner.

A meeting of this magnitude would not be possible without the active cooperation of many program planners and staff workers. The program committee set the scope and direction of the scientific program early in the planning. A subset of the members of that committee worked behind the scenes and recruited two organizers for each mini-conference. They were D. Hitlin, N.G. Deshpande, D.E. Soper, R.C. Hwa, S.L.

Adler, and H. Lubatti, in order of the mini-conferences listed for the parallel sessions. The organizers, whose names appear on the page before this preface, did the most important job in the program, as described earlier; to them many thanks are due. J. Gunion and H. Lubatti assisted the members of the local organizing committee in sorting the contributed papers for the parallel sessions. N.G. Deshpande and G. Eilam further gave time to undertake certain administrative tasks for the Meeting. The local staff members, Sharon Finke, Katherine Wilson, and Laila Zakis, deserve all the credits for the smooth running of the Meeting. Each of them has a special talent and disposition, as well as skills to manage their respective areas in the enterprise; and by being willing to go beyond the call of duty, they were able to make the joint effort more than the sum of the parts. How often does a physicist attending a conference get to catch his return flight but by the grace of the personal chauffering of a staff member when there was not enough time even to call and wait for a taxi? I might add as a final remark that this has been a most interesting opportunity for me to learn about the sociological/professional behavior of high-energy physicists, which exhibits a distinct difference in both the means and the variances between the theorists and experimentalists.

Rudolph C. Hwa

Eugene, Oregon

December, 1985

TABLE OF CONTENTS

Preface   vii

PLENARY SESSIONS

Status of the Standard Model	*R. D. Peccei*	1
Intermediate Vector Bosons and Jets at the CERN pp̄ Collider	*J.-M. Gaillard*	25
Unified String Theories	*D. J. Gross*	49
Results From UA1	*J. Rohlf*	66
Phenomenology of Collider Physics	*V. Barger*	95
Rare Decays, Monopoles, Neutrino Masses, Cygnus X3 Phenomena	*D.H. Perkins*	113
Difference Equations as the Basis of Fundamental Physical Theories	*T.D. Lee*	135
e^+e^- Highlights 1985	*K. Berkelman*	150
Accelerator Physics in SSC Design	*A.W. Chao*	170
Particle Physics: Past and Future	*S. Weinberg*	186

PARALLEL SESSIONS

I. ELECTROWEAK INTERACTIONS

Kobayashi-Maskawa Matrix Element	*W.Y. Keung*	207
Experimental Study of Sigma Beta Decay and its Implications	*R. Winston*	215
Recent Results on D Meson Decays from the Mark III	*D. H. Coward*	219
Lifetime of Charmed Particles	*M. Bosman*	232
Nonleptonic Decays of Charmed Mesons and Baryons	*H. Y. Cheng*	240
New Results on τ Lepton from PEP	*K.K. Gan*	248
PETRA and PEP Measurements of the Average B Hadron Lifetime	*G. J. Baranko*	258
A Review of Recent Results on B Decays	*A.H. Jawahery*	267
CP Violation in Kaon Decays	*M. B. Wise*	279
Lattice Calculation of Weak Matrix Elements	*A. Soni*	287
A Review of Experiments Measuring ε'/ε	*G. D. Gollin*	299
Toponium-Z^0 Interference	*P. J. Franzini*	310
Higgs Boson Spectrum from Infrared Fixed Points	*C.N. Leung*	319
Electroweak Interference in e^+e^- Collisions	*J. Moromisato*	327

Recent Results from Deep Inelastic υ N Experiments   *F.S.Merritt*   335
A Study of the Weak Neutral-Current in
  Deep-Inelastic Neutrino-Nucleon Scattering   *F. E. Taylor*   343

II. EXPERIMENT AND PHENOMENOLOGY BEYOND THE STANDARD MODEL

Strong New Limits on Gluino and Scalar Quark
  Masses   *R.M. Barnett*   353
Search for Anomalous Single Photon Production
  at PEP   *R.J. Wilson*   362
Neutrino Production of Like-Sign Dimuons   *K. Lang*   370
Same-Sign Dilepton Production by Neutrinos   *M.J. Murtagh*   378
Underground Muons from Cygnus X-3   *L. E. Price*   389
Neutrino Mass and Related Problems   *B. Kayser*   397
Search for Heavy Neutrino Production at PEP   *G. Feldman*   411
Review of Neutrino Mass Measurements   *T.J. Bowles*   418
Neutrino Oscillations: An Experimental Review   *M. H. Shaevitz*   426
Double Beta Decay: Recent Experimental Results   *M. S. Witherell*   437
Detecting the Higgs   *J. F. Gunion*   445
Search for Supersymmetric Particles at PEP
  and PETRA   *S.L. Wu*   471
Review of Recent Rare K Decay Experiments   *M. E. Zeller*   495

III. QUANTUM CHROMODYNAMICS

The Status of Perturbative QCD   *J. C. Collins*   505
Testing QCD in Deep Inelastic Lepton Scattering   *J. Carr*   512
QCD at the CERN $p\bar{p}$ Collider   *W. Scott*   520
The Status of Perturbative QCD (An Update)   *F. Halzen*   529
Testing QCD with Two Photons   *D. O. Caldwell*   542
Status of α_S Determination in e^+e^- Annihilation   *R.Y. Zhu*   552
Simulation of Hadronic Reactions   *F. E. Paige*   566
Application of Monte Carlo Renormalization Group
  Methods in Non-Abelian Lattice Gauge Theories   *F. Karsch*   575
Lattice Gauge Theory with Fermions: A Progress
  Report   *J. B. Kogut*   584
Lattice Gauge Theory with Special Processors
  and Super Computers   *N. H. Christ*   593
Recent Results From ARGUS   *R. Davis*   602

IV. HADRON PHYSICS

Study of Minimum-Bias-Trigger Events at $\sqrt{s}$ = 0.2 - 0.9 TeV with Magnetic and Calorimetric Analysis at the CERN Proton-Antiproton Collider   *G. Piano-Mortari*   615

Small-x Behavior and Minijets in QCD   *A. H. Mueller*   634

A Two-Component Model for Soft Hadronic Interactions at High Energy   *T. K. Gaisser*   643

Gluons and the Rising Total Cross Section   *B. Margolis*   648

Recent Results on J/ψ Decays from the Mark III   *W. J. Wisniewski*   657

Partial Wave Analysis of KKPi System in D and E/IOTA Region   *S. Protopopescu*   671

New Results at $\sqrt{s}$ = 200 and 900 GeV from the CERN Collider   *P. Carlson*   680

Is Cygnus X-3 Strange?   *L. McLerran*   689

Formation and Signatures of Quark-Gluon Plasma   *R. C. Hwa*   699

The Status of Skyrmions   *E. Braaten*   707

A Dependent Effects in Particle Production   *D. D. Reeder*   715

V. THEORY BEYOND THE STANDARD MODEL

Higher Dimensions, Supersymmetry, Strings   *P. G. O. Freund*   727

Superstring Phenomenology   *P. Candelas*   737

The Heterotic String   *J. A. Harvey*   753

The One-Loop Effective Lagrangian of the Superstring   *B. A. Ovrut*   760

Making Sense of Anomalous Gauge Theories   *R. Jackiw*   772

Torsion and Geometrostasis in Covariant Superstrings   *C. Zachos*   779

Ambitwistors (and Strings?)   *J. Isenberg*   787

Infinity Cancellation for O(32) Open Strings   *P. H. Frampton*   797

The Thermodynamics of Superstrings   *M. J. Bowick*   805

Toward A Covariant String Field Theory   *S. Raby*   813

Fermion Families in Superstring Theory   *M. Visser*   829

E_6 Symmetry Breaking In Superstring Models   *G. C. Segrè*   835

On the Status of Superstring Models   *C. R. Nappi*   843

Radiative Corrections in String Theories   *S. Weinberg*   850

VI. DETECTORS AND ACCELERATORS

Review of Recent Progress in the Development of Čerenkov Ring Imaging Detectors	*D. Leith*	863
Uranium-Liquid Argon Calorimetry: Preliminary Results from the DO Tests	*B. Cox*	887
Some New Development in Instrumentation	*D. F. Anderson*	904
Improvement Program of the UA2 Detector	*J.-M. Gaillard*	912
Silicon Microstrip Detectors	*M. Turala*	922
Scintillating Fiber Detectors	*R. C. Ruchti*	932
Transition Radiation: A Review	*J. R. Hubbard*	941
Initial Operation of the Fermilab Antiproton Source	*S. D. Holmes*	949
The Status of SLC	*S. Ecklund*	958
Progress with HERA	*P. Schmüser*	968
Experimentation at p$\bar{\text{p}}$ Colliders	*D. B. Cline*	976
Very High Energy Polarized Beams for the SSC	*Y. Makdisi*	992
Accelerators for the Study of Many Particle Systems	*J. R. Alonso*	1001
A Bismuth Germanate Electromagnetic Calorimeter: CUSB-II	*P. Franzini &* *J. Lee-Franzini*	1009

List of other papers presented	1021
List of participants	1029
Author Index	1039

STATUS OF THE STANDARD MODEL

R. D. Peccei

Deutsches Elektronen-Synchrotron DESY
2000 Hamburg 52
Fed. Rep. Germany

ABSTRACT

I illustrate by means of a variety of recent examples how well the standard SU(3)xSU(2)xU(1) model works. Among the topics discussed are: W and Z physics; some aspects of perturbative QCD; theoretical and experimental constraints and prospects for Higgs boson detection; and CP violation within the Kobayashi-Maskawa framework.

1. PREMISES

A report on the standard SU(3)xSU(2)xU(1) model of the strong and electroweak interactions these days can follow two roads. Either it tries to be encyclopedic, and details the extensive evidence that exists supporting the model in a variety of different physical contexts, or it picks and chooses some significant recent results, which exemplify again how well the model works. This report follows the second route. The examples I have chosen, for illustration, are a matter of personal taste, although I believe they fairly represent what might be considered highlights in the field this year. It will be noticed that no sharp distinction exists anymore between strong and weak interaction tests. Hadronic interactions are used to extract properties of the weak bosons and, conversely, the $p_\perp$ distributions of these bosons are used to test QCD. The only sad note, in all this unity, is that the standard model in 1985 works all too well! Alas, there is not much that can be done about this, until more detailed and higher energy experiments find some real traces of disagreement. In 1985, the anomalies of 1984 appear to have been only statistical fluctuations!

2. W AND Z PHYSICS

The first topic which I would like to discuss is the progress made in determining the properties of the weak intermediate bosons in the collider experiments at CERN.

2.1 W and Z Masses

The standard electroweak theory of Glashow, Salam and Weinberg /1/ makes precise predictions for the masses of the intermediate vector bosons, in terms of low energy parameters. It is convenient /2/ to adopt a definition of the Weinberg angle in terms of the W and Z masses:

$$\sin^2 \theta_W = 1 - M_W^2/M_Z^2 \tag{1}$$

Then the W mass, including radiative corrections, can be expressed as /3/:

$$M_W^2 = \frac{\pi \alpha}{\sqrt{2} \; G_\mu \; \sin^2 \theta_W (1-\Delta r)} \tag{2}$$

where G_μ is the Fermi constant determined from μ decay, in which certain electromagnetic contributions are explicitly included /4/. Numerically one has /4/ $G_\mu = (1.16638 \pm 0.00002) \times 10^{-5} \text{GeV}^{-2}$. The Weinberg angle, defined by Eq. (1), can be extracted from radiatively corrected ν deep inelastic scattering, with the result /5/

$$\sin^2 \theta_W = 0.217 \pm 0.014 \tag{3}$$

Finally, the quantity $(1-\Delta r)$ is theoretically calculable and has the value /6/, for $m_t = 40$ GeV, $m_H = m_Z$

$$1 - \Delta r = 0.9304 \pm 0.0020 \tag{4}$$

The dependence of (4) on m_t and m_H is mild and will be commented upon below.

Using the above results, and a precise value for the fine structure constant, gives for the W and Z masses the predictions

$$M_W = 83.0 \; ^{+2.9}_{-2.7} \; \text{GeV}; \qquad M_Z = 93.8 \; ^{+2.4}_{-2.2} \; \text{GeV} \tag{5}$$

The main error in Eqs. (5) is due to the error in $\sin^2 \theta_W$ from Eq. (3). The most recent UA1 and UA2 values, as reported by Di Lella /7/, are in perfect agreement with the above predictions

$$M_W = 83.1 \; ^{+1.3}_{-0.8} \pm 3 \quad \text{GeV} \quad \text{UA}_1$$

$$M_W = 81.2 \pm 1.1 \pm 1.3 \quad \text{GeV} \quad \text{UA}_2 \tag{6}$$

$$M_Z = 93.0 \pm 1.6 \pm 3 \quad \text{GeV} \quad \text{UA}_1$$

$$M_Z = 92.5 \pm 1.3 \pm 1.5 \quad \text{GeV} \quad \text{UA}_2$$

(the first error is statistical, the second is systematic), but the errors in both the theoretical prediction and experiment are too large to test significantly the radiative corrections. I should note that calculating $\sin^2 \theta_W$, using Eq. (1), from the collider data one cancels most of the systematic error. The UA1 and UA2 average /7/

$$\overline{\sin^2 \theta_W} = 0.218 \pm 0.023 \qquad\qquad (7)$$

agrees very nicely with that obtained from the low energy experiments and the precision of the determination is quite comparable to that in Eq. (3).

It is of obvious theoretical interest to eventually be able to check the radiative corrections embodies in Δr. It turns out, however, that most of the 7 % change in $(1-\Delta r)$ in Eq. (4) is theoretically rather trivial, coming essentially from the effects of the running of α to the W mass scale. The interesting variation in Δr, coming from the properties of the Higgs sector or the existence of widely split fermion doublets, is at the 1 % level. For instance /6/, changing m_H from 10^2 GeV to 10^3 GeV changes Δr by 0.009. To be able to test the electroweak theory to this level necessitates two very precise measurements (cf. Eqs. (1) and (2)). One of these will be provided by SLC and LEP, through the measurement of the Z^0 mass to better than one part per mil ($\delta M_Z / M_Z \sim 10^{-3}$). The other will require either a comparable measurement of the W mass, ($\delta M_W / M_W \sim 10^{-3}$), or an extremely accurate measurement of $\sin^2 \theta_W (\delta \sin^2 \theta_W \sim 2.5 \times 10^{-3})$ by measuring the forward-backward asymmetry at the Z^0 peak to 2 parts per mil. This latter measurement, although very difficult, appears to be feasible at LEP /8/. Interestingly enough, measuring M_W to 100 MeV also appears feasible at LEPII /9/. Let me briefly comment on how this may be done.

Four ways have been suggested /9/ for measuring the W mass in the process $e^+e^- \to W^+W^-$. They involve:

i)   Measuring the threshold dependence of $\sigma(e^+e^- \to W^+W^-)$.
ii) Measuring the endpoint in the electron spectrum in $W \to e\nu$ decays.
iii) Measuring the jet-jet invariant mass arising from hadronic decays of the W.
iv) Measuring the $e\nu$ invariant mass in $W \to e\nu$ decays.

The statistical and systematic errors for all these four methods, in one year of running at LEPII, lie in the 100 MeV range. Let me illustrate this for case iv). Here one selects events in which one W decays hadronically and the other leptonically. Having determined the W axis from the jet-jet analysis, as shown in Fig. 1, then all the remaining kinematics is fixed. At LEPII, in contrast to the collider, $M_{e\nu}$ and not only the transverse mass $M^T_{e\nu}$ is determined. Fig. 2 shows a Monte Carlo reconstruction /9/ of $M_{e\nu}$ using an integrated luminosity of 100 pb^{-1}. The statistical error here is ± 55 MeV and there is a systematic shift of the input W mass of 80 MeV.

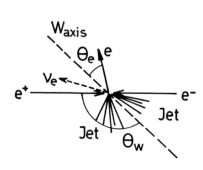

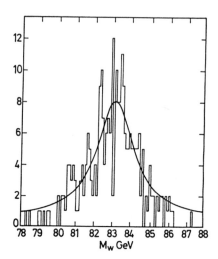

Fig. 1: Kinematic reconstruction of $M_W = M_{e\nu}$

Fig. 2: Monte Carlo reconstruction of $M_{e\nu}$, from Ref. 9

2.2 Neutrino Counting

In the standard model, a precise measurement of the Z^0 width gives a determination of the number of neutrino species and, inferentially, of the number of generations. Unfortunately, the present uncertainty in mass resolution, and the scant number of events, preclude a direct measurement of Γ_Z at the collider. Nevertheless, one can infer the number of neutrino species N_ν by using a bit of theory /10/. What is measured at the collider is the ratio

$$R = \frac{\sigma_Z \, B(Z \to e^+ e^-)}{\sigma_W \, B(W \to e\nu)} \tag{8}$$

of the production of Z's and W's, multiplied by their decay branching ratio into electrons. The average value for R determined by UA1 and UA2 is /7/

$$\overline{R} = 0.125 \pm 0.023 \tag{9}$$

and at 90 % confidence level R > 0.096. Let me display the explicit dependence of R on N_ν. One has

$$R = \frac{\sigma_W}{\sigma_Z} \frac{\Gamma(Z^0 \to e^+ e^-)}{\Gamma(W \to e\nu)} \frac{\Gamma_W^{st}}{\Gamma_Z^{st} + (N_\nu - 3) \, \Gamma(Z \to \overline{\nu}\nu)} \tag{10}$$

Using some theoretical imput, therefore, one may compute N_ν from the measured value of R. The cross section ratio can be calculated in QCD rather accurately /11/ (see below) and one finds $\sigma_W/\sigma_Z = 0.30 \pm 0.02$. The ratio of leptonic widths is also well known since it follows from the low energy couplings of the Z^0 and W ($\Gamma_Z/\Gamma_W \simeq 0.37$). Finally, the total widths Γ_W^{st}, Γ_Z^{st} in the standard model do depend on m but one has, nearly, $\Gamma_W^{st} \simeq \Gamma_Z^{st}$. It is obvious, therefore, from (10) and the experimental result (9) that there is not much room for extra neutrinos.

To be specific, using the choice of parameters of Deshpande et al. /10/ ($M_W = 83$ GeV, $M_Z = 94$ GeV, $\sin^2 \theta_W = 0.22$, $m_t = 40$ GeV) one has $\Gamma(Z^0 \to e^+e^-)/\Gamma(W \to e\nu) = 0.368$ and $\Gamma_W = 2.82$ GeV, $\Gamma_Z = 2.83$ GeV. Then the collider measurement of R /7/ implies $N_\nu = 1.3 \pm 2.7$, while using the 90 % confidence limit on R one has

$$N_\nu < 5.4 \pm 1.0 \tag{11}$$

Given the very little amount of theory imputed, this is quite impressive. The collider is already catching up with well known nucleosynthesis bound /12/ $N_\nu \simeq 3 - 4$!

2.3   Weak Boson Production and QCD

The total production cross section, as well as the $P_\perp$ distribution of the weak bosons, at the collider can be reliably computed in perturbative QCD. To lowest order in α_S the weak boson production proceeds by quark - antiquark annihilation. In $0(\alpha_S)$, however, both produced gluons and quark-gluon processes must be included, as shown schematically in Fig. 3 /13/.

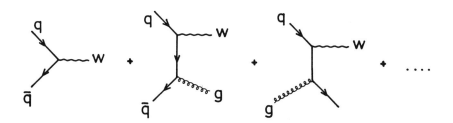

Fig. 3: Lowest order contributions to W production in hadronic collisions

Using the full theoretical machinery of perturbative QCD, prediting σ_W and σ_Z has passed from being an ancient and honorable art /14/ to a science /11/. To a very good approximation, one finds that one can include the $O(\alpha_s)$ corrections by just multiplying the lowest order parton result by an overall factor (K-factor).

Schematically, one can write for the production cross sections

$$\sigma_{W/Z} = K(Q^2) \, N_{W/Z} \int dx_1 dx_2 \, \delta(x_1 x_2 - \tau) \, \{q(x_1, M^2)\overline{q}(x_2, M^2) + 1 \leftrightarrow 2\}$$

(12)

Here $\tau = M^2/S$, with M being the boson mass; Q^2 is a dynamical scale associated with the K-factor; $N_{W/Z}$ is the appropriate weak vertex factor squared and the quark distributions have been evolved to M^2. In evaluating this formula to predict $\sigma_{W/Z}$ there are two sources of uncertainty:

1) One needs to know the parton densities for all x, evolved to M^2 — which is a large scale. Fortunately, at the CERN collider the dominant values of x_i in (12) are $x_1 \simeq x_2 \simeq \sqrt{\tau} \simeq 0.15$. For this values, valence-valence collisions dominate and the relevant densities are quite well known. Going up in energy, as will happen with Tevatron, is less favorable from this point of view.

2) The scale Q^2 is not really determined until $O(\alpha_s^2)$ terms are computed. The natural choice for Q^2 would be $Q^2 \simeq M^2$, although one could envisage $Q^2 \simeq \langle P_\perp^2 \rangle_{W/Z}$, which is much less than M^2. At any rate, assuming that $Q^2 \simeq M^2$, one expects considerably smaller K-factors than in Drell-Yan processes, where the pair invariant mass is much smaller than the W/Z mass. Typically /11/ $K \simeq 1.3 - 1.4$ here.

The results of the recent calculation of Altarelli, Ellis and Martinelli /11/are presented in Table I. The lower error in the table is due to effects of variation in the parton density, keeping $K(Q^2 = M^2)$ fixed. The larger error takes $Q^2 = \langle P_\perp^2 \rangle_{W/Z}$, thereby increasing the value of K.

Table I   Predictions for W and Z production at the CERN collider

$\sqrt{s}$ GeV	σ_W (M_W = 83 GeV) nb	σ_Z (M_Z = 94 GeV) nb
540	$4.2 \, ^{+1.3}_{-0.6}$	$1.3 \, ^{+0.4}_{-0.2}$
630	$5.3 \, ^{+1.6}_{-0.9}$	$1.6 \, ^{+0.5}_{-0.3}$

Using $B(W \to e\nu) = 0.089$; $B(Z \to e^+ e^-) = 0.032$ one can compare these results with the values of $\sigma \cdot B$ measured at the collider /7/. This is done graphically in Fig. 4, and one sees that the agreement is fine.

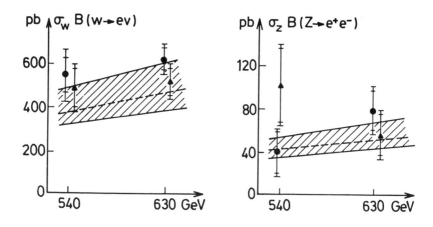

Fig. 4: Comparison of UA₁ ● and UA₂ ▲ data on (σ.B) with the QCD calculations of Ref. 11.

More exclusive quantities, than the total production cross section, can also be computed. An example of this is the W $p_\perp$ distribution. The calculation of this distribution is easy for $p_\perp \gg M_W$, but
for $p_\perp \simeq M_W$ one must worry about all orders in α_s. The point is that
one encounters large logarithms, of the type $\alpha_s^n(p_\perp^2) \ln{}^m M_W^2 / p_\perp^2$, which
cannot be ignored. Fortunately, one has been able to resum terms of
this type /15/ and one obtains an eikonal-like representation for the
differential $p_\perp$ distribution /11/

$$\frac{d\sigma}{dp_\perp^2} = \int \frac{d^2b}{4\pi} e^{-i\,\vec{p}_\perp \cdot \vec{b}} \sigma_o(1+A) e^{S(b)} + Y(p_\perp) \qquad (13)$$

Here σ_o is the lowest order cross section and A contains the multiplicative $O(\alpha_s)$ corrections. The non singular $O(\alpha_s)$ corrections are in
$Y(p_\perp)$, while the dangerous terms which have been resumed are in $S(b)$
which is now known to $O(\alpha_s^2)$ /16/. Although $S(b)$ affects the shape of
the $p_\perp$ distribution, it does not contribute to the total cross section
since $S(0) = 1$.

In Fig. 5, I show the $p_\perp$ distribution for the W boson determined
by the UA₂ collaboration /7/, compared to the QCD prediction computed
from Eq. (13) by Altarelli, Ellis, Greco and Martinelli /11/. The fit
is obviously excellent. Note that the 1983 spectacular UA₂ events A, B,
C with more data now appear to be consistent with theory. Very similar
fits have been shown by the UA₁ collaboration /17/.

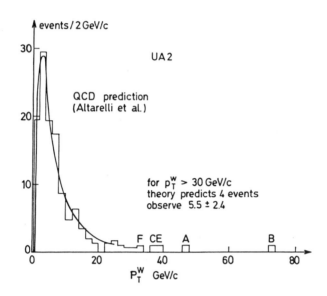

events/2 GeV/c

UA 2

QCD prediction
(Altarelli et al.)

for $p_T^W > 30$ GeV/c
theory predicts 4 events
observe 5.5 ± 2.4

F CE   A   B

P_T^W GeV/c

Fig. 5:   Transverse momentum distribution of the W boson compared to
the QCD calculation of Ref. 11

3.   QCD NEWS

In this section I want to discuss some additional QCD tests for
which new or more refined data has become available this year. I begin
by looking again at some results coming out of the CERN collider.

3.1   Jets at the Collider

The two jet cross section at the collider arises from a combina-
tion of many subprocesses: $q\bar{q} \to q\bar{q}$; $qg \to qg$; $gg \to q\bar{q}$; etc. The $\hat{t}$ channel gluon exchange which enters in most of these subprocesses gives a
typical Rutherford angular dependence:

$$\frac{d\sigma}{d\cos\theta} \sim \frac{\alpha_s^2}{\hat{s}(1 - \cos\theta)^2} \qquad (14)$$

This characteristic dependence, which was already apparent in the early running of the collider /18/, can be clearly seen in Fig. 6, taken
from a recent UA1 publication /19/. In fact, the inclusion of small
scale breaking effects, both in $\alpha_s(\hat{t})$ as well as in the evolution of
the structure functions, seems to improve the fit. For the di jet mass
range $\hat{s} = m_{2j} = 150 - 250$ GeV, initial state gluons are quite important. One finds /19/ that the qq to qg to gg subprocesses are in the

ratio of 36 : 52 : 12.
Further, the gluonic
contributions are domi-
nant at small values
/20/.

Gluons are not
only important in the
initial state in the
collider. There is
clear evidence now
for their "presence"
in the final state,
as three jet events
are clearly distin-
guishable in the data
/21/. Both the UA$_1$
and the UA$_2$ collabo-
rations have attempted
to extract from the
ratio of 3-jet to 2-
jet processes a value
for α_s. This is not an
easy task and at pre-
sent the inferred re-
sults should be consi-
dered only as semiquan-
titative.

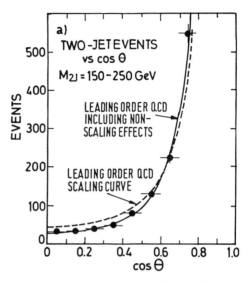

Fig. 6: Rutherford behaviour of the
2-jet cross section

If c_{3j} and c_{2j} are the fraction of 3-jet and 2-jet events one
expects, then the cross section ratio is just

$$\frac{\sigma_{3j}}{\sigma_{2j}} = \frac{c_{3j}}{c_{2j}} \frac{\alpha_s^3}{\alpha_s^2} \tag{15}$$

Thus, if the jet fractions are equal and if the 3-jet and 2-jet processes depended on the same scale, the cross section ratio would directly measure α_s. Unfortunately the fractions c_{3j} and c_{2j} depend
crucially on the subprocess and on the experimental cuts one is imposing, so that a careful analysis is needed before one can extract α_s
from the cross section ratio. Furthermore, the typical scale appropriate for a 3-jet process is not the same as that for a 2-jet process, so that it is also not possible to simply cancel the α_s factors
in (15). Finally, it should be commented that although c_{3j} and c_{2j} are
calculable in perturbative QCD, their full calculation is not completed. Virtual processes of higher order need yet to be included, to properly determine the appropriate scale to evaluate numerator and denominator in (15).

To give a feeling of some of the ambiguities one encounters in the present analysis, I show in Fig. 7 a plot of σ_{3j}/σ_{2j} from the UA$_1$

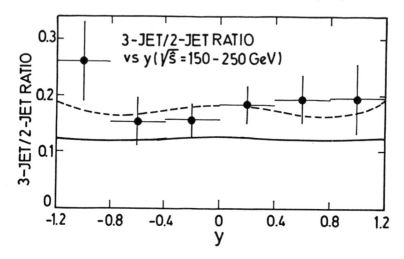

Fig. 7:   Ratio of 3-jet to 2-jet cross section from Ref. 19

collaboration /19/, along with two theoretical fits. The solid line corresponds to a QCD fit in which the 2-jet and 3-jet scales are taken to be the same, while for the dotted line one has $<q^2>_{3j} \simeq 4/9<q^2>_{2j}$. Clearly the latter curve gives a better fit to the data, for the choice of α_s taken. However, by increasing α_s the solid line could also be brought in agreement with the data. Hence, roughly speaking, the difference between the solid and dashed line gives one an idea on the uncertainty in α_s. This is borne out by the value obtained by the collaboration /19/, after their analysis

$$\alpha_s (4000 \text{ GeV}^2) = 0.16 \ \underline{+0.02} \ \underline{+0.03} \tag{16}$$

where the first error is statistical and the second is systematic. A similar result was obtained by the UA$_2$ collaboration, although their value for α_s /22/ is somewhat larger, taking the K factors the same:

$$\alpha_s \frac{K_3}{K_2} = 0.23 \ \overset{+}{-} \ 0.01 \ \overset{+}{-} \ 0.04 \tag{17}$$

The discrepancy, in my opinion, reflects nothing more than the uncertainties in the analysis.

3.2   Measurement of $F_L(x)$

One of the nicests tests of QCD which became available this year concerns the longitudinal structure function $F_L(x)$ in deep inelastic scattering. This structure function is a combination of F_2 and F_1,

$$F_L = F_2 - 2 \times F_1 \tag{18}$$

and is a measure of the spin of the constituents of the proton. In the parton model, where only spin 1/2 quarks contribute to the scattering, F_L vanishes. This is the famous Callan Gross relation /23/. However, in QCD, where also gluons enter, $F_L(x)$ acquires a non zero value. In fact, one predicts /24/

$$F_L(x;q^2) = \frac{\alpha_s(q^2)}{2\pi} x^2 \int_x^1 \frac{d\xi}{\xi^2} \left\{ \frac{8}{3} F_2(\xi;q^2) + 16(1-\frac{x}{\xi})\xi \ G(\xi;q^2) \right\} \tag{19}$$

where G is the gluon density function in the proton.

Unfortunately F_L is very difficult to extract experimentally and up to recently the available data did not allow for a significant test of (19). This situation has changed this year thanks to the new, high statistics, CDHSW data. Having more than half a million ν and $\bar{\nu}$ events they can extract F_L by studying the y distribution of

$$\frac{d\sigma^{\bar{\nu}}}{dx\,dy} - (1-y)^2 \frac{d\sigma^\nu}{dx\,dy}$$

Their results, presented by Feltesse /25/ are shown in Fig. 8, along with a QCD fit. The quality of the data and the theoretical agreement are impressive.

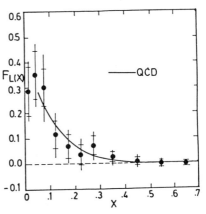

Fig. 8: CDHSW data on $F_L(x)$, from Ref. 25.

3.3 Hard Photons and QCD

Another topic in which considerable experimental and theoretical progress has been achieved this year concerns direct photon production in hadronic reactions. Aurenche, Baier, Fontannaz and Schiff /26/ have completed a computation of higher order corrections for this process (see Fig. 9 for the relevant graphs). Interestingly enough, they find

that, by optimizing the scales at which $\alpha_s(q^2)$ and the structure functions are evaluated, their results are very stable and lead to K factors very near to unity. Aurenche et al. /26/ have compared their calculations with a host of different prompt photon data, over a wide ener-
gy range from $\sqrt{s}$ = 19 GeV to $\sqrt{s}$ = 630 GeV. For $p_\perp \geq$ 4 GeV, they find

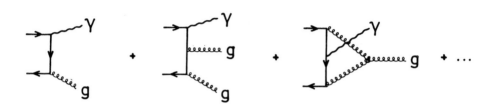

Fig. 9:   Processes leading to hard γ production

very nice fits to the data, using for structure functions those of
Duke and Owens, SetI with $\Lambda_{\overline{MS}}$ = 200 MeV /27/. I show some of these fits
in Figs. 10.

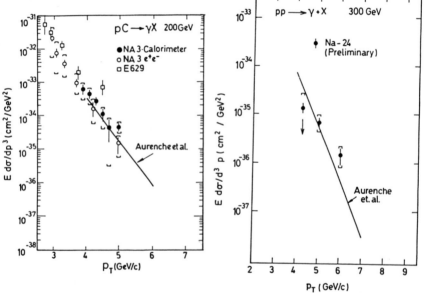

Fig. 10a: pC → γX at
$\sqrt{s}$ = 19.4 GeV,
from Ref. 28

Fig. 10b: pp → γX at
$\sqrt{s}$ = 23.8 GeV,
from Ref. 29

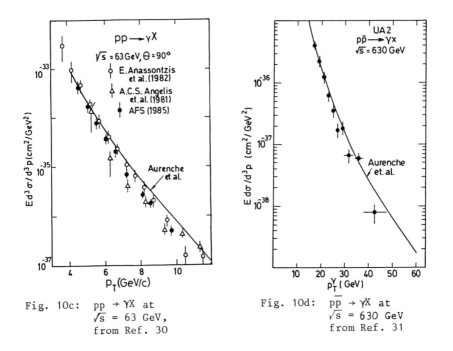

Fig. 10c: pp → γX at
√s̄ = 63 GeV,
from Ref. 30

Fig. 10d: p̄p → γX at
√s̄ = 630 GeV
from Ref. 31

Aurenche, Baier, Douiri, Fontannaz and Schiff /32/ have also computed the $O(\alpha_S)$ corrections to Compton scattering $\gamma q \rightarrow \gamma q$. Their

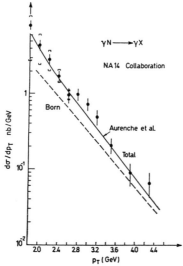

Fig. 11: Deep inelastic Compton scattering γN → γX data from Ref. 33
at E_γ = 100 GeV

results are in good agreement with the new NA 14 data /33/. Indeed, as shown in Fig. 11, the $O(\alpha_s)$ corrections improve the fit.

4. HIGGS BOSONS

The Higgs boson although an integral part of the standard model, is its most ephemeral entity. It is there to preserve renormalizability, although it could be obviated if some sort of dynamical symmetry breaking of SU(2)xU(1) obtained. Naturally, its theoretical and experimental investigation is of fundamental concern.

4.1 An Experimental Bound on m_H?

The mass of the Higgs boson, m_H, is not predicted by the standard model. Furthermore, since the coupling of the Higgs to fermions is proportional to the fermion's mass, Higgs bosons are difficult to produce. As a result, obtaining any kind of bound on Higgs bosons is a difficult business. Perhaps the most promising way to look for, relatively light, Higgs bosons is through the Wilczek mechanism /34/, in which a heavy quarkonia decays into a Higgs boson plus a photon. Since this rate is proportional to the mass squared of the heavy quark, it is clear that Υ decays are the most reasonable hunting ground, at present, for Higgs bosons.

The ratio between Υ decays into a Higgs boson and a photon and its decay into μ pairs is given by /34/:

$$\frac{\Gamma(\Upsilon \to H\gamma)}{\Gamma(\Upsilon \to \mu^+\mu^-)} = \frac{G_F \, m_b^2}{\sqrt{2}\,\pi\alpha} \left\{ 1 - \frac{m_H^2}{m_\Upsilon^2} \right\} \simeq \frac{M_\Upsilon^2/M_W^2}{8 \sin^2\theta_W} \left\{ 1 - \frac{m_H^2}{M_\Upsilon^2} \right\} \tag{20}$$

For m_H not too near the kinematic boundary, this ratio is of the order of 2 %, and hence is within experimental reach. Recent CUSB data, presented by J. Lee Franzini at the Autun Meeting /35/, appears to exclude, at the 90 % confidence limit, Higgs bosons lighter than about 4.5 GeV. This is seen in Fig. 12. However, the Wilczek formula Eq. (20) has large QCD corrections and the issue is unsettled.

It has been known for a long time that the QCD corrections to the rate for quarkonia to decay into lepton pairs are rather large /36/. It turns out that the QCD corrections to decays of quarkonia into a Higgs and a photon, calculated by Vysotosky /37/ are even larger. Taking into account of both corrections one has

$$\frac{\Gamma(\Upsilon \to H\gamma)}{\Gamma(\Upsilon \to \mu^+\mu^-)} = R_{Wilczek} \frac{\left\{ 1 - (40/3\pi)\,\alpha_s\, F(m_H^2/M_\Upsilon^2) \right\}}{\left\{ 1 - (16/3\pi)\,\alpha_s \right\}} \tag{21}$$

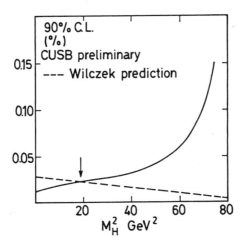

Fig. 12: Comparison of the CUSB data, Ref. 35, with the Wilczek prediction

The function $F(m_H^2/M_T^2)$ is explicitly given in Vysotosky's paper /37/. However, for $m_H \lesssim 0.8\ M_T$, $F \simeq 1$. Because the corrections in Eq. (21) are so large, one probably cannot trust them. But even proceeding naivly, expanding in α_s, one sees that these corrections vitiate the CUSB bound. Imagining that α_s is small in Eq. (21), one has for the ratio, for light m_H:

$$R = R_{Wilczek} \left\{ 1 - \frac{8\ \alpha_s(M_T^2)}{\pi} \right\} \tag{22}$$

If $\alpha_s(M_T^2) = 0.15$ then the square bracket above is 0.62. For $\alpha_s(M_T^2) = 0.20$, the Wilczek prediction is reduced by 50 %. Given that the data in Fig. 12 is barely below the original Wilczek prediction, it is clear that including QCD corrections removes the bound on m_H altogether. Of course, if one could experimentally go well below the Wilczek prediction in T decays, then even taking into account of the QCD corrections, one could rule out sufficiently light Higgs bosons.

4.2   Toponium as a Higgs Detector

If the top quark mass is really in the range $m_t = 40 \pm 10$ GeV, suggested by the UA_1 collaboration /38/, then toponium will give a very strong bound on the Higgs mass. Recalling that the Wilczek rate grows with the square of the quarkonia mass (cf. Eq. (20)), one sees that for toponium near 80 GeV the rate for decay into $H\gamma$ is of the order of the $\mu^+\mu^-$ rate. Even large QCD corrections will not affect this qualitative fact and one should be able to rule out (or discover!) Higgs bosons with masses within 5 - 10 GeV of the kinematic limit.

This matter has been studied carefully recently, in occasion of the LEP Jamboree /39/. First of all, provided the top quark has a mass less than half the Z^0 mass, its discovery at LEP should be straightforward.(Because of the much larger energy spread, the SLC is not in such a favorable condition.) Z^0 decays should determine the top mass within one GeV. Then at LEP, doing a scan of the relevant 2 GeV region, in 80 MeV steps, one should be able to determine the existence of toponium, by means of topological cuts, in about 2 weeks /39/. Having found toponium, then a Hγ signal is relatively easy to detect, even with rather small luminosity. This is shown in Fig. 13, for the case of a 70 GeV toponium, where the integrated luminosity needed for a 3σ Hγ signal is plotted versus the Higgs mass. Since at LEP an integrated luminosity of around 10 pb^{-1} per month is expected, it is clear that, if toponium is at 70 GeV, then the discovery of a 60 GeV Higgs boson will require only about one month running.

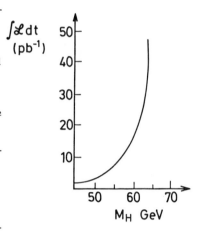

Fig. 13: Integrated luminosity needed for Higgs boson detection, if toponium has a mass of 70 GeV

4.3   Theoretical Guesstimates on m_H

Although, as I have already mentioned, the Higgs mass is not predictable in the standard model, predicting m_H is a favorite theoretical pastime. The results obtained, in general, reflect the prejudices put in. Two examples will suffice. Beg Panagiotakopoulos and Sirlin /40/ by requiring stability of the Higgs self coupling λ in the perturbative renormalization group equations, find that this coupling should be bound by the U(1) coupling constant squared. This implies, immediately, a bound on the Higgs mass:

$$m_H \lesssim 2\sqrt{2} \cot\theta_W M_W < 130 \text{ GeV} \tag{23}$$

Kubo, Sibold and Zimmermann /41/, on the other hand, solve the renormalization group equations under the assumptions that all couplings are driven by just one coupling. This reduction of couplings means that, effectively, all couplings are driven by the largest gauge coupling. Neglecting the SU(2)xU(1) couplings altogether and all Yukawa couplings, except that of the t quark, they find that either

$$m_H = 61 \text{ GeV}; \quad m_t = 81 \text{ GeV} \tag{24a}$$

or

$$m_H > 40 \text{ GeV}; \quad m_t < 81 \text{ GeV} \qquad\qquad (24b)$$

if this coupling reduction holds.

Less dependent on particular dynamical assumptions is the study of the Higgs sector on the lattice. This has been undertaken by a number of authors recently /42/, most notably by Montvay and collaborators. What has been investigated is the pure SU(2) Higgs model in the presence of gauge fields and most particularly what happens when the Higgs self coupling λ becomes very large. What Montvay finds /42/ is that as $\lambda \to \infty$ the ratio of the Higgs mass to the W mass remains of $O(1)$. Furthermore, his numerical results seem to show λ independence, over a wide range of λ, suggesting perhaps that λ may be an irrelevant variable. These results hold in a strong coupling region for the gauge coupling and need to be extrapolated to small g to make direct connection with physics. When this extrapolation is attempted Montvay finds that the m_H/M_W ratio grows. A preliminary value obtained this way is that $m_H \simeq 6 \ M_W$ /42/.

Although these lattice investigations are really just beginning, they appear extremely interesting theoretically. Obviously, they still need considerable refinement. For instance, at the moment, no U(1) factor is included at all. Nevertheless, if these more refined lattice calculations continued to converge on such a large value for the Higgs mass, this would put the Higgs boson out of reach experimentally until the advent of the SSC!

5. KM MATRIX, CP VIOLATION AND ALL THAT

As a last topic of discussion I want to consider CP violation in the standard model. In particular I want to examine critically whether the usual explanation of CP violation, through the appearance of a phase in the Kobayashi Maskawa mixing matrix, is still tenable, or whether finally one is forced to go beyond the standard model. My answer (unfortunately?) will be that everything is still compatible with the standard model, although the model is being challenged.

5.1 Quark Mixings

The mass matrices for the quarks and leptons are beyond prediction in the standard model and so are the mixings among the quarks. However, the model does predict that the mixing matrix (KM matrix) is unitary. Thus, even though very little is known about the top quark, quite a lot is known about the mixing matrix elements V_{tb}, V_{ts} and V_{td}. In particular, the comparatively recent discovery of a very long B lifetime coupled to stringent bounds on the ratio of $b \to u$ to $b \to c$ transitions have provided important information for the structure of the KM matrix, establishing that $V_{us} \gg V_{cb} \gg V_{ub}$. Wolfenstein /43/ has given a handy approximate parametrization of the KM matrix, which takes these new facts into account and is easy to remember:

$$
V_{KM} = \begin{vmatrix} 1 - \frac{1}{2}\lambda_W^2 & \lambda_W & \lambda_W^3\, A(\rho - i\eta) \\ -\lambda_W & 1 - \frac{1}{2}\lambda_W^2 & \lambda_W^2\, A \\ \frac{\lambda^3}{3}\, A(1 - \rho - i\eta) & -\lambda_W^2\, A & 1 \end{vmatrix} \tag{25}
$$

where

$$
\lambda_W \simeq \sin\theta_c \simeq 0.23; \quad A \simeq 1; \quad \rho^2 + \eta^2 \leqslant 0.25 \tag{26}
$$

5.2   CP Violation - ε Parameter

The smallness of V_{cb} and V_{ub}, along with a possible relatively light top quark /38/, have brought the standard model explanation for the CP parameter ε into question . (In fact, originally, Glashow Ginsparg and Wise /44/ used ε and the smallness of V_{cb}, V_{ub} to get a lower bound on m_t.) Recall that ε, in the standard model, is given by the imaginary part of the box graph shown in Fig. 14, in which all the charge 2/3 quarks enter. For the imaginary part, the t quark contribution is crucial and this is surpressed because V_{td} and V_{ts} are small (c.f. Eq. (25)).

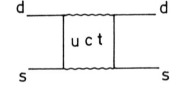

Fig. 14:   Box graph whose imaginary part contributes to ε

Instead of expressing ε in terms of the KM matrix elements it is perhaps more useful to detail its dependence directly on other measured quantities /45/. Let me denote the B lifetime by

$$
\tau_B = \beta \times 10^{-12}\ \text{sec} \tag{27a}
$$

and the ratio of the b $\rightarrow$ u to b $\rightarrow$ c leptonic decays by

$$
R = \frac{\Gamma(b \rightarrow u e \nu)}{\Gamma(b \rightarrow c e \nu)} \tag{27b}
$$

Experimentally the world average of all existing experiments, computed by D. Haidt /46/, gives for β:

$$\beta = 1.0 \pm 0,19 \tag{28a}$$

J. Lee Franzini /35/ quotes a 90 % confidence limit for R of

$$R < 0.03 \tag{28b}$$

while Thorndyke /47/ has a more conservative limit:

$$R < 0.04 \tag{28c}$$

The dominant dependence of ε on these two parameters and on the top quark mass scales as

$$\varepsilon \sim \frac{1}{\beta} \sqrt{R} \, m_t^2 \ . \tag{29}$$

Hence a longer B lifetime, a tighter bound on the $b \to u$ to $b \to c$ ratio and a smaller value for m_t all conspire to make ε smaller. Clearly, at some point, the standard model explanation for ε would then cease to be tenable. Even now, the situation is somewhat fluid, since a determination of ε requires besides the above experimental information also a theoretical calculation for the matrix element of $(\overline{d}\gamma_\mu(1-\gamma_5)s)^2$ between K and $\overline{K}$ states. This calculation is also quite uncertain.

It has been customary to denote by B the ratio of the above matrix element to that computed via vacuum insertion. Various approaches have been followed to compute B, ranging from bag model calculations to lattice calculations, with results ranging from about − 0.4 to +2.5 for B. Two values of B appear special: B = 1, which is just the vacuum insertion approximation and B = 1/3, which is a result that follows from current algebra /48/ (see below). At any rate, the predicted value for ε is proportional to B and thus, whether one considers the standard model to be in trouble or not depends in part on the value for B assumed.

Buras /49/ has displayed this interrelation, between experimental and theoretical input for ε, in a nice way. He computes, as a function of β, R and m_t, what is the minimum value of B necessary to fit in the standard model the experimental value for ε. As an illustration, I display in Table II, some of his results /49/ for a selected range of β, R and m_t.

Table II Minimum values of B needed to fit ε in the standard model

m_t (GeV)	β	0.7	1.0	1.3	
25		0.45	0.75	1.08	
40		0.23	0.41	0.63	R = 0.03
55		0.14	0.24	0.40	
25		0.86	1.42	2.01	
40		0.47	0.82	1.22	R = 0.01
55		0.29	0.53	0.82	

Clearly if $\beta = 1$, R = 0.03 and m_t = 40 GeV a value of B = 1/3 would just be slightly unacceptable. However, if R = 0.01, with $\beta = 1$ and m_t = 40 GeV, really B must be near unity, for the standard model not to be in trouble. I would characterize the situation as tantalyzing, but not yet critical. Obiously, theoretically, it is important to pin down the value of B.

5.3 B Parameter Controversy

Given the above discussion, it is particularly important to examine the theoretical basis of predictions for the B parameter which yield a small value. Of particular importance, in this respect, is the current algebra prediction of Donoghue, Golowich and Holstein /48/ for B, B $\simeq$ 1/3. Their result is rather easy to understand and appears to require very little theoretical input. What these authors realized is that, in the limit of good chiral $SU(3)_L \times SU(3)_R$, the 4-quark operator that enters in the K-$\overline{\text{K}}$ matrix element (and therefore in B) is related to that which describes the $K^+ \to \pi^+\pi^0$ decay. Both of these operators transform as $(27_L, 1_R)$. Using an effective chiral Lagrangian the amplitudes for both of these processes are then fixed, save for an overall normalization constant. This constant can, however, be determined from the experimental value for $A(K^+ \to \pi^+\pi^0)$ and it is this procedure which gives B $\simeq$ 1/3.

Because the logic above is so simple, considerable credence was given to a small value for B. However, recently, Bijnens, Sonoda and Wise /50/ have brought into question the validity of the chiral limit for the evaluation of B. They have calculated $O(m_\pi^2 \ln m_K^2)$ corrections to the chiral limit of Donoghue et al. /47/ and found that, for the case of the $\Delta s = 2$ operators, these corrections are larger than the zero order result! Thus the chiral limit inference that B $\sim$ 1/3 may be flawed.

This situation has been made even more difficult to interpret by a very recent paper of Pich and de Rafael /51/. These authors have reexamined this issue from a completely different point of view - that of QCD sum rules - and find again that B $\simeq$ 1/3! Let me briefly discuss their idea, so as to give a flavor of their calculations. What one needs to calculate is the K-$\overline{\text{K}}$ matrix element of the $\Delta s = 2$ operator $0 = (\bar{d}\gamma_\mu(1-\gamma_5)s)^2$. What one does, instead, is to calculate the matrix element of an appropriate effective chiral operator, with the same transformation properties as 0. To actually get a number, of course, one must know the proportionality constant relating 0 to this effective operator. To be specific, de Rafael and Pich /51/ replace 0 by

$$0 \to \mathcal{L}_{\text{chiral}}^{\text{eff}} = G_{\Delta s=2} ([f_\pi^2 U\partial_\mu U^+]_{23})^2 \qquad (30)$$

where U is an SU(3) matrix of Goldstone boson fields, but $G_{\Delta s=2}$ is unknown. To get a value of B they need to fix $G_{\Delta s=2}$.

Donoghue et al. /47/ fixed $G_{\Delta s=2}$ by relating it to the amplitude for $K^+ \to \pi^+\pi^0$, determined experimentally. Pich and de Rafael /51/, on the other hand, do this by comparing the behaviour of two different two point functions in the fashion of QCD sum rules. What they consider are the two point functions of the operators 0 and of $\mathcal{L}_{eff}$, $\Delta(q^2)$ and $\Delta_{eff}(q^2)$, and compare integrals over their spectral functions:

$$\int dq^2 \text{ Im } \Delta(q^2) = \int dq^2 \text{ Im } \Delta_{eff} (q^2) \qquad (31)$$

For the integral on the left-hand side above they use the answer obtained from the QCD short distance behaviour. For the right-hand side, they make use of resonance saturation in the chiral model. Matching the results gives a value for $G_{\Delta s=2}$ and therefore B. What Pich and de Rafael find in this way is

$$B = 0.33 \pm 0.09 \qquad (32)$$

It appears rather amazing to me that this calculation should agree so well with the simple chiral result of Donoghue et al./48/. Thus, it could well be that the agreement is fortuitous. However, Guberina, Pich and de Rafael /52/, have made an analogous, but in detail different, calculation for the operator that enters in $K^+ \to \pi^+\pi^0$ and computed the amplituds $A(K^+\to\pi^+\pi^0)$. Their result $A_{GPR} = 1.8 \times 10^7$ sec^{-1} is in excellent agreement with the experimental value ($A_{exp} = 1.7 \times 10^7$ sec^{-1}). This would argue that their methods are reliable and so that one should trust also their result (32).

5.4   CP Violation - ε'/ε

The situation with ε'/ε is even more uncertain than that with ε. Now, even though some of the dependence on the specific size of the KM matrix elements is milder, there is a new, uncertain, hadronic matrix element to estimate /53/:

$$\langle\pi\pi| \; 0_{Penguin} \; |K\rangle \sim B' \qquad (33)$$

Also B' estimates can vary by more than a factor of 2, so that the theoretical uncertainty in $\varepsilon'/\varepsilon \sim B'/B$ can be really quite large. Two typical ranges, for $m_t = 40$ GeV, $\beta = 1$, which appear in the recent literature, are

$$10^{-3} \leq \varepsilon'/\varepsilon \leq 15 \times 10^{-3} \qquad \text{/Ref. 54/} \qquad (34a)$$

$$2 \times 10^{-3} \leq \varepsilon'/\varepsilon \leq 8 \times 10^{-3} \qquad \text{/Ref. 49/} \qquad (34b)$$

Thus the wonderful experimental limits /55/

$$\varepsilon'/ \varepsilon = (-4.6 \pm 5.3 \pm 2.4) \times 10^{-3} \qquad \text{/Chicago-Saclay/}$$

$$\varepsilon'/ \varepsilon = (1.7 \pm 8.2) \qquad \times 10^{-3} \qquad \text{/BNL-Yale/}$$

badly need a more accurate theory prediction, to really push the standard model.

6.   CONCLUDING REMARKS

I conclude with pretty much the observation I made at the beginning: SU(3)xSU(2)xU(1) works remarkably well! Some hope does exist for finding some discrepancy in this beautiful edifice. For one thing, the Higgs sector is essentially unknown. Here toponium and the lattice calculations may begin to shed some light. It is possible that the CP violation parameter ε may need some new physics for its explanation. But that would require $R \leq 0.01$, m_t being really light and that the theoretical ambiguities in B were finally resolved! Also, it is clear that a measurmend of ε'/ε at the 10^{-3} level would seriously impact the model, expecially if some of the theoretical ambiguities in B',B were under better control.

My personal conclusion is that the real question should be shifted from: does the standard model work? to, why is the standard model a description of nature? The list of unanswered questions in this latter case is rich and deep. A partial sampling includes:

Why SU(3)xSU(2)xU(1)?
Why chiral fermions?
Why is $(\sqrt{2}\ G_F)^{-1/2} \sim 10^3\ \Lambda_{QCD}$?
Why do the fermions replicate?
What fixes the fermion masses and mixing?

REFERENCES

/1/ S.L. Glashow, Nucl. Phys. 22 (1961) 579; A. Salam in Elementary
    Particle Theory, ed. by N. Svartholm (Almquist and Wiksells, Stockholm 1969); S. Weinberg, Phys. Rev. Lett. 19 (1967) 1264.
/2/ A. Sirlin, Phys. Rev. D22 (1980) 971.
/3/ W. Marciano, Proceedings of the Fourth Topical Conference on Proton-Antiproton Collider Physics, Bern 1984.
/4/ A. Sirlin, Phys. Rev. D29 (1984) 89.
/5/ W. Marciano and A. Sirlin, Nucl. Phys. B189 (1981) 442;
    C. Llewellyn Smith and J. Wheater, Phys. Lett. 105B (1981) 486.
/6/ W. Marciano and A. Sirlin, Phys. Rev. D22 (1980) 2695;
    W. Marciano Ref. 3
/7/ L. di Lella, Proceedings of the International Europhysics Conference
    on High Energy Physics, Bari 1985.
/8/ G. Altarelli et al., report of the Precision Studies at the Z$^{\circ}$Working Group for the LEP Jamboree. To be published in the Jamboree
    Proceedings.
/9/ G. Barbiellini et al., report of the High Energy Working Group for
    the LEP Jamboree. To be published in the Jamboree Proceedings.
/10/ F. Halzen and K. Mursula, Phys. Rev. Lett. 51 (1983) 857;
    K. Hikasa, Phys. Rev. D29 (1984) 1939; N.G. Deshpande, E. Eilam,
    V. Barger and F. Halzen, Phys. Rev. Lett. 54 (1985) 1757.

/11/ G. Altarelli, R.K. Ellis, M. Greco and G. Martinelli, Nucl. Phys.
B246 (1984) 12; G. Altarelli, R.K. Ellis and G. Martinelli,Zeit.
Phys. C27 (1985) 617.
/12/ K. Olive, D. Schramm, G. Steigman, M. Turner and J. Yang, Ap. J.
246 (1981) 547
/13/ G. Altarelli, Phys. Rept. 81C (1982) 1.
/14/ C. Quigg, Rev. Mod. Phys. 49 (1977) 297 and references therein.
/15/ Yu.L. Dokshitzer, D.I. Dyakonov and S.I. Troyan, Phys. Rept. 58C
(1980) 269; G. Parisi and R. Petronzio, Nucl. Phys. B154 (1979) 427;
J.C. Collins and D.E. Soper, Nucl. Phys. B197 (1982) 446;
J.C. Collins, D.E. Soper and G. Sterman, Nucl. Phys. B250 (1985)
199.
/16/ J. Kodaira and L. Trentadue, Phys. Lett. 123B (1983) 335;
C.T.H. Davies and W.J. Stirling, Nucl. Phys. B244 (1984) 337.
/17/ M. Levi, in Proceedings of the 5th Topical Workshop on Proton-
Antiproton Collider Physics, St. Vincent 1985, ed. M. Greco
(World Scientific, Singapore 1985).
/18/ G. Arnison et al., Phys. Lett. 136B (1984) 294; P. Bagnaia et al.,
Zeit. Phys. C20 (1983) 117.
/19/ W. Scott, in Proceedings of the International Symposium on Physics
of Proton-Antiproton Collision, Tsukuba 1985, ed. by Y. Shimizu
and K. Takikawa; CERN-EP / 85-98
/20/ C. Rubbia, in Quarks Leptons and Beyond, ed. H. Fritzsch, R.D. Peccei,
H. Saller and F. Wagner (Plenum Press, N.Y. 1985)
/21/ W. Scott, Ref. 19; F. Pastore, in Proceedings of the 5th Topical
Workshop on Proton-Antiproton Collider Physics, St. Vincent 1985,
ed M. Greco (World Scientific, Singapore 1985).
/22/ P. Jenni, private communication.
/23/ C.G. Callan and D.J. Gross, Phys. Rev. Lett. 22 (1969) 156.
/24/ A.J. Buras, Rev. Mod. Phys. 52 (1980) 199.
/25/ J. Feltesse, Proceedings of the International Europhysics Conference
on High Energy Physics, Bari 1985.
/26/ P. Aurenche, R. Baier, M. Fontannaz and D. Schiff, Bielefeld preprint forthcoming.
/27/ D.W. Duke and J.F. Owens, Phys. Rev. D27 (1984) 508.
/28/ M. Cohen, Orsay Thesis, LAL 85/14.
/29/ K. Pretzl, Proceedings of the XVI Symposium on Multiparticle Dynamics,
Kyriat Anavim, 1985.
/30/ E. Anassontzis et al., Zeit. Phys. C13 (1982) 277; A.C.S. Angelis
et al., Phys. Lett. 98B (1981) 115; T. Akesson et al., AFS Contribution to Bari Conference 1985.
/31/ P. Hansen, Proceedings of the XVI Symposium on Multiparticle Dynamics, Kyriat Anavim 1985.
/32/ P. Aurenche, R. Baier, A. Douiri, M. Fontannaz and D. Schiff, Zeit.
Phys. C24 (1984) 309.
/33/ P. Astbury et al., Phys. Lett. 152B (1985) 419.
/34/ F. Wilczek, Phys. Rev. Lett. 39 (1977) 1304.
/35/ J. Lee Franzini, Proceedings of the IV Meeting of Physics in Collision, Autun 1985.
/36/ R. Barbieri, R. Gatto, R. Kögerler and Z. Kunszt, Phys. Lett. 57B
(1975) 455.

/37/ M. Vysotosky, Phys. Lett. 97B (1980) 159.

/38/ G. Arnison et al., Phys. Lett. 147B (1984) 493.

/39/ W. Buchmüller et al., report of the Toponium Working Group for the LEP Jamboree. To be published in the Jamboree Proceedings

/40/ M.A. Beg, C. Panagiotakopoulos and A. Sirlin, Phys. Rev. Lett. 52 (1984) 883.

/41/ J. Kubo, K. Sibold and W. Zimmermann, Nucl. Phys. to be published.

/42/ I. Montvay, DESY 85-005; I. Montvay, in Proceedings of the Conference on Advances in Lattice Gauge Theory; W. Langguth and I. Montvay, Phys. Lett. to be published; J. Jersák, C.B. Lang, T. Neuhaus and G. Vones, Aachen preprint PITHA 85/05.

/43/ L. Wolfenstein, Phys. Rev. Lett. 51 (1983) 1945.

/44/ S.L. Glashow, P. Ginsparg and M. Wise, Phys. Rev. Lett. 50 (1983) 1415.

/45/ A. Buras, W. Slominski and H. Steger, Nucl. Phys. B238 (1984) 529.

/46/ D. Haidt, private communication.

/47/ E. Thorndyke, Proceedings of the 1985 International Symposium on Lepton and Photon Interactions at High Energy, Kyoto 1985.

/48/ J. Donoghue, E. Golowich and B. Holstein, Phys. Lett. 119B (1982) 412.

/49/ A. Buras, Proceedings of the International Europhysics Conference on High Energy Physics, Bari 1985.

/50/ J. Bijnens, H. Sonoda, M. Wise, Phys. Rev. Lett. 53 (1984) 2367.

/51/ A. Pich and E. de Rafael, Phys. Lett. 158B (1985) 477.

/52/ B. Guberina, A. Pich and E. de Rafael, Phys. Lett. to be published.

/53/ F. Gilman and M. Wise, Phys. Lett. 83B (1979) 83; B. Guberina and R.D. Peccei, Nucl. Phys. B163 (1980) 289.

/54/ P. Langacker, Proceedings of the Aspen Winter Conference Series, Aspen 1985.

/55/ Chicago-Saclay, R.K. Bernstein et al., Phys. Rev. Lett. 54 (1985) 1631; Brookhaven-Yale, J.K. Black et al., Phys. Rev. Lett. 54 (1985) 1628.

INTERMEDIATE VECTOR BOSONS AND JETS AT THE CERN pp̄ COLLIDER

Jean-Marc Gaillard

Laboratoire de l'Accélérateur linéaire

91405 Orsay, France

ABSTRACT

We present recent results from the analysis of data collected up to the end of 1984 by the UA1 and UA2 experiments at the CERN pp̄ Collider. These results, which correspond to a total integrated luminosity of about 400 nb^{-1}, cover two physics subjects: the production of jets of particles at high transverse momentum, and the production and decay of the weak intermediate bosons, $W^{\pm}$ and Z^0. All experimental results are found to be in good agreement with the predictions of the Standard SU(3) × SU(2) × U(1) Model.

1. FOREWORD

The first pp̄ collisions in the CERN SPS Collider were observed in July 1981[1] at an energy $\sqrt{s}$ = 546 GeV. At the end of that same year the first physics run provided the first clear evidence for the production of large-transverse-momentum (p_T) hadron jets in hadron-hadron collisions[2,3].

A year later, a second physics run with an integrated luminosity 200 times larger led to the discovery of the W boson[4,5]. In the spring of 1983, with an eightfold increase of the integrated luminosity, the Z^0 was also discovered[6,7].

The energy of the Collider was increased to $\sqrt{s}$ = 630 GeV for the following run during the autumn of 1984. At the end of that run, the UA1 and UA2 experiments had collected data corresponding to total integrated luminosities of 399 nb^{-1} and 452 nb^{-1}, respectively, since the start of the Collider operations.

I have been asked to review the results obtained so far by the UA1 and UA2 experiments on jets and the W and Z intermediate bosons. I have found it necessary to make a selection of the topics to be covered by a single talk. For the jet physics the review will be limited to the production properties of jets and their comparison with QCD predictions. In the case of the W and Z vector bosons, only the leptonic decays have been observed. The emphasis will be put on the electronic modes for which a total of about 262 W → eν and 34 Z^0 → e^+e^- are now available from the two experiments. The W and Z muonic decay results of UA1 are consistent with the electronic mode information, but less accurate. They will not be discussed in the present report. The UA1 observation of W → τν followed by τ → hadrons will be presented by J. Rohlf.

2. LARGE-TRANSVERSE-MOMENTUM JET PRODUCTION

2.1 Introduction

The unambiguous identification of jets at the CERN p$\bar{\text{p}}$ Collider[2,3] has evidenced in a direct way the prediction that the hard scattering of hadron constituents should result in the production of two hadronic jets having the same large transverse momenta as the scattered partons[8]. The high centre-of-mass energy of the $\bar{\text{p}}$p Collider gives the possibility for detailed measurements of hadronic jet production and their fragmentation properties in an energy domain where hard processes can be separated clearly from the soft hadronic interactions.

A schematic representation of the hard collision between two partons a + b → c + d is shown in Fig. 1; $F_a(x)$ and $F_b(x)$ are the structure functions of the partons a and b. The fragmentation of the scattered partons c and d into hadrons is described by the functions D(z).

The contribution of the elementary subprocess, a + b → c + d, to the total cross-section can be written as

$$d^3\sigma/(dx_1\, dx_2\, d\cos\theta^*) = [F_a(x_1,Q^2)/x_1]\, [F_b(x_2,Q^2)/x_2]\, [\pi\alpha_s^2(Q^2)/2\hat{s}]\, |M|^2\, , \qquad (1)$$

where $\alpha_s(Q^2)$ is the running QCD coupling constant and $\hat{s}$ is the centre-of-mass energy of the subprocess. The matrix element $|M(t,s,u)|^2$ depends, in fact, only upon $\cos\theta^*$, where θ^* is the scattering angle of the subprocess. The total cross-section is the incoherent sum of all the contributions from the different subprocesses.

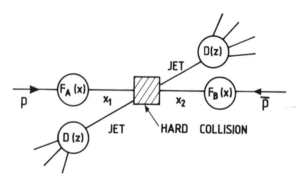

Fig. 1 Hard scattering between two partons. F(x) = structure function; D(z) = fragmentation function.

The dominance of the two-jet production process is now a well-established experimental observation.

More detailed studies of the 1983 data have been pursued by UA1 and UA2. The data of the 1984 run with a larger Collider energy and a big increase of the integrated luminosity have provided additional possibilities of testing the QCD predictions.

2.2   Inclusive Cross-section

The results of the measurements[9-11] of the inclusive jet production cross-section, $d^2\sigma/d\eta\, dp_T$, at $\sqrt{s}$ = 546 GeV and for the interval of pseudo-rapidity $0 < |\eta| < 0.7$ are shown in Fig. 2 as a function of the jet transverse momentum. The additional systematic errors, not shown by the error bars of Fig. 2, are about 65% for the UA1 data and 45% for the UA2 data. Also shown in Fig. 2 is a band of QCD predictions[12], whose width serves to illustrate the uncertainties in the theory arising mostly from different parametrizations of the structure functions.

In spite of the large experimental and theoretical uncertainties, the agreement between data and theory is remarkable, especially because the theoretical curves of Fig. 2 are not a fit to the data but represent an absolute prediction.

Most systematic uncertainties cancel when comparing the result at two energies. The UA2 Collaboration[13] has analysed the two sets of data corresponding to $\sqrt{s}$ = 546 GeV (1983) and $\sqrt{s}$ = 630 GeV (1984) and the inclusive cross-section distributions are shown in Fig. 3a. The cross-section is systematically larger at 630 GeV than at 546 GeV as expected since x_1 (x_2) decreases with increasing c.m. energy for a fixed p_T. The ratio of the two cross-sections as a function of p_T is shown in Fig. 3b together with the QCD prediction. As noted above, both the theoretical and experimental systematic uncertainties largely cancel in that ratio.

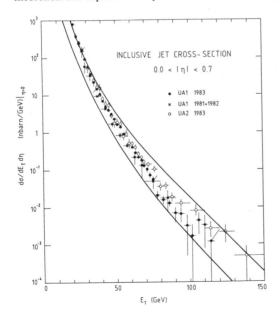

Fig. 2 Inclusive jet production cross-section at $\sqrt{s}$ = 546 GeV.

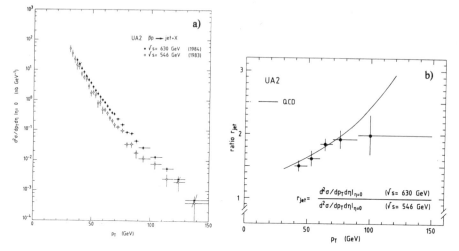

Fig. 3 Comparison between $\sqrt{s} = 630$ GeV and $\sqrt{s} = 546$ GeV:
a) inclusive jet-production cross-sections;
b) ratio of the inclusive jet production cross-sections; the data are compared with QCD prediction.

The UA2 Collaboration has also analysed the two-jet mass distribution in the 1984 data. Indications from the 1983 UA2 data of a structure at a mass of about 150 GeV for the events with two jets only have not been confirmed by the new data with two and a half times more statistics.

2.3 Angular Distribution of Parton–Parton Scattering

The study of the jet angular distribution offers the possibility of measuring directly the angular distribution of parton–parton scattering. However, the inclusive cross-section is the incoherent sum of the contributions of different subprocesses: gluon-gluon, quark–gluon, and quark–antiquark scattering. A great simplification is due to the fact that all the subprocesses have almost the same $\cos \theta^*$ dependence in the case of vector gluons:

$$d\sigma/d \cos \theta^* \simeq (1 - \cos \theta^*)^{-2}.$$

In this approximation, where the singularity dominates, the total cross-section factorizes and can be written as

$$d^3\sigma_{tot}/dx_1 \, dx_2 \, d \, \cos \theta^* = [F(x_1)/x_1] \, (\pi\alpha_s^2/2\hat{s}) \, |M|^2 \, [F(x_2)/x_2] \,, \qquad (2)$$

where

$$F(x) = G(x) + (4/9) \left\{ \sum_q [Q(x) + \bar{Q}(x)] \right\}. \qquad (3)$$

Here $G(x)$, $Q(x)$, and $\bar{Q}(x)$ are the structure functions of the gluons, the quarks, and the antiquarks, respectively.

The adequacy of the factorization approximation and of the structure-function parametrization have been verified by UA2 and UA1 measurements. For example, in Fig. 4 the combined structure function F(x) as measured by UA1 [14] and UA2 [15] is compared with the combination of Eq. (3) as it is obtained from neutrino-experiment measurements [16] evolved from $Q^2 \simeq 20 \, GeV^2$ to $Q^2 = 2000 \, GeV^2$.

The angular distribution of the jet pairs obtained by UA1 [14] in 1982 is compared in Fig. 5 with the predictions of QCD and Abelian scalar theories. The data agree with vector-gluon QCD predictions. Similar results have been obtained by UA2 [15] over a more limited angular range.

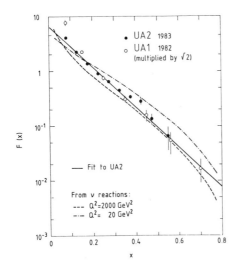

Fig. 4 Effective structure function. The full line represents an exponential fit to the data. The dashed lines are computed from neutrino deep inelastic scattering.

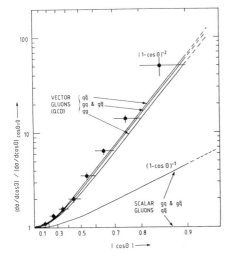

Fig. 5 Angular distribution of jet pairs compared with QCD and Abelian scalar theories.

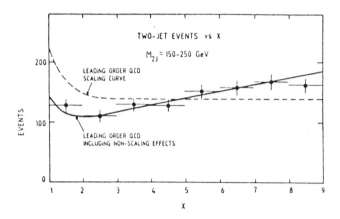

Fig. 6 Evidence for non-scaling effects in the jet-pair distribution

The above approximations do not take into account the variation of Q^2 over the angular range, $Q^2 = -\hat{t} = (\hat{s}/2)(1 - \cos \theta^*)$, which introduces small non-scaling effects, since α_s and $F(x)$ are also Q^2 dependent. Evidence for these non-scaling effects has been obtained from the analysis of the 1983 UA1 data[17], as shown in Fig. 6. The variable $\chi = (1 + \cos \theta^*)/(1 - \cos \theta^*)$ used in Fig. 6 is such that in the scaling approximation $d\sigma/d\chi = constant$ for $\chi > 2$. Non-scaling effects are clearly required in order to fit the data.

2.4 Three-Jet Events and α_s Determination

In the previous section we have seen that the features of two-jet final states are well described by the leading term in the perturbative expansion of the parton–parton cross-section in powers of the strong coupling constant α_s. The next term of the expansion implies the existence of three-jet final states in which a bremsstrahlung gluon has been radiated. The observation and a detailed study of the properties of these final states are a crucial test of the theory.

The theoretical expression to lowest order of the n-jet cross-section can be written as:

$$\sigma_n^{LO} = (\alpha_s)^n (\hat{s})^{-1} \sum_{i,j} \int F_i(x_i) F_j(x_j) |T_n^{ij}|^2 \Phi_n (dx_i/x_i)(dx_j/x_j) , \qquad (4)$$

where $\hat{s} = x_1 x_2 s$, Φ_n is the n-body phase-space factor, and T_n^{ij} is the QCD matrix element. As a result of the bremsstrahlung nature of the gluon radiation, T_n^{ij} diverges when the mass of one of the parton pairs in the initial or final state approaches zero. These divergences are cancelled by non-leading contributions to the cross-sections associated with lower parton multiplicities in the final state. The experimental configurations close to the diverging limits are avoided in the data selection by requiring the jets to be well separated from the incoming partons and from each other. The latter condition ensures also a good definition of the jets.

Owing to several additional factors the situation is more complex than that indicated by Eq. (4):

i) Higher order contributions cause the n jet cross-section to be larger than the leading-order expression by a factor K_n. In principle, the value of K_n can be calculated, but this has not yet been done. Experience with other processes suggests that K_n may take values between 1 and 2. Lacking precise calculations of K_n, the two-jet and three-jet data are compared with expressions proportional to $K_2\alpha_s^2$ and $K_3\alpha_s^3$, respectively, with the result that the ratio is sensitive to the quantity $\alpha_s K_3/K_2$ instead of simply α_s.

ii) The Q^2 dependence of the structure functions and of α_s cause sizeable non-scaling effects as seen in Section 2.3. In particular, the strong coupling constant α_s can be written as:

$$\alpha_s = 12\pi/(33 - 2N_f) \ln (Q^2/\Lambda^2) \tag{5}$$

with N_f equal to 5. For a given process, the definition of Q^2 is somewhat arbitrary. Different definitions of Q^2 will yield different values of the scale parameter Λ, but may not change significantly the measured value of α_s provided that the same Q^2 range is used to evaluate the two-jet and three-jet cross-ssections.

iii) Individual jets can be lost either by merging several jets into one or by not being detected.

The detailed analysis of the UA1 and UA2 Collaborations[17,18] are reported elsewhere. The results are

$$\alpha_s K_3/K_2 = 0.162 \pm 0.024 \text{ (stat.)} \pm 0.03 \text{ (syst.)}$$
$$\alpha_s K_3/K_2 = 0.23 \ \pm 0.01 \ \ \text{(stat.)} \pm 0.04 \text{ (syst.)} \tag{6}$$

for UA1 and UA2, respectively.

The large systematic errors aim at taking into account the uncertainties due to the effects mentioned above. The UA2 analysis uses the same Q^2 scale for two-jet and three-jet events, whereas UA1 assumes a different Q^2 scale for the two types of processes, which reduces the $\alpha_s K_3/K_2$ value from 0.226 (same Q^2 scale) to 0.162 (preferred UA1 result). The quoted systematic errors do not include the uncertainty due to the ambiguity in the choice of the Q^2 scale.

The results from $e^+ e^-$ experiments extrapolated to the Collider Q^2 range yield values of about 0.14 for α_s with large systematic variations. The difference between this value and the UA2 result for $\alpha_s K_3/K_2$ may be due to the effect of higher order corrections or to a different Q^2 scale in two-jet and three-jet processes. The latter point of view is advocated by UA1 in choosing their preferred value.

The above discussion shows that conceptual rather than instrumental limitations preclude a more accurate and more reliable measurement using the Collider data: a calculation of K_3/K_2 and a deeper understanding of the relevant Q^2 scales are the next questions to be addressed.

3. PHYSICS OF THE INTERMEDIATE VECTOR BOSONS

3.1 Introduction

The analysis of the data collected by the UA1 and UA2 experiments at $\sqrt{s}$ = 546 GeV in 1982 and 1983 has established the existence of the $W^{\pm}$ and Z^0 intermediate bosons[4-7]. The study of their properties has shown good agreement with the expectations of the Standard Model. However, a few phenomena — possible hints for new physics — were observed, such as the excess of $Z \rightarrow ee\gamma$ decays and events with an electron, large missing transverse energy, and hard jets[19]. The analysis of the data collected in 1984 at $\sqrt{s}$ = 630 GeV with an increase by a factor of 2.5 of the integrated luminosity has not confirmed these observations. Therefore the interpretation of these phenomena in terms of conventional processes has become more likely.

The larger statistics available with the 1984 run has made it possible to confront the experimental results with the theoretical predictions in a quantitative way, especially for the $W \rightarrow e\nu$ channel with data from both UA1 and UA2. The $W \rightarrow \mu\nu$, $Z^0 \rightarrow \mu\mu$, and $W \rightarrow \tau\nu$ decays have been observed by UA1.

3.2 Electron Identification

In a search for high-p_T electrons at the Collider the main background results from hadronic jets consisting of one or more high-p_T π^0's and one charged particle, which could be either a charged jet fragment or an electron from photon conversion. Such a configuration is hard to distinguish from a genuine electron in the UA1 and UA2 detectors.

As a consequence, both experiments use somewhat similar electron-identification criteria, which aim at a high rejection of jets while maintaining a reasonably high efficiency for electrons. These criteria require the energy deposition in the calorimeter to match that expected from an isolated electron, which implies that a very large fraction of events containing electrons inside or near a jet are rejected. In the following, we describe the main cuts used to select an inclusive electron sample.

i) The transverse energy E_T associated with the energy cluster in the calorimeter is required to exceed a given threshold. For the data discussed here this threshold is set at 15 GeV for both experiments.

ii) The shower energy leakage into the hadronic compartment of the calorimeter is required to be small.

iii) In the UA1 experiment the longitudinal development of the electromagnetic shower, measured over four samples, is required to be compatible with that expected for an electron. In the UA2 experiment, the lateral shower profile is required to be small, using the small cell size of the calorimeters.

iv) The presence of a charged particle track pointing to the energy cluster in the calorimeter is required.

v) In the UA1 experiment the momentum measurement is used to ensure that the track transverse momentum be larger than 7 GeV/c (or compatible with 15 GeV/c within 3σ). The UA2 apparatus has magnetic field only in the two forward regions $20° < \theta < 37.5°$ with respect to the beams. In these regions the track momentum is required to match the particle energy, as measured in the calorimeter, within 4σ.

vi) A distinctive feature of the UA2 detector is the presence of a «preshower» counter (a 1.5 radiation length thick converter, followed by a proportional chamber) in front of the calorimeters. This counter is used to verify that the electromagnetic shower is initiated in the converter and that its position in space matches accurately with the measured charged-particle track, as expected for electrons.

vii) Finally, an explicit isolation criterion is applied. In UA1 this is done by imposing an upper limit on the amount of transverse energy (typically less than 3 GeV) associated with charged particles and calorimeter cells contained in a cone of $\sim 40°$ half-angle around the electron track. In UA2 this cone has a half-angle of typically $\sim 15°$.

The combination of all these cuts is estimated to be $\sim 75\%$ efficient for isolated electrons in both experiments.

3.3   Neutrino Identification

The presence of a non-interacting high-p_T neutrino in the final state is characteristic of $W \rightarrow e\nu$ decays. Since a large fraction of the total collision energy is carried by particles at very small angles, which cannot be detected because they remain inside the machine vacuum pipe, only the missing transverse momentum, $\vec{p}_T^{\,miss}$, can be reliably measured. For events containing an electron candidate, $\vec{p}_T^{\,miss}$ is identified with the neutrino transverse momentum, $\vec{p}_T^{\,\nu}$.

In the UA1 experiment, for events containing an electron candidate of transverse momentum $\vec{p}_T^{\,e}$, $p_T^{\,\nu}$ is defined as

$$\vec{p}_T^{\,\nu} = -\vec{p}_T^{\,e} - \sum_i \vec{p}_T^{\,i}, \tag{7}$$

where $\vec{p}_T^{\,i}$ is a vector with magnitude equal to the energy deposited in the i-th cell of the calorimeter, and directed from the event vertex to the estimated impact point on the cell. The sum is extended over all calorimeter cells. However, the central part of the UA1 detector ($|\eta| < 1.5$, where η is the pseudorapidity) has imperfect calorimetry inside a wedge of $\pm 4°$ with respect to the vertical plane containing the beam axis. For this reason, the measurement of $\vec{p}_T^{\,\nu}$ is unreliable whenever $\vec{p}_T^{\,\nu}$ is close to this region. After rejecting such events by a $\pm 15°$ cut , the $p_T^{\,\nu}$ resolution becomes almost Gaussian.

In the case of the UA2 detectors, there is no particle detection at angles $\theta < 20°$ to the beams. Furthermore, the two forward regions ($20° < \theta < 40°$ to the beams) provide only partial detection because of incomplete azimuthal coverage (due to twelve toroid coils) and incomplete hadronic calorimetry. This results in non-Gaussian tails in the $\vec{p}_T^{\,miss}$ resolution. The probability of losing one of the jets in a two-jet event varies between $\sim 10\%$ at $p_T = 15$ GeV/c, to $\sim 2\%$ at $p_T = 40$ GeV/c.

For each event containing an electron candidate, the UA2 definition of $\vec{p}_T^{\,\nu}$ is

$$\vec{p}_T^{\,\nu} = -\vec{p}_T^{\,e} - \sum_j \vec{p}_T^{\,j} - \lambda\vec{P}, \tag{8}$$

where the sum extends to all reconstructed jets with $p_T^{\,j} > 5$ GeV/c, and the vector $\vec{P}$ is the total transverse momentum carried by the system of all other particles not belonging to the jet. The factor λ, of the order of 1.5, is an empirical correction factor which takes into account the non-linearity of the calorimeter response to low-energy particles. Its value is determined by applying the condition $\langle \vec{p}_T^\nu \rangle = 0$ to the sample of $Z \to e^+ e^-$ events observed in UA2.

3.4 The final $W \to e\nu$ Event Samples

Figure 7 shows, for UA1 and UA2, the distribution of the events containing at least one electron candidate with $p_T^e > 15$ GeV/c in the (p_T^e, p_T^ν) plane. In the high-p_T^e region ($p_T^e \gtrsim 25$ GeV/c), signals from $W \to e\nu$ ($p_T^\nu \approx p_T^e$) and $Z \to e^+ e^-$ ($p_T^\nu \approx 0$) are clearly visible above the background of misidentified hadrons, which is dominant at low p_T^e.

These distributions demonstrate clearly that the missing transverse momentum p_T^ν is measured more accurately in UA1 than in UA2, for the reasons given in the previous section. For the UA1 data the cloud of low-p_T^ν background events dies off for $p_T^e > 15$ GeV/c, whereas in the case of UA2 it extends further. For both experiments the $W \to e\nu$ signal at large $p_T^e \cong p_T^\nu$ is clearly visible.

The final UA1 $W \to e\nu$ sample is defined by requiring $p_T^\nu > 15$ GeV/c. In the full data sample, this condition is satisfied by 172 events. Background contributions to this sample are listed in Table 1. The background from misidentified hadrons is estimated from the shape of the p_T^ν resolution. After rejecting events for which either $\vec{p}_T^e$ or $\vec{p}_T^\nu$ points in the direction of the vertical axis within $\pm 15°$, 148 events are left. The p_T^e distribution of these events is shown in Fig. 8a.

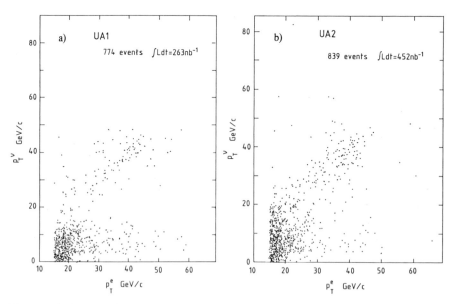

Fig. 7 Distribution of the events containing at least one electron candidate in the p_T^e, p_T^ν plane: a) UA1, 1984 sample; b) UA2, full sample.

Table 1

W → eν event samples and backgrounds

	UA1	UA2
p_T^e threshold (GeV/c)	15	25
Number of events	172	119
Hadronic background	5.3 ± 1.9	5.6 ± 1.7
Z^0 → e (detected) e (undetected)	–	4.4 ± 0.8
W → $\tau\nu_\tau$, τ → $e\nu_e\nu_\tau$	9.1 ± 0.5	1.7 ± 0.2
W → $\tau\nu_\tau$, τ → $\nu_\tau\pi^0$ + hadrons	2.7 ± 0.4	–
W → eν signal	155 ± 13	107 ± 11

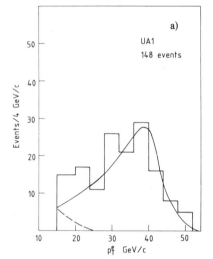
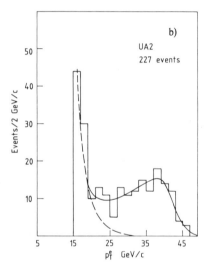

Fig. 8 a) The p_T^e distribution for electron candidates in events having p_T^ν > 15 GeV/c (UA1); b) The p_T^e distribution for electron candidates satisfying the requirement ϱ_{opp} < 0.2 (UA2). Broken curves: estimated hadronic background. Full curves: expected distributions for values of m_W as given by Eqs. (11).

The final UA2 W → eν sample is defined by applying a topological cut which minimizes the background from misidentified hadrons accompanied by a jet at opposite azimuth. As a measure of the fraction of the electron transverse momentum balanced by jets at opposite azimuth, the quantity

$$\varrho_{opp} = -\vec{p}_T^e \sum_j \vec{p}_T^j /|\vec{p}_T^e|^2 \qquad (9)$$

is defined, where the sum extends over all reconstructed jets (if any) with $p_T^j > 3$ GeV/c separated in azimuth from $\vec{p}_T^e$ by at least 120°. Most W → eν decays belong to the category of events with large p_T imbalance ($\varrho_{opp} \approx 0$), whereas for misidentified high-p_T hadrons values of ϱ_{opp} near unity are expected. Figure 8b shows the p_T^e distribution of all electron candidates satisfying $\varrho_{opp} < 0.2$. Although the Jacobian peak structure is very apparent, the hadronic background is still dominant for $p_T^e < 20$ GeV/c. For this reason the final UA2 W → eν sample consists only of the 119 events having $p_T^e > 25$ GeV/c. Background contributions to this sample are listed in Table 1. The background from misidentified hadrons is estimated from a sample of high-p_T π^0's which have the opposite jet outside the detector acceptance.

We note (see table 1) that the contribution from W → $\tau\nu_\tau$, followed by τ → e$\nu_e\nu_\tau$, is larger in UA1 than in UA2 because of the lower p_T^e threshold used to define the final sample. However, the contribution from Z → e^+e^- decays with one electron outside the detector acceptance, is larger in UA2 because the probability of detecting both electrons is only ~ 60% in UA2, while it is close to 100% in UA1.

3.5   Cross-sections for inclusive W Production

The cross-sections for inclusive W production, followed by the decay W → eν, σ_W^e, are computed in a straightforward way from the number of observed W → eν signal events.

The results obtained by the UA1 and UA2 experiments are listed separately for $\sqrt{s} = $ 546 and 630 GeV in Table 2. The quoted systematic uncertainties arise mainly from the uncertainties on the total luminosity [± 15% in UA1, ± 8% in UA2 which benefits from the measurement of the total cross-section by UA4[20] in the same intersection].

These results are consistent with the corresponding theoretical predictions[21], also given in Table 2. These predictions have large uncertainties arising from uncertainties in the structure functions and higher order QCD corrections.

In contrast, the ratio r = $\sigma_W^e(630$ GeV$)/\sigma_W^e(546$ GeV$)$ has negligible systematic errors and small theoretical uncertainties. The average of the two measurements, r = 1.11 ± 0.15, agrees with the theoretical prediction[21], r = 1.26 ± 0.02.

Table 2

W → eν cross-sections

	σ_W^e (nb) $\sqrt{s}$ = 546 GeV	$\sqrt{s}$ = 630 GeV	$r = \dfrac{\sigma_W^e (630 \text{ GeV})}{\sigma_W^e (546 \text{ GeV})}$
UA1	0.55 ± 0.08 ± 0.09	0.63 ± 0.05 ± 0.09	1.15 ± 0.19
UA2	0.50 ± 0.09 ± 0.05	0.53 ± 0.06 ± 0.05	1.06 ± 0.23
Theory[21]	$0.36 \begin{smallmatrix} +0.11 \\ -0.05 \end{smallmatrix}$	$0.47 \begin{smallmatrix} +0.14 \\ -0.08 \end{smallmatrix}$	1.26 ± 0.02

(The second error is the systematic uncertainty.)

3.6   Determination of the W Mass

To extract a value of the W mass, m_W, from the W → eν event samples, both experiments define for each event a transverse mass m_T

$$m_T^2 = 2p_T^e p_T^\nu (1 - \cos \Delta\Phi) \tag{10}$$

with the property $m_T \leqslant m_W$, where $\Delta\Phi$ is the azimuthal separation between $\vec{p}_T^e$ and $\vec{p}_T^\nu$. A Monte Carlo simulation is used to generate m_T distributions for different values of m_W, and the most probable value of m_W is found by a maximum–likelihood fit to the experimental distributions. This technique has the advantage that m_T is rather insensitive to the W transverse motion, contrary to other variables such as p_T^e or p_T^ν.

In the UA1 experiment a sample of 86 events for which both p_T^e and p_T^ν exceed 30 GeV/c is selected from the 148 events of Fig. 4a. The m_T distribution for this sample, which is practically background-free, is shown in Fig. 9a.

In UA2, the m_T distribution of the entire W → eν sample (119 events) is used (Fig. 9b). The backgrounds discussed in Section 2.3 have a negligible effect on the mass determination, because they are predominantly at small m_T, and the best fit value of m_W depends mainly on the upper edge of the distribution.

The results of the fit are:

(UA1) $m_W = 83.5 \, {}^{+\,1.1}_{-\,1.0}$ (stat.) ± 2.8 (syst.) GeV/c^2

(UA2) $m_W = 81.2 \pm 1.1$ (stat.) ± 1.3 (syst.) GeV/c^2 . $\tag{11}$

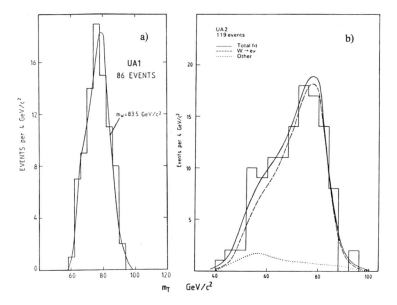

Fig. 9 Electron–neutrino transverse mass distribution for the UA1 (a) and UA2 (b) W → eν event samples. The curves represent best fits to the data.

The systematic errors reflect the uncertainty on the absolute energy scale of the calorimeters, which is $\pm 3\%$ in UA1 and $\pm 1.6\%$ in UA2. These errors are quoted separately because they cancel in the ratio m_W/m_Z. An additional systematic error of ± 0.5 GeV/c^2 in the UA2 result, mainly due to uncertainties in the measurement of p_T^e, has been added in quadrature to the statistical error.

The smaller systematic uncertainty of the UA2 experiment [see Eqs. (11)] results from a better control of the calorimeter calibration, which was performed for all calorimeter cells using beams of known energies before the start of the experiment. This calibration has then been repeated periodically for a fraction of the calorimeter modules.

The expected p_T^e and m_T distributions for the m_W values given by Eqs. (11) are shown in Figs. 8 and 9.

A fit to the m_T distributions using the W width, Γ_W, as a second free parameter, provides a way to obtain an upper limit on Γ_W. The results from the two experiments are $\Gamma_W < 6.5$ GeV/c^2 (UA1) and < 7 GeV/c^2 (UA2) at the 90% confidence level.

3.7   Charge Asymmetry in the Decay $W \to e\nu$

At the energies of the CERN $p\bar{p}$ Collider, W production is dominated by $q\bar{q}$ annihilation involving at least one valence quark or antiquark. As a consequence of the $V - A$ coupling, the helicity of the quarks (antiquarks) is -1 $(+1)$ and the W is almost fully polarized along the $\bar{p}$ beam.

Similar helicity arguments applied to $W \to e\nu$ decay predict that the leptons (e^- or ν_e) should be preferentially emitted opposite to the direction of the W polarization, and antileptons (e^+ or $\bar{\nu}_e$) along it. More precisely, the angular distribution of the charged lepton in the W rest frame has the form dn/d $\cos\theta^* \propto (1 + q \cos\theta^*)^2$, where θ^* is the $e^+ (e^-)$ angle in the W rest frame, measured with respect to the $\bar{p}$ direction and q is the sign of the charge of the lepton. This is true only if the W transverse momentum, p_T^W, is zero. For $p_T^W \neq 0$ the initial parton directions are not known and the Collins–Soper convention is used to define θ^*.

A further complication arises from the fact that p_L^ν is not measured, and the condition that the invariant mass of the $e\nu$ pair be equal to m_W gives two solutions for p_L^ν or p_L^W. The UA1 analysis retains only the events for which one solution is unphysical and the charge sign is unambiguously determined (75 events). Figure 10a shows the $\cos\theta^*$ distribution from UA1, corrected for the detector acceptance. The expected $(1 + q \cos\theta^*)^2$ form agrees well with the data.

In the UA2 experiment only the charge-averaged $\cos\theta^*$ distribution can be measured, because there is no magnetic field over most of the solid angle. The p_L^ν solution corresponding to the smaller value of $|p_L^W|$ is chosen. The experimental data agree well with the form $1 + \cos^2\theta^*$ expected in this case (see Fig. 10b). From the $W \to e\nu$ events detected in the two forward regions, where a magnetic field is present, a charge asymmetry measurement is obtained, $A = 0.43 \pm 0.17$. This value is in good agreement with the result of Monte Carlo calculation, $A = 0.53 \pm 0.06$, which assumes a $V - A$ decay matrix element.

None of those measurements can distinguish between $V - A$ and $V + A$. Such a separation would require a measurement of the lepton helicity.

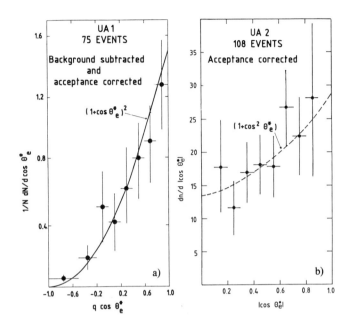

Fig. 10 a) The UA1 distribution of the product $Q \cos \theta^*$ for electrons from $W \rightarrow e\nu$ decay, where $Q = +1(-1)$ for e^+ (e^-), and $\theta^* = 0$ along the $\bar{p}$ beam; b) The $|\cos \theta^*|$ distribution, as measured by UA2.

3.8 Longitudinal Momentum of the W

Figure 11 shows the distribution of the fractional beam momentum carried by the W-boson, $x_W = 2p_L^W/\sqrt{s}$, where the smaller $|p_L^W|$ value is chosen, for the total UA1 sample. This distribution, shown separately for $\sqrt{s} = 546$ and 630 GeV, is expected to reflect the structure functions of the annihilating partons. From the relations

$$x_W = x_p - x_{\bar{p}} \tag{12}$$

$$m_W^2/s = x_p x_{\bar{p}} \quad (\text{if } p_T^W \ll m_W) \tag{13}$$

one can extract the fractional momentum x_p ($x_{\bar{p}}$) of the parton contained in the incident p ($\bar{p}$). Using the events with an unambiguous determination of the charge sign (118 events in the UA1 sample), x_p ($x_{\bar{p}}$) can be identified with the fractional momentum of a u($\bar{d}$) quark for a W^+, and a d($\bar{u}$) quark for a W^-.

Figures 12a and 12b show the resulting u- and d-quark x-distributions, which agree with the expectation from the structure functions as parametrized by Eichten et al.[23].

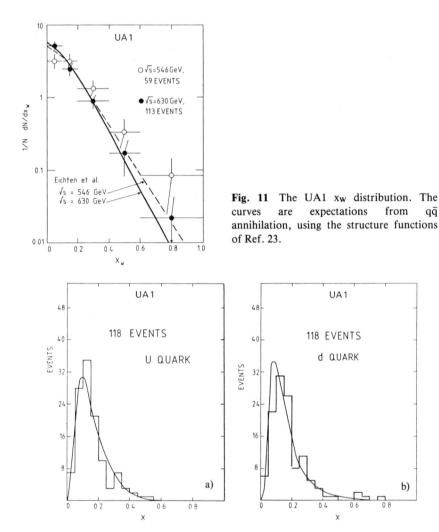

Fig. 11 The UA1 x_W distribution. The curves are expectations from $q\bar{q}$ annihilation, using the structure functions of Ref. 23.

Fig. 12 The x-distributions for (a) u-quarks, and (b) d-quarks, as measured by UA1. The curves are expectations from the structure functions of Ref. 23.

3.9 Transverse Momentum of the W

The W transverse momentum p_T^W is obtained by adding the measured vectors $\vec{p}_T^e$ and $\vec{p}_T^\nu$. In the case of the UA2 sample, the topological cut discussed in Section 2.3 may reject high-p_T^W events if they contain jets emitted opposite to the decay electron in a plane normal to the beams. For this reason the cuts $p_T^e > 25$ GeV/c and $\varrho_{opp} < 0.2$ [see Eq. (3)], used to define the W $\rightarrow e\nu$ sample, are replaced by the cuts $p_T^e > 15$ GeV/c, $p_T^\nu > 25$ GeV/c, and $m_T >$

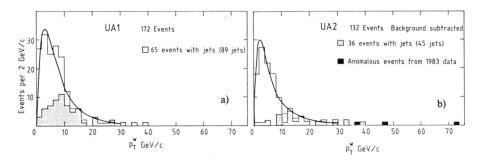

Fig. 13 The p_T^W distribution, as measured by UA1 (a) and UA2 (b). The shaded regions represent events which contain at least one jet with $E_T > 5$ GeV/c. The three events shown as dark areas in (b) are the anomalous UA2 events discussed in Ref. 19. The curves are QCD predictions from Ref. 21.

20 GeV/c². Figure 13 shows the p_T^W distributions for UA1 and UA2. The two sets of data agree well with each other and with QCD predictions[21].

The events containing at least one jet with $E_T > 5$ GeV are shown as dashed areas in Fig. 13. As expected, they correspond to W bosons produced at high values of p_T^W. In UA2, 28% of the W's are observed in association with at least one jet, whereas this fraction is 38% in UA1. This difference reflects the larger acceptance of the UA1 detector.

For the jets of the UA1 sample, the angular distribution $dn/d|\cos \theta^*|$, where θ^* is the angle between the jet and the beam direction in the W jet(s) rest frame, is shown in Fig. 14. This distribution is in agreement with the $(1 - |\cos \theta^*|)^{-1}$ expected for t-channel quark exchange.

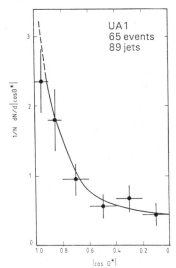

Fig. 14 Distribution of $|\cos \theta^*|$ for jets produced in association with W $\rightarrow$ eν events. The angle θ^* is the jet angle in the W-jet(s) rest frame with respect to the beam axis. Only jets with $E_T > 5$ GeV are considered. The curve is a QCD prediction.

The UA2 Collaboration[19] published the observation of three anomalous events from the 1983 data sample. These events could be interpreted as W → eν produced in association with very hard jets. Two of these events, because of their very large p_T^e values ($p_T^e > 50$ GeV/c) and the very hard jets accompanying the (eν) pair, could not be easily interpreted as originating from known QCD processes. No additional event of this type was observed in the 1984 data, making the interpretation of these events much more likely in terms of standard QCD processes.

3.10   The Z → e⁺e⁻ Event Samples

The electron identification criteria used for W → eν being about 75% efficient in both UA1 and UA2 experiments, their application to both electrons from Z^0 decay would result in the loss of about 50% of the Z^0 → e⁺e⁻ event samples. The selection of electron-pair candidates is therefore performed by using the less stringent selection criteria that both energy clusters should be compatible with an electron from calorimeter information alone, and that at least one cluster should satisfy the full electron identification criteria. The distributions of the invariant mass, m_{ee}, of the electron-pair candidates are shown in Fig. 15. Above a threshold of 20 GeV/c², both distributions show a rapidly falling continuum at mass values of less than 50 GeV/c², and a well-separated peak near $m_{ee} \approx 90$ GeV/c². The events in this peak are interpreted as Z → e⁺e⁻ decays (18 events in UA1, 16 in UA2). Background estimates give less than 0.3 event under the Z peak for both distributions of Fig. 12.

The measured values of the Z mass, m_Z, are obtained from the m_{ee} distributions by a maximum likelihood fit of a Breit–Wigner shape distorted by the experimental mass resolution. The UA1 result, based on 14 well-measured e⁺e⁻ pairs, is

$$m_Z = 93.0 \pm 1.4 \text{ (stat.)} \pm 3.2 \text{ (syst.)} \text{ GeV/c}^2 . \qquad (14)$$

The corresponding UA2 result, based on thirteen well-measured e⁺e⁻ pairs, is

$$m_Z = 92.5 \pm 1.3 \text{ (stat.)} \pm 1.5 \text{ (syst.)} \text{ GeV/c}^2 . \qquad (15)$$

In both cases, the systematic error reflects the uncertainty in the absolute calibration of the calorimeter energy scale.

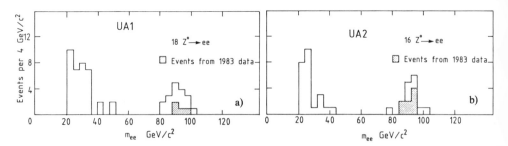

Fig. 15 Invariant mass distribution of electron pairs measured in the UA1 (a) and UA2 (b) experiments. The shaded events are Z → e⁺e⁻ decays observed in the 1982–83 data samples.

3.11   The Z Width and the Number of Neutrino Species

Within the context of the Standard Model, the value of the Z width, Γ_Z, is related to the number of fermion doublets for which the decay $Z \rightarrow f\bar{f}$ is kinematically allowed. Under the assumption that for any additional fermion family only the neutrino is significantly less massive than $m_Z/2$, we can write

$$\Gamma_Z = \Gamma_Z \text{ (three families)} + 0.18 \,\Delta N_\nu \,, \tag{16}$$

where Γ_Z is in GeV/c^2, and ΔN_ν is the number of additional neutrino species.

A value of Γ_Z can be obtained, in principle, by the same maximum likelihood fit as used to determine m_Z. However, in the present experiments the expected width for three fermion families ($\sim 2.8 \text{ GeV}/c^2$) is comparable with the experimental mass resolution. As an example, in the UA2 sample the r.m.s. deviation of the 13 mass values from the value of m_Z given by Eq. (15) is 3.58 GeV/c^2, which is very similar to the average of the measured errors $\langle\sigma\rangle = 3.42 \text{ GeV}/c^2$. Under these circumstances the determination of Γ_Z depends critically on a precise knowledge of both the measurement errors and the shape of the experimental resolution. In the present experiments, given also the small event samples available, this method does not lead to reliable results.

A model-dependent method[24], which does not depend on the mass resolution, consists in measuring the ratio $R = \sigma_Z^e/\sigma_W^e$, which is related to Γ_W and Γ_Z by the equation

$$R = (\sigma_Z/\sigma_W)(\Gamma_Z^{ee}/\Gamma_Z)(\Gamma_W/\Gamma_W^{e\nu}) \,, \tag{17}$$

where $\sigma_Z(\sigma_W)$ is the inclusive cross-section for Z(W) production, and $\Gamma_Z^{ee}(\Gamma_W^{e\nu})$ is the partial width for the decay $Z \rightarrow e^+e^-$ ($W \rightarrow e\nu$). While Γ_Z^{ee} and $\Gamma_W^{e\nu}$ are given directly by the Standard Model, the ratio σ_Z/σ_W can be calculated in the framework of QCD, with the property that most theoretical uncertainties cancel out. A recent extimate[21] gives $\sigma_Z/\sigma_W = 0.30 \pm 0.02$.

The measured values of R are

$$R = 0.108 \begin{array}{c} + 0.025 \\ - 0.026 \end{array} \tag{18}$$

from UA1, and

$$R = 0.136 \begin{array}{c} + 0.041 \\ - 0.033 \end{array} \tag{19}$$

from UA2, where in both cases the errors are dominated by statistics because the value of the integrated luminosity cancels out. Both values agree with the expectation[25] $R = 0.11 \pm 0.01$, based on three fermion families. Taking the average of the UA1 and UA2 results we obtain:

$$R_{av} = 0.118 \pm 0.020 \,. \tag{20}$$

By making the additional assumption that the masses of the charged fermions of any new family are sufficiently heavy not to affect the value of Γ_W, it becomes possible to determine Γ_Z and ΔN_ν using Eqs. (16) and (17). The lower limit $R_{av} > 0.094$ (90% confidence level) corresponds to the limit $\Gamma_Z < 3.17 \pm 0.31 \text{ GeV}/c^2$, which gives $\Delta N_\nu < 2.6 \pm 1.7$, where the errors reflect the theoretical uncertainties.

3.12 Radiative Z Decays

Among the four $Z \to e^+e^-$ events recorded by UA1 in 1983, one event of the type $Z \to e^+e^-\gamma$ was observed[26]. This event contained a 39 GeV photon separated in space by $14° \pm 4°$ from an electron of 9 GeV. A similar event, with a 24 GeV photon separated in space by $31° \pm 1°$ from an electron of 11 GeV, was also present among the eight events recorded by UA2[27]. The probability of observing these $ee\gamma$ configurations, or less likely ones, as a result of internal bremsstrahlung, in a total sample of 12 events was estimated to be ~ 0.6%.

From the analysis of the 1984 data, the total $Z \to e^+e^-$ (or $e^+e^-\gamma$) sample has increased from 12 to 34 events. However, no new event compatible with $Z \to e^+e^-\gamma$ decay has been observed. As a consequence, the probability of the internal bremsstrahlung hypothesis has increased from ~ 0.6% to ~ 4%, a tolerable level.

3.13 Comparison with the Standard Model

In order to compare the measurements of m_W and m_Z with the predictions of the Standard Model, we must use suitably renormalized and radiatively-corrected quantities[28]. We shall use the scheme where $\sin^2 \theta_W$ is defined as[29]

$$\sin^2 \theta_W = 1 - (m_W/m_Z)^2 , \tag{21}$$

which leads to the following predictions

$$m_W^2 = A^2/[(1 - \Delta r) \sin^2 \theta_W] \tag{22}$$

$$m_Z^2 = 4A^2/[(1 - \Delta r) \sin^2 2\theta_W] , \tag{23}$$

where $A = (\pi\alpha/\sqrt{2}G_F)^{1/2} = (37.2810 \pm 0.0003)$ GeV/c^2, using the measured values of α and G_F. The quantity Δr reflects the effect of one-loop radiative corrections, and has been computed to be[29] $\Delta r = 0.0696 \pm 0.020$ for a mass of the top quark $m_t = 36$ GeV/c^2 and assuming that the mass of the Higgs boson, m_H, is equal to m_Z.

Using Eqs. (22) and (23) we can extract two values of $\sin^2 \theta_W$ from the values of m_W and m_Z measured in each experiment. We then combine them to obtain

$$\sin^2 \theta_W = 0.216 \begin{array}{c} + 0.004 \\ - 0.005 \end{array} \text{(stat.)} \pm 0.014 \text{ (syst.)} \tag{24}$$

from UA1; and

$$\sin^2 \theta_W = 0.226 \pm 0.005 \text{ (stat.)} \pm 0.008 \text{ (syst.)} \tag{25}$$

from UA2. In both cases the systematic error reflects the uncertainty on the mass scale from the uncertainty of the calorimeter calibration.

Within errors, the two values of $\sin^2 \theta_W$ agree with each other, and also with the value $\sin^2 \theta_W = 0.220 \pm 0.008$ obtained from an average of low-energy data[30] after applying radiative corrections. The low-energy data include recent results from the CDHS[31] and CCCFRR[32] experiments.

By using Eq. (21) it is possible to measure $\sin^2 \theta_W$ with no systematic error from the mass scale. The results are

$$\sin^2 \theta_W = 0.194 \pm 0.031 \tag{26}$$

from UA1, and

$$\sin^2 \theta_W = 0.229 \pm 0.030 \tag{27}$$

from UA2. The weighted average of these two results is

$$\sin^2 \theta_W = 0.212 \pm 0.022 , \tag{28}$$

which represents a less precise measurement than the results given by Eqs. (24) and (25), owing to the limited event samples available at present.

By using the $\sin^2 \theta_W$ definition given by Eq. (21) we have implicitly assumed that the ϱ parameter, defined as[33)]

$$\varrho = m_W^2/(m_Z^2 \cos^2 \theta_W) , \tag{29}$$

is equal to 1. We can test this assumption by combining Eqs. (22) and (29) to obtain

$$\varrho = m_W^2/[m_Z^2(1 - B^2/m_W^2)] , \tag{30}$$

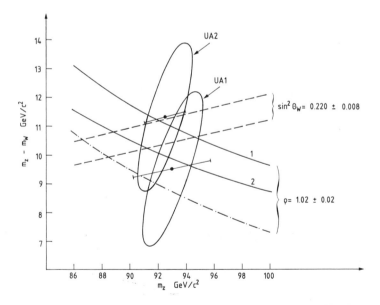

Fig. 16 68% confidence level contours in the plot of m_Z-m_W versus m_Z taking into account the statistical errors only. The error bars applied to the centre of the ellipses represent the translations allowed by the magnitude of the systematic errors. Curve 1 (2) is the prediction of the Standard Model with (without) radiative corrections. The region between the two dashed lines is the region allowed by the low-energy measurement of $\sin^2 \theta_W$; that between curve 1 and the dash-dotted line is allowed by the low-energy measurement of ϱ.

where $B^2 = A^2/(1 - \Delta r)$. The UA1 result is

$$\varrho = 1.028 \pm 0.037 \text{ (stat.)} \pm 0.019 \text{ (syst.)} \tag{31}$$

and the corresponding UA2 result is

$$\varrho = 0.996 \pm 0.033 \text{ (stat.)} \pm 0.009 \text{ (syst.)} . \tag{32}$$

These two results agree with each other and with the value $\varrho = 1.02 \pm 0.02$ (see the compilations of Refs. (29) and (34)]. We recall that the value $\varrho = 1$ corresponds to the minimal Standard Model with only one isodoublet of complex Higgs fields.

The sensitivity of the measurements of m_W and m_Z to the radiative corrections is illustrated in Fig. 16, which shows the 68% confidence level contours from the two experiments in the plot of $m_Z - m_W$ versus m_Z. Also shown in Fig. 16 are the ranges of $\sin^2 \theta_W$ and ϱ allowed by the low-energy measurements. Within the present statistical and systematic errors, the need for radiative corrections in the Standard Model cannot be demonstrated.

4.   CONCLUSIONS

After three years of very sucessful operation of the Collider, the analysis of the data collected by the UA1 and UA2 experiments allows very meaningful comparisons with the theoretical predictions.

The agreement between the measured W and Z properties, with, for example, a total sample of more than 250 W $\rightarrow$ eν decays, and the predictions of the SU(3) $\times$ SU(2) $\times$ U(1) Standard Model is remarkable. The unexplained effects which had been observed by the UA1 and UA2 collaborations in the 1982–83 data have not been confirmed by the analysis of the three times larger 1984 data samples.

The jet results are in agreement with the QCD predictions, although, in that case, the uncertainties, both theoretical and experimental, are an important limiting factor. In particular, theoretical calculations of higher order terms in the jet production are urgently needed.

In the immediate future (1985–86), the integrated luminosities collected by UA1 and UA2 will be increased by a factor of 2 to 3 with respect to the present ones. In 1987, after a shutdown of a little more than a year, the Collider will resume operation with the new antiproton collector (ACOL) providing a factor of 10 increase in luminosity. The UA1 and UA2 detectors will also be upgraded. By the end of 1988 an integrated luminosity of about 10 pb^{-1} should have been collected by each experiment.

ACKNOWLEDGEMENTS

I am grateful to P. Bagnaia, D. Froidevaux, L. Di Lella and J. Sass, whose reviews I have used in preparing this report. I should also like to thank the Scientific Reports Editing and Text Processing Sections for their remarkable competence and efficiency.

References

1) The staff of the CERN proton–antiproton project, Phys. Lett. **107B**, 306 (1981).

2) Banner, M. et al., Phys. Lett. **118B**, 203 (1982).

3) Arnison, G. et al., Phys. Lett. **123B**, 115 (1983).

4) Arnison, G. et al., Phys. Lett. **122B**, 103 (1983).

5) Banner, M. et al., Phys. Lett. **122B**, 476 (1983).

6) Arnison, G. et al., Phys. Lett. **126B**, 398 (1983).

7) Bagnaia, P. et al., Phys. Lett. **129B**, 130 (1983).

8) Feynman, R.P., Photon Hadron Interactions (Benjamin, New York, 1972);
Berman, S.M. and Jacob, M., Phys. Rev. Lett. **25**, 1683 (1970);
Berman, S.M., Bjorken, J.D. and Kogut, J.B., Phys. Rev. **D4**, 3388 (1971).

9) Arnison, G. et al., Phys. Lett. **132B**, 214 (1983).

10) Bagnaia, P. et al., Z. Phys. **C20**, 117 (1983).

11) Bagnaia, P. et al., Phys. Lett. **138B**, 430 (1984).

12) Horgan, R. and Jacob, M., Nucl. Phys. **B179**, 441 (1981).
Furmanski, W. and Kowalski, H., Nucl. Phys. **B224**, 523 (1983).
Kunszt, Z. and Pietarinen, E., Phys. Lett. **132B**, 453 (1983).
Antoniou, N.G. et al., Phys. Lett. **128B**, 257 (1983).
Humpert, B., Z. Phys. **C27**, 257 (1985).
Humpert, B., Phys. Lett. **140B**, 105 (1984).

13) Appel, J.A. et al., Phys. Lett. **160B**, 349 (1985).

14) Arnison, G. et al., Phys. Lett. **136B**, 294 (1984).

15) Bagnaia, P. et al., Phys. Lett. **144B**, 283 (1984).

16) Abramowicz, H. et al., Z. Phys. **C12**, 289 (1982); **C17**, 283 (1983); **C25**, 29 (1984).

17) Arnison, G. et al., Phys. Lett. **158B**, 494 (1985).

18) The UA2 Collaboration, Results on three-jet events in the UA2 detector at the CERN p̄p Collider, presented by P. Bagnaia at the 5th Int. Conf. on Physics in Collisions, Autun, France, July 1985.

19) Bagnaia, P. et al., Phys. Lett. **139B**, 105 (1984).

20) Bozzo, M. et al., Phys. Lett. **147B**, 392 (1984).

21) Altarelli, G. et al., Nucl. Phys. **B246**, 12 (1984).
Altarelli, G., Ellis, R.K. and Martinelli, G., Z. Phys. **C27**, 617 (1985).

22) Collins, J.C. and Soper, D.E., Nucl. Phys. **B193**, 381 (1981); **B194**, 445 (1982); **B197**, 446 (1982).

23) Eichten, E. et al., Rev. Mod. Phys. **56**, 579 (1984).

24) Halzen, F. and Mursula, K., Phys. Rev. Lett. **51**, 857 (1983).
Hikasa, K., Phys. Rev. **D29**, 1939 (1984).

25) Deshpande, V.G. et al., Phys. Rev. Lett. **54**, 1757 (1985).

26) Arnison, G. et al., Phys. Lett. **135B**, 250 (1984).

27) Bagnaia, P. et al., Z. Phys. **C24**, 1 (1984).

28) Sirlin, A., Phys. Rev. **D22**, 971 (1980).
Marciano, W.J., Phys. Rev. **D20**, 274 (1979).
Veltman, M., Phys. Lett. **91B**, 95 (1980).
Antonelli, F. et al., Phys. Lett. **91B**, 90 (1980).

29) Marciano, W.J. and Sirlin, A., Phys. Rev. **D29**, 945 (1984).

30) Marciano, W.J. and Sirlin, A., Nucl. Phys. **B189**, 442 (1981).
For a more recent review, see J. Panman, preprint CERN–EP/85–35 (1985).

31) Abramowicz, H. et al., Z. Phys. **C28**, 51 (1985).

32) Reutens, P.G. et al., Phys. Lett. **152B**, 404 (1985).

33) Ross, D. and Veltman, M., Nucl. Phys. **B95**, 135 (1975).
Hung, P.Q., and Sakurai, J.J., Nucl. Phys. **B143**, 81 (1978).

34) Kim, J. et al., Rev. Mod. Phys. **53**, 211 (1980).
For more recent reviews, see: Langacker, P., Proc. 22nd Int. Conf. on High-Energy Physics, Leipzig, 1984, eds. Meyer, A. and Wieczorek, E. (Akademie der Wissenschaften der DDR, Zeuthen, 1984), p. 215.
Pullia, A., Proc. Fifty Years of Weak Interaction Physics, eds. Bertin, A., Ricci, R.A. and Vitale, A. (Italian Physical Society, Bologna, 1984), p. 333.

UNIFIED STRING THEORIES

David J. Gross[*]

Joseph Henry Laboratories
Princeton University
Princeton, New Jersey 08544

I.   INTRODUCTION

High energy physics is, at present, in an unusual state. It has
been clear for some time that we have succeeded in achieving many of
the original goals of particle physics. We have constructed theories
of the strong, weak and elecromagnetic interactions and have understood
the basic constituents of matter and their interactions. The "standard
model" has been remarkably successful and seems to be an accurate and
complete description of physics, at least at energies below a Tev.
Indeed, as we have heard in the experimental talks at this meeting,
there are at the moment no significant experimental results that cannot
be explained by the color gauge theory of the strong interactions (QCD)
and the electroweak gauge theory. New experiments continue to confirm
the predictions of these theories and no new phenomenon have appeared.

This success has not left us sanguine. Our present theories
contain too many arbitrary parameters and unexplained patterns to be
complete. They do not satisfactorily explain the dynamics of chiral
symmetry breaking or of CP symmetry breaking. The strong and electroweak interactions cry out for unification. Finally we must ultimately
face up to including quantum gravity within the theory. However, we
theorists are in the unfortunate situation of having to address these

[*]Research supported in part by NSF Grant PHY80-19754.

questions without the aid of experimental clues. Furthermore extrapolation of present theory and early attempts at unification suggest
that the natural scale of unification is 10^{16} Gev or greater, tantalizingly close to the Planck mass scale of 10^{19} Gev. It seems very likely
that the next major advance in unification will include gravity. I do
not mean to suggest that new physics will not appear in the range of
Tev energies. Almost all attempts at unification do, in fact, predict
a multitude of new particles and effects that could show up in the Tev
domain (Higgs particles, Supersymmetric partners, etc.), whose discovery and exploration is of the utmost importance. But the truly new
threshold might lie in the totally inaccessible Planckian domain.

In this unfortunate circumstance, when theorists are not provided
with new experimental clues and paradoxes, they are forced to adopt new
strategies. Given the lessons of the past decades it is no surprise
that much of exploratory particle theory is devoted to the search for
new symmetries. However it is not enough simply to dream up new symmetries, one must also explain why these symmetries are not apparent,
why they have been heretofore hidden from our view. This often
requires both the discovery of new and hidden degrees of freedom as
well as mechanisms for the dynamical breaking of the symmetry.

Some of this effort is based on straightforward extrapolations of
establisshed symmetries and dynamics, as in the search for grand unified theories (SU_5, SO_{10}, E_6,...), or in the development of a predictive theory of dynamical chiral gauge symmetry breaking (technicolor,
preons, ...). Ultimately more promising, however, are the suggestions
for radically new symmetries and degrees of freedom.

First there is supersymmetry, a radical and beautiful extension
of space-time symmetries to include fermionic charges. This symmetry
principle has the potential to drastically reduce the number of free
parameters. Most of all it offers an explanation for the existence
of fermionic matter, quarks and leptons, as compelling as the argument
that the existence of gauge mesons follows from local gauge symmetry.

An even greater enlargement of symmetry, and of hidden degrees of
freedom is envisaged in the attempts to revive the idea of Kaluza and
Klein, wherein space itself contains new, hidden, dimensions. These

new degrees of freedom are hidden from us due to the spontaneous compactification of the new spatial dimensions, which partially breaks many of the space-time symmetries of the larger manifold. Although strange at first, the notion of extra spatial dimensions is quite reasonable when viewed this way. The number of spatial dimensions is clearly an experimental question. Since we would expect the compact dimensions to have sizes of order the Planck length there clearly would be no way to directly observe many (say six) extra dimensions. The existence of such extra dimensions is not without consequence. The unbroken isometries of the hidden, compact, dimensions can yield a gravitational explanation for the emergence of gauge symmetries (and, in supergravity theories, the existence of fermionic matter). A combination of supergravity and Kaluza-Klein thus has the potential of providing a truly unified theory of gravity and matter, which can provide an explanation of the known low energy gauge theory of matter and predict its full particle content.

Attempts to utilize these new symmetries in the context of ordinary QFT, however, have reached an impasse. The problems one encounters are most severe if one attempts to be very ambitious and contemplate a unified theory of pure supergravity (in, say, 11 dimensions), which would yield the observed low energy gauge group and fermionic spectrum upon compactification. First of all we do not have a satisfactory quantum theory of gravity, even at the perturbative level. Einstein's theory of gravity, as well as its supersymmetric extensions, is nonrenormalizable. We know that that means that there must be new physics at the Planck length. We are clearly treading on thin ice if we attempt to use this potentially inconsistent theory as the basis for unification.

Even if we ignore this issue, and focus on the low energy structure of such theories, it appears to be impossible to construct realistic theories without a great loss of predictive power. The primary obstacle is the existence of chiral fermions (i.e. the fact that the weak interactions are V-A in structure). In order to generate the observed spectrum of chiral quarks and leptons it appears to be necessary to retreat from the most ambitious Kaluza-Klein program, which would uniquely determine the low energy gauge group as isometries of some compact space and introduce gauge fields by hand. Furthermore the supergravity theories ubiquitously produce a world which would have an intolerably large cosmological constant. Finally no realistic and compelling model has emerged. This brings us to string theories which offer a way out of this impasse.

II. STRING THEORIES

String theories offer a way of realizing the potential of super-symmetry, Kaluza-Klein and much more. They represent a radical departure from ordinary quantum field theory, but in the direction of increased symmetry and structure. They are based on an enormous increase in the number of degrees of freedom, since in addition to fermionic coordinates and extra dimensions, the basic entities are extended one dimensional objects instead of points. Correspondingly the symmetry group is greatly enlarged, in a way that we are only beginning to comprehend. At the very least this extended symmetry contains the largest group of symmetries that can be contemplated within the framework of point field theories--those of ten-dimensional supergravity and super Yang-Mills theory.

The origin of these symmetries can be traced back to the geometrical invariance of the dynamics of propagating strings. Traditionally string theories are constructed by the first quantization of a classical relativistic one dimensional object, whose motion is determined by requiring that the invariant area of the world sheet it sweeps out in space-time is extremized. In this picture the dynamical degrees of freedom of the string are its coordinates, $X_\mu(\sigma,\tau)$ (plus fermionic coordinates in the superstring), which describe its position in space time. The symmetries of the resulting theory are all consequences of the reparametrization invariance of σ,τ parameters which label the world sheet. As a consequence of these symmetries one finds that the free string contains massless gauge bosons. The closed string automatically contains a massless spin two meson, which can be identified as the graviton whereas the open string, which has ends to which charges can be attached, yields massless vector mesons which can be identified as Yang-Mills gauge bosons.

String theories are inherently theories of gravity. Unlike ordinary quantum field theory we do not have the option of turning off gravity. The gravitational, or closed string, sector of the theory must always be present for consistency even if one starts by considering only open strings, since these can join at their ends to form closed strings. One could even imagine discovering the graviton in the attempt to construct string theories of matter. In fact this was the course of events for the dual resonance models where the graviton (then called the Pomeron) was discovered as a bound state of open strings. Most exciting is that string theories provide for the first time a consistent, even finite, theory of gravity. The problem of ultraviolet divergences is bypassed in string theories which contain no short distance infinities. This is not too surprising considering the extended nature of strings, which softens their interactions. Alternatively one notes that interactions are introduced into string theory by allowing the string coordinates, which are two dimensional fields, to propagate on world sheets with nontrivial topology that describe strings splitting and joining. From this first quantized point of view one does not introduce an interaction at all, one just adds handles or holes to the world sheet of the free string. As long as reparametrization invariance is maintained there are simply no possible counterterms. In fact all the divergences that have ever appeared in string theories can be traced to infrared divergences that are a consequence of vacuum instability. All string theories contain a massless partner of the graviton called the dilaton. If one constructs a string theory about a trial vacuum state in which the dilaton has a nonvanishing vacuum expectation value, then infrared infinities will occur due to massless dilaton tadpoles. These divergences however are just a sign of the instability of the original trial vacuum. This is the source of the divergences that occur in one loop diagrams in the old bosonic string theories (the Veneziano model). Superstring theories have vanishing dilaton tadpoles, at least to one-loop order. Therefore both the superstring and the heterotic string are explicitly finite to one loop order and there are strong arguments that this persists to all orders!

String theories, as befits unified theories of physics, are incredibly unique. In principle they contain no freely adjustable parameters and all physical quantities should be calculable in terms of h, c, and m_{planck}. In practice we are not yet in the position to exploit this enormous predictive power. The fine structure constant α, for example, appears in the theory in the form $\alpha \exp(-D)$, where D is the aforementioned dilaton field. Now the value of this field is undetermined to all orders in perturbation theory (it has a "flat potential"). Thus we are free to choose its value, thereby choosing one of an infinite number of degenerate vacuum states, and thus to adjust α as desired. Ultimately we might believe that string dynamics will determine the value of D uniquely, presumably by a nonperturbative mechanism, and thereby eliminate the nonuniqueness of the choice of vacuum state. In that case all dimensionless parameters will be calculable. Even more, string theories determine in a rather unique fashion the gauge group of the world and fix the number of space-time dimensions to be ten.

Finally and most importantly, string theories lead to phenomeno-logically attractive unified theories, which could very well describe the real world.

III. CONSISTENT STRING THEORIES

The number of consistent string theories is extremely small, the number of phenomenologically attractive theories even smaller. First there are the closed superstrings, of which there are two consistent versions. These are theories which contain only closed strings which have no ends to which to attach charges and are thus inherently neutral objects. At low energies, compared to the mass scale of the theory which we can identify as the Planck mass, we only see the massless states of the theory which are those of ten dimensional supergravity. One version of this theory is non-chiral and of no interest since it could never reproduce the observed chiral nature of low energy physics.

The other version is chiral. One might then worry that it would suffer from anomalies, which is indeed the fate of almost all chiral supergravity theories in ten dimensions. Remarkably the particular supergravity theory contained within the chiral superstring is the unique anomaly free theory in ten dimensins. It however contains no gauge interactions in ten dimensions and could only produce such as a consequence of compactification. This approach raises the same problems of reproducing chiral fermions that plagued field theoretic Kaluza-Klein models and has not attracted much attention.

Open string theories, on the other hand, allow the introduction of gauge groups by the time honored method of attaching charges to the ends of the strings. String theories of this type can be constructed which yield, at low energies, N=1 supergravity with any $SO(N)$ or $Sp(2N)$ Yang-Mills group. These however, in addition to being somewhat arbitrary, were suspected to be anomalous. The discovery by Green and Schwarz, last summer, that for a particular gauge group--SO_{32}--the would be anomalies cancel, greatly increased the phenomenological prospects of unified string theories.

The anomaly cancellation mechanism of Green and Schwarz can be understood in terms of the low energy field theory that emerges from the superstring, which is a slightly modified form of d=10 supergravity. One finds that the dangerous Lorentz and gauge anomalies cancel, if and only if, the gauge group is SO_{32} or $E_8 \times E_8$. The ordinary superstring theory cannot incorporate $E_8 \times E_8$. The apparent correspondence between the low energy limit of anomaly free superstring theories and anomaly free supergravity theories provided the motivation that led to the discovery of a new string theory, by J. Harvey, E. Martinec, R. Rohm and myself, whose low energy limit contained an $E_8 \times E_8$ gauge group --the heterotic string. The heterotic string is a closed string theory that produced by a stringy generalization of the Kaluza-Klein mechanism of compactification, gauge interactions. These are determined by consistency to be $E_8 \times E_8$ or Spin $32/Z_2$. It is of more than academic interest to construct this theory since its phenomenological prospects are much brighter.

IV.   THE HETEROTIC STRING

Previously known string theories are the bosonic theory in 26 dimensions (the Veneziano model) and the fermionic, superstring theory in ten dimensions (an outgrowth of the Ramond-Neveu-Schwarz string). The new string theory is constructed as a chiral hybrid of these. To see how this is possible let us recall how string theories are con- structed.

Free string theories are constructed by first quantization of an action given by the invariant area of the world sheet swept out by the string, or by its supersymmetric generalization. For the bosonic string the action is

$$S = -\int d\tau d\sigma \ \frac{1}{4\pi\alpha'} \ [\eta_{\alpha\beta} g^{ab} \partial_a X^\alpha \partial_b X^\beta], \tag{1}$$

where $X^\alpha(\sigma,\tau)$ labels the space time position of the string, embedded in some D dimensional manifold ($\alpha=1,2,\ldots,D$), with σ,τ labeling the world sheet that the string sweeps out. It appears to be possible to consruct consistent string theoreis as long as the above two dimensional sigma-model is conformally invariant. For the moment we take the big space to be flat, so that η_{ab} is the Minkowski metric. This is essentially a choice of vacuum for the quantum string theory. In order to describe the real world however one will be interested in non flat D dimensional manifolds. The reparametrization invariance of the action (in σ,τ) permits one to choose the metric of the world sheet to be conformally flat and in which the timelike parameter of the world sheet, τ, is identified with light cone time. In this light cone gauge the theory reduces to a two-dimensional free field theory of the physical degrees of freedom--the transverse coordinates of the string, subject to constraints. This procedure is valid however only in the critical dimension of 26 for the bosonic string and 10 for the fermionic string.   In other dimensions of space time the existence of conformal anomalies imply that the conformal degree of freedom of the internal metric does not decouple. If it is ignored there is a breakdown of world sheet reparametrization invariance.

In the critical dimension the physial degrees of freedom, being massless two-dimensional fields, can be decomposed into right and left movers, i.e. functions of $\tau-\sigma$ and $\tau+\sigma$ respectively. If we consider only closed strings then the right and left movers never mix. This separation is maintained even in the presence of string interactions, as long as we allow only orientable world sheets on which a handedness can be defined. This is because the interactions between closed strings are constructed, order by order in perturbation theory, by simply modifying the topology of the world sheet on which the strings propagate. In terms of the first quantized two-dimensional theory no interaction is thereby introduced; the right and left movers still propagate freely and independently as massless fields. Thus there is in principle no obstacle to constructing the right and left moving sectors of a closed string in a different fashion, as long as each sector is separately consistent, and together can be regarded as a string embedded in ordinary space-time. This is the idea behind the construction of the heterotic string, which combines the right movers of the fermionic superstring with the left movers of the bosonic string. It is necessarily a theory of closed and orientable strings, since one can clearly distinguish an orientation on such a string. In some sense the heterotic string is inherently chiral; indeed we do not have the option, present in other closed string theories, of constructing a left-right symmetric theory.

The physical degrees of freedom of the right-moving sector of the fermionic superstring consist of 8 transverse coordinates $X^i(\tau-\sigma)$ ($i=1,\ldots,8$) and 8 Majorana-Weyl fermionic coordinates $S^a(\tau\ \sigma)$. The physical degrees of freedom of the left-moving sector of the bosonic string consist of 24 transverse coordianes, $X^i(\tau+\sigma)$ and $X^i(\tau+\sigma)$ ($i=1,\ldots8$, $l=1,\ldots16$). Together they comprise the physical degrees of freedom of the heterotic string. The eight transverse right and left-movers combine with the longitudinal coordinates to describe the position of the string embedded in ten dimensional space. The extra fermionic and bosonic degrees of freedom parametrize an internal space.

The light cone action that yields the dynamics of these degrees of freedom can be derived from a manifestly covariant action, and one can easily quantize it. The only new feature that enters is the compactification and quantization of the extra 16 left-moving bosonic coordinaes. It is this compactification, on a uniquely determined 16 dimensional compact space, that leads to the emergence of Yang-Mills interactions.

The extra 16 left-moving coordinates of the heterotic string can be viewed as parametrizing an "internal" compact space T. This interpretation should not be taken too literally; in fact one can equally well represent these degrees of freedom by 32 real fermions. for consistency we take T to be a flat compact manifold, i.e. a 16 dimensional torus. Since closed strings contain gravity in their low energy limit we would expect that a compactified string theory will contain massless vector mesons associated with the isometries of the compact space. For T this would yield the 16 gauge bosons of $U(1)^{16}$. A remarkable feature of closed string theories is that for special choices of the compact space there will exist extra massless gauge bosons, which are in fact massless solitons. They combine with the Kaluza-Klein gauge bosons to fill out the adjoint representation of a simple Lie group whose rank equals the dimension of T. In the case of the heterotic string the structure of T is so constrained by requirements of consistency that only two choices are possible. These produce the gauge bosons of $G = Spin(32)/Z_2$ or $E_8 \times E_8$.

The heterotic string theory has, by now, been developed to the same stage as other superstring theories. Interactions have been introduced and shown to preserve the symmetries and consistency of the theory, radiative corrections calculated and shown to be finite.

V.   STRING PHENOMENOLOGY

In order to make contact beween the string theories and the real world one is faced with a formidable task. These theories are formulated in ten flat space-time dimensions, have no candidates for fermionic matter multiplets, are supersymmetric and contain an unbroken large gauge group--say $E_8 \times E_8$. These are not characteristic features of

the physics we observe at energies below a Tev. If the theory is to
describe the real world one must understand how six of the spatial
dimensions compactify to a small manifold leaving four flat dimensions,
how the gauge group is broken down to $Su_3 \times SU_2 \times U_1$, how supersymmetry is
broken, how families of light quarks and leptons emerge, etc. Much of
the recent excitement concerning string theories has been generated by
the discovery of a host of mechanisms, due to the work of Witten and of
Candelas, Horowitz and Strominger, and of Dine, Kaplonovsky, Nappi,
Seiberg, Rohm, Breit, Ovrut, Segre, and others, which indicate how all
of this could occur. The resulting phenomenology, in the case of the
$E_8 \times E_8$ heterotic string theory is quite promising.

The first issue that must be addressed is that of the compactification of six of the dimensions of space. The heterotic string, as
described above, was formulated in ten dimensional flat spacetime.
This however is not neccessary. Since the theory contains gravity
within it the issue of which spacetime the string can be embedded in is
one of the string dynamics. That the theory can consistently be constructed in perturbation theory about flat space is equivalent to the
statement that ten dimensional Minkowski spacetime is a solution of the
classical string equations of motion. Such a solution yields the background expectation values of the quantum degrees of freedom. We can
then ask are there other solutions of the string equations of motion
that describe the string embedded in, say, four dimensional Minkowski
spacetime times a small compact six dimensional manifold?

At the moment we do not possess the full string functional equations of motion, however one can attack this problem in an indirect
fashion. One method is to deduce from the scattering amplitudes that
describe the string fluctuations in ten dimensional Minkowski space an
effective Lagrangian for local fields that describe the string modes.
Restricting one's attention to the massless modes, the resulting Lagrangian yields equations which reduce to Einstein's equations at low
energies, and can be explored for compactified solutions. Another
method is to proceed directly to construct the first quantized string
about a trial vacuum in which the metric n_{ab} (as well as other string
modes) have assumed background values. In this approach one starts

with the action of Equation (1), or its supersymmetric generalization, but allows $n_{ab}(x)$ to be the metric of a curved manifold. A consistent string theory can be developed as long as the two dimensional field theory of the coordinates $X^{\alpha}(\sigma,\tau)$ is conformally invariant. This is a nontrivial requirement, since the theory described by (1) is an interacting nonlinear σ-model. The condition that the two dimensional theory be conformal invariant is equivalent to demanding that the string equations of motion are satisfied. Thus one can search for alternative vacuum states by looking for σ-models (actually supersymmetric σ-models), for which the relevant β functions (which are local functions of the metric $n_{ab}(x)$ and its derivatives) vanish. In addition one must check that the anomaly in the commutators of the stress energy tensor is not modified. Given such a theory one can construct a consistent string theory and if $n_{ab}(x)$ describes a curved manifold the string will effectively be embedded in this manifold.

Remarkably there do exist a very large class of conformally invariant supersymmetric σ-models, that yield solutions of the string classical equations of motion to all orders and describe the compactification of ten dimensions to a product of four dimensional Minkowski space times a compact internal six dimenional manifold. These compact manifolds are rather exotic mathematical constructs (they are Kahler and admits a Ricci flat metric--i.e. they have SU_3 holonomy) and are called "Calabi-Yau" manifolds. In general they have many free parameters (moduli) which, among the rest, determine their size. Once again, this is an indication of the enormous vacuum degeneracy of the string theory, at least when treated perturbatively, and leads to many (at the present stage of our understanding) free parameters. This abundance of riches should not displease us, at the moment we would like to know whether there are any solutions of the theory which resemble the real world, later we can try to understand why the dynamics picks out a particular solution.

In the case of the heterotic string it is not sufficient to simply embed the string in a Calabi-Yau manifold. One must also turn on an SU_3 subgroup of $E_8 \times E_8$ gauge group of the string. This is because the internal degree of freedom of the heterotic string consist of right-moving fermions, which feel the curvature of space-time, and left-moving coordinates which know nothing of the space-time curvature but are sensitive to background gauge fields. Unless there is a relation between the curvature of space and the curvature (field strength) of the gauge group there is a right left mismatch which gives rise to anomalies. Therefore one must identify the space-time curvature with the gauge curvature (embed the spin connection in the gauge group). One does this by turning on background gauge fields in an SU_3 subgroup of one of the E_8's, thereby breaking it down to E_6 (or possibly O_{10} or SU_5).

These Calabi-Yau compactifications, produce for each manifold K, a consistent string vacuum, for which the gauge group is no larger than $E_6 \times E_8$ and N=1 supersymmetry is preserved. Furthermore there now exist massless fermions which naturally form families of quarks and leptons. Recall that after Kaluza-Klein compactification the spectrum of massless chiral fermions is determined by the zero modes of the Dirac operator on the internal space. Since, for heterotic string, the gauge and spin connections are forced to be equal one can count the number of chiral fermions by geometrical arguments. The massless fermions fall into 27's of E_6. This is good, E_6 is an attractive grand unified model and each 27 can incorporate one generation of quarks and leptons. The number of generations is equal to half the Euler character of the manifold (which counts the number of "handles" it has), and is normally quite large. If there exists a discrete symmetry group, Z, which acts freely on K, one can consider the smaller manifold K/Z, whose Euler character is reduced by the dimension of Z. By this trick, and after some searching, manifolds have been constructed with 1,2,3,4,... generations. It seems that to be realistic we must restrict attention to manifolds with three, or perhaps four, generations.

The compactification scheme also produces a natural mechanism for the breaking of E_6 down to the observed low energy gauge group. If K/Z is multiply connected one can allow flux of the unbroken E_6 (or of the E_8, for that matter) or to run through it, with no change in the vacuum energy. The net effect is that when we go around a hole in the manifold through which some flux runs we must perform a nontrivial gauge transformation on the charged degrees of freedom. These noncontractible Wilson loops act like Higgs bosons, breaking E_6 down to the largest subgroup that commutes with all of them. By this mechanism one can, without generating a cosmological constant, find vacua whose unbroken low energy gauge group is, say $Su_3 \times SU_2 \times U_1 \times$ (typically, an extra U_1 or two). Moreover there exists a natural reason for the existence of massless Higgs bosons which are weak isospin doublets (and could be responsible for the electroweak breaking at a Tev), without accompanying color triplets. Many of the successful features of grand unified models, such as the prediction of the weak mixing angle, carry over, and many of the unsuccessful predictions, such as quark lepton mass ratios, do not.

Of course it is also necessary to break the remaining N=1 supersymmetry. For this purpose the extra E_8 gauge group might be useful. Below the compactification scale it yields a strong, confining gauge theory like QCD, but without light matter fields. In general this sector would be totally unobservable to us, consisting of very heavy glueballs, which would only interact with our sector with gravitational strength at low energies. However there could very well exist in this sector a gluino condensate which can serve as source for supersymmetry breaking.

Thus the heterotic string theory appears to contain, in a rather natural context, many of the ingredients necessary to produce the observed low energy physics. I do not mean to suggest that there are not many problems and unexplained mysteries. There exists the danger (common to many grand unified models, especially suprrsymmetric ones) of too rapid proton decay, there is no deep understanding of why the cosmological constant, so far zero, remains zero to all orders, and when supersymmetry is broken, at least by the mechanism discussed above, the theory tends to relax back to ten dimensinal flat space. Nonetheless, the early successes are very reassuring and they give one the feeling that there are no insuperable obstacles to deriving all of low energy physics from the $E_8 \times E_8$ heterotic string theory.

VI. OUTLOOK

I do not want to leave the impression that string theory has brought us close to the end of particle physics. Quite the opposite is the case. Not only are there many unsolved problems and deep mysteries that need to be understood before one can claim success, in addition we have only begun to probe the structure of these new theories. I prefer, therefore, to conclude with a list of open problems.

VI.1 What is String Theory?

We do not fully understand the deep symmetry principles and symmetries that underly string theories. To date these theories have been constructed in a somewhat adhoc fashion and often the formulism has produced, for reasons that are not totally understood, structures that appear miraculous.

VI.2 How Many String Theories Are There?

Do there exist more consistent theories than the known five? Do there exist fewer, in the sense that some of the ones we know already are perhaps different manifestations (different vacua?) of the same theory?

VI.3 String Technology

This is not a question but a program of development of the techniques for performing calculations within string theory, including control of multiloop perturbation theory and the construction of manifestly covariant and supersymmetric methods of calculation. In addition one needs to develop, in a manifestly covariant approach, a useful second quantized formulation of the theory--string field theory.

VI.4 What is the Nature of String Perturbation Theory

Does the perturbative expansion of the string theory converge? If not, when does it give a reliable asymptotic expansion? How can one go beyond perturbation theory?

VI.5 String Phenomenology

Here there are many issues that remain to be resolved. They can all be included in the question--can one construct a totally realistic model which agrees with observation and why is it picked out?

VI.6 What Picks the Correct Vacuum?

This is one of the greatest mysteries of the theory, which seems to have an enormous number of acceptable vacuum.states. Why then don't we live in ten dimensional flat space? How does the value of the dilaton field get fixed and thereby the dilaton acquire a mass? Does the vanishing of the cosmological constant survive the physical mechanism that lifts the vacuum degeneracy?

VI.7 What Is the Nature of High Energy Physics?

By this I mean what does physics look like at energies well above the Planck mass scale? This is a question that is addressable, in principle, for the first time and might be of more than academic interest for cosmology. Does the string undergo a transition to a new phase at high temperatures and densities? Can one avoid in string theory the ubiquitous singularities that plague ordinary general relativity?

VI.8 Is There a Measurable, Qualitatively Distinctive, Prediction of String Theory

String theories can make many "postdictions" (such as the calculation of mass ratios of quarks and leptons, Higgs masses, gauge couplings, etc.). They can also make many new predictions (such as the masses of the various supersymmetric partners). These would be sufficient to establish the validity of the theory, however one could imagine conventional field theories coming up with similar pre or post dictions. It would be nice to predict a phenomenon which might be accessible at observable energies and is uniquely characteristic of string theory.

REFERENCES
   I have made no attempt to give detailed references to the papers
in this rapidly growing field. An incomplete set of references to
recent work is given below.

Reviews:
Unified String Theories (Proceedings of the String Workshop at Santa
Barbara, World Scientific, 1986).
Superstrings (Reprint Volume, edited by J. Schwarz, World Scientific,
1985).

Superstring Anomalies
M.B. Green and J.H. Schwarz, Phys. Lett. 149B (1984) 117;
L. Alvarez-Gaume and E. Witten, Nucl. Phys. B234 (1983) 269.

Heterotic String
D.J. Gross, J.A. Harvey, E. Martinec and R. Rohm, Phys. Rev. Lett. 54
(1985) 502, Nucl. Phys. B256 (1985) 253, and to be published.

String Phenomenology
P. Candelas, G. Horowitz, A. Strominger and E. Witten, Nucl. Phys. B258
(1985) 46; E. Witten, Nucl. Phys. B258 (1985) 75.

RESULTS FROM UA1[*]

Aachen, Annecy, Birmingham, CERN, Harvard, Helsinki, Imperial College, Kiel, Queen Mary College, NIKHEF, Padua, Paris, Riverside, Rome, Rutherford, Saclay, Vienna, Wisconsin

presented by
J. Rohlf
(Harvard University)

1. INTRODUCTION

New results are reported from the UA1 experiment on proton-antiproton collisions at high energies. The data sample recorded on tape is shown in Table 1. This talk is divided into the following parts: measured properties of W and Z^0 particles (section 2); and analysis of events with large missing energy, including search for the decay $W \rightarrow \tau\nu$ (section 3).

Table 1. UA1 data sample collected thus far.

year	c.m. energy	luminosity (integrated)
1982-3	546 GeV	136 nb^{-1}
1984	630 GeV	263 nb^{-1}
	total:	399 ± 60 nb^{-1}

2. W AND Z^0 PROPERTIES

The observed numbers of events with intermediate vector bosons are given in Table 2. The difference in the raw number of decays into electron and muon channels is accounted for by acceptance, mainly due to the trigger. The event selection to find W and Z^0 particles is

[*] Invited talk at the American Physical Society, Division of Particles and Fields, Eugene, Oregon, August 13, 1895

straight-forward[1]; the selection cuts for the electron channel are:

1) High transverse energy (E_T) electromagnetic
   calorimeter cluster, $E_T > 15$ GeV
2) High transerse momentum (p_T) track, $p_T > 7$ GeV
3) Isolation of the electron candidate, requiring less than
   10% extra energy in a cone of size $\Delta R = 0.4$
   ($\Delta R^2 = \Delta\phi^2 + \Delta\eta^2$)
4) Missing transverse energy (ΔE_M) greater than 15 GeV
   for W selection, or a second electromagnetic cluster for
   Z^0 selection.

Selection cuts for the muon channel are similar except that electromagnetic shower cuts are replaced with track cuts in the muon chambers, and more severe cuts on the quality of the track are made to remove backgrounds due to kaon and pion decays[1].

Table 2. Observed numbers of W and Z^0 events (total data sample)

channel	events
$W \to e\nu$	172
$W \to \mu\nu$	47
$Z^0 \to e^+e^-$	18
$Z^0 \to \mu^+\mu^-$	10

The mass distribution for events with two electromagnetic clusters is shown in Figure 1. The contribution from the decays $Z^0 \to e^+e^-$ (18 events) is clearly separated from jet backgrounds, which cause electron misidentification and end at about 50 GeV. Figure 2 shows the Z^0 mass distribution, after a calorimeter fiducial cut to select the events with the cleanest showers, together with the results of a breit-wigner fit to the Z^0 mass. We obtain the value,

$$m_Z = 93.0 \pm 1.4 \pm 3.2 \text{ GeV.}$$

Measurement of the Z^0 mass is already dominated by systematic error.

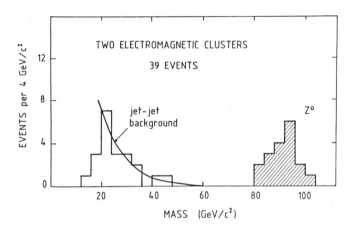

Figure 1. Two-cluster invariant mass distribution

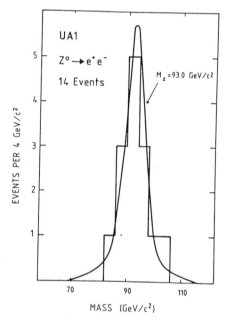

Figure 2. e^+e^- invariant mass distribution.

The dimuon mass spectrum is shown in Figure 3. As is the case in the electron channel, the contribution from the Z^0 is clearly separated from the continuum background. However, there are two major differences between the electron and muon analyses: 1) the muon identification allows us the possibility of going lower in lepton momentum and thus, sensitivity to dimuon masses down to about 10 GeV, and 2) the mass resolution in the Z^0 region is much worse for the muon channel.

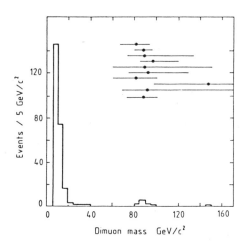

Figure 3. Dimuon invariant mass distribution.

For the selection of $W \to e\nu$ events, we demand a single electron and missing transverse energy greater than 15 GeV. The electron E_T and ΔE_M distributions are shown in Figure 4; events with showers near the edges of the detector have been removed. The expected contribution for the decay $W \to e\nu$ is shown as a solid curve. The shaded part of the histogram shows the expected backgrounds from electron misidentification and from $W \to \tau\nu$ decays. These bacgrounds make up about 10% of the total number of events, and are concentrated at low E_T.

The W transverse mass distribution (m_T) is shown in Figure 5a. In order to get an essentially background free sample of $W \to e\nu$ decays for the purpose of fitting the W mass, we make the more stringent cuts:

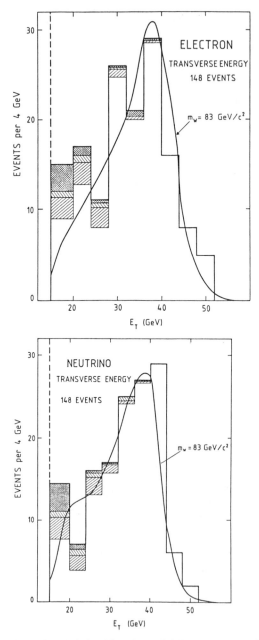

Figure 4. a) electron E_T distribution from W decays, expected backgrounds are shaded; b) missing transverse energy distribution

p_T (electron) > 30 GeV   and ΔE_M > 30 GeV. The events that pass these cuts are the events which best determine the position of the jacobian peak of the W (see Figure 5b). We obtain the mass value:

$$m_W = 83.5\ ^{+\ 1.1}_{-\ 1.0}\ \pm 2.8\ \ \text{GeV}.$$

The W and Z^0 measurements are summarized in Table 3 and derived quantities are given in Table 4. There is good agreement with results reported by the UA2 Collaboration[2-3].

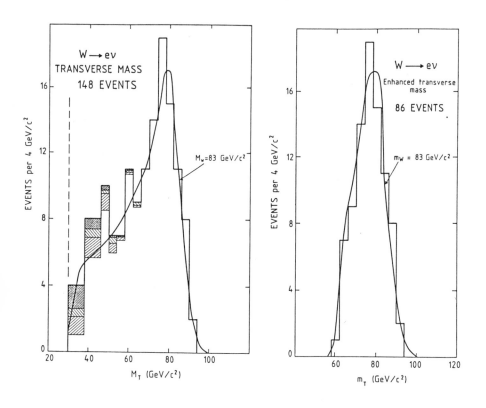

Figure 5. a) transverse mass of electron and missing energy, b) same but with 30 GeV cuts for the electron and the neutrino.

Table 3. UA1 W and Z^0 measurements

$$m_W \qquad 83.5 \,{}^{+\,1.1}_{-\,1.0} \pm 2.8 \;\; \text{GeV}$$

$$m_Z \qquad 93.0 \pm 1.4 \pm 3.2 \;\; \text{GeV}$$

$$\Gamma_W \qquad < 6.5 \;\; \text{GeV} \;\; (90\% \text{ CL})$$

$$\Gamma_Z \qquad < 8.1 \;\; \text{GeV} \;\; (90\% \text{ CL})$$

$$\sigma \cdot B_{W \to e\nu} \qquad 0.55 \pm 0.08 \pm 0.09 \;\; \text{nb}, \quad \sqrt{s} = 546 \;\text{GeV}$$
$$0.63 \pm 0.05 \pm 0.09 \;\; \text{nb}, \quad \sqrt{s} = 630 \;\text{GeV}$$

$$\sigma \cdot B_{Z \to ee} \qquad 40 \pm 20 \pm 6 \;\; \text{pb}, \quad \sqrt{s} = 546 \;\text{GeV}$$
$$79 \pm 21 \pm 12 \;\; \text{pb}, \quad \sqrt{s} = 630 \;\text{GeV}$$

Table 4. Standard Model parameters

$$\sin^2\theta_W \equiv 1 - m_W^2/m_Z^2 \qquad 0.19 \pm 0.03$$

$$\sin^2\theta_W \equiv (38.5 \;\text{GeV})^2 / m_W^2 \qquad 0.213 \,{}^{+\,0.005}_{-\,0.006} \pm 0.015$$

$$\rho \equiv \sec^2\theta_W \; m_W^2/m_Z^2 \qquad 1.04 \pm 0.04$$

$$R \equiv \sigma \cdot B_{W \to e\nu} / \sigma \cdot B_{Z \to ee} \qquad 9.3 \,{}^{+\,2.6}_{-\,1.6}$$

The Feynman-x distribution of W particles (x_W) is shown in Figure 6. A slightly softer x distribution is observed at higher c.m. energy, as is expected from proton structure functions[5]. Figure 7 shows the fraction of proton momentum (x) carried by quarks or antiquarks which make W's. The peak at about x=0.15 corresponds to the case where the W is produced nearly at rest in the laboratory frame. There is no difference observed in the structure of the antiproton compared with the proton.

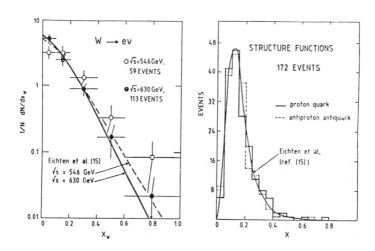

Figure 6. a) Feynman-x distribution of W particles (left);
b) fraction of proton (antiproton) momentum carried by quarks (antiquarks) which make W particles

The p_T distribution of the W particle is shown in Figure 7. There is excellent agreement with the latest QCD calculations[6]. The shaded portion of the histogram indicates those events which have visable jets[7]. All events with W transverse momentum greater than 20 GeV have clearly visable jets balancing the W p_T. For the events with jets, we plot the angular distribution of the jet, with respect to the beam axis, in the rest frame of the W (Figure 8). The data show the characteristic gluon bremsstrahlung angular distribution as expected from QCD. The mass of the W-jet system for events with 1 jet is shown in Figure 9. No significant structure is observed.

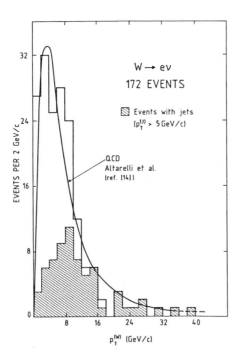

Figure 7. W transverse momentum distribution

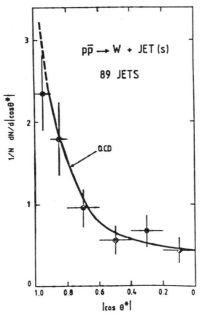

Figure 8. Angular distribution of the jet with respect to the beam axis in the rest frame of the W, for events with a W and a jet

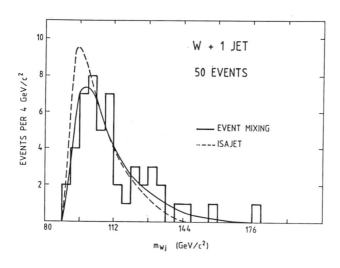

Figure 9. Mass of the (W + jet) for events with a W and a jet

The angular distribution of the electron (positron) with respect to the proton (antiproton) direction in the rest frame of the W is shown in Figure 10. A clear forward-backward asymmetry due is observed. The decay angular distribution is consistent with maximum polarization of the W at production and maximum parity violation in its decay. Some more details may be found in reference 4.

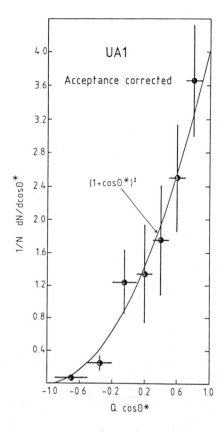

Figure 10. Angular distribution of the electron (positron) with respect to the proton (antiproton) direction in the rest frame of the W

3. MISSING ENERGY EVENTS

The missing energy analysis is motivated by observation in the 1983 data of a few events containing a single jet of large transverse energy[8] The background to these events from QCD jets (taking into account the detector resolution) is small in the UA1 experiment, provided that the missing E_T is large and isolated. The higher statistics of the 1984 data have given us a greater understanding of proton-antiproton collisions, as viewed with the UA1 detector, and has enabled us to more deeply investigate the possible conventional sources for these events.

The jet energy resolution, for the UA1 detector estimated by monte carlo techniques, is shown in Figure 11.

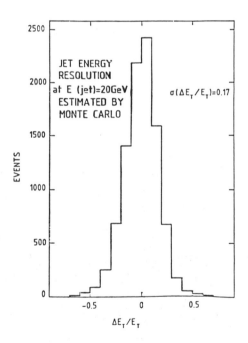

Figure 11. Energy resolution expected for a jet of energy 20 GeV; the resolution ($\Delta E/E$) is about 17%.

The balance of transverse momentum observed in two-jet events is consistent with our expectation based on the resolution shown in Figure 11. This is shown in Figure 12, where we histogram the missing transverse energy squared divided by the sum of the scalar E_T (x). We shall be concerned with a search for a possible excess of events with large missing transverse energy, beyond what may be explained by detector resolution effects.

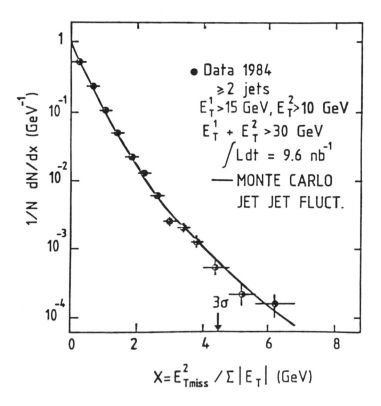

Figure 12. Transverse momentum balance observed in two-jet events compared with monte carlo expectation (solid line) based on jet resolution calculations.

3.a) Search for the Decay, $W \to \tau \nu$

The decay chain $W \to \tau \nu$, $\tau \to \nu$ + hadrons, has a pure monojet signature. Observation of this decay mode would complete the leptonic decay modes of the W into known leptons and is interesting in itself. However, it is also of special interest to isolate this decay because it poses the potentially largest background in the search for new processes giving large missing energy.

The four standard deviation[8] (4σ) monojet selection contains the following cuts:

1) $\Delta E_M > 4\sigma$ and $\Delta E_M > 15$ GeV,

2) one (and only one) jet with $E_T > 12$ GeV,

3) at least one central detector track with $p_T > 1$ GeV, within
   a distance $\Delta R = 0.4$ of the calorimeter jet axis

4) jet-jet fluctuation veto demanding no calorimeter jet with
   $E_T > 8$ GeV and no CD jet with $p_T > 5$ GeV coplanar with the monojet
   or the missing energy direction (± 30 deg.)

5) event validation by scanning to remove beam halo, cosmic rays,
   beam gas interactions, double interactions, reconstruction
   problems, vertical jets, leakage / punch-through, high p_T muons
   (recalulate ΔE_M)

A total of 29 events (including six from 1983) pass the monojet cuts[9-11]. The jet E_T and ΔE_M distributions are shown in Figure 13a-b. A scatter-plot of jet E_T vs. ΔE_M is shown in Figure 14.

A rather extensive search was made for measured quantities which could be used to separate the tau events from other jets. Two simple quantities were constructed for this purpose. The first we denote by the variable F, which is defined to be the fraction of the jet energy contained in a cone of size $\Delta R = 0.4$ centered on the calorimeter jet axis.

$$F = \frac{\Sigma E_T \text{ (in cone } \Delta R = 0.4)}{\Sigma E_T \text{ (in cone } \Delta R = 1.0)}$$

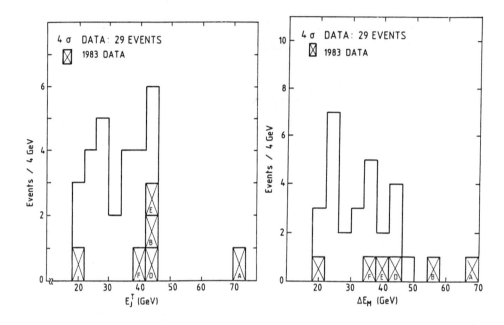

Figure 13. a) Jet transverse energy for monojet events (left);
b) missing transverse energy for monojet events (right)

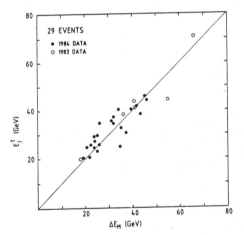

Figure 14. Scatter-plot of jet transverse energy (vertical axis) *vs.*
missing transverse energy (horizontal axis) for monojet events.

The distribution of the variable F is shown in Figure 15 for our 29 monojet events, compared with the measured distribution from balanced jet events and a monte carlo distribution for tau decays.

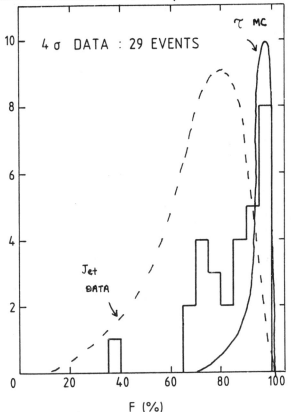

Figure 15. F distribution for monojet events compared with jet data and tau monte carlo.

The second variable we use is the matching ($\Delta\phi$ and $\Delta\eta$) of the largest p_T track with the direction of the jet axis as measured in the calorimeter.

$$\Delta R \equiv (\Delta\phi + \Delta\eta)^{1/2}$$

The ΔR distribution for the 29 monojets is shown in Figure 16 together with the jet data and the tau monte carlo.

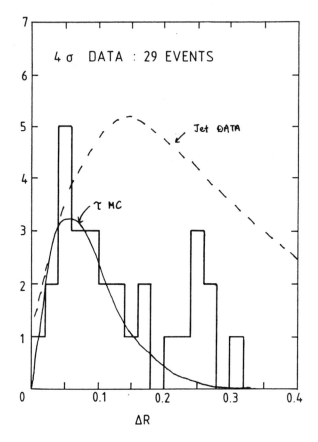

Figure 16. Matching between the highest p_T track and the jet axis

Figure 17 shows the distributions of F and ΔR separated by observed charged particle jet multiplicity for $p_T > 1$ GeV (n). Some jets fit the tau hypothesis very well, having a large value of F, a small value of ΔR, and charged particle multiplicity of 1 or 3.

We may perform a relative likelihood for each event to be a tau jet based on our Monte Carlo simulation of F, ΔR, and n. The tau likelihood distribution is shown in Figure 18a, compared with the distribution expected for backgrounds (Figure 18b, discussed in more detail later). One of the tau candidates is shown in Figure 19.

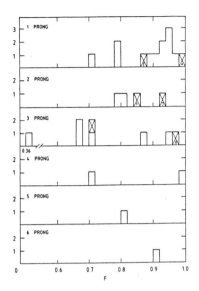

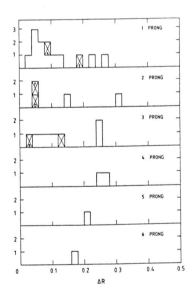

Figure 17. a) F distribution for various observed jet multiplicites (top); b) same for ΔR (bottom).

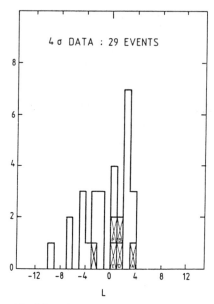

Figure 18a. Tau likelihood distribution for monojet data (arbitrary horizontal scale)

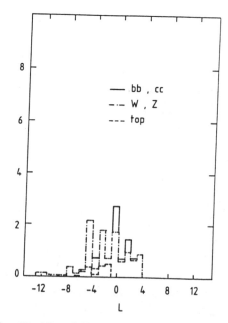

Figure 18b. Tau likelihood distribution for monte carlo backgrounds of Table 5 (excluding $W \rightarrow \tau \rightarrow$ hadrons)

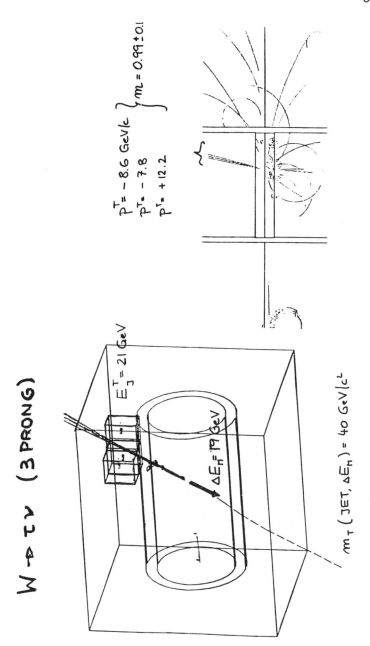

Figure 19. Observed tau candidate event

To further evaluate backgrounds, we must consider all known sources of missing energy. These ocurr when a W or Z^0, real or virtual, is produced. The main processes are listed in Table 5. Detailed monte carlo calculations were performed using data as input whenever possible. For this, we used a technique of mixed events, taking the X of Table 5 from our data and superimposing a theoretical W or Z^0 decay.

Table 5. Possible conventional sources of monojets

W with hard gluon, W decay overlaps jet	$p\bar{p} \rightarrow W + X$
	$\rightarrow e\nu$
	$\rightarrow \mu\nu$
	$\rightarrow \tau\nu$
	$\rightarrow e\nu\nu$
	$\rightarrow \mu\nu\nu$
	$\rightarrow \nu + hadrons$
Z^0 with hard gluon, Z^0 decay into neutrinos (or taus)	$p\bar{p} \rightarrow Z^0 + X$
	$\rightarrow \nu\nu$
	$\rightarrow \tau\tau$
heavy flavor pairs, one jet has leading neutrino	$p\bar{p} \rightarrow c\bar{c} + X$
	$b\bar{b} + X$
	$t\bar{t} + X$
heavy flavors from W	$p\bar{p} \rightarrow W + X$
	$\rightarrow c\bar{s}$
	$\rightarrow t\bar{b}$
heavy flavors from Z^0	$p\bar{p} \rightarrow Z^0 + X$
	$\rightarrow c\bar{c}$
	$\rightarrow b\bar{b}$
	$\rightarrow t\bar{t}$

We define a clean tau sample with the likelihood cut (see Figure 18) of
L > 2. This cut is somewhat arbitrary, and we shall address our sensitivity
to it later. Figure 20 shows the jet E_T distribution for the best tau
candidate events (L > 2) together with the expected background
contributions. The number of monojets observed with tau likelihood
greater than 2 is 9 events[12]. The expected physics contributions from
the processes listed in Table 5 (other than tau) is 1.7 ± 0.4 events (see
Figure 18b). We observe a tau signal of 7.3 ± 3.0 events which may be
compared to our monte carlo expectation of 10.6 ± 0.5 events, assuming
electron-tau universality. The tau acceptance including the trigger, event
selection, and tau branching ratio into hadrons is about 4.6%. We calculate
a cross section times branching ratio for W → τν to be:

$$\sigma \cdot B_{W \to \tau\nu} = 0.42 \pm 0.18 \pm 0.06 \ nb.$$

The ratio of branching ratios for W decay into tau divided by W decay into
electron is:

$$\frac{B\,(W \to \tau\nu)}{B\,(W \to e\nu)} = 0.7 \pm 0.3$$

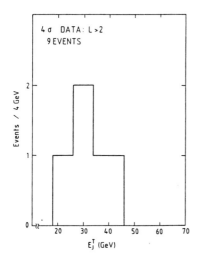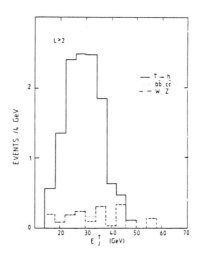

Figure 20. a) Jet transverse energy distribution for tau candidate events
(left), compared with b) expected backgrounds of Table 5 (right)

3.b) Inclusive Search for Events with Large Missing Energy

We may remove most of the tau candidate events with a likelihood cut complimentary to that used to select taus. After making an anti-tau cut, L < 2, we are left with 20 events. The question is then, can we explain the remaining events by the processes listed in Table 5? The jet E_T and ΔE_M distributions are shown in Figures 21-2, together with the result of background calculations of the processes listed in Table 5.

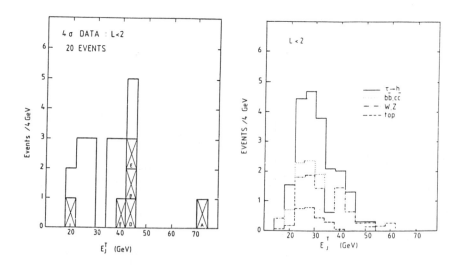

Figure 21. a) Jet transverse energy distribution for non-tau candidates (left) and b) expected contributions from Table 5 (right)

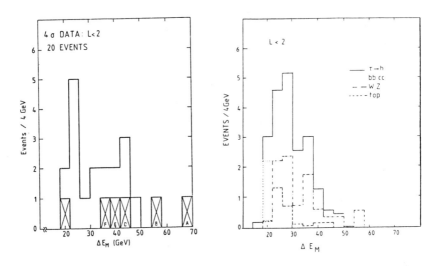

Figure 22. a) Missing transverse energy distribution for non-tau candidates (left) and b) expected contributions from Table 5 (right)

Figure 23 shows a scatter-plot of tau likelihood *vs.* jet E_T for all 29 monojet events. The tau events are expected in the region L > 0 and E_T < 40 GeV. The region L < 0 and E_T < 40 GeV is where the QCD backgrounds dominate. There is a cluster of events (5 from 1983 and 4 from 1984) which tend to be somewhat tau-like but have large jet E_T. The effect is not nearly as striking with the 1984 data alone as it was with the 1983 data alone; however, there still remain a few extra narrow jets of high transverse energy. If they were all taus, then we would not understand their energy spectrum.

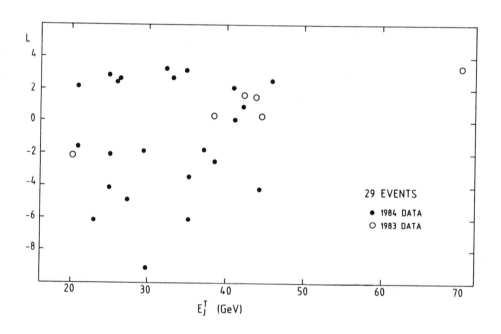

Figure 23. Tau likelihood (vertical axis) *vs.* jet transverse energy

We have also looked for events with large missing transverse energy and more than one jet by changing only selection cut 2 above to two or more jets with $E_T > 12$ GeV. The results are summarized in Table 6. We find that backgrounds expected from jet fluctuations are small if the missing transverse energy is isolated, as was observed in the single jet case. We expect a background from conventional sources (see Table 5) of about 3 events for the isolated case. We observe 3 events; however, they have higher transverse energies than predicted (Figure 24). One of these events is shown in Figure 25.

Table 6. Events with two or more jets and large ΔE_M

	4σ	4σ isolated
Jet fluctuations	16.9	0.
Physics	11.2	3.0
TOTAL	28.1	3.0
DATA	31	3

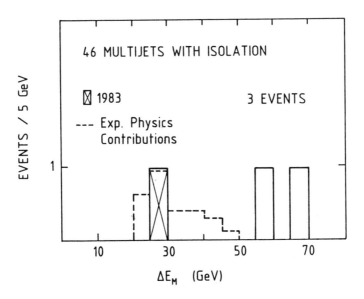

Figure 24. Missing energy distribution for events with two or more jets and isolated ΔE_M

92

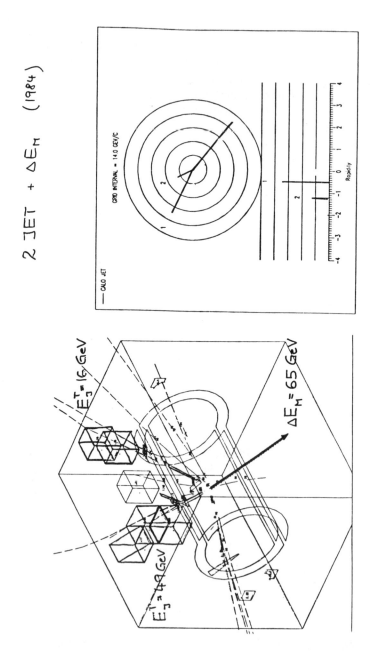

Figure 25. Observed multi-jet event with large isolated missing transverse energy

4. SUMMARY

The measured W and Z^0 properties from electron and muon channels are in excellent agreement with standard model expectations. This includes the mass, width, and cross section measurements as well as the production and decay properties.

The decay $W \rightarrow \tau\nu$ has been observed.

The missing energy events (non-tau) are not yet completely understood. A detailed monte carlo study has been performed of known physical processes which give energetic neutrinos. In absolute rate we may account for the observed number of events with large missing transverse energy. However, the data have a few more narrow (tau-like) jets at large E_T than expected. In addition, two multi-jet events are found at large E_T where backgrounds are predicted to be small.

References and Footnotes

1. G. Arnison *et al.* (UA1 Collaboration), Phys. Lett. **122B**, 103 (1983); **126B**, 398 (1983); **129B**, 273 (1983); **134B**, 469 (1984); **147B**, 241 (1984).
2. M. Banner *et al.* (UA2 Collaboration), Phys. Lett. **122B**, 476 (1983); P. Bagnaia *et al.*, Phys. Lett. **129B**, 130 (1983); Z. Phys. C: Particles and Fields **24**, 1 (1984).
3. See talk of J.-M. Gaillard (this conference) for a detailed comparison of UA1 and UA2 W and Z^0 measurements.
4. G. Arnison *et al.* (UA1 Collaboration), CERN-EP/85-108, submitted to Nuovo Cimento Letters; G. Arnison *et al.*, paper in preparation.
5. E. Eichten *et al.*, Rev. Mod. Phys. **56**, 579 (1984).
6. G. Altarelli *et al.*, Nucl. Phys. **B246**, 12 (1984); G. Altarelli *et al.*, Z. Phys. C. : Particles and Fields **27**, 617 (1985).
7. For some details of the jet algorithm used, see G. Arnison *et al.* (UA1 Collaboration), Phys. Lett. **123B**, 115 (1983).
8. G. Arnison *et al.* (UA1 Collaboration), Phys. Lett. **139B**, 115 (1984).

9. The cuts used here are more stringent than that of reference 8, due mainly to the isolation cuts. All of the highest E_T events of 1983 pass the tighter cuts except event C.

10. A new hardware trigger was added in 1984 to select events with imbalance greater than 17 GeV and a jet greater than 15 GeV. This had no effect in the 4σ sample for events with large ΔE_M because all of these events satisfy other hardware triggers. This trigger did select about 20% more events at low ΔE_M compared with the 1983 triggers.

11. While the electromagnetic energy calibration is relatively well understood from the W sample, a study is being undertaken to further check the hadronic energy calibration. If we reduce the hadronic energy scale, some events at low ΔE_M would be removed.

12. Event A of 1983 (reference 8) was an event with a very high transverse momentum muon in a jet. Since this event is very unlikely to be a tau, we exclude it from the tau sample and include it in the non-tau sample.

PHENOMENOLOGY OF COLLIDER PHYSICS

VERNON BARGER

Physics Department, University of Wisconsin
Madison, Wisconsin 53706
U.S.A.

ABSTRACT

QCD shower calculations of W, Z production are shown to be well approximated by the $q\bar{q} \to Wg$ subprocess with a low-p_T cutoff chosen to reproduce the total cross section through order α_s. A similar approximation is applied to heavy-quark production from $2 \to 3$ parton subprocesses in $p\bar{p}$ collisions. The results agree with CERN dimuon events for a $B^0 - \bar{B}^0$ mixing parameter $\epsilon \gtrsim 0.1$, which is larger than Standard Model expectations and at the e^+e^- experimental limit. Some aspects of searches for supersymmetry and a fourth generation of leptons and quarks are addressed.

1. INTRODUCTION

Tremendous strides have been made in understanding the physics of high energy collisions in the short time that data have been available from experiments at the CERN $p\bar{p}$ collider. To a considerable extent theoretical progress has been in the refinement of QCD calculations for Standard Model processes, a necessary first step in determining whether signals of new physics exist. In this report, efforts to describe the $p\bar{p}$ collider data by Standard Model processes and to possibly uncover evidence for new phenomena in and beyond the Standard Model will be described.

2. QCD CALCULATIONS

Three different QCD calculational methods have been pursued:

(i) **perturbative** calculation of low-order subprocesses. In some cases of interest perturbative calculations exist only in tree order, so regularization of soft and collinear singularities (*e.g.* by p_T cutoffs) is required. QCD-evolved initial parton distributions from deep inelastic scattering are input. For recent perturbative applications see, *e.g.*, Refs. 1, 2.

(ii) **impact-parameter** summation of multiple gluon emission to all orders in leading log approximation. This approach, which has been applied only to weak boson and Drell-Yan lepton pair production, can predict the p_T of the weak boson or lepton pair, but cannot provide individual jet information. Calculations of this type can be found, *e.g.*, in Ref. 3.

(iii) **Monte Carlo shower** simulation of multiple gluon and light-quark emissions from incident partons. The shower method gives predictions at the parton level to all orders of all observables. The evolution of the initial partons is generated directly by the showers and includes the effects of transverse momenta from parton emissions that are not included in Altarelli-Parisi evolution. Integrations over unresolved soft emissions cancel virtual divergences. Shower calculations are made, *e.g.*, in Refs. 4, 5.

None of these approaches include soft interactions in the underlying event, such as occur in minimum bias events. Such soft-physics effects can occasionally be important as, for example, in the summed E_T of all hadrons or in lepton isolation cuts.

The subsequent discussion focuses on the approaches (i) and (iii) which give more comprehensive predictions.

3. W AND Z PRODUCTION

The by now rather mundane measurements of W and Z boson production provide an important first test of QCD calculations. The shower calculation[5] of $p\bar{p} \to W^+$ production starts from the hard-scattering $u\bar{d} \to W^+$ cross section σ_{BORN}, which gets multiplied by probabilities for successive emission of partons in the shower, as the virtual mass-squared t of the parton lines are evolved backwards from the hard scale $-t = Q^2 \approx M_W^2$ to the initial scale $-t = Q_0^2$. The individual branching probabilities

$$dP(t, z) = \frac{dt}{-t} \frac{\alpha_s(-t)}{2\pi} dz \, P_{A/B}(z) , \qquad (1)$$

are determined by the Altarelli-Parisi splitting functions $P_{A/B}$ for parton transitions $B \to A$, with arguments z given by the ratio of subprocess c.m. energy-squared before and after an emission. The shower cross section is

$$\sigma^{\text{shower}} \simeq \int dx_+ \, dx_- \, x_+ \, u(x_+, Q^2) \, x_- \, \bar{d}(x_-, Q^2) \, K \sigma_{\text{BORN}} \qquad (2)$$

where, somewhat symbolically,

$$q(x, Q^2) = \int \prod_j dP(t_j, z_j) \, q(x_N, Q_0^2) \,. \qquad (3)$$

The integral represents a Monte Carlo average over many shower configurations. Differential probabilities for various parton configurations follow from selection of events according to σ^{shower}. A jet algorithm is used to convert partons into jets. The K factor[6]

$$K = 1 + \frac{16\pi^2}{9} \frac{\alpha_s(Q^2)}{2\pi} \qquad (4)$$

represents the $O(\alpha_s)$ virtual gluon correction to the total cross section ($K \simeq 1.3$ for $Q^2 = M_W^2$). It may be noted that the evolved quark distributions from Drell-Yan showers $q_{SH}(x, Q^2)$ in Eq. (3) differ from the deep inelastic distributions $q_{DIS}(x, Q^2)$ by a correction of order α_s due to differences in kinematics.[7]

3.1 W Transverse Momentum

Shower QCD results[5] for $p_T(W)$ are compared with the UA1 data[8] in Fig. 1. A sharp spike predicted at low p_T (dashed curve) gets smeared by measurement resolution; with a realistic Gaussian smearing of standard deviation $\sigma = 3$ GeV on each component of $\vec{p}_T$, the predicted $p_T(W)$ distribution is in good accord with the data. The tail of the distribution at $p_T(W) \gtrsim 15$ GeV is independent of this smearing and is given by the $O(\alpha_s)$ perturbative calculation.

3.2 Jet Properties

By combining partons into jets using an algorithm similar to that used experimentally for defining jets of hadrons, the shower calculations can be compared with jets produced in association with the W. Fig. 2 shows the results[5] for the jet polar angle θ^* in the $W +$ jet rest frame, showing the expected peaking at $|\cos \theta^*| = 1$ from the $O(\alpha_s)$ subprocess.

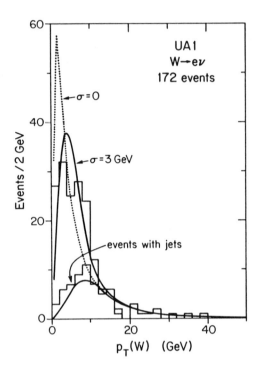

Fig. 1.   Shower Monte Carlo results for the $p_T(W)$ distribution compared with preliminary UA1 $W \to e\nu$ data.

The predicted fractions of events with n jets[5]

$$
\begin{array}{ccccc}
 & n = 0 & n = 1 & n = 2 & n = 3 \\
\sigma(n)/\sum \sigma(n) = & 0.6 & 0.3 & 0.06 & 0.01
\end{array}
\tag{5}
$$

are compared with the measured multiplicity distributions from UA1[8] in Fig. 3. We note that $\sigma(n \geq 2 \text{ jets})$ is an order of magnitude smaller than $\sigma(n \leq 1 \text{ jet})$. The predicted jet fractions depend critically on the jet E_T threshold in the jet algorithm, which is taken here to be 5 GeV, and high multiplicities may be affected by hadronic contributions from the underlying event.

3.3   Poor Man's Shower Approximation

A simplified description of the dominant features of W production can be obtained *solely* from the $O(\alpha_s)$ diagrams as follows.[9] We regularize the

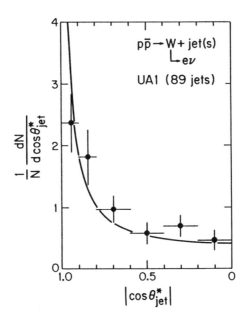

Fig. 2. Shower calculations for the distribution of jet polar angle θ^* with respect to the beam axis in the W+jet rest frame compared with preliminary UA1 data.

divergence at $p_T = 0$ with a cutoff chosen such that the $O(\alpha_s)$ diagrams give the proper total cross section

$$\sigma_{\mathrm{TOT}}(W) = \sigma(u\bar{d} \to W^+ g) + \sigma(\text{crossed Compton graphs}) = K\,\sigma_{\mathrm{BORN}}\,. \quad (6)$$

The p_T-dependence at high p_T is properly given by the $O(\alpha_s)$ diagrams and $\langle p_T(W)\rangle$ is also correctly predicted. In this $O(\alpha_s)$ approximation hard final-state partons will give rise to one-jet events, while soft contributions will give no jet events.

Equation (6) implies that the $u\bar{d} \to W^+$ Born contribution is being cancelled by the contributions from virtual gluon graphs and the $O(\alpha_s)$ parton emission graphs with p_T below the cutoff. Equation (6) is empirically realized with a sharp cutoff $p_T \geq 1.3$ GeV or with a smooth cutoff $f = 1 - \exp(-p_T^2/4)$ such that

$$\sigma_{\mathrm{TOT}}(W) = \int f K\,\sigma(2 \to 2)\,. \quad (7)$$

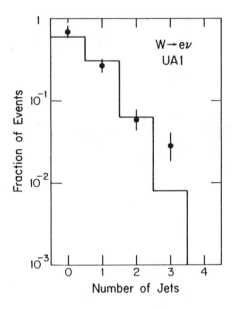

Fig. 3. Shower predictions for fractions of events with n jets (shown as a historgram) compared with preliminary UA1 data.

Taking this cutoff prescription and folding the $O(\alpha_s)$ subprocess with evolved quark distributions, the calculated $W + 1$-jet observables (such as $p_T(W)$, $\cos\theta^*_{\text{jet}}$, and $\sigma(1\text{-jet})/\sigma(0\text{-jet})$) are essentially indistinguishable from the full shower calculations and this $O(\alpha_s)$ approximation thereby accounts for the major features of W production; we infer that similar simplifying approximations can be usefully applied to heavy-quark production where the application of shower methods is exceedingly complicated.

4. DRELL-YAN DIMUON PRODUCTION

The QCD subprocesses for Drell-Yan production of $\mu^+\mu^-$ pairs correspond to those for weak boson production, and calculations similar to those described above can be performed. However, comparison of the predictions with data is not as straightforward since ψ, Υ, and heavy-quark pair production with semileptonic decays also lead to low-mass dimuons. Figure 4 shows expected Drell-Yan and heavy-quark dimuon cross sections without acceptance cuts, versus the dimuon mass. In the $\mu^+\mu^-$ case the heavy-quark contributions are larger than the Drell-Yan contributions for $m(\mu^+\mu^-) < 20$ GeV.

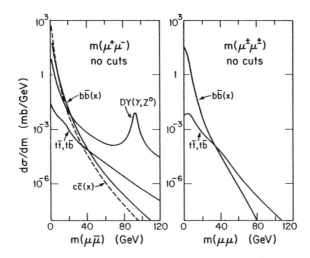

Fig. 4. Predicted Drell-Yan and heavy-quark dimuon cross sections, without acceptance cuts, versus dimuon mass.

Isolation requirements on the muons help to separate the sources; in $\mu^+\mu^-$ events of Drell-Yan origin the muons will generally be isolated, since the accompanying hadron jets will not often overlap the muons, whereas muons from heavy-quark decay will often lie in or near jets. Following the UA1 procedure,[10] the hadronic p_T is summed within cones $\Delta R = \left[(\Delta\phi)^2 + (\Delta\eta)^2\right]^{1/2} < 0.7$ around each muon and an isolation parameter $s = \left\{(\sum_1 p_T)^2 + (\sum_2 p_T)^2\right\}^{1/2}$ is defined, where subscripts 1, 2 refer to cones about muons 1 and 2. Dimuon events with $s < 3$ GeV are "isolated", and those with $s > 3$ GeV are "non-isolated". In our calculations[11] we sum the p_T of partons within these cones and add a soft-hadron contribution from the underlying event, assumed to have Gaussian distribution with $\sigma = 1.5$ GeV (which fits the hadronic p_T around the muons in UA1 data[10] on $W \to \mu\nu$ and $Z \to \mu\mu$, as well as minimum bias events). We impose the most recent UA1 acceptance cuts $p_T(\mu) > 3$ GeV, $|\eta(\mu)| < 2$ and $m(\mu\mu) > 6$ GeV.

Figure 5(a) compares the observed $m(\mu^+\mu^-)$ distribution for *isolated* muons with Drell-Yan shower predictions (at $\sqrt{s} = 540$–630 GeV, for integrated luminosity 0.38 pb^{-1}, reported geometry and track acceptance factors 0.45 × 0.58 and acceptance cuts above). The high-mass tail, for $m(\mu\mu) > 12$ GeV, agrees with the Drell-Yan expectation. At lower $m(\mu\mu)$ contributions from $b\bar{b}$ and Υ, Υ', and Υ'' production are also likely to be important; estimates of these contributions are also shown.

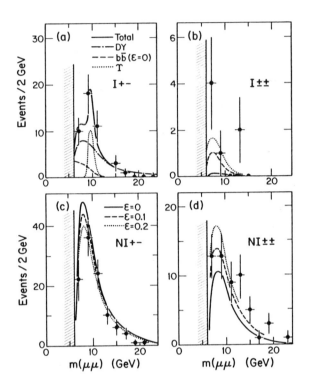

Fig. 5. Predicted dimuon mass distributions compared with preliminary UA1 data.

5. DIMUONS FROM HEAVY-QUARK PRODUCTION

The lowest-order hard scattering subprocesses for hadroproduction of heavy quarks Q are $q\bar{q}, gg \rightarrow Q\bar{Q}$. Perturbative or shower calculations of higher-order contributions can be calculated at each invariant mass $M(Q\bar{Q})$ in essentially the same way as for W, Z or Drell-Yan production.

In a perturbative approach based *solely* on the contributions from the $2 \rightarrow 3$ subprocesses $q\bar{q}, gg \rightarrow Q\bar{Q}g$ and $gq \rightarrow Q\bar{Q}q$, a smooth p_T-cutoff $f(p_T)$ on the divergences at $p_T = 0$ is chosen such that the $2 \rightarrow 3$ cross sections reproduce the folded $2 \rightarrow 2$ cross section at each $M(Q\bar{Q})$

$$\sum \int f \sigma(2 \rightarrow 3) = \sum \sigma(2 \rightarrow 2).$$ (8)

Here we neglect the K-factor corrections since the overall normalization also depends on other parameters. The $2 \rightarrow 3$ mechanism then gives $Q\bar{Q}$ events both with and without a recognizable extra jet, as determined by the jet algorithm.

Predictions for lepton final states from heavy-quark decays depend on the $Q \rightarrow$ meson $(Q\bar{q})$ fragmentation, the Q decay matrix, and the semileptonic branching fractions. These properties are taken from analyses of heavy-quark production in e^+e^- collisions. The $b \rightarrow B$ fragmentation is harder than $c \rightarrow D$ and the V-A spectrum of $b \rightarrow \ell$ is harder than $c \rightarrow \ell$. As a consequence, more leptons of b origin survive the acceptance cuts, by about a factor of 4 over charm.

With the UA1 acceptance cuts almost all dimuon events with *non-isolated* muons should be due to two primary heavy-quark parents. For no $B^0 - \bar{B}^0$ mixing, the dominant sources are

$$\mu^- \longleftrightarrow\!\!\times\!\!- \begin{array}{c} b\bar{b} \\ \hline \bar{c}c \end{array} -\!\!\times\!\!\longleftrightarrow \mu^+ \tag{9a}$$

$$\mu^- \longleftrightarrow\!\!\times\!\!- b\bar{b} -\!\!\times\!\!\longrightarrow \bar{c} -\!\!\times\!\!\longrightarrow \mu^- . \tag{9b}$$

In unlike-sign events, each muon is from the primary semileptonic decay of a heavy quark, whereas like-sign events involve one secondary muon from the $b \rightarrow c$ cascade decay. With $B^0 - \bar{B}^0$ mixing there are additional like-sign contributions

$$\mu^- \longleftrightarrow\!\!\times\!\!- b\bar{b} \longrightarrow b -\!\!\times\!\!\longrightarrow \mu^- . \tag{9c}$$

5.1   NO MIXING

For the UA1 integrated luminosity acceptance cuts and dimuon detection efficiency, we predict[11] the following breakdown of non-isolated and isolated dimuon events.

	$+-$		$\pm\pm$
	8	$\Upsilon, \Upsilon', \Upsilon''$	—
isolated	29	DY	—
	8	$b\bar{b}$	0.3
non-isolated	6	DY $+\Upsilon$	—
	92	$b\bar{b}$	32
	36	$c\bar{c}$	—

Comparison of the predicted event rates with the preliminary UA1 data[11] are given in the following table:

	$+-$	$\pm\pm$	
isolated	45	0.3	THEORY
	(44 ± 7)	(7 ± 3)	EXPR
non-isolated	134	32	THEORY
	(106 ± 10)	(55 ± 8)	EXPR

The predicted same-sign rates are considerably below the data. This discrepancy suggests that $B^0 - \bar{B}^0$ mixing effects may be present at an appreciable level.[12,13]

5.2 $B^0 - \bar{B}^0$ Mixing

The mixing parameter ϵ_s is the fractional probability for $B^0(b\bar{s}) \to \bar{B}^0(\bar{b}s)$ decay transitions, and similarly ϵ_d describes $B^0(b\bar{d})$ mixing. In the Standard Model the t-exchange box diagrams are dominant and the value of the mixing parameter depends on the product $B f_B^2 m_t^2 U_{t(s,d)}^2$, where B is the deviation from the vacuum saturation approximation, f_B is the B-decay constant, m_t is the t-quark mass and U is the weak-current matrix. A recent determination[14] of f_B finds $f_B < f_\pi$, in which case $B f_B^2 \lesssim 0.02$ is expected. For $m_t = 40$ GeV this corresponds to mixing parameters $\epsilon_s \lesssim 0.25$ and $\epsilon_d << \epsilon_s$ (see, e.g., Ref. 15). Ignoring b-baryon production and assuming meson production with semileptonic decay to contribute in the ratio $b\bar{u} : b\bar{d} : b\bar{s} = 1 : 1 : 0.5$, the overall probability for a single B to give a wrong-sign muon through mixing is $\epsilon = 0.4\epsilon_d + 0.2\epsilon_s \lesssim 0.05$. From dimuons produced in high-energy e^+e^- collisions, the Mark II collaboration[16] have set an upper bound $\epsilon \leq 0.12$ (90% CL) for $B^0 - \bar{B}^0$ mixing. The predicted dimuon rates with mixing can be easily obtained from those without mixing, from the formula

$$\sigma(i,\epsilon) = [(1-\epsilon)^2 + \epsilon^2]\,\sigma(i,0) + 2\epsilon(1-\epsilon)\,\sigma(j,0)\,, \qquad (10)$$

where $i, j = US, LS$, and $i \neq j$. For the *non-isolated* events the ratio of like-sign to unlike-sign dimuon cross sections for various ϵ values are

ϵ	$R = \sigma(LS)/\sigma(US)$
0	0.24
0.1	0.34
0.2	0.44
EXPR	0.46 ± 0.10

The $\epsilon = 0.1$ value would be a compromise between the e^+e^- upper limit of $\epsilon <$ 0.12 and the $p\bar{p}$ dimuons which suggest large $B^0 \leftrightarrow \bar{B}^0$ mixing. However, it is conceivable that fragmentation is different at PEP and SPS collider energies and that the value $\epsilon = 0.2$ is realized in $p\bar{p}$ collisions.

Figs. 5(c), (d) compare predicted dimuon mass distributions for non-isolated events with the UA1 data. The principal effect of mixing is to change the normalization of the like-sign dimuons.

More spectacular changes due to mixing occur in the *isolated* $\pm\pm$ mass distribution, as shown in Fig. 5(b). Mixing increases the number of expected events from 1 event for $\epsilon = 0$ to 4 events for $\epsilon = 0.1$ (7 events for $\epsilon = 0.2$). However, the enhancement in the 12–14 GeV mass bin is not accounted for.

To summarize, these QCD calculations of dimuons from $b\bar{b}$ and $c\bar{c}$ sources are in broad accord with the UA1 data, provided that there is $B^0 - \bar{B}^0$ mixing of order $\epsilon \gtrsim 0.1$, which is close to the Mark II upper limit and above Standard Model expectations.

6. SEARCH FOR SUPERSYMMETRY

When monojets with large missing p_T (denoted by $\not{p}_T$) were first reported by the UA1 collaboration,[17] it appeared that the signal was considerably in excess of Standard Model backgrounds. This caused considerable excitement in the theoretical community and numerous new physics models were proposed to explain these $\not{p}_T$ events, among which supersymmetry was a leading contender. However, with the data accumulated from the subsequent collider run at $\sqrt{s} = 630$ GeV, the evidence for the monojet signal has become far less compelling. The expected Standard Model monojet backgrounds[18] are comparable to the observed signal, though the background may fall faster with increasing $p_T(j)$ than the data,[17] leaving a possible excess of 7 monojet events at $p_T(j) > 30$ GeV. Should this be a real signal for a new physics source, the associated cross section would be $\sigma \lesssim 25$ pb. This upper limit can be used to place lower bounds on sparticle masses.

6.1 SUSY Scenarios

One favorite new physics scenario for $\not{p}_T$ events was the production of squark pairs.[19] In this particular scenario $m_{\tilde{g}} > m_{\tilde{q}}$ is assumed and the $\not{p}_T$ arises from $\tilde{q} \to q\tilde{\gamma}$ decays. For the calculated monojet cross sections to be less than 25 pb the squark masses must be at least 60 GeV. Then the p_T (monojet) distribution is much harder than that observed and $\sigma(2j)/\sigma(1j) > 2$, whereas UA1 finds more monojets than dijets. Thus prospects for early discovery of squarks at the collider have dimmed.

The $\tilde{g}\tilde{g}$ scenario[20] with $\tilde{g} \to q\bar{q}\tilde{\gamma}$ decays also is incompatible with the present experimental situation on $\not{p}_T$ events and the gluino mass must be above 60 GeV.

Only the heavy-squark/light-gluino production scenario,[21,22] with the $\tilde{g}$ long-lived,[22] remains compatible with a possible monojet excess for $p_T(j)$ near 40 GeV. The $\tilde{q} \to q\tilde{\gamma}$ decay gives a Jacobian distribution with $p_T(j)$ peaked at $\frac{1}{2}m_{\tilde{q}}$, monojets dominate over dijets, and the cross section is of the right order. However, until the experimental situation is clarified, this scenario will also remain in a state of limbo.

6.2 Z Decays to Sparticles

The decays of Z^0 bosons are a potential source of sparticles[23,24] if the $\tilde{\omega}$ and $\tilde{\mu}$ masses are less than $\frac{1}{2}M_Z$. Suppose that gauge fermion ($\tilde{\omega}^\pm$) and/or slepton ($\tilde{\ell}$) decays are kinematically allowed. Then $\mu^+\mu^- + \not{p}_T$ events result from the decays

$$Z \to \tilde{\omega}\bar{\tilde{\omega}} \qquad\qquad Z \to \tilde{\mu}\bar{\tilde{\mu}}$$
$$\tilde{\omega} \to \tilde{\mu}\nu \text{ or } \mu\tilde{\nu} \qquad \tilde{\mu} \to \mu\tilde{\gamma} \tag{11}$$

These dimuon signals are readily observable at the CERN $p\bar{p}$ collider if the $\tilde{\omega}$ and/or $\tilde{\mu}$ masses are below 40 GeV. The contributions of these sparticle sources are dominantly at large $p_T(\mu\mu) \approx \not{p}_T \gtrsim 10$ GeV and large $m(\mu\mu) \gtrsim 10$ GeV, whereas no dimuon events have been reported in this kinematic region. The dilepton data may thereby be used to set significantly improved lower bounds on the $\tilde{\omega}$ and $\tilde{\ell}$ masses.

7. FOURTH FAMILY

There is increasing theoretical interest in the possible existence of a fourth generation of leptons and quarks, which we denote by the symbols

$$\begin{pmatrix} \nu_L \\ L \end{pmatrix} \begin{pmatrix} a \\ v \end{pmatrix}$$

A fourth generation of quarks could conceivably be relevant to CP violation.[25]

In some superstring theories[26] the number of generations is even. In renormalization-group equations for the evolution of couplings, infrared fixed points place upper limits on heavy-fermion masses, typically[27]

$$m_L \lesssim 90 \text{GeV} , \qquad m_Q \lesssim 150 \text{GeV} . \tag{12}$$

Searches can be made[28] at the CERN $p\bar{p}$ collider for fourth-generation states of mass $\lesssim 60$ GeV and even higher mass fermions could be produced at the Fermilab Tevatron.

7.1 Heavy Charged Lepton

The $W \to L\nu_L$ decay mode with subsequent $L \to \nu_L u\bar{d}$, $\nu_L c\bar{s}$ decays is a promising channel[29] to search for a charged heavy lepton $m_L \lesssim 50$ GeV with a light neutrino ν_L. The signal is large $\not{p}_T$ and jets. The L decay will give two narrow jets that are reasonably well separated in azimuth. In addition, there may be an extra jet from a hard gluon produced in association with the W. The $W \to \tau\nu_\tau$ background is identifiable from the narrow τ hadronic jet; the smaller $Z \to \nu\bar{\nu}$ background is dominantly one jet.

7.2 Heavy Neutral Lepton

A heavy neutrino ν_4 produced via $Z \to \nu_4\bar{\nu}_4$ may be detectable through its charged current decays in flight[30] (e.g., $\nu_4 \to e\bar{e}\nu_e$). Recently searches have been made[31] at PEP for such events produced in e^+e^- collisions via virtual Z^0. By searching for secondary vertices between 0.2 mm and 10 cm from the interaction point, a range of neutrino masses and mixing matrix elements $\sum_\ell |U_{4\ell}|^2$ is excluded up to a neutrino mass of 14 GeV.

7.3 Neutrino Counting

An effect of any further light particles (such as neutrinos, heavy neutral or charged leptons, and supersymmetric particles) is to increase the W and Z widths from their Standard Model values. The change in Γ_Z/Γ_W in turn changes the W/Z production ratio

$$\begin{aligned} R &\equiv \frac{\left[\sigma(W^+) + \sigma(W^-)\right] B(W \to e\nu)}{\sigma(Z^0) \, B(Z \to e\bar{e})} \\ &= \frac{\sigma(W^+) + \sigma(W^-)}{\sigma(Z^0)} \frac{\Gamma_Z}{\Gamma_W} \frac{\Gamma(W \to e\nu)}{\Gamma(Z \to e\bar{e})} . \end{aligned} \tag{13}$$

The predicted R value for three generations (with $m_t = 40$ GeV), is[32] $R = 8.9 \pm 0.5$, where allowance has been made for theoretical uncertainties in the evolved quark distributions. The combined UA1 and UA2 measured value is $R = 8.0^{+1.8}_{-1.2}$. Up to two extra light neutrinos are allowed at the 1σ experimental upper bound; appreciable contributions to the Z width from light supersymmetric particles are excluded.[32]

7.4 Heavy-quark Mixing

CP violation in $K_L \to 2\pi$ decay is conventionally explained by the phase in the 3×3 Kobayashi-Maskawa matrix. However, if the current limit[33] of $\Gamma(b \to u)/\Gamma(b \to c) < 0.04$ becomes more stringent, it may then be impossible to account for observed magnitude of the CP impurity parameter ϵ from the box diagram contributions of u, c, and t quarks alone. A fourth family could be important in explaining both ϵ and ϵ'/ϵ in the kaon system and modify expectations for $D^0 - \bar{D}^0$ and $B^0 - \bar{B}^0$ mixing.

With four-quark generations, there are three new angles plus two new phases. Generally it is expected that cross-generational transitions will be suppressed, as in the first matrix:

$$|U| = \begin{array}{c} \\ u \\ c \\ t \\ a \end{array} \begin{array}{cccc} d & s & b & v \\ \left(\begin{array}{cccc} 1- & \theta & & \\ \theta & 1- & \theta^2 & \\ & \theta^2 & 1- & \theta^3 \\ & & \theta^3 & 1- \end{array} \right) \end{array} \tag{14}$$

However, a flavor-flip solution has been proposed[34] of the approximate form

$$|U| = \begin{array}{c} \\ u \\ c \\ t \\ a \end{array} \begin{array}{cccc} d & s & b & v \\ \left(\begin{array}{cccc} 1- & \theta & & \\ \theta & 1- & & \\ & & \theta & 1- \\ & & 1- & \theta \end{array} \right) \end{array} \tag{15}$$

in which $t \to v$ and $a \to b$ are the dominant fourth-generation weak couplings.

The production and decay of t quarks via $W \to t\bar{b}$ and $t \to be\nu$ gives $e + 2$-jet events with distinctive transverse mass characteristics. With the flavor flip matrix, a possible alternative would be $W \to t\bar{v}$ with $t \to ve\nu$ or $\bar{v} \to \bar{c}e\nu$ leptonic decays. With $m_v > 20$ GeV the absolute rate of $e + 2$-jet events is lower than for $W \to t\bar{b}$. Moreover, the transverse masses peak at lower values, as shown in Fig. 6. The present UA1 evidence for a t quark of mass 40 GeV disfavors the flavor flip possibility.

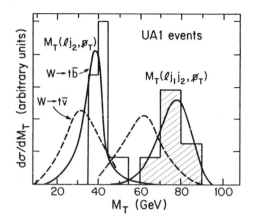

Fig. 6.   Distributions for e+two-jet events from $W \rightarrow t\bar{\nu}$ decay compared with corresponding results for $W \rightarrow t\bar{b}$.

8. CONCLUSION

The early indications of anomalous events in CERN $p\bar{p}$ collider experiments provided a strong stimulus for calculations of QCD backgrounds and for new physics phenomenology. The experimental case for monojets with large p_T is no longer compelling and we are left without a smoking gun to indicate supersymmetry particles at present energies. A useful product of the p_T phenomenology are lower mass bounds for squarks and gluinos of at least 60 GeV and a better understanding of SUSY signals. In the dimuon events the promise of new discovery remains. Perhaps the high observed value of the like-sign/unlike-sign dimuon ratio is due to $B^0 - \bar{B}^0$ mixing. Finally, searches for members of a fourth family remain of great interest in $p\bar{p}$ collider experiments.

ACKNOWLEDGMENTS

I wish to thank R.J.N. Phillips, J. Ohnemus, W.-Y. Keung, T. Gottschalk, and J. Hilgart for valuable assistance in the preparation of this report. This research was supported in part by the University of Wisconsin Research Committee with funds granted by the Wisconsin Alumni Research Foundation, and in part by the U.S. Department of Energy under contract DE-AC02-76ER00881.

REFERENCES

1. Ellis, S. D., Kleiss, R., and Stirling, W. J., Phys. Lett. 154B, 435 (1985).

2. Barger, V., Baer, H., Hagiwara, K., Martin, A. D., and Phillips, R.J.N., Phys. Rev. D29, 1923 (1984); Halzen, F. and Hoyer, P., Phys. Lett. 154B, 324 (1985); Schmitt, I., Sehgal, L. M., Tholol, H., and Zerwas, P., Phys. Lett. 139B, 99 (1984); Kunszt, Z., Pietarinen, F., and Reya, E., Phys. Rev. D21, 733 (1980); Collins, J. C., Soper, D. E., and Sterman, G., Oregon Report OITS 292 (1985), unpublished.

3. Altarelli, G., Ellis, R. K., and Martinelli, G., Phys. Lett. 151B, 57 (1985); Altarelli, G., Ellis, R. K., Greco, M., and Martinelli, G., Nucl. Phys. B246, 12 (1985); Halzen, F., Martin A. D., and Scott, D. M., Phys. Rev. D25, 754 (1982).

4. Gottschalk, T., Caltech Report CALT-68-1241 (1985, unpublished); Paige, F., ISAJET Version 5 (1985, unpublished); Sjöstrand, T., Fermilab Report No. 85/23-T, (1985, unpublished); Odorico, R., Nucl. Phys. B228, 381 (1983), Phys. Rev. D31, 49 (1985); Fox, G. C., in *Proceedings of the 1981 SLAC Summer Institute*, edited by Mosher, A., (1981), p. 181.

5. Barger, V., Gottschalk, T., Ohnemus, J., and Phillips, R.J.N., Wisconsin Report PH/247, (1985).

6. Altarelli, G., Ellis, R. K., and Martinelli, G., Nucl. Phys. B157, 461 (1979); Kubar-André, J., and Paige, F., Phys. Rev. D19, 221 (1979); Abad, J. and Humpert, B., Phys. Lett. 78B, 627 (1978), 80B, 433 (1979).

7. Gottschalk, R., reports at Oregon DPF Workshop (1985).

8. UA1 Collaboration: Arnison, S. *et al.*, Phys. Lett. 122B, 103 (1983), 126B, 398 (1983), 134B, 469 (1984), 147B, 241 (1984); Rubbia, C., report at Kyoto Conference (1985).

9. Barger, V. and Phillips, R.J.N. (unpublished).

10. UA1 Collaboration: Arnison, G. *et al.*, CERN Report No. EP/85-19; Eggert, K., report at Madison Conference (1985); Rubbia, C., report at Kyoto Conference (1985).

11. Barger, V. and Phillips, R.J.N. Wisconsin Report PH/266 (1985).

12. Barger, V. and Phillips, R.J.N., Phys. Lett. 143B, 259 (1984); Wisconsin Report PH/239 (1985).

13. Ali, A., report at KEK Conference (1985); Van Eijk, R., report at Madison Conference (1985); Ali, A. and Jarlskog, C., Phys. Lett. 144B, 266 (1984).

14. Suzuki, M., LBL Report 19983 (1985). See also Claudson, M., Harvard preprint HUTP-81/A016; Golowich, E., Phys. Lett. 91B, 271 (1980).

15. Pakvasa, S., Phys. Rev. $\underline{D28}$, 2915 (1983); Brown, T. and Pakvasa, S., Phys. Rev. $\underline{D31}$, 1661 (1985); Paschos, E. and Türke, U., Nucl. Phys. $\underline{B243}$, 29 (1984); Chau, L.-L. and Keung, W.-Y., Phys. Rev. $\underline{29}$, 592 (1983); Buras, A. J., Slominski, W., and Steger, H., Nucl. Phys. $\underline{B238}$, 529 (1984); Gilman, F. G. and Hagelin, J., Phys. Lett. $\underline{113B}$, 443 (1984); Bigi, I. I. and Sanda, A. J., Phys. Rev. $\underline{29}$, 1393 (1984).

16. Mark II Collaboration: Schaad, T., *et al.*, SLAC Report, 1985; see also Bartel, W. *et al.*, Phys. Lett. $\underline{146B}$, 437 (1984).

17. UA1 Collaboration: Arnison, G. *et al.*, Phys. Lett. $\underline{139B}$, 115 (1984); Rohlf, J., report to this conference; Rubbia, C., report to Kyoto Conference (1985).

18. Cudell, J. R., Halzen, F. and Hikasa, K., Phys. Lett. $\underline{157B}$, 447 (1985); Altarelli, G., Ellis, R. K., and Martinelli, G., CERN Report No. TH.4015, 1985; Ellis, S. D., Kleiss, R. and Stirling, W. J., CERN Report No. TH.4144, 1985; Glover, E.W.N. and Martin, A. D., Durham Report DTP/85/10; Odorico, R., Bologna Report IFUB 85/9.

19. Ellis, J. and Kowalski, H., Nucl. Phys. $\underline{B246}$, 189 (1984); Barger, V., Hagiwara, K., and Keung, W.-Y., Phys. Lett. $\underline{145B}$, 147 (1984); Allan, A. R., Glover, E.W.N., and Martin, A. D., Phys. Lett. $\underline{146B}$, 247 (1984); Haber, H. E. and Kane, G. L., Phys. Lett. $\underline{142B}$, 212 (1984); Glück, M., Reya, E., and Roy, D. P., Phys. Lett. $\underline{155B}$, 284 (1985); Delduc, F., Navelet, H., Peschanski, R., and Savoy, C. A., Phys. Lett. $\underline{155B}$, 173 (1985).

20. Reya, E. and Roy, D. P., Phys. Lett. $\underline{141B}$, 442 (1984), Phys. Rev. Lett. $\underline{52}$, 881 (1984), Dortmund Report 85/1, 1985; Ellis, J. and Kowalski, H., Phys. Lett. $\underline{142B}$, 441 (1984).

21. Herraro, M. J., Ibáñez, K., López, C., and Yndurain, F., Phys. Lett. $\underline{132B}$, 199 (1983), $\underline{145B}$, 430 (1984); Barnett, R. M., Haber, H. E., and Kane, G. L., Phys. Rev. Lett. $\underline{54}$, 1983 (1985); DeRújula, A. and Petronzio, R., CERN Report No. TH.4070/84, 1984; Herzog, F. and Kunszt, Z., Bern Report BUTP-85/8, 1984; Barger, V., Jacobs, S., Woodside, J., and Hagiwara, K., Wisconsin Report PH/232 (1985).

22. Barger, V., Hagiwara, K., Keung, W.-Y., and Woodside, J., Phys. Rev. Lett. $\underline{53}$, 641 (1984); Barger, V., Hagiwara, K., Keung, W.-Y., and Woodside, J., Phys. Rev. $\underline{31}$, 528 (1985).

23. Barger, V., Keung, W.-Y., and Phillips, R.J.N., Phys. Rev. Lett. $\underline{55}$, 166 (1985).

24. Baer, H. and Tata, X., Phys. Lett. 155B, 278 (1985); Baer, H., Ellis, J., Nanopoulos, D. V., and Tata, X., Phys. Lett. 153B, 265 (1985).

25. He, X.-G. and Pakvasa, S., University of Hawaii Report No. UH-511-533-85; Gronau, M. and Schechter, J., SLAC Report No. SLAC-PUB-3451, 1985; Türke, U., Paschos, E. A., Usler, H., and Decker, R., Dortmund Report No. DO-TH 84/26; Hayashi, T., Tanimoto, M., and Wakaizumi, S., Kure Report No. KTCP-8501, 1985.

26. Candelas, P., Horowitz, G. T., and Witten, E., Princeton Report 85-0512.

27. Bagger, J., Dimopoulos, S., and Masso, E., SLAC-PUB-3437, 1984, SLAC-PUB-3587, 1985; Cvetic, M. and Preitschoff, C. R., SLAC-PUB-3685, 1985; Goldberg, H., Northeastern Report NUB 2680, 1985; Paschos, E., Z. Phys. C26, 235 (1984); Halley, J., Paschos, E., and Usler, H., Dortmund Report No. DO-TH 84/24; Hill, C., Phys. Rev. 24, 691 (1981); Pendelton, B. and Ross, G., Phys. Lett. 98B, 291 (1981).

28. Barger, V., Baer, H., Hagiwara, K., and Phillips, R.J.N., Phys. Rev. D30, 947 (1984).

29. Barger, V., Baer, H., Martin, A. D., Glover, E.W.N., and Phillips, R.J.N., Phys. Lett. 133B, 449 (1983), Phys. Rev. D29, 2020 (1984).

30. Thun, R., Phys. Lett. 134B, 459 (1984); Barger, V., Keung, W.-Y., and Phillips, R.J.N., Phys. Lett. 141B, 126 (1984).

31. Gilman, F. J. and Rhie, S. H., Phys. Rev. D32, 324 (1985); Feldman, G. J. et al., SLAC-PUB-3581, 1985; Ash, W. W. et al., SLAC-PUB-359, 1985.

32. Deshpande, N. G., Eilam, G., Barger, V., and Halzen, F., Phys. Rev. Lett. 54, 1757 (1985).

33. Chen, A., et al., Phys. Rev. Lett. 52, 1084 (1984).

34. Pakvasa, S., Sugawara, H., and Tuan, S. F., Z. Phys. C4, 53 (1980).

35. UA1 Collaboration: Arnison, G., et al., Phys. Lett. 147B, 493 (1984); Norton, A., report at New Particles '85 Conference, Madison, 1985.

RARE DECAYS, MONOPOLES, NEUTRINO MASSES, CYGNUS X3 PHENOMENA

D.H. Perkins

Department of Nuclear Physics
University of Oxford
Keble Road
Oxford
ENGLAND

This review will cover the following topics:-

- Proton Decay Experiments
- Monopole searches
- Neutrino masses (including double β-decay)
- Neutrino oscillations
- Cygnus X3 phenomena.

At the present time, there is no convincing evidence for proton decay, neutrinoless beta decay, monopoles, finite neutrino masses or neutrino oscillations. The underground muon data purportedly correlated with the binary Cygnus X3 is very confused and hardly makes a consistent story.

So, most of what I say will be in a sense negative: lower limits on lifetimes for proton decay and double beta decay, upper limits on neutrino masses and mixing. Of course everyone likes to see signals, but their absence can also be highly significant. Recall that the experiments of Michelson, Morley and Miller were among the most important in science in this century.

1) PROTON DECAY EXPERIMENTS.

Let us first remind ourselves of the reasons that have prompted proton decay searches. First is that given by Lee & Yang in 1955, following the extension of the principle of local gauge invariance to non-Abelian fields by Yang and Mills in 1954. If the baryon number B were absolutely conserved, then in the context of local gauge invariance, one expects that a new long-range field coupled to B should exist, just as conservation of electric charge implies the existence of a field (the electromagnetic field) coupled to charge.

There are stringent limits on the coupling of a possible long-range field coupling to baryon number, following as a by-product of experiments carried out over the last 60 years to check the Equivalence Principle of general relativity. The most accurate experiments show that the ratio of gravitational to inertial mass is indeed the same in different materials (Al and Pt) to an accuracy of 1 in 10^{12}. A given mass of platinum contains fewer nucleons, by about 3 parts in 10^4, than the same mass of aluminium, on account of the differences in binding energy per nucleon and of the neutron-proton mass ratio. This result can be expressed as a limit to $K_B/K < 10^{-9}$, where K is the Newtonian constant and K_B is the analogous coupling of the baryon number field. While not proving that such a field does not exist, it is clear that a new field with such a weak coupling would make enormously more difficult the problems that we already have, in seeking to unite the different fundamental interactions.

A second reason for believing that protons are unstable at some level was given in 1966 by Sakharov, following the discovery of CP violation in 1964. In the context of the big bang model, Sakharov emphasized three conditions needed to account for the baryon-antibaryon asymmetry of the universe;- baryon number violation, CP violation (unequal production of left-moving quarks and right-moving antiquarks) and an "arrow of time" (non-equilibrium expansion). Sakharov emphasized the inevitability of proton decay (conversion of baryon to lepton plus mesons) and took for his energy scale the Planck mass, estimating a proton lifetime of 10^{50} years or more.

In grand unified theories, characterized by a mass $M_X \sim 10^{15}$ GeV for the GUT bosons (where the electroweak and strong couplings, α_g, merge together), the proton lifetime via X-exchange is

$$\tau = \frac{A \, M_X^4}{\alpha_g^2 M_p^5}$$

where A is a dimensionless number of order unity and contains the matrix elements for the conversion of a quark pair to a lepton and quark $(Q+Q \to X \to l^+ + \bar{Q})$. In minimal SU(5), $A \simeq 1$ and $\tau(p \to e^+\pi^\circ) = 10^{29\pm1}$ years. This prediction is in definite disagreement with the recent result (Bionta *et al* 1985) from the IMB water Cerenkov experiment, yielding

$$\tau(p \to e^+\pi^\circ) > 2.10^{32} \text{ years.}$$

Other versions of GUT make less exact predictions for the lifetime and cannot be excluded. Those GUTS incorporating supersymmetry involve supersymmetric Higgs exchange and therefore predict decay to the heaviest possible quarks and leptons $i.e.$ the decay modes $p \to \mu^+ K^\circ$, $\bar{\nu}_r K^+$.

Table 1 summarizes the results on total event rates in the different detectors: the tracking-type detectors consisting of steel plates and proportional counters (Kolar Gold Field, KGF), streamer tubes (NUSEX) or flash chambers triggered by Geiger planes (FREJUS); and the water Cerenkov detectors (Kamioka, IMB and HPW). The event rates observed are in good agreement with those calculated by Gaisser et al (1983) assuming they are due entirely to interactions of atmospheric neutrinos. (The uncertainty in expected rates is of order 30%).

Practically all the experiments report proton decay candidates, but no-one actually claims to have observed a clear signal. A small fraction of neutrino reactions (up to 5%) can simulate some modes of proton decay rather closely. Roughly 10% of all the neutrino reactions are charged-current reactions with at least 2 prongs and visible energy 0.9 ± 0.3 GeV. They should be characterized by the equivalence of the net secondary momentum and visible energy, $\Sigma p_s \simeq \Sigma E_s$, with however fluctuations associated with

Table 1: **Event Rates in Proton Decay Dectors.**

Group	KGF	NUSEX	FREJUS	KAMIOKA	IMB
Method	Fe/PWC	Fe/ST	Fe/FC	Water – Cerenkov	
$\neq$ Events	17	31	19	29	401
Kiloton years	0.22	-	0.29	0.27	3.77
Rate/Kty	77±19	151±28	97±25	123±23	142±7
Prediction (Gaisser) of Neutrino event rate	85	132	132	99	132

Fermi motion ($p_f \sim 200$ MeV/c) and scattering/absorption of the secondary hadrons in the parent nucleus or along their trajectories. On the other hand, many nucleon decays should result in ≥ 2 prong events, with a "back-to-back" configuration, that is $\Sigma p_s \simeq 0$, again with the variations due to Fermi motion and secondary scattering.

Two different approaches have been made to the estimation of neutrino background for event configurations simulating proton decay. Fig. 1 shows IMB results on the asymmetry, A, versus Cerenkov energy E_c. The anisotropy is defined as the vector sum of unit vectors drawn from the effective event vertex to each of the hit photo-multipliers, divided by the number of hits. For a single straight track, A will be ~ 0.7 (the cosine of the Cerenkov

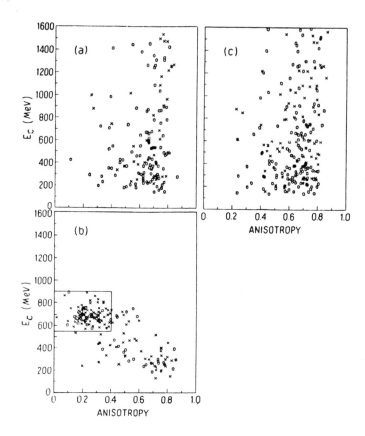

Fig.1 Plot of anistropy A versus Cerenkov energy E_c in IMB experiment.

(a) 169 contained events in 204 day run in underground detector

(b) simulation of a sample of proton decay evants in the mode $p \rightarrow \mu^+\pi^o$

(c) simulation of 204 days of atmospheric neutrino interactions.

angle, 42°), while for two back-to-back tracks, $A \simeq 0$. Fig 1 shows, in (a), the observed values for a sample of 169 events; in (b), the values expected from a computer simulation of the decay $p \rightarrow \mu^+\pi^\circ$, taking account of Fermi motion and scattering; and in (c), the distribution for simulated neutrino events, corresponding to the rates expected for the period of observation (204 days). The neutrino event simulation is based on the topologies of neutrino interactions observed in accelerator experiments using bubble chambers (BNL Ne/H data, ANL D_2 data, Gargamelle freon data). Clearly, the observed underground event distribution is compatible with the expected neutrino background. In particular, no events are observed which could correspond to the decay $p \rightarrow e^+\pi\circ$, giving the limit previously quoted. From this experiment, the 90%CL limits for various decay modes are:-

Table 2: **IMB Limits.**

	($\tau_{min} \times$ branching ratio)
$p \rightarrow e^+\gamma, \mu^+\gamma, e^+\pi^\circ, \mu^+\pi^\circ$	$2\text{-}3.10^{32}$ years.
$n \rightarrow e^+\pi^-, \mu^+\pi^-$	$5\text{-}8.10^{31}$ years.
$p \rightarrow \mu^+\rho^\circ, \mu^+\omega, \nu\rho^+, \nu K^+, \mu K^\circ$	$0.8\text{-}3.10^{31}$ years.

The strength of the tracking calorimeter—type detectors is, of course, that they have much better spatial resolution that the water devices (1cm versus 1m), and can resolve multiprong events and detect non-relativistic secondaries. Their disadvantages are equally clear. They are more expensive ton for ton, and the interpretation of event topology is complicated by nuclear interactions, scattering and absorption in the iron plates. Fig. 2 shows a proton decay candidate from the NUSEX experiment, with the two interpretations: proton decay $p \rightarrow \mu^+K^\circ$, or a neutrino reaction $\nu N \rightarrow \mu\pi N$, with back-scattering of the pion. While, in my opinion, the simulation of neutrino interactions in water from the configurations of accelerator neutrino interactions in deuterium and heavy liquids is dubious at best, the Monte-Carlo simulation of neutrino reactions in iron is virtually impossible. The NUSEX collaboration (alone amongst the experiments on proton decay) have therefore exposed a module of their underground detector to a low energy accelerator (ν_μ) beam. From analysis of 400 interactions they can directly measure the probability that the neutrino background can contain an event like that in Fig. 2. This *measured* background is 0.15 events.

(a)   NUSEX PROTON DECAY CANDIDATE

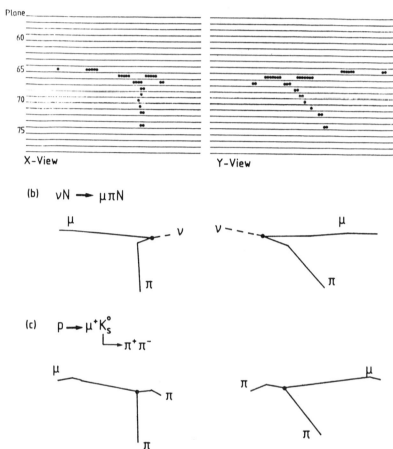

Fig.2 (a)   The orthogonal views of a NUSEX event that is a candidate for proton decay.

(b)   Interpretation of event in terms of a neutrino interaction $\nu\,N \to \mu\pi\,N$, with large-angle scattering of the pion.

(c)   Interpretation according to the decay $p \to \mu^+ K_s^o,\, K_s^o \to \pi^+\pi^-$.

What are the future possibilities for the proton decay experiments? First, one has to remark that proton decay, if it is ever observed, will be a most important happening in its own right, as well as a crucial test of grand unified theories. Second, it is clear that, for particular decay modes—those in the first line of Table 2—the water Cerenkov method has no equal and appears not to be background limited. Therefore, in time the lifetime limits for these modes can be pushed to 10^{33} years and beyond.

If multibody decays to non-relativistic particles are important, then some improvements are certainly possible in the tracking detectors. The Soudan II 1 kiloton detector will be operating within two years. It will be an advance on previous detectors, since it has a honeycomb steel geometry (so secondaries are not lost as in planar steel sheets) and employs drift tube counters which measure ionization (and hence track direction) as well as track coordinates.

Finally, second generation detectors have been proposed, including an ambitious scheme for a 10kton liquid argon calorimeter in the Gran Sasso tunnel. So, the subject is still very much alive and will continue to be so at least to the end of the century.

Table 3: **Resolution of Tracking Detectors.**

Experiment	Mass	Counter dimensions	Resolution	Steel
KGF	140t.	10 x 10cm (PWC	3cm	Planar
NUSEX	150t.	1 x 1cm (ST)	5mm	Planar
FREJUS	900t.	0.5 x 0.5cm (FT)	2mm	Planar
SOUDAN II	1200t.	ϕ=14mm Drift tubes	2mm	Honeycomb
(1987)		+ ionization.		

2) MONOPOLE SEARCHES.

Two methods have been used to search for GUT monopoles, expected to be of mass $\sim 10^{16}$GeV and of cosmic velocities $\beta > 10^{-4}$. First, induction experiments use SQUID magnetometers to detect the change in magnetic flux $\Delta\phi$ when a monopole traverses a (superconducting) loop, where

$$\Delta\phi = 4\pi g$$

and the monopole charge g is an integer times the Dirac charge

$$g = n(\hbar c/2e)$$

SQUIDS are certainly sensitive enough to detect such a flux change, and the method has the great advantage that it is independent of monopole mass or velocity—but has the disadvantage that the sensitive area is restricted. Present limits on monopole flux, with detectors of area up to $1m^2$, are $F < 10^{-11}$ monopoles $cm^{-2}sr^{-1}s^{-1}$.

The second type of experiment relies on monopole interactions with matter. For velocities $\beta > 10^{-2}$, the monopole acts like an electric charge $g\beta$ and the usual Bethe-Bloch ionization loss formula applies. For $10^{-4} < \beta < 10^{-3}$, the energy-loss mechanism is expected to be by atomic excitation via interaction with the magnetic moments of atomic electrons (the Drell effect). De-excitation is by photon emission (e.g. fluorescence), or ionization if the energy is transferred to molecules of low ionisation potential (Penning effect). For example in a helium-methane mixture,

$$He \overset{M}{\underset{\rightarrow}{}} He^* + CH_4 \rightarrow He + CH_4^+ + e^-.$$

A large number of experiments have been carried out with scintillation or proportional counters, with areas upto $1000m^2sr$. It is usual to have at least 2 counter layers and require the time delay appropriate to a slow monopole traversal. Present flux limits by these techniques are $F < 10^{-13} - 10^{-14}$ monopoles $cm^{-2}sr^{-1}s^{-1}$, from the UCSD and Texas A and M experiments.

There is a third class of experiments which requires for their interpretation assumptions about monopole interactions with matter. As an example, it has been proposed that even a slow monopole could attach to a heavy nucleus (Al) in traversing the earths' crust and therefore leave "fossil" tracks in mica, and this approach leads to very low ($F < 10^{-17}$) flux limits (Price, 1984). At the other extreme, it has been proposed that monopoles may calatyze nucleon decay (Rubakov 1982, Callan 1982), leading to flux limits from proton decay experiments of $F < 10^{-14}$ (depending somewhat on monopole velocity and cross-section). It goes without saying that these two competing processes are mutually exclusive.

In summary, the most stringent present limits on GUT monopole fluxes are in the region of $F < 10^{-13} - 10^{-14} cm^{-2} st^{-1} s^{-1}$, that is a factor $10^2 - 10^3$ larger than the "Parker bound" (Turner et al 1982) of $F < 10^{-15}$ for the maximum galactic flux which could be tolerated by the need to maintain the observed galactic magnetic field. This shortcoming will be remedied by forthcoming experiments with large area ionization detectors.

3) NEUTRINO MASSES (DIRECT).

Improved limits on the ν_τ mass from observations at DESY and PEP on the process $e^+e^- \rightarrow \tau^+\tau^-$ have been reported this year from the ARGUS ($m(\bar{\nu}_\tau) < 70$ MeV) and HRS ($m(\nu_\tau) < 82$ MeV) experiments.

A new limit on the ν_μ mass has been reported from SIN, from a precision measurement of the muon momentum in $\pi^+ \to \mu^+\nu_\mu$ decay, using CPT and the most exact value of pion mass (m_{π^-}) from mesic atom spectroscopy. They obtain $m(\nu_\mu) < 0.25$ MeV.

The situation on the ν_e mass has not changed. The ITEP result (Lubimov *et al*) from analysis of the Kurie plot in tritium β-decay is given as $m(\bar{\nu}_e) = 33 \pm 1.1$ eV. The analysis of this data, and the assumptions about energy resolution functions, have not gained general acceptance but the result has not yet been disproved in independent experiments. A number of new experiments, including one with an atomic tritium beam, are beginning to take data and results are expected soon.

Earlier this year, Simpson (1985) reported a kink in the tritium Kurie plot at 1 keV electron energy, (*i.e.* 17 keV from the end point) which was interpreted as evidence for a massive (17keV) neutrino in addition to one of low mass, with a mixing coefficient of 3%. This seems to have been disproved in 4 experiments on S^{35} decay (Q=168 keV). Cosmologists have all heaved a great sigh of relief.

4) DOUBLE β-DECAY.

Let us first recall that neutrinoless double β-decay (DBD) is impossible for massless neutrinos.

In the V-A theory and with a Dirac neutrino $(m_\nu = 0)$, the β-decay and inverse β-decay processes are

$$n \to p + e^- + \bar{\nu}^{RH} \qquad (i)$$

$$\nu^{LH} + n \to p + e^- \qquad (ii)$$

where ν and $\bar{\nu}$ are distinct particles. In the Majorana picture $\nu \equiv \bar{\nu}$ but even so reaction (ii) is forbidden by helicity:-

$$2n \to n + p + \bar{\nu}^{RH} \equiv n + p + \nu^{RH} \not\to 2p + 2e^- \qquad (iii)$$

Neutrinoless DBD is possible if (a) neutrinos are Majorana particles; (b) the interaction is not pure V-A (that is it has a (V+A) admixture of amplitude η; and/or (c) the neutrino mass is finite (so that both LH and RH components exist). In the context of gauge theories, it has however been shown (Kayser, these proceedings) that condition (b) is not sufficient and condition (c) as well as (a) must hold. Since the RH helicity admixture for a massive neutrino in the V-A theory introduces a factor $1 - v/c \simeq m_\nu^2/2E_\nu^2$, the rate of 0ν DBD varies as the mass squared.

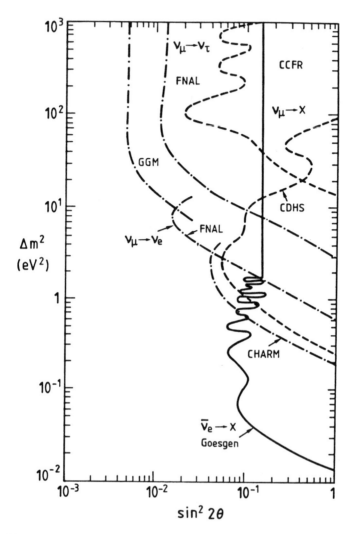

Fig.3 Limits on neutrino oscillations, from accelerator and reactor experiments. The plot is of Δm^2 (the difference in mass squared of the mass eigenstates) versus $\sin^2 2\theta$ where θ is the mixing angle. Regions to the right of the various curves are excluded at 90% CL. (After Boehm 1984).

Nobody has yet detected 0ν DBD *in vivo* (*i.e.* in a physical, non-geochemical experiment). One searches for a line in the electron spectrum at an energy (for the two electrons) corresponding to the Q-value in (iii). The best limit from the Milano group (Fiorini 1984), for the transition $Ge^{76} \rightarrow Se^{76} + 2e^-$ is $\tau > 2.10^{23}$ years, corresponding to $m_{\nu_e} < 3.2$ eV. Because of theoretical uncertainties on nuclear matrix elements, the number is uncertain by a factor of 2.

It is of course only in conflict with the ITEP result ($m = 33$ eV) if the neutrino is a Majorana particle.

5) NEUTRINO OSCILLATIONS.

Last year, an experiment at the Bugey reactor (Cavaignac *et al* 1984) reported a difference in the positron spectra from the reactions

$$\bar{\nu}_e + p \rightarrow e^+ + n$$

at two distances (13.6m and 18.3m) from the reactor core, which was interpreted as evidence for oscillations, with a mass difference $\Delta m^2 \sim 0.1$ eV^2 and a mixing parameter $\sin^2 2\theta \sim 0.3$.

Since then, the Goesgen reactor experiment has been performed at 3 distances (37.9m, 45.2m, 64.7m); they find no evidence for oscillations and this result is in conflict with that at Bugey.

On the accelerator front, nothing much has changed, and the limits (see Fig. 3) are much as last year. A recent PS BEBC experiment (Wisconsin, Athens, Padova, Pisa) with a baseline of 0.82Km and $< E_\nu > = 1.5$ GeV reports $\sin^2 2\theta < 10^{-2}$ and $\Delta m^2 < 0.14$ eV^2. Simultaneous fits to all the accelerator and reactor data submerge the Bugey result and find no evidence for a closed region in the Δm^2 *vs.* $\sin^2 2\theta$ plot.

6) CYGNUS X3 PHENOMENA.

Cygnus X3 is an X-ray binary discovered in 1966, distant $\geq 30,000\,ly$, with right ascension $\alpha = 307.6°$ and declination $\delta = 40.8°$ (*i.e.* it is in the Northern Sky and only dips below the northern horizon by $\sim 10°$ at 50° latitude). It is invisible in the optical region (absorption) but detected at radio, infra-red, X-ray and γ-ray wavelengths. The X-ray and IR data show a clear period of 4.8 hours between sharp minima. This period is assumed to be related to orbital motion and eclipse of a compact, active star (N star, pulsar ...) by a close and diffuse companion. Accurate measurements (Van der Klis *et al* 1981) of the X-ray period give a value $P_0 = 0.1996830$ days, which is lengthening by about $1.2.10^{-9}$ parts per revolution.

124

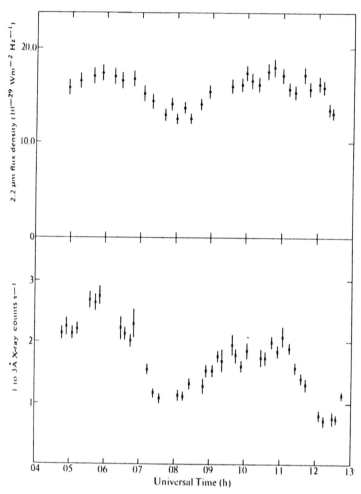

Fig.4 Results of Becklin *et.al.* (1973) of photon fluxes from Cygnus X3 observed during a 9hr period on July 9, 1973, in (top) infra-red ($\lambda = 2.2\mu$ m) and (bottom) X-rays ($\lambda = 1 - 3\text{Å}$). The IR observations were with the Hale 200" telescope, the X-rays from the Copernicus satellite.

Fig. 4 shows typical results in the IR and X-ray signals, in a 9 hour period (Becklin *et al* 1973). It is apparent that the X-ray intensity changes from period to period. The most prominent feature of Cygnus X3 is indeed its extreme variability—factors of 10 in X-ray emission on a timescale of months/years and 20-30% fluctuations on a timescale of seconds. This variability is observed in an extreme form for occasional violent radio bursts.

We are mostly concerned here with γ-ray and underground muon signals. The low energy γ's (~ 1 GeV) have been detected in shower counters carried in balloons and satellites. High energy γ's (1-10^3 TeV) will produce extensive showers in the atmosphere, and have been observed by (a) twin or multiple ground based mirrors detecting the Cerenkov light emitted as the shower particles traverse the atmosphere (b) ground arrays to detect charged particles in the shower. Underground muons have been detected in large proton decay experiments. From about 30 experiments on γ-rays and muons, slightly more than half claimed a signal originating from Cygnus X3; the others had not seen anything, or were not claiming a significant effect. Part of this result is surely because some of the experiments had poor angular resolution or statistics, but there is equally no question that the photon flux from Cygnus fluctuates very widely.

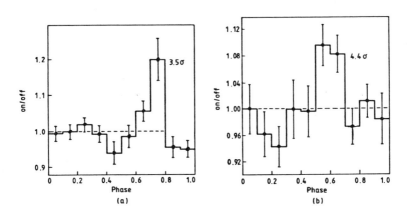

Fig.5 Results from twin mirror detectors of Cerenkov light from air showers of order 1TeV energy. The ratio of signals (on source)/(off source) is shown as a function of Cygnus X3 phase.

(a)   Results of Danaher *et.al.* (1981) of 95 30-minute scans (10mins off, 10mins on, 10mins off source) that is outside and inside a 2.5° aperture.

(b)   Results of Lamb *et.al.* (1982), with similar angular resolution. In both graphs, the off-source mean is shown by the dashed line. The enhancements at phase 0.7-0.8 and 0.5-0.7 are given as 3.5σ and 4.5σ respectively.

Fig. 5 shows typical atmospheric Cerenkov results (Danaher *et al* 1981, Lamb *et al* 1982) showing enhancements at the 3-4σ level in the phase plot, at phases of 0.6-0.8. Fig. 6 shows results from the Kiel air shower array (Samorski and Stamm (1983)). This consists of 28 1m^2 scintillators at ground level and is sensitive to shower sizes $N > 10^5$, that is to incident γ energies of 2000-20000 TeV. The angular resolution (from timing of the shower front) is $\leq 1°$ and they accepted events coinciding with the direction of Cygnus (within 1.5$°$ in δ and 2$°$ in α). The total observation time was 3800 hours. The phase plot shows a large peak at ϕ=0.35 using an early version of the ephemeris. Using the Van der Klis ephemeris, like the other experiments, this would move the peak to ϕ=0.25. These results have been confirmed, with poorer statistical weight because of poor timing/angular resolution, by the Haverah Park experiment (Lloyd-Evans *et al* 1983).

The high energy photon spectrum from Cygnus X3 deduced from the air-shower data is found to be much harder, $f(E)dE \sim E^{-2}dE$, than the general cosmic ray primary spectrum, $f(E)dE \sim E^{-3}dE$.

Now I come to the underground muon data, from the Soudan I, Nusex and Frejus experiments. Fig. 7 shows Soudan I data (Marshak *et al* 1985), for muons pointing within a 3$°$ half-angle cone of Cygnus (off-source data are shown as the horizontal line). Fig. 7(a) shows single muons. The χ^2-probability that the observed deviations about the mean in the phase plot are random is $\sim$10% (χ^2=27.7 for 19 degrees of freedom). However, if one asks in addition the probability that 4 out of 5 bins between ϕ=0.65 and 0.9 should show an excess of 60$\pm$17 events, the random probability drops to $\sim 10^{-3}$. In an attempt to select intervals of time when Cygnus was "active", pairs of muons pointing to Cygnus and occuring within a 30 minute interval were also plotted—see Fig. 7(b). The random probability of the enhancement (ϕ=0.65-0.9) in the phase plot is given as 3.10^{-4}. The data is integrated over a period of $\sim$1 year.

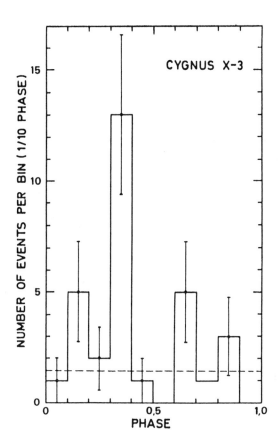

Fig.6 Results of Samorski and Stamm (1983) on air shower events of $> 10^5$ particles recorded
by the Kiel Air-shower array (28 1m^2 scintillators). The plot is of events within $\pm$
1.5 $^\circ$ in declination and right ascension of Cygnus X3 over 3800 hours of observation, in terms of the orbital phase. Dashed line is off-source background (1.44 $\pm$ .04
events/bin). Typical primary energies are 2000-20,000 TeV.

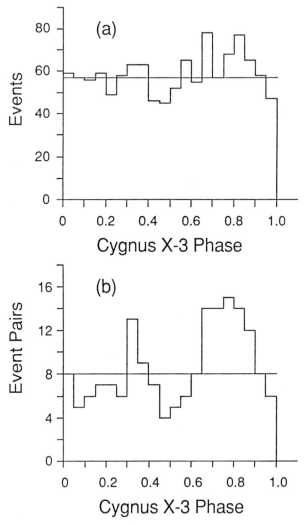

Fig.7 Soudan I results (Marshak *et.al.* 1985) on underground muons (minimum energy ∼ 1TeV at sea-level) recorded in proton decay detector. The phase plots are for muons pointing within 3° half-angle of Cygnus X3.

(a) plot for single muons: off source background shown by dotted line.

(b) plot for pairs of muons recurring within 30 minute interval. The phase plotted is for the mean of the two muon arrival times.

Fig. 8(a) shows the Nusex data (collected over 2.4 years) on muons in the Mont Blanc experiment (Battistoni *et al* 1985). The phase plot includes muons inside a $10° \times 10°$ box in δ and α, centred on Cygnus X3. There is a large peak at ϕ=0.7-0.8. Comparison with off-source background yields a probability that this occurs by random fluctuation as 10^{-4}. Fig. 8(b) shows the distribution in δ and α of the muons in the phase peak. The angular resolution of the detector is given as of order $\pm 1°$, but no muons are observed inside a $2° \times 2°$ box centred on Cygnus. The very wide angular spread of the muons in the phase peak is very strange: in the words of the authors "we are unable to find an easy interpretation of this effect". If the large angular spread is due to some scattering process, it is difficult to understand why it is not larger for the Soudan I experiment, with much lower energy muons (see Table 5).

Finally, Fig. 9 shows the results from the Frejus experiment (Barloutaud *et al* 1985), again inside a $10° \times 10°$ bin. Although there is some enhancement at ϕ=0.6-0.7, the authors do not claim any signal.

In order to discuss those Cygnus-related effects, I show in Table 4 the on-source/off-source ratios for a sample of the air shower experiments (integrated over all phases). Clearly, some of these experiments show a significant excess when pointing at Cygnus, and one can believe that they have seen it. The discrepancy between the Kiel and Leeds experiments can be attributed to the poor angular resolution in the latter case.

Table 4. On-source/off-source Ratios for EAS.

Atmospheric Cerenkov ($E_0 \sim 1$ TeV)	ON/OFF	S.D.
Danaher *et al* (1981)	1.20±.05	3.75σ
Lamb *et al* (1982)	1.04±.01	4σ
Ground air shower arrays ($E_0 = 10^3 - 10^4$ TeV)		
Samorski and Stamm (1983)	2.15 ±0.4	5σ
Lloyd-Evans *et.al.* (1983)	1.044 ±.026	1.7σ

For the underground muons in Table 5 the case is quite different. When integrated over time, there is no significant excess of muons when pointing at the source. This would not, of itself, preclude a significant signal in the phase plot when it is divided into fine bins, but it clearly reduces the credibility of the claims for Soudan I and Nusex to have seen a signal. If Cygnus X3 is generating a muon signal in the phase plot, this has to be

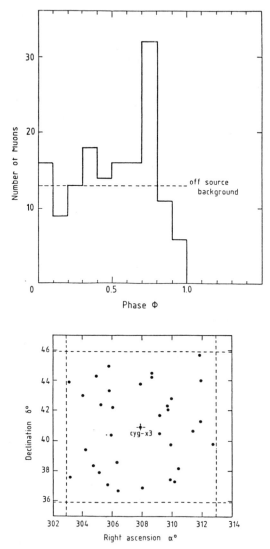

Fig.8 Nusex data (Battistoni *et.al.* 1985) on underground muons ($E_{min} \sim$ 5TeV) in Mont Blanc tunnel, within a 10° × 10° bin in declineation and right ascension centred on Cygmus X3. Upper diagram is the phase plot, the lower diagram the angular distribution of events in the phase peak at 0.7-0.8. The angular resolution in the experiment is of order 1° × 1°.

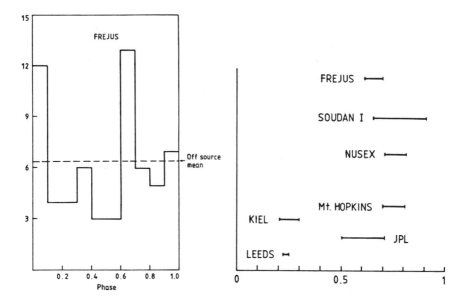

Fig.9 Phase plot of muons in Frejus proton decay experiment, as reported at the Bari conference, for muons within a 10° × 10° bin centred on Cygnus X3. (Berger *et.al.* Bari Conf. 1985).

Fig.10 Positions of peaks in Cygnus X3 phase plots for the Kiel and Haverah Park air-shower experiments (Samorski and Stamm (1983), Lloyd-Evans *et.al.* (1983)), from atmospheric Cerenkov air-shower detectors (JPL, Lamb *et.al.* (1982), Mt. Hopkins (Danaher *et.al.* (1984)) and three proton decay experiments (Frejus, Soudan I, Nusex).

seen as an excess in the on-source/off-source ratio, and until such time as such a clear excess is detected, one has to be very cautious. The fact that phase peaks occur in all three experiments, where they are also visible in the atmospheric Cerenkov data, is at present an interesting curiosity. Fig.10 shows where the peaks are seen in different experiments.

Table 5. On Source/Off Source ratios for Muons.

Experiment	Vertical depth	Minimum Muon energy	On/Off Ratio
Soudan I	1800 mwe	0.6 TeV	Single μ 1.03±.03
			μ pair 1.00±.09
Nusex	5000 mwe	5 TeV	1.16± 0.10
Frejus	5000 mwe	5 TeV	0.93±.09

The reason for the excitement over a possible muon signal associated with Cygnus X3 derives from the following arguments.

First, a correlation in direction and in phase with the X-ray period implies that the radiation from Cygnus to Earth must be neutral and with velocity β very close to that of light : $(1 - \beta) < 0.5$ hours/30,000 years $= 10^{-9}$. For this reason (and no other, as far as I know) it is assumed that the air showers correlated with Cygnus are photon-initiated. However, if the muon signals are taken seriously, then the intermediary radiation cannot be photons, because the underground muon fluxes given by Soudan I and Nusex are about equal to the photon fluxes at the same energy from the air shower data. A calculation of the muon flux produced by photons in the atmosphere (by pair production, or via pion production and decay) requires that the muon flux at energy E should be $< 10^{-3} - 10^{-4}$ of the photon flux at the same energy.†

Neutrinos from Cygnus X3, interacting in the rock and and producing muons in the detectors, are equally excluded because there should be no dependence on zenith angle or depth, (whereas there is a factor $\sim$50 difference in muon flux in the Soudan I and Nusex experiments). In any case, the neutrino fluxes required would be impossibly large.

---

† The existence of energetic γ-rays from Cygnus X3 seems to be established. Since such gammas will certainly produce muons in the atmospheric, some underground muon signal, at the level of $\sim 10^{-4}$ of the photon signal, is expected in any case. However, it seems most unlikely that the underground experiments would ever achieve this level of sensitivity.

Neutrons as an intermediary are also excluded, because their finite lifetime implies neutron energies $> 10^6$ TeV to survive the journey, and the flux required would be several times that of all cosmic rays above this energy.

Assuming a typical source particle energy $E_0 \sim 10$ TeV, the requirement $1 - \beta < 10^{-9}$ implies $\gamma > 3.10^4$ and thus a rest-mass $M_0 \lesssim 1$ GeV and a proper lifetime $\tau_0 > 10^8$ sec. Such neutral particles should have been produced at, and detected near, accelerators long ago. This implies a new form of matter, whimsically christened "cygnets". Theorists have speculated on strange neutral quark matter $(u, d, s$ quarks), $\Lambda\Lambda$ bound states, etc. Clearly, there may yet be new physics in the stars.

What is needed most of all in the future is **vast improvements in the experiments.** First, the signal/background ratios can be improved with better angular resolution in the detectors. The limiting factor is, of course, multiple Coulomb scattering in the rock overburden. For depths greater than 1800 mwe, the rms scattering angle for an $E^{-3.7}$ spectrum of muons (one index steeper than the cosmic ray spectrum) is ~ 6 mrad at 1800 mwe (Soudan I) and ~ 5.5 mrad at 6000 mwe. By contrast, the angular resolution of the Soudan I detector is 25 mrad, so there is a clear order of magnitude to be gained here.

Second, because of the episodic nature of the signal, it will be important to correlate the different underground experiments in real time. Thirdly, it should be possible to correlate results from ground-based (EAS) arrays with the underground data. It is also clear that underground detectors of much larger area are needed to improve the statistical

significance of the results. Finally, a more realistic statistical analysis of the data is necessary. In this respect, it should be noted that in the assessment of the significance of non-uniform phase plots, part of the contribution to χ^2 comes from regions where the flux is below the background level, as well as positive phase peaks, and this is clearly unphysical.

REFERENCES

Barloutaud, R. *et.al.* Report by Ch. Berger at Int.Conf. on High Energy Physics, Bari (1985)

Battistoni, G. *et.al.* submitted to Phys. Lett. B

Becklin, E.E. *et.al.* Nature **245** 302 (1973)

Bionta, R.M. *et.al.* Phys. Rev. Lett. 54, 22 (1985)

Boehm, F. Proc. 5th Workshop on Grand Unification, Brown University, Providence RI (World Scientific) (1984)

Fiorini, E. *et.al.* Proc. 5th Workshop on Grand Unification, Brown University, Providence RI (World Scientific) (1984)

Gaisser, T. *et.al.* Phys.Rev.Lett. **51** 223 (1983)

Lamb, R.C. *et.al.* Nature **296** 543 (1982)

Lee. T.D. and C.N. Yang, Phys.Rev. **98** 1501 (1955)

Lloyd-Evans J. *et.al.* Nature **305** 784 (1983)

Marshak, M.L. *et.al.* Phys.Rev.Lett. **54**, 2079 (1985)

Price, P.B. CERN EP/84-28 (1984)

Sakharov, A. JETP Letters **5**, 24 (1967)

Samorski, M. and W. Stamm, Astr.J. **268** 217 (1983)

Simpson, J.J. Phys.Rev.Lett. **S4**, 1891 (1985)

Turner, M.S. *et.al.* Phys.Rev. **D26** 1296 (1982)

Van der Klis, M. and J.M. Bonnet-Bidand, Astron. & Astroph. **95** L5 (1981).

DIFFERENCE EQUATIONS AS THE BASIS
OF FUNDAMENTAL PHYSICAL THEORIES

T. D. Lee

Columbia University, New York, N. Y. 10027

1. TIME AS A DYNAMICAL VARIABLE

In this talk I would like to regard time as a dynamical variable. As we shall see, this new theory is formulated in terms of difference equations, instead of the usual differential equations. We will first review briefly the classical theory of this new mechanics, and then go over to the quantum theory.

1.1 Classical Mechanics

Take the simplest example of a one-dimensional particle of unit mass moving in a potential $V(x)$. In the usual continuum mechanics the action is

$$A(x(t)) = \int_0^T \left[\tfrac{1}{2} \dot{x}^2 - V(x) \right] dt \ , \tag{1}$$

where $x(t)$ can be any smooth function of the time t. Keeping fixed the initial and final positions, say x_0 and x_f, at $t = 0$ and T, we determine the orbit of the particle by the stationary condition

$$\frac{\delta A}{\delta(x(t))} = 0 \tag{2}$$

which leads to Newton's equation

$$\ddot{x} = -dV/dx \ . \tag{3}$$

This research was supported in part by the U.S. Department of Energy.

In the above, x is the dynamical variable and t is merely a parameter. Next, we shall see how this customary approach may be modified in the discrete version.

Let the initial and final positions of the particle be the same

$$x_0 \text{ at } t = 0 \quad \text{and} \quad x_f \text{ at } t = T . \tag{4}$$

In the discrete mechanics we restrict the usual smooth path to a "discrete path" $x_D(t)$, which is continuous but piece-wise linear.

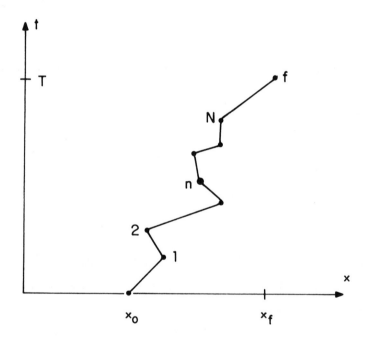

Figure 1 .

As shown in the above figure, such a path is characterized by its vertices $n = 1, 2, \cdots, N$, each of which carries a space-time position x_n and t_n. For convenience, we shall label these n so that they are arranged in a time-ordered sequence

$$0 < t_1 < t_2 < t_3 < \cdots < t_N < T . \tag{5}$$

Nearest neighboring vertices are then linked by straight lines to form the discrete path $x_D(t)$, which appears as a one-dimensional lattice with n as lattice sites. In the following, we shall keep the site-density

$$\frac{N}{T} \equiv \frac{1}{\ell} \tag{6}$$

fixed, and regard ℓ as a fundamental constant of the theory. The action integral (1) evaluated on such a discrete path $x_D(t)$ is

$$A_D = A(x_D(t)) = \tfrac{1}{2} \sum_n \left[\frac{(x_n - x_{n-1})^2}{t_n - t_{n-1}} - (t_n - t_{n-1}) \, \overline{V}(n) \right] \tag{7}$$

where

$$\overline{V}(n) = \frac{1}{x_n - x_{n-1}} \int_{x_{n-1}}^{x_n} V(x)\, dx \tag{8}$$

is the average of $V(x)$ along the straight line between x_{n-1} and x_n.

Because the path $x_D(t)$ is completely specified by its vertices $n(x_n, t_n)$ a variation in $x_D(t)$ is equivalent to a variation in all the positions of its vertices

$$d\left[x_D(t)\right] = \prod_n \left[dx_n\right]\left[dt_n\right] . \tag{9}$$

Correspondingly, the dynamical equation (2) becomes the difference equations

$$\frac{\partial A_D}{\partial x_n} = 0 \tag{10}$$

and

$$\frac{\partial A_D}{\partial t_n} = 0 . \tag{11}$$

We see that in this new mechanics the roles of x_n and t_n are quite similar. Both appear as dynamical variables. For each x_n or t_n we have one difference equation, (10) or (11). The former gives Newton's law on the lattice and the latter gives the conservation of energy

$$E_n \equiv \frac{1}{2}\left(\frac{x_n - x_{n-1}}{t_n - t_{n-1}}\right)^2 + \overline{V}(n) = E_{n+1} \quad . \tag{11}$$

In the usual continuum mechanics, conservation of energy is a consequence of Newton's equation. Here, these two equations (10) and (11) are independent. If we regard the total momentum of the system as the sum of the particle-momentum and the impulse generated by the potential, then Newton's equation (10) is equivalent to the conservation of momentum. Because the action A_D is stationary under a variation in x_n and in t_n for every n, the discrete theory retains the translational invariance of both space and time, and that leaves the conservation laws of energy and momentum intact.

When $\ell \to 0$, the site density $\to \infty$ and the discrete path $x_D(t)$ can assume the form of any smooth path $x(t)$: consequently the discrete mechanics approaches the usual continuum mechanics. However, we are interested in $\ell \neq 0$, in which case the discrete mechanics differs from the continuum theory.

1.2 Nonrelativistic Quantum Mechanics

When we go over from classical to quantum mechanics, in the usual continuum theory the particle can take on any smooth path $x(t)$; each path carries an amplitude e^{iA} where $A = A(x(t))$ is the same action integral (1). In Feynman's path integration formalism, the matrix element of e^{-iHT} in the usual continuum quantum mechanics is given by

$$\langle x_f | e^{-iHT} | x_0 \rangle = \int e^{iA(x(t))} d[x(t)] \quad , \tag{12}$$

in which all paths $x(t)$ have the same end-points (4) and

$$H = -\frac{1}{2}\frac{\partial^2}{\partial x^2} + V(x) \quad . \tag{13}$$

Sometimes it is more convenient to consider the analytic continuation of T to $-iT$. The operator e^{-iHT} becomes then e^{-HT}, and its matrix element is given by

$$\langle x_f | e^{-HT} | x_0 \rangle = \int e^{-\mathcal{A}(x(t))} d[x(t)] \tag{14}$$

where

$$\mathcal{A}(x(t)) = \int_0^T \left[\frac{1}{2}\dot{x}^2 + V(x) \right] dt \quad . \tag{15}$$

In the corresponding discrete theory, we again restrict the particle to move only along the discrete path $x_D(t)$. By using (7) and (9), we see that (12) becomes

$$\int e^{iA_D} \prod_n \left[dx_n \right] \left[dt_n \right] \quad . \tag{16}$$

Likewise, (14) and (15) become

$$< x_f \mid G_N(T) \mid x_0 > \equiv \int e^{-\mathcal{A}_D} \prod_{n=1}^{N} \left[dx_n \right] \left[dt_n \right] \tag{17}$$

where

$$\mathcal{A}_D = \mathcal{A}(x_D(t)) \quad . \tag{18}$$

When the vertices $n = 1, 2, \cdots$ are arranged in a time-ordered sequence (5), by using (15) and (18) we see that the discrete action $\mathcal{A}_D$ is given by

$$\mathcal{A}_D = \sum_{n=1}^{N+1} \left[\frac{(x_n - x_{n-1})^2}{2(t_n - t_{n-1})} + (t_n - t_{n-1}) \, \overline{V}(n) \right] \tag{19}$$

with $x_{N+1} = x_f$ and $t_{N+1} = T$, as shown in Figure 1.

In the integration over $\prod_n \left[dt_n \right]$, whenever t_i appears larger than, say, t_{i+1} , we should re-link the vertices so that the newly linked ones are in a time-ordered sequence. Alternatively, we may re-label them so that (5) remains valid; such a re-labeling of vertices clearly does not change the discrete path $x_D(t)$. [This is necessary because in the usual nonrelativistic continuum mechanics the path $x(t)$ is a single-valued function of t.]

In the quantum version of the discrete mechanics it is more convenient to regard the constraint (6) as a condition on the average site-density. This can be most easily arranged by considering an ensemble sum over N :

$$\mathcal{G}(T, \ell) \equiv \sum_{N=0}^{\infty} \frac{1}{N!} \left(\frac{1}{\ell} \right)^N G_N(T) \tag{20}$$

where $G_N(T)$ refers to the matrix defined by (17). One may readily verify that this Green's function satisfies

$$\frac{\partial}{\partial(1/\ell)}\, \mathcal{G}(T, \ell) = \int_0^T \mathcal{G}(\tau, \ell)\, \mathcal{G}(T-\tau, \ell)\, d\tau, \qquad (21)$$

from which it follows that for large T the operator $\mathcal{G}(T, \ell)$ becomes

$$\mathcal{G}(T, \ell) \sim e^{-\mathcal{H}\ell T} \qquad (22)$$

where $\mathcal{H}$ is Hermitian. When $\ell \to 0$, $\mathcal{H}$ reduces to the continuum Hamiltonian H, given by (13). The analytic continuation of $\mathcal{G}(T, \ell)$ from T to iT leads at large T to the unitary operator $e^{-i\mathcal{H}\ell T}$, which is the S-matrix of the theory. Therefore, the unitarity of the S-matrix is maintained in the new mechanics.

2. RELATIVISTIC QUANTUM FIELD THEORY

As an example, let $\phi(x)$ be a scalar field in the usual continuum theory with x denoting the space-time coordinates. In the path integration formulation the operator e^{-HT} is given by, similar to (14),

$$e^{-HT} = \int e^{-\mathcal{A}} \left[d\phi(x) \right] \qquad (23)$$

where H is the Hamiltonian operator, $\mathcal{A}$ the usual continuum action in the euclidean space and T the total "time" interval. (Here, as in (14)–(15), "time" refers to the euclidean time.) Because in the usual continuum theory the space-time coordinates x are parameters, and only $\phi(x)$ are dynamical variables, the functional integration in (23) is over $\left[d\phi(x) \right]$, not $\left[dx \right]$.

In the discrete version, we impose a constraint on the (average) number N of experiments that can be performed within any given space-time volume Ω, with $N/\Omega \equiv \ell^{-4} =$ fundamental constant. Each measurement determines the field $\phi(i)$ as well as the space-time position $x(i)$ with $i = 1, 2, \cdots, N$. The i will be referred to as lattice sites, as illustrated by Figure 2(a).

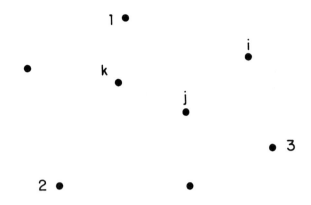

Figure 2(a).

As we shall see, the Green's function (23) will be replaced by

$$\int e^{-A_D} \left[dx(i) \right] \left[d\phi(i) \right] . \tag{24}$$

Because $\phi(i)$ and $x(i)$ are all dynamical variables, in the discrete theory we in-tegrate over $\left[d\phi(i) \right]$ as well as $\left[dx(i) \right]$. The latter integration makes it ob-vious that rotational and translational symmetries can be maintained in the dis-crete theory.

To simulate the local character of the usual continuum theory, each site in the discrete theory is coupled only to its neighboring sites, as illustrated in Figure 2(b). The whole volume is then divided into triangles if the dimension of $x(i)$ is $d = 2$, tetrahedra if $d = 3$, 4-simplices when $d = 4$, etc. An example of such a simplicial lattice when $d = 2$ appears in Figure 2(b).

We give the algorithm[2] of linking an arbitrary distribution of sites into a simplicial lattice for any dimension d: select any group of $d + 1$ sites. Consider the hypersphere (in the d-dimensional euclidean space) whose surface passes

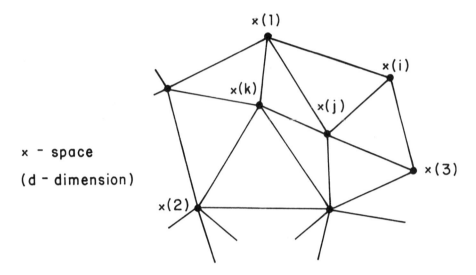

x – space

(d – dimension)

Figure 2(b).

through these $d + 1$ sites. If the interior of the sphere is empty of sites, link these sites to form a d - simplex; otherwise, do nothing. Proceed to another group of $d + 1$ sites, and repeat the same steps. The d - simplices thus formed never intersect each other, and the sum total of their volumes fills the entire space.

Each site i carries, in addition to its space-time coordinates $x(i)$, also a $\phi(i)$. Viewed in the x - ϕ space, the lattice forms a d - dimensional surface represented by $\phi_D(x)$, called the "discrete" function; it is continuous but piecewise flat within each d - simplex as illustrated in Figure 2(c).

The discrete action $\mathcal{A}_D$ in (24) can be readily evaluated by using the usual continuum action $\mathcal{A}(\phi(x))$, but restricting $\phi(x)$ to the discrete function:

$$\mathcal{A}_D \equiv \mathcal{A}(\phi_D(x)) \ . \tag{25}$$

For example, if

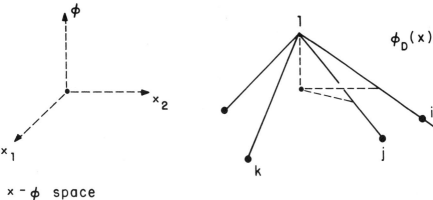

x - φ   space

(d + 1   dimension)

Figure 2(c)

$$\mathcal{A}(\phi(x)) = \int \left[\tfrac{1}{2}(\vec{\nabla}\phi)^2 + V(\phi) \right] dx \qquad (26)$$

where dx is the d-dimensional volume element in the x-space, then setting
$\phi(x)$ to be the discrete function $\phi_D(x)$ we find

$$\mathcal{A}_D = \mathcal{A}(\phi_D(x)) = \tfrac{1}{2} \sum_{\ell_{ij}} \lambda_{ij} \left[\phi(i) - \phi(j) \right]^2 + \sum_i \omega_i V(\phi(i)) \qquad (27)$$

where the first sum is over all links ℓ_{ij} and the second over all sites i , ω_i is
the volume of the Voronoi cell that is dual to the site i , and[3]

$$\lambda_{ij} = -\frac{1}{d^2} \sum \frac{1}{V(ij)} \vec{\tau}(i) \cdot \vec{\tau}(j) \qquad (28)$$

in which the sum extends over all d-simplices V(ij) that share the link ℓ_{ij} .
In V(ij) , each vertex, say k , faces a (d-1)-dimensional simplex $\tau(k)$. In
(28), V(ij) denotes also the volume of the d-simplex and $\vec{\tau}(i)$ is the outward

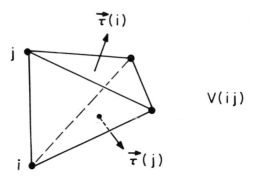

$\vec{\tau}(i)$

j

$V(ij)$

i $\vec{\tau}(j)$

Figure 3.

normal vector of $\tau(i)$ times its $(d-1)$-dimensional volume, as illustrated in Figure 3. As in the previous section, mathematically the discrete theory can be regarded as a special case of the usual continuum theory: one in which $\phi(x)$ is restricted to those continuous but piece-wise flat functions $\phi_D(x)$ with a fixed average density of vertices (i.e., lattice sites). Because the site density is an invariant, rotational and translational invariances can both be preserved in the discrete theory.

Since the discrete surface, described by $\phi_D(x)$, is characterized by the positions $\phi(i)$ and $x(i)$ of its vertices, a variation over the functional space $\phi(x)$ in the usual continuum theory becomes

$$\left[d\,\phi_D(x) \right] = \prod_i \left[d\,\phi(i) \right] \left[dx(i) \right] . \tag{29}$$

Correspondingly (23) becomes (24). As $x(i)$ changes, the linking algorithm keeps track of how these vertices should be linked, so that the discrete action $\mathcal{A}_D$ is extensive; i.e., $\mathcal{A}_D$ is proportional to the overall space-time volume Ω when Ω is large. Thereby, the unitarity of the S-matrix can be established, as before.

In the usual continuum theory, the equation of motion is given by the partial differential equation

$$\frac{\delta A(\phi(x))}{\delta \phi(x)} = 0 \quad . \tag{30}$$

Here in the discrete version it is replaced by the difference equations

$$\frac{\partial A_D}{\partial \phi(i)} = 0 \quad \text{and} \quad \frac{\partial A_D}{\partial x(i)} = 0 \quad ; \tag{31}$$

the former is the field equation on the lattice and the latter expresses the conservation law of the energy-momentum tensor.

3. LATTICE GRAVITY

The usual Einstein action in general relativity is

$$A(S) = \int_S \sqrt{|g|} \, R \, dx \tag{32}$$

where S is a d-dimensional smooth continuous surface, $|g|$ is the absolute value of the determinant of the matrix of the metric tensor $g_{\mu\nu}$ on S, R is the scalar curvature and dx is the d-dimensional volume element in the space-time coordinate x.

For lattice gravity, we consider first a (random) lattice $\overset{o}{L}$ in a flat d-dimensional eucliden space R_d. Label each site by $i = 1, 2, \cdots$. For every linked pair of sites i and j there is a link-length $\overset{o}{\ell}_{ij}$.

Consider now an arbitrary variation

$$\overset{o}{\ell}_{ij} \to \ell_{ij} \quad . \tag{33}$$

Correspondingly, each d-simplex, say $\overset{o}{\tau}$, in $\overset{o}{L}$ remains a d-simplex τ with the same vertices, but different link-lengths. These new link-lengths ℓ_{ij} are assumed to satisfy all simplicial inequalities, so that each d-simplex τ, by itself, can still be realized in a flat d-dimensional space R_d. In general the entire new lattice cannot fit into R_d. This then defines[4] a d-dimensional non-flat lattice surface L.

Sometimes, it is convenient to embed L in a flat space R_N. This is possible if

$$N = d + n$$

is sufficiently large; in that case

$$\ell_{ij}^2 - [\vec{r}(i) - \vec{r}(j)]^2 \quad \text{in} \quad R_N \tag{34}$$

with $\vec{r}(i)$ the cartesian N-dimensional position vector of the ith site in R_N.

Next we wish to evaluate Einstein's action (32) when S is restricted to the lattice surface L. At first sight, it might appear difficult because the metric g_{ij} would change discontinuously from simplex to simplex, the Christoffel symbol would then acquire δ-functions, and the scalar curvature δ'-functions. Since Einstein's action is nonlinear in g_{ij}, one might expect the resulting expression to be totally unmanageable. It turns out that this is not so.

It can be shown that the Einstein formula (32) evaluated on any d-dimensional lattice space L gives the discrete action[5,6]

$$A(L) \equiv \int_L \sqrt{|g|}\, \mathcal{R}\, dx \tag{35}$$

$$= 2 \sum_s s\, \epsilon_s \tag{36}$$

where dx is the d-dimensional volume element, s is the volume of the $(D-2)$ simplex, ϵ_s is Regge's deficit angle around s and the sum extends over all s in the lattice. [See ref. 7 for the definition of ϵ_s.] The right-hand side of (36) is precisely the formula of Regge calculus.[7]

In Regge's original approach, he considered the discrete action as an approximation to Einstein's continuum action. Here we are reversing the role and regarding the discrete action $A(L)$ as more fundamental. It is therefore satisfying to realize that Regge's action is identical to Einstein's action, but evaluated on L.

The quintessence of Einstein's theory of general relativity lies in its invariance under a general coordinate transformation

$$x \rightarrow x' \tag{37}$$

that leaves ds^2 unchanged. Since the action for the lattice space L is the discrete action

$$A_D \equiv A(L) = \int_L \sqrt{|g|} \; \mathcal{R} \; dx = 2\sum_s s \, \epsilon_s \; , \tag{38}$$

the discrete theory clearly remains invariant under the coordinate transformation (37). Thus, the entire apparatus of coordinate invariance in the usual continuum theory automatically applies to the lattice theory as well. In addition, as we shall see, the lattice theory enjoys still another totally new class of symmetries which does not exist in the usual continuum theory. Aesthetically, this adds greatly to the appeal of lattice gravity. For physical applications, when the link-length ℓ is small, our general formula (38) insures that all known tests of general relativity are automatically satisfied. Furthermore, by keeping ℓ nonzero, we see that the lattice action A_D per volume possesses only a finite degree of freedom. The normal difficulty of ultra-violet divergence that one encounters in quantum gravity disappears in the lattice theory. All these suggest that the lattice theory with a nonzero ℓ may be more fundamental. The usual continuum theory is quite possibly only an approximation.

To amplify the aforementioned symmetry properties, let us consider any lattice L. From (38), we see that the discrete action A_D, through its right-hand side, is a function of the link-lengths ℓ_{ij}:

$$A_D = A_D(\ell_{ij}) \; . \tag{39}$$

We may also characterize the lattice by other means of parametrization. We assume all the lattice sites i to lie on a D-dimensional smooth enveloping surface S, with $z_\mu(i)$ as the coordinates of the site i on S, where $\mu = 1, 2, \cdots, d$. Thus, ℓ_{ij} can also be determined by giving S and $z_\mu(i)$. Hence, we can also express A_D as a function of the enveloping surface S and the site positions on S:

$$A_D = A_D(S, z_\mu(i)) \; . \tag{40}$$

Thus, we can have new symmetry transformations:

(i) fix $z_\mu(i)$, but vary $S \to S'$,

and (ii) fix S, but vary $z_\mu(i) \to z'_\mu(i)$.

These symmetries are exact if ℓ_{ij} are unchanged; they can be approximate even if ℓ_{ij} does change, provided that the link-lengths are sufficiently small and $\sqrt{|g|}\; R\, dx$ remains the same on the enveloping surface.

In the usual continuum theory, the physical space-time points and the underlying four-dimensional manifold are the same. Here, they are distinct; the former is related to measurements, while the latter is purely a mathematical artifice (like the choice of gauge in the usual continuum theory of a spin 1, or 2 , field).

4. CONCLUDING REMARKS

By regarding space and time as dynamical variables, a fundamental length ℓ can be introduced which removes all ultraviolet divergences, and therefore may make quantization of gravity possible. As we have shown, such a discrete theory can also be viewed as the mathematical limit of the usual continuum theory, but with a fixed density of lattice sites. Because this is an invariant constraint, the discrete theory shares the same symmetries of the usual continuum theory. In this way, we have succeeded in the creation of theories with finite degrees of freedom, but which retain all the good properties of the usual continuum theory. We suggest that this discrete formulation might be more fundamental. If so, our basic physical laws should be expressed in terms of difference equations, and the usual differential formulation would be only an approximation.

References

1. See, e.g., T. D. Lee, proceedings of the International School of Subnuclear Physics, Erice, 1983, and the references given there.

2. N. H. Christ, R. Friedberg and T. D. Lee, Nucl.Phys. B202, 89 (1982).

3. R. Friedberg and T. D. Lee, unpublished.

4. For physical application to general relativity, in order to maintain the quasi-local character of the discrete action, we must link only neighboring sites. Thus, when the new link-lengths ℓ_{ij} are too large, the sites have to be re-linked. Details will be given elsewhere.

5. R. Sorkin, Phys.Rev. D12, 385 (1975).

6. R. Friedberg and T. D. Lee, Nucl.Phys. B242, 145 (1984).

7. T. Regge, Nuovo Cimento 19, 558 (1961).

e^+e^- HIGHLIGHTS 1985

Karl Berkelman

Laboratory of Nuclear Studies, Cornell University

Ithaca, NY 14853

ABSTRACT

A dozen or so detectors at DESY, SLAC, and Cornell
have been producing results on a wide variety of
physics topics, year in year out: searches for
everything from glueballs to winos, weak-
electromagnetic asymmetries, weak decays of τ leptons
and D, F, and B mesons, tests of perturbative QCD in
the e^+e^- total annihilation cross section, in gluon
bremsstrahlung, and in quarkonium annihilation rates,
soft QCD studies of spectroscopy and fragmentation,
and so on. This year is no exception. You will
forgive me if I don't do justice to all of this work,
but concentrate on a few areas in which I think the
most significant recent progress has been made.

1. SEARCHES

1.1 The xi(2230)

The news is that the peak at 2.2 GeV in $K\bar{K}$ seen several years ago
by the Mark III collaboration[1] in radiative ψ decays is still there.
There were some doubts raised by inconsistencies in the original data
and by the fact that it did not show up in the data of the Orsay DM2
group. But now Mark III has extensive new ψ data[2] which show the
$\xi(2230)$ with 4 standard deviation significance in both K^+K^- and $K^o\bar{K}^o$.
What it is is still a mystery. Suggestions include Higgs

(ruled out for most models), glueball, high-spin $s\bar{s}$ meson, or four-quark state.

1.2 The zeta(8300)

Here the news is that the peak reported last year by the Crystal Ball[3] at 1.07 GeV in the single photon inclusive spectrum from upsilon decays is not really there. Although it was originally seen with 4 and 3 standard deviation significance in two independent data samples, later runs by the Crystal Ball,[4] CUSB,[5] ARGUS,[6] and CLEO[7] collaborations showed no effect. The best upper limit, from both the Crystal Ball and CUSB, is a branching ratio of 0.08%, well below the 0.5% originally reported. The CUSB data at higher photon energies actually ule out the upsilon radiative decay to a standard neutral Higgs of mass below about 5 GeV, at least if you ignore QCD corrections in the cross section calculation.

1.3 Monojets

Last year UA1[8] reported five events of $p\bar{p}$ annihilation into a single jet of transverse momentum greater than 40 GeV/c and mass apparently in the 3 to 10 GeV range. This year the monojets are still with us, but are less distinctive. They may be somewhat improbable examples of $W \rightarrow \tau\nu$ or of QCD background, or they may still be new physics. Anyway, they provoked a flurry of theoretical activity. In the scenario suggested by Glashow and Manohar[9] they come from the decay of a Z° to two neutral Higgs or Higgsinos or heavy neutrinos, one stable (or decays into light neutrinos in the latter case) the other producing a quark-antiquark pair which becomes the monojet. The calculated branching ratio of about 3% agrees with the rate observed.

Although the beam energies at PEP and PETRA are well below the Z° mass, there should be enough of a tail to the Z° Breit-Wigner to give a rate for $e^+e^- \rightarrow Z^\circ$ which would imply about a dozen monojet events in a 100 pb^{-1} run. Table I shows the results from five experiments.

Table I. Monojet searches

Expt.	Ref.	Cand's	Bkgd	Br upper limit
HRS	10	1	3.3	1.5%
Mark II	11	2	2	0.7%
MAC	12	11	13.2	0.5%
JADE	13	0	0	0.6%
CELLO	14	0	0	1.2%

The measured upper limits for the Z° to monojet branching ratio (assuming a mass between 3 and 10 GeV), are all well below the predicted 3%. If the monojets are real, they do not come from Z° decays.

1.4 Supersymmetry partners

There are several kinds of e^+e^- reactions which could give evidence for the existence of selectrons and photinos (see Fig. 1). The experimental signatures are as follows:

A1 (if $\tilde{e} \rightarrow e\tilde{\gamma}$) -- an electron pair with missing momentum,

A2 (if $\tilde{e}$ is stable) -- higher apparent muon pair yield,

B -- a single electron only observed,

C1 (if $\tilde{\gamma}$ is stable) -- a radiated photon only,

C2 (if $\tilde{\gamma} \rightarrow \tilde{\gamma}G$) -- a photon pair with missing momentum.

Most experiments at PETRA and PEP have searched for such events, but none have been found above background. The implied limits on selectron and photino masses are related and are model dependent. As an example I show in Fig. 2 the regions excluded under the assumptions of a stable photino and equal masses for right and left handed selectrons. For photino masses above 10 GeV the best lower limit on the selectron mass is about 22 GeV (95% conf.), set by JADE[15] and CELLO[16] using A1 and A2. For lower photino masses the best limits

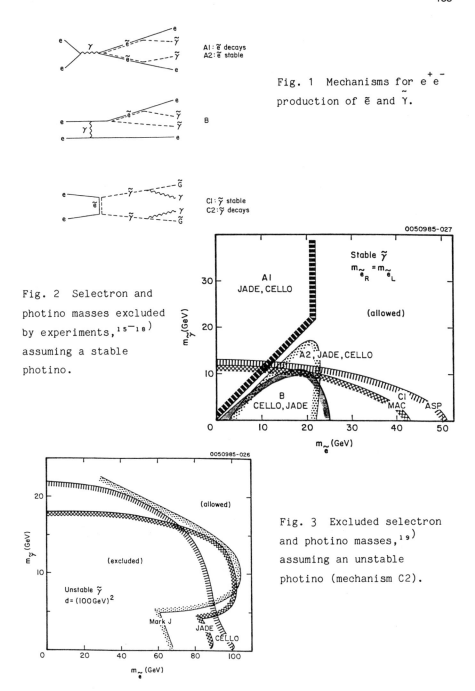

Fig. 1 Mechanisms for e^+e^- production of $\tilde{e}$ and $\tilde{\gamma}$.

Fig. 2 Selectron and photino masses excluded by experiments,[15−18] assuming a stable photino.

Fig. 3 Excluded selectron and photino masses,[19] assuming an unstable photino (mechanism C2).

come from MAC[17] and ASP[18] using process C: $m_{\tilde{e}} > 51$ GeV at $m_{\tilde{\gamma}}=0$. ASP is a new experiment at PEP designed for this search; it uses a lead glass shower detector system with a veto system sensitive down to 21 mrad. For the case of an unstable photino, decaying radiatively into the lightest supersymmetry partner (goldstino?), the best limits are from PETRA experiments[19] and are shown in Fig. 3. Some other recently measured limits are shown below.

Table II. Limits on masses of supersymmetric partners

Particle	Expt.	Ref.	m_{min}
$\tilde{\mu}$	MkJ,JADE	20, 21	20 GeV
$\tilde{\tau}$	MkJ	20	17 GeV
$\tilde{W}/\tilde{H}^{\pm}$	MkJ,JADE	20, 22	22.5 GeV

2. WEAK DECAYS

2.1 The tau lepton

The present situation in τ decays was reviewed by K. K. Gan in one of the parallel sessions.[23] Here I give only a brief summary.

a) There are new data on the weak-electromagnetic interference in e^+e^- from JADE and CELLO.[24] All data are nicely in accord with the standard prediction based on $\sin^2\theta_w = 0.22$.

b) There are many measurements of the τ lifetime, all from e^+e^- experiments. The world average is now 3.0 ± 0.2 ps, to be compared with the predicted 2.8 ± 0.2 ps.

c) A new more accurate measurement of the leptonic branching ratio, $Br(\tau \to \ell\nu\bar{\nu}) = 0.178 \pm 0.005$ (Average of e and μ), was reported by the MAC collaboration,[25] in excellent agreement with theory.

d) CLEO has made a new measurement[26] of the Michel parameter describing the shape of the electron and muon spectra in the leptonic decays of the tau: $\rho = 0.70 \pm 0.10 \pm 0.03$, to be compared with 0.75, 0, and 0.375 expected for V-A, V+A, and V or A.

e) New upper limits on the tau-neutrino mass come from ARGUS (70 MeV, using $\tau \to A_1 \nu$)[27] and from HRS (85 MeV, using $\tau \to 5\pi^{\pm}(\pi^{\circ})\nu$).[28]

f) Branching fractions have been accurately measured for presumably all of the significant decay modes of the tau. Where theoretical predictions exist, the agreeement is excellent. The only trouble is that the measured branching fractions sum to only 93%. No one has a plausible explanation for the missing 7%. Except for this one could say that the τ lepton is perfectly understood.

2.2 The D meson

There are two motives in studying heavy quark decays. (1) We want to check the standard electroweak model of six quarks and six leptons. No discrepancies have been found. (2) We want to measure the unknown parameters of the model: the three angles and one phase of the Kobayashi-Maskawa matrix generalization[29] of the Cabibbo theory. The decays of states containing only u, d, s, and c are sensitive mainly to the well-known θ_1. The decays involving the transitions b → c and b → u are sensitive to the angles θ_2 and θ_3. In order to extract information on b quark decays we must understand the mechanisms of heavy quark decay in mesons. To do this we study charmed meson decays, where the relevant K-M angle is known.

2.2.1 Branching ratios from SPEAR.

The new data on D decays come mainly from Mark III at SPEAR,[30] running on the $\psi''(3770)$ D-factory. Besides having smaller statistical errors than earlier data, the new measurements use a much more reliable normalization. Instead of dividing observed rates by the total D event rate inferred from the shape of the ψ'' peak, Mark III uses the events in which one D decay is reconstructed, as a pure unbiased event sample for studying the other D. This only has to be done for a few higher-rate modes to serve as a normalization for all modes. The new branching ratios are listed in

156

the table. Where they overlap, the new numbers are about a factor of
two higher than the old numbers: that is, the ψ'' cross section implied
by the Mark III double-tag data is half of that assumed in normalizing
the old data. This will have the effect of halving all other reported
D production cross sections, too.

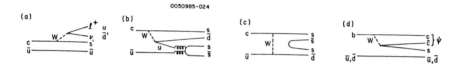

Fig. 4 Diagrams for (a) spectator decay of a D°, (b,c) the decay
D° → K̄°φ, and (d) the decay B → ψX.

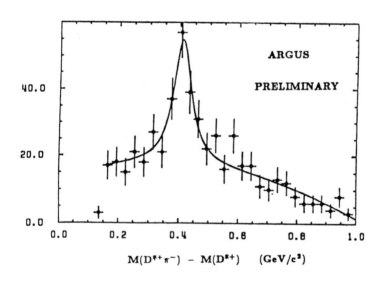

Fig. 5 D*π - D* mass difference spectrum.[33]

Table III. % Branching ratios for D decays, from Mark III

Semileptonic modes

$D^+ \to \bar{K}^0 e^+ \nu$ $9.3 \pm 2.2 \pm 0.3$

       $K^- \pi^+ e^+ \nu$ $4.1 \pm 0.9 \pm 0.1$

       $\bar{K}^0 \mu^+ \nu$ 14.6 ± 6.1 (preliminary)

$D^0 \to K^- e^+ \nu$ $3.2 \pm 0.5 \pm 0.1$

       $K^- \pi^0 e^+ \nu$ $0.9 \pm 0.4 \pm 0.1$

       $\bar{K}^0 \pi^- e^+ \nu$ $3.0 \pm 1.4 \pm 0.2$

       $K^- \mu^+ \nu$ 4.9 ± 1.2 (preliminary)

Cabibbo favored hadronic modes

$D^+ \to \bar{K}^0 \pi^+$ $3.5 \pm 0.5 \pm 0.4$

       $K^- \pi^+ \pi^+$ $11.3 \pm 1.3 \pm 0.8$ (double-tag)

       $\bar{K}^0 \pi^+ \pi^0$ $14.1 \pm 2.8 \pm 2.1$ (double-tag)

       $\bar{K}^0 \rho^+$ $12.2 \pm 2.8 \pm 1.9$

       $\bar{K}^{*0} \pi^+$ $3.0 \pm 1.9 \pm 1.7$

       $\bar{K} \pi^+ \pi^+ \pi^-$ $8.5 \pm 1.3 \pm 1.1$

       $K^- \pi^+ \pi^+ \pi^0$ $7.2 \pm 1.4 \pm 1.6$

$D^0 \to K^- \pi^+$ $5.1 \pm 0.4 \pm 0.4$ (double-tag)

       $\bar{K}^0 \pi^0$ $2.2 \pm 0.4 \pm 0.2$

       $\bar{K}^0 \eta$ $1.8 \pm 0.8 \pm 0.3$

       $\bar{K}^0 \omega$ $3.8 \pm 1.5 \pm 1.0$

       $K^- \pi^+ \pi^0$ $18.5 \pm 1.3 \pm 1.6$ (double-tag)

       $K^- \rho^+$ $13.7 \pm 1.3 \pm 1.5$

       $K^{*-} \pi^+$ $7.1 \pm 1.6 \pm 1.3$

       $\bar{K}^{*0} \pi^0$ $2.1 \pm 0.9 \pm 0.6$

       $\bar{K}^0 \pi^+ \pi^-$ $7.6 \pm 0.8 \pm 0.7$

       $\bar{K}^0 \rho^0$ $1.3 \pm 0.4 \pm 0.3$

       $K^{*-} \pi^+$ $7.3 \pm 1.2 \pm 0.9$

       $\bar{K}^0 K^+ K^-$ $1.5 \pm 0.5 \pm 0.3$

       $K^- \pi^+ \pi^+ \pi^-$ $11.5 \pm 0.8 \pm 0.8$ (double-tag)

       $\bar{K}^0 \pi^+ \pi^- \pi^0$ $13.7 \pm 2.5 \pm 3.2$

Cabibbo suppressed hadronic modes

$D^+ \rightarrow \bar{K}^\circ K^+$ $(0.317 \pm 0.086 \pm 0.048) \times (\rightarrow \bar{K}^\circ \pi^+)$

$\pi^- \pi^+ \pi^+$ $(0.042 \pm 0.016 \pm 0.010) \times (\rightarrow K^- \pi^+ \pi^+)$

$K^- K^+ \pi^+$ $(0.059 \pm 0.026 \pm 0.009) \times (\rightarrow K^- \pi^+ \pi^+)$

$\phi \pi^+$ $(0.084 \pm 0.021 \pm 0.011) \times (\rightarrow K^- \pi^+ \pi^+)$

$\bar{K}^{*\circ} K^+$ $(0.048 \pm 0.021 \pm 0.011) \times (\rightarrow K^- \pi^+ \pi^+)$

$D^\circ \rightarrow \pi^- \pi^+$ $(0.033 \pm 0.010 \pm 0.006) \times (\rightarrow K^- \pi^+)$

$K^- K^+$ $(0.122 \pm 0.018 \pm 0.012) \times (\rightarrow K^- \pi^+)$

$\pi^+ \pi^- \pi^\circ$ $1.11 ^{+0.43}_{-0.35} \pm 0.18$

$\pi^+ \pi^- \pi^+ \pi^-$ $1.47 ^{+0.67}_{-0.49} \pm 0.19$

Thanks to the factor of two from the renormalization as well as the increase in the number of modes measured, we now have 87% of all D decays accounted for. Summed over all observed modes, the ratio of Cabibbo suppressed to favored hadronic branching fractions is 0.08. This rich mine of information will keep theorists busy for years, with the result (I hope) that we will really understand the mechanisms of decay of mesons containing heavy quarks.

2.2.2 A nonspectator mode. One particular mode deserves more comment. The decay $D^\circ \rightarrow \bar{K}^\circ \phi$ cannot occur through the simple spectator graph (Fig. 4a), since the final state does not contain the spectator $\bar{u}$ quark. You either have to (Fig. 4b) annihilate a u from the W with the spectator $\bar{u}$ to produce an $s\bar{s}$ pair, which is color suppressed and Zweig-violating, or (Fig. 4c) exchange a W with the spectator and then pop an $s\bar{s}$ pair. We know that the simple spectator graph cannot be dominant, since it implies equal lifetimes for the charged and neutral D, but it would be nice to see a decay which cannot be spectator. The ARGUS group[31] reports a signal for $e^+ e^- \rightarrow D^\circ X$, $D^\circ \rightarrow \bar{K}^\circ \phi$, with the $\bar{K}^\circ$ detected through its decay to two charged pions and the ϕ detected in $K^+ K^-$. They calculate a branching ratio for $D^\circ \rightarrow \bar{K} \phi$ of 1.4 ± 0.5%. CLEO[32] confirms it with a branching ratio of 0.9 ± 0.3 ± 0.3%. Mark III[30] has an upper limit of 2.5%. Although the results appear consistent, there is controversy over how to treat the background from $D^\circ \to \bar{K}^\circ K^+ K^-$, where the $K^+ K^-$ do not form a ϕ. Mark III takes very seriously the rather large $K^+ K^-$ contribution below the ϕ; ARGUS ignores it; CLEO tries to fit it. I think the apparent $\bar{K}^\circ \phi$ signal is real, but it is not clear how large it is. It is certainly much smaller than the 13.7% rate for the spectator allowed $D^\circ \to K^- \rho^+$.

2.2.3 **The D**.** Let me mention another D result from ARGUS here, although it has nothing to do with weak decays. ARGUS[33] has reported a new charmed meson, called for now the D**(2421). They have a peak of $143\ ^{+43}_{-29}$ events in the $D^* \pi$ - D^* mass difference spectrum (Fig. 5), seen in the production and decay chain

$$e^+ e^- \to D^{*\circ} X$$
$$\searrow D^{*+} \pi^-$$
$$\searrow D^\circ \pi^+$$
$$\searrow K^- \pi^+ \text{ or } K^- \pi^+ \pi^+ \pi^- \text{ or } K^- \pi^+ \pi^\circ \text{ (\& ch.conj.)}$$

The measured mass is 2421 ± 7 MeV, as expected for the P states of $c\bar{u}$; the width is 73 ± 23 MeV.

2.3 The B meson

Two measurements are made in B decays to extract information about the K-M angles θ_2 and θ_3: the ratio of decay rates to noncharm and charmed decays, [b→u]/[b→c], is measured at CESR and DORIS; and the b lifetime is measured at PEP and PETRA.

2.3.1 **Noncharm/charmed ratio.** Actually, all we have is an upper limit on [b→u]/[b→c]. It is obtained by looking at the momentum spectrum of electrons or muons in the semileptonic decays of the B and noting that if the B decays to a noncharm hadron state plus $\ell\nu$, the lepton momentum can be higher than if the B decays to a charmed hadron state (typically more massive) plus $\ell\nu$. Until this conference the (90% conf.) upper limit was 4% from CLEO[34] and 5.5% from CUSB,[35]

which you could perhaps combine to give a 3% limit. The good news is
that the data are getting better: there are new results from CLEO[36]
and ARGUS,[37] which if analyzed as before, would give 1% and 4%
limits. The bad news is that it is now clear that the extraction of a
limit from the measured spectrum by fitting to a theoretical model is
flawed by the uncertainties in the model. This was made obvious when
the more accrurate CLEO data did not fit well the popular Altarelli
model[38] with the usually assumed value of the parameter describing
the Fermi momentum of the quarks in the meson. Letting this parameter
float in the fit (there is no reason to take it as known) weakens the
[b→u]/[b→c] upper limit to 4%. A similar limit is obtained by fitting
only the spectrum above the maximum momentum available in charmed
final states. Significant progress in pushing down this limit (or
actually measuring [b→u]/[b→c]) by this method depends on better
understanding of the theory. If this ratio is actually as low as 1%,
it will imply that θ_3 is too small to allow the CP violation in K°
decays to be expained by the phase in the K-M matrix.

There is another way to get evidence for the b→u coupling.
CLEO[39] has searched for exclusive noncharm decays. The upper limits
(90% conf.) on several low-multiplicity modes are given below;
combinatoric background makes the higher multiplicities less useful:

$$B^- \rightarrow \rho^\circ \pi^- \qquad\qquad < 2 \times 10^{-4}$$

$$\rho^\circ A_1^- \qquad\qquad 8$$

$$\rho^\circ A_2^- \qquad\qquad 4$$

$$\rho^\circ X^- \qquad\qquad 2 \qquad (1.0 < M_X < 1.6 \text{ GeV}, \ \Gamma_X = 100 \text{ MeV})$$

$$\overline{B}^\circ \rightarrow \pi^+ \pi^- \qquad\qquad 2$$

$$\pi^+ A_1^- \qquad\qquad 24$$

$$\pi^+ A_2^- \qquad\qquad 20$$

$$\pi^+ X^- \qquad\qquad 7 \qquad (1.0 < M_X < 1.6 \text{ GeV}, \ \Gamma_X = 100 \text{ MeV})$$

Although these limits are as much as two orders of magnitude smaller
than for the corresponding low multiplicity charmed modes, one does
not know how the relative probability of various multiplicities
depends on the masses.

2.3.2 <u>Lifetime.</u> Over the past few years there has been a steady accumulation of b lifetime measurements from most of the PEP and PETRA detectors. To get an enriched sample of b jets some use leptons with high p_T relative to the jet axis as a signature, others use an event shape criterion to pick up the broader b jets. The lifetime is measured either from the lepton impact parameter or using vertices. All the data are consistent. Last year at the Leipzig conference the measurements averaged about 1.5 ± 0.3 psec; measurements reported at this conference[40] bring the mean down to 1.06 ± 0.17 psec.

Two B meson decays have been observed in an event recorded by the CERN WA75 experiment,[41] using emulsion, silicon microstrips, and a magnetic spectrometer: $B^- \rightarrow D^0 \mu \nu$ after about 0.08 psec, and $B^0 \rightarrow D^- X^+$ after about 0.5 psec. Although the first decay time is rather short compared to the mean life measured in the $e^+ e^-$ experiments, one such event is not enough to challenge those measurements.

With only an upper limit for [b→u]/[b→c] and no measurement of the K-M phase parameter δ, one cannot derive values for the angles θ_2 and θ_3, only an allowed region in θ_2 vs. θ_3 which is different for each assumed value of δ. Since useful b decay first became available two years ago[42] the measured lifetime has decreased and become more accurate, and the [b→u]/[b→c] limit has perhaps decreased. These trends cause the allowed values for θ_2 to increase (0.05 ± 0.02) and for θ_3 to decrease (<0.04).

2.3.3 <u>B to ψ decay.</u> Up to now our knowledge of B decay branching ratios has been confined to the inclusive semileptonic modes and a few exclusive low-multiplicity modes with D or D* plus a few pions, all compatible with the spectator mechanism and color mixing suppression. An interesting possiblility is that the B could decay to the ψ, as suggested back in 1979 by Fritzsch.[43] This mode presumably proceeds by the spectator graph with $s\bar{c}$ produced at the W vertex (which should happen about 10% of the time) and the $\bar{c}$ combining with the c from the b (if the colors match). If color mixing is suppressed, the match will occur only 1/9 of the time, so naively the branching ratio

for B → ψX should be of the order of 1%. CLEO[44] and ARGUS[45] have
now seen such decays; the evidence for the ψ can be seen in the mass
spectrum of e^+e^- or $\mu^+\mu^-$ pairs from B's produced at the T(4S) B-factory
resonance. The branching ratio for B → ψX measurements are 1.1 ± 0.2
± 0.2 (CLEO) and 1.4 ± 0.6 (ARGUS). The theoretical uncertainties
(e.g., how often should a $c\bar{c}$ form a ψ?), do not allow us to conclude
that soft gluons do or do not cancel the color mixing suppression.
CLEO has four candidates for exclusive decays to ψK* (none for ψK);
ARGUS has four exclusive decay candidates in various modes (with an
estimated background of one event).

2.3.4 <u>The B* meson.</u> Here again, let me insert a topic that doesn't
belong in weak decays. The CUSB group[46] has seen the vector state of
the B system, the B*, radiatively decaying to the B. In an energy
scan above the T(4S) resonance they see evidence for an almost
monoenergetic (doppler broadened) peak in the inclusive photon
spectrum, corresponding to a B* - B mass difference of 53 ± 2 ± 4 MeV,

Fig. 6 Inclusive photon energy
spectrum for events containing
a lepton.[46]

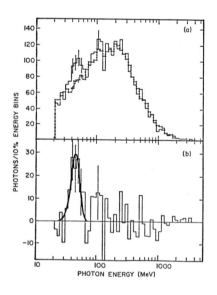

in agreement with most predictions. That it is associated with B's is confirmed by the fact that requiring a lepton in the event enhances the signal (Fig. 6). In the range of e^+e^- energy covered in the scan, one is presumably producing $B\bar{B}^*$ or $B^*\bar{B}$ or $B^*\bar{B}^*$; CUSB measures an average of 1.5 ± 0.2 B^* per $b\bar{b}$ event.

3. JET FRAGMENTATION

3.1 Survey

For quark jets of energies up to 20 GeV, e^+e^- experiments are our best source of information on fragmentation topologies, inclusive particle momentum spectra and abundances, two-particle correlations, and so on. Table IV indicates the wide range of data available,[47] a # indicating the presence of data and a number indicating that the inclusive spectrum data cover a wide enough range of momenta to permit calculation of the integrated abundance. There are single-particle data now on pseudoscalar and vector mesons, octet and decuplet baryons (the latter rather sparse), as well as charmed particles. Two-particle correlations have been measured for pions, kaons, protons, and lambdas. There are also extensive data from CLEO and ARGUS on upsilon decays, supposedly telling us something about gluon jets.

Table IV. Data on single particle inclusive production
(average number per event) and two-particle correlations

	10 GeV CLEO[ARGUS]	Mark II	HRS	TPC	JADE	TASSO
		--------29 GeV--------			-----34 GeV----	
$\pi^{\pm}$	[5.8±.3]		#	10.7±.6		10.9±.3
π°	3.0±.7			5.3±.7	6.1±.3	6.1±2.0
η					.64±.15	
$K^{\pm}$	[.89±.17]	#	#	1.35±.13		1.74±.15
K°	.92±.12	1.27±.15	1.44±.10	1.22±.15	1.47±.17	1.48±.05
ρ°	.50±.09	#	#		.98±.15	#
ϕ	.08±.02		.08±.01	.08±.02		
$K^{*\pm}$	.45±.08				.87±.18	
$K^{*\circ}$	.38±.09			.49±.08		
p	[.21±.06]		#	.60±.08	#	.68±.06
Λ	.066±.010	.21±.02	.20±.02	.20±.02	.23±.06	.31±.04
Ξ^{-}	.005±.001			.025±.012		.026±.012
Δ^{++}						<.10
$\Sigma^{*\pm}$	[.003±.001]				<.09	
$\Xi^{*\circ}$	[.002±.001]					
D	# [#]					
D*	# [#]	#	#		#	#
F	# [#]		#			#
Λ_{c}	# [#]					
$\pi\pi$	#			#		
$K\bar{K}$	#			#		
πK				#		
$p\bar{p}$	#			#		
$\Lambda\bar{\Lambda}$	#	#	#	#		#
$\Lambda\bar{p}$				#		#

3.2 Charmed baryons

Obviously, I don't have time to discuss the implications of all of these measurements. Let me just pick a couple of interesting items. CLEO and ARGUS have now produced the first fragmentation function for the charmed baryon Λ_c. CLEO[48] detects the Λ_c in the $\Lambda\pi\pi\pi$ mode; ARGUS[49] prefers the $pK^-\pi^+$ mode. Since the branching ratios are not very well known, I plot the two fragmentation function results (Fig. 7) with arbitrary scales chosen to make the magnitudes compatible. The spectrum peaks at an x $(=p/p_{max})$ of about 0.7, somewhat higher than for the D or F, and inconsistent with the prediction of deGrand.[50]

3.3 Baryon production mechanisms

The TPC has produced an impressive quantity of very nice fragmentation data. As an illustration of the kind of sophisticated study one can now do, let me mention their investigation of the mechanism for baryon production using proton-antiproton correlation data.[51] There are three models for making baryons in jets, schematically defined in Fig. 8. The usual kinds of plots of two-particle correlations, say the distribution of the angle between p and $\bar{p}$ or the distribution in the difference of p and $\bar{p}$ azimuths with respect to the jet axis, do not distinguish among models; all models predict that the p and $\bar{p}$ tend to be in the same jet with a slight preference for the same azimuth. However, the observation of a correlation between the $p\bar{p}$ axis and the jet axis in the $p\bar{p}$ center of mass (Fig. 9) rules out the cluster model, in which baryon-antibaryon pairs come from essentially isotropic heavy meson decays. If we look at a transverse momentum correlation (defined for components out of the event plane, to minimize the effect of gluon bremsstrahlung), the data (Fig. 10) favor the case in which a meson is popped between the proton and antiproton (the "popcorn" model), thus ruling out the simple diquark model.

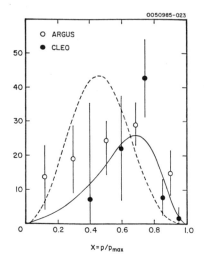

Fig. 7 Fragmentation function
for $e^+e^- \to \Lambda_c X$.[48-49]

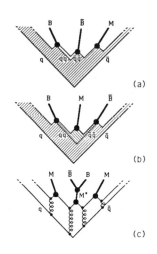

Fig. 8 Models for baryon-antibaryon
production: (a) stable diquarks,
(b) "popcorn", and (c) cluster.

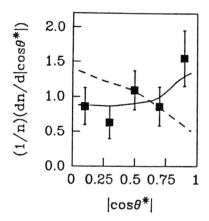

Fig. 9 Measured angular distribution of the $p\bar{p}$ axis with
respect to the jet axis in the
$p\bar{p}$ c.m.s.[51] compared with
the cluster model (dashed), and
other models (solid).

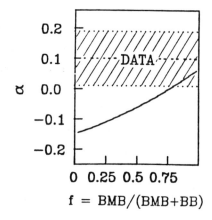

Fig. 10 Comparison of measured
$\alpha = \langle p_{T,p} \cdot p_{T,\bar{p}} \rangle / \langle p_T^2 \rangle$ (for p_T
out of the event plane)[51] with
the "popcorn" model as a function
of the probability of producing a
meson between the qq and $\overline{qq}$.

REFERENCES

1. Hitlin, D. (Mark III), Proceedings of the 1983 International Symposium on Lepton and Photon Interactions at High Energies (Cornell, 1983).
2. Wisniewski, W., (Mark III), presented at this conference.
3. Trost, H.J. (Crystal Ball), XXII International Conference on High Energy Physics (Leipzig, 1984).
4. Bloom, E.D. (Crystal Ball), SLAC-PUB-3686 (May, 1985).
5. Franzini, P. (CUSB), presented at this conference.
6. Albrecht, H. et al. (ARGUS), Phys. Lett. 154B, 452 (1985), and Davis, R., presented at this conference.
7. Peterson, D. (CLEO), invited talk at International Conference on Hadron Spectroscopy, Maryland 1985.
8. Arnison, G. et al. (UA1), Phys. Lett. 139B, 115 (1984).
9. Glashow, S.L. and Manohar, A., Phys. Rev. Lett. 54, 526 (1985).
10. Akerlof, C. et al. (HRS), Phys. Lett. 156B, 271 (1985).
11. Feldman, G.F. et al. (Mark II), Phys. Rev. Lett. 54, 2289 (1985).
12. Ash, W.W., et al. (MAC), Phys. Rev. Lett. 54, 2477 (1985).
13. Bartel, W. et al. (JADE), Phys. Lett. 155B, 288 (1985).
14. Behrend, H.J. et al. (CELLO), DESY 85-061 (July, 1985).
15. Bartel, W. et al. (JADE), Phys. Lett. 152B, 383 (1985).
16. CELLO collaboration, contributed paper to International Europhysics Conference on High Energy Physics (Bari, 1985).
17. Fernandez, E. et al. (MAC), Phys. Rev. Lett. 54, 1118 (1985).
18. Wilson, R. (ASP), presented at this conference.
19. Behrend, H.J. et al. (CELLO), Phys. Lett. 123B, 127 (1983), and reference 16;
    Bartel, W. et al. (JADE), Phys. Lett. 139B, 325 (1984);
    Adeva, B. et al. (Mark J), MIT technical report no. 139 (1984);
20. Adeva, B. et al. (Mark J), Phys. Lett. 152B, 433 (1985).
21. Bartel, W. et al. (JADE), Phys. Lett. 152B, 392 (1985).
22. Bartel, W. et al. (JADE), preprint (July, 1985).
23. Gan, K.K., presented at this conference.

24. CELLO collaboration, contributed paper to International Symposium on Lepton and Photon Interactions at High Energies (Kyoto, 1985); Bartel, W. et al. (JADE), contributed paper to International Europhysics Conference on High Energy Physics (Bari, 1985).

25. Read, A. (MAC), presented at this conference.

26. Behrends, S. et al. (CLEO), CLNS 85/660 (Aug., 1985).

27. Albrecht, H. et al. (ARGUS), DESY 85-054 (June, 1985).

28. Abachi, S. et al. (HRS), Purdue U. preprint PU-85-535 (June, 1985).

29. Kobayashi, M. and Maskawa, T., Prog. Theor. Phys. $\underline{49}$, 652 (1973).

30. Coward D. (Mark III), presented at this conference, and Baltrusaitis, R.M. et al., Phys. Rev. Lett. $\underline{55}$, 150 (1985).

31. Albrecht, H. et al. (ARGUS), Phys. Lett. $\underline{158B}$, 525 (1985).

32. Avery, P. et al. (CLEO), preprint (Aug., 1985).

33. Albrecht, H. et al. (ARGUS), preprint (July, 1985), and Davis, R., presented at this conference.

34. Chen, A. et al. (CLEO), Phys. Rev. Lett. $\underline{52}$, 1084 (1984).

35. Klopfenstein, C. et al. (CUSB), Phys. Lett. $\underline{130B}$, 444 (1983).

36. Jawahery, A. (CLEO), presented at this conference.

37. ARGUS collaboration, preprint (July, 1985).

38. Altarelli, G. et al., Nucl. Phys. $\underline{B208}$, 365 (1982).

39. CLEO collaboration, preprint (Aug., 1985).

40. Baranko, G., review of HRS, JADE, MAC, Mark II, and TASSO results presented at this conference.

41. Albanese, J.P. et al. (WA75), Phys. Lett. $\underline{158B}$, 186 (1985).

42. Stone, S., Proceedings of the 1983 International Symposium on Lepton and Photon Interactions at High Energies (Cornell, 1983).

43. Fritzsch, H., Phys. Lett. $\underline{86B}$, 164 (1979), and $\underline{86B}$, 343 (1979).

44. Haas, P. et al. (CLEO), Phys. Rev. Lett. $\underline{55}$, 1248 (1985).

45. Davis, R. (ARGUS), presented at this conference.

46. Han, K. et al. (CUSB), Phys. Rev. Lett. $\underline{55}$, 36 (1985).

47. Oddone, P., LBL-19195 (Dec., 1984), and Izen, J.M., DESY 84-104 (Oct., 1984) for earlier references, plus the following more recent reports:

Behrends, S. et al. (CLEO), Phys. Rev. D. 31, 2161 (1985);
Albrecht, H. et al. (ARGUS), Phys. Lett. 150B, 235 (1985), and
153B, 343 (1985), and Davis, R., presented at this conference;
de la Vaissiere, C. (Mark II), Phys. Rev. Lett. 54, 2071 (1985);
Yamamoto, H. et al. (DELCO), Phys. Rev. Lett. 54, 522 (1985);
Derrick, M. et al. (HRS), Phys. Lett. 158B, 519 (1985), and Phys.
Rev. Lett. 54, 2568 (1985), and Purdue U. preprint PU-85-537
(July, 1985);
Abachi, S. et al. (HRS), Purdue U. preprint PU-85-536 (1985);
Aihara, H. et al. (TPC), Phys. Rev. Lett. 54, 274 (1985);
Bartel, W. et al. (JADE), DESY 85-029 (Apr., 1985);
TASSO collaboration, contributed paper to the International
Europhysics Conference on High Energy Physics (Bari, 1985).

48. Bowcock T. et al. (CLEO), Phys. Rev. Lett. 55, 923 (1985).

49. Orr, R.S. (ARGUS), presented at the International Europhysics
Conference on High Energy Physics (Bari, 1985).

50. DeGrand, T.A., Phys. Rev. D 26, 3298 (1982).

51. Aihara, H. et al. (TPC), Phys. Rev. Lett. 55, 1047 (1985).

ACCELERATOR PHYSICS IN SSC DESIGN*

Alexander W. Chao
SSC Central Design Group†
c/o Lawrence Berkeley Laboratory
1 Cyclotron Road
Berkeley, California USA 94720

ABSTRACT

An elementary description of the accelerator physics considerations encountered in the design of the Super-conducting Super Collider is presented. An attempt has been made to introduce the terminology and the basic physics issues from a user's point of view.

1. THE END PRODUCT

The end product of a storage ring collider can be summarized by three parameters: the type of colliding particles, the particle energy, and the luminosity. For the SSC, these parameters are chosen to be

$$\text{particle type} = pp$$
$$\text{particle energy } E = 20 \text{ TeV} \tag{1}$$
$$\text{luminosity } \mathscr{L} = 10^{33} \text{ cm}^{-2}\text{sec}^{-1}$$

We begin with a discussion on luminosity. Consider a certain type of high energy physics events of interest with cross section Σ. Obviously the counting rate $\mathscr{R}$ of these events in a collider is proportional to Σ. The proportionality constant is the luminosity, i.e.,

$$\mathscr{R} = \mathscr{L}\Sigma. \tag{2}$$

Consider two beam bunches with N particles each colliding head-on. Let the bunches have a round gaussian transverse distribution with rms size σ, as shown in Fig. 1. Let f be the frequency of collisions occurring at the collision point under consideration. Luminosity is given by[1]

$$\mathscr{L} = N^2 \, f/4\pi \, \sigma^2. \tag{3}$$

---

*Talk presented at 1985 DPF annual meeting, Eugene, Oregon.
†Operated by Universities Research Association for the Department of Energy.

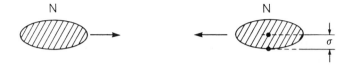

Fig. 1. Two colliding bunches.

The collision frequency is related to the revolution frequency f_{rev} by

$$f = f_{rev} \, B, \tag{4}$$

where B is the number of bunches in each beam.
These parameters for the present SSC design are[2]

$$\begin{aligned}
N &= 1.4 \times 10^{10} \\
f_{rev} &= 3000 \text{ sec}^{-1} \\
\sigma &= 7 \text{ } \mu m \\
B &= 10000
\end{aligned} \tag{5}$$

With a circumference of 100 km, the spacing between bunches in each beam is 10 m.

We now ask what happens to a proton as it passes through the collision point. The probability that a proton actually collides with a proton in the on-coming bunch is $P = N \Sigma_{tot}/4\pi \, \sigma^2$, where Σ_{tot} is the total cross section of collision. Assuming $\Sigma_{tot} = 137$ mb, the probability of collision is found to be 3×10^{-10} per crossing, which is very small. The protons basically just pass through the on-coming beam without actual collisions.

However, there are two effects caused by beam crossings to be considered. The first is on the lifetime of the beam. With $N = 1.4 \times 10^{10}$, about 4 protons will find a partner to collide with per crossing. The beam lifetime due to pp collisions is 3.3×10^9 crossings, which corresponds to 1.8×10^5 sec, or about 50 hours, since $f_{rev} = 3000$ sec^{-1} and there are 6 collision points per revolution.

The second, more important, effect caused by beam crossings is the perturbation on particle motion due to elastic scattering by the collective Coulomb field associated with the on-coming bunch. This perturbation is referred to as the beam-beam interaction,[3] which is described next.

2. BEAM-BEAM EFFECT

The beam-beam interaction constitutes one of the main limiting
effects on the luminosity of a storage ring collider. To achieve a
high luminosity, one needs a high beam intensity and a small beam
area. These requirements, however, must be made so that the beam-beam
effect is not made untolerably strong.

Consider a test particle in beam bunch 1 that passes through beam
bunch 2 with a transverse displacement x from center, as shown in
Fig. 2. In the relativistic limit, the electric field seen by the
test particle points in the radial direction perpendicular to its
direction of motion. Applying Gauss' law yields

$$E_r = (2Ne/\ell) \ [1 - \exp(-x^2/2\sigma^2)]/x, \tag{6}$$

where ℓ is the length of the beam bunch.

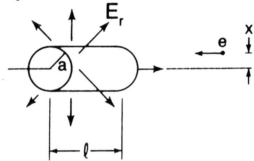

Fig. 2. The beam-beam encounter seen by a test charge.

In addition to the electric field, there is a magnetic field
B_Θ of the same strength (in cgs units) as the electric field.
The Lorentz forces due to the electric and the magnetic fields add to
give a force twice that due to the electric field alone. For the SSC
with   = 15cm, the electric field is found to be 17 MV/m and the magnetic field is 0.5 kG evaluated at the edge of the beam, $x = 1.6\sigma$,
where the fields are maximum.

The beam-beam interaction imposes limitation on luminosity not
because it is extraordinarily strong but because it is extraordinarily
nonlinear. The linear part of the force acts like a quadrupole magnet, whose effects can be compensated by adjusting the strengths and
arrangements of the neighboring quadrupoles.

Figure 3 illustrates a comparison between a quadrupole magnet
force and the beam-beam force. The linear quadrupole force provides
transverse focussing which confines particles to execute simple harmonic oscillations, called the betatron oscillations.[4] The slope
of the force directly relates to the betatron oscillation tune υ,
which is defined to be the betatron oscillation frequency divided by

f_{rev}. Any additional linear force will cause a shift of the tune value. In particular, the linear part of the beam-beam force gives rise to a beam-beam tune shift given by[3]

$$\xi = N \ r_0 \ \beta^*/4\pi \ \sigma^2\gamma, \tag{7}$$

where β^* is the β-function[4] at the collision point, γ is the relativistic factor, $r_0 = e^2/mc^2 = 1.53 \times 10^{-18}$m is the classical radius of a proton. We will discuss more on the β-function later. As mentioned, the linear beam-beam tune shift is not difficult to compensate.

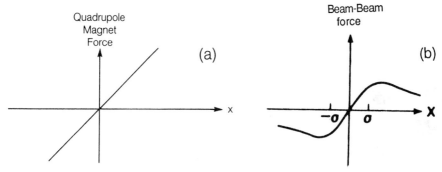

Fig. 3.   (a) quadrupole magnet force as a function of the transverse displacement x, and (b) the same for the beam-beam force. The quadrupole linear force gives rise to the betatron tune. The dashed line is the linear part of the beam-beam force. It gives rise to a beam-beam tune shift.

The nonlinear part of the beam-beam force, on the other hand, is not so easy to deal with. As shown in Fig. 3(b), the beam-beam nonlinearity starts around $x > \sigma$. It turns out that the strength of the beam-beam nonlinearity is specified by the same quantity that specifies the linear part of the beam-beam force, namely the beam-beam tune shift ξ. (This is an important and nontrivial observation. It led to the discovery of low-β^* insertion,[5,6] which will be discussed later for the SSC.) To control the beam-beam nonlinear effects, it is therefore necessary to limit ξ. One of the fundamental constants in the design of a storage ring collider is in fact the maximum tolerable ξ in the presence of the nonlinear beam-beam perturbation. The conventional wisdom is that this maximum allowed for proton storage rings is somewhere between 0.003 and 0.005. For the SSC, we have

$$\xi = 0.0017 \text{ per head-on crossing.} \tag{8}$$

The fact that this is smaller than what is believed to be achievable
will be explained later. Note that since there are 6 crossing points,
the total beam-beam tune shift is 6 x 0.0017 = 0.01. Inserting the
SSC parameters into Eq.(8), we find the needed β^*, β-function at
the collision points, is 1.0 m.

3.   LOW-β^* INSERTION

The β-function[4] has several physical meanings. The ones
relevant to us here are

- β-function is a function of position around the storage ring.
- the equilibrium beam size at position s is proportional to $\sqrt{\beta(s)}$.
- the sensitivity of particle motion to perturbations at position
  s is specified by $\beta(s)$. The larger $\beta(s)$ is, the more sensitive is
  particle motion to perturbations.

At the collision point, we want to have a small beam size to
enhance the luminosity. We also want to desensitize particle motion
to the nonlinear beam-beam force. Both require small β^*.
The value of $\beta^* = 1$ m is to be compared with the average
β-function in the storage ring which is determined by a convenient
and economic spacing between quadrupole magnets in the lattice design.
For the SSC, the average β-function is approximately 200 m. This
means there needs to be a special lattice insertion consisting of a
sequence of quadrupole magnets to focus the β-function from an average value of 200 m down to 1 m at the collision points. This special
insertion is called the low-β^* insertion; it strongly enhances the
luminosity of storage ring colliders. The price to pay is that it
also strains the optics of the storage ring, as will be discussed in
section.[5]

4.   CROSSING ANGLE

Figure 4 is a sketch of the beam trajectories and lattice design
around a collision point for the SSC. The two beams are designed to
cross at a small angle α to avoid multiple head-on collisions in the
region ±65 m around the collision point. Without a crossing angle,
there will be 26 head-on collisions per beam-beam crossing with bunch
spacing of 10 m, yielding a clearly excessive beam-beam tune shift.
Even with a crossing angle, the beam-beam encounter includes 26
long range interactions, a few of them are illustrated in Fig. 5. The
crossing angle must be chosen large enough to reduce the long range
beam-beam effects. It must also be small enough so that the two beams
can use common quadrupole magnets, i.e., the two triplets nearest to
and on each side of the collision point.

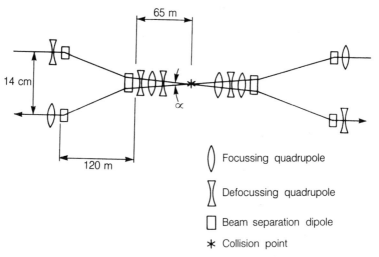

Focussing quadrupole

Defocussing quadrupole

Beam separation dipole

∗ Collision point

Fig. 4. Beam trajectories and a possible lattice design around a
collision point for the SSC.

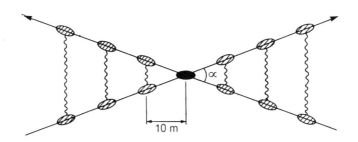

Fig. 5. Long range beam-beam encounters.

There are a few effects of a crossing angle. The first is that
the effective overlapping of the colliding bunches is reduced by a
factor $1/(1+f^2)$, where

$$f = \sigma_z \, \alpha/2\sigma, \qquad (9)$$

with σ_z the rms bunch length. Figure 6 shows the bunch overlapping as a function of f. To avoid loss of luminosity, f needs to
be smaller than 1.

large f   small f

Fig. 6. Effect of crossing angle on the overlapping of the two beam
distributions and thus its effect on luminosity.

The second effect is that the beam-beam interaction with a crossing angle will drive a whole set of potentially harmful resonances,
called the synchro-betatron resonances.[7] The strength of these
resonances is proportional to f. We like to have small f also for
this reason.

The third effect is that the long range beam-beam force introduces
a beam-beam tune shift term in addition to that due to the direct collision. The total tune shift is given by

$$(\text{total}) = (\text{direct}) + (\text{long range}), \qquad (10)$$

where (direct) is given by Eq.(7). The long range contribution is
related to the direct contribution by[2]

$$(\text{long range}) = g \ (\text{direct}), \qquad (11)$$

where

$$g = 2n/(\beta^*\alpha/\sigma)^2$$

with n the number of long range encounters per crossing (n = 26 for
SSC).

Assuming σ_z = 7 cm, σ = 7 μm, α = 50 μrad for the SSC, we find
f = 0.25 and g = 1.0. These values look quite acceptable. In particular, the total beam-beam tune shift is 0.0034, which is close to
what is believed to be the beam-beam stability limit mentioned before.

5. CHROMATIC OPTICAL ABERRATIONS

There are several limitations in reducing β* indefinitely in
order to gain luminosity. Practical limitation on the strength of the
low-β* insertion quadrupole magnets is one example. Here we will
discuss another limitation, i.e., the chromatic aberration of the
storage ring optics which draws a substantial contribution from the
low-β* insertions.

At the start, a storage ring is composed of bending magnets and quadrupole magnets--bending magnets to guide the trajectory of particles and quadrupoles to provide the focussing. The β-function around the storage ring is sketched in Fig. 7. The insertion has produced a small β*, but it also produces a large β-function, which for the SSC is $\hat{\beta}$ = 4000 m, at the insertion quadrupole magnets.

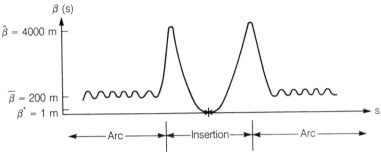

Fig. 7. Sketch of the β-function in SSC.

If the particles in the beam do not have any energy spread, a storage ring consisting of only bending and quadrupole magnets will satisfactorily produce the desired behavior shown in Fig. 7 and there will be no optical restriction on indefinitely reducing β*. The motion of particle is linear and is perfectly stable.

The difficulty arises when the beam has a finite energy spread. To see that, consider the effect of a quadrupole magnet on the motion of an off-momentum particle with energy error $\delta = \Delta E/E$. The kick angle is given by

$$\Delta x' = K\ x$$

$$= K_0\ x/(1+\delta)$$

$$= K_0 x\ (1 - \delta + \delta^2 - \ ...),\qquad (12)$$

where K_0 is the quadrupole gradient seen by an on-momentum particle. The factor $1/(1+\delta)$ represents the rigidity in the kick to the off-momentum particle under consideration.

The rigidity factor is expanded in Eq.(12) into a power series to show its nonlinear behavior in delta. It is very nonlinear even it may not look like so. This is especially the case if β* is small (and thus $\hat{\beta}$ is large) because significant chromatic aberrations comes from the insertion quadrupoles, which are strong to begin with and their effects are strongly enhanced due to the very large $\hat{\beta}$ there.

In a storage ring consisting of only bending and quadrupole magnets, therefore, particle motion is linear in x but nonlinear in δ. Figure 8(a) shows schematically the stability aperture diagram under this condition. In the aperture diagram, the maximum stable betatron amplitude is called the dynamic aperture. The maximum stable energy width is called the momentum aperture. Figure 8(a) shows an infinite dynamic aperture and a very small momentum aperture in the case when the ring consists of only bending and quadrupole magnets. The momentum aperture is too small to accommodate the energy spread of the beam particles.

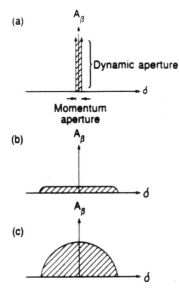

Fig. 8. Aperture diagram for a storage ring that consists of (a) dipole and quadrupole magnets only, (b) dipole, quadrupole and sextupole magnets, and (c) same as (b) but with special sextupole arrangements. A_β is the betatron amplitude, δ is the relative energy error. Shaded regions indicate region of stable motion.

6. SEXTUPOLES

To control the chromatic aberrations, sextupole magnets are installed in the storage ring in addition to bending and quadrupole magnets. Sextupoles have the property that they act like quadrupoles when the beam passes through them off-centered horizontally. We recall that a particle with $\delta \neq 0$ has its closed orbit displaced horizontally by an amount $\eta\delta$, where η is the horizontal dispersion function.[4] The kick given by a sextupole is therefore

$$\Delta x' = S (x + \eta\delta)^2$$
$$= S x^2 + 2 S \eta x \delta + S \eta^2 \delta^2. \tag{13}$$

The third term in Eq.(13) is not too significant. The middle term, which is the reason sextupoles are installed, is made to cancel to first order the chromatic nonlinearities due to the quadrupoles.

Unfortunately we still are left with the first term in Eq. (13). It produces a serious side effect due to its nonlinear nature in x. As a result, although we have removed to a large extent the chromatic aberrations, we have introduced new nonlinearities in the betatron motion, which substantially suppress the dynamic aperture. The situation is sketched in Fig. 8(b). The achieved stable region is still not acceptable.

7. SEXTUPOLE SCHEMES

It is possible to improve the situation substantially by properly choosing the locations and strengths of the sextupoles. The idea is to make their x-nonlinearities cancel among themselves. There are a few schemes to do that; the simplest is the achromat scheme.[8]

In the achromat scheme, sextupoles are arranged in pairs in which two sextupoles of equal strength are spaced by a -1 transformation apart in the betatron motion, as shown in Fig. 9. It is easy to show that if an on-momentum particle enters the first sextupole with coordinate and slope of (x_0, x_0'), it will exit the second sextupole with $(-x_0, -x_0')$, independent of the existence of the sextupoles. The nonlinear effects of the sextupoles thus cancel each other as far as the betatron motion [the first term in Eq.(13) is concerned. On the other hand, the middle term is still active, yielding the needed control over chromatic aberrations.

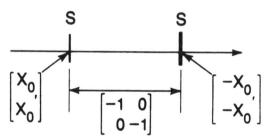

Fig. 9.   A sextupole pair used in the achromat scheme.

Figure 8(c) shows the aperture diagram when sextupoles are arranged to minimize their betatron nonlinear effects. The arrangement does not affect the momentum aperture much but it increases the dynamic aperture substantially. The achromat arrangement have been adopted by the SSC design.

8. MAGNET FIELD ERRORS

We have so far discussed two sources of optical aberrations due to low-β^* and sextupoles, but there is another source, i.e., that due to the magnet field errors. Figure 10 is a sketch of two possible designs of the SSC superconducting bending magnets. One is a superferric design that reaches a field of 3 tesla. The other is a $\cos\theta$ design that gives about 6 to 6.5 tesla field.

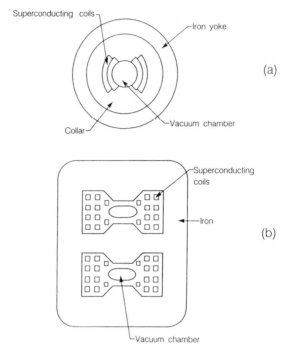

Fig. 10. Sketches of two SSC superconducting bending magnet designs. (a) the cosΘ design and (b) the superferric design.

The magnet field is determined by the placement of the super-conducting coils. Field errors are produced due to design (systematic errors), or due to construction (random errors). The field error is specified by a set of error coefficients a_n and b_n defined by

$$\Delta(B_x + iB_y) = \sum_n (b_n + ia_n)(x + iy)^n. \qquad (14)$$

To compare the nonlinearity due to magnet field errors to that of the beam-beam force, note that the beam-beam force deviates from linearity at a transverse distance of the order of the beam size at the interaction point (of the order of 7μm) while a magnet field nonlinearity has the characteristic distance of the magnet coil or gap size (of the order of a few centimeters).

Extensive effort has been made on the magnet designs to make the error coefficients small. Table I gives a recent set of values of the random field error coefficients for the SSC cosΘ and the superferric magnet designs.[9] As a rough rule, a value larger than about 1 unit in the table means the multipole field error has the potential of causing instabilities in particle motion. However, closer studies must be performed to evaluate individual cases.

The magnet field errors are to be included in the optical aber-rations just as the chromatic and sextupole aberrations discussed before. The stability aperture diagram for the $\cos\theta$ design, for example, is shown in Fig. 11 using results obtained by particle tracking.[10] The stable region is sufficient for the SSC beam to operate. In particular, the achieved dynamic aperture is about 5 mm and the momentum aperture is ±0.15%.

Table I.   A recent set of random multipole error coefficients of the $\cos\theta$ and the superferric magnet designs.[9]   The unit of a_n and b_n is 10^{-4}cm^{-n}.

coefficient	superferric	$\cos\theta$
a_2	1.1	0.63
b_2	1.0	2.15
a_3	1.3	0.69
b_3	0.8	0.35
a_4	0.8	0.14
b_4	0.4	0.59
a_5	0.7	0.16
b_5	0.4	0.059
a_6	0.7	0.034
b_6	0.5	0.076
a_7		0.030
b_7		0.016
a_8		0.0064
b_8		0.021
a_9		0.0056
b_9		0.0030
a_{10}		0.0012
b_{10}		0.0071

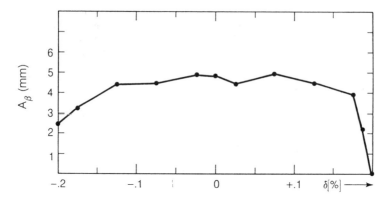

Fig. 11. The aperture diagram achieved by a cosθ magnet design.

9. SYNCHROTRON RADIATION

Synchrotron radiation has been an important effect in electron circular accelerators. SSC will be the first proton storage ring in which synchrotron radiation plays a noticeable role. This is particularly because the radiation is deposited in a cryogenic environment. The following table gives the property of the radiation for the superferric and the cosθ magnets.

B (tesla)	typical photon energy	radiation power
3	132 eV	3.9 KW
6.5	286 eV	8.5 KW

Synchrotron radiation power has to be removed by the cryogenic system. In addition to sychrotron radiation, there are other sources of heat generation:
- The collision points are heated by the colliding protons. With 4 protons in each bunch colliding per crossing, the associated energy release is about 700 watts. With 6 collision points, the total heat generation is 4.2 kW.
- The vacuum chamber inevitably has discontinuous joints. As beam bunches pass by these locations, parasitic energy losses occur (see next section). This contributes to a heating of about 4.5 kW for each ring.[2]
- Particles collide with residual gas molecules in the vacuum chamber, causing particle losses and generating heat. The heat load is estimated to be 0.2 kW per ring.

The above add up to a total heat load of 31 kW for the cosθ magnet design. About 22 kW of this total load is deposited in the cryogenic part of the 2 storage rings. With a cooling efficiency of 1/500, this means the power needed to remove this heat load is 22kW x 500 = 11 MW.

In addition to the radiation power, synchrotron radiation causes the beam emittance to shrink slowly.[1,2] The damping time of the beam size for the 6 tesla case is about 30 hours. This is a beneficial effect since it enhances the luminosity.

10. COLLECTIVE EFFECTS

In the SSC, there are several phenomena that depend on the beam intensity. Beam-beam effect is one of them. Another example occurs when particles within a single bunch Coulomb scatter with one another, yielding either particle losses from the bunch bucket[11] or a growth in the beam emittance.[12]

A discontinuous vacuum chamber is another important source of collective effects. Figure 12 illustrates what happens as a beam bunch passes by a cavity-like discontinuity in vacuum chamber. The bunch and the cavity interact in Fig. 12(b). A wake of electromagnetic field is generated. As the bunch leaves the region, Fig. 12(c), the wake field is trapped by the cavity. The bunch therefore has lost an energy that is equal to the energy stored in the wake field. The wake field can be either longitudinal or transverse. The longitudinal wake field retards beam motion and is directly related to the parasitic energy loss discussed in the previous section.

The wake field can also cause instabilities if it is strong enough. The strength of wake field is specified by basically two quantities. The longitudinal wake field is specified by the longitudinal impedance designated by Z_n/n. The transverse wake field is specified by the transverse impedance Z_t. Adding up contributions from all vacuum chamber discontinuities envisioned for the SSC, the estimated imped-ances for the superferric and the $\cos\theta$ designs are given below:[13]

	superferric	$\cos\theta$
Z_n/n	0.35Ω	0.35Ω
Z_t	75 Ω/m	50 Ω/m

The parasitic loss of 4.5 kW per beam is obtained from the longi-tudinal impedance. As to the transverse impedance, its dominating effect is to cause an instability called single bunch transverse mode coupling instability, or sometimes called the fast head-tail instabil-ity.[14,15] To overcome this instability, the beam needs to have a minimum rms energy spread given by

$$\sigma_E/E = Ne^2 c\beta Z_t/(45\ ER\alpha), \tag{15}$$

where N is the number of particles per bunch, β is the β-function at the location of the transverse impedance, R is the storage ring radius and α is an optics parameter called momentum compaction factor.[4]

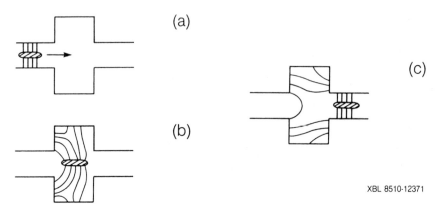

XBL 8510-12371

Fig. 12. Wake field due to beam-cavity interaction. (a) before the beam passes the cavity. (b) during beam passage. (c) after passage.

In order to provide sufficient room for the beam to operate, it is necessary to have a momentum aperture of at least $\pm 4\ \sigma_E/E$. Substituting parameters in Eq. (15), we find that the momentum aperture needed to stabilize the beam is

$$\pm\ 4\ \sigma_E/E = \pm\ 0.4 \times 10^{-3} \text{ for the } \cos\theta \text{ design}$$

$$\pm\ 1.3 \times 10^{-3} \text{ for the superferric design.} \qquad (16)$$

The needed momentum apertures for both designs are within the achieved values discussed in section 8 and Fig. 11.

11. RECAP

We started with the end product, in particular the luminosity, requirements. Beam dynamics is introduced by asking what happens to particle motion at the collision points. The beam-beam interaction causes a beam-beam tune shift which specifies the strength of the perturbation.

The beam-beam interaction in SSC has a long range contribution. Its strength is controlled by the crossing angle. The long range force also contributes to the beam-beam tune shift.

To reduce the beam-beam effect while maintaining a high luminosity, a low β^* is introduced by installing low-β^* insertions around the collision points. These low-β^* insertions introduce strong chromatic aberrations that complicate the optics. To compensate for these chromatic aberrations, sextupoles are introduced in the lattice. The sextupoles have their own nonlinear effects. This is taken care of by arranging them according to an achromat scheme so that their nonlinear effects cancel among themselves.

Synchrotron radiation begins to become a significant design consideration in the SSC. The power radiated must be removed by the cryogenic system. We also discussed the collective effects. Wake fields and impedances are introduced. The instability effects are found to be not very crucial for the SSC.

REFERENCES

1. M. Sands, SLAC-121/UC-28 (1970).

2. SSC Reference Design Studies for US Department of Energy, May 1984.

3. For a review, see for example, "Nonlinear Dynamics and the Beam-Beam Interaction," AIP Conf. Proc. No. 57, Brookhaven, 1979, edited by M. Month and J.C. Herrera.

4. E.D. Courant and H.S. Snyder, Ann. Phys. 3, 1 (1958).

5. K.W. Robinson and G.-A. Voss, Cambridge Electron Accel. report CEAL-TM-149 (1965) unpublished.

6. P.L. Morton and J.R. Rees, IEEE Trans. Nucl. Sci. NS-14, 630 (1967).

7. A. Piwinski, DESY Report 77/18 (1977).

8. K.L. Brown and R.V. Servranckx, Summer School on High Energy Accelerators, Stony Brook/Brookhaven, 1983, AIP Proc No. 127, page 62.

9. SSC Aperture Task Force, SSC-SR-1013 (1985).

10. Clustered Interaction Region Study Group, SSC-SR-1014 (1985).

11. C. Bernardini, G.F. Corazza, G.Di Giugno, G. Ghigo, J. Haissinski, P. Marlin, R. Querzoli, and B. Touschek, Phys. Rev. Lett. 10, 407 (1963).

12. A. Piwinski, Proc. 9th Int. Conf. on High Energy Accelerators, 1974, page 405.

13. J. Bisognano, SSC-25 (1985) IEEE Trans. Nucl, Sci., NS-32, No. 5, 2392 (1985).

14. R. Talman, CERN/ISR-TH/811-17 (1981), unpublished.

15. R.D. Kohaupt, DESY Report M-80/19 (1980), unpublished.

"Particle Physics: Past and Future"

Steven Weinberg
Theory Group, Physics Department
University of Texas
Austin, TX   78712

In thinking over what I might say here about the history of
particle physics, it struck me that our specialty carries with it
a somewhat a different sense of history than do other areas of
science. Our friends in mathematics or biology or even in
condensed matter physics have before them an unlimited number of
problems, and they will never run out of things to worry about.
In that sense the history of most other fields of science is a
little like political history. On the other hand, we who work on
particles and fields have really only one problem, which is to
learn the laws of nature that underly all other sciences. For
this reason it makes sense to describe the history of our field in
terms of progress toward a definite specific goal. Political
history used to be written in this way, the goal being taken as
civilization or a liberal society or what have you. The historian
Herbert Butterfield even coined a term for historians who write in
that style, as having a "Whig" view of history. The worst thing
apparently that a historian can say about another historian these
days is to say that he or she is Whigish. However, it seems to me
that in our field, perhaps as nowhere else in the history of
science (or of anything else) it really is appropriate to take a
Whig view, and judge ourselves according to how we are moving
toward our ultimate goal. At any rate, this will be a Whigish
talk.

I don't mean to imply that we always know how to judge our
progress, nor that we move steadily in one direction. In fact, it
seems to me that a good deal of the history of particle physics in
the last thirty years or so has been a story of oscillation
between two broad lines of approach to the underlying laws, two

approaches that might briefly be called quantum field theory and S-matrix theory. Speaking more loosely, we oscillate between a view of nature as a world of fields, in which the particles are just quanta of the fields, and as a world of particles, of which the fields are just macroscopic coherent states. I am not talking here about how we view our existing theories, which is not so important, but about the strategy that we think will work best in making future progress toward the laws of nature. Very recently this oscillation between S-matrix theory and quantum field theory seems to have entered a new phase, in the revival of string theories. I will talk about this a little later, but first I want to remind you of the historical context of these recent developments.

Quantum field theory of course began as a theory of photons. One of the first papers on quantum mechanics, the one by Born, Heisenberg and Jordan in 1926, showed how Einstein's photons would emerge from the application of quantum mechanics to the electromagnetic field. The formalism was used by Dirac a year later in his theory of spontaneous emission, and then further developed by Fermi. However, the idea of a quantum field theory of everything, electrons and protons as well as photons, really begins with a pair of papers in 1929 by Heisenberg and Pauli.

This grand new view of nature ran into trouble almost immediately. In 1930 Oppenheimer and Waller discovered the ultraviolet divergencies, that have been on our minds ever since. And then a little bit later in the 1930's, it began to appear that quantum electrodynamics was not working very well in describing the development of electromagnetic cosmic-ray showers at high energy. This was a red herring; the anomalies were actually caused by the production of mesons, but mesons were not known at that time, and it appeared like a breakdown of quantum electrodynamics. The conventional wisdom about quantum field theory in the 1930's was that it provides a good low energy approximation, which can be used up to energies of the order of 100 MeV or so, but it should not be taken seriously beyond that.

Quantum field theory had a first revival in 1947, with the development of Lorentz invariant methods of calculation, the solution of the problem of infinities in quantum electrodynamics, and the thrilling calculation of quantities like the magnetic moment of the electron and the Lamb shift to many decimal places, with perfect agreement with experiment. I was not doing physics then, and I can only say that I envy the physicists who were.

However, as exciting as that break-through of the late '40's was, it did not take long before there began another period of disillusionment about quantum field theory. It was found almost right away that the renormalization theory which took care of the infinities of quantum electrodynamics simply didn't work as applied to the weak interactions in the only theory which was then extant, the Fermi theory of weak interactions. Also, for strong interactions it seemed clear that no perturbation theory would work at all. If you have a chance, take a look at Fermi's 1952 Silliman Lectures at Yale. Fermi surveyed the state of particle physics as of that moment, a few years after the grand revival of the late 1940's. It's really very depressing. He considers one process after another, and in each case does approximate calculations and continually admits that there is very little reason to believe in the validity of the approximations.

There is another reason for the renewed disillusionment with quantum field theory that is a little more indirect, but still I think important. Quantum field theory has built into it the idea of an elementary particle as being something fundamentally different from an atomic nucleus or a blackboard eraser. An elementary particle is taken to be any of the particles whose field appears in the Lagrangian of the theory. As long as the universe seemed to consist of a small number of species of elementary particles, the photon, electron, proton, neutron, and neutrino, it really was imaginable that one could have a field theory that would include all these particles' fields. However, it soon became clear that if the proton is going to be called an elementary particle, and the neutron is going to be called an elementary particle, then you have to call hundreds of things elementary particles. You really can't make a case for the proton being any more elementary than the ρ meson or the second excited state of the lambda hyperon. So it seemed pointless to explore quantum field theories in which a few particles that happened to be discovered first appeared as elementary particles.

Out of this sense of disillusionment with quantum field theory in the 1950's, there at last developed an alternative approach. In its initial version, it was called dispersion theory, because it was based on the earlier dispersion theory of electromagnetic propagation. Chew, Gell-Mann, Goldberger, Low, Nambu, Thirring, Toll, Wheeler, and others developed the idea that the analyticity of the S-matrix would provide relations between the real and imaginary parts of scattering amplitudes, while the unitarity of the S-matrix would provide relations between imaginary parts of scattering amplitudes and the squares of their absolute values, and in that way one would have a closed system of equations, which one might either solve perturbatively or perhaps in a self consistent manner, and would completely free us from the conceptual and mathematical difficulties of quantum field theory.

In a sense this represented the return to a philosophical preconception that had played an honorable role in physics, the idea that not only should physics aim at the prediction of observables, but that everything in our equations should be directly observable. Quantum field theory manifestly doesn't satisfy this sort of logical positivism. In quantum field theory we deal with objects like the quantum field of the electron, which we are never going really to measure; they are just mathematical tools for calculating the S-matrix. It was in a way an act of purification to throw out quantum field theory, and just make a theory of observables: masses and S-matrix elements. Wheeler in

1937 and Heisenberg and Møller in the 1940's had already taken steps toward an S-matrix theory of this sort, but now in the 1950's it seemed that dispersion theory would provide the dynamical background for such a theory.

Unfortunately, unitarity connects every physical process to every other process, so the S-matrix program required that one learn how to apply the fundamental principles of analyticity and unitarity and Lorentz invariance and all, not just to simple elastic scattering amplitudes, but to multiparticle amplitudes where any number of particles come into a collison and any number of particles go out. A full-fledged dynamical S-matrix theory began to be developed, particularly in Berkeley, and particularly by a group around Chew, including Mandelstam, Stapp and others. I remember that while I was at Berkeley during this period, S-matrix theory appeared the best hope of progress in physics. I didn't get into it myself, but influenced by S-matrix theory, I tried to see how the basic facts about electromagnetism and gravitation could possibly be understood in an S-matrix context, and showed for example that in any S-matrix theory of mass zero spin one particles the couplings of these photons would have to be those dictated by gauge invariance, and in any theory of mass zero spin two particles, the couplings of these gravitons would have to be those dictated by general covariance, as also remarked independently by Feynman. I will come back to this point a little later.

Despite the great hopes that were entertained at the beginning for S-matrix theory, it in turn foundered. By the mid-1960's, I think few physicists had still any great hopes for the S-matrix program. There are a number of reasons for this, that are instructive to recall. One is that there had never been developed any well-motivated approximation scheme for solving problems numerically, any rationale for simplifying equations in any way so that you could actually get solutions. There was no small parameter in the theory. Another problem was that in order to say what you mean by the requirement of analyticity for something more complicated than a forward scattering amplitude, one must understand the theory of many complex variables, and deal with cuts as well as poles in more than one complex variable at a time. It may be that an adequate understanding of this sort of mathematics is beyond the capabilities of the human brain; I was able to confirm quite convincingly that it's beyond me.

Finally, S-matrix theory had really very little to show in the way of experimental success. Dispersion theory provides useful information about forward scattering amplitudes, but beyond that, the grand S-matrix program did not make much contact with experiment. In particular, although the S-matrix theorists devoted a great deal of attention to pion scattering, they did not anticipate the key point that pion interactions at low energy are rather weak, which was revealed by the applications of current algebra starting with the Adler-Weisberger sum rule in the mid-1960's. The S-matrix program finally died, not in the sense that it was proved wrong, but rather that physicists gave up trying to make any progress with it.

Before the S-matrix program finally died, it had a curious last burst of activity. Theorists who had tried very hard to develop an S-matrix theory for complicated multiparticle amplitudes, and had found that it was impossible to formulate even what the axioms were, began instead to take a more modest view and simply try to guess model scattering amplitudes, which would have most of the properties that were required by the fundamentals of quantum mechanics. In particular Veneziano in 1968 wrote down a toy amplitude for the scattering of two spinless particles, which in a remarkable way embodied the kind of properties that we always thought would come out of an S-matrix theory. It had Regge asymptotic behavior, because of the exchange of an infinite number of particles of higher and higher spin in the cross channel, and when one looked at the scattering amplitude produced by the forces produced by this infinite number of exchanges, sure enough you found as particles in the direct channel just the ones that were producing the forces in the first place; which is why these were called dual models. Great effort went into the exploration of these dual models, starting with the paper of Veneziano. They were very rapidly generalized to multiparticle processes, by Bardacki and Ruegg, Goebel and Sakita, Chan and Tsun, and Koba and Nielsen in 1969, and it was then pointed out by Nambu, Nielsen, and Susskind that the problem to which the dual scattering amplitudes were the solution was the problem of the motion of a relativistic string. That is, the infinite number of particles whose exchange produces the Regge trajectory are in one to one correspondence with the modes of vibration of a relativistic string. For the Veneziano model in its original form the string is open, with free ends which travel at the speed of light. The strings have a certain tension, and since this was all in the context of strong interaction physics, this string tension was imagined to be about a pion mass per fermi, or roughly $(100 \text{ MeV})^2$. There were other models developed; another model called the Virasoro–Shapiro model was found to be based on a closed string, with a string tension that again was taken to be about $(100 \text{ MeV})^2$. Spinors were incorporated into string theories by Neveu and Schwarz, Ramond, and others, and it was this formalism that led Wess and Zumino in 1974 to their development of supersymmetry.

It's really remarkable, now looking back at the period of the late 1960's and early 1970's, how much work went into these string theories without the slightest encouragement from experiment. In fact, the string theories incorporated features that were not only not confirmed by experiment, but were in gross contradiction with experiment. One of these features was that these theories contained massless particles, and given the background of these theories, these were taken to be massless strongly interacting particles, which clearly would have already been observed. In the open string theory, there was a massless spin one particle, and in the closed string theory massless particles of spin two and also spin zero. Another problem was that when one tried to take into account the unitarity corrections to the S—matrix elements, it was found that these theories really only made sense in 26 dimensions, or if you included fermions in the theory, then in 10 dimensions. That was quite embarrasing. Schwarz and Scherk in 1974 suggested that these theories should not be thought of as describing the strong interactions; rather, the mass zero spin two particle that appeared in the closed string theory should be identified as the graviton. In other words, the string tension would not be of order $(100 \text{ MeV})^2$, but would be something like the square of the Planck mass, 10^{38} GeV^2. One does not assume general covariance or the Equivalence Principle here, but these theories are Lorentz covariant, and thus as I mentioned earlier, any mass zero spin two particle in them would have to interact like the graviton of General Relativity. However, this proposal received little attention, I think above all because it was embarrasing to be working on theories that made sense only in 10 or 26 dimensions.

But perhaps even more influential than any failure of the dual models or string theory was a second revival of quantum field theory, which began at just about the time of the development of string theory. I needn't dwell on this, as I suppose it's familiar to everyone here. First, it was shown by t'Hooft and others that the earlier suggestion of a spontaneously broken renormalizable gauge theory of electroweak interactions would indeed work mathematically. Very soon thereafter, the discovery of neutral currents and the measurement of their properties showed that not only <u>could</u> an electroweak theory be constructed along these lines, but the already constructed SU(2) x U(1) theory was in fact the correct theory of electroweak interactions. At the same time, there was also the development at last of a theory of strong interactions, based on much the same Yang—Mills mathematical structures, in which instead of the gauge symmetry being spontaneously broken, it provides a trapping mechanism which hides the underlying constituents, the gluons and quarks, from our view. We had in the early 1970's a fully developed standard model of weak, electromagnetic, and strong interactions, which apparently was capable of describing everything to do with accessible particle physics. Exciting experiments then later in the 70's gave a final confirmation to these theories, showing that in fact the neutral currents did behave in the way they were supposed to, the quarks and leptons really are what we think they are (except that there are more of them), and the W and the Z are really there.

Once again it became possible to think of a quantum field theory as being a fundamental theory of nature. In particular it made sense once again to really talk about particles as being elementary, because we no longer had to think of the proton or the neutron or the ρ meson as elementary particles. The elementary particles had been reduced to a fairly manageable set; leptons, quarks, gluons, W, Z, the photon, and maybe a few Higgs bosons. One thing that especially excited me, was the fact that at last in these theories we could understand in a natural way why symmetries like parity and strangeness and isotopic spin were symmetries of some interactions and not of others. All the mysterious facts that my generation of physicists had had to learn as empirical rules when we were graduate students suddenly began to make sense rationally.

Coleman has recently published a collection of his lectures given at Erice, whose preface contains an eloquent statement of enthusiasm for the revival of quantum field theory. The lectures span the period from 1966 to 1979, and of this period Coleman says, "This was a great time to be a high-energy theorist, the period of the famous triumph of quantum field theory. And what a triumph it was, in the old sense of the word: a glorious victory parade, full of wonderful things brought back from far places to make the spectator gasp with awe and laugh with joy."

I certainly agree with Coleman. Nevertheless, by the end of this period, by the beginning of the 1980's, there began to be felt some sense of dissatisfaction with quantum field theory. For instance, I was asked in 1981 to give a talk at the 50th anniversary of the Lawrence Berkeley Laboratory, and took the occasion to suggest that it might be time for another turn away from quantum field theory. If I may quote myself, "These have been exciting times. Quantum field theory is riding very high and one might be forgiven for a certain amount of complacency with it. But perhaps we will now see another swing away from quantum field theory. Perhaps that swing will be back in the direction of something like S—matrix theory."

Part of my reason for this remark was that I was at Berkeley, and I wanted to say nice things about S—matrix theory. However, I gave two serious reasons for reservations about the future of quantum field theory.

One reason is that theorists had failed to make further progress in explaining or predicting the properties of elementary particles, beyond the progress that was already well in hand by the early 1970's. The standard model has many loose ends; mass ratios, coupling constants, a whole menu of quarks and leptons, and we have simply not succeeded in explaining it. Several attractive ideas were tried: grand unification, technicolor, preons, supersymmetry, Kaluza—Klein theories, and all that, but despite so much clever mathematical work, almost nothing has come out in the way of concrete numbers that could be compared with experiments. Perhaps the only success in that hard quantitative sense was the grand—unification prediction of $\sin^2\theta$, a prediction that does seem to be quite robust, and also to agree with experiment. It soon became clear that in trying to make the next step beyond the standard model we would probably have to understand physics at or near the Planck scale, partly because that's where whatever grand unified group there might be would break down, and also because after all gravity exists, and gravity becomes a strong interaction at the Planck scale. Unfortunately, throughout the 1970's most of us saw no hope for a quantum field theory of gravity.

The second reason that I gave in 1981 for being skeptical about the future of quantum field theory is that we could understand its successes in the low energy range, up to a TeV or so, without having to believe in quantum field theory as a fundamental theory. There is a folk theorem (a term of Wightman, meaning something which is generally known to be true although it hasn't been proved), that says that any theory which satisfies the axioms of S-matrix theory, and contains only a finite number of particles with mass below some M, will at energies below M look like a quantum field theory involving just these particles. That is, quantum field theory by itself has no content; it is just a way of calculating the most general scattering amplitudes that obey the axioms of S-matrix theory. Of course, one might argue that the quantum field theories that we have developed, like quantum chromodynamics, are not just any old quantum field theories, but simple, even beautiful, field theories. However, another folk theorem tells us that in the effective field theory, that essentially any theory reduces to at sufficiently low energy, the non-renormalizable interactions are all suppressed by powers of the underlying fundamental mass scale, which one might imagine is the Planck mass scale. Thus, the physics we see at accessible energies should be described by a renormalizable effective field theory, and we know that the interactions in such theories are always limited in number and complexity. To see anything else we would have to do experiments at the Planck scale, except that a few interactions though very weak may be detectable for special reasons, like for example the special circumstance that gravity adds up coherently, so that although the gravitational interaction is fantastically weak, we can still measure its effect for macroscopic bodies.

To summarize: quantum field theory has not done much for us lately, and what it had done for us earlier we can understand without having to believe that quantum field theory is in any sense fundamental.

When I made these remarks in 1981, I had no idea what direction S-matrix theory would take as a possible replacement for quantum field theory. In fact, Schwarz had been forcefully advocating string theory as the only hope for a quantum theory of gravity ever since his 1974 work with Scherk, but he was ignored by almost everyone (myself included). In 1980 Green and Schwarz proved the space-time supersymmetry of superstring theory. In the following two years they developed a new supersymmetric formalism for superstrings and used it to invent the Type II superstring theories and to prove their finiteness at one loop. At the 1982 Solvay Conference in Austin, Zumino mentioned this work of Green and Schwarz as the natural candidate for a finite theory of gravitation. Then in 1983, at the Fourth Workshop on Grand Unification, Witten gave an influential talk about d=10 superstring theory, and some of the rest of us began to take such theories seriously as a promising approach to quantum gravity. The time was ripe for this suggestion, as it had not been in 1974, partly because we were all so frustrated with everything else, and also because several years of work on Kaluza-Klein theories had made us comfortable with the idea that spacetime might really have more than 4 dimensions, with all but 4 wrapped up in a compact manifold of very small circumference.

Witten in his talk also pointed out the theoretical obstacle, having to do with a hexagon anomaly, that impeded this development, and later with Alvarez-Gaumé discovered the cancellation of this anomaly in one example of a superstring theory, that unfortunately seemed phenomenologically unpromising. Then the great breakthough came last year when Green and Schwarz demonstrated that there were a few potentially realistic superstring theories, with very specific gauge groups, in which the anomalies that Witten had worried about cancelled. This started an explosion of interest in string theory, which has not yet even peaked.

I suppose that this should be scored as a victory for the S-matrix approach. String theory grew out of S-matrix theory, but in a sense it has some of the features of both S-matrix theory and quantum field theory — the experts have not yet settled down in their view of what string theory really is. Indeed, this is one of the things that makes the theory hard to learn; not everyone will not tell you the same thing about what it is you're supposed to be learning.

On one hand there is a view of string theory, which takes seriously that these are theories of strings. Instead of quantum fields that are time-dependent functions of the position in space of a particular particle, you have quantum fields that are time-dependent functionals of the configuration in space of a moving string. This second quantized quantum field theory of strings unfortunately does not yet exist, but many of the leading experts in this area are working very hard to develop it. The particles I may remind you are the normal modes of these strings, so when you calculate an S-matrix element (which in the end is what you always have to do) you imagine a string in a particular normal mode colliding with another string in another normal mode, and perhaps two strings joining together to make a single string, and then that single string breaking apart to be two other strings, which finally wind up in two other normal modes. Calculations are not actually done that way. The description I just gave is often presented in public talks about string theories by the experts, but as far as I can tell they don't actually do calculations that way. It's too hard, and the formalism hasn't been developed yet.

There is another approach to string theory, which is the one that almost everyone actually uses. In this other approach, one first starts with the observation that a string moving through space sweeps out a two dimensional surface in space-time. You can describe the string by giving the space-time coordinates x^μ as functions of two parameters. One parameter σ tells you where along the string you are, and the other parameter τ tells you how long the string has had to move. So the string theory can be regarded as a quantum field theory in two dimensions, the "fields" being taken as the d quantities $x^\mu(\sigma,\tau)$, (where d = 4 or 10 or 26 or whatever) with perhaps some spinors $\psi(\sigma,\tau)$ as well. The interpretation of this two-dimensional field theory in terms of physical processes in d spacetime dimensions has to be accomplished by asking what sort of quantum averages in the two-dimensional world have the unitarity and Lorentz transformation properties that we require for the S-matrix in d-dimensional spacetime, so we have a curious blend here of quantum field theory and S-matrix theory.

It is very natural to write the Lagrangian for the two-dimensional theory so that it is independent of the choice of the parameters σ and τ, which of course requires the introduction of a two-by-two metric tensor. As emphasized by Polyakov, it turns out then that the old string theory is not only a generally covariant two-dimensional field theory, but is invariant as well under conformal transformations, in which the metric is multiplied with an arbitrary function of the two-dimensional coordinates. This may sound like I'm getting into technicalities here, but the addition of conformal invariance to 2-dimensional general covariance and d-dimensional Lorentz invariance has an overwhelmingly important consequence: the string Lagrangian must be that of a free field theory in two dimensions, with the non-triviality of the S-matrix in flat d-dimensional spacetime arising not from interaction terms in the Lagrangian but from the non-trivial topology of the Riemann surface described by the 2x2 metrics. So here we have the realization of an old dream of what a fundamental theory ought to be: interactions are not something we insert in a more-or-less arbitrary way into a Lagrangian (and might if we wished leave out altogether) but are inevitable consequences of the nature of the theory's degrees of freedom.

The differences between these two views of string theory can be illustrated by considering how each would deal with a "one-loop" calculation of the S-matrix for 2-particle scattering in a closed string theory. In the 2-dimensional field theory approach, one imagines the 2-dimensional space to form a torus, and carries out a free-field quantum average of a product of four "vertex functions" of position on the torus, one for each incoming or outgoing particle. On the other hand, in the second quantized string theory approach, one imagines two closed strings in different normal modes approaching, joining to form one string, then breaking up again into two closed strings, then joining again to form one closed string, and finally breaking apart again to form two closed strings in definite normal modes. This gives one more of a sense of the physical reality of strings, but it is not a very elegant way to describe a torus.

I have not been entirely impartial here in drawing the contrast between the two leading approaches to string theory. My preference for the 2-dimensional field theory approach may be due in part to the fact that this is the only approach that I have so far been able to learn. Certainly one should try to understand all possible approaches. Also, it may be, as often argued, that the second-quantized field theory of strings offers the best hope for an understanding of non-perturbative effects. Nevertheless, since string theory is supposed to be better than quantum field theory, it does not seem clear to me that the best strategy is to make string theory look as much as possible like field theory, only with strings instead of particles, which I take is the spirit of the second-quantized approach.

In the last few minutes, I want to take up the question that has doubtless been on the minds of all those of you who have not yet become string mavens. The question is: Is it safe to ignore string theory, and hope that it will go away? For a theorist, the question takes the form whether it is necessary to learn all about automorphic functions, Riemann surfaces, Virasoro algebras, and all that, or just bypass all this effort and wait for the next fashion in theoretical physics. For the experimentalist, the question is whether it is worthwhile beginning to think of possibly testing these theories?

In trying to answer these questions, I must say right away that there is not the slightest shred of experimental evidence for string theory. The same was also true of the other theories that we developed in our desperate attempt to go beyond the standard model, in particular for supersymmetry and Kaluza-Klein theories, which have now been incorporated into superstring theory. Never has so much brilliant mathematics been done by physicists with so little encouragement from experiment. Furthermore, just as for supersymmetry and Kaluza-Klein theories, the string theories have still not settled down so that they could make very definite predictions which could be tested experimentally. However in this respect I think string theories are really different from the Kaluza-Klein and supersymmetry theories. Supersymmetry is a symmetry like Lorentz invariance; it allows a tremendous variety of possible dynamical theories. Likewise, Kaluza-Klein theory is just a general idea, that there might be some higher number of dimensions which are compactified; again, this idea allows a great many specific theories. String theories, on the other hand, are very rigid. There are almost no string theories at all, and of the few possibilities only one at present (the "heterotic" superstring of Gross, Harvey, Martinec, and Rohm) seems at all promising phenomenologically. As I discussed earlier, what you're really doing in string theory is studying two-dimensional conformal gravity (actually supergravity), which is a free field theory, so that the only interactions are those that arise from the topology of two dimensional manifolds. Also, the topology of a two dimensional manifold is completely specified by the number of handles that you put on it (assuming it an orientable closed surface). And so there's nothing you can tinker with in these theories; they're either right or wrong as they stand. We don't have any experimental evidence for string theories, and I can't really tell the experimentalists what they should look for, but the string theorists in the next few years should be able to come up with definite predictions, which can then be tested. Already as we heard this morning from Nappi and Segré, there is an indication that any realistic string theory when compactified down to four dimensions will probably contain at least an extra $U(1)$, so that it is worth looking for one more gauge symmetry in addition to the $SU(3)xSU(2)xU(1)$ of the standard model. But the precise features of this extra $U(1)$ are certainly not yet predicted. The biggest gap that will have to be crossed before such predictions can be made is in understanding the dynamics of the compactification from 10 to 4 spacetime dimensions. Candelas, Horowitz, Strominger, Witten, and others have made some progress in understanding the general features of this compactification, but much remains to be done.

I have remarked that there's no evidence for string theories, but there are other criticisms of a more fundamental nature. Georgi has argued to me that it really would be very unlikely that string theory should provide a fundamental theory of gravity and everything else, in part because after all string theories developed out of the original guess by Veneziano of a scattering amplitude for strong interaction meson-meson scattering. Why in the world should we believe that mathematical structures that grew out of the attempt to understand the strong interactions should be applicable not to the strong interactions but to everything, including gravity? I don't agree with this argument, because I don't agree with its historical basis. The string theories and the dual models which preceded them did not grow out of an attempt to understand the detailed empirical facts of the strong interactions; in fact, they never were much good at that. They grew out of an attempt to find some solution to the problem of constructing scattering amplitudes that satisfy the basic axioms of analyticity, unitarity, and so on. And these are of course the same problems that we're all solving, whether at 100 MeV or at the Planck scale. There's a good chance that the way of satisfying these fundamental S-matrix principles that's provided by string theory is unique, at least if we want to include gravitation. The string theorists of the late '60's and early '70's guessed that the kind of dual model that they were proposing would turn out (when suitably unitarized) to be the more or less unique solution of the axioms of S-matrix theory. And maybe it is, not as applied only to the strong attractions, but as applied to everything, including gravitation. In fact, maybe we have really now the answer to the problem that I mentioned at the beginning. Maybe we really do know the laws of nature, and the only thing that is left is to work hard for the next few years, and try to figure out how the ten dimensions get compactified to four, and then find out what low-mass quarks and leptons and gauge bosons are left over, calculate their mass ratios, check to see whether they agree with experiment, and then go home.

Relying on a general sense that nothing ever in this life works out the way we want it to, ("we" here means theorists), my guess is that there are many surprises that will be provided to us both by imaginative theorists and by enterprising experimentalists before we finally get to the solution to our problem. Nevertheless, I would argue that it is not safe to ignore string theories and wait for the next change in fashion. These theories are much too promising and too beautiful for us not to take them very seriously and explore their consequences for as long as it takes.

(Supported in part by NSF Grant 8304629 and the Robert A. Welch Foundation.)

I

ELECTROWEAK INTERACTIONS

Session Organizers: D. Hitlin and J.L. Rosner

KOBAYASHI-MASKAWA MATRIX ELEMENT

Wai-Yee Keung

Department of Physics, University of Illinois-Chicago, IL60680

ABSTRACT

We review the origin of the CP nonconservation. The implication of a rather long-lived b-quark is studied. Its phenomenological consequences to m_t, rare K decay and B-$\bar{\text{B}}$ mixing etc. are given in the context of the Standard Model of 6 flavors. A convenient parametrization of the quark mixing angles is proposed. Finally, we show the relation between the elusive "Penguin" mechanism and the ϵ'/ϵ measurement.

1. CP NONCONSERVATION

The asymmetry of the interference patterns among the particle channel and the anti-particle channel could result in CP violation. The transition amplitude for a path j can usualy be factorized into two terms: (1) the generic coupling g_j which stands for the product of coupling constants or mixing angles and becomes complex conjugated under CP inversion; and (2) the dynamical factor A_j, which comes from either

(i) the time evolution term $\exp(-iE_j t)$ or
(ii) the final state interaction or the Feynman intergral $\int dQ.../(q^2-m^2+i\epsilon)$,

and remains unchanged under CP inversion. The resultant amplitude is given by summing up all contributions from different paths j .

$$\text{Amplitude} \quad (i \to f) = \sum_j g_j A_j$$

$$\text{Probability} \quad (i \to f) = \sum_j \sum_k g_j g_k A_j^* A_k^*$$

For the CP conjugated process,

$$\text{Probability} \quad (\bar{i} \to \bar{f}) = \sum_j \sum_k g_j^* g_k A_j A_k^*$$

$$\text{Prob.}(i \to f) - \text{Prob.}(\bar{i} \to \bar{f}) = 2 \sum \text{Im}(\, g_j g_k^*) \, \text{Im} \, (A_j A_k^*)$$

Therfore, in order to manifest CP Nonconservation, it is necessary that two or more paths with unparallel couplings g_j and unparallel dynamical factors A_j in the complex plane.

1.1 Particle-Antiparticle Oscillation

K-$\bar{\text{K}}$ ocsillation arises from the the box diagram with W and quark exchanges. It includes 9 paths of different combinations of intermediate quark states, i.e. $i=u\bar{u}$, $c\bar{c}$, $u\bar{c}$, $u\bar{t}$ ..). Mixing effect does not imply CP Nonconservation. Instead, CP Nonconservation happens if Prob.$(K\rightarrow\bar{K})$ $\neq$ Prob.$(\bar{K}\rightarrow K)$. Typical couplings are

$$g_{(k=u\bar{c})} = V_{cd}^* V_{cs} V_{ud}^* V_{us} \ , \ g_{(k=c\bar{c})} = V_{cd}^* V_{cs} V_{cd}^* V_{cs} \ .$$

If there were only two families (four flavors u, d, s and c), unitarity requires

$$V_{cd}^* V_{cs} = - V_{ud}^* V_{us}$$

so that $g_{(k=c\bar{c})}$ is parallell to $g_{(i=u\bar{c})}$ etc. It becomes clear that 3 families are necessary for CP nonconservation in the Standard Model. The other criterion $(A_k \not\parallel A_j)$ is satified here because some dyanmical factors, e.g. $A_{(k=u\bar{u})}$ but not $A_{(k=c\bar{c})}$, acquire imaginary absorptive parts from the intermediate $u\bar{u}$ states. It is commonly defined in the literature[1]:

$$M_{KR} = \Sigma\ g_j\text{Re}\ A_j \ , \ \Gamma_{KR} = -2\ \Sigma\ g_j\text{Im}\ A_j$$

$$\text{Amplitude}\ (R \rightarrow K) = \Sigma\ g_j A_j = M_{KR} - \tfrac{1}{2}i\ \Gamma_{KR} \ .$$

The time-evolution eigenstates K_L and K_S have asymmetric components in K and $\bar{K}$:

$$K_{L,S} = [2(1+|\epsilon|^2)]^{-1}[(1+\epsilon)\ K \mp (1-\epsilon)\ \bar{K}]$$

Here

$$\left(\frac{1-\epsilon}{1+\epsilon}\right)^2 = \frac{M_{KR}^* - i\tfrac{1}{2}\Gamma_{KR}^*}{M_{KR} - i\tfrac{1}{2}\Gamma_{KR}} \ \text{or} \ \frac{\Sigma g_j^* A_j}{\Sigma g_j A_j}$$

The real part of ϵ explicilty breaks the CP symmetry. It is measured by:

$$\frac{\text{Br}\ (K_L\rightarrow \pi^- e^+\nu)\ -\text{Br}\ (K_L\rightarrow \pi^+e^-\nu)}{\text{Br}\ (K_L\rightarrow \pi^-e^+\nu)\ +\text{Br}\ (K_L\rightarrow \pi^+e^-\nu)} = 2\ \text{Re}\,(\epsilon) = 0.0033$$

1.2 Time Evolution

The imaginary part of ϵ will not give explicit CP asymmetry in the $K_{L,S}$ compositions as an global phase can be rotated away. However, through the unparrell time propagators $\exp(-m_{L,S}\tau i)$, $\text{Im}(\epsilon)$ can also give rise to the CP nonconservation if a common final state is monitored. For example, the rate difference between $K^0\rightarrow 2\pi$ and $\bar{K}^0\rightarrow 2\pi$ has a term proportional to

$$\text{Im}\ \epsilon\ \text{Im}\ \exp\,(-i\Delta m\tau - \Gamma_S\tau/2) \ .$$

When combining with the explicit CP asymmetric term from $\text{Re}\,(\epsilon)$, i.e.

$$-\text{Re } \epsilon \text{ Re}[1 - \exp(-i\Delta m\tau - \Gamma_S\tau/2)],$$

we obtain the overall asymmetry as[1]

$$|\epsilon| \exp(-\Gamma_S\tau/2)\cos(\Delta m\tau - \phi) - \text{Re }(\epsilon)$$

Here ϕ is the argument of the complex ϵ. Experimentally, we are able to measure the time evolution profile deviated from the CP conservation for the process $K \to 2\pi$. It is known that[2]

$$|\epsilon| = 0.00227$$

$$\phi \sim 45°$$

This implies that both the real part and the imaginary part contribute about equally for $K \to 2\pi$. Nonetheless, in the heavy quark system, e.g. $B°-\bar{B}°$, the explicit CP violating factor Re(ϵ) could be very small but not the Im(ϵ), , then we may observe CP violation in an exclusive mutual channel for the $B-\bar{B}$ system[3].

1.3 Direct Decay

So far we oversimplify the situation. In fact, there is another intrinsic and direct CP violation for $K \to 2\pi$ and $\bar{K} \to 2\pi$, where both the final state configurations $\Delta I = 1/2$ or $3/2$ with different phase shifts also pick up different weak mixing angles. Especially, the penguin diagram $\Delta I = \frac{1}{2}$ has an imaginary part dominantly given by the virtual top quark contribution in the Kabayahi-Maskawa representation. Parameter ϵ' characterizes this effect[1].

$$\frac{\text{Br }(K \to \pi^+\pi^-) - \text{Br }(\bar{K} \to \pi^+\pi^-)}{\text{Br }(K \to \pi^+\pi^-) + \text{Br }(\bar{K} \to \pi^+\pi^-)} = 2 \text{ Re } \epsilon'$$

Also,

$$\frac{\text{Amp }(K_L \to \pi^+\pi^-)}{\text{Amp }(K_S \to \pi^+\pi^-)} = \eta^{+-} = \epsilon + \epsilon'$$

$$\frac{\text{Amp }(K_L \to \pi^0\pi^0)}{\text{Amp }(K_S \to \pi^0\pi^0)} = \eta^{00} = \epsilon - 2\epsilon'$$

ϵ' is small partially because of the $\Delta I = \frac{1}{2}$ rule which is still a mystery in the theory. Latest experiments[4] measured the ratio of the ratios of the above expressions and obtained:

YALE-BNL $\quad -0.0065 \leq \epsilon'/\epsilon \leq 0.0099$
UC-FNAL-SACLAY $\epsilon'/\epsilon = -0.0046 \pm 0.0053 \pm 0.0024$

2. QUARK MIXING MATRIX

Nuclear beta decay and hyperon beta decay tell us the magnitude of V_{ud} and V_{us}. They are

$$V_{ud} = 0.97$$
$$V_{us} = 0.23 \quad .$$

Recent measurement[5] of the b-quark lifetime of the order of one picosecond and the limit[6] of $\Gamma(b\rightarrow u)/\Gamma(b\rightarrow c)$, which gives the upper bound of the amplitude ratio $R=|V_{ub}/V_{cb}|$, impose informative constraints[7] on V_{cb} and V_{ub}:

$$V_{cb} = s_3 + s_2\, e^{i\delta}$$
$$|V_{cb}| \sim 0.06$$
$$V_{ub} = s_1 s_3 \leq 0.01 \quad .$$

They are pretty small compared to V_{us}. We have a hierachy situation that the amplitude is weaker for much off diagonal transition. The fact of a rather long-lived b-quark implies that the top quark mass could not be too light[7,8] in the Standard Model. Otherwise it is difficult to saturate the ϵ value of the CP violation in K_L-K_S mixing. However, the lower limit of m_t depends strongly on the theoretical uncertain parameter B_K which measures

$$B_K = -3< K \,|\,[\bar{d}\,\gamma_\mu(1-\gamma_5)\,s]^2\,|\bar{K}>/(4f_K^2 M_K) \quad .$$

The current estimations[9] of this parameter vary from 0.33 to about 1. A light t quark of 40 Gev is only compatible with the Standard Model for B_K about unity. Fig. 1 shows the allowed regions of s_2 and s_3 by fitting the observed ϵ value (ϵ' effect is ignored at the present moment). We also observe that further constraint upon $R=|V_{ub}/V_{cb}|$ will push the lower bound of m_t upward.

Fig. 1 Allowed regions of s_2 and s_3.

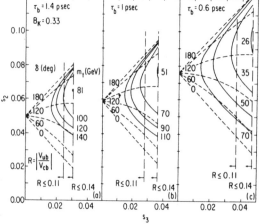

2.1 RARE DECAY $K^+ \to \pi\nu\bar{\nu}$ and B^0-$\bar{B}^0$ MIXING

With the b-quark lifetime as an input, we can better estimate an interesting rare decay $K^+ \to \pi\nu\bar{\nu}$. in the Standard Model. Fig. 2 shows the possible rate. The allowed regions are within the banana-like curves for different b lifetime.

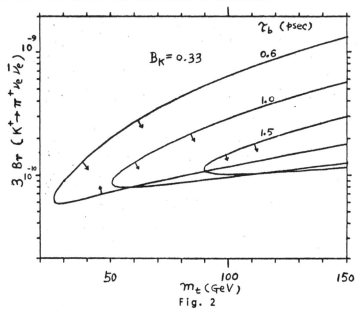

Fig. 2

About the B^0-$\bar{B}^0$ mixing, the long lifetime provides enough oscillation before the decay, so large mixing is expected. For B_s (B_d), the mixing probability[7] lies in the range 0.1-0.9 (0.01-0.1).

3. REPARAMETRIZATION

There are some shortcomings of the conventional Kobayashi-Maskawa parametrization.
(a) It easily gives the wrong impression that CP violation could be large in the heavy quark sector, becuse of the large imaginary parts of some quark mixing amplitudes, e.g. $V_{tb} = -e^{i\delta}$ However, the CP violating effect is always proportional a small factor $s_2 s_3 s_\delta$ after some algebraic manipulations.

(b) V_{cb} does not tell s_2 or s_3 explicitly.

We present[10] a better parametrization in terms of angles

$$s_x = V_{us} = 0.23 \quad (10^{-1})$$
$$s_y = V_{cb} \sim 0.06 \quad (10^{-2})$$

$$s_z = |V_{ub}| < 0.01 \quad (\ 10^{-3}\ )$$
$$V_{ub} = s_z\ e^{-i\phi}\ .$$

The matrix elements can be expanded to the order of 10^{-3} in the real parts and 10^{-5} in the imaginary parts as follows:

$$V = \begin{pmatrix} c_x & s_x & s_z e^{-i\phi} \\ -s_x - s_y s_z e^{i\phi} & c_x & s_y \\ s_x s_y - s_z e^{i\phi} & -s_y - s_x s_z e^{i\phi} & 1 \end{pmatrix}$$

This parametrization has the imaginary parts minimized less than 10^{-3} already. It is easy to remember the upper right corner, then the rest can be derived from unitarity immediately. The measurable quantity $|V_{cb}|$ from b-lifetime is directly related to the angle s_y. Furthermore, the universal CP violation factor can be shown to be

$$X_{CP} = s_x s_y s_z\ s_\phi$$

4. ABOUT ϵ'/ϵ

We can further pursuit the value of ϵ' in the Standard Model.

$$\epsilon'/\epsilon = B_K'(s_y\ s_z/s_x)\ [\ \ln\ (m_t^2/m_c^2)\ -\ \dots]/\sqrt{2}$$

Here we encounter another uncertain hadronic parameter about the penguin amplitude.

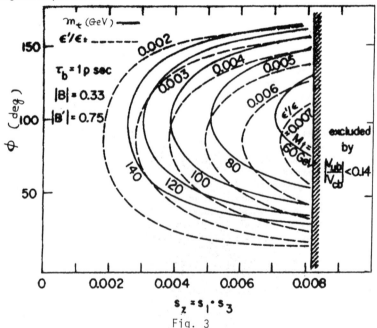

Fig. 3

$$B_K' = \sqrt{2} < 2\pi \ (I=0) \ |Q_6| \ K^0 > / (1 \ GeV^3)$$

which ranges[11] from

0.27	Vacuum Insertion
0.77	Chiral Lagrangian
-0.9 ~ 1.2	Bag Model

If B_K and B_K' are reliably known, we can refine our previous analysis by fitting ϵ with various possible values of ϵ'. We observe[12] that the contours of constant m_t and ϵ'/ϵ define a map of the allowed regions of the mixing angles, i.e. s_y and s_z (See Fig. 3). Therefore, all angles can be determined with the measurement of m_t and ϵ'/ϵ provided B_K and B_K' are known without uncertainty. On the other hand, for given ϵ'/ϵ, m_t is also bound above as shown in Fig. 4. It occurs because the m_t dependent term in ϵ'/ϵ formula tends to a constant at large m_t. So $s_y s_z$ is bound below. To avoid oversaturate ϵ, m_t has an upper limit.

ACKNOWLEDGEMENT

I wish to thank L.-L. Chau and H.-Y. Cheng for the collaboration. This work was supported by the Department of Energy under contract DE-FG02-87ER40173.

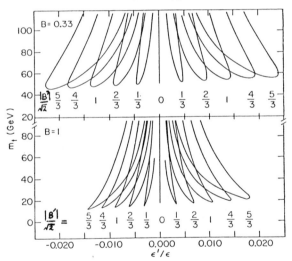

Fig. 4   Allowed regions (inside the curves of m_t and ϵ'/ϵ). See Ref. 12 for the conditions of the signs.

REFERENCE

1. C.G. Wohl et al. (Particle Data Group), Mod. Rev. Phys. **51** , S1 (1984).
2. For example, see E.Commins and P. Bucksbaum, "Weak Interaction of Leptons and Quarks", Cambridge Univ. Press (1983).
3. I.I. Bigi and A.I. Sanda, Phys.Rev. **D29** , 1393 (1984).
4. R. Bernstein et al., Phys. Rev. Lett. **54** , 1631 (1985); J.K. Black et al., Phys. Rev. Lett. **54** , 1628 (1985).
5. E. Fernandez et al., Phys. Rev. Lett. **51** ,1022 (1983); N.S. Lockyer et al. *ibid 51* , 1316 (1983).
6. C. Klopfenstein et al., Phys. Lett. **130B** , 444 (1983).
7. L.-L. Chau and W.-Y. Keung, Phys. Rev. **29** , 592 (1984).
8. P. Ginsparg, S. Glashow and M. Wise, Phys. Rev. Lett. **50** , 1415 (1983).
9. J. F. Donoghue, E. Golowich and B.R. Holstein, Phys. Lett. **119B** , 412 (1982);J. Bijnens, H. Sonoda, M.B. Wise, Phys. Rev. Lett. **53** , 2367 (1984); A. Pich and E. deRafael, Phys. Lett. **158B** , 477 (1985); A.J. Buras and J.-M. Gerard, MPI-PAE/PTh 40/85 (1985).
10. L.-L. Chau and W.-Y. Keung, Phys. Rev. Lett. **53** , 1802 (1984).
11. J.F. Donoghue, Phys. Rev. **30** , 1499 (1984); T.N. Pham, Phys. Lett. **145B** , 113 (1984).
12. L.-L. Chau, H.-Y. Cheng and W.-Y. Keung, Phys. Rev. **32** , 1837 (1985).

EXPERIMENTAL STUDY OF SIGMA BETA DECAY AND ITS IMPLICATIONS

Roland Winston

The Enrico Fermi Institute and Department of Physics
The University of Chicago, 5640 S. Ellis Avenue
Chicago, Illinois 60637, U. S. A.

ABSTRACT

The high statistics data sample of polarized Σ^- beta decays from Fermilab Experiment 715 permits a critical test of the Cabibbo theory. In addition, the induced form factors f_2 (weak magnetism) and g_2 (pseudo tensor) can be extracted with sufficient sensitivity to compare with calculations of SU(3) breaking effects.

1. INTRODUCTION

Baryon beta decay is described by the Cabibbo theory[1] which assumes that: 1) the weak vector current and the electromagnetic current consist of the appropriate components of a common octet current of SU(3): 2) the axial-vector current is the same component of another octet current; and 3) there is a universal suppression factor for strangeness changing transitions. Although SU(3) symmetry is known to be broken (e.g., the baryon mass differences), no significant effects have yet been seen in semi-leptonic decays[2].

2. THEORY

For the semileptonic decay $\Sigma^- \rightarrow n + e^- + \bar{\nu}_e$, the matrix element can be written as

$$M = \frac{G_F}{\sqrt{2}} < n \; |j^\mu| \; \Sigma^- > \bar{u}(e^-)\gamma_\mu \; (1 + \gamma_5) \; u \; (\bar{\nu}_e) \tag{1}$$

where G_F is Fermi weak coupling constant. We assume the leptonic current is V-A. The hadronic current in its most general form can be written as

$$<n \ |J^\mu|\Sigma^-> \ = \sin\theta_c \bar{u}(n) \left\{ f_1(q^2)\gamma\mu + \frac{f_2(q^2)}{M_{\Sigma^-}} \sigma^{\mu\nu}q_\nu + \frac{f_3(q^2)}{M_{\Sigma^-}} q^\mu \right.$$

$$\left. + [g_1(q^2)\gamma\mu + \frac{g_2(q^2)}{M_{\Sigma^-}} \sigma^{\mu\nu} q_\nu + \frac{g_3(q^2)}{M_{\Sigma^-}} q^\mu]\gamma_5 \right\} u(\Sigma^-) \qquad (2)$$

where $\sin\theta_c$ is Cabibbo angle and q^2 is the momentum transfer squared.

The form factors are real if time reversal symmetry is assumed. With Cabibbo theory, the form factors at $q^2 = 0$ of the baryon octet semileptonic decays are related to each other by SU(3) Clebsch-Gordon coefficients. In the case of Σ^- beta decay, in the SU(3) symmetry limit, $f_2(0)/f_1(0)$ is $\mu_n + \frac{\mu_p}{2}$ where μ_n and μ_p are the neutron and proton anomolous magnetic moments respectively. The ratio $g_1(0)/f_1(0)$ is F-D where D and F are symmetric and antisymmetric couplings of two SU(3) octets to form a third. A recent measurement[2] gave F = 0.477 ± 0.012 and D = 0.756 ± 0.011. The g_2 is second class. The contributions from f_3 and g_3 are negligible because they are proportional to electron mass.

The q_2 dependence of the form factors is formulated[2] as

$$f_1(q^2) = f_1(0) \ (1 + 2q^2/M_V^2) \qquad (3)$$

$$g_1(q^2) - g_1(0) \ (1 + 2q^2/M_A^2) \qquad (4)$$

with $M_V = 0.97$ GeV/c_2 and $M_A = 1.25$ GeV/c^2.

Decay product angular distribution asymmetries in the center of mass can be expressed as

$$\frac{dN_i}{d\Omega} = (1 + \alpha_i P \cdot \Omega) \qquad (5)$$

where $i = n, \ e^-, \ \bar{\nu}_e$. P is the Σ^- polarization vector. In the case that P points in the y direction, it can be written as $1 + \alpha_i P\cos\theta_y$.

Differential decay rate and α_i's have been calculated by several authors. In our analysis we have used the calculations by Garcia et al.[3]

3. THE SIGN OF g_1/f_1

For Σ^- leptonic decay, the Cabibbo theory makes the distinctive prediction of a *negative* g_1/f_1 (i.e., a V + A hadronic matrix element) which is opposite in sign to neutron beta decay.[4] Thus a V - A interaction on the quark level gives a "V + A" matrix element on the hadron level. Consequently the electron asymmetry from polarized Σ^- beta decay would be expected to be large and negative. An entirely analogous phenomenon occurs in nuclear beta decay where a "V - A" interaction on the nucleon level can give "V + A" matrix elements in specific decays;

e.g., in tritium beta decay ($H^3 \rightarrow He^3 + e^- + \bar{\nu}_e$). Consequently, the electron asymmetry from polarized H^3 decay would be $\simeq -1$ even though the decay of the "constituent" neutron is $\simeq$ V-A and free neutron decay has a very small electron asymmetry.

It was therefore disconcerting that experimental measurements prior to 1984 favored a small, positive electron asymmetry.[5] The situation changed dramatically one year ago when Fermilab experiment E-715 reported a precise determination of the electron asymmetry performed with a polarized Σ^- beam.[6] The new value $\alpha_e = -0.56 \pm 0.12$ is in striking agreement with the Cabibbo theory. In contrast with the earlier low energy polarized Σ^- experiments with limited statistics, the high fluxes of polarized Σ^- from the Fermilab hyperon beam (e.g., $\simeq 1 kHz$ Σ^- at $\simeq 22\%$ polarization) enabled E-715 to collect $\simeq 5 \times 10^4$ beta decays. Moreover, the ability to produce polarized Σ^- in both vertical and horizontal planes and to reverse the sign of polarization gave the hyperon beam experiment a degree of control over systematic effects not available to the earlier low-energy measurements.

4. FORM FACTOR ANALYSIS

4.1 Determination of g_1

The most precise value of this ratio is obtained by analyzing the data in the Dalitz plot variables (electron and neutron kinetic energy in the Σ^- center of mass) *provided* the g_2 form factor is fixed, say at zero. We find that this fit is relatively insensitive to radiative corrections and to f_2, somewhat sensitive to q^2 (neglecting q^2 dependence shifts $|g_1/f_1|$ up by 0.045) and very sensitive to g_2. In fact, one determines essentially the combination $(g_1 + 0.24g_2)$. By combining the Dalitz plot with the asymmetries we are able to fit for g_1 and g_2 separately; this analysis will be discussed below. Making the conventional assumption that $g_2 = 0$, E-715 obtains a *preliminary* value which is reasonably consistent with the electron asymmetry as well as the two previous hyperon beam experiments with unpolarized decays.[2,7]

4.2 Determination of f_2

The weak magnetism form factor is sensitive primarily to the electron spectrum. It is therefore of critical importance to allow for radiative corrections in the decay matrix element and for radiation in material traversed by the electron in the spectrometer. One obtains consistent results using radiative corrections of Toth et al[8] and of Garcia et al.[3] The $\simeq 4\%$ radiation length traversed by electons in the apparatus was both calculated and calibrated with mono-energetic electrons. The preliminary result for f_2/f_1 is close to but somewhat smaller in magnitude than the SU(3) value. In this connection one may mention that a large class of models that correlate both magnetic moment and semi-leptonic data[9] give values in the range $-0.7 < f_2/f_1 < -1.3$.

4.3   Determination of g_2

By combining the Dalitz plot analysis with the asymmetries from polarized decays, the data is no longer sensitive only to the form factor combination $g_1 + 0.24g$. In fact it is possible to fit for g_1 and g_2 separately. In this analysis one uses the distribution in the Dalitz plot as well as the electron, neutron, and neutrino asymmetries. The electron asymmetry has been published.[6] Then the current values for all three asymmetries are:

$$\alpha_e = -0.53 \pm 0.14; \quad \alpha_n = 0.58 \pm 0.15, \quad \alpha_\nu = -0.27 \pm 0.07 \tag{6}$$

where the errors include an estimate of systematic effects. Since the analysis of the data is in progress, these values should also be regarded as a report in progress. The preliminary finding is that g_2 is very small, but that it may be resolvable from zero. In this connection we note that a recent bag model calculation by Carson, Oakes and Willcox[9] predicts a small non-zero value $g_2 \simeq -0.1$.

5.   CONCLUSIONS

Recent high statistics data from polarized Σ^- beta decay confirms that the Cabibbo theory accounts remarkably well for all observables in hyperon beta decay. The precision of the data may be sufficient to begin to see differences in the induced form factors from their SU(3) symmetric values.

ACKNOWLEDGEMENTS

I am grateful to A. Bohm and S. Y. Hseuh for discussions and assistance in preparing this paper. Work was supported by the U.S. Department of Energy under Contract No. DE-FG02-84ER 13178.

REFERENCES

1.   Cabibbo, N., Phys. Rev. Lett. 10, 531 (1963).

2.   Bourquin, M. et al., Z. Phys. C12, 307 (1982).

3.   Garcia, A. and Kielankowski, P., The Beta Decay of Hyperons, New York: Springer-Verlag (1985).

4.   Cabibbo, N. and Chilton, F., Phys. Rev. B137, 1628 (1965).

5.   Keller, P. et al., Phys. Rev. Lett. 48, 971 (1982).

6.   Hseuh, S.Y. et al., Phys. Rev. Lett. 54, 2399 (1985).

7.   Tanenbaum, W. et al., Phys. Rev. D12, 1871 (1975).

8.   Toth, K., Margaritisz, T. et al., Cern Preprint TH3169 (1975).

9.   Bohm, A., private communication.

10.   Carson, L.J., Oakes, R.J. and Willcox, C.R., ANL Preprint 85-99 (1985).

RECENT RESULTS ON D MESON DECAYS FROM THE MARK III

David H. Coward

Stanford Linear Accelerator Center
Stanford University, Stanford, California 94305

Representing the MARK III Collaboration

1. INTRODUCTION

The MARK III Collaboration recently completed the analysis of a number of decay modes of charged and neutral D mesons produced in electron-positron collisions near the peak of the $\psi(3770)$ resonance at SLAC's SPEAR storage ring. The mesons were produced nearly at rest in pairs, either D^+D^- or $D^0\bar{D}^0$, at a center-of-mass energy below the threshold for DD^* production. The unique kinematics of the production allow us to isolate the charmed meson signal clearly and unambiguously. The data were collected with the MARK III Spectrometer, a large solid angle magnetic detector which has been described in detail elsewhere.[1] Our data sample corresponds to an integrated luminosity of approximately 9.3 inverse picobarns.

New results will be presented on the absolute branching ratios of D mesons into hadronic final states, branching ratios for three body decays via pseudoscalar-vector intermediate states, and branching ratios for Cabibbo allowed and Cabibbo suppressed decays. Inclusive and exclusive branching ratios for the semi-leptonic decays of D mesons will be presented, as well as the first measurement of the vector form factor in the decay $D^0 \rightarrow K^- e^+ \nu$, evidence for interference in D^+ decays, and new information on the contributions of W exchange diagrams to D^0 decays.

2. ABSOLUTE BRANCHING RATIOS TO HADRONIC FINAL STATES

Since charmed D mesons are produced in pairs in our data sample, we can make a unique identification of the charm of a single D meson through the reconstruction of the hadronic decay of the D or $\bar{D}$. These reconstructed events form our single-tag sample, containing 3435 D^0 and 1729 D^+ mesons. We then study the decay of the recoil $\bar{D}$ or D meson, and determine their absolute

220

branching ratios into hadrons, electrons or muons. These events with both D's reconstructed give us our double-tag sample. Here and throughout this paper, we adopt the convention that reference to a particle state also implies reference to its charge conjugate. Because of the good mass resolution of the MARK III detector, we have a very clean sample of tagged charm events. In Fig. 1 we show mass plots for 105 D^+ and 367 D^0 double-tag combinations into the listed hadronic final states.

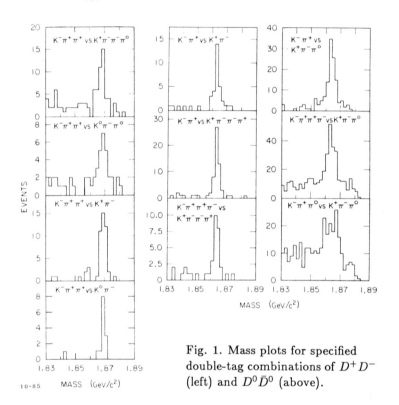

Fig. 1. Mass plots for specified double-tag combinations of D^+D^- (left) and $D^0\bar{D}^0$ (above).

The method of analysis is as follows. Let D_i be the ith decay mode of a D meson. The efficiency of tagging the particular decay in the detector is ϵ_i. B_i is the branching ratio for the ith final state, and $N_{D\bar{D}}$ is the number of produced $D\bar{D}$ pairs. The number of tags are given by the following equations:

$$N_{D_i} = 2\,N_{D\bar{D}}\,\epsilon_i\,B_i \qquad \text{single tags}$$
$$N_{D_jD_k} = 2\,N_{D\bar{D}}\,\epsilon_j\,B_j\,\epsilon_k\,B_k \qquad \text{double tags.}$$

The tagging efficiencies, ϵ_i, are determined by Monte Carlo simulation programs.

The numbers of events for particular decay modes, N_{D_i} and $N_{D_j D_k}$, are selected (cf Fig. 1) from the single-tag and double-tag samples. The equations are fit to the D^0 and D^+ data independently, yielding the B_i's and $N_{D\bar{D}}$.

The fits are performed on the reaction $e^- e^+ \to X\bar{X} \to$ hadronic final states, with $M_X \equiv M_{\bar{X}}$. These results do not depend on a measurement of the luminosity or on the knowledge of the shape and magnitude of the $D\bar{D}$ production cross section. The results also are nearly free from systematic errors, being dependent only on the systematic errors in the calculated tagging efficiencies. The absolute branching ratios obtained from the fits are given in the far right column of Table 1.

Table 1

Cabibbo Allowed D^0 and D^+ Branching Ratios

Decay Mode	$\sigma \cdot B$ (nb)	$B_{s.t.}$ (%)	B_{fit} (%)
$K^-\pi^+$	.237±.009±.013	4.9±0.4±0.4	5.1±0.4±0.4
$\bar{K}^0\pi^0$	.108±.020±.010	2.2±0.4±0.2	
$\bar{K}^0\eta$	.088±.039±.012	1.8±0.8±0.3	
$\bar{K}^0\omega$	.187±.073±.047	3.8±1.5±1.0	
$K^-\pi^+\pi^0$	.978±.065±.137	20.1±1.9±3.0	18.5±1.3±1.6
$\bar{K}^0\pi^+\pi^-$	.372±.030±.031	7.6±0.8±0.7	
$K^-\pi^+\pi^-\pi^+$	.566±.027±.061	11.6±1.0±1.4	11.5±0.8±0.8
$\bar{K}^0\pi^+\pi^-\pi^0$	.666±.113±.153	13.7±2.5±3.2	
$\bar{K}^0\pi^+$	.126±.012±.009	3.5±0.5±0.4	4.0±0.6±0.4
$K^-\pi^+\pi^+$	.399±.017±.028	11.1±1.4±1.2	11.3±1.3±0.8
$\bar{K}^0\pi^+\pi^0$	.714±.142±.100	19.8±4.6±3.2	14.1±2.8±2.1
$\bar{K}^0\pi^+\pi^+\pi^-$	.305±.031±.030	8.5±1.3±1.1	
$K^-\pi^+\pi^+\pi^0$	.260±.040±.054	7.2±1.4±1.6	7.5±1.5±1.6

The average luminosity for the entire run at 3.768 GeV $(9325 \pm 466 \text{ nb}^{-1})$ is determined from wide angle Bhabhas and muon pairs in our detector. From this value and the number of produced $D\bar{D}$ pairs $(22700 \pm 1600 \pm 1660 \ D^0\bar{D}^0$ and $16800 \pm 2000 \pm 1600 \ D^+\bar{D}^+)$, we obtain cross sections for D^0 and D^+ production at the $\psi(3770)$: $\sigma_{D^0} = 4.9 \pm 0.3 \pm 0.4$ nb and $\sigma_{D^+} = 3.6 \pm 0.4 \pm 0.4$ nb. In Table 1 we include values of cross section times branching ratio, $\sigma \cdot B$, for a larger number of decays from a different single-tag analysis. Branching ratios for these single-tag decays, $B_{s.t.}$, are obtained by using the number of produced $D\bar{D}$ pairs given by the fits. The agreement between the branching ratios determined by the two methods is excellent.

A comparison between this experiment and other experiments performed earlier at SPEAR shows that the $\sigma \cdot B$ values for these experiments agree within statistical errors. Thus the differences in the production cross sections between these experiments (see Table 2) are probably the causes of many of the discrepancies between our branching ratios and those of earlier experiments.

Table 2
Comparison of Energies and Cross Sections at the $\psi(3770)$

	LGW	MARK II	C.B.	MARK III
$E_{c.m.}$ (GeV)	3.774	3.771	3.771	3.768
σ_{D^0} (nb)	11.5±2.5	8.0±1.0±1.2	6.8±1.2	4.9±0.3±0.4
σ_{D^+} (nb)	9.0±2.0	6.0±0.7±1.0	6.0±1.1	3.6±0.4±0.4

3. THREE BODY HADRONIC DECAYS

We now turn to an analysis performed to isolate any pseudoscalar-vector (PV) substructure in the three body decays $D^0 \to \bar{K}^0\pi^+\pi^-$, $D^0 \to K^-\pi^+\pi^0$, and $D^+ \to \bar{K}^0\pi^+\pi^0$. Each mode is fit to a sum of interfering Breit-Wigner amplitudes (for $K^*(892)$ and ρ) and a constant amplitude for three body phase space. Appropriate phase space factors and decay angular distributions are included for the PV channels. Fits are performed using a maximum likelihood technique. As an illustration of the analysis, the Dalitz plot and $\pi^+\pi^0$ two body projection for the decay $D^0 \to K^-\pi^+\pi^0$ are shown in Fig. 2. Our results on the PV decays are summarized in Table 3. The branching ratios for the PV intermediate states

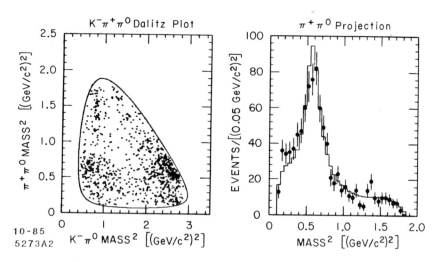

Fig. 2. (a) $D^0 \to K^-\pi^+\pi^0$ Dalitz plot. (b) $\pi^+\pi^0$ projection. The solid curve is the fit.

for $K^-\pi^+\pi^0$ and $\bar{K}^0\pi^+\pi^0$ are normalized to the double-tag fits, while the ratios for $\bar{K}^0\pi^+\pi^-$ come from the single-tag analysis using the number of produced $D^0\bar{D}^0$ pairs given by the fit. The agreement within errors of the two independent determinations of $B(K^{*-}\pi^+)$ in the $K^-\pi^+\pi^0$ and $\bar{K}^0\pi^+\pi^-$ final states provides a good check of the systematic errors in the analysis.

Table 3

Pseudoscalar-Vector Decays of D^0 and D^+

Decay Mode	Fraction (%)	$\sigma \cdot B$ (nanobarns)	B (%)
$D^0 \to K^-\pi^+\pi^0$			
$K^-\rho^+$	74.0± 6.9	.72±.07±.11	13.7±1.3±1.5
$K^{*-}\pi^+$	12.9± 3.4	.38±.09±.08	7.1±1.6±1.3
$\bar{K}^{*0}\pi^0$	7.6± 3.9	.12±.05±.03	2.1±0.9±0.6
non-resonant	5.5± 5.3	.05±.04±.03	1.0±0.8±0.6
$D^0 \to \bar{K}^0\pi^+\pi^-$			
$K^{*-}\pi^+$	63.9± 8.8	.36±.05±.04	7.3±1.2±0.9
$\bar{K}^0\rho^0$	16.8± 5.9	.06±.02±.01	1.3±0.4±0.3
non-resonant	19.3± 9.3	.07±.03±.02	1.5±0.7±0.3
$D^+ \to \bar{K}^0\pi^+\pi^0$			
$\bar{K}^0\rho^+$	86.5±10.4	.62±.14±.09	12.2±2.8±1.9
$\bar{K}^{*0}\pi^+$	7.0± 5.9	.15±.09±.09	3.0±1.9±1.7
non-resonant	6.5± 6.8	.04±.04±.03	0.9±0.8±0.6

From Tables 1 and 3, we extract the following ratios, with common systematic errors removed:

$$B(D^0 \to \bar{K}^0\pi^0)/B(D^0 \to K^-\pi^+) = 0.45 \pm 0.08 \pm 0.05$$
$$B(D^0 \to \bar{K}^{*0}\pi^0)/B(D^0 \to K^{*-}\pi^+) = 0.30 \pm 0.14 \pm 0.08$$
$$B(D^0 \to \bar{K}^0\rho^0)/B(D^0 \to K^-\rho^+) = 0.09 \pm 0.03 \pm 0.02$$

4. CABIBBO-SUPPRESSED HADRONIC DECAYS

In Table 4, we list the branching ratios for a number of Cabibbo suppressed decay modes. Some of these branching ratios have been published[2] as ratios relative to Cabibbo allowed decays. They have been converted to the values given in Table 4 through the use of the $B_{s.t.}$ values given in Table 1.

Table 4

Cabibbo Suppressed D^0 and D^+ Branching Ratios

Decay Mode	B (%)
D^0	
$K^- K^+$	$0.60 \pm 0.10 \pm 0.08$
$\pi^- \pi^+$	$0.16 \pm 0.05 \pm 0.03$
$\bar{K}^0 K^0$	≤ 0.62
$(K^0 K^- \pi^+)_{nonres}$	≤ 1.80
$\bar{K}^0 K^{*0}$	≤ 0.83
$K^{*-} K^+$	$1.02 \pm 0.47 \pm 0.21$
$\pi^- \pi^+ \pi^0$	$1.11^{+0.43+0.18}_{-0.35-0.18}$
$\pi^- \pi^+ \pi^- \pi^+$	$1.47^{+0.61+0.19}_{-0.49-0.19}$
D^+	
$\pi^+ \pi^0$	≤ 0.53
$K^+ \bar{K}^0$	$1.11 \pm 0.34 \pm 0.21$
$\pi^+ \pi^+ \pi^-$	$0.47 \pm 0.19 \pm 0.12$
$\phi \pi^+$	$0.93 \pm 0.26 \pm 0.17$
$\bar{K}^{*0} K^+$	$0.53 \pm 0.24 \pm 0.14$
$(K^- K^+ \pi^+)_{nonres}$	$0.66 \pm 0.30 \pm 0.12$

5. SEMILEPTONIC BRANCHING RATIOS

For completeness, we mention our recently published results from our measurement of the inclusive electron spectra from D^0 and D^+ decays.[3]

$$B(D^+ \rightarrow e^+ + X) = 17.0 \pm 1.9 \pm 0.7 \ \%$$

$$B(D^0 \rightarrow e^+ + X) = 7.5 \pm 1.1 \pm 0.4 \ \%$$

$$B(D^+ \rightarrow e^+ + X)/B(D^0 \rightarrow e^+ + X) = 2.3^{+0.5}_{-0.4} \pm 0.1$$

If we neglect the Cabibbo-suppressed semileptonic branching ratios, then the ratio of the D^+ to D^0 lifetimes equals the ratio of their respective semileptonic branching ratios.[4]

Using the single-tag sample (which contains a tag with a definite charm signature), we have reconstructed the recoil D's into the following exclusive final states:

$$D^0 \rightarrow K^- e^+ \nu, \ K^- \pi^0 e^+ \nu, \ \bar{K}^0 \pi^- e^+ \nu \qquad D^+ \rightarrow \bar{K}^0 e^+ \nu, \ K^- \pi^+ e^+ \nu$$

We require (1) correct multiplicity and total charge, (2) particle identification by time-of-flight (TOF), (3) lepton identication by TOF, shower, and muon systems, and (4) no additional gamma with energy greater than 0.100 GeV. The exclusive reconstruction process gives a very clean sample. In Table 5 we show the number of reconstructed events and branching ratios for several exclusive D meson decay modes. An examination of the $K\pi e\nu$ events shows that all are consistent with a decay through a $K^*e\nu$ intermediate state. Calling all $K\pi$ combinations K^*, and correcting for the unobserved $\bar{K}^0\pi^0e^+\nu$ state, we sum the $Ke\nu$ and $K^*e\nu$ channels and obtain:

$$B(D^+ \to \bar{K}^0e^+\nu + \bar{K}^{*0}e^+\nu) = 15.5 \pm 2.6 \pm 0.3 \text{ \%}$$

$$B(D^0 \to K^-e^+\nu + K^{*-}e^+\nu) = 7.1 \pm 1.6 \pm 0.2 \text{ \%}$$

The agreement between the inclusive and exclusive decay rates shows that there is little room left for other decay channels, and is consistent with the expected Cabibbo suppressed contribution of order $\tan^2\theta_c \approx 0.05$.

Table 5

Exclusive Semileptonic Branching Ratios

Channel	Events	Bkd. Events	B(%)
$K^-e^+\nu$	49	2.4	3.2±0.5±0.1
$K^-\pi^0e^+\nu$	4	0.0	0.9±0.5±0.1
$\bar{K}^0\pi^-e^+\nu$	5	0.0	3.0±1.4±0.2
$K^{*-}e^+\nu$			3.9±1.5±0.2
$\bar{K}^0e^+\nu$	19	0.6	9.3±2.2±0.3
$K^-\pi^+e^+\nu$	21	0.0	4.1±0.9±0.1
$\bar{K}^{*0}e^+\nu$			6.2±1.4±0.4

6. MEASUREMENT OF THE VECTOR FORM FACTOR

The decay $D^0 \to K^-e^+\nu$ can be calculated assuming that the matrix element is dominated by a single pole, where the $(c\bar{s})$ F^* is the lowest lying vector meson with the correct quantum numbers. If the mass of the electron is set to zero, then the matrix element for the decay can be written

$$M = G_F\cos\theta_c f_+(t)(P_D + P_K)^\mu \bar{u}(\nu_e)\gamma_\mu(1 - \gamma_5)v(e^+).$$

In the rest frame of the D^0, where $t \equiv (P_D - P_K)^2 = m_D^2 + m_K^2 - 2m_DE_K$ and E_K is the kaon energy, we can integrate over the lepton variables and the direction of the kaon and obtain the kaon energy spectrum:

$$W(x_K) = |f_+(t)|^2 [x_K^2 - 4\lambda^2]^{3/2}$$

where $x_K = 2E_K/m_D$, $\lambda = m_K/m_D$, and $f_+(t)$ is the form factor associated with the vector part of the current. The simplest prediction for $f_+(t)$ is $f_+(t) = f_+(0)\, m_{F^*}{}^2/(m_{F^*}{}^2 - t)$. The K^- detection efficiency is flat over the low and middle t range, but falls off at large t. This falloff is due to the decay of low momenta K^-'s ($\leq 0.2\ GeV/c$) in the detector. In Fig. 3, we show the efficiency-corrected kaon energy spectrum for the 49 observed events, and the fit, calculated using the simple pole form for $f_+(t)$. The best fit value for M_{F^*} is $2.1^{+1.7}_{-0.4}$ GeV/c^2. The agreement is good, suggesting that the single vector exchange is an adequate description of the physical process.

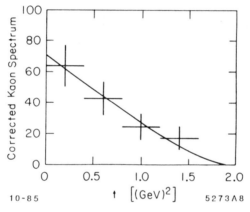

Fig. 3. Efficiency-corrected kaon energy spectrum $(W(x_K))$ and fit assuming the simple pole form for $f_+(t)$.

7. EVIDENCE FOR D$^+$ INTERFERENCE

The Cabibbo allowed spectator diagram for D^0 and D^+ decay is shown in Fig. 4(a). Fig. 4(b) shows the W exchange diagram for D^0 decay. Evidence for the existence of this diagram will be given in the next section. For certain other decay modes, two versions of the spectator diagram may exist for D^+ decays, as illustrated for a particular final state in Fig. 4(c).[5] In the presence of strong color clustering, the two distinct diagrams would result in identical final states. The distructive interference of the amplitudes for these two diagrams would reduce the D^+ hadronic width and thus lengthen its lifetime relative to the D^0. We search for this type of interference by comparing particular ratios of branching ratios.

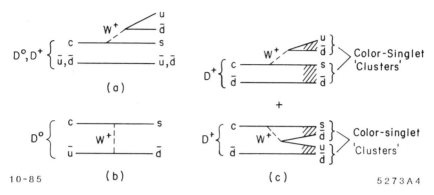

Fig. 4. Possible decay diagrams for D^0 and D^+ mesons.

The Cabibbo allowed decay $D^+ \to \bar{K}^0\pi^+$ is shown in the two diagrams of Fig. 4(c). The Cabibbo suppressed decay $D^+ \to \pi^0\pi^+$ goes by diagrams similar to those in Fig. 4(c), while the suppressed decay $D^+ \to \bar{K}^0K^+$ goes by a diagram similar to Fig. 4(a). We measure $B(D^+ \to \pi^0\pi^+)/B(D^+ \to \bar{K}^0\pi^+) \leq 0.15$ at 90% C.L. This number is based on fluctuating the one observed event to 4.2 events. The observed upper limit is consistent with the SU(3) prediction[6] of $\frac{1}{2}\tan^2\theta_c \approx 0.03$. Both decays in the ratio are reduced by the interference of diagrams like those in Fig. 4(c). On the other hand, in the ratio $B(D^+ \to \bar{K}^0K^+)/B(D^+ \to \bar{K}^0\pi^+) = 0.317 \pm 0.086 \pm 0.048$, only the decay in the denominator is reduced by the interference effect. Since the first ratio is significantly smaller than the second, this is evidence that D^+ interference is operative. We would expect the ratio $B(D^+ \to \bar{K}^{*0}K^+)/B(D^+ \to \bar{K}^{*0}\pi^+)$, for which we measure $0.18 \pm 0.14 \pm 0.11$, to behave like the ratio $B(D^+ \to \bar{K}^0K^+)/B(D^+ \to \bar{K}^0\pi^+)$. However, the accuracy of the measurement of the $B(D^+ \to \bar{K}^{*0}\pi^+)$ branching ratio is so poor that we must wait for a more accurate determination from the analysis of the three body decay $D^+ \to K^-\pi^+\pi^+$.

8.   EVIDENCE FOR W EXCHANGE DIAGRAMS

The D^0 can decay, in principle, into $K^0\bar{K}^0$, $K^0\bar{K}^{*0}$, $\bar{K}^0K^{*0}$ and $\bar{K}^0\phi$ final states. The $\bar{K}^0\phi$ final state, produced through a diagram similar to that shown in Fig. 4(b), is Cabibbo allowed, but thought to be helicity suppressed at the $W\bar{u}d$ vertex. The other three decay channels are Cabibbo suppressed. In addition, $D^0 \to K^0\bar{K}^0$ is also SU(3) forbidden. Since W exchange occurs only for D^0 decays (the similar modes available to the D^+ go by W annihilation and are Cabibbo suppressed), its existence could be another of the mechanisms contributing to the non-equality of the D^0 and D^+ lifetimes.

We see one event from the decay $D^0 \to K^0 \bar{K}^0$, which leads to the ratio of branching ratios:

$$B(D^0 \to K^0 \bar{K}^0)/B(D^0 \to K^- \pi^+) \leq 0.11 \text{ at } 90\% \text{ C.L.}$$

We have looked in the final state $(K^0 K^- \pi^+$ or $\bar{K}^0 K^+ \pi^-)$ from D^0 decay, and have seen the distributions shown in Fig. 5. The all neutral final state, which comes only from W exchange and is therefore the more interesting, can be analyzed, leading to the following ratio of branching ratios:

$$B(D^0 \to K^{*0} \bar{K}^0 + \bar{K}^{*0} K^0)/B(D^0 \to K^- \rho^+ + K^{*-} \pi^+) \leq 0.034 \text{ at } 90\% \text{ C.L.}$$

The less interesting charged final state includes contributions from both spectator and non-spectator diagrams. Our analysis leads to the following ratio of branching ratios:

$$B(D^0 \to K^{*-} K^+ + K^{*+} K^-)/B(D^0 \to K^- \rho^+ + K^{*-} \pi^+) = 0.05 \pm 0.03.$$

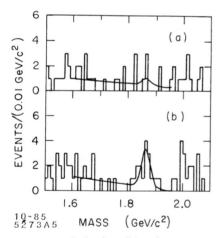

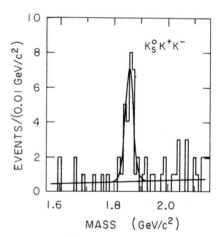

Fig. 5. (a) $\bar{K}^{*0} K^0 + \bar{K}^0 K^{*0}$
(b) $K^{*-} K^+ + K^{*+} K^-$.

Fig. 6. $K_S^0 K^+ K^-$ mass distribution and fit. The background is derived from off-momentum events.

Now we turn to our search for the decay $D^0 \to \bar{K}^0 \phi$. In Fig. 6 we show a mass plot for the decay $D^0 \to K_S^0 K^+ K^-$, where the K_S^0 decays to $\pi^+ \pi^-$. We observe a direct signal of 25.2 ± 5.4 events. We identify kaons by dE/dx and TOF cuts, K_S^0's by $\pi^+ \pi^-$ vertex cuts, and the D^0's by the requirement that the absolute value of the D momentum be not more than 0.050 GeV from the nominal value. The background shape is determined from off-momentum D^0's; D^0's whose momenta are between 0.060 and 0.110 GeV/c from the nominal value. A cut on the D^0 invariant mass of $\pm$ 0.040 GeV/c^2 selects 28 events, of which 4.8 $\pm$ 2.4 are background. If we then plot those 28 events as a function of the K^+K^- mass, we find 4 events below, 11 events within, and 13 events above the ϕ region (1.019 $\pm$ 0.015 GeV/c^2). These events are plotted in Fig. 7.

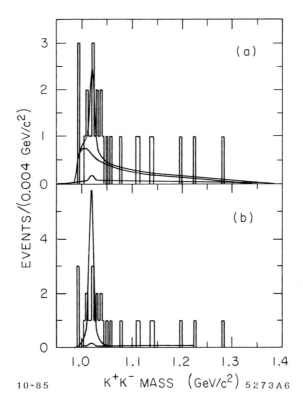

Fig. 7. (a)K^+K^- mass in $K^0_S K^+K^-$, with fit including background, $\bar{K}^0\phi$ and $\bar{K}^0\delta^0$. (b) As in (a) except for upper limit assuming only $\bar{K}^0\phi$ and background.

A number of processes have been considered that might contribute backgrounds to $\bar{K}^0K^+K^-$, and specifically into the ϕ region as defined above. Except for the decay mentioned below, we have found no processes that can feed large contributions into the ϕ region. We have no experimental information about the decay $D^0 \to \bar{K}^0\delta^0$. However, a Monte Carlo simulation of this decay, using the

Flatte parameterization[7] of the δ^0, leads to a peak at low K^+K^- mass and a long tail into the high mass region. In Fig. 8 we show the Dalitz plot for the 28 events in the $\bar{K}^0 K^+ K^-$ state, and 400 Monte Carlo events each in the $\bar{K}^0\delta^0$ and $\bar{K}^0\phi$ final states. The latter Monte Carlo shows the strong angular distribution expected from the pseudoscalar-vector decay. Additional Monte Carlo efficiency studies indicate no significant distortion of the K^+K^- mass distribution near threshold, for pseudoscalar or vector parents. The Monte Carlo results suggest that the high mass tail cannot be ignored but needs to be extrapolated into the low mass region. Our conclusion is that the observed distribution favors the ϕ and a low mass K^+K^- enhancement.

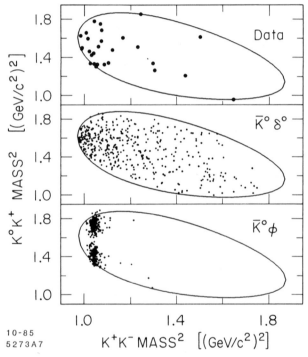

Fig. 8. Dalitz plots for $K_S^0 K^+ K^-$ data, and $\bar{K}^0\delta^0$ and $\bar{K}^0\phi$ Monte Carlo predictions.

In Fig. 7, we compare the data with two different fits to signal plus background. From a fit that includes contributions from $\bar{K}^0\delta^0$ and non-resonant background (Fig. 7a), we obtain 5.2 ± 3.3 events from the decay $D^0 \rightarrow \bar{K}^0\phi$. This translates to a branching ratio $B(D^0 \rightarrow \bar{K}^0\phi) = 0.7 \pm 0.5 \pm 0.2\%$. Alternatively, we obtain an upper limit ignoring both the low mass events and the high mass tail but including a contribution from a non-resonant background (Fig. 7b):

$$B(D^0 \to \bar{K}^0\phi) \leq 2.5\% \text{ at } 90\% \text{ C.L.}$$

This result is based on 17.5 events (including contributions from systematic errors).

9. CONCLUSIONS

We summarize the new results from the MARK III collaboration:

(a) We have made high statistics measurements of the absolute D meson branching ratios. These are independent of cross section measurements.

(b) We have presented measurements of $D \to (e + X)$ and the ratio of the D^+ to D^0 lifetimes. The error in the lifetime ratio is comparable with the error in the ratio of the world averages of the direct D^+ and D^0 lifetimes.

(c) We have presented new measurements of $D \to Ke\nu$ and $D \to K^*e\nu$, and the f_+ vector form factor.

(d) We have made measurements of a large number of D branching ratios. We can account for about 85% of all D^+ and D^0 decays.

(e) We have presented evidence for D^+ interference.

(f) We have presented evidence for the existence of non-spectator diagrams. There appear to be other structures present in the $\bar{K}^0K^+K^-$ final state at high and low K^+K^- mass that are inconsistent with background, and feed into the low K^+K^- mass region.

10. ACKNOWLEDGEMENTS

I would like to thank Hai-Yang Cheng for a number of discussions concerning the interpretation of our data, and G. T. Blaylock, D. Coffman, and R. H. Schindler for help in the preparation of this paper. This work was supported in part by the U.S. National Science Foundation and the U.S. Department of Energy under Contracts No. DE-AC03-76SF00515, No. DE-AC02-76ER01195, No. DE-AC03-81ER40050, and No. DE-AM03-76SF00034.

11. REFERENCES

[1] D. Bernstein et al., Nucl. Instrum. Methods 226, 301 (1984).
[2] R. M. Baltrusaitis et al., Phys. Rev. Lett. 55, 150 (1985).
[3] R. M. Baltrusaitis et al., Phys. Rev. Lett. 54, 1976 (1985).
[4] A. Pais and S. B. Treiman, Phys. Rev. D 15, 2529 (1977).
[5] B. Guberina et al., Phys. Lett. 89B, 111 (1979).
[6] R. L. Kingsley et al., Phys. Rev. D 11, 1919 (1975).
[7] S. M. Flatte, Phys. Lett. 63B, 224 (1976).

LIFETIME OF CHARMED PARTICLES

M. Bosman

Max-Planck-Institut für Physik und Astrophysik, Munich, FRG

ABSTRACT

Experimental aspects of recent measurements of charmed particle lifetimes are discussed and the theoretical relevance of the results is briefly reviewed.

1. INTRODUCTION

In the standard electroweak theory, the flavour changing transition among quarks and leptons are described by their coupling to the $W^{\pm}$ bosons. The assumption of a universal weak coupling for all fundamental fermions links the free c quark decay time to the μ decay time. In the absence of Cabibbo suppression, the predicted lifetime is $\tau_c \sim 8 \cdot 10^{-13}$sec. We will report here on lifetime measurements of charmed mesons and baryons.

2. EXPERIMENTAL RESULTS

Both fixed target and e^+e^- collider experiments have contributed to the measurement of charmed particle lifetimes. In fixed target experiments, the ratio of charm production cross-section to the total inelastic cross-section varies from 1:10 for neutrinos to 1:1000 for hadrons. High resolution devices (emulsion [1 μm], silicon microstrip counters [5 μm], bubble chambers [20-50 μm]) have been used to identify charmed particles by their decay topology: the distance of closest approach of a decay track to the production vertex (impact parameter) is typically 30 μm for a particle which lives 10^{-13}s. The lifetime can be extracted after correction for the possible bias introduced in the sample by the selection criteria. In e^+e^- experiments, the ratio of charmed events to all hadronic events of 1/3 is more favourable. Kinematic cuts alone enable the isolation of a clean sample of charmed events. High resolution drift chambers (σ ~ 100 μm) close to the beam pipe have been used for lifetime measurements. In this case the precision in the determination of the impact parameter is 200–400 μm, which is larger than the average measured effect.

2.1 D°,D± Lifetimes

Experiments which have presented new results or whose data samples are the most significant will be reviewed. A compilation of all the data can be found in Table 1.

2.1.1 <u>Bubble chamber experiments.</u> The NA27 [1] experiment at CERN has taken 760 K pictures from interactions of 360 GeV π⁻ and p in the LEBC bubble chamber (20 μm bubble size). 230 charm decays have been found by requiring at least one track with 50 μm impact parameter or a visible jump in ionization. The events used for the lifetime determination are unique fits to the D hypothesis (3c fits) allowing for missing neutrals (0c fits) only if the kinematic solutions are within 10% of each other. Those events were taken from the c3,v4,c5 topologies where the scanning efficiency was shown to be high and uniform (95±5%). The lifetimes of the 29 D° and 32 D⁺ are corrected for the maximum and minimum detectable lifetime in each decay configuration (Fig. 1a). The minimum detectable lifetime is the important correction in this case : ~ 30% for the D°, and 10% for the D⁺. No systematic error could be found by varying the cuts so only the statistical error is quoted (Table 1).

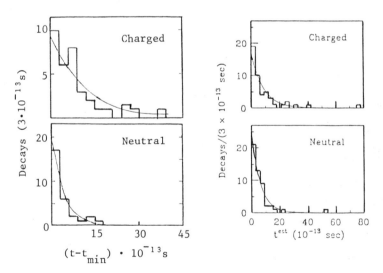

Fig. 1 D°,D⁺ corrected lifetime distribution for NA27 a) and SHF b).

The SLAC Hybrid Facility[2] (SHF) has recorded 690 K interactions of 20 GeV γ in a bubble chamber (40–55 μm bubble size). 100 charm decays were found in the topologies v2,c3,v4,c5 by requiring 2 tracks with an impact parameter of 110 μm and 40 μm, respectively, and a minimum decay length of

0.6 mm. These cuts ensure a high and constant efficiency as measured with K° and Λ (97$\pm^2_4$%). To be able to use decays with missing neutrals, the D momentum was estimated by the relation $(1/p_D)^{\text{est.}} = \alpha \cdot m_{\text{vis}}/m_D \cdot 1/p_{\text{vis}}$ where $\alpha = 1.10\pm0.02$. The uncertainty in α arises from the errors on the branching ratios. 50 D° and 48 D^+ have been used for the lifetime measurement (Fig. 1b). The minimum detectable lifetime correction is of the order of 50%. A larger systematic error is quoted for the D^+ due to the possible contamination by F^+ and Λ^+_c (Table 1).

2.1.2 An experiment using a <u>silicon microstrip vertex detector</u>, the NA11[3] experiment at CERN, has recorded 4.6 million 200 GeV π^- Be interactions triggering on a single electron. A telescope of silicon microstrip counters, which provides a precision of 10 μm on the impact parameter, has been used to fully reconstruct 54 D mesons in the decay modes $K\pi$, $K\pi\pi$, $K\pi\pi\pi$. The kaon was required to be identified in the Cerenkovs. The minimum detectable lifetime corrections are similar to the SHF case. No systematic error could be detected by varying the cuts; so only the statistical error is quoted (Table 1).

2.1.3 <u>Emulsion experiments.</u> The emulsion experiments have resolution of 1 μm and can detect very short decay paths (20 μm) but maintaining a good scanning efficiency over long distances is difficult.

WA58[4] at CERN has analysed 9000 interactions of γ's of 20-70 GeV/c^2 in an emulsion of $\sim$ 7 mm effective thickness. Charged decays were looked for by following tracks from the production vertex up to the end of the emulsion : the efficiency for finding decays is estimated to be constant (94%). For neutral decays a volume scan is performed in a cone of $\pm15^\circ$: the efficiency for finding decays drops rapidly down to 0 at 2 mm. An efficiency curve was determined by studying γ conversions. It was then used to weight the events. The average potential decay length is $\sim$ 3.5 mm and hence a large correction for the maximum detectable lifetime is needed. 43 D° and 23 D^+ are selected, distributed about equally between the 3c and 0c fit categories. For the 0c fits the momentum of the D is averaged over the different kinematical solutions. The systematic error is estimated by using the higher and lower momentum solutions for the ambiguous decays (Table 1).

The Fermilab experiment E531[5] has studied 2500 γ interactions in their 50 mm thick emulsion. A scanback method was used with the aid of a thin emulsion sheet placed behind the emulsion. The scanning efficiency as a function of distance was found to be fairly flat for $\gamma \to e^+e^-$. Although this check could be made with more statistics. 57 D° events and 11 D^+, all D^+ being ambiguous with F and Λ_c, have been used for the lifetime determination. The last update on the published results was at the '83 Como conference[6] (Table 1).

2.1.4 <u>e^+e^- experiments.</u> Many e^+e^- experiments have now measured the D° lifetime. They obtain a clean D° signal with typically 10-15% background by selecting D° from the $D^{*+} \to D^\circ + \pi^+$ decay chain and applying a cut on z_{D°. The z_{D° cut reduces also the number of D°'s coming from B's to a few percent. Mark II[7], from 205 pb^{-1}, has presented a lifetime measurement from 74 D° decays in $K\pi$ and $K\pi\pi^\circ$ (Fig. 2). Preliminary results have been presented at this conference by HRS[8] (31 D° → Kπ from 74 pb⁻¹), CLEO[9] (130 D° → Kπ from 74 pb⁻¹), TASSO[10] (8 D° → Kπ). The DELCO[11] collaboration has used instead of the z_{D^0} cut its particle identification capability and required the K or the π to be identified. 240 candidates have been collected in 150 pb⁻¹ but the statistical significance is however limited by the resolution on the impact parameter of the order of 500 μm. For the D⁺ the situation is different : the equivalent decay cascade is D*⁺ → D⁺+π° with a very low momentum π°. The Mark II[7] collaboration using this decay cascade obtains a signal of 23 D⁺ → Kππ on top of a 40% background. HRS[8] has shown at this conference preliminary results from an inclusive D⁺ → Kππ signal of 175 events on top of 3 times as much background.

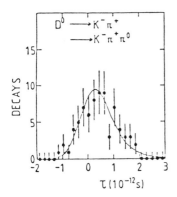

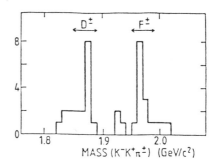

Fig. 2 Lifetime distribution from the Mark II data.

Fig. 3 Effective mass of K⁺K⁻π± for tracks forming a secondary vertex (NA11).

2.1.5 Summary of D° and D⁺ lifetimes. A compilation of the D° and D⁺ lifetime measurements is given in Table 1. To merge the information, average and combined errors have been calculated for each experiment; individual measurements have been weighted by the inverse square of the fractional error[6]. The average D° lifetime is $\tau_{D^0} = 4.3 \pm 0.3 \cdot 10^{-13}$ sec. The values of the various experiments scatter reasonably around the average value. Two experiments have a relative error smaller than 20% and carry the most weight: E531 and SHF. SHF is also the one which deviates the most and increases the average value by $0.4 \cdot 10^{-13}$ sec. The average D⁺ lifetime is $\tau_{D^+} = 9.2 \pm^{0.9}_{0.8} \cdot 10^{-13}$ sec. The experiment which deviates the most is WA58, an experiment where large systematic corrections have to be applied. Their result reduces the lifetime by $0.3 \cdot 10^{-13}$ sec. The averaged lifetimes of the D° and D⁺ have now errors smaller than 10%. A substantial improvement would require experiments with a sample of a few hundred events and measurement precision of a few microns to reduce both the statistical and systematic errors. The ratio of the D⁺ to the D° lifetime deviates significantly from 1 : $\tau_{D^+}/\tau_{D^0} = 2.1 \pm 0.3$. This value is affected by the low SHF ratios of $1.4 \pm 0.3 \pm^{0.2}_{0.1}$.

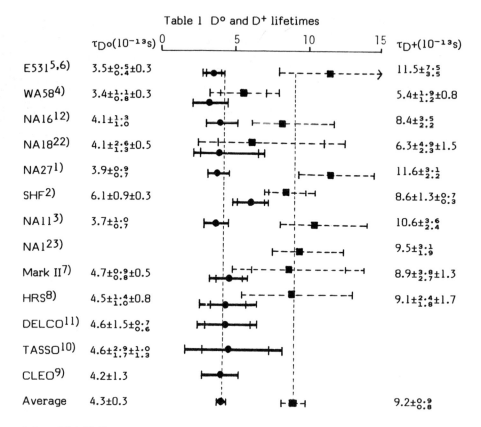

Table 1 D^0 and D^+ lifetimes

$\tau_{D^0}(10^{-13}s)$		$\tau_{D^+}(10^{-13}s)$
E531[5,6]	$3.5\pm^{0.5}_{0.4}\pm0.3$	$11.5\pm^{7.5}_{3.5}$
WA58[4]	$3.4\pm^{1.1}_{0.8}\pm0.3$	$5.4\pm^{1.9}_{1.2}\pm0.8$
NA16[12]	$4.1\pm^{1.3}_{1.0}$	$8.4\pm^{3.5}_{2.2}$
NA18[22]	$4.1\pm^{2.6}_{1.3}\pm0.5$	$6.3\pm^{4.9}_{2.3}\pm1.5$
NA27[1]	$3.9\pm^{0.9}_{0.7}$	$11.6\pm^{3.1}_{2.2}$
SHF[2]	$6.1\pm0.9\pm0.3$	$8.6\pm1.3\pm^{0.7}_{0.3}$
NA11[3]	$3.7\pm^{1.9}_{0.7}$	$10.6\pm^{3.6}_{2.4}$
NA12[3]		$9.5\pm^{3.1}_{1.9}$
Mark II[7]	$4.7\pm^{0.9}_{0.8}\pm0.5$	$8.9\pm^{3.8}_{2.7}\pm1.3$
HRS[8]	$4.5\pm^{1.4}_{1.0}\pm0.8$	$9.1\pm^{2.4}_{1.8}\pm1.7$
DELCO[11]	$4.6\pm1.5\pm^{0.7}_{0.6}$	
TASSO[10]	$4.6\pm^{2.9}_{1.7}\pm^{1.0}_{1.3}$	
CLEO[9]	4.2 ± 1.3	
Average	4.3 ± 0.3	$9.2\pm^{0.9}_{0.8}$

2.2 F^+ Lifetime

The F meson contains an s quark ; its production cross–section can be expected to be suppressed compared to the D. A clean sample of F should not contain decays that can be interpreted as a D or a Λ_c. To resolve the ambiguities, one needs fully constrained decays and in most cases particle identification. No unambiguous F candidates have been found by NA16[12], NA27[13], WA58[4], SHF[2]. E531[5,6] has reported 8 F candidates incompatible at the 90% CL with Cabibbo–favoured D decays. Their average mass is $1994\pm15\pm25$ MeV/c², the systematic error of 25 MeV/c² is due to the possible bias introduced in half of the sample by the requirement that the mass of the candidate should be bigger than 2000 MeV/c². The NA11[14] experiment has collected 12 fully reconstructed F decays to KKπ with both kaons identified (Fig. 3). Their average mass is 1972 ± 2 MeV, a value for the F mass now confirmed by many experiments[24]. Three events in that sample are ambiguous in mass with Λ_c. The HRS[15] has presented at this conference the first attempt at a F^+ lifetime measurement in e^+e^- experiments: they have shown a signal of about 16 F → φπ decays (m_F 1963 ± 4 MeV) on top of about 16 events background. The summary of the results is presented in Table 2. The average F lifetime is $\tau_{F^+} = 3.0\pm^{0.9}_{0.5}10^{-13}$sec. Those results are based on small statistics and the purity of the samples is still of concern.

Table 2 F^+ lifetime

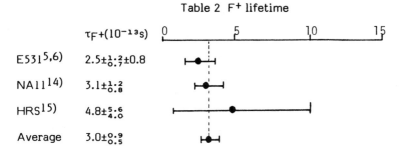

	$\tau_{F^+}(10^{-13}s)$	0	5	10	15
E531[5,6]	$2.5\pm^{1.2}_{0.7}\pm0.8$				
NA11[14]	$3.1\pm^{1.2}_{0.8}$				
HRS[15]	$4.8\pm^{5.6}_{4.0}$				
Average	$3.0\pm^{0.9}_{0.5}$				

2.3 Charmed Baryon Lifetime

The Λ_c search presents the same difficulties as the F's. NA11[14], NA16[12] and SHF[2] have found no unambiguous candidates. To try to select a clean sample, WA58[4] has used events with 2 visible decays and require the Λ_c decay to involve a Λ^o or a proton. 10 candidates are found: three are 3c fits, only one containing 3 charged prongs. E531[5,6] has selected 8 candidates by requiring an identified baryon in the decay and a kinematic fit to the Λ_c mass : 5 events are 3c fits, 3 of them containing 3 charged prongs. Preliminary results from NA27[16] show 10 Λ_c candidates, all 3 charged prong events with kinematic fit to the Λ_c mass, no conflict with particle identification and not ambiguous with Cabibbo-favoured D decays. Ambiguities with F's are still under study. A summary of the results is given in Table 3. The average value, $\tau_{\Lambda_c^+} = 1.8\pm^{0.4}_{0.3}\cdot10^{-13}$sec, is lower than the value presented at Leipzig[17], and is strongly influenced by the preliminary value of NA27. The same concern as for the F case is valid here : small statistics sample and possible contamination.

Table 3 $\Lambda_c^{\pm}$ lifetime

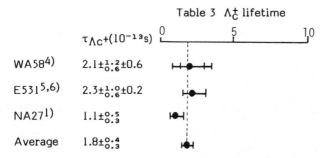

	$\tau_{\Lambda_c^+}(10^{-13}s)$	0	5	10
WA58[4]	$2.1\pm^{1.2}_{0.6}\pm0.6$			
E531[5,6]	$2.3\pm^{1.0}_{0.6}\pm0.2$			
NA27[1]	$1.1\pm^{0.5}_{0.3}$			
Average	$1.8\pm^{0.4}_{0.3}$			

Evidence for the charmed strange baryon $\Xi_c^{\pm}$ (csu)[18] at a mass of 2460±15 MeV/c² has been presented by the WA62[19] experiment which used an incident Σ^- beam and triggered on strangeness -2 or -3 states. The effective $\Lambda K^-\pi^+\pi^-$ mass distribution shows a narrow peak of 53 events on top of 59 events background. With proportional chambers of 140 μm resolution, the precision of the measurement of the decay path of the Ξ_c^+ is 5 mm. They measure $\tau(\Xi_c^+) = (4.8^{+2.1}_{-1.6}\pm^{2.0}_{.0})\bullet10^{-13}$ sec.

3. DISCUSSION OF THE RESULTS

It seems now well established that the D° and D^+ lifetimes are different : $\tau_{D^+}/\tau_{D^\circ} = 2.1\pm0.3$. This ratio can also be inferred from the ratio of semi-leptonic branching ratios. The Mark III[20] collaboration has reported $Br(D^+ \to e^+X)/Br(D^\circ \to e^+X) = \tau_{D^+}/\tau_{D^\circ} = 2.3^{+0.5}_{-0.4}\pm0.1$ in agreement with the ratio from the direct measurement. The leptonic width of the D mesons being negligible and the semileptonic partial width of the D's being equal for Cabibbo-favoured decays, a difference in lifetimes implies a difference in the hadronic widths of the 2 states. Various theoretical explanations[21] have been proposed. Destructive interferences of identical final states could suppress the D^+ decay. W-exchange diagram, present at the Cabibbo-allowed level only in the D° case, could enhance the D° decays, this process being however helicity suppressed. In addition, possible hard and soft gluon effects and colour suppression complicate the picture. The study of exclusive D decay modes and other charmed particle lifetimes will help to disentangle the different effects. The case of the F is interesting because, similar to W-exchange diagram for the D°, the W annihilation process is Cabibbo allowed for the F. The observation of $\tau_F \sim \tau_{D^\circ} < \tau_{D^+}$ supports the importance of these diagrams. But we have seen that the situation of the F^+ lifetime measurement is not very satisfactory yet and better data are needed to make quantitative comparisons. The charmed baryons' case is also interesting. The Λ_c (cdu) can decay at the Cabibbo-allowed level by W exchange and it is neither helicity nor colour suppressed. In addition both annihilation and spectator diagrams have identical quarks in the final state and hence interference effects must be considered. There is no exchange diagram at the Cabibbo-favoured level for the Ξ_c^+ (csu). The observation that $\tau_{\Lambda_c} < \tau_{\Xi_c^+}$ would again support the importance of exchange diagrams. But the same caveat is true here : more precise are needed.

Acknowledgements

I would like to acknowledge useful discussions with my colleagues and, in particular, V. Castillo, M. Iori and V. Lüth. Finally, I like to thank Carole Ponting for typing this report.

REFERENCES

1) Aguilar-Benitez, M. et al., CERN-EP/85-130 (1985), submitted to Zeitschr. Phys. C;
Iori, M., presentation at the Int. Europhysics Conf. on High Energy Physics, Bari, 1985.

2) Abe, K. et al., SLAC-PUB-3722 (1985), submitted to Phys. Rev. D.

3) Bailey, R. et al., Zeitschr. Phys. C., Particles & Fields 28, 357 (1985).

4) Adamovich, M.I. et al., Phys. Lett. 140B, 119 and 123 (1984);
   Castillo, V., private communication.

5) Ushida, N. et al., Phys. Rev. Lett. 48, 844 (1982); Phys. Rev. Lett. 51, 2362 (1983).

6) Reay, N.W., Int. Conf. on Physics in Collision III (Como) 1983.

7) Gladney, L., Ph.D. Thesis, Stanford 1985.

8) Blockus, D, presentation at this conference.

9) Hempstead, M., presentation at this conference.

10) Strom, D., presentation at this conference.

11) Yamamoto, H. et al., SLAC-PUB-3628 (1985), submitted to Phys. Rev. D.

12) Aguilar-Benitez, M. et al., Phys. Lett. 122B, 312 (1983).

13) Aguilar-Benitez, M. et al., CERN-EP/85-02 (1985), submitted to Phys. Rev. D.

14) Bailey, R. et al., Phys. Lett. 139B, 320 (1984);
    Kwan, S., presentation at the Int. Europhysics Conf. on High Energy Physics, Bari, 1985.

15) Jung, C., presentation at this conference.

16) Nowak, A., presentation at the Int. Europhysics Conf. on High Energy Physics, Bari, 1985.

17) Klanner, R., Proc. of the 22nd Int. Conf. on High Energy Physics, Leipzig, 1984, ed. by Meyer, A. and Wieczorek, E.

18) Porter, F.C. et al., Particle Data Group, LBL 18834 (1985).

19) Biagi, S.F. et al., Phys. Lett. 122B, 455 (1983); Phys. Lett. 150B, 230 (1985).

20) Coward, D., presentation at this conference.

21) For a review see, for example, Rück, R., Habilitationsschrift, München 1983.

22) Badertscher, A. et al., Phys. Lett. 123B, 471 (1983).

23) Albini, E. et al., Phys. Lett. 110B, 339 (1982).

24) Particle Data Group, Rev. Mod. Phys. 56 (1984).

NONLEPTONIC DECAYS OF CHARMED MESONS AND BARYONS

Hai-Yang Cheng

Physics Department, Brandeis University
Waltham, MA 02254

ABSTRACT

Two-body decays and the lifetimes of charmed mesons and baryons
are discussed. Existence of final-state interactions and nonperturbative effects is established and emphasized. Various phenomenological models for charmed-meson decays are tested by comparing
with the new Mark III measurements.

1.   TWO-BODY DECAYS OF CHARMED MESONS

There are two main difficulties in the study of exclusive
nonleptonic decays of mesons: one is the final-state interaction, the
other is the pre-asymptotic nonperturbative effect. The final-state
interaction is particularly complicated for charmed mesons because it
arises from resonance effects which would not only modify the phase
shifts of isospin amplitudes but also induce inelasticities. On the other
hand, at least some nonperturbative corrections are empirically of
order[1] $4\pi^2 f_p^2/m_p^2$, which becomes less important only for heavy mesons,
such as beauty and truth mesons.

New Mark III measurements do shed light on our understanding of
charmed-meson decays. The quark-diagram formulation provides a framework in which experimental results can be analyzed in a model-independent
way. All nonleptonic weak decays of mesons can be classified according
to six different quark diagrams[2,3]: the color-enhanced (-suppressed)
spectator diagram a (b), the W-exchange (annihilation) c (d), and two

penguin diagrams e and f. In terms of quark-diagram amplitudes, the Cabibbo-allowed $D \to K\pi$ decay amplitudes are given by

$$A(D^\circ \to K^-\pi^+) = V_{ud}V^*_{cs}(a+c) \qquad Br = (4.9\pm0.4\pm0.4)\%$$
$$A(D^\circ \to \bar{K}^\circ\pi^\circ) = V_{ud}V^*_{cs}(b+c)/\sqrt{2} \qquad (2.2\pm0.4\pm0.2)\% \qquad (1)$$
$$A(D^+ \to \bar{K}^\circ\pi^+) = V_{ud}V^*_{cs}(a+b) \qquad (3.5\pm0.5\pm0.4)\%$$

Branching ratios reported by the Mark III Collaboration[4] are also given in (1). Because of the improved error bars, it is easily seen that the data (1) cannot be fitted if a, b, c are real[5,6]. This implies the importance of the final-state interaction which would induce complexity to decay amplitudes. The final-state interaction is, however, more complicated for charmed-meson decays since it involves resonance effects and multiple channel problems. Color suppression of $D^\circ \to \bar{K}^\circ\pi^\circ$ is now partially relieved because $K^-\pi^+$ mode can be converted into $\bar{K}^\circ\pi^\circ$ through final-state interactions. A fit to (1) can be found[5] with

$$(a+c)/(a+b) = 2.8 , \qquad \delta_{1/2} - \delta_{3/2} = 78^\circ \qquad (2)$$

Based on QCD-corrected weak Hamiltonian the relative magnitudes of quark-diagram amplitudes a, b, c can be evaluated using the vacuum insertion approximation. The results are[5]*

$$b/a = 0.48[3(2c_+ - c_-)/(2c_+ + c_-)] , \quad c/a = -0.05[3(2c_+ - c_-)/(2c_+ + c_-)]$$
$$(a+c)/(a+b) = 0.64 \quad \text{(without QCD correction)} \qquad (3)$$
$$1.45 \quad \text{(with QCD correction)}$$

The W-exchange c is suppressed due to the smallness of the form factor at large momentum transfer. Hence neither of the solutions (3) is consistent with the experimental result (2). The naive picture has to be modified and this generally calls for a nonperturbative effect. Among many improved models I will just mention two "orthogonal" models: (i) Gluon effects in the W-exchange. W-exchange is enhanced by the radiative gluon which is emitted from the initial light quark and creates the final quark-antiquark pair. An estimate[7] gives $c/a = 0.8 \sim 1.0$. Since $b/a = -0.30$ is not modified, it turns out that $(a+c)/(a+b) = 2.6 \sim 2.9$, which is consistent with (2). (ii) Final-state soft-gluon effects. It has been pointed out that a contribution may be received from the four-quark operators consisting of color-octet currents through multiple soft gluon exchanges between outgoing quarks and antiquarks[8]. Denoting the new contribution by the parameter ϵ, the experimental value (2) can be

---

* With QCD correction we use $c_+ = 0.69$ and $c_- = 2.09$.

fitted provided that $\epsilon = -0.6$. Since c/a is not affected by this mechanism, so we have*

$$c/a = 0.03 \quad , \quad b/a = -0.63 \quad \Rightarrow \quad (c_-/c_+)_{eff} = 5.2 \qquad (4)$$

we note that $\epsilon = -0.6$ is close to the value -1 as determined from $K^+ \to \pi^+ \pi^+ \pi^-$[8].

We thus have two extreme models that in the model (i) W-exchange is important while in the second model the amplitude b plays an essential role. The observed $D \to K\pi$ data can be explained by these two models only if the final-state interaction is included. Models (i) and (ii) are indistinguishable in $D \to K\pi$ decays, but can be tested in other channels:

(a) Mark III data give[10]

$$\frac{\Gamma(D^+ \to \bar{K}^0 K^+)}{\Gamma(D^+ \to \bar{K}^0 \pi^+)} = (0.317 \pm 0.086 \pm 0.048) \Rightarrow R_1 = \left| \frac{a - d + e}{a + b} \right|^2_{exp} = 6.4 \pm 1.7 \qquad (5)$$

Neglecting the penguin contribution e for the moment, we find that R_1 is predicted to be 3.0 and 7.3 respectively in models (i) and (ii). A detailed calculation shows that the penguin contribution to $D^+ \to \bar{K}^0 K^+$ is destructive and small[11]. Therefore, a small amount of penguin contributions improves the agreement of the model (ii) with experiment.

(b) The quark-diagram amplitudes of $D^0 \to \bar{K}^0 \eta_8$, $\bar{K}^0 \eta_0$ are given by

$$A(D^0 \to \bar{K}^0 \eta_8) = (b - c + 2\lambda c)/\sqrt{6} \quad , \quad A(D^0 \to \bar{K}^0 \eta_0) = (b + 2c - \lambda c)/\sqrt{3} \qquad (6)$$

where λ is a suppression factor associated with the pair production of strange quarks which vanishes in the SU(3) limit. Including the $\eta - \eta'$ mixing and phase-space corrections, it follows that

$$R_2 = \frac{\Gamma(D^0 \to \bar{K}^0 \eta)}{\Gamma(D^0 \to \bar{K}^0 \pi^0)} = \begin{cases} \text{model (i)} : 0.12 \text{ for } \lambda = 0, \quad 0.03 \text{ for } \lambda = 1/3 \\ \text{model (ii)}: 0.4 \quad \text{(insensitive to } \lambda) \end{cases} \qquad (7)$$

Hence the model (i) predicts a very small R_2 for a reasonable value of λ . The experimental ratio[4] $R_2 = 0.8 \pm 0.4 \pm 0.2$ favors the second model in which the color-suppressed spectator contribution is enhanced.

For $P_c \to VP$ decays there are two new features which we would like to mention. First, the W-exchange or annihilation is no longer subject to form factor suppression. Secondly, owing to the purity of quark contents in ϕ and ω mesons, some $P_c \to \phi P, \omega P$ decays consist of only one type of quark-diagram amplitudes (The "hairpin" diagram is suppressed by the OZI rule.). Consequently, it becomes possible to "see" the W-exchange or annihilation directly in $P_c \to VP$ decays. Indeed, $D^0 \to \phi \bar{K}^0$, which can proceed only through the W-exchange has already been observed

---

* This is a type of $\underline{6}$ dominance model. The Bauer-Stech's model[9] also corresponds to $\epsilon = -2/3$, i.e. the effective color suppression is infinite.

by ARGUS and CLEO collaborations[12].

From the following channels

$$A(F^+ \to \phi\pi^+) = V_{ud} V^*_{cs} a' \qquad Br = (3.3 \pm 1.1)\%[13], \ 4.4\%[14], \ 11\%[15]$$
$$A(D^+ \to \phi\pi^+) = V_{ud} V^*_{cd} b' \qquad (0.93 \pm 0.26 \pm 0.16)\%[4],[10] \qquad (8)$$
$$A(D^+ \to \bar{K}^{*0}\pi^+) = V_{ud} V^*_{cs}(a'+b') \qquad (4.2 \pm 2.5 \pm 2.4)\% \ \text{preliminary}[4]$$

one can determine the relative sign and magnitudes of a' and b', where primed amplitudes are for the case when the vector meson comes from the charmed-quark decay. Standard model calculation based on QCD-corrected weak Hamiltonian yields

$$b'/a' = -0.66[3(2c_+ - c_-)/(2c_+ + c_-)] = 0.4 \qquad (9)$$

Due to the QCD correction and the structure constant f_{ijk} in the form factor $\langle P^i | A^j_\mu | V^k \rangle$, the relative sign is predicted to be positive. This is ensured by the V-spin symmetry argument[16]: If V-spin were exact, $D^+ \to \bar{K}^{*0}\pi^+$ would receive contributions only from 0_- operator. Using $Br(F^+ \to \phi\pi^+) = 4\%$ we obtain $|b'/a'| \sim 0.8$. Furthermore, we learn from $Br(D^+ \to \bar{K}^{*0}\pi^+)$ that b' is antiparallel to a'. This means that the QCD correction is not operative in D $\to$ VP decays! This is really striking because the 6 domin-ance model, which is favored by the data of D$\to$ PP (and likely D$\to$ VV), definitely does not work in the VP case! In fact, a fit of (9) to b'/a' ~ -0.8 gives $(c_-/c_+)_{eff} \sim 0.9$. Similar conclusions are also obtained for the W-exchange, for which we will not go into further details.

2. LIFETIME DIFFERENCE BETWEEN D° AND D$^+$

The large number of D branching ratios measured by Mark III can now account for 85% of the D° and D$^+$ decays[4]. As a result, the lifetime difference between D° and D$^+$ should be understandable at the level of two-body decays since three- and four- body decays are dominated by quasi-two-body channels. The longer lifetime of D$^+$ compared to D° may be attributed to (a) suppression of D$^+$ rate. Due to identical-particle effects in the Cabibbo-allowed decays of D$^+$, only a few channels are open to D$^+$. More precisely, there are 16 channels for Cabibbo-favored D° decays, but only 4 channels for D$^+$. Further, all Cabibbo-allowed D$^+$ decay amplitudes are always of the form (a+b) with destructive interfer-ence[9]. In particular, the destructive interference is very severe in the VP case (recall that b'/a'~ -0.8). (b) enhancement of D° rate. The W-exchange diagram, which does not exist in D$^+$ decays, may enhance the

D° decay rate. A detailed analysis indicates, however, that although the W-exchange is important in $D \to VP$ decays, it gives a destructive contribution to $D \to VP$ rate! For $D \to 2P$, $2V$ decays, the D° partial rate is boosted a great deal in the first model, but not much in the model (ii).

Since for $D \to PP$, model (ii) is favored by data, this implies that D° and D^+ lifetime difference is attributed mainly to the large suppression of the D^+ rate. Even for the first model, the D^+-rate suppression is still a sizable effect apart from the enhancement of D° rate. Overall, the lifetime difference between D° and D^+ cannot be solely due to the nonspectator contribution as proposed in models with the soft-gluon emission or with the presence of gluons in the initial D state[17]. Finally, we remark that in the quark picture the Pauli-interference[18] corresponds to the aforementioned D^+- rate suppression mechanism. This interference effect is enhanced by nonperturbative corrections.

3. CHARMED-BARYON NONLEPTONIC DECAYS

To solve the s-/p-wave ratio problem in hyperon decays, a popular approach is to include all linear corrections $R(q)$ to the usual soft-pion results

$$\langle B\, \pi^\alpha(q) \mid H_w \mid Y \rangle = -i\sqrt{2}\langle B \mid [Q_\alpha^5,\, H_w] \mid Y \rangle / f_\pi + P(q) + R(q) \qquad (10)$$

where $R(q)$ vanishes in the soft-pion limit. Examples of $R(q)$ are meson-pole (or, factorizable term) and excited baryon-pole contributions. To be more specific we write symbolically[19]

$$A = A^{com} + A^{pole} + A^{pole(*)} + A^{fact}$$
$$B = B^{com} + B^{pole} + B^{pole(*)} + B^{fact} + B^{cont} \qquad (11)$$

where (*) denotes contributions from $\frac{1}{2}^-$ low-lying baryon resonances for s waves and from excited $\frac{1}{2}^+$ baryon poles for p waves. The last three terms of Eq.(11) vanish in the soft-pion limit. B^{cont} is a new nonpole contribution found in chiral perturbation theory [20]. Several authors[21] have shown the importance of these baryon-pole terms. Taking into account these linear corrections, Bonvin[22] concluded that individual s- and p-wave amplitudes can be reproduced to an accuracy of about 10%.

As in hyperon decays, exclusive nonleptonic decays of charmed-baryon has been studied in the current-algebra approach. However, only A^{com}, B^{pole}, and factorizable terms were considered in earlier calculations; contributions from parity-violating matrix elements $b_{\alpha\beta}$ were ignored[23,24].

Nevertheless, the SU(4) symmetry is badly broken and significant $b_{\alpha\beta}$ could be induced. This was studied in Refs. 19 and 25 within the framework of the MIT bag model. It was found in Λ_c^+ decays $b_{\alpha\beta}/a_{\alpha\beta}$ ($a_{\alpha\beta}$ being parity-conserving baryon matrix elements) ranges from 0.1 to 0.4 . Although parity-violating matrix elements do modify significantly some of the s- and p-wave amplitudes, their corrections to the charmed-baryon nonleptonic decay rate are generall small[19,25]. Since branching ratios of two-body decays are measured indirectly, it is more sensible to compare the rate ratios with experiment. Theoretical predictions are[19]

$$R_1 = \Gamma(\Lambda_c^+ \to \Lambda\pi^+)/\Gamma(\Lambda_c^+ \to p\bar{K}^0) = 3.7 \qquad (0.57\pm0.35)$$
$$R_2 = \Gamma(\Lambda_c^+ \to \Delta^{++}K^-)/\Gamma(\Lambda_c^+ \to p\bar{K}^0) = 0.33 \qquad (0.41\pm0.36) \qquad (12)$$

Experimental values are given in brackets. While R_2 is in good agreement with data, R_1 is too large by a factor of 7. The difficulty with R_1 might be cured by including the linear corrections $A^{pole(*)}$ and $B^{pole(*)}$ which thus far have not been computed or by the nonperturbative effect discussed in the next section.

4. LIFETIME OF CHARMED-BARYONS

W-exchange plays an essential role in inclusive nonleptonic decays of charmed baryons since it is not necessarily subject to helicity and color suppressions. According to nonrelativistic model calculations[26], W-exchange dominates the inclusive decay rates of Λ_c^+ and A^0 and it explains the observed lifetime pattern of A^+ and Λ_c^+. At a first sight, it seems to be surprising that the lifetime of charmed baryons can be understood without invoking nonperturbative effects. This is re-examined in the relativistic quark model[27] and it is found that W-exchange is overestimated in the nonrelativistic model by at least a factor of four. This is analogous to the Pauli-interference effect in D^+ inclusive

decays which would be overestimated in the nonrelativistic model if the Fermi motion of quarks were not taken into account[28]. However, the W-exchange contribution estimated in the MIT bag model does not suffice to explain the observed width of Λ_c^+; it must be enhanced substantially by some nonperturbative effects[27]. Thus the nonspectator contribution dominates the inclusive decay rate of Λ_c^+ and A^0 owing to nonperturbative corrections.

The nonperturbative effect, if exists, should also manifest itself in two-body decays of Λ_c^+. In fact, it has been argued[23] that the large discrepency between theory and experiment for R_1 (Eq.(12)) is improved by including soft-gluon corrections with $\epsilon = -3/4$. We then have a parallel picture for understanding charmed-meson and -baryon decays.

5. CONCLUSIONS AND DISCUSSIONS

It is known that the naive model based on the QCD-corrected weak Hamiltonian fails to explain the bulk of the experimental data for two-body decays of charmed mesons. From the Mark III data of $D \to K\pi$, the existence of final-state interactions (more precisely, resonance effects) and nonperturbative corrections is now well established. Any viable model has to accommodate and elucidate these two necessary ingredients. It is shown that the nonperturbative effect which enhances the color-suppressed spectator amplitude is preferred by the Mark III measurements on other $D \to PP$ decays. That is to say the $\underline{6}$ dominance is favored over W-exchange dominance by data. It is thus striking that experimental results for $P_c \to VP$ indicate that QCD correction is not operative there and effective coefficients c_- and c_+ are such that $(c_-/c_+)_{eff} \sim 1$. This means the nonperturbative effect in VP modes compensates QCD correction so that the effective theory is the one without hard-gluon corrections. The model-independent result $b'/a' \sim -0.8$ implies a large destructive interference in Cabibbo-allowed $D^+ \to VP$ decays. This together with the destructive interference (due to QCD correction) in other Cabibbo-favored D^+ modes and the identical-particle effect in D^+ decays accounts for the main part of the D°-D^+ lifetime difference.*

For charmed-baryon two-body decays we have shown the importance of parity-violating matrix elements for some s- and p-wave amplitudes. For inclusive decays the nonperturbative effect is crucial for understanding the lifetime of Λ_c^+ and A°. On the experimental side, more data are needed and the measurement of the decay asymmetry, if feasible, is urged. On the theoretical side, other linear corrections to soft-meson results need to be elaborated. A new framework beyond current algebra is required to understand the quasi-two-body decays, namely, $B_c \to B+V$.

---

* Due to the lack of space, many other topics such as the ratio of $D^\circ \to K^+K^-$ to $\pi^+\pi^-$ and nonresonant three-body decays are not discussed here.

I wish to thank Dr. Ling-Lie Chau for many discussions on the subject and Dr. David H. Coward for explaining to me the Mark III data. This work was supported in part by DOE contract DE-AC03-ER03230 .

REFERENCES

1. Khoze, V.A. and Shifman, M.A., DESY 83-105 (1983).
2. Chau, L.L., Phys. Rep. C95, 1 (1983).
3. Gorn, M., Nucl. Phys. B191, 269 (1981).
4. Mark III Collaboration, D. Coward in these proceedings.
5. Chau, L.L. and Cheng, H.Y., Brookhaven preprint (1985).
6. Kamal, A.N., SLAC-PUB-3443 and SLAC-PUB-3593 (1985).
7. Chernyak, V.I. and Zhitnitsky,A.R., Nucl. Phys. B201, 492 (1982); Phys. Rep. C112, 173 (1984).
8. Deshpande, N., Gronau, M., Sutherland, D., Phys. Lett. 90B, 431 (1980); Deshpande, N. and Sutherland, D., Phys. Lett. B183,367(1981).
9. Bauer, M. and Stech, B., Phys. Lett. 152B, 380 (1985); Stech, B., Heidelberg preprint HD-THEP-85-8 (1985).
10. Baltrusaitis, R., et al., Phys. Rev. Lett. 54, 150 (1985)..
11. Chau, L.L. and Cheng, H.Y., Brookhaven preprint (1985).
12. Albrecht,H., et al., DESY 85-048 (1985);paper contributed to 1985 Lepton and Photon Conference at Kyotoby the CLEO Collaboration.
13. Derrick, M., et al., Phys. Rev. Lett. 54, 2569 (1985).
14. Chen, A., et al., Phys. Rev. Lett. 51, 643 (1983).
15. Althoff, M., et al., Phys. Lett. 136B, 130 (1984).
16. Einhorn, M.B. and Quigg, C., Phys. Rev. D12, 2015 (1975); Bigi, I.I., Z. Phys. C6, 83 (1980).
17. Fritzsch, H. and Minkowski, P., Phys. Lett. 90B, 455 (1980); Bernreuther, W., Nachtman, O. and Stech, B., Z. Phys. C4, 257 (1980); Rosen, S.P., Phys. Rev. Lett. 44, 4 (1980); Bander, M., Silverman, D. and Soni, A., Phys. Rev. Lett. 44, 7 (1980).
18. Kobayashi, T. and Yamazaki., Prog. Theor. Phys. 65, 775 (1981); Peccei,R.D. and Rückl, R., in Proc. Ahrenshoop Symp. (1981).
19. Cheng, H.Y., to appear in Z. Phys. C (1985).
20. Georgi, H. and Manohar, M., Nucl. Phys. B234, 189 (1984); Donoghue, J.F., Golowich, E. and Lin, Y-C. R., UMHEP-223 (1985).
21. Le Yaouanc, A., et al., Nucl. Phys. B149, 321 (1979); Milosević,M., Tadić, D. and Trampetić, J., Nucl. Phys. B207, 461 (1982); Pham, T.N., Phys. Rev. Lett. 53, 326 (1984).
22. Bonvin, M., Nucl. Phys. B238, 241 (1982).
23. Guberina, B., Tadić, D., Trampetić, J., Z. Phys. C13, 251 (1982).
24. Ebert, D. and Kallies, W., Phys. Lett. 131B, 183 (1983); Hussain,F. and Scadron, M.D., Nuovo Cimento 79A, 248 (1984).
25. Ebert, D. and Kallies, W., Yad. Fiz. 40, 1250 (1984).
26. Barger, V., Leveille, J.P., Stevenson, P.M., Phys. Rev. Lett. 44, 226 (1980); Rückl, R., Phys. Lett. 120B, 449 (1983).
27. Cheng, H.Y., to appear in Z. Phys. C (1985).
28. Altarelli, G. and Maiani, L., Phys. Lett. 118B, 414 (1982).

NEW RESULTS ON τ LEPTON FROM PEP

K.K. Gan

Purdue University, W. Lafayette, IN 47906

ABSTRACT

This is a review of the new results on τ lepton from PEP. The results include the precise measurements of the electroweak parameters and lifetime, the first observation of the five-charged-particle decay, new limits on the τ neutrino mass, and measurements of all the major exclusive decay branching ratios with improved precision. The inclusive one- and three-charged-particle topological branching ratios have also been remeasured with high accuracy. All the results are consistent with the predictions of the standard model, except that the sum of the exclusive one-charged-particle decay branching ratios falls significantly below the inclusive measurement. Possible explanations of this discrepancy are discussed.

1. INTRODUCTION

The τ lepton has been a subject of extensive study since its discovery in 1975.[1] All measurements[2] indicate that it is a sequential lepton in the standard gauge theory[3] of electromagnetic and weak interactions. Most of the experiments were performed at a low center-of-mass energy where the τ lepton is not well separated from the low multiplicity hadronic events. The high statistics data sample collected at the higher energy PEP e^+e^- storage ring provides more precise tests of the predictions of the standard model.

In this model, the coupling strength of the τ lepton to the weak charged and neutral currents is of the same as that for the other two charged leptons, e and μ. This lepton universality for the weak charged current can be tested by measuring the τ lifetime. For the weak neutral current, the lepton universality can be tested by measuring the forward-backward asymmetry in the angular distribution of the $\tau^+\tau^-$ pairs. The precise results presented in this review provide stringent tests on the lepton universality.

The τ decay is a good laboratory for studying the hadronic weak current. The high lepton mass allows the τ to decay to a variety of final hadronic states and the V$-$A coupling of the hadronic weak current limits the quantum numbers to a restricted class of resonances such as ρ, K^*, A_1, ρ', etc. Many individual branching ratios can be predicted[4] using the conserved-vector-current (CVC)[5] and the partially-conserved-axial-current (PCAC)[6] hypotheses, together with the sum rules.[7-8] The predictions can be tested with the new and precise measurements of the branching ratios presented in this review.

The results on the branching ratios discussed here are those from the PEP and PETRA experiments, plus new results on the leptonic branching ratios from MARK III at SPEAR. In comparing the sums of the exclusive branching ratios with the inclusive measurements, I do not use any old SPEAR and DORIS results, except the measurement of the $\tau \to \pi\nu$ decay branching ratio,[9] which has not been measured at PEP or PETRA with higher precision. Many of the earlier measurements have large systematic uncertainties (see Section 8). I assume that the errors of all measurements are not correlated when computing the weighted averages. Each measurement is weighted with the systematic and statistical errors combined in quadrature.

2. CHARGE ASYMMETRY

In the standard model,[3] the reaction $e^+e^- \to \tau^+\tau^-$ can proceed through both the electromagnetic and weak neutral currents. The interference between the two currents produces a forward-backward asymmetry in the differential cross section. The standard model predicts the differential cross section to be

$$\frac{d\sigma}{d\cos\theta} = \frac{\pi\alpha^2}{2s}R_{\tau\tau}[1 + \cos^2\theta + \frac{8}{3}A_{\tau\tau}\cos\theta],$$

where $R_{\tau\tau}$ and $A_{\tau\tau}$ are related to the vector and axial-vector weak coupling constants by $R_{\tau\tau} = 1 + 2g_v^e g_v^\tau Re\chi + (g_v^{e^2} + g_a^{e^2})(g_v^{\tau^2} + g_a^{\tau^2})|\chi|^2$, $A_{\tau\tau} = \frac{3}{2R_{\tau\tau}}g_a^e g_a^\tau Re\chi$, with $\chi = \frac{1}{4\sin^2\theta_w \cos^2\theta_w} \frac{s}{s-M_Z^2+iM_Z\Gamma_Z}$. In the standard model, $g_a^e g_a^\tau = g_a^2 = \frac{1}{4}$ and $g_v^e g_v^\tau = g_v^2 = \frac{1}{4}(1 - 4\sin^2\theta_w)^2 \simeq 0.0036$, for[10] $\sin^2\theta_w = 0.22 \pm 0.01$. Using these values, together with the Z_0 mass[10] of $M_z = (93 \pm 2)$ GeV/c^2, the model predicts $R_{\tau\tau} = 1.00$ and $A_{\tau\tau} = -5.9\%$ at $\sqrt{s} = 29$ GeV.

The measurements of $A_{\tau\tau}$ from the PEP experiments[11-12] are summarized in Table 1. The weighted average of the asymmetries is in good agreement with the standard model prediction. The weak coupling constant computed from this weighted average is consistent with the lepton universality of $g_a^2 = \frac{1}{4}$. The results are also in good agreement with the PETRA measurements.[13]

Table 1. Measurements of the charge asymmetry and axial-vector coupling constant.

Detector	$A_{\tau\tau}(\%)$	$g_a^e g_a^\tau$
HRS	$-5.2 \pm 1.7 \pm 0.5$	$0.23 \pm 0.08 \pm 0.03$
MAC	$-5.5 \pm 1.2 \pm 0.5$	$0.22 \pm 0.05 \pm 0.03$
MARK II	-4.2 ± 2.0	$0.19 \pm 0.09 \pm 0.02$
Average	-5.1 ± 0.9	0.22 ± 0.04
Theory	-5.9	0.25

3. LIFETIME

Measurement of the τ lifetime provides a direct study of the coupling strength of the τ to the charged weak current. In the standard model, the τ

decay $\tau \to e\nu\nu$ proceeds in perfect analogy to the μ decay $\mu \to e\nu\nu$. Assuming $\mu - \tau$ universality of the weak coupling and that the τ neutrino is massless, then the τ lifetime is related to the μ lifetime by

$$\tau_\tau = (\frac{m_\mu}{m_\tau})^5 \tau_\mu B(\tau \to e\nu\nu).$$

With the world average measurement (see next section) of the electron branching ratio $B(\tau \to e\nu\nu) = (17.8 \pm 0.6)\%$, the predicted[4] lifetime is $\tau_\tau = (2.85 \pm 0.10) \times 10^{-13} s$.

The HRS and MAC experiments have both submitted preliminary results on the τ lifetime to this conference. The results, together with other measurements,[14-15] are summarized in Fig. 1. All results are in agreement and the weighted average[16] of $\langle \tau_\tau \rangle = (2.80 \pm 0.20) \times 10^{-13} s$ is in excellent agreement with the theoretical prediction and confirms $\mu - \tau$ universality of the weak coupling to the level of 4%. This may be compared to the 0.8% on the $e - \mu$ universality coming from the study of pion decay.[17]

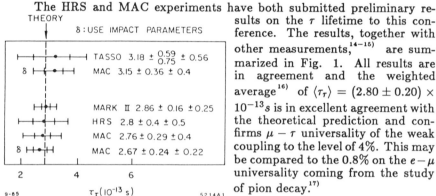

Fig. 1. Measurements of the τ lifetime.

4. e AND μ DECAYS

In the standard model, the μ decay branching ratio of the τ lepton is the same as the e decay branching ratio except for a small suppression[4] coming from the phase space factor, $B_\mu/B_e = 0.97$. Prior to this conference, the PDG[9] world averages of the two branching ratios yielded $B_\mu/B_e = 1.12 \pm 0.10$. If B_μ were indeed greater than $0.97B_e$, there would be an interesting implication on the minimal standard model with one Higgs doublet. The minimal model could be extended to include another Higgs doublet.[18] This extension allows the τ lepton to decay through the $H^\pm$ propagators as well as the standard $W^\pm$ bosons, and increases the ratio of the branching ratios to

$$\frac{B_\mu}{B_e} = 0.97 + 0.0086 \cdot \frac{\cot^4 \alpha}{M_H^4},$$

where $\cot \alpha$ is the ratio of the vacuum expectation values of the two Higgs doublets and M_H is the mass of the Higgs in GeV/c^2. Therefore precise measurements of the leptonic decay branching ratios provide a test of the minimal standard model. Since the leptonic decay modes account for about 40% of the τ decay, precise measurements of these two major decay modes also allow for more accurate determination of the sum of the exclusive decay branching ratios and thus facilitate the search for any unexpected decay modes. Furthermore, precise measurements enable more definite theoretical predictions of other decay branching ratios to be made since they are related to the leptonic branching ratios.

The MARK III and MAC collaborations have reported new and precise results[19] on the leptonic branching ratios. The results, together with the PETRA measurements,[20-21] are summarized in Table 2. All results are in good agreement and the weighted average yields $B_\mu/B_e = 0.97 \pm 0.05$, in excellent agreement with the prediction of the minimal standard model. More accurate measurements of the leptonic branching ratios are needed before one can draw any conclusion on the existence of a charged Higgs.

Table 2. Measurements of the leptonic branching ratios.

Detector	$B_e(\%)$	$B_\mu(\%)$
MARK III	$18.2 \pm 0.7 \pm 0.5$	$18.0 \pm 1.0 \pm 0.6$
MAC	$17.4 \pm 0.8 \pm 0.5$	$17.7 \pm 0.8 \pm 0.5$
CELLO	$18.3 \pm 2.4 \pm 1.9$	$17.6 \pm 2.6 \pm 2.1$
TASSO	$20.4 \pm 3.0 \,^{+1.4}_{-0.9}$	$12.9 \pm 1.7 \,^{+0.7}_{-0.5}$
PLUTO	$13.0 \pm 1.9 \pm 2.9$	$19.4 \pm 1.6 \pm 1.7$
Average	17.8 ± 0.6	17.3 ± 0.6

5. ρ AND K^* DECAYS

The ρ and K^* decays of the τ lepton allow studies of the hadronic weak current. The ρ branching ratio can be calculated[4] by using CVC to relate the coupling strength of the ρ to the weak charged-vector-current and the electromagnetic neutral-vector-current. Gilman and Rhie[22] used the measurement of the $e^+e^- \to \gamma^* \to \rho$ cross section to calculate the electromagnetic coupling and predicted that $B_\rho/B_e = 1.23$. The Cabibbo-suppressed K^* decay is related[4] to the Cabibbo-favoured ρ decay by

$$B_{K^*}/B_\rho = \tan\theta_c \cdot PS \cdot \frac{g_{K^*}^2}{g_\rho^2}.$$

The factor PS corrects for the difference in the available phase spaces. The relationship between g_{K^*} and g_ρ, the coupling strengths of the ρ and K^* to the vector current, depends on whether the SU(3) symmetry is exact or broken. If the SU(3) symmetry is exact, then $g_{K^*}^2 = g_\rho^2$ and the prediction is $B_{K^*}/B_\rho = 0.038$. On the other hand, if the symmetry is broken, then the Das-Mathur-Okubo[8] sum rules give $g_{K^*}^2/m_{K^*}^2 = g_\rho^2/m_\rho^2$ and the prediction becomes $B_{K^*}/B_\rho = 0.052$.

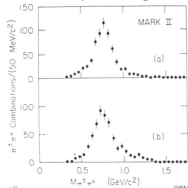

Fig. 2. $\pi\pi^0$ mass spectrum for events that have (a) two photons with good fit to the π^0 hypothesis, (b) one single energetic photon with energy of 2 GeV or more.

The MARK II collaboration[23] has reported new results on ρ (see Fig. 2) and K^*. The results are $B_\rho = (22.3 \pm 0.6 \pm 1.4)\%$ and $B_{K^*} = (1.3 \pm 0.3 \pm 0.3)\%$. The ρ branching ratio is in good agreement with the CELLO measurement[24] of $B_\rho = (22.1 \pm 1.9 \pm 1.6)\%$ and leads to a weighted average of $B_\rho = (22.2 \pm 1.3)\%$, in excellent agreement with the CVC prediction of $B_\rho = (21.9 \pm 0.7)\%$. Using this weighted average of B_ρ, together with the MARK II result on B_{K^*}, one obtains $B_{K^*}/B_\rho = 0.059 \pm 0.018$. Within these limited statistics, the result favours a broken SU(3) symmetry. Higher statistics are required before one can draw a definite conclusion.

6. 3π AND 4π DECAYS

The existence of an axial-vector state, the A_1 meson, predicted by the quark model has been known for many years. However, some puzzling features pertaining to the A_1 mass and width remain. The main difficulty with the study of the A_1 produced in hadronic collisions is the uncertainty in the subtraction of the Deck background, arising from the diffractive dissociation of the incident particles. There is no such problem in the $\tau \to A_1\nu$ decay and so it provides more direct information on the properties of the A_1.

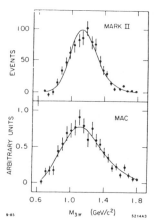

Fig. 3. Invariant mass distribution of the 3π system.

The MAC and MARK II collaborations have both reported preliminary results on the A_1 to this conference. Fig. 3 shows the invariant mass spectrum of the 3π system measured by the two experiments. The results of the fits to the mass distributions are summarized in Table 3, together with the PDG[9] world average measurements from the hadronic data. The MAC's measurement of the A_1 mass is somewhat lower than the MARK II, although the two mass distributions appear to peak near the same value. The discrepancy is due to the different mass fitting functions used.[25] Exactly what is the proper fitting function for extracting the mass and width will remain controversial for years to come. Both experiments find that the mass of the A_1 is lower than the PDG world average. The width of the two measurements are in good agreement with each other but substantially larger than the hadron result. The results conflict with the hadronic data, but more work is needed to understand how to extract the properties of the A_1 from the data.

The MAC and MARK II experiments both find that the 3π system is consistent with the $\tau \to A_1\nu \to \rho\pi\nu$ decay. The MARK II group has performed a spin-parity analysis on the 3π system and set an upper limit of 20% for a non-$\rho\pi$ resonance contribution at 95% CL. The MARK II experiment also finds that a 1^+s wave $\rho\pi$ resonance is favoured, and sets 95% CL upper limits on 1^+d and 0^-p wave contributions of 30% and 20%, respectively.

Table 3. Measurements[26] of the mass and width of the A_1.

Detector	$M_{A_1}(MeV/c^2)$	$\Gamma_{A_1}(MeV/c^2)$
MARK II	1230 ± 20	540 ± 90
MAC	$1098 \pm 7 \pm 20$	$525 \pm 10 \pm 20$
PDG	1275 ± 30	315 ± 45

The MAC[15] and MARK II groups have also measured the 3π and $3\pi\pi^0$ decay branching ratio of the τ lepton by classifying the three-charged-particle final states into those that contain zero and one or more energetic photons. The results on the branching ratios from the two experiments, together with those from CELLO,[24] are summarized in Table 4. All the measurements are in good agreement. There is no firm

Table 4. Measurements of the 3π and $3\pi\pi^0$ decay branching ratios of the τ lepton.

Detector	$B_{3\pi}(\%)$	$B_{3\pi\pi^0}(\%)$
MAC	8.1 ± 0.8	5.2 ± 0.8
MARK II	$7.8 \pm 0.5 \pm 0.8$	$4.7 \pm 0.5 \pm 0.8$
CELLO	$9.7 \pm 2.0 \pm 1.3$	$6.2 \pm 2.3 \pm 1.7$
Average	8.1 ± 0.6	5.0 ± 0.6

prediction for the 3π branching ratio due to the theoretical uncertainty in calculating the axial-current decay. The $3\pi\pi^0$ decay branching ratio can be calculated using CVC together with the measurements of the $e^+e^- \to \pi^+\pi^-2\pi^0$ cross section and is found[22] to be $\sim 4.9\%$. Therefore the weighted average of the measured $3\pi\pi^0$ branching ratios is in agreement with the CVC prediction.

7. 5π AND 6π DECAYS

The decay of the τ lepton to five charged particles has been reported by the HRS collaboration.[27] Ten events were observed with an estimated background of less than 0.5 events. This results in a measured branching ratio of $B_5 = (0.13 \pm 0.04)\%$. The MARK II collaboration[28] has also reported the observation of the five-charged-particle decay and measures a branching ratio of $B_5 = (0.16 \pm 0.08 \pm 0.04)\%$. The weighted average of the two measurements is $B_5 = (0.14 \pm 0.04)\%$. The HRS collaboration also finds that five of the events contain photons which are consistent with coming from a single π^0 decay. With this interpretation, they measure $B(\tau \to 5\pi\nu), B(\tau \to 6\pi\nu) = (0.067 \pm 0.030)\%$. There are no firm theoretical predictions on these branching ratios.[22]

The upper limit on the τ neutrino mass has been improved several times over the last year. The MARK II collaboration[28] used the end-point of the 5π mass spectrum to reduce the previous 95% CL upper limit from 143 MeV/c^2 to 125 MeV/c^2. The HRS collaboration[29] used both 5π and 6π decay modes and has set a new limit of 84 MeV/c^2. The Argus collaboration[30] used the end-point of the 3π momentum spectrum to place a new limit of 70 MeV/c^2. These results exclude the $\nu_\tau \to \mu e \nu_e$ decay.

8. TOPOLOGICAL BRANCHING RATIOS

The topological decay branching ratios of the τ lepton to one and three charged particles, denoted by B_1 and B_3 respectively, have been measured by HRS[31] and MAC[15] with very high precision. The results, along with other measurements,[21,24,32] are summarized in Fig. 4. All results are in good agreement and the weighted average is $B_3 = (13.2 \pm 0.2)\%$. This is substantially below the old SPEAR/DORIS[33] average of $B_3 = (32 \pm 4)\%$. The discrepancy is due to the large systematic problems in the early low energy measurements. At low energies, the τ events are more spherical and less well separated from the low multiplicity hadronic events. Moreover, the measurements were performed before other properties of the τ such as the mass and leptonic branching ratios were well measured. It is for these reasons that I avoid using any old SPEAR/DORIS results on branching ratios.

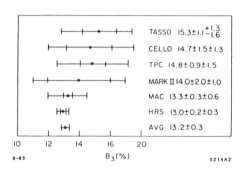

Fig. 4. Measurements of B_3.

9. COMPARISON OF THE INCLUSIVE AND EXCLUSIVE BRANCHING RATIOS

With the major exclusive decay branching ratios of the τ lepton remeasured with high precision, it is interesting to compare the sums of the exclusive branching ratios with the inclusive measurements and so to search for any unexpected decay modes. For the three-charged-particle final states, the sum (see Table 5) of exclusive decay modes is equal to the inclusive measurement of $B_3 = (13.2 \pm 0.2)\%$. This is expected as the $3\pi\nu$ and $3\pi\pi^0\nu$ branching ratios are measured by classifying the three-charged-particle final states into those that contain zero and one or more energetic photons and thus unitarity is ensured. For the one-charged-particle final states, the sum of the exclusive decay modes is more than three standard deviations below the inclusive measurement of $B_1 = (86.7 \pm 0.3)\%$.

It is unlikely that the discrepancy between the inclusive and exclusive measurements of B_1 is due to the statistical fluctuations in some of the measurements

Table 5. Experimental and theoretical values of the branching ratios.[34]

Decay Mode	Branching Ratio(%) Experiment	Theory
1-prong:		
$e\nu\nu$	17.8 ± 0.6	17.8
$\mu\nu\nu$	17.3 ± 0.6	17.3
$\pi\nu$	10.3 ± 1.2	10.8
$K\nu$	0.6 ± 0.2	0.7
$\rho\nu$	22.2 ± 1.3	21.9
$K^*\nu$	1.3 ± 0.4	1.1
$\pi 2\pi^0\nu$	8.1 ± 0.6	
$\pi 3\pi^0\nu$	1.0 ± 0.1	1.0
$\pi 4\pi^0\nu + \pi 5\pi^0\nu$	0.14 ± 0.04	
Total	78.7 ± 2.1	
3-prong:		
$3\pi\nu$	8.1 ± 0.6	
$3\pi\pi^0\nu$	5.0 ± 0.6	4.9
$3\pi 2\pi^0\nu + 3\pi 3\pi^0\nu$	0.14 ± 0.04	
Total	13.2 ± 0.8	
5-prong:		
$5\pi\nu + 5\pi\pi^0\nu$	0.14 ± 0.04	

because the theoretical predictions and experimental results agree very well. As an example one can consider the possibility that several of the major decay branching ratios with one charged particle in the final states are low and their actual values correspond to $B_e = 19.0\%$. Then the sum of the one-charged-particle branching ratios would be $(83.5 \pm 2.1)\%$. However, this would require

5 of the measured branching ratios to be one standard deviation below their actual values. Therefore statistical fluctuations are an unlikely explanation for the discrepancy. Note that the measurement of the τ lifetime, assuming $\mu - \tau$ universality, predicts an electron branching ratio of $B_e = (17.5 \pm 1.3)\%$, which is on the low side. More precise measurement of the τ lifetime will provide a better independent check. Also note that the uncertainties in the $\pi\nu$ and $\rho\nu$ branching ratios now dominate the overall error in the exclusive measurement of B_1. Therefore more precise measurement of these two decay modes will provide a better understanding of the discrepancy.

It is also unlikely that the discrepancy can be accounted for by the expected but not yet observed decay modes, such as $\tau \to K^*K\nu, \eta\pi\pi^0\nu, Q_1\nu, Q_2\nu$. These decay modes are expected to have a total contribution[35] of $\sim 1.5\%$. There is another conventional decay mode, $\tau \to \rho'(1600)\nu \to \pi\pi^0\nu$, which could potentially have large contribution. However, as can be seen in Fig. 2, there is no event in the ρ' region. Therefore these not yet observed decay modes are probably too small to account for the discrepancy.

Truong[36] had previously studied the discrepancy with less well measured branching ratios and pointed out that the discrepancy could be due to the not yet observed axial current hadronic decays of the τ lepton. The hypothesis was proposed because the axial current branching ratio was significantly below the vector current branching ratio, ignoring the small Cabibbo-suppressed contribution. The new results listed in Table 5 show that the axial current contribution of $(26.5 \pm 1.7)\%$ is approximately equal to the vector current contribution of $(28.2 \pm 1.4)\%$ and thus do not support this hypothesis.

A possible explanation with τ decay involving supersymmetry particles can be ruled out. If the process $\tau \to \widetilde{W}\tilde{\nu} \to W\tilde{\gamma}\tilde{\nu}$ occurs, then $\tilde{\gamma}$ and $\tilde{\nu}$ will carry away a large fraction of the energy and force the virtual W to decay mostly to one-charged-particle final states and thus provide a possible explanation. However, the large fraction of missing energy carried away by the invisible particles will result in an enhancement in the lower end of the one-charged-particle momentum spectrum. This is an effect not observed in the data.[37]

Could the discrepancy be due to a new particle that couples strongly to the one-charged-particle final states being produced ? This is unlikely as all the new particle searches at PEP[38] and PETRA[13] produce negative results. It should be noted, however, that all the new particle searches were performed in kinematic regions where there is no τ signal. Therefore the searches are insensitive to new particles that mimic the τ decays. However, all the measurements of the τ cross section show that the τ lepton is a point-like particle with $R_{\tau\tau} = 1.00$. Moreover, a new particle could alter the $1+\cos^2\theta + A\cos\theta$ angular distribution. Fortunately, this new particle hypothesis can easily be tested by remeasuring the B_1 and B_3 at lower energies at SPEAR and CESR/DORIS. Both branching ratios should be independent of the beam energy unless a new particle is being produced. A more difficult test is to compare the τ lifetime of the one-charged-particle and three-charged-particle final states using the impact parameter technique. One can also use $\frac{dE}{dx}$ (TPC) to study the 1-prongs in the events of 1-3 topology, selected with the 3-prong as a tag, and search for a new heavy and stable particle.

Gilman and Rhie[22] suggest that the τ lepton could have unconventional decays such as $\tau \to S\nu$, where S is a virtual particle and couples mostly to one-charged-particle final states. This is an explanation that cannot be ruled out with the present data.

10. SUMMARY

In summary, the electroweak parameters and the τ lifetime have been measured with high precision and the results are consistent with lepton universality. The five-charged-particle decay of the τ lepton has been observed and the new limit on the τ neutrino mass is now below the μ mass. Major decay branching ratios of the τ lepton have been remeasured with high accuracy and the results are consistent with the standard model predictions. The inclusive charged particle topological branching ratios have also been remeasured with very high precision. However, the sum of the exclusive decay branching ratios with one charged particle in the final states is significantly below the inclusive measurement. The best way to resolve the discrepancy is to use a technique similar to that employed to measure the $3\pi\nu$ and $3\pi\pi^0\nu$ branching ratios, which are measured by classifying the three-charged-particle final states into those that contain zero and one or more energetic photons. For the one-charged-particle final states, one can classify them as the decay of the τ lepton into $e\nu, \mu\nu, \pi\nu, K\nu, \pi\pi^0\nu, K^*\nu, \pi 2\pi^0\nu, \pi 3\pi^0\nu$, etc. This will ensure unitarity and so may show up the "missing" decay modes.

The author wishes to thank D.L. Blockus, M. Derrick, J.M. Dorfan, F.J. Gilman, D. Koltick, M.L. Perl, A.L. Read, W.B. Schmidke, J.G. Smith, and J.P. Venuti for useful discussions. This work was supported in part by the U.S. Department of Energy under Contract No. DE-AC02-76ER01428.

REFERENCES

1. M.L. Perl et al., Phys. Rev. Lett. 35, 1489 (1975).
2. M.L. Perl, Ann. Rev. Nucl. Part. Sci. 30, 299 (1980).
3. S.L. Glashow, A. Salam, and S. Weinberg, Rev. Mod. Phys. 52, 515 (1980).
4. Y.S. Tsai, Phys. Rev. D 4, 2821 (1971); H.B. Thacker and J.J. Sakurai, Phys. Lett. 36B, 103 (1971).
5. R.P. Feymann and M. Gell-Mann, Phys. Rev. 109, 193 (1958).
6. J. Bernstein, S. Fubini, M. Gell-Mann, and W. Thirring, Nuovo Cimento 17, 757 (1960).
7. S. Weinberg, Phys. Rev. Lett. 18, 507 (1967).
8. T. Das, V.S. Mathur, and S. Okubo, Phys. Rev. Lett. 18, 761 (1967).
9. Particle Data Group, Rev. Mod. Phys. 56, No. 2, Part II (1984).
10. G. Arnison et al., Phys. Lett. 126B, 398 (1983); P. Bagnaia, Phys. Lett. 129B, 130 (1983).
11. M.E. Levi et al., Phys. Rev. Lett. 51, 1941 (1983); E. Fernandez et al., Phys. Rev. Lett. 54, 1620 (1985).
12. K.K. Gan, Ph.D. Thesis, Purdue University, 1985 (unpublished).
13. M. Davier, in Proc. of the 12th SLAC Summer Inst. in Part. Phys., ed. P. McDonough (SLAC, Stanford, CA, 1984), p. 349.
14. J.A. Jaros, in Proc. of the 12th SLAC Summer Inst. in Part. Phys., ed. P. McDonough (SLAC, Stanford, CA, 1984), p. 427.

15. E. Fernandez et al., Phys. Rev. Lett. 54, 1624 (1985).

16. In the three MAC measurements of the τ lifetime, only the one with the smallest error is used in calculating the weighted average since the errors of the three measurements are correlated.

17. D.A. Bryman et al., Phys. Rev. Lett. 50, 7 (1983).

18. Y.S. Tsai, private communication.

19. R.M. Baltrusaitis et al., SLAC-PUB-3732, 1985; W.W. Ash et al., SLAC-PUB-3745, 1985 (both submitted to Phys. Rev. Lett.).

20. H.J. Behrend et al., Phys. Lett. 127B, 270 (1983); Ch. Berger et al., Z. Phys. C 28, 1 (1985).

21. M. Althoff et al., Z. Phys. C 26, 521 (1985).

22. F.J. Gilman and S.H. Rhie, Phys. Rev. D 31, 1066 (1985).

23. J.M. Yelton et al., SLAC-PUB-3579, 1985 (submitted to Phys. Rev. Lett.).

24. H.J. Behrend et al., Z. Phys. C 23, 103 (1984).

25. Due to the limited space, details of their fits cannot be discussed here. See their upcoming publications for a full discussion.

26. The errors quoted for the MARK II results are statistical only. The systematic uncertainties are expected to be larger.

27. I. Beltrami et al., Phys. Rev. Lett. 54, 1775 (1985).

28. P.R. Burchat et al., Phys. Rev. Lett. 54, 2489 (1985).

29. S. Abachi et al., Purdue University Report No. PU-85-541, 1985 (to be published).

30. H. Albercht al., DESY 85-054, 1985 (submitted to Phys. Lett. B).

31. C. Akerlof et al., Phys. Rev. Lett. 55, 570 (1985).

32. C.A. Blocker et al., Phys. Rev. Lett. 49, 1369 (1982); H. Aihara et al., Phys. Rev. D 30, 2436 (1984).

33. J. Burmester et al., Phys. Lett. 68B, 297 (1977); W. Bacino et al., Phys. Rev. Lett. 41, 13 (1978); R. Brandelik et al., Phys. Lett. 73B, 109 (1978).

34. All the experimental branching ratios listed in the table are from the results presented in this review except for the $\tau \to K\nu$ and $\pi 3\pi^0\nu$ decay modes. The value for the $K\nu$ decay is given by G.B. Mills et al., Phys. Rev. Lett. 52, 1944 (1984). For the $\pi^\pm 3\pi^0\nu$ decay, the theoretical prediction of Gilman and Rhie[22] is used instead of the experimental measurement which has large uncertainty. I assume that the $\pi 2\pi^0\nu$ and $3\pi\nu$ decay modes are dominated by the A_1 and, for the 5π and 6π decay modes, the branching ratios into one, three or five charged π final states are equal.

35. An educated guess given by W. Ruckstuhl, in Proc. of the 12th SLAC Summer Inst. in Part. Phys., ed. P. McDonough (SLAC, Stanford, CA, 1984), p. 466.

36. T.N. Truong, Phys. Rev. D 30, 1509 (1984).

37. For example, see Fig. 24 of Ref. 12.

38. R. Prepost, in Proc. of the 12th SLAC Summer Inst. in Part. Phys., ed. P. McDonough (SLAC, Stanford, CA, 1984), p. 488.

PETRA AND PEP MEASUREMENTS OF THE AVERAGE B HADRON LIFETIME

Gregory J. Baranko

Physics Department
University of Wisconsin - Madison[1]

ABSTRACT

Measurements of the average B hadron lifetime by six e^+e^- experiments at PETRA and PEP are reviewed. The results are also expressed in the context of the Kobayashi-Maskawa flavor mixing matrix.

1.   INTRODUCTION

In the past few years, experiments at PETRA and PEP have come to play a major role in the lifetime measurements of the τ lepton and hadrons containing a charm or bottom quark. Although the precision wire chambers employed in these detectors still cannot match the resolution often found in devices used by fixed target experiments (bubble chambers, emulsion and silicon strip detectors), there is a great advantage in that the e^+e^- annihilation process used to produce the weakly decaying particles is a well understood, easily modelled reaction with relatively low and readily predictable backgrounds. The heavy quark production rate is also significant, comprising almost half of the total hadronic cross section. Furthermore, the method of data acquisition and analysis applied to lifetime measurements in the e^+e^- experiments is less affected by the kinds of acceptance and scanning biases that can plague the fixed target results.

One of the e^+e^- lifetime measurements that has relied substantially on these advantages has been the average bottom hadron, B, lifetime. Six experiments at PETRA and PEP have now reported on this study: JADE,[2] MAC,[3] MARK II,[4] DELCO,[5] TASSO,[6] and the HRS.[7] The measurement of the B lifetime is interesting because it provides information on the strengths of flavor transitions involving the b quark and the resulting consequences that these couplings have for other phenomena associated with flavor mixing in weak decays. This almost direct inference from B hadron decays to b quark properties is fostered by the assumption that nonspectator contributions are less important to the description of B decays than for lighter hadrons.[8] If this is true, then the B lifetime can be estimated[9] using the spectator model in an analogy to the muon lifetime, τ_μ, as:

$$\tau_B = \tau_\mu \cdot \left[\frac{m_\mu}{m_b} \right]^5 \cdot BR(B \rightarrow X\ell\nu) \cdot x_{KM}^b \qquad (1)$$

where $$x_{KM}^b = \frac{1}{k_c |U_{cb}|^2 + k_u |U_{ub}|^2}$$

and $BR(B \rightarrow X\ell\nu)$ is a nominal, semileptonic branching fraction for a B hadron. The flavor mixing factor, x_{KM}^b, is expressed in terms of the Kobayashi-Maskawa matrix elements[10] U_{qb} with mass and QCD correction factors k_q (of order 1) for the transitions to the lighter quarks, $q = u$ and c.[11]

Measurements of inclusive lepton production at CESR have shown that $BR(B \rightarrow X\ell\nu) = (11.8 \pm 0.4 \pm 0.7)\%$.[11-13] If there were no K-M suppression, $x_{KM}^b = 1$, then one would expect $\tau_B \simeq 1.1 \cdot 10^{-15}$ s, using $m_b = 5$ GeV/c^2. However, x_{KM}^b may increase the lifetime for weaker b quark couplings to u and c quarks, or small U_{cb} and U_{ub}. Finally, measurements at CUSB,[12] CLEO,[13] and ARGUS[14] have shown that:[11]

$$R_B = \frac{\Gamma(B \rightarrow X_u \ell\nu)}{\Gamma(B \rightarrow X_c \ell\nu)} = \frac{k_u}{k_c} \cdot \frac{|U_{ub}|^2}{|U_{cb}|^2} < 4 \% \text{ at } 95 \% \text{ C.L.} \qquad (2)$$

Thus, measuring τ_B primarily determines $|U_{cb}|$. From the constraints that these and other experiments place on the K-M matrix elements, along with a unitarity requirement, several interesting predictions may be made, such as: a lower bound on the top quark mass,[15] a lower bound on Re(ε'/ε),[16] and the amount of mixing and CP violation in neutral D and B mesons.[17] Obtaining more precise values for the matrix elements may also provide timely insight into the problem of family replication and the spectrum of heavier leptons and quarks that might yet be discovered. Taking advantage of such a predictive opportunity would give experimentalists and theorists alike an important head start on the physics to come from the next generation of colliders.

2. METHODS

This past year's work at PETRA and PEP has brought new and statistically improved measurements of the average B hadron lifetime. Although the present data samples are not yet large enough to allow these experiments to reconstruct B hadrons in explicit decay channels, the B's have special properties that can be exploited to identify their decay products and tag $b\bar{b}$ events. Their relatively large mass, decay multiplicity, and energy fraction (E_B/E_{Beam}) produce events with larger than average sphericities and, in semileptonic decays, leptons with a large transverse momentum, p_t, relative to the event axis. Charm particles resulting from B decays also are expected to be at lower energy fractions than when produced from primary charm quarks.

Using these event characteristics, all of the experiments first enrich a data sample with $b\bar{b}$ events. JADE, MAC, MARK II, and DELCO do this by requiring an identified lepton with high p_t, typically 1 - 1.5 GeV/c. TASSO's enrichment relies on the higher sphericity of b quark jets compared to lighter quark jets. The sphericity of each of the jets in an event is calculated in a frame boosted towards the B rest frame. The enrichment is made by requiring the product of the two sphericities to be large. Finally, the HRS observes the low energy fraction $D^{*\pm}$ from the B decay, relying on the special kinematics of the transition $D^{*+} \to D^0\pi^+$ with $D^0 \to K^-\pi^+$ (including charge conjugate decays) to improve the signal's significance. They require $.2 < E_D^{*\pm}/E_{Beam} < .4$ and estimate that 23% of these $D^{*\pm}$ candidates are from B decays.

Since the B hadrons are not explicitly reconstructed, their decay distance is parameterized by observing displacements of their decay products, in an inclusive manner, from the B production point. All of the methods assume that the B is produced at the e^+e^- interaction point, IP. This IP is found only on a run by run basis, usually by averaging the vertex position of Bhabha events. In all but the HRS analysis, the impact parameters of selected charged tracks in the $b\bar{b}$ enriched samples are used to measure the B lifetime. The impact parameter is defined as the track's distance of closest approach to the IP. This distance is projected into the plane transverse to the beam line, where the spatial resolution for track reconstruction is best. In order to obtain a relation between the B decay distance and the impact parameter, δ, of a track, the B hadron is assumed to travel along the direction of the event axis. The direction is calculated using either the thrust or sphericity algorithms. In the JADE, MAC, and MARK II analyses, only the impact parameters of high p_t electrons and muons are used. DELCO observes electrons identified by their Cherenkov counters. TASSO's method uses any good quality charged track with $p > 1$ GeV/c.

In the impact parameter method, the average lifetime for the mixture of B hadrons produced at the PETRA and PEP experiments is extracted by comparing the δ distribution with its Monte Carlo prediction for varying B lifetimes. These calculations include background sources for the candidate B decay products and a complete detector simulation. Both approaches to the measurement, using either the high p_t lepton or selected charged tracks, have particular merit. The lepton analyses take advantage of the better studied semileptonic B decays and suffer less from systematic errors associated with modelling the decays in the Monte Carlo. However, using several charged tracks per event gives more sensitivity to the B decay point, due to the higher multiplicity of B decays compared to lighter quarks. There is also less sensitivity to the uncertainty in the B production point, as these errors per track tend to cancel for several tracks in the same event.

In the HRS analysis, the B lifetime is measured by observing the lifetimes of D^o candidates from the low energy fraction $D^{*\pm}$. A three term maximum likelihood fit is made to the proper time distribution of the D^o decays. One term with 54% probability is taken as a zero lifetime background described by a Gaussian shape. A second term with

23% probability describes the direct charm contribution by an exponential distribution convoluted with a Gaussian resolution. The third term, with 23% probability, is an exponential for the B lifetime convoluted with an exponential for the D^o lifetime, and again convoluted with a Gaussian. A D^o lifetime of $4.5 \cdot 10^{-13}$ s was used.

3. RESULTS

The values for τ_B obtained by the six experiments are shown in Table 1. The first error quoted is statistical; the second is considered as systematic. The integrated luminosities that were available for these measurements, L, were taken at a center of mass energy W = 29 GeV for the PEP experiments and at several energies up to a maximum of 46.7 GeV for the PETRA experiments. In the applicable analyses, the table gives the average impact parameter, $\bar{\delta}$, with its statistical error, and the experimental resolution per track for δ, σ_δ, including the contribution from the beam size. Also shown in the table are the number of candidate tracks from B decays in the $b\bar{b}$ enriched samples, N_B, and the estimated fraction of the tracks that are considered as the B signal, f_B. This fraction includes both the direct decay $B \to X\ell\nu$ and the cascade process $B \to C \to X\ell\nu$. For TASSO,

TABLE 1

AVERAGE B HADRON LIFETIME MEASUREMENTS

EXP	$\bar{W}$[GeV]	L[pb^{-1}]	N_B	f_B	σ_δ[µm]	$\bar{\delta}$[µm]	τ_B[10^{-12} s]
JADE	35	63	99	.75	570	326±56	$1.8 \begin{smallmatrix} +0.5 \\ -0.4 \end{smallmatrix} \pm 0.4$
MAC	29	210	505	.58	530	70±22	$0.81 \pm 0.28 \pm 0.17$
MARK II	29	220	272	.64	200	80±17	$0.85 \pm 0.17 \pm 0.21$
DELCO	29	214	113	.83	400	288±48	$1.47 \begin{smallmatrix} +0.30 \\ -0.26 \end{smallmatrix} \pm 0.29$
TASSO	44	25	1530	.30	380	91±17	$1.57 \pm 0.32 \begin{smallmatrix} +0.37 \\ -0.34 \end{smallmatrix}$
HRS	29	97	21	.23	N.A.	N.A.	$0.97 \begin{smallmatrix} +0.97 \\ -0.53 \end{smallmatrix}$

WEIGHTED AVERAGE 1.11 ± 0.16

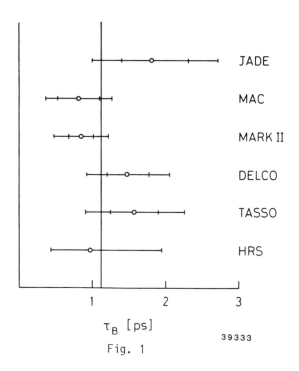

$$\tau_B \ [\text{ps}]$$

Fig. 1

39333

the data are from the analysis using the vertex detector, although τ_B is the value obtained by combining results from their old and new detector configurations. The distribution of these measurements is shown in Figure 1. The first error bar away from the measured value is statistical. The systematic error is added linearly onto the statistical error for the second bar. Only the statistical error is shown for the preliminary HRS measurement.

JADE first used the impact parameter method to make an upper limit on τ_B with a small data sample. The present result is based on about 7 times more data and is consistent with their previous upper limit within the quoted errors. JADE also uses their muon tagged event sample to make a δ distribution for all selected charged tracks that are weighted by the event shape and p_t of the muon. Fitting this distribution gives[18] $\tau_B = (1.7 \pm 0.6 \pm 0.4) \cdot 10^{-12}$ s, in agreement with their lepton-only analysis. The updated MAC measurement[19] is based on a data sample that was increased by 30 % this past year and is somewhat lower than their previously reported value. MARK II has also compared the result from their updated lepton-only method[20] with an alternative analysis[21] that observes the displacement of jet vertices in the same sample of lepton tagged events. Using this analysis, they obtain $\tau_B = (1.25 \; {}^{+\,0.26}_{-\,0.19} \; {}^{+\,0.35}_{-\,0.39}) \cdot 10^{-12}$ s. DELCO[22] has now doubled the data sample used in their original publication. The new value is also consistent with the original measurement. TASSO[23] has increased their data sample with the vertex detector to about twice the sample used in publication for this detector configuration. The new value combines their updated result using the vertex detector with the previously reported result using only the large central detector. Although the HRS result has a large statistical error, it uses a valuable new approach to this measurement, as it brings in very different systematic errors.

Although the B lifetime results are comparable, it should be noted that a given measurement is an average over all produced B hadrons, and may depend slightly on the particular selection procedures. For example, in the lepton method, the B lifetime is a weighted average depending not only on the production rates of the different B hadrons, but also depending on their semileptonic branching ratios. For inclusive hadronic analyses, only the weighting due to the production rates would enter. However, these differences are not expected to be large.[20] The HRS result, using the $D^{*\pm}$, may also be more dependent on the B^o_d lifetime, assuming that spectator decays dominate. In order to obtain an average τ_B for the four, lepton-only analyses, TASSO's inclusive charged track method, and the HRS result, the measurements are assumed to have Gaussian errors. Then taking statistical and systematic errors in quadrature, the weighted average is $\tau_B = (1.11 \pm 0.16) \cdot 10^{-12}$ s.

4. FLAVOR MIXING

The relation (1) between the B lifetime and the K-M elements U_{qb} can be rewritten as[11]

$$|U_{cb}| = [\; \frac{BR(B \to X\ell\nu)}{\tau_B} \cdot \frac{K_{cb}}{1 + R_B} \;]^{1/2} \tag{3}$$

where $K_{cb} = (2.35 \pm 0.13) \cdot 10^{-14}$ s and R_B is from (2). Using

BR = 11.8 %, $R_B = 0$, and the average for τ_B gives $|U_{cb}| = 0.050 \pm 0.005$. An upper limit for $|U_{ub}|$ can also be obtained from the expression[11] $|U_{ub}| = |U_{cb}| \cdot [R_B/2.26]^{1/2}$. The above values give $|U_{ub}| < 0.007$ at 95% C.L. Due to uncertainties in the parameters of the relationship between τ_B and the matrix elements, for example the mass of the b quark in K_{cb} and model dependencies in the determination of R_B, an additional error of ± 0.005 for $|U_{cb}|$ should be considered.

From the value of $|U_{cb}|$, it follows that $k_c \simeq 0.4$ and $X_{KM}^b \simeq 10^3$ in relation (1). This is a large factor (highly suppressed transition), even when compared to the Cabibbo suppression of strange particle weak decays. Thus, for a unitary K-M matrix, a strong coupling between the top and bottom quarks may be expected. The long B lifetime may also allow future experiments to tag these hadrons by the displacements of their decay products. This could provide an important experimental passage to more detailed studies of the weak interaction. For example, the ability to tag B hadrons and the existence of strong couplings amoung the heavier quarks may enhance the experimental success of top and Higgs physics.

Acknowledgements

I would like to thank all of the contributing collaborations from PEP and PETRA for their assistance in collecting the results presented in this paper. I thank Prof. Sau Lan Wu for her support and encouragement in these studies. I have benefited from discussions and material from Dr. Harsh Venkataramania. The University of Wisconsin students on TASSO helped correct the text. It is also a pleasure to thank the DPF organizers and hosts for an enlightening conference.

References

1. Supported by the US Department of Energy contract DE-AC02-76ER00881 and the US National Science foundation grant INT-8313994 for travel.
2. JADE Collab., W. Bartel et al., Phys.Lett. 114B, 71 (1982); JADE Collab., P. Steffen, contribution to the Int. Conf. High Energy Physics, Leipzig (1984).

3. MAC Collab., E. Fernandez et al., Phys.Rev.Lett. 51, 1022 (1983).

4. MARK II Collab.,N.S.Lockyer et al., Phys.Rev.Lett 51, 1316(1983).

5. DELCO Collab., D.E. Klem et al., Phys.Rev.Lett 53, 1873 (1984).

6. TASSO Collab., M.Althoff et al., Phys.Lett. 149B, 524 (1984).

7. HRS Collab., contribution to this conference. Talk by D.R. Rust.

8. R. Ruckl, Weak Decays of Heavy Mesons, CERN-TH.4013/84 and CERN-Preprint (1983).

9. M.K. Gaillard and L. Maiani, Proc. Summer Institute on Quarks and Leptons, Cargese, 1979, Plenum Press, 433 (1980).

10. M. Kobayashi and K. Maskawa, Prog. Theor. Phys. 49, 652 (1973).

11. J. Lee-Franzini, invited talk at the Europhysics Study Conference on Flavor Mixing in Weak Interactions, Erice, March (1984).

12. C. Klopfenstein et al., Phys.Lett. 130B, 444 (1983).

13. A. Chen et al., Phys.Rev.Lett. 111B, 1084 (1984).

14. D.B. MacFarlane, Presentation at the XXth Recontre de Moriond, Les Arcs, France 1985.

15. P.H. Ginsparg, S.L. Glashow and M.B. Wise, Phys.Rev.Lett. 50, 1415 (1983).

16. F.J. Gilman and J.S. Hagelin, Phys.Lett. 126B, 111 (1983). Ginsparg and M.B. Wise, Phys.Lett. 127B, 265 (1983).

17. L.L. Chau, W.Y. Keung, M.D. Tran, Phys.Rev. D27, 2145 (1983). E.A.Paschos, B.Stech and U.Tuerke, Phys.Lett. 128B, 240 (1983).

18. JADE Collab., J. Spitzer, private communication.

19. MAC Collab., W.T. Ford, contribution to Aspen Winter Physics Conference (1985). Also U. of Colorado preprint COLO-HEP-87.

20. MARK II Collab., Contribution to the Physics in Collision IV Conference, Santa Cruz, California, August 1984. See also report by J.A. Jaros, SLAC-PUB-3519.

21. L.J. Golding, Ph.D. Thesis, LBL (1985).

22. DELCO Collab., D.E. Klem, private communication.

23. TASSO Collab., Contribution to European Physical Society Meeting on High Energy Physics, Bari, Italy (1985). Bari Conference paper P07-34.

A REVIEW OF RECENT RESULTS ON B DECAYS

Abolhassan Jawahery

Physics Department, Syracuse University

Syracuse, New York 13210

1. INTRODUCTION:

Weak decays of b-flavored mesons provide a valuable tool for studying properties of the standard model. In the standard model, the b quark is a member of the third weak iso-spin doublet (t,b'), where the b' is related to the mass eigenstates (d,s,b) via the Kobayashi-Maskawa mixing matrix V_{ij},

$$\begin{vmatrix} c_1 & s_1 c_3 & s_1 s_3 \\ -s_1 c_2 & c_1 c_2 c_3 - s_2 s_3 e^{i\delta} & c_1 c_2 s_3 + s_2 c_3 e^{-i\delta} \\ s_1 s_2 & -c_1 s_2 c_3 - c_2 s_3 e^{i\delta} & -c_1 s_2 s_3 + c_2 c_3 e^{i\delta} \end{vmatrix}$$

where $c_i = \cos(\theta_i)$ and $s_i = \sin(\theta_i)$. θ_1, θ_2, θ_3 and δ are the four parameters of the KM matrix.[1] In this framework the b quark can couple to the u or c quarks with relative coupling strengths of $V_{ub} = s_1 s_3$ and $V_{cb} = c_1 c_2 s_3 + s_2 c_3 e^{-i\delta}$. A major goal of the experimental study of B decays is to determine the ratio $|V_{ub}/V_{cb}|$. The measurement of the b life time, $\tau_b \propto (|V_{ub}|^2 + f_{ps}|V_{cb}|^2)$, combined with the measurement of the ratio of the partial widths, $\Gamma(b \to u l \nu)/\Gamma(b \to c l \nu) = (|V_{ub}|)^2/(f_{ps}|V_{cb}|)^2$ would determine V_{ub} and V_{cb}. In the above relations, f_{ps} is a phase-space factor.

On the experimental front much of the information about b quarks is extracted from studies of B meson decays. The B meson, which is composed of a b quark and a light anti-quark, is believed to decay predominantly through the so-called spectator mechanism. In this scheme the B meson decay is described simply as the decay of the b to a c or u quark and a virtual W and the subsequent formation of final state particles from fragments of the virtual W and the spectator light quark (Fig. 1a). Studies of D decays have shown that the spectator diagram alone is not sufficient in describing the experimental data. It is therefore expected that other processes such as W-exchange and annihilation diagrams (Figs. 1 b, and c) also contribute at some level to the decay of the B meson.[2]

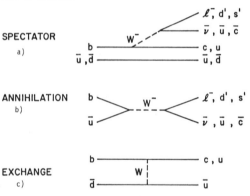

Fig. 1: B meson decay diagrams

In this report, I will discuss the experimental status of the determination of $|V_{ub}/V_{cb}|$, measurement of the color mixed process $B \to \psi X$, observation of the decay $B \to \phi X$ and recent results on B^0-$\bar{B}^0$ mixing. CLEO, CUSB and ARGUS groups have conducted B decay studies using data acquired from electron-positron annihilations at the $\Upsilon(4S)$ resonance at CESR and DORIS-II. The $\Upsilon(4S)$, which is a $b\bar{b}$ bound state, is above the kinematic threshold for decay into $B\bar{B}$, and thereby provides a valuable factory for B meson production.[3]

2. DETERMINATION OF $|V_{ub}/V_{cb}|$

CLEO measurements of the yield of K's, D^0's and D^{*+}'s have shown
the dominance of b→c in B meson decays.[4] However, because theoretical
evaluations of these processes suffer from large uncertainties one can
not set a realistic limit on b→u transitions using these measurements.
A sensitive indicator of b→u is the end-point of the lepton spectrum.
In the decay B→Xlν the end point of the lepton spectrum is sensitive to
the mass of the hadronic system m_x. For b→clν the minimum value of m_x
is the D mass at 1.860 GeV/c^2, while for b→ulν the hadronic system can
be substantially lighter. Thus, the latter results in a harder lepton
momentum distribution than that expected for b→clν.

In order to apply this method, one requires a model of the
semileptonic decays of the B meson. A description of the spectator
model, known as the Altarelli model,[5] has been used to obtain the
lepton momentum spectrum for B→X_clν and B→X_ulν, where X_c and X_u are
hadronic systems for b→c and b→u transitions respectively. This model
includes lowest order QCD corrections to the spectator diagram. Effects
due to the bound state structure of the meson are taken into account by
attributing a Fermi momentum $\vec{P}$, distributed as $|\vec{P}|^2\exp(-P^2/2P_f^2)$, to
the b quark in the B rest frame and folding this motion with the
spectrum from the quark decay. In addition to the spectrum for primary
leptons, one requires the lepton momentum distribution from the
semileptonic decays of secondary D's. This is determined by applying
the Altarelli model to the semileptonic decays of D's and folding the
resulting momentum spectrum with the measured momentum distribution of
D's from B decay.

Recently the CLEO group has used its entire data sample at the
$\Upsilon(4S)$, which consist of 73 pb^{-1} collected during 1983 and 1985,
combined with 31 pb^{-1} continuum data to obtain the momentum spectra of
electrons and muons from B decays. The spectra are shown in Fig. 2.
Theoretical curves are computed for P_f=0.150 GeV/c. The solid curve is

the spectator model calculation for B→X$_c$lν. Dash-dot curves show separately the distributions for primary leptons from B→X$_c$lν and secondary leptons from B→D→Xlν and the dashed curve is a representation of B→X$_u$lν for R=Γ(b→u)/Γ(b→c)=.2.

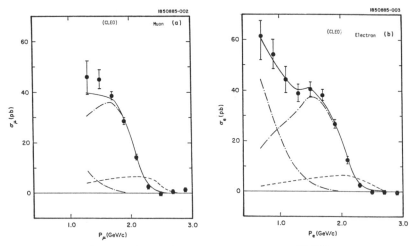

Fig. 2: Lepton momentum spectrum (CLEO).

The experimental spectrum is consistent with 100% b→clν. A fit to the B→X$_c$lν spectrum yields the semileptonic branching ratios of the B meson, B(B→Xμν)=.108±.004 and B(B→Xeν)=.112±.004.

The spectrum was also fit to a function which was a linear combination of spectra for B→X$_c$lν and B→X$_u$lν and secondary leptons from D decays. This analysis yields R=Γ(b→u)/Γ(b→c)=-.05±.02 which at 90% confidence level corresponds to Γ(b→u)/Γ(b→c)<0.01. Note that this result is significantly lower than the previous limits of .04 and .055 reported by CLEO[6] and CUSB[7] groups. However, the significant deviation of the above value of R from zero indicates possible inadequacy of the theoretical description of the lepton spectrum. By varying parameters of the model, the CLEO group finds that the spectrum is most sensitive to the value of P$_f$, which is the least well understood parameter of the model. The best fit to the spectrum corresponds to $P_f=0.250\pm0.035$ GeV/c, giving $\Gamma(b\rightarrow u)/\Gamma(b\rightarrow c)<.03$ at 90% confidence level. A similar method has been used by the ARGUS group to set a limit on R (Fig. 3).

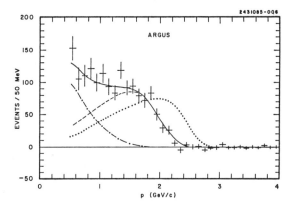

Fig. 3: Electron momentum spectrum (ARGUS).

A summary of the results from all three groups is given in Table 1.

Table 1

Upper limit on $R=\Gamma(b\rightarrow u)/\Gamma(b\rightarrow c)$ (90% c.l)

ARGUS	0.04
CUSB[7]	0.055
CLEO	0.03

The upper bound of 3% on the ratio of the partial widths $\Gamma(b\rightarrow u l\nu)/\Gamma(b\rightarrow c l\nu)=(|V_{ub}|)^2/(0.45|V_{cb}|)^2$ leads to a limit on $|V_{ub}/V_{cb}|<.12$ at 90% confidence level. Using this limit in conjunction with the measurements of the b-lifetime[8] $\tau_b=(1.11\pm.16)\times10^{-12}$ and relations,

$$\Gamma(b\rightarrow c)=(1+R)^{-1}\cdot(B(B\rightarrow Xl\nu)/\tau_b)$$

$$\text{and } \Gamma(b\rightarrow c)=(G_F^2 m_b^5/192\pi^3)\cdot(p|V_{bc}|^2)$$

one finds the values of $|V_{cb}|=.05\pm.004$ and $|V_{ub}|<.006$. The limit on $|V_{ub}|=s_1s_3$ combined with the value of $s_1=.231\pm.003$ restrict the value of $s_3<0.06$.

It should be noted that the current limit on R is strongly dependent on the value of the free parameter P_f and increases for larger values of P_f. As a consequence, further improvement in this measurement requires more accurate theoretical analysis of the problem.

In attempting to find examples of b→u transitions, the CLEO group has searched for two body non-charm final states of the B meson. They find no clear signal above the background and set limits on branching ratios of B to several non-charm exclusive final states (Table 2).

Table 2

Upper limits on B to exclusive non-charm final states

Decay modes	Upper limit on branching ratio (%)
$B^0 \to \pi^+\pi^-$	.02
$B^- \to \rho^0\pi^-$	.02
$B^0 \to \pi^\pm A_1$	.24
$B^0 \to \pi^\pm A_2$	.20
$B^- \to \rho^0 A_1^-$	.08
$B^- \to \rho^0 A_2^-$	.04

Because no reliable theoretical evaluations of these processes have yet been done one can not, at this point, relate these limits to the ratio of b→u to b→c transitions.

3 OBSERVATION OF THE DECAY B→ψX

The CLEO and ARGUS groups have recently detected the decay B→ψx. Because it relates to the so-called color mixed diagrams in B decay, the observation of this decay is of considerable importance.[9] The spectator diagram for the B→ψx transition is shown in Fig. 4.

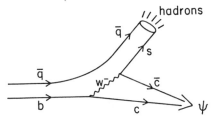

Fig. 4: Color mixed diagram for B→ψX

In the spectator scheme, the production of ψ in B decay is a result of an interference between the c$\bar{\text{s}}$ from the virtual W decay and the c$\bar{\text{q}}$ system. In the absence of QCD corrections, the color mismatch between the two systems would result in a suppression of 1/9 in the transition rate.[1] It is however argued that soft gluon exchanges could eliminate this explicit color mismatch.[10]

Both the CLEO and ARGUS groups have detected ψ's via ψ→μ$^+$μ$^-$ and ψ→e$^+$e$^-$. They find a clear ψ signal in the invariant mass specrta of μ$^+$μ$^-$ and e$^+$e$^-$ from Υ(4S) data (Figs.5 and 6). Having studied their continuum data sample, the CLEO group has determined that the continuum contribution to the ψ signal is negligible. They have also determined that the non-resonant background in the dilepton mass spectrum can be accounted for by a combination of the known processes B→Xlν, D→Xlν, B→D→Xlν and fake leptons. The inclusive branching ratio of B→ψX is measured to be 1.37±.6% (ARGUS) and 1.07±.18±.22% (CLEO).

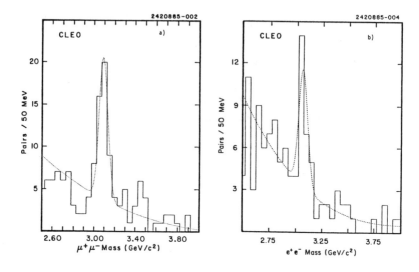

Fig. 5: Invariant mass distribution of a) $\mu^+\mu^-$ and b) e^+e^- (CLEO)

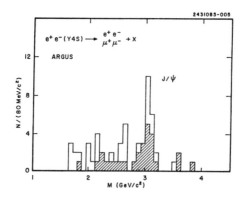

Fig 6: Invariant mass distribution of $\mu^+\mu^-$ and e^+e^- (ARGUS)

    This result is consistent with an evaluation of the spectator diagram with full color suppression (.6 to 2.4%).[11] However, it is pointed out that given the theoretical uncertainties in computing the rate for B→ψX, one can also accommodate this result in models with no color suppression. It is clear that more theoretical work is needed in this area.

In addition to measuring the inclusive branching ratio of B→ψX, both groups have looked for exclusive decays B→ψK(nπ). The invariant mass distributions of the combinations are shown in Fig. 7.

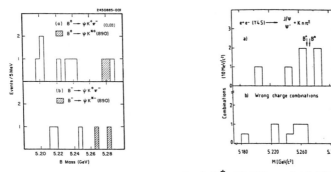

Fig: 7. Invariant mass spectra of a)ψK* (CLEO) and b)ψK(nπ) (ARGUS)

The CLEO group finds four candidate events consistent with B→ψK*. The ARGUS data shows four events at the B mass with an estimated background of 1 event.

4. OBSERVATION OF THE DECAY B→φX

Observation of the decay B→φX is reported by the CLEO group. A possible scheme for φ production in B decays is shown in Fig. 8. In this picture φ's are produced as the decay products of secondary F's in B decay.

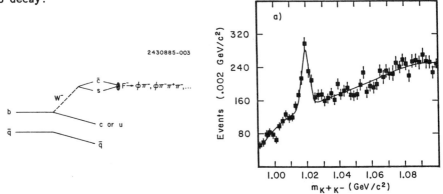

Fig. 8: A possible Scheme for B→φX   Fig. 9: K⁺K⁻ invariant Mass (CLEO).

The invariant mass of K^+K^- pairs for the $\Upsilon(4S)$ data, shown in Fig. 9, exhibits a clear signal at the ϕ mass. After subtracting the continuum contribution and correcting for the efficiencies, they find $B(B \to \phi X) = .025 \pm .006 \pm .005$. Using the observed decay modes of D to ϕ and B to D,[4] they estimate $B(B \to FX).B(F \to \phi X) < .025$ at 90% confidence level.

5. $B^0 - \bar{B}^0$ MIXING

Particle-antiparticle mixing has only been observed in the K^0 system. In the standard model, analogous to the K^0 system, a B^0 can transform to a $\bar{B}^0$ through the so-called box diagram (Fig. 10).[11]

Fig. 10: Box diagram $B \to \bar{B}$ 0050185-003

Mixing is described by the parameter r defined as:

$$r = (\delta m/\Gamma)^2 / [2 + (\delta m/\Gamma)^2],$$
$$\delta m/\Gamma \propto Bf^2 m_t^2 |V_{t(s,d)}|^2,$$

Where $\delta m = m_{B^0} - m_{\bar{B}^0}$ and Γ is the total width of B^0. The KM matrix elements $V_{ts} = |-c_1 s_2 s_3 + c_2 s_3 e^{i\delta}|$ and $V_{td} = s_1 s_2$ determine the coupling of t to s and d quarks. These relations indicate that mixing may be larger in the $B(b\bar{s})$ system as compared with $B(b\bar{d})$.

The CLEO group have searched for like-sign dileptons in the $\Upsilon(4S)$ decays as a signal for mixing in the B^0 ($b\bar{d}$) system. The mixing parameter in this case is defined as $y = (N_{l^-l^-} + N_{l^+l^+})/N_{l^+l^-}$. Note that $y = 2r/(1+r^2)$ is zero for no mixing and is one for complete mixing.[13]

The dominant sources of the observed dileptons in the $\Upsilon(4S)$ decays are a) parallel decays where both B's decay as $B \rightarrow Xl\nu$, b) cascade decays where $B \rightarrow Dl\nu$ is followed by $D \rightarrow Xl\nu$ and b) hadrons faking a lepton signal. Comparing the observed number of dileptons with that expected from the above sources in the absence of mixing, they find no excess of like-sign dileptons and set upper limit on the B^0-$\bar{B}^0$ mixing.

Because the $\Upsilon(4S)$ decays to both B^+B^- and $B^0\bar{B}^0$ pairs, the limit on the $B^0\bar{B}^0$ mixing would depend on the parameter $R_B=(f_0/f_\pm) \cdot (B(B^0 \rightarrow Xl\nu)/B(B^\pm \rightarrow Xl\nu)^2$, where $f_0=.4$ and $f_\pm=.6$ are fractions of neutral and charged B's produced in the $\Upsilon(4S)$ decays. In Fig. 11, the 90% confidence level upper limit on $B^0\bar{B}^0$ mixing is plotted as a function of the ratio $B(B^0 \rightarrow Xl\nu)/B(B^\pm \rightarrow Xl\nu)$. For equal semileptonic branching ratios y is less than .3.

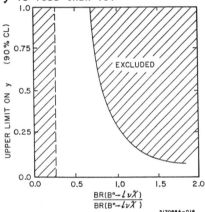

Fig. 11: Upper limit on B^0-$\bar{B}^0$ mixing

6. CONCLUSION

Recent studies of the lepton momentum spectrum from semileptonic decays of B's have shown that the experimental limit on $R=\Gamma(b \rightarrow u)/\Gamma(b \rightarrow c)$ is subject to the uncertainties in the theoretical description of the B decays. Any improvement in the current limit on $R<.03$ requires more accurate theoretical understanding of the probelm. The inclusive branching ratio for the process B→ψX is measured. This allows various models of the color mixed processes to be tested. The decay B→φX is detected and its branching ratio is measured to be B(B→φX)=.025±.006±.005. The CLEO data has shown that if B(B⁰→Xlν)=B(B±→Xlν), mixing in the B(bd̄) is less than 30%.

I would like to thank M. Hempstead, J. Kandaswamy, R. Namjoshi and G. C. Moneti for help in preparation of this manuscript.

REFERENCES:
1) M. Kobayashi and T. Maskawa, Prog. Theor. Phys. 652(1973)
2) For a detailed review see R. Ruckl, "Weak Decays of Heavy Mesons", CERN-TH.4013/84 and CERN-Preprint (1983).
3) D. Andrews et al., Phys. Rev. Lett. 45, 219(1980)
4) A. Brody et al., Phys. Rev. Lett., 48, 1070(1982)
   J. Green et al., Phys. Rev. Lett. 51, 347(1983)
   S.E. Csorna et al., Phys. Rev. Lett.,54, 1894(1985)
5) G. Altarelli et al., Nucl. Phys. B208, 365(1982)
6) A. Chen et al., Phys. Rev. Lett. 52, 1084(1984)
7) C. Klopfenstein et al., Phys. Lett., 130B, 444(1983)
8) W. T. Ford, "Charm and Heavy Flavor Lifetimes", Talk given at Aspen Winter physics Conference, 1985.
9) H. Fritzsch, Phys. Lett. 86b, 164(1979)
10) K. Jaganathan, Private Communication
11) J. H. Kuhn and R. Ruckl Phys. Lett. 135B, 477(1984)
    D. H. Cox et al., University of Alabama Preprint (1985)
12) For a review see A. J. Buras et al., MPI-PAE/PTH 7/84
13) P. Avery et al., Phys. Rev. Lett. 53, 1309(1984)

CP VIOLATION IN KAON DECAYS

Mark B. Wise

California Institute of Technology, Pasadena, CA 91125

CP violation has been observed in the neutral kaon system. Nonzero values for the quantities

$$\eta_{+-} \equiv \frac{\langle \pi^+\pi^- | H^{|\Delta S|=1} | K_L \rangle}{\langle \pi^+\pi^- | H^{|\Delta S|=1} | K_S \rangle}, \qquad \eta_{00} \equiv \frac{\langle \pi^0\pi^0 | H^{|\Delta S|=1} | K_L \rangle}{\langle \pi^0\pi^0 | H^{|\Delta S|=1} | K_S \rangle}, \qquad (1)$$

indicate that CP is violated. Experimentally η_{+-} and η_{00} are about 10^{-3}. The parameters η_{+-} and η_{00} depend on CP violation in the first order weak ($\sim G_F$) kaon decay amplitudes

$$\langle \pi\pi(I = 0) | H^{|\Delta S|=1} | K^0 \rangle = iA_0 e^{i\delta_0}, \qquad (2a)$$

$$\langle \pi\pi(I = 2) | H^{|\Delta S|=1} | K^0 \rangle = iA_2 e^{i\delta_2}. \qquad (2b)$$

In the phase convention A_0 real an imaginary part for A_2 is a signal for CP violation. CP violation in kaon decay amplitudes is usually characterized by

$$\varepsilon' \equiv \frac{i}{\sqrt{2}} e^{i(\delta_2 - \delta_0)} \frac{\text{Im}A_2}{A_0} \simeq \frac{1}{\sqrt{2}} e^{i\pi/4} \frac{\text{Im}A_2}{A_0}. \qquad (3)$$

η_{+-} and η_{00} also depend on CP violation in second order weak ($\sim G_F^2$) $K^0 - \bar{K}^0$ mixing. The states K_S and K_L are

$$K_S \equiv \frac{1}{[2(1 + |\varepsilon|^2)]^{1/2}} [(1 + \varepsilon)K^0 - (1 - \varepsilon)\bar{K}^0] \qquad (4a)$$

$$K_L \equiv \frac{1}{[2(1 + |\varepsilon|^2)]^{1/2}} [(1 + \varepsilon)K^0 + (1 - \varepsilon)\bar{K}^0]. \qquad (4b)$$

The quantity ε determines the deviation of K_S and K_L from CP eigenstates and is a measure of CP violation. In the convention A_0 real

$$\varepsilon \cong \frac{e^{i\pi/4}}{\sqrt{2}} \frac{\mathrm{Im}\, M_{12}}{m_{K_S} - m_{K_L}}, \tag{5}$$

where

$$M_{12} = \langle \bar{K}^0 | H^{|\Delta S|=2} | K^0 \rangle / \langle \bar{K}^0 | K^0 \rangle. \tag{6}$$

The CP violation parameters η_{+-} and η_{00} can be expressed simply in terms of ε and ε'

$$\eta_{+-} = \varepsilon + \varepsilon' \tag{6a}$$

$$\eta_{00} = \varepsilon - 2\varepsilon'. \tag{6b}$$

The CP violation observed at the present time is consistent with $\varepsilon' = 0$. The CP violation parameter ε is sensitive to physics beyond the standard model associated with a very large mass scale M_x. More explicitly, if the effective Hamiltonian for $K^0 - \bar{K}^0$ mixing contained the SU(3) × SU(2) × U(1) invariant operator

$$O = \frac{i}{M_x^2} (\bar{c}_L \gamma^\mu u_L + \bar{s}_L \gamma^\mu d_L)(\bar{s}_R \gamma_\mu d_R) + h.c. \tag{7}$$

then the experimental value for ε demands that

$$M_x \gtrsim 10^8\, GeV. \tag{8}$$

Physics associated with a mass scale of order $10^8\, GeV$ could not appreciably influence first order weak amplitudes like A_0 and A_2. So if the CP violation observed in ε were due to new physics beyond the standard model associated with a tremendously large mass scale M_x then $\varepsilon' \cong 0$. Measurement for a nonzero value for ε' would be very important since it would eliminate this possibility.

Since CP violation is sensitive to new physics associated with a very large mass scale it is a good place to look for deviations from the standard model. In the standard model with minimal particle content CP can be violated in only two ways. The charged W-bosons coupled to the quarks through the weak current

$$J_\mu^{(+)} = \frac{g_2}{2\sqrt{2}}(\bar{u}, \bar{c}, \bar{t})\gamma_\mu(1 - \gamma_5)U\begin{bmatrix} d \\ s \\ b \end{bmatrix}. \tag{9}$$

Here U is a 3×3 unitary matrix that arises from the diagonalization of the quark mass matrices. By adjusting the phases of the quark fields U can be written as

$$U = \begin{bmatrix} c_1 & -s_1 c_3 & -s_1 s_3 \\ s_1 c_2 & c_1 c_2 c_3 - s_2 s_3 e^{i\delta} & c_1 c_2 s_3 + s_2 c_3 e^{i\delta} \\ s_1 s_2 & c_1 s_2 c_3 + c_2 s_3 e^{i\delta} & c_1 s_2 s_3 - c_2 c_3 e^{i\delta} \end{bmatrix}, \tag{10}$$

where $c_i \equiv \cos\vartheta_i$, $s_i \equiv \sin\vartheta_i$, $i\in\{1,2,3\}$ and the angles ϑ_1, ϑ_2 and ϑ_3 lie in the first quadrant. In general the phase δ cannot be removed from the matrix U by a redefinition of the quark fields so it is a source of CP violation[1]. The other source of CP violation in the standard model is the strong interaction vacuum angle $\bar{\vartheta}$. However the upper limit on the electric dipole moment of the neutron implies that[2].

$$\bar{\vartheta} \lesssim 10^{-9}. \tag{11}$$

Strong interaction CP violation is much too small to be the source of the CP violation observed in the neutral kaon system

If any of the angles ϑ_i are zero then it is possible to remove the phase δ from the matrix U be a redefinition of the quark fields. Therefore the CP violating quantities ε and ε' are proportional to the sines of all the mixing angles in addition to that of the phase δ. Experimental information on neutron β decay and semileptonic hyperon decays give (for small s_3)

$$s_1 \cong 0.22. \tag{12}$$

The weak mixing angles ϑ_2 and ϑ_3 are constrained by experimental information on B meson decays. Since the b-quark is heavy compared with the typical strong interaction scale the rate for semileptonic B-meson decay can be approximated by the decay rate of the free b-quark

$$\Gamma(B \to e\bar{\nu}x) \cong \Gamma(b \to c e\bar{\nu}) + \Gamma(b \to u e\bar{\nu}) \tag{13}$$

where

$$\Gamma(b \to c\,e\,\bar{\nu}) = |U_{cb}|^2 \frac{G_F^2 m_b^5}{192\,\pi^3} f(m_c/m_b) \tag{14a}$$

$$\Gamma(b \to u\,e\,\bar{\nu}) = |U_{ub}|^2 \frac{G_F^2 m_b^5}{192\,\pi^3} \, . \tag{14b}$$

and

$$f(x) = 1 - 8x^2 + 8x^6 - x^8 - 24x^4 \ln x \, . \tag{15}$$

With $m_b = 4.8\,GeV$ and $m_c = 1.5\,GeV$ these eqs. imply

$$\Gamma(b \to c\,e\,\bar{\nu}) = 4.3|U_{cb}|^2 \times 10^{13} s^{-1} \tag{16a}$$

$$\Gamma(b \to u\,e\,\bar{\nu}) = 8.7|U_{ub}|^2 \times 10^{13} s^{-1} \, . \tag{16b}$$

These can be compared with experiment through

$$\Gamma(b \to c\,e\,\bar{\nu}) = Br(b \to c) \frac{Br(B \to e\,\bar{\nu}x)}{\tau_B} \tag{17a}$$

$$\Gamma(b \to u\,e\,\bar{\nu}) = Br(b \to u) \frac{Br(B \to e\,\bar{\nu}x)}{\tau_B} \, . \tag{17b}$$

Using the experimental result

$$Br(B \to e\,\bar{\nu}x) = 11.6\% \, ,$$

and writing

$$Br(b \to u) = 0.04y \, ,$$

the above imply

$$|U_{cb}|^2 \cong 3 \times 10^{-3}(10^{-12}s/\tau_b) \tag{18a}$$

$$|U_{ub}|^2 \cong 5 \times 10^{-5}y(10^{-12}s/\tau_b) \, . \tag{18b}$$

We also know that $y \lesssim 1$ and $\tau_B \cong 10^{-12}s$ so the above imply that s_2 and s_3 are quite small. To leading nontrivial order in the small angles the above become

$$\{s_2^2 + s_3^2 + 2s_2 s_3 c_\delta\} = 3 \times 10^{-3}(10^{-12}s/\tau_B) \tag{19a}$$

$$\{s_3^2\} = 1y \times 10^{-3}(10^{-12}s/\tau_B) \, . \tag{19b}$$

Our goal is to compare standard model predictions for ε and ε' with experiment. To do this we must compute $H^{|\Delta S|=2}$ and $H^{|\Delta S|=1}$ and then take the relevant matrix elements of these effective Hamiltonians.

The effective Hamiltonian for $K^0 - \bar{K}^0$ mixing has been computed in the leading logarithmic approximation by successively treating the W-boson, t-quark, b-quark and c-quark as heavy and integrating them out of the theory. The result, to leading nontrivial order in the large masses is[3]

$$H^{|\Delta S|=2} = \frac{G_F^2}{16\pi^2} s_1^2 (\bar{s}_\alpha \gamma_\mu (1 - \gamma_5) d_\alpha)(\bar{s}_\beta \gamma^\mu (1 - \gamma_5) d_\beta) m_c^2$$

$$\bullet \; \{\eta_1 c_2^2 (c_1 c_2 c_3 - s_2 s_3 e^{-i\delta})^2 + \eta_2 s_2^2 (c_1 s_2 c_3 + c_2 s_3 e^{-i\delta})^2 (m_t / m_c)^2$$

$$+ \; 2\eta_3 s_2 c_2 (c_1 c_2 c_3 - s_2 s_3 e^{-i\delta})(c_1 s_2 c_3 + c_2 s_3 e^{-i\delta})\ln(m_t^2 / m_c^2)\} + h.c. \qquad (20)$$

Here η_1, η_2 and η_3 are QCD correction factors. They are roughly independent of the top quark mass and have the values $\eta_1 \cong 0.7$, $\eta_2 \cong 0.6$ and $\eta_3 \cong 0.4$ (using $m_c = 1.5\,GeV$, $m_b = 4.8\,GeV$, $M_W \cong 85\,GeV$, $\Lambda_{QCD} = 0.1\,GeV$ and $\alpha_s(\mu^2) = 1$).

In eq. (20) long distance contributions the the effective Hamiltonian have been neglected since they do not contain a factor of a heavy quark mass squared. A typical long distance contribution is the time ordered product of two effective Hamiltonians for $|\Delta S| = 1$ weak nonleptonic kaon decays. The contribution of this operator might well dominate $\mathrm{Re}\,M_{12}$. Its contribution to $\mathrm{Im}\,M_{12}$ involves CP violation in $|\Delta S| = 1$ amplitudes and is expected to be of order $20\varepsilon'$. The twenty comes from the fact that these amplitudes are not necessarily suppressed by the $\Delta I = \frac{1}{2}$ rule. Anticipating a modest improvement in the present limit on $|\varepsilon'/\varepsilon|$ I shall neglect the long distance contributions to $\mathrm{Im}\,M_{12}$. Then to leading nontrivial order in the small angles

$$\varepsilon \cong \frac{-s_1^2 B G_F^2 f_\pi^2 m_K^2 m_c^2}{16\sqrt{2}\,\pi^2 (m_{K_S} - m_{K_L})} s_2 s_3 s_\delta [-\eta_1 + \eta_3 \ln(m_t^2 / m_c^2)$$

$$+ \; \eta_2 (m_t^2 / m_c^2)(s_2^2 + s_2 s_3 c_\delta)] e^{i\pi/4} . \qquad (21)$$

Here f_π is the pion decay constant and B is defined by

$$\langle \bar{K}^0 | \int d^3x \, O'(x) | K^0 \rangle = B f_\pi m_K^3$$

$$O' = (\bar{s}_\alpha \gamma^\mu (1 - \gamma_5) d_\alpha)(\bar{s}_\beta \gamma^\mu (1 - \gamma_5) d_\beta) . \qquad (22)$$

Under chiral $SU(3)_L \times SU(3)_R$, O' transforms as $(27_L, 1_R)$. The effective Hamiltonian for $|\Delta S| = 1$, $|\Delta I| = 3/2$ weak nonleptonic decays is

$$H\begin{vmatrix}\Delta S|=1\\\Delta I|=3/2\end{vmatrix} = \frac{-G_F}{2\sqrt{2}} s_1 c_2 c_3 CO \,, \tag{23a}$$

where

$$O = (\bar{s}_\alpha \gamma_\mu (1 - \gamma_5) d_\alpha)(\bar{u}_\beta \gamma^\mu (1 - \gamma_5) u_\beta) + (s_\alpha \gamma_\mu (1 - \gamma_5) u_\alpha)(\bar{u}_\beta \gamma^\mu (1 - \gamma_5) d_\beta)$$

$$- (\bar{s}_\alpha \gamma_\mu (1 - \gamma_5) d_\alpha)(\bar{d}_\beta \gamma^\mu (1 - \gamma_5) d_\beta) \,, \tag{23b}$$

and C is a QCD correction factor which for the parameters used previously has the value 0.4. The operator O also transforms as $(27_L, 1_R)$ and so chiral perturbation theory relates the $K^+ \to \pi^+ \pi^0$ matrix element of O to the $K^0 - \bar{K}^0$ matrix element of O'[4]

$$\langle \pi^+ \pi^0 | H\begin{vmatrix}\Delta S|=1\\\Delta I|=3/2\end{vmatrix} | K^+ \rangle = \frac{-3iG_F}{8} Cm_K^3 s_1 c_1 c_3 B \times \{1 + \frac{55}{6} \frac{m_K^2}{(4\pi f_\pi)^2} \ln(\frac{m_K^2}{\mu^2}) + \cdots \}.$$

$$\tag{24}$$

Unfortunately the chiral perturbation expansion does not seem well enough behaved to trust the leading order prediction (which is $|B| \simeq 0.37$).

The expression for ε can be used to get information on the phase δ. For example, if $c_\delta < 0$,

$$|s_2 s_3 s_\delta| > \frac{0.06 |0.37/B|}{\{-0.7 + 0.4 \ln(m_t^2/m_c^2) + 0.6(\sqrt{3} + \sqrt{y})^2 10^{-3}(10^{-12} s/\tau_\beta)(m_t^2/m_c^2)\}} \,. \tag{25}$$

(For $c_\delta > 0$ there is a more stringent bound.) To get a feeling for what this implies note that for $|0.37/B| = 1/2$, $\tau_B = 10^{-12} s$, $m_t = 45 \, GeV$ and $y = 1$ the above implies

$$|6 s_2 s_3 s_\delta| \gtrsim 5 \times 10^{-3} \,. \tag{26}$$

Given the constraints on s_2 and s_3 from B decays it is clear that s_δ must be of order unity for the CP violation in the standard model to account for the measured value of ε. An improvement in our ability to compute B and in our knowledge of m_t, $\Gamma(b \to u)/\Gamma(b \to c)$ and τ_B could show that the standard model with minimal particle content is not compatible with the observed CP violation in kaon decays.

The other CP violation parameter ε' depends on the effective Hamiltonian for $|\Delta S| = 1$ weak nonleptonic decays. Since the CP violating phase δ only enters in the coupling of the heavy quarks CP violation only arises from Penguin type diagrams.

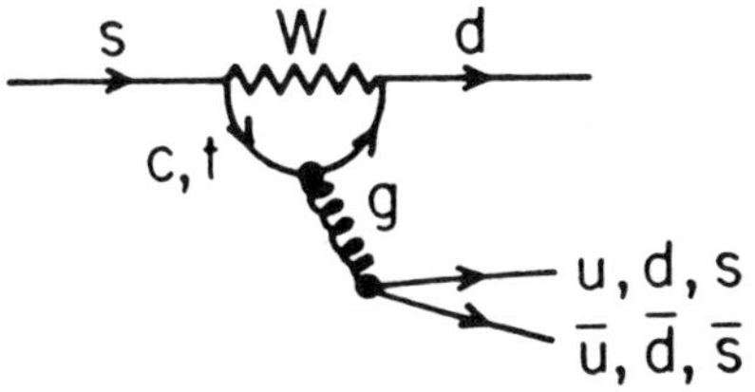

Fig. 1

These diagrams give rise to operators in the effective Hamiltonian for $|\Delta S| = 1$ weak nonleptonic decays that have a $(V - A) \times (V + A)$ chiral structure. For example,

$$Q_6 = (\bar{s}_\alpha \gamma_\mu (1 - \gamma_5) d_\beta)[(\bar{u}_\beta \gamma^\mu (1 + \gamma_5) u_\alpha) + (\bar{d}_\beta \gamma^\mu (1 + \gamma_5) d_\alpha) + (\bar{s}_\beta \gamma^\mu (1 + \gamma_5) s_\alpha)] \ .$$

The operator Q_6 is purely $\Delta I = 1/2$ and gives an imaginary part to A_0. The redefinition of kaon fields to comply with the phase convention A_0 real induces an imaginary part to A_2. It has been suggested that the $(V - A) \times (V + A)$ chiral structure of Q_6 causes it to have large matrix elements[5] of order $1 \, GeV^3$. If this is the case then[6]

$$|\varepsilon'/\varepsilon| \cong |6s_2 s_3 s_6 \delta| \left| \frac{\langle \pi\pi(I = 0)| \int d^3x \, Q_6(\vec{x})|K^0\rangle}{1 \, GeV^3} \right| |\tilde{C}_6/0.1| \ . \tag{27}$$

In eq. (27) $\tilde{C}_6$ is related through mixing angles to the coefficient of Q_6 in the effective Hamiltonian for $|\Delta S| = 1$ weak decays and has the value 0.1 in the leading logarithmic approximation (using the same strong interaction parameters as before and $m_t = 30 \, GeV$). Recall that for $|0.37/B| = 1/2$, $\tau_B = 10^{-12}s$, $m_t = 45 \, GeV$ and $y = 1$ we had

$$|6s_2 s_3 s_\delta| \gtrsim 5 \times 10^{-3} \ .$$

An improvement in our ability to compute the hadronic matrix elements $\langle \bar{K}^0| O'|K^0\rangle$ and $\langle \pi\pi(I = 0)| \int d^3x \, Q_6(\vec{x})|K^0\rangle$ could bring the standard model into conflict with experimental limits on ε'/ε.

REFERENCES

1. M. Kobayashi and T. Maskawa, Prog. Theor. Phys. 49, 652 (1973).

2. V. Baluni, Phys. Rev. D19, 227 (1979).

    R. Crewther, P. Di Vecchia, G. Veneziano and E. Witten, Phys. Lett. 88B, 123 (1979).

3. F. Gilman and M. Wise, Phys. Lett. 93B, 129 (1980); Phys. Rev. D27, 1128 (1983).

4. H. Bijnens, H. Sonoda and M. Wise, Phys. Rev. Lett. 53B, 2367 (1985);

    J. Donoghue, E. Golowich and B. Holstein, Phys. Lett. 119B, 412 (1982).

5. M. Shifman, A. Vainstein and V. Zakharov, JETP Lett. 22, 55 (1975); Nucl. Phys. B120, 316 (1977);

    J. Donoghue, E. Golowich, B. Holstein and N. Ponce, Phys. Rev. D23, 1213 (1981);

    A. Soni, in the Proceedings of the Oregon DPF meeting (1985).

6. F. Gilman and M. Wise, Phys. Lett. 83B, 83 (1979); Phys. Rev. D20, 2392 (1979).

LATTICE CALCULATION OF WEAK MATRIX ELEMENTS[*]

C. Bernard, T. Draper,[†] G. Hockney, and A. Soni

University of California, Los Angeles, CA 90024

USA

ABSTRACT

We present the first results from a small ($6^3 \times 10$) lattice lattice of a calculation of non-leptonic weak matrix elements. The $\Delta I = 1/2$ rule is studied as a test case. For a lattice meson of mass $\simeq$ kaon mass we find a significantly enhanced $\Delta I = 1/2$ amplitude and a $\Delta I = 3/2$ amplitude compatible with zero within our statistics. The dominance of the $\Delta I = 1/2$ amplitude appears to be due to a class of graphs called the eye graphs. Qualitatively similar results are found whether or not the charm quark is integrated out _ab_ _initio_. We also report preliminary results on other weak matrix elements.

A long standing problem in low energy hadronic physics has been the calculation of non-leptonic weak matrix elements. Prominent examples here are the $\Delta I=1/2$ rule, whose origin has remained obscure, and the CP violation parameters ε and ε' in neutral kaon decays. In fact, accurate calculations of the relevant hadronic matrix elements,

---

[*]Presented by A. Soni.

[†]University of California, Irvine, CA 92717.

coupled with existing experimental measurements of ϵ and ϵ', could provide non-trivial tests of the standard model. Lattice Monte-Carlo (MC) techniques offer a unique opportunity for performing such calculations directly from the fundamental theory. However, the efforts of the past few years suggest that practical difficulties often seriously limit attainable accuracies with these methods. It thus seems reasonable to begin by studying effects for which even qualitative results can be physically significant.[1-3] The $\Delta I = 1/2$ rule, i.e. the empirical statement that $\Delta I = 1/2$ amplitudes are enhanced over $\Delta I = 3/2$ amplitudes by a factor of ~ 20, is one such effect. Because the enhancement is so large an inaccuracy of order 50% (which is not unusual in current MC calculations) need not mask its origin. The same machinery can, of course, also be used to compute other nonleptonic weak interaction matrix elements. Here we report the first results from our lattice calculation.

To relate matrix elements amenable to a MC calculation to the experimentally measured $K \to \pi\pi$ amplitudes, three key theoretical ingredients are used: the operator product expansion and renormalization group (OPE/RG), lattice weak coupling perturbation theory, and chiral perturbation theory[4] (CPTh). The OPE/RG is required because the characteristic scale for weak interactions is mass $M_W \sim 80$ GEV of the W boson, while the lattice ultraviolet cutoff (dictated by low energy hadronic physics and computer time) is $\pi/a \sim 3$ GeV in our calculation (a is the lattice spacing). Thus the W field has to be integrated out from the weak Hamiltonian. When performed in the continuum this procedure leads to two four-quark operators $(O_\pm)$ in which the charm quark appears as an explicit field.[5] One can also go further by integrating out the charm quark (the penguin approach).[6-8] One then has six four-quark operators Q_1-Q_6: Q_1-Q_4 are LL; Q_5, Q_6 are LR. Since the charm mass is not very large the reliability of this approximation is uncertain, but we study it anyway for purposes of comparison and because it may be useful for coarse- grained lattices. Note that for matrix elements relevant to CP violation in the standard model the top quark, at least, must be integrated out by a penguin-like approach (since $m_t \gg$

a^{-1}).

In order to use the standard results for the OPE/RG from the continuum literature, a lattice weak coupling calculation, relating matrix elements of continuum operators to their lattice counterparts, is required. Such a relation has the following generic appearance[9] (sum on $j, j \neq i$):

$$
\begin{aligned}
O_i^{cont} = & [1 + (g^2/16\pi^2) Z_{1i}(r, \mu a)] O_i^{latt} \\
& + (g^2/16\pi^2) Z_{2ij}(r, \xi a) O_j^{latt} + (g^2/16\pi^2) Z_3(r, \mu a) \\
& \times \bar{s}\gamma_\mu(1-\gamma_5) t^a d[\bar{u}\gamma_\mu t^a u + \bar{d}\gamma_\mu t^a d + \bar{s}\gamma_\mu t^a s]
\end{aligned} \tag{1}
$$

Here Z's are finite renormalization constants, r is the Wilson parameter, and μ is the continuum renormalization point. The $O(g^2)$ terms can be very important even for weak coupling because naive $O(g^0)$ LL operators can mix at this order with LR operators whose matrix elements are often much larger. Note, however, that the calculations to $O(g^2)$ should be sufficient unless new operators which appear only at $O(g^4)$ [10] (such as $\bar{s}\sigma^{\mu\nu}F_{\mu\nu}d$) turn out to have anomalously large matrix elements.

We thus wish to calculate matrix elements of the type $\langle\pi\pi|\bar{s}\Gamma_1 u\bar{u}\Gamma_2 d|K\rangle$, where Γ_i represents an arbitrary Dirac matrix. This is a four-point function and practical considerations make it difficult to evaluate on the lattice. An approximation scheme for light meson masses, namely CPTh, allows one to reduce one of the pion fields and relate the $K \to \pi\pi$ matrix element to a suitable linear combination of $K \to \pi$ and $K \to 0$ matrix elements.[4]

After replacing the remaining mesons by their interpolating fields, Green's functions such as $\langle 0|\bar{d}\gamma_5 u(x) \bar{s}\Gamma_1 u\bar{u}\Gamma_2 d(0) \bar{u}\gamma_5 s(y)|0\rangle$ result. Wick contraction yields Fig. 1a (the "figure-eight" graph)[11] and Fig. 1b (the "eye" graph). Despite its quark loop the eye graph should not be eliminated in the quenched approximation since it originates simply from a W boson correction to a single valence quark line (see Ref. 1). Furthermore, since it includes all possible gluon corrections, the eye graph cannot be eliminated by normal ordering the operator.

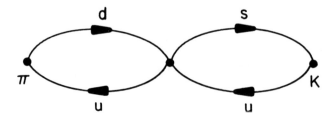

Fig. 1(a). The "figure-eight" graph.

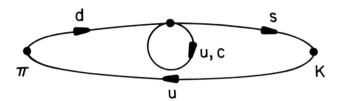

Fig. 1(b). The "eye" graph.

The inclusion of the eye graphs introduces a technical complication[1] which requires the "exponentiation" of the kaon (i.e. putting a kaon source term into the action).[12] This is equivalent to using a modified source term in the Gauss-Seidel algorithm. The eye graphs then become tractable to usual (quenched) MC, requiring only about twice the time of a hadron mass calculation.

The background gauge fields were generated by the standard Metropolis MC technique with 12 hits per site on a lattice of size 6^3 x 10 with periodic boundary conditions at $\beta = 6/g^2 = 5.7$. Two independent sets of eight configurations were used which evolved to equilibrium from different starting points. One thousand passes were used for thermalization; 500 passes separated each configuration used. Each of the configurations was copied in the time direction to form the background in which 6^3 x 17 quark lattices were embedded. The quarks had periodic boundary conditions in the spatial direction and "free"[13] (Neumann) boundary conditions in the time direction.

(See Ref. 3 for details.) By varying the fermion lattice size, we checked that the time boundary effects were no more than a few percent of the final results.

The quark propagators (with Wilson r = 1) were calculated for seven values of the hopping constant, k = .094, .123, .150, .155, .162, .164 and .165; k_c was found to be .171 ± .002.[14] Table 1 shows

K	am_q	am_M	af_M	am_M^2/m_q
.150	~.34	1.09±.05	.22±.01	~3.4
.155	~.26	.90±.07	.19±.01	~3.1
.162	~.15	.64±.11	.15±.02	~2.7
.164	~.12	.57±.12	.14±.04	~2.8
.165	~.10	.54±.11	.14±.05	~2.9

Table 1.

the pseudoscalar meson mass (m_M) and the decay constant f_M (normalized to f_π = 132 MeV) for k ≥ .150. With $a^{-1} \simeq 1$ GeV,[15] the meson with k = .162, .164 or .165 has roughly the mass of a kaon, but note that the lattice meson is made of degenerate quarks, unlike the physical kaon. Note that af_M and (for k > .155) am_M^2/m_q are fairly constant, indicating compatibility with chiral behavior. The decay constant is found by calculating the matrix element of the axial vector current,[16] i.e. from $\langle 0|A_\mu(x)|M(p)\rangle = if_M e^{ip \cdot x} p_\mu$. We use $am_q = \ell n[1+.5(k^{-1}- k_c^{-1})]$ to find the quark masses.

Fig. 2 shows the $\Delta I = 1/2$ and the $\Delta I = 3/2$ amplitudes for K → π as a function of k both with and without the penguin approach. When the charm quark was included as an explicit field we used k =.094, the corresponding $c\bar{c}$ pseudoscalar (i.e. η_c) has $am_M \simeq 3$. As the meson mass gets lighter, the 3/2 amplitude decreases and for k = .162

-.165, i.e.($m_M \sim m_K$), it is compatible with zero within statistics. On the other hand, the 1/2 amplitude is seen to increase for lighter mesons. For $m_M \sim m_K$, the 1/2 amplitude is much larger than the 3/2 amplitude. It is interesting that the results from both approaches are in rough agreement.[17]

Table 2 shows a ratio of the eye graph to the figure-eight graph contribution for the K-π matrix elements of some four-quark

K	am_M	EYES / EIGHTS			$a\, A^{latt}(K_S \to \pi\pi)$
		$\langle O_+ \rangle$	$\langle O_- \rangle$	$\langle Q_6 \rangle$	
.150	1.09±.05	-2	-2	-24	.9± .1
.155	.90±.07	-3	-5	-26	1.7± .2
.162	.64±.11	≯20	20	-34	5.6± .2
.164	.57±.12	23	10	-37	9.5±1.9
.165	.54±.11	18	7	-40	13.2±5.6

Table 2.

operators. The ratio eyes/eights becomes large as the meson mass gets lighter. Since the eye graph is purely $\Delta I = 1/2$ while the figure-eight is a mixture of $\Delta I = 1/2$ and $\Delta I = 3/2$, the $\Delta I = 1/2$ rule seems to be due to the dominance of eye graphs over figure-eights. We note that the eye graph contributions in Table 2 are the result of large cancellations ($\simeq 80\%$) between the up and the charm quark loops.

CPTh would also require the K → 0 amplitude (as well as K → π) in order to predict the K → ππ amplitude.[4] Our calculations of K → 0 show that the contribution of this amplitude is $\lesssim 15\%$ of that of K → π and thus can be ignored at this qualitative stage of the project.

There is a feature of Fig. 2 that is disturbing. From CPTh one expects both amplitudes to vanish as m_M goes to zero. The $\Delta I = 1/2$ amplitudes in the figure show no such behavior. It is possible that

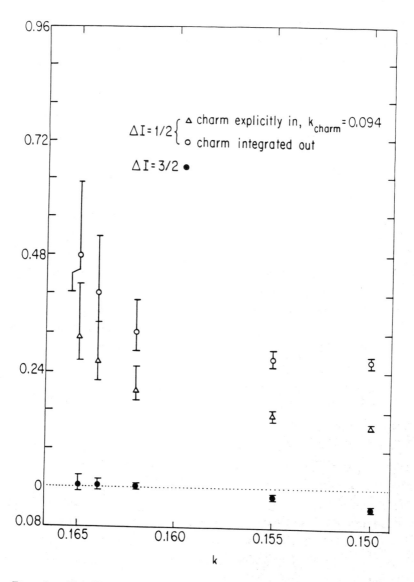

Fig. 2. The K → ππ amplitudes vs. hopping constant k. The ΔI = 3/2 amplitude is essentially unchanged whether or not charm is integrated out.

this is a lattice artifact, due to bad chiral properties of Wilson fermions at the not-very weak coupling of β = 5.7. However, the fact that both GIM subtracted and non-subtracted operators have similar mass dependence may be an indication that we are seeing the continuum physics of this rather heavy mass range: There is no sign in the eye graphs of spurious, mass-independent contributions (such as occur for $\langle \bar{\psi}\psi \rangle$).

Since CPTh behavior is not observed, the $K \to \pi\pi$ rate cannot be reliably computed from our lattice $K \to \pi$ amplitude. However, for what it is worth, we show in Table 2 the calculated (explicit charm) $\Delta I = 1/2$ $K \to \pi\pi$ reduced amplitude,[3,18] A^{latt}, which should be compared with the experimental value, $A^{expt} = 2.25$ GeV. Although the error bars are large, the lack of CPTh behavior is again obvious (in CPTh, A is independent of meson mass), and the magnitude of A^{latt} is quite large for k = .162 to .165. Clearly, better control of the chiral behavior is needed: future projects will use lower masses, weaker coupling, and various values of Wilson r, including r = 0 (Kogut-Susskind fermions).

We have also looked at $\langle \pi | Q_{5,6} | K \rangle$. With the penguin approach and CPTh, this matrix element can be related[19] to ϵ'/ϵ. Thus $|\epsilon'/\epsilon|^{latt} \simeq 15.6 \ |\xi^{latt} \ s_2 s_3 s_\delta c_2|$ where $s_2 s_3 s_\delta c_2$ are KM angles and

$$\xi^{latt} \equiv \langle \pi^+ | Im(\tilde{C}_5 Q_5 + \tilde{C}_6 Q_6) | K^+ \rangle / \langle \pi^+ | C_+ O_+^{\Delta I=1/2} + C_- O_- | K^+ \rangle \tag{2}$$

where C's are the Wilson coefficients.[7] From Table 3 we see that for k ⟩ .162, ξ^{latt} is fairly independent of k and $\simeq 3.5$. Thus this ratio appears to scale as CPTh suggests and is larger than estimates[19] based on the bag model. Further, we find very good agreement (within ~ 2%) between ξ^{latt} and the corresponding quantity calculated with charm included explicitly. However, since the individual amplitudes do not show CPTh behavior, the relation of ϵ'/ϵ to ξ^{latt} is at this point untrustworthy. We must also emphasize that the matrix elements for the non-GIM subtracted operators which enter here have a systematic uncertainty originating from possible mixing[20] at higher orders in g^2 with $(r/a) \ \bar{s}\sigma_{\mu\nu} F^{\mu\nu} d$. Note that even for the left-right operators $Q_{5,6}$, the eye graphs completely dominate over figure-eights (Table 1).

K	am_M	ξ^{latt}	$a^4 M_{LL}$	$a^4 M_{LR}$
.150	1.09±.05	5.7±.3	−.109±.006	.60±.12
.155	.90±.07	4.4±.1	−.056±.013	.52±.19
.162	.64±.11	3.7±.5	$-.003^{+.056}_{-.011}$	$.38^{+.21}_{-.09}$
.164	.57±.12	3.6±.6	$.006^{+.078}_{-.016}$	$.41^{+.19}_{-.11}$
.165	.54±.11	3.5±.5	$.007^{+.094}_{-.016}$	$.45^{+.19}_{-.13}$

Table 3.

Finally, we have examined the K^0-$\bar{K}^0$ matrix elements of the $\Delta S = 2$ LL,LR operators,[21] i.e. $M_{LL,LR} \equiv \langle K^0 | \bar{s}\gamma_\mu(1-\gamma_5)d\bar{s}\gamma^\mu(1\pm\gamma_5)d | \bar{K}^0 \rangle$. (See Table 3.). For $am_M \gtrsim .9$, the values of M_{LL} are quite consistent with vacuum saturation (computed using the lattice values of f_M); however, for $m_M \simeq m_K$, our values for M_{LL} are at this point compatible with zero with very large statistical errors. In contrast, the LR matrix element, M_{LR}, seems to show little variation with meson mass. We find that M_{LR} is 2 to 5 times more than implied by vacuum saturation.

To summarize, this work illustrates that lattice MC techniques can be very useful for attacking the old but very difficult problem of the non-leptonic decays of strange partcles. Even at this exploratory stage, we find a useful qualitative understanding of the origin of the $\Delta I = 1/2$ rule. The numerical values of some other quantities such as M_{LL} and ξ^{latt} appear interesting but more definite statements must await better control of systematics such as mass dependence (CPTh). The theoretical machinery that we have set up will now be used on a bigger (i.e. $12^3 \times 16$ gauge) lattice requiring > 2000 Cray X-MP hours. The goal is a level of accuracy of ~ 30% in our calculations.

We are grateful to W. Celmaster, F. Gilman, H. D. Politzer, and
M. Wise for useful discussions. The research of C.B., T.D., and G.H.
was supported by the NSF and the research of A.S. was supported by
the DOE Outstanding Junior Investigator Program. The computing was
done at the DOE Livermore MFE Computing Network.

REFERENCES

1. For a progress report on this project, see C. Bernard, in Gauge
Theory On A Lattice (C. Zachos et al., eds.), 1984, p. 85. See
also Refs. 2 and 3.

2. T. Draper, "Lattice Evaluation of Strong Corrections to Weak
Matrix Elements--ΔI = 1/2 Rule," Ph.D. Thesis (UCLA), August,
1984.

3. C. Bernard, G. Hockney, A. Soni, and T. Draper, UCLA/85/TEP/14,
invited paper presented at the Conference, "Advances in Lattice
Gauge Theories," Tallahassee, Florida, April 10-13, 1985.

4. C. Bernard, T. Draper, A. Soni, H. D. Politzer, and M. B. Wise,
UCLA/85/TEP/14, CALT 68-1211, to be published in Physical Review
D.

5. M. K. Gaillard and B. W. Lee, Phys. Rev. Lett. 33, 108 (1974);
G. Altarelli and L. Maiani, Phys. Lett. 52B, 351 (1974).

6. M. A. Shifman, A. I. Vainshtein, and V. I. Zakharov, Nucl. Phys.
B120, 316 (1977).

7. M. B. Wise, SLAC-PUB-227 (1980). We follow this reference in
defining Q_1-Q_6.

8. F. J. Gilman and M. B. Wise, Phys. Rev. D20, 2392 (1979);
B. Guberina and R. D. Peccei, Nucl. Phys. B163, 289 (1980);
C. T. Hill and G. Ross, Nucl. Phys. B171, 141 (1980);
J. Donoghue, E. Golowich, B. Holstein, and N. Ponce, Phys. Rev.
D23, 1213 (1981).

9. The constants Z_{1i} and Z_{2ij}, which are the only non-zero constants for four-quark operators with all flavors distinct, have
been calculated by G. Martinelli (Phys. Lett. 141B, 395

(1984)). We have repeated this calculation and agree with his numerical values for Z_{1i} and Z_{2ij}, but disagree by a few signs and factors of 2 in the definitions of the off-diagonal operators O_j^{latt}. We have also calculated the additional graphs necessary when the operators contain quark-antiquark pairs which may contract with each other ("eye" graphs). For $r = 1$, we find $Z_3 = \pm(.062\pm.001-(8/3) \ln(a\mu))$ for $O_\pm$ without GIM subtraction. For details see C. Bernard, T. Draper and A. Soni (in prepaation).

10. M. A. Shifman, A. I. Vainshtein, and V. I. Zakharov, JETP Lett. 22, 55 (1975); C. T. Hill and G. Ross; op. cit.

11. N. Cabibbo, G. Martinelli, and R. Petronzio, Nucl. Phys. B244, 381 (1984); and R. C. Brower, M. B. Gavela, R. Gupta, Maturana, Phys. Rev. Lett. 53. 1318 (1984) have reported lattice calculations of weak matrix elements retaining only figure-eight graphs.

12. Similar methods have been used previously, see e.g. S. Gottlieb, P. B. Mackenzie, H. B. Thacker, and D. H. Weingarten, Phys. Lett. 134B, 346 (1984); C. Bernard, T. Draper, K. Olynyk, and M. Rushton, Phys. Rev. Lett. 49, 1076 (1982).

13. C. Bernard, T. Draper, and K. Olynyk, Phys. Rev. D27, 227 (1983).

14. One of the sixteen configurations was eliminated because it had an extraordinary low m_M over several time slices and was therefore very sensitive to finite size limitations. Including it would not change the qualitative conclusions reached here but would increase the statistical errors significantly for $k = .162$ and $.164$. See Ref. 3 for details.

15. This estimate is based on S. Otto (private communication); D. Barkai, K. J. M. Moriarty, and C. Rebbi, Phys. Rev. D30, 1293 (1984).

16. Previously (Refs. 3 and 4) we used other methods for calculating f_M which depended either on a knowledge of the quark mass or on chiral perturbation theory and thus seem less reliable than the method we now use. We include perturbative corrections to A_μ: B. Meyer and C. Smith, Phys. Lett. 123B, 62 (1983);

298

G. Martinelli and Z. Yi-Cheng, ibid., 433; R. Groot, J. Hoek, and J. Smit, Nucl. Phys. <u>B237</u>, 111 (1984).

17. The penguin approach is sensitive to the choice of renormalization point. We choose $\mu = (\Lambda_{\overline{MS}}/\Lambda_{latt}) \, a^{-1}$ (see Ref. 3). Other, perhaps somewhat less reasonable, choices produce variations of $\sim \pm 50\%$. The "explicit charm" method suffers from the fact that our lattice cutoff is not $\gg M_{charm}$. Replacing k_{charm} by .123 ($am_M \simeq 2$ instead of $\simeq 3$), reduces the $\Delta I = 1/2$ amplitude by almost a factor of 2.

18. The values for A^{latt} in Table 2 differ from those given in Ref. 3 because of the new method for computing f_M. See Ref. 16.

19. F. J. Gilman and J. S. Hagelin, Phys. Lett. 133B, 443 (1983); J. F. Donoghue et al. op. cit.; P. H. Ginsparg and M. B. Wise, Phys. Lett. <u>127B</u>, 265 (1983); M. B. Wise, Caltech preprint 68-1179 (1984).

20. M. Bochiccio, L. Maiani, G. Martinelli, G. Rossi, and M. Testa, INFN preprint #452 (1985). Our measurement of $K \to 0$ seems to show that mixing with two-quark operators which are total divergences (such as $\bar{s}d$ and $\bar{s}\not{D}d$) is small. See Ref. 4.

21. Note that only figure eights contribute to these quantities. Lattice calculations of M_{LL} also appear in N. Cabibbo et al. R. C. Brower et al., op. cit.

A REVIEW OF EXPERIMENTS MEASURING ε'/ε

George D. Gollin

Joseph Henry Laboratories, Princeton University
Princeton, New Jersey 08544

ABSTRACT

Recent experiments setting limits on the magnitude of the ratio of two CP nonconservation parameters ε'/ε are discussed. Gauge theory calculations suggest that $\varepsilon'/\varepsilon > .005$, somewhat at odds with the results $\varepsilon'/\varepsilon = -.0046 \pm .0058$ from a Chicago-Saclay measurement and $\varepsilon'/\varepsilon = .0017 \pm .0082$ from a Yale-BNL experiment.

1. INTRODUCTION

Experiments performed over the twenty-two years since the discovery of CP violation have failed to find evidence for any CP non-invariance outside of the K^0-$\bar{K}^0$ system. CP nonconservation in $K_L \to \pi\pi$ decays and in the charge asymmetry for $K_L \to \pi\mu\nu$ and $\pi e\nu$ decays can be accounted for by an imbalance in the $\Delta S=\pm 2$ transitions $K^0 \leftrightarrow \bar{K}^0$. Gauge theories of the weak interactions suggest that a $\Delta S=1$ contribution to CP violation also should exist. This contribution arises from interference between the $\Delta I=1/2$ and $\Delta I=3/2$ $K_L \to \pi\pi$ amplitudes and might produce a measurable difference in the ratios $\Gamma(K_L \to \pi^0\pi^0)/\Gamma(K_S \to \pi^0\pi^0)$ and $\Gamma(K_L \to \pi^+\pi^-)/\Gamma(K_S \to \pi^+\pi^-)$. Defining η = Amplitude $(K_L \to \pi\pi)$/ Amplitude $(K_S \to \pi\pi)$, one has $\eta_{00} = \varepsilon - 2\varepsilon'$ and $\eta_{+-} = \varepsilon + \varepsilon'$, or $\eta_{00}/\eta_{+-} \approx 1 - 3\varepsilon'/\varepsilon$. The ratio ε'/ε is indicative of the relative amounts of $\Delta S=1$ and $\Delta S=2$ CP violation in K_L decay.

Calculations[1]) using the Standard Model tend to find $\varepsilon'/\varepsilon \gtrsim .005$ while those based on models with CP violation in the Higgs sector[2]) yield larger values. There is still a fair amount of uncertainty in predictions of a lower bound on $|\varepsilon'/\varepsilon|$. Before the current round of experiments discussed below, a Princeton group[3]) measured $|\eta_{00}/\eta_{+-}|$ = 1.03±.07 from data which included 124 $K_L \to \pi^0\pi^0$; an Aachen-CERN-Torino group[4]) found $|\eta_{00}/\eta_{+-}|$=1.00±.06 based on 167 $K_L \to \pi^0\pi^0$. These experiments were performed more than a year before the publication of Kobayashi's and Maskawa's 1973 paper[5]) discussing CP violation in a six-quark Standard Model. To test this new framework for CP violation, recent experiments have been performed by a Chicago-Saclay[6]) collaboration at Fermilab and a Yale-BNL[7]) collaboration at Brookhaven. More ambitious than the earlier measurements, the Fermilab experiment recorded 3152 $K_L \to \pi^0\pi^0$ while the Brookhaven experiment obtained 1122 $K_L \to \pi^0\pi^0$ after background subtractions.

2. EXPERIMENTAL CONSIDERATIONS

Since $\Gamma(K_L \to \pi\pi)/\Gamma(K_S \to \pi\pi) = |\eta|^2$, and since the phases of the complex numbers ϵ' and ϵ are approximately equal[8]), one may write

$$\frac{\Gamma(K_L \to \pi^0\pi^0)}{\Gamma(K_S \to \pi^0\pi^0)} \Bigg/ \frac{\Gamma(K_L \to \pi^+\pi^-)}{\Gamma(K_S \to \pi^+\pi^-)} \approx 1 - 6\frac{\epsilon'}{\epsilon}. \tag{1}$$

A determination of the four $K \to \pi\pi$ rates yields a value for ϵ'/ϵ. Because experiments generally record substantially more $K_S \to \pi\pi$ and $K_L \to \pi^+\pi^-$ than $K_L \to \pi^0\pi^0$, the statistical error on a measurement of ϵ'/ϵ is roughly $17\%/\sqrt{N}$. Here N is the number of $K_L \to \pi^0\pi^0$ events obtained; the 17% arises from the factor of 6 multiplying ϵ'/ϵ in equation 1. The level of precision needed to test current ideas about CP violation requires experiments to measure ratios of kaon decay rates to an accuracy of a few percent. As a result, detectors must be designed to minimize systematic uncertainties as well as to record large numbers of $K \to \pi\pi$ events. Some possible sources of systematic error are the following:

1. Imperfect knowledge of K_L and K_S fluxes.
2. Different acceptances for K_L and K_S decays.
3. Different reconstruction efficiencies for $\pi^0\pi^0$ and $\pi^+\pi^-$ final states.
4. Contamination from K_S produced by neutron, diffractive K_L and inelastic K_L interactions in the detector.
5. Changes in apparatus resolution and calibration during data taking.
6. Background subtractions.

The Chicago-Saclay group produced a pair of side-by-side K_L beams of $\sim 6 \times 10^5$ K_L per 850 millisecond spill by striking a 30 cm beryllium target with 6×10^{12} 400 GeV protons. The 5mr production angle beams contained about 10 times as many neutrons as kaons. K_S were produced by attenuating one of the beams with 29 inches of beryllium absorber, then passing the surviving K_L through a 40 inch carbon regenerator. The Yale-BNL experimenters obtained a single beam of K_L from interactions of 28 GeV protons in a 20 cm copper target. Their zero-degree beam of 3×10^6 K_L per one second spill of $3 \times 10''$ protons contained about 30 times as many neutrons as kaons; K_S were produced by inserting an 80 cm graphite regenerator into the K_L beam periodically. Both experiments used coherently regenerated K_S to calculate the denominators of equation 1, correcting for K_S contributions from incoherent and inelastic processes.

3. THE CHICAGO-SACLAY EXPERIMENT AT FERMILAB

The Chicago-Saclay group consisted of R.H. Bernstein, G.J. Bock, D. Carlsmith, D. Coupal, J.W. Cronin, myself, Wen Keling, K. Nishikawa, H.W.M. Norton, B. Winstein, B. Peyaud, R. Turlay, and A. Zylberstejn. The detector is shown below in Figure 1.

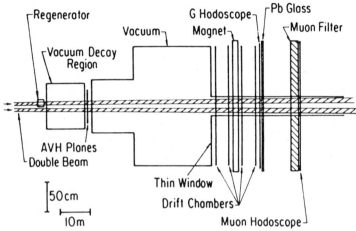

Figure 1. Chicago-Saclay experiment

A pair of neutral beams entered from the left in the figure. K_L decays from the unobstructed beam were detected by the drift chamber spectrometer and 804-block lead glass array. Most decays in the beam passing through the regenerator were $K_S \to \pi\pi$; the regenerator and upstream 29-inch absorber (not shown) switched beams after every accelerator pulse. The apparatus simultaneously recorded K_L and K_S decays to one of the two relevant final states. Changing from $\pi^0\pi^0$ to $\pi^+\pi^-$ data mode required removal of the A scintillator and a 0.1 radiation length lead sheet between A and V.

The charged ($\pi^+\pi^-$) mode trigger demanded signals in V and H, in-time signals in some of the drift chambers, and hits in the upper and lower halves of the G hodoscope. Some data were taken with a lead curtain in front of the G array; large signals in G or signals in the muon hodoscope served to veto $K \to \pi e \nu$ and $\pi \mu \nu$ decays. Events were reconstructed by calculating the invariant mass, total energy, and momentum for pairs of tracks. The charged mode mass resolution was 4.5 MeV/c^2. Coherently regenerated K_S were selected by requiring the apparent change in P_T^2, the square of the kaon's transverse momentum after passing through the regenerator to be less than 300 (MeV/c^2). P_T^2 resolution was about 30 (MeV/c)^2; the mean reconstructed kaon momentum was 70 GeV/c.

The neutral mode trigger required one photon to convert in the lead sheet between the A and V scintillators. Hits corresponding to the two tracks, split horizontally by the analyzing magnet, were registered by the G hodoscope. A coarse calculation of the kaon invariant mass based on the pattern of energy deposition in the lead glass vetoed some events with missing photons. Information from the glass array was used offline to determine the event's mass and the distance from decay vertex to lead glass. Extrapolation of the e^+/e^- tracks back to this point yielded the transverse position of the vertex. A kaon's P_T^2 was measured by determining the angle between the line from the target to the regenerator and the line from the regenerator through the decay vertex to the center-of-energy in the glass array. Mass and P_T^2 resolutions were 6.5 MeV/c^2 and 400 (MeV/c)^2 respectively; the mean reconstructed kaon momentum was about 90 GeV/c. K_S were required to have P_T^2 less than 2500 (MeV/c)^2.

The ratio of the number of reconstructed $\pi\pi$ decays in the K_L beam and regenerated beam, as a function of proper time and kaon energy is

$$\frac{\Gamma(K_L \text{ beam})}{\Gamma(\text{regenerated beam})} = \frac{|ne^{-\Gamma_L\tau/2}|^2 \times K_L \text{ flux} \times \text{acceptance}}{A|\rho e^{-\Gamma_S\tau/2} + ne^{-\Gamma_L\tau/2}|^2 \times K_L \text{ flux} \times \text{acceptance}}. \quad (2)$$

n is $\varepsilon - 2\varepsilon'$ for $\pi^0\pi^0$ decays and $\varepsilon + \varepsilon'$ for $\pi^+\pi^-$ decays. Γ_L and Γ_S are K_L and K_S decay rates, respectively, and τ is the proper time for the kaon. $\tau=0$ corresponds to the downstream end of the regenerator. ρ is the regeneration amplitude, typically ten times greater than n, and A is an attenuation factor due to scattering in the carbon regenerator and upstream beryllium absorber. The twin beam technique makes the ratio insensitive to the K_L flux; terms involving A are eliminated by comparing the ratios for charged and neutral mode data. Hardware failures and inefficiencies caused by noise, halo muons, and so forth interfere equally with events from both beams.

Since the detector's fiducial decay volume is about 3 τ_S but only .005 τ_L long, the spatial distributions of K_S and K_L decays differ appreciably. The Chicago-Saclay group used a Monte Carlo calculation to correct for the minor acceptance variations between the up- and downstream regions of the decay volume, permitting all the K_L decays (including those upstream of the regenerator) to contribute to the numerator of equation 2. The copious $K_L \rightarrow 3\pi^0$ signal provided a check on neutral mode Monte Carlo assumptions while semileptonic decays tested the charged mode simulation. Figure 2 shows $K_L \rightarrow 2\pi^0$ data for different kaon momenta.

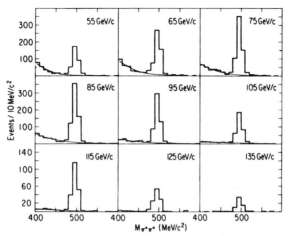

Fig. 2. Mass distributions for $K_L \to 2\pi^0$ candidates
with $P_T^2 < 2500$ $(MeV/c)^2$

The 8% background under the K_L peak is due to misidentified $K_L \to 3\pi^0$ decays as well as neutron interactions which produced pairs of π^0's. Backgrounds under the $K_S \to 2\pi^0$, $K_L \to \pi^+\pi^-$, and $K_S \to \pi^+\pi^-$ mass peaks were considerably less. The major background to coherently regenerated $K_S \to \pi^0\pi^0$ came from inelastically and diffractively produced K_S. Figure 3 shows distributions of P_T^2 for $K_S \to \pi^0\pi^0$; a clear coherent spike at small P_T is visible for each energy range. The smooth background from incoherent K_S is subtracted to yield the total coherent $K_S \to \pi^0\pi^0$ sample.

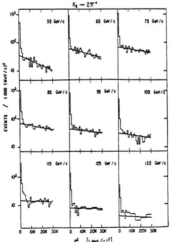

Fig. 3. P_T^2 distributions for $K_S \to 2\pi^0$ candidates with mass
within 20 MeV/c^2 of 498 MeV/c^2

The event sample which remains after all subtractions is described in Table 1. Note that the decay volumes used in $K_L \to \pi^+\pi^-$ and $K_L \to \pi^0\pi^0$ reconstruction have different lengths.

Table 1. Event totals and corrections.

Mode	Events after subtraction	Non-$\pi\pi$ background	Incoherent $K \to \pi\pi$
$K_L \to \pi^0\pi^0$	3152 ± 61	266.0 ± 13.0	90.7 ± 9.5
$K_S \to \pi^0\pi^0$	5663 ± 84	35.3 ± 5.9	825.7 ± 28.7
$K_L \to \pi^+\pi^-$	10638 ± 106	324.9 ± 18.0	42.5 ± 6.5
$K_S \to \pi^+\pi^-$	25751 ± 163	44.8 ± 6.7	439.2 ± 21.0

If n_{+-} and n_{00} were known, equation 2 would allow extraction of $\rho(P_K)$, the regeneration amplitude as a function of K_L momentum, from either the neutral or charged data samples. The Chicago-Saclay analysis extracts ε'/ε by determining the value of ε' which gives the best fit when data from both modes are used to measure $\rho(P_K)$. The group believes that most systematic errors stem from uncertainties in the amount of background to subtract. The estimated error caused by the Monte Carlo correction for the differences in K_L and K_S acceptances is small compared to background uncertainties. Their final result is $\varepsilon'/\varepsilon = -.0046 \pm .0053 \pm .0024$. The first error is statistical, the second systematic.

4. THE YALE-BNL EXPERIMENT AT BROOKHAVEN

The Yale-BNL group consisted of J.K. Black, S.R. Blatt, M.K. Campbell, H. Kasha, M. Mannelli, M.P. Schmidt, C.B. Schwarz, R.K. Adair, R.C. Larsen, L.B. Leipuner, and W.M. Morse. The detector is shown in Figure 4.

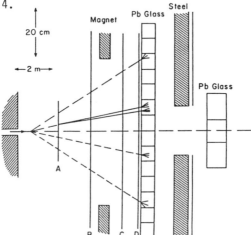

Fig. 4. The Yale-BNL experiment

A single cylindrical beam entered from the left in the figure. Decays were detected by PWC planes A, B, C, and D, scintillator at A, B, and D, and the upstream array of 208 lead glass blocks. An 80 cm graphite regenerator (not shown) was placed in the beam every two hours for 300 pulses, one meter upstream of the shielding wall. The neutral beam profile was Gaussian with σ=5.2 mm in the decay volume; $K \to \pi^+\pi^-$ and $K \to \pi^0\pi^0$ were recorded simultaneously, either through K_S decays when the regenerator was in place or through K_L decays when the beam was unobstructed. A 1 mm thick lead sheet containing a horizontal gap 4 cm high between two scintillator arrays at A converted one photon for neutral decays.

The $\pi^+\pi^-$ trigger demanded hits in scintillators at A, B, and D consistent with two tracks passing through the slot in the lead sheet. By requiring signals in opposite quadrants in the D hodoscope, events with a large P_T imbalance were rejected. All vetoes were common to both charged and neutral triggers to avoid systematic effects due to different dead times and sensitivities to noise for the $\pi^+\pi^-$ and $\pi^0\pi^0$ triggers. Hits in the scintillators downstream of the steel muon filter reduced the trigger rate from $K_L \to \pi\mu\nu$. Signals in a ring of counters shielded by lead on the perimeter of the B array served to reject $K_L \to 3\pi^0$ decays with photons aimed wide of the lead glass. The experiment's data acquisition program attempted to track-find events and discarded those it was unable to reconstruct. About three percent of the hardware $\pi^+\pi^-$ triggers satisfied the tracking requirements and were written to tape. These were later reconstructed offline; mass resolution was 15 MeV/c^2. $K \to \pi^+\pi^-$ were required to have P_T^2 less than 900 (MeV/c)^2. Typical reconstructed kaon energies were in the range 8-12 GeV for charged decays.

The $\pi^0\pi^0$ trigger demanded one photon's conversion in the lead sheet at plane A, a single struck counter at B, two struck counters at D, and at least 1 GeV in upper and lower halves of the lead glass. More than 5 GeV total energy in the glass array was also required in the trigger. The data acquisition program looked for tracks and isolated depositions of energy in the lead glass. Events without good tracks or with fewer than four energy clusters were discarded. Five percent of the hardware $\pi^0\pi^0$ triggers satisfied the online program's requirements and were written to tape. The offline reconstruction algorithm assumed that kaons were not scattered out of the beam by the regenerator. The decay vertex was chosen to be in the beam at the point closest to the extrapolation into the decay volume of the trajectory of the converted photon. Mass resolution was 30 MeV/c^2; P_T^2 resolution was roughly 1000 (MeV/c)^2. Typical reconstructed kaon energies for neutral decays were 10-14 GeV.

The ratio of reconstructed $\pi^0\pi^0$ to $\pi^+\pi^-$ decays vs. proper time is

$$\frac{\Gamma(oo)}{\Gamma(+-)} = \frac{|T_{L_{00}}|^2 e^{-\Gamma_L \tau} \times K_L \text{ flux} \times \pi^0\pi^0 \text{ acceptance}}{|T_{L_{+-}}|^2 e^{-\Gamma_L \tau} \times K_L \text{ flux} \times \pi^+\pi^- \text{ acceptance}} \qquad (3)$$

for an unobstructed K_L beam and

$$\frac{\Gamma(oo)}{\Gamma(+-)} = \frac{A|T_{Loo}e^{-\Gamma_L\tau/2} + T_{Soo}\rho e^{-\Gamma_S\tau/2}|^2 \times K_L \text{ flux} \times \pi^0\pi^0 \text{ acceptance}}{A|T_{L+-}e^{-\Gamma_L\tau/2} + T_{S+-}\rho e^{-\Gamma_S\tau/2}|^2 \times K_L \text{ flux} \times \pi^+\pi^- \text{ acceptance}} \qquad (4)$$

for a regenerated beam. T_{Loo} is the $K_L \to 2\pi^0$ transition amplitude while T_{Soo}, T_{Li-}, and T_{S+-} describe the other three decays; $T_L/T_S = n$. The remaining terms in equations 3 and 4 are the same as those appearing in equation 2. ρT_S is roughly thirty-five times greater than T_L; simultaneous detection of $\pi^+\pi^-$ and $\pi^0\pi^0$ final states eliminates the ratios' sensitivities to beam flux, regeneration amplitude, and attentuation in the regenerator. By switching frequently between K_L and K_S, the experimenters hoped to minimize the sensitivity of the double ratio (equation 1) to drifts in lead glass response, PWC efficiency, and so forth. The Yale-BNL analysis did not rely on a Monte Carlo calculation to correct for K_L/K_S acceptance differences. Instead, it determined the double ratio in bins of kaon energy and decay position, extracted ε'/ε, then averaged the values obtained. Figure 5 shows mass distributions for K_L and K_S decays. dQ is defined as the difference between reconstructed mass and 498 MeV/c^2. Events in the shaded regions contributed to ε'/ε.

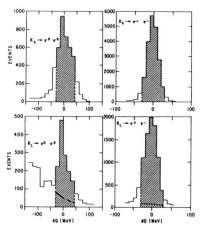

Fig. 5. Mass distributions for $K \to \pi\pi$ candidates

The 21% background under the $K_L \to 2\pi^0$ peak comes from $K_L \to 3\pi^0$ decays and neutron interactions. The contribution to the $K_S \to \pi\pi$ signals from incoherent K_S production is $(1.5\pm0.5)\%$ in the neutral mode and about 0.2% in the charged mode. The Yale-BNL data sample is described in Table 2.

Table 2. Event totals and corrections

Mode	Events after subtraction	Background (excluding incoherent K_S)	Incoherent $K \to \pi\pi$
$K_L \to \pi^0 \pi^0$	1122	239 ± 41	
$K_S \to \pi^0 \pi^0$	3267	40 ± 7	50 ± 17
$K_L \to \pi^+ \pi^-$	8081	599 ± 123	
$K_S \to \pi^+ \pi^-$	20921		42

After averaging over energy-momentum bins, this group finds $\varepsilon'/\varepsilon = .0017$ ±.0072±.0043. The first error is statistical, the second systematic. The dominant systematic uncertainty comes from subtraction of backgrounds under the $K_L \to \pi^0 \pi^0$ mass peak.

5. COMPARISON OF THE TWO EXPERIMENTS

The different beam and detector configurations chosen by the two groups give rise to different sets of systematic uncertainties in the two results. The K_L beam energy at Fermilab is about eight times higher than at Brookhaven. The BNL beam is a thin cylinder while each beam at FNAL has an area of 10 square inches inside the decay volume. The Chicago-Saclay group records K_L and K_S decays to a chosen final state simultaneously while the Yale-BNL collaboration selects K_L or K_S and runs charged and neutral decays simultaneously. The analysis of the Fermilab experiment included a Monte Carlo to correct for the different K_L and K_S vertex distributions while the Brookhaven analysis discarded the decays upstream of the regenerator, avoiding reliance on an acceptance simulation.

The resolution of lead glass photon detectors improves with increasing energy. Since both groups attributed most of their systematic uncertainty to background subtractions, narrower $K_L \to 2\pi^0$ mass distributions and more powerful rejection of $K_L \to 3\pi^0$ background produced smaller errors in the Chicago-Saclay measurement. The Yale-BNL group found their lead glass resolution to be ~ 13%/$\sqrt{E}$ for typical photon energies of a few GeV while the Chicago-Saclay group obtained (2+6/$\sqrt{E}$)% for E ~ 22 GeV.

The coherent regeneration power of carbon decreases[6]) with kaon energy as $E_K^{-1.2}$. As a result, the ratio of coherent to incoherent K_S regeneration in the Fermilab data is about a tenth as great as in the BNL data. However, the uncertainties in the incoherent $\pi\pi$ backgrounds listed in Table 1 are small compared to the errors in the amounts of coherent K_S decay. As a result, the decreased regeneration power at Fermilab energies is unimportant.

By extrapolating the trajectory of the converted photon back to its intersection with the narrow beam in $K \to \pi^0 \pi^0$, the Yale-BNL group determines a longitudinal (Z) decay coordinate to an accuracy of 12 cm or about one fifth of a K_S lifetime. Z, and consequently the mass, are miscalculated for kaons which scattered in the regenerator. Since the coherent regeneration amplitude is so large at Brookhaven energies, this is not a significant problem. The Chicago-Saclay collaboration measures the vertex Z position solely on the basis of leadglass information to an accuracy of 1.8 m or about two-fifths of a K_S lifetime. Extrapolation of the converted photon's path back to this Z plane yields the transverse position of the vertex. This method of reconstruction permits correct determination of mass and P_T^2 for kaons which have scattered in the regenerator, important at the higher Fermilab energies where coherent regeneration is reduced.

Reconstruction of $\pi^0\pi^0$ final states depends on the performance of the groups' lead glass arrays while detection of $\pi^+\pi^-$ decays requires accuracy and efficiency of the tracking chambers. The responses of lead glass and wire chambers to stray muons, neutrons, photons, rf noise, temperature variations, and so forth are rather different. Introduction of a two interaction length regenerator into a beam composed largely of neutrons changes the experimental environment downstream of the regenerator. Drifts in phototube gains, wire chamber efficiencies and ADC responses also influence the reconstruction efficiency for $\pi\pi$ states. The Yale-BNL collaboration switched between K_L and K_S modes frequently to minimize sensitivity to slow variations in detector parameters. Their method did not cancel all effects related to the presence or absence of a regenerator. Any decrease in photon reconstruction efficiency during K_S running due to neutron interactions, for example, would need to be measured and taken into account. These sorts of problems -- changes in $K_S \rightarrow \pi^0\pi^0$ detection efficiency not cancelled by changes in $K_L \rightarrow \pi^0\pi^0$ or $K_S \rightarrow \pi^+\pi^-$ efficiencies -- are thought to be small compared to the quoted errors. The Chicago-Saclay experiment recorded a single final state of both K_L and K_S decays, switching infrequently between charged and neutral modes. Noise rates, changing detector parameters, and hadron debris from the regenerator affected both K_L and K_S, and largely cancelled in the ratio of equation 2.

The Yale-BNL analysis extracted ε'/ε for decay data binned in kaon momentum and energy. Since geometrical acceptances are identical for K_L and K_S with the same energies and decay vertices, their result did not require a Monte Carlo acceptance simulation. To allow the use of K_L decays from the region upstream of the regenerator, the Chicago-Saclay analysis used an apparatus simulation to correct the K_L and K_S data samples for the differences in acceptance. Reanalysis of the data using bins of momentum and decay vertex yields $\varepsilon'/\varepsilon = -.0044 \pm .0062$, in excellent agreement with the more precise result of $-.0046 \pm .0053$. Errors are statistical.

The systematic uncertainties associated with both groups' results are smaller than the quoted statistical errors. The systematic errors of the Yale-BNL group are about twice as large as those of the Chicago-Saclay collaboration.

6. FUTURE PROSPECTS

The Chicago-Saclay group has evolved into a Chicago-Fermilab-Princeton-Saclay collaboration. A new experiment, whose goal is an ε'/ε measurement to a precision of .001, is underway. The new detector uses the same double beam technique as the previous experiment. Data from a 1985 run are being analyzed; more data taking is scheduled for 1987. Changes include increased acceptance and $K_L \to 3\pi^0$ veto power. A CERN-Dortmund-Edinburgh-Orsay-Pisa-Siegen group[9] (NA31) is performing a very different sort of ε/ε' measurement using a liquid argon photon calorimeter, tracking chambers, and a hadron calorimeter. NA31 uses a single beam produced by collisions of protons with a moveable target. By changing the longitudinal position of the target, the mix of K_L and K_S can be adjusted. The experiment has no analyzing magnet, relying instead on calorimetry and tracking chambers to determine $\pi^+\pi^-$ momenta and trajectories. The NA31 group simultaneously records $\pi^+\pi^-$ and $\pi^0\pi^0$ decays; all information about the neutral final states is obtained from the photon calorimeter. Their goal is a precision of .001 in ε'/ε. The compensation for systematic effects caused by rate-dependent reconstruction efficiencies and drifts in calorimetry calibration will present a significant challenge in the analysis of the CERN data.

REFERENCES

1. Gilman, F.J. and Hagelin, J.S., Phys. Lett. B **133**, 443 (1983).

2. Deshpande, N.G. Phys. Rev. D **23**, 2654 (1981).

3. Banner, M. et al., Phys. Rev. Lett. **28**, 1597 (1972).

4. Holder, M., et al., Phys. Lett. B **40**, 141 (1972).

5. Kobayashi, M., and Maskawa, T., Progr. Theor. Phys. **49**, 652 (1973).

6. Bernstein, R.H., et al., Phys. Rev. Lett. **54**, 1631 (1985). More detailed information is available in Bernstein, R.H., Ph.D. Thesis, University of Chicago (1984) (unpublished).

7. Black, J.K. et al., Phys. Rev. Lett., 1628 (1985). Black, J.K., Ph.D. Thesis, Yale University (1984) (unpublished) and Schwarz, C.B., Ph.D. Thesis, Yale University (1985) (unpublished) contain more detailed information.

8. See K. Kleinknecht, Ann. Rev. Nucl. Sci. **26**, 1 (1976) for a review of CP violation phenomenology.

9. Cundy, D. et al., CERN/SPSC/81-110, SPSC/P174, Proposal for approved CERN experiment NA31 (1981).

TOPONIUM-Z^0 INTERFERENCE*

Paula J. Franzini

Stanford Linear Accelerator Center
Stanford University, Stanford, California, 94305

ABSTRACT

A study of interference of the Z^0 boson and toponium states is presented. The simple case of the Z^0 mixing with one $t\bar{t}$ state is discussed in detail. Effects of mixing with the full $t\bar{t}$ spectrum, of the smearing due to beam spread, and of different potentials, are then shown.

1. INTRODUCTION

Why do we expect toponium-Z^0 mixing to be of interest? From the absence of flavor-changing neutral currents in B decay, we are confident that the bottom quark must have an as-yet-unobserved partner. Experimentally, $m_t < 23$ GeV is excluded, while UA1 data suggests a top quark of mass between 30 and 50 GeV. It appears quite possible that $t\bar{t}$ bound states will have masses near that of the Z^0 (93 GeV), and thus vector $(J^{PC} = 1^{--})$ $t\bar{t}$ states (henceforth V) could be nearly degenerate with the Z^0. We expect the effects of $V - Z$ mixing to be seen soon, at both SLC and LEP.

I first present a few ways of understanding the nearly complete destructive interference of the Z boson with one V state. Then, after a brief review of toponium spectroscopy, I discuss the mixing of the Z with the full spectrum of toponium states (when the Z and V are nearly degenerate); I show the effects of finite beam width on the cross-sections and asymmetries. I then display the striking effects that remain if the Z is relatively far away from the V ($10 - 20$ GeV), and conclude by contrasting the effects of the Richardson potential, the Cornell potential, and a non-standard Higgs sector.

This talk is based on work done with Fred Gilman and Gregory Athanasiu.[1-3]

---

* Work supported by the Department of Energy, contract $DE - AC03 - 76SF00515$.

2. MIXING OF THE Z^0 WITH A SINGLE $t\bar{t}$ STATE

I begin with a qualitative argument to show that the interference is indeed destructive. To be specific we consider the process $e^+e^- \to \mu^+\mu^-$(other final states are discussed analogously; see Ref. 1). This process occurs predominantly as $e^+e^- \to Z_0 \to \mu^+\mu^-$, while another contribution is $e^+e^- \to Z_0 \to V_0 \to Z_0 \to \mu^+\mu^-$ (for now, we neglect the small contributions due to γ couplings). The first has an amplitude proportional to the propagator $1/(s - M_{Z_0}^2 + i\Gamma_{Z_0}M_{Z_0})$, and therefore to $1/i\Gamma_{Z_0}$ on the peak of the Z_0 resonance. If, for simplicity, I choose the Z_0 and V_0 resonances to be degenerate, the amplitude from the second contribution is similarly proportional to $1/(i\Gamma_{Z_0}i\Gamma_{V_0}i\Gamma_{Z_0})$. Thus we have a relative minus sign between these two amplitudes, i.e., destructive interference.

I can extend this argument by replacing the Z_0 propagator by the iterated series

$$\frac{1}{s - M_{Z_0}^2} + \frac{1}{s - M_{Z_0}^2} \cdot \left(a \cdot \frac{1}{s - M_{V_0}^2} \cdot a \cdot \frac{1}{s - M_{Z_0}^2} \right) \tag{1}$$

$$+ \frac{1}{s - M_{Z_0}^2} \cdot \left(a \cdot \frac{1}{s - M_{V_0}^2} \cdot a \cdot \frac{1}{s - M_{Z_0}^2} \right)^2 + \cdots = \frac{s - M_{V_0}^2}{(s - M_{Z_0}^2)(s - M_{V_0}^2) - a^2} \, .$$

Here, and often in what follows, $M_{Z_0}^2$ is used as a shorthand for the full expression $M_{Z_0}^2 - i\Gamma_{Z_0}M_{Z_0}$, and a is the $Z - V$ coupling.

What does this expression tell us? For energies a few GeV away from a V_0 resonance, $(s - M_{Z_0}^2)(s - M_{V_0}^2)$ is large compared to a^2; as expected, we recover the Z_0 propagator. When we are sitting on the V_0 resonance we get zero for the amplitude—*complete destructive interference.*

Strictly speaking, the amplitude only vanishes if we make some simplifying assumptions:

(1) I have ignored the fact that $e^+e^- \to \mu^+\mu^-$ can also proceed via a virtual photon. This is a good approximation, since the photon, by definition, contributes an R-value of about[†] one, while the R-value on the Z_0 peak is 200. (I note here that on the Z_0 peak, the Z amplitude is imaginary while that of the photon is real, so that there is no $\gamma - Z$ interference. However, in general we must compute $Z\gamma V$ mixing. The effect of the photon is small enough to be negligible, except in the determination of the asymmetry parameters.)

(2) I have implicitly assumed that the width of the V_0 is zero. The expression $s - M_{V_0}^2$ really stands for $s - M_{V_0}^2 + iM_{V_0}^2\Gamma_{V_0}$ which can only be zero (for a physically allowed value of s) if $\Gamma_{V_0} = 0$. This is also a good approximation, since the expected width of a $t\bar{t}$ $1S$ state (here, and throughout this section, I use the Richardson potential to estimate $t\bar{t}$ properties) is ≈ 100 keV, compared to $\Gamma_Z = 2.7$ GeV.

---

† since R-value is defined in terms of the QED cross-section at the electron mass scale.

(3) Finally, I have ignored the "direct" couplings of the V_0, that is, the V_0 coupling to fermions through the photon instead of through the Z_0. This approximation is analogous to, and comparable in magnitude with, the second one.

2.1 Mass-Mixing Approach

Now I would like to present another way of analyzing this problem. The pure states V_0 and Z_0 are nearly degenerate, with mass-squared matrix

$$\mathcal{M}_0^2 = \begin{pmatrix} M_{V_0}^2 - i\Gamma_{V_0}M_{V_0} & \delta m^2 \\ \delta m^2 & M_{Z_0}^2 - i\Gamma_{Z_0}M_{Z_0} \end{pmatrix}. \tag{2}$$

δm^2 is the $Z_0 - V_0$ coupling, given by

$$\delta m^2 = \frac{(g_V)_t}{\frac{2}{3}e} \left[2\sqrt{3}\sqrt{4\pi\alpha(M_Z)} \frac{2}{3}\sqrt{M_{V_0}} |\Psi(0)| \right], \tag{3}$$

where $\alpha(M_Z)$ is the fine structure constant evaluated at the relevant mass scale and $\Psi(0)$ is the wave function of the $t\bar{t}$ system at the origin. The quantity in brackets is the $\gamma - V$ coupling, which is multiplied by the ratio of the weak charge of toponium to its electromagnetic charge to get the $Z - V$ coupling.

There are two ways one can deal with this mass matrix. It can be diagonalized, yielding a matrix with physical masses on the diagonal. The amplitude for the process $e^+e^- \to \mu^+\mu^-$ is then just the sum of the amplitudes for this process occurring via each of the physical particles, separately. This can be written as:

$$A = (g_Z \quad g_V)_I \left[s - \begin{pmatrix} M_Z^2 - iM_Z\Gamma_Z & 0 \\ 0 & M_V^2 - iM_V\Gamma_V \end{pmatrix} \right]^{-1} \begin{pmatrix} g_Z \\ g_V \end{pmatrix}_F. \tag{4}$$

Here g_V and g_Z are the rotated, or physical, couplings. Equivalently, we have

$$A = (g_{Z_0} \quad g_{V_0})_I \left[s - \begin{pmatrix} M_{Z_0}^2 - iM_{Z_0}\Gamma_{Z_0} & \delta m^2 \\ \delta m^2 & M_{V_0}^2 - iM_{V_0}\Gamma_{V_0} \end{pmatrix} \right]^{-1} \begin{pmatrix} g_{Z_0} \\ g_{V_0} \end{pmatrix}_F \tag{5}$$

where we have rewritten $(\vec{g}_0\mathcal{U})(\mathcal{U}^{-1}(s - \mathcal{M})^{-1}\mathcal{U})(\mathcal{U}^{-1}\vec{g}_0)$, canceling out the unitary transformations $\mathcal{U}$ that rotate couplings and masses from one basis to the other.

In this non-diagonalized basis, if we set $g_{V_0} = 0$, $\Gamma_{V_0} = 0$, and $s = M_{V_0}^2$, we have

$$A = (g_{Z_0} \quad 0)_I \begin{pmatrix} 0 & -\delta m^2 \\ -\delta m^2 & M_{V_0}^2 - M_{Z_0}^2 \end{pmatrix} \begin{pmatrix} g_{Z_0} \\ 0 \end{pmatrix}_F = 0. \tag{6}$$

So in this formalism also, it is easy to see the complete destructive interference.

If we diagonalize the mass matrices, we find that the physical masses and widths are shifted away from their original values. This is not an important effect for the M_Z and M_V, which get shifted equally and oppositely by at most 4 MeV. Similarly, Γ_Z is $\Gamma_{Z_0} - \Delta\Gamma$ and Γ_V is $\Gamma_{V_0} + \Delta\Gamma$; $\Delta\Gamma$ can be as big as 20 MeV—irrelevant for the Z, but very important for the V, which has an unmixed width of 100 keV. This maximal $\Delta\Gamma$ is achieved when the Z and V are degenerate; when they are, e.g., 2 GeV apart, $\Delta\Gamma$ drops to 5 MeV (these $\Delta\Gamma$ are for the $1S$ state).

We have then, for the R-value, the following expression:

$$R = .1365 \frac{\alpha^2(M_Z)}{\alpha^2(m_e)} s^2 \left(\frac{s - M_{V_0}^2}{(s - M_{V_0}^2)(s - M_{Z_0}^2) - \delta m^2} \right)^2 \qquad (7)$$

where .1365 comes from combinations of θ_W.[*] In Fig. 1 I show the results of our calculation: the solid line is exact (we include the V_0's direct couplings and width, and the photon term; we deal with the cross-sections for various helicity combinations separately); the dashed line is the result of ignoring the above parenthetical effects; the effect of the Z_0 alone is shown for comparison (dotted line). The two graphs differ only in scale.

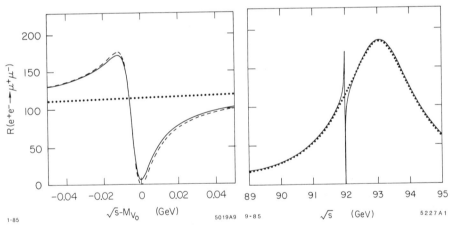

Figure 1. $R(e^+e^- \to \mu^+\mu^-)$ for one toponium state mixing with the Z_0.

3.   WHAT WE WILL SEE: MANY STATES, SMEARING, AND ALL THAT

I begin with a brief review of heavy quarkonia, with particular reference to toponium. These systems are well-described by non-relativistic potential models; for the c and b quark systems, a wide range of successful forms have been proposed.

---

[*] The ratio of α^2's in Eq. (7) is about 1.15. See previous footnote.

That they all work is not too surprising, as they approximately coincide in the range $.1 < R < 1$ fm, where the RMS radii of the observed $c\bar{c}$ and $b\bar{b}$ states lie. However, $t\bar{t}$, due to its large mass, will have a smaller radius, and test a region where the potentials differ. As examples, I choose:

1. Cornell:[4]

$$V(r) = \frac{-.48}{r} + \frac{r}{5.4756(\text{GeV})^{-2}} \qquad (8)$$

—a combination of Coulomb at short range and linear confinement—and

2. Richardson:[5] the single dressed gluon exchange amplitude (in momentum space)

$$\widetilde{V}(q^2) = \frac{3\alpha_s(q^2)}{3q^2} \qquad (9)$$

interpolated with a linear potential (in coordinate space) at large distances.

These two potentials give rather different level spectra (shown in Fig. 2) and wavefunctions; for example, $\psi(0)_{1S}$ is three times larger for Cornell than for Richardson (the Cornell potential, unlike Richardson's, does not incorporate asymptotic freedom, and is more singular at short distances). Since both the bare V widths and those acquired from mixing go as $|\psi(0)|^2$, these numbers increase by a factor of ten for Cornell—resulting in, for example, a maximal width from mixing of .2 GeV.

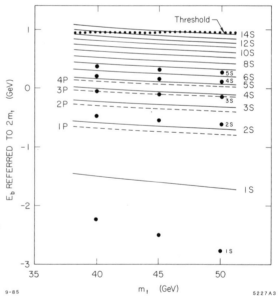

Figure 2. Binding energies of toponium, versus top mass, for Richardson (solid lines, S states; dashed, P states) and Cornell (dots–S states only) potentials. The threshold for open top production, calculated using the Richardson potential, is also shown (dotted line).

9-85

m_t (GeV)

5227A3

It is then a simple extension of the formalism developed in Sec. 2 to deal with the mixing of the Z_0 with the (about) thirteen toponium states we expect below the open top threshold. We obtain cross-sections such as the one shown in Fig. 3. The height of the "spikes" are given by

$$C * s^2 \left| \frac{1}{s - M_{Z_0}^2 + iM_{Z_0}\Gamma_{Z_0} - \frac{\delta m^2}{s - M_{V_0}^2}} \right|^2 . \tag{10}$$

The peak of a given V_0 resonance occurs when the real part of the denominator vanishes, at a value of s very close to $M_{V_0}^2$. The height of the peak is thus $C * s^2/(m_{Z_0}\Gamma_{Z_0})^2$. The height of the Z_0 peak can also be gotten from Eq. (10), by dropping the V_0 mixing term; maximizing, we obtain the exact expression found for the spike (s equals the relevant mass squared). This explains why all the peaks, including that of the Z_0, are on the same gently rising curve. (In Fig. 3 the P-states are ignored; they cause similar spikes, but are unobservably narrow, as their coupling, and hence acquired width, is suppressed relative to the S-states.)

Of course, real machines, such as SLC and LEP, will not resolve these very narrow spikes; we must convolute the curves with a Gaussian (with width related to the beam spread) in order to approximate what will be measured. In Fig. 4, I show R for the Z alone, and Fig. 3 convoluted with Gaussians appropriate to $\sigma_{\mathrm{beam}} = 40$ MeV and 100 MeV. LEP is expected to run (without wigglers) at the former beam width; SLC is expected to achieve the latter, and perhaps with special effort, the former.

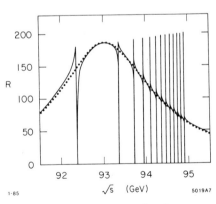

Figure 3. $R(e^+e^- \to \mu^+\mu^-)$ for several toponium states mixing with the Z (Richardson potential, $m_t = 47$ GeV). The dotted line is the Z_0 alone.

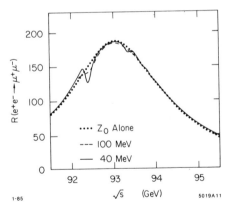

Figure 4. $R(e^+e^- \to \mu^+\mu^-)$, smeared, for various expected beam widths.

316

I next remark that even for a V relatively far away from the Z, the enhancement due to mixing should be quite noticeable. The height of the peak does not decrease, though its width does. The smeared height is therefore greatly reduced, but should be compared to the also much reduced background due to the Z.

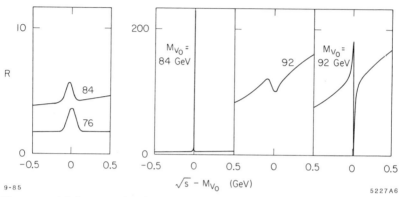

Figure 5. $R(e^+e^- \to \mu^+\mu^-)$ smeared and not, for $M_{V_0} = 76$, 84 and 92 GeV.

I now present smeared polarization and forward-backward asymmetries for various values of M_{V_0}. These are found by calculating the cross-sections (for each individual helicity configuration), smearing them, and then taking the appropriate differences and ratios. Since the asymmetries also crucially depend on the $ZV\gamma$ interference, the results do not seem to have a simple qualitative explanation. In Fig. 6 I show the asymmetries; the effects are in fact more striking for V moderately far away from Z.

All the results I have shown so far used the Richardson potential. I shall briefly show the effects of using the Cornell potential, and the Richardson potential combined with a non-standard Higgs sector. Consider the 2-Higgs model of Glashow, Weinberg and Paschos,[6] where one Higgs couples to up-type quarks, and one to down-type. There is a neutral-Higgs (H_0) exchange contribution to the toponium potential, where the H_0 coupling is enhanced by the vacuum-expectation-value ratio ξ/η (ξ being the VEV of the Higgs coupling to down type quarks and η to up-type). The extra contribution is an attractive Yukawa, in momentum space

$$-\left(\frac{\xi}{\eta}\frac{gm_t}{2M_W}\right)^2 \frac{1}{m_H^2 + q^2} \qquad \text{or} \qquad -\left(\frac{\xi}{\eta}\frac{gm_t}{2M_W}\right)^2 \frac{e^{-rm_H}}{4\pi r} \qquad (11)$$

in coordinate space. This addition has the effects of increasing the wavefunctions at the origin, since it pulls in the wavefunctions, and of lowering states (increasing binding energies); it changes the level spacings, since it affects the lowest lying states the most. Finally, if the Higgs term is strong enough[*] it has a very curious

---

[*] that is, ξ/η equals about 5, if we are using the Cornell potential, or 10, for Richardson.

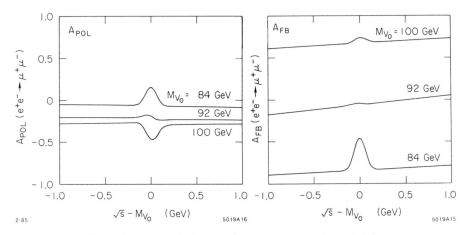

Figure 6. A_{pol} and A_{fb} for three different values of M_{V_0}.

effect—it causes the 2S state to lie below the 1P. This effect does not happen for any standard quarkonium potential, and is related[7] to the fact that $\Delta V(r) < 0$ for the Higgs potential and not so for any standard quarkonium potential. In Fig. 7, I show $R(e^+e^- \to \mu^+\mu^-)$, smeared ($\sigma_{\text{beam}} = 40$ MeV), for Richardson alone, Cornell alone, and Richardson with Higgs[†]. Note the qualitative similarity between the second and third figure.

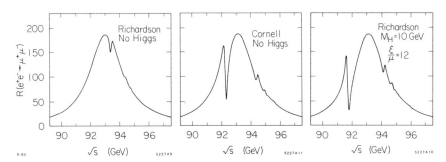

Figure 7. Effects of varying quarkonium potential.

In summary, we have seen that toponium and the Z_0 almost completely destructively interfere. Toponium states pick up a large width from mixing—the 1S state, with a bare width of 100 keV, can acquire a width of as much as 20 MeV

---

† the parameters have been chosen to be dramatic; they are all but excluded by $B\bar{B}$ mixing[3]

(using the Richardson potential). While the beam widths of machines such as SLC and LEP will greatly blur the sharp spikes that we find, effects will be visible as wiggles in cross-sections and asymmetry parameters. The exact potential for toponium (and thus exactly what we will see) is not very well known. The Higgs (in a 2-Higgs model) can have noticeable effects, but it may be hard to distinguish these effects from those of different potentials; the 2S-1P level inversion is a possible qualitative difference, if the Higgs couplings are rather large.

REFERENCES

1. P. J. Franzini and F. J. Gilman, Phys. Rev. **D32**, 237 (1985). See this paper for a complete set of references.

2. Similar work to Ref.1 has been done by S. Güsken, J. H. Kühn, and P. M. Zerwas, SLAC PUB 3580; J. H. Kühn, and P. M. Zerwas, Phys. Lett. **154B**, 448 (1985); L. J. Hall, S. F. King, and S. R. Sharpe, Harvard preprint HUTP-85/A012 (1985) (unpublished).

3. G. G. Athanasiu, P. J. Franzini, and F. J. Gilman, SLAC PUB 3648 (1985).

4. E. Eichten et. al., Phys. Rev. **D17**, 3090 (1978); **D21**, 203 (1980).

5. J. L. Richardson, Phys. Lett. **82B**, 272 (1979).

6. S. Glashow and S. Weinberg, Phys. Rev. **D15**, 1958 (1977), E. A. Paschos, Phys. Rev. **D15**, 1966 (1977).

7. A. Martin, CERN preprint TH4060/84 (1984) (unpublished).

Higgs Boson Spectrum From Infrared Fixed Points

CHUNG NGOC LEUNG

Fermi National Accelerator Laboratory

P. O. Box 500, Batavia, Illinois 60510

ABSTRACT

The fixed point structure of the renormalization group equations for the scalar quartic couplings in the one and two-doublet models is studied. Masses of the physical Higgs bosons can be determined by the infrared fixed points of the quartic coupling constants. The existence of these fixed points in the two-doublet model requires the presence of a heavy fourth generation in which quarks are coupled to both doublets. Otherwise, the potential can become quartically unstable at low energies for arbitrary initial stable values of the coupling constants.

Despite the successes of the standard model, the Higgs sector remains a mystery. The main reason is that the mass of the physical Higgs boson m_H depends on the quartic coupling constant λ in the scalar potential, which is a free parameter of the model. Many attempts [1-5] have been made to constrain or to predict λ (and hence m_H) all of which involve extra assumption(s). Pendleton and Ross[3] first suggested the interesting possibility that low energy physics may be dictated by the infrared (IR) fixed point structure of the renormalization group equations (RGE). If the RGE for the various couplings in a theory possess stable IR fixed points, the couplings will be swept towards these fixed points when evolving from high energy (e.g., a unification scale Λ_U) to low energy (e.g., the weak interaction scale $\Lambda_w \sim M_w$), irrespective of their initial values. Consequently predictions for low energy parameters can be obtained without knowledge of the symmetry conditions at Λ_U. Implicit in this approach is the assumption that a desert exists and perturbation theory is valid throughout the desert.

The work[6] I am going to describe was done in collaboration with Chris Hill and Sumathi Rao. We study numerically the IR fixed point structure of RGE for the scalar quartic couplings in the standard model and its extension to two Higgs doublets. The masses of the physical Higgs bosons are determined from the fixed points of the quartic coupling constants.

In the standard model with one Higgs doulbet, the physical Higgs (mass)^2 is $m_H^2 = \lambda \nu^2$, where $\nu/\sqrt{2} = 175$ GeV is the vacuum expectation value and λ is the quartic coupling constant in the potential

$$V(\phi) = \mu^2 \phi^\dagger \phi + \frac{\lambda}{2}(\phi^\dagger \phi)^2 \,. \tag{1}$$

In the two-doublet model the potential is

$$
\begin{aligned}
V(\phi_1, \phi_2) = {} & \mu_1^2 \phi_1^\dagger \phi_1 + \mu_2^2 \phi_2^\dagger \phi_2 + \frac{\lambda_1}{2}(\phi_1^\dagger \phi_1)^2 + \frac{\lambda_2}{2}(\phi_2^\dagger \phi_2)^2 \\
& + \lambda_3 (\phi_1^\dagger \phi_1)(\phi_2^\dagger \phi_2) + \lambda_4 |\phi_1^\dagger \phi_2| \\
& + \frac{\lambda_5}{2}[(\phi_1^\dagger \phi_2)^2 + h.c.] \,.
\end{aligned}
\tag{2}
$$

There are four physical Higgs bosons, one charged, one neutral pseudoscalar and two neutral scalars. Their (mass)^2 are, respectively,

$$
\begin{aligned}
m_\pm^2 &= -\frac{1}{2}(\lambda_4 + \lambda_5)\nu^2 \\
m_p^2 &= |\lambda_5|\nu^2 \\
m_1^2 &= \frac{1}{2}\eta_+ \nu^2 \\
m_2^2 &= \frac{1}{2}\eta_- \nu^2 \,,
\end{aligned}
\tag{3}
$$

where

$$
\begin{aligned}
\eta_\pm = {} & (\lambda_1 cos^2\beta + \lambda_2 \sin^2\beta) \\
& \pm [(\lambda_1 \cos^2\beta - \lambda_2 \sin^2\beta)^2 + (\lambda_3 + \lambda_4 + \lambda_5)^2 \sin^2 2\beta]^{1/2}
\end{aligned}
\tag{4}
$$

and

$$tan\beta = \frac{\nu_2}{\nu_1} \,. \tag{5}$$

Here $\nu_1/\sqrt{2}(\nu_2/\sqrt{2})$ is the vacuum expectation value of $\phi_1(\phi_2)$ and $\nu^2 = \nu_1^2 + \nu_2^2$.

In order to have a residual $U(1)_{EM}$ symmetry so that the photon remains massless, λ_4 must be negative. Then the rquirement that the potential energy of the vacuum be bounded below necessarily implies the following conditions in tree-approximation:

$$\lambda_1 > 0,$$

$$\lambda_2 > 0, \tag{6}$$

$$\text{and} \quad \sqrt{\lambda_1 \lambda_2} > -\lambda_3 + |\lambda_4| + |\lambda_5| \, .$$

The RGE for the gauge, Yukawa and quartic couplings are given in Ref. 6. The RGE are numerically integrated from $\Lambda_U = 10^{15}$ GeV to $\Lambda_w = 100$ GeV assuming large initial values for the Yukawa and quartic couplings (by large couplings we do not mean the saturation of unitarity bounds, $g^2 \sim 4\pi$, but rather g^2 "of order unity"). For simplicity the Yukawa couplings are assumed to be diagonal[7]. The Yukawa couplings of the light fermions (lighter than the t-quark) are set to be zero since they have negligible effects on the evolution of the coupling constants except for the counting in the gauge coupling beta-functions. Fixed points of the scalar quartic couplings are universal values attained from a sample of random (but satisfying Eq. (6)) initial values.

We now discuss the results. Consider first the one-doublet model with three fermion generations. The evolution of the quartic coupling λ and the Yukawa coupling of the t-quark g_t is shown in Fig. 1a. For sufficiently large initial values of g_t λ reaches a fixed point. For smaller initial g_t it terminates on the dashed line of Fig. 1a. This can be turned into a relationship between m_H and m_t, the mass of the t-quark, which is shown in Fig. 1b. Curiously, m_H lies around 170 GeV for a large range of m_t. If more heavy fermions are present, m_H tends to increase.

In the two-doublet model fixed points exist only if both doublets are coupled to heavy quarks. This can be understood from the structure of the RGE. Suppose all Yukawa couplings are negligible. Then the RGE for λ_1 (and similarly λ_2) can be written as (g_2 and g_1 are the gauge couplings for SU(2) and U(1), respectively)

$$16\pi^2 \mu \frac{\partial \lambda_1}{\partial \mu} = 12[\lambda_1 - \frac{1}{8}(3g_2^2 + g_1^2)]^2 + 2\lambda_3^2 + 2(\lambda_3 + \lambda_4)^2 + 2\lambda_5^2$$

$$+ \frac{9}{16}(g_2^4 + g_1^4) + \frac{3}{8}g_2^2 g_1^2 \, . \tag{7}$$

Notice that the right hand side is always positive and thus no fixed point exists. Consequently $\lambda_1(\lambda_2)$ always decreases from its initial value when evolved to lower energy. In fact it decreases so fast that it can exit the stability region becoming negative and eventually negative infinite, thereby causing other couplings to diverge. This suggests the interesting possibility that a theory which is perturbatively weak at high energies can become non-perturbatively strong at lower energies. This is reminiscent of technicolor and might be applied to generate the breaking of the electroweak symmetries. Coupling ϕ_1 (or ϕ_2) to a heavy quark with Yukawa coupling f introduces to the RGE of λ_1 (or λ_2) the terms $12\lambda_1 f^2 - 12 f^4$. The f^4-term is negative and hence a fixed point can exist if f is sufficiently large. Coupling to heavy lepton with equal Yukawa coupling introduces the similar terms but 3 time smaller, hence less effective in driving $\lambda_1(\lambda_2)$ to its fixed point.

Let us consider then the following Yukawa coupling scheme consistent with the natural suppression of off diagonal neutral couplings[8]

$$L_{yuk.} = \bar{Q}_L U u_R \phi_1^c + \bar{Q}_L D d_R \phi_2 + \bar{L}_L \mathcal{L} \, e_R \phi_2 + h.c. \, , \tag{8}$$

in which at least one of each charged fermion species has a large Yukawa coupling (corresponding to coupling scheme I in Ref. 6). This necessarily implies the existence of a fourth generation. Here $Q_L(L_L)$ is the left-handed quark (lepton) doublet, u_R, d_R, and e_R are, respectively, the right-handed up-quarks, down-quarks, and charged leptons, and U, D and $\mathcal{L}$ are Yukawa coupling matrices. Generation indices have been suppressed. Fixed points for the quartic couplings exist in this case and are shown in Fig. 2. Notice that the fixed point for λ_4 is negative (corresponding to a massless photon) for arbitrary initial values. It is somewhat remarkable that the renormalization group fixed point will select the physically interesting vacuum!

The masses of the physical Higgs particles can be determined from Eq. 3 with the fixed point values of the quartic couplings. These are shown in Table 1. The masses of the neutral scalars $m_{1,2}$ depend on the unknown vacuum expectation value ratio ν_2/ν_1. However, they lie within a finite region (as shown in Table 1) for ν_2/ν_1 ranging from 0 to ∞. (A recent renormalization group analysis[9] constrains $(\nu_2/\nu_1)^2$ to be less than 60. This does not affect the range shown in Table 1). The mass of the pseudoscalar m_p depends on λ_5 whose fixed point is zero (since the right hand side of the RGE is proportional to λ_5) but is never reached in the finite running time, although it tends to be small. The values of m_p shown in Table 1 are for typical values of λ_5 at Λ_w.

Also shown in Table I is the dependence of the Higgs masses on m_t. There is essentially no difference between $m_t = 0$ and $m_t = 50$ GeV, whereas a heavier

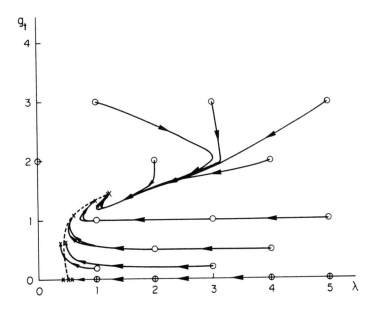

Fig. 1a. Flow of λ and g_t towards fixed points in the standard one-doublet model.

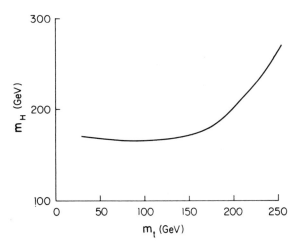

Fig. 1b. Relation between the Higgs mass and the t-quark mass in the one-doublet model.

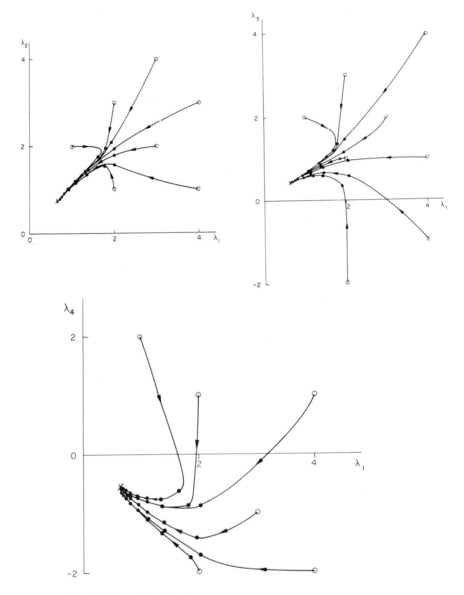

Fig. 2 Flow of $\lambda_1, \lambda_2, \lambda_3,$ and λ_4 towards fixed points.

Table I. Fixed point masses (in GeV) of the physical Higgs particles (N = number of generations).

	m_t	$m_\pm$	m_1	m_2	m_p
	0	188	$190 - 231$	$0 - 105$	"61"
$N = 4$	50	186	$188 - 225$	$0 - 105$	"59"
	172	136	$164 - 216$	$0 - 131$	"25"
	0	142	$144 - 191$	$0 - 106$	"21"
$N = 5$	50	140	$143 - 186$	$0 - 105$	"21"
	135	118	$133 - 174$	$0 - 110$	"13"

t-quark tends to lower the masses. Similar effects occur with the addition of extra heavy generations as seen in the case with N=5. All the masses lie in an interesting range which is accessible to experiment in the near future.

Other distinct Yukawa coupling schemes are also considered in Ref. 6. We do not have the space-time to discuss them here.

ACKNOWLEDGEMENTS

I wish to thank C. Hill and S. Rao for a most enjoyable collaboration. The hospitality of the Aspen Center for Physics where part of this talk was prepared is also gratefully appreciated.

REFERENCES

[1]   A.D. Linde, JETP Lett. **23,** 64(1976); S. Weinberg, Phys. Rev. Lett. **36,** 294 (1976).

[2]   B.W. Lee, C. Quigg and H.B. Thacker, Phys. Lett. **38,** 883 (1977); Phys. Rev. **D16,** 1519 (1977); M.Veltman, Acta Phys. Pol. **B8,** 475 (1977); D.Dicus and V. Mathur, Phys. Rev. **D7,** 3111 (1973).

[3]   B. Pendleton and G. Ross, Phys. Lett. **98B,** 291 (1981).

[4]   N. Cabibbo, *et al.*, Nucl. Phys. **B158,** 295 (1979); M. Machacek and M. Vaughn, Phys. Lett. **103B,** 427 (1981).

[5] L. Maiani, G. Parisi and R. Petronzio, Nucl. Phys. **B136,** 115 (1978);
R. Dashen and H. Neuberger, Phys. Rev. lett. **50,** 1897 (1983); D.J.E. Callaway, Nucl. Phys. **B233,** 189 (1984); M.A.B Bég, C. Panagiotakopoulos and
A. Sirlin, Phys. Rev. Lett. **52,** 883 (1984); E. Ma, Phys. Rev. **D31,** 1143
(1985); K. S. Babu and E. Ma, UH-511-552-85 (1985).

[6] C. T. Hill, C. N. Leung and S. Rao, Fermilab-Pub-85/56-T (to be published
in Nucl. Phys. B).

[7] E. A. Paschos, Z. Phys. **C26,** 235 (1984); J. W. Halley, E. A. Paschos and
H. Usler, DO-TH 84/24 (1984).

[8] S. Weinberg, Phys. Rev Lett. **37,** 657 (1976); S. Glashow and S. Weinberg,
Phys. **D5,** 1958 (1977).

[9] J. Bagger, S. Dimopoulos and E. Masso, SLAC-PUB-3587 (1985).

ELECTROWEAK INTERFERENCE IN e^+e^- COLLISIONS

Jorge H. Moromisato
Dept. of Physics, Northeastern University
Boston, Massachusetts 02115

1. INTRODUCTION

The basic process in e^+e^- collision is the anihilation of the colliding particles into a virtual photon, and its immediate decay into either a lepton pair or a quark-antiquark pair: $e^+e^- \to \gamma \to f\bar{f}$. But we also have the corresponding weak process: $e^+e^- \to Z^0 \to f\bar{f}$, where $f\bar{f}$ stands for fermion pair. At the relatively low energy of PEP and PETRA this second process is largely suppressed but its effects can still be observed mainly by its interference with the electromagnetic process. This interference gives rise to a Forward-Backward asymmetry in the angular distribution of the fermion pair. This paper will be concerned mainly with the current experimental results on $\mu\mu$, $\tau\tau$, ee, $c\bar{c}$, and $b\bar{b}$ asymmetries, from PEP and PETRA.

2. THEORETICAL PREDICTIONS.

To lowest order (Born approximation) the cross section for the process $e^+e^- \to f\bar{f}$, is given in the Standard Model[1] by:

$$\frac{d\sigma}{d\Omega} = \frac{\alpha^2}{4s} Q_f^2 \left[A_0(1 + cos^2\theta) + A_1 cos\theta \right] \tag{1}$$

where
$$A_0 \equiv 1 + \frac{2Re\chi}{-Q_f} g_V^e g_V^f + |\frac{\chi}{Q_f}|^2 (g_V^{e\,2} + g_A^{e\,2})(g_V^{f\,2} + g_A^{f\,2}) \tag{2}$$

$$A_1 \equiv \frac{4Re\chi}{-Q_f} g_A^e g_A^f + 8\,|\frac{\chi}{Q_f}|^2 g_V^e g_V^f g_A^e g_A^f \tag{3}$$

and
$$\chi \equiv \frac{1}{sin^2 2\theta_w} \frac{s}{s - M_Z^2 + iM_Z\Gamma_Z} \tag{4}$$

or
$$\chi \equiv \frac{\rho G_\mu}{2\pi\alpha\sqrt{2}} M_Z^2 \frac{s}{s - M_Z^2 + iM_Z\Gamma_Z} \tag{4a}$$

The standard notation is followed throughout, e.g: 's' is the c.m. energy squared, 'α' is the fine structure constant, 'ρ' is the ratio of strenghts of the neutral coupling to the charge coupling, which in the Standard Model is set to equal to 1.0, 'G_μ' is the Fermi constant determined from μ decay data. The total cross section is, integrating (1): $\sigma = \frac{\alpha^2}{4s}\left(2\pi\frac{8}{3}A_0\right) = \frac{4}{3}\frac{\pi\alpha^2}{s}A_0 = \sigma_{QED}\,A_0$ thus A_0 is just the ratio of the total cross section to the lowest order QED cross section, and is usually known as $R_{f\bar{f}}$:

$$R_{f\bar{f}} \equiv A_0 = 1 + \frac{2Re\chi}{-Q_f}\,g_V^e\,g_V^f + \qquad O(\chi^2) \qquad (5)$$

The Forward-Backward asymmetry is defined as: $A_{f\bar{f}} \equiv \frac{\sigma_F - \sigma_B}{\sigma}$, and we have, in terms of the weak coupling constants:

$$A_{f\bar{f}} \equiv \frac{3}{2}\frac{Re\chi}{-Q_f}\frac{g_A^e\,g_A^f}{R_{f\bar{f}}} + \qquad O(\chi^2) \qquad (6)$$

The $O(\chi^2)$ terms in eqs. (5) and (6) contain the factor g_V which in the lepton pair case is almost zero, and can be neglected. The situation is different in the case of quark pairs as can be seen from Table 1.

Table 1
Weak coupling constants in the Standard Model

fermion	Q_f	g_A	g_V ($sin^2\theta_W = 0.22$)
e, μ, τ	-1	$-\frac{1}{2}$	-0.04
u, c, t	$\frac{2}{3}$	$\frac{1}{2}$	0.19
d, s, b	$-\frac{1}{3}$	$-\frac{1}{2}$	-0.35

3. RADIATIVE CORRECTIONS

It is now customary to separate the radiative corrections, to the so called 'one loop' level, into three parts[2]:

I. QED corrections to the photon exchange diagram (fig. 1),
II. QED corrections to the Z^0 exchange diagram (fig. 2),
III. Weak self field, or vacuum polarization of the Z^0 (fig. 3).

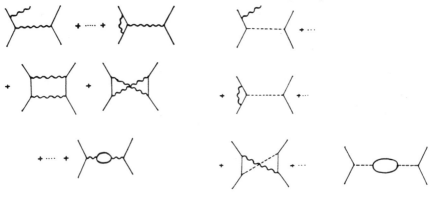

Fig. 1 Reduced QED diagrams   Fig. 2 Z^0 exchange   Fig. 3 Z^0 self energy.

Correction I, also known as 'Reduced QED' is the largest of all, it lowers by $\sim 1/3$ the magnitude of the asymmetry effect at PEP energies, i.e. these QED processes produce a FB asymmetry of $\sim 2\%$ which is opposite to the one from the interference term. Correction II is only about 1/5 as large, but has the same sign, as correction I. And correction III is very close in magnitude to correction II but has the opposite sign; i.e. II and III tend to cancel each other. Usually only the Reduced QED correction (I) is applied to the data, using Monte Carlo techniques[3], thus the experimental numbers can be compared directly to the Born approximation values.

To lowest order the expresion for the FB asymmetry is given by eq. (6) but there are two different expressions for the term $'\chi'$: eqs (4) and (4a). These two expressions are supposedly equivalent but in fact eq. (4a) gives a numerical value for χ which is about 7% larger than for eq. (4). Two arguments have been put forward to explain this discrepancy; a) the numerical value of χ, to lowest order, depends on the specific parametrization (or renormalization parameters) chosen, and, b) The constant G_μ, which is determined from measurements of the muon decay rate, already contains some of the weak radiative correction, the largest of them being the weak vector self energy[4], which happens to be a $\sim 7\%$ effect. These two explanations may or may not be equivalent but the bottom line is that the various theoretical calculations[5] of the one loop corrected FB asymmetry are essentially the same regardless of which of the two expressions they start with.

4. EXPERIMENTAL RESULTS: LEPTON SECTOR.

4.1 Results on $e^+e^- \to \mu^+\mu^-$

This is the process in which the most accurate measurement of the FB asymmetry have been performed. Experiments at PEP and at PE-TRA have masured this quantity and their results are all in reasonable agreement with the Standard Model prediction. The highest statistics data come from the MAC experiment whose result are shown in Fig. 4. Most of the reported results have been corrected only for the reduced QED radiation, which, as explained before, gives the same result, to within the present theoretical uncertainty, as a 'complete radiative correction' (i.e. one including the three types mentioned in Section 3). A couple of experiments use a 'Full QED' correction, namely type I plus type II, which means that their data

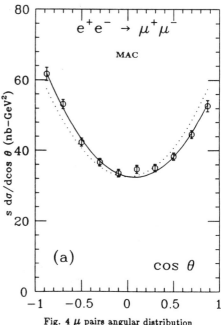

Fig. 4 μ pairs angular distribution

contain the effect of the Z^0 selfenergy. Those results should, therefore, be compared to theoretical numbers which are about 7% larger in numerical value than the one given by eq. (4). The experimental results shown in Table 3 contain either the reduced QED or the complete radiative correction, and have been modified, whenever necessary, by this 7% factor, in order to make them compatible.

Table 2

FB Asymmetry in $e^+e^- \to \mu^+\mu^-$

Experiment	$\sqrt{s}$ (Gev)	N_{events}	$A_{\mu\mu}$ (%)	Theory
HRS[HR1]	29.0	5057	-4.9 ± 1.6	-5.9
MAC[M1]†	29.0	16058	$-5.9. \pm .8 \pm .2$	-5.9
MARK II[MK1]	29.0	5312	-7.1 ± 1.7	-5.9
PEP (comb)	29.0	26427	$-5.9 \pm .67$	-5.9
CELLO[C1]†	34.2	387	-6.0 ± 6.4	-8.5
JADE[J1]	34.4	3405	$-11.1 \pm 1.8 \pm 1.0$	-8.7
MARK J[MJ1]	34.6	3658	-11.7 ± 1.9	-8.8
PLUTO[P1]	34.7	1550	$-13.4 \pm 3.1 \pm < 1$	-8.9
TASSO[T1]†	34.5	2673	$-9.1 \pm 2.3 \pm .5$	-8.7
PETRA$_{Low}$(comb)	34.5	11673	-11.2 ± 1.1	-8.8
CELLO[C2]	43.9	477	-18.3 ± 5.9	-15.6
JADE[J1]	43.0	571	$-20.1 \pm 4.3 \pm 1.7$	-14.9
MARK J[MJ2]	39.8	341	-10.9 ± 5.7	-12.0
	44.6	673	-16.7 ± 4.0	-16.0
TASSO[T2]	43.7	483	-19.0 ± 5.0	-15.4
PETRA$_{High}$(comb)	43.3	2545	-17.3 ± 2.2	-15.1

† Uses Full QED correction; result reduced by 7%. See text.

Table 3 shows that PETRA's numbers are somewhat larger, in absolute value, than expected (about two $\sigma's$, for their combined 'low energy' data).

4.2 Results on $e^+e^- \to \tau^+\tau^-$

The efficiency for τ pair reconstruction is lower than for μ pairs since only the decay products of the $\tau's$ are detected, and not all of the decay channels are used. This account for the somewhat lower statistics and higher systematic errors of the FB asymmetry measurements. The current experimental results are presented in Table 4.

Table 3

FB Asymmetry in $e^+e^- \to \tau^+\tau^-$

Expt	$\sqrt{s}$ (Gev)	N_{events}	$A_{\tau\tau}$ (%)	Theory
HRS[HR3]	29.0	2518	-6.1 ± 2.4	-5.9
MAC[M2]	29.0	10155	$-5.5 \pm 1.2 \pm .5$	-5.9
MARK II[MK1]	29.0	3714	-4.2 ± 2.0	-5.9
PEP (comb)	29.0	16387	$-5.3 \pm .95$	-5.9
CELLO[C3]	34.2	434	-10.3 ± 5.2	-8.5
JADE[J2]	34.6	1998	$-6.0 \pm 2.5 \pm 1.0$	-8.8
MARK J[MJ1]	34.6	860	-7.8 ± 4.0	-8.8
TASSO[T3]	34.4	350	$-4.9 \pm 5.3 \pm 1.5$	-8.7
PETRA$_{Low}$(comb)	34.5	3642	-6.9 ± 1.0	-8.7
CELLO[C2]	43.9	397	$-15.4 \pm 5.2 \pm 1.1$	-15.6
JADE[J3]	43.0	800	$-18.8 \pm 4.6 \pm 1.0$	-14.8
PETRA$_{High}$(comb)	43.1		-18.3 ± 2.2	-15.0

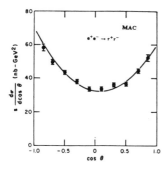

Fig. 5 τ pairs angular distribution.

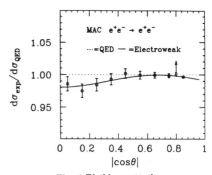

Fig. 6 Bhabha scattering.

4.3   Result on $e^+e^- \to e^+e^-$ (Bhabha Scattering)

For Bhabha scattering the contribution from the space like diagrams, Fig. 7, which is absent in the other leptonic final states processes, dominates the cross section and all but swamps out the electroweak effect.

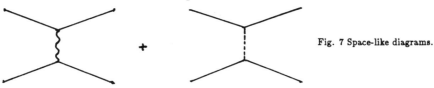

Fig. 7 Space-like diagrams.

In a high statistics experiment one can see some deviation from QED in the angular distribution of these events. Fig. 6 shows recent results from the MAC experiment, in which the weak interaction effect is clearly present. Results from Bhabha Scattering have been presented by other experiments as well[HR3][MK1][C4][MJ3][T1]

4.4   Electroweak Parameters from Lepton Pair asymmetry.

From the combined PEP results on μ pair asymmetry, and using eq. (6) with the Z^0 mass fixed at $M_Z = 93$ Gev, we find for the axial coupling: $g_A^e g_A^\mu = .5 \pm .05$. Alternatively, if we assume that $\sin^2\theta_W = .22$, we can determine with good precision the mass of the Z^0: $M_Z = 93 \pm 5$ Gev. A remarkable feat considering that the c.m. energy at PEP, or at PETRA, is so far below the Z^0 mass shell.

5.   EXPERIMENTAL RESULTS: THE QUARK SECTOR.

5.1 Results on $e^+e^- \to c\bar{c}$

The majority of results on charm quark asymmetry come from experiments using reconstructed D^* as a tag for the charm pair creation process. The technique used to reconstruct D^* was pioneered by Mark II[6] and takes advantage of the fact that the Q value in the decay $D^{*+} \to D^0 + \pi^+$ is very low, the mass difference $(M_{D^{*+}} - M_{D^0})$ being equal to 145 Mev. Thus, even though the individual mass resolution may be poor, the resolution on their mass difference is much better, allowing for the selection of a rather clean sample of c quark events (fig. 8a).

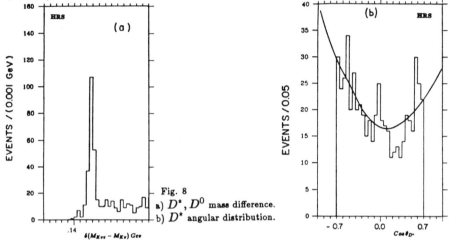

Fig. 8
a) D^*, D^0 mass difference.
b) D^* angular distribution.

Fig. 8b shows the angular distribution of the D^{*+}, whose direction is very much the same as the initial c quark states, as seen by the HRS Experiment at PEP.

Another procedure to isolate charmed quark events is by using lepton tagging of heavy flavors. in this method the semileptonic decay of heavy quark mesons is identified by the high transverse momentum of the prompt muons or electrons.

Table 4 gives the current results on charm asymmetry. The combined average for the charm axial coupling constant is $g_A^c = 0.49 \pm 0.1$.

Table 4
FB Asymmetry in $e^+e^- \rightarrow c\bar{c}$

Expt	$\sqrt{s}$	$A_{c\bar{c}}$ (%)	Theory	g_A^c	Method
HRS[HR4]	29.0	-14 ± 5	-9.5	.76 $\pm$ .26	D^*, D^+, D^0
MAC[M4]	29.0	-5 ± 11	-9.5	.82 $\pm$ 1.8	μ tag
TPC[TP1]	29.0			.95 $\pm$ .53	μ, e tags
JADE[J5]	34.4	-14 ± 9	-12.9	.54 $\pm$ .35	D^*
MARK J[MJ4]	34.5	-16 ± 9	-13.5	.59 $\pm$ .33	μ tag
TASSO[T4]	34.4	-10 ± 9	-12.9	.39 $\pm$ .35	D^*, e tag

5.2 Results on $e^+e^- \rightarrow b\bar{b}$

Since no meson containing a b quark can be used in the same way as the D^* for charm, all the measurements of b quark asymmetry have been done using the above mentioned lepton tagging technique. This measurement benefits from two important advantages over the charm case: the effect is twice as large and the selection of a b enriched sample is more efficient because the decay lepton tends to have a larger transverse momentum. However the cross section for $b\bar{b}$ production is only one fourth of the charm cross section and this fact tends to even out the score. The best measurement of the b quark FB asymmetry is the one by the JADE Collaboration ,using inclusive muon events at $\sqrt{s} = 34.6 Gev$, which is shown in Fig. 9.

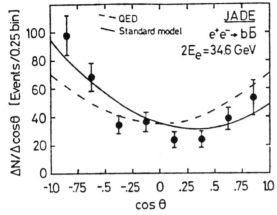

Fig. 9 b quarks angular distibution

Table 5 gives the present experimental results from PEP and PETRA. The combined average for the axial coupling constant is $g_A^b = 0.49 \pm 0.1$.

Table 5
FB Asymmetry in $e^+e^- \to b\bar{b}$

Expt	$\sqrt{s}$	$A_{b\bar{b}}$ (%)	Theory	g_A^b	Method
TPC[TP1]	29.0			$-.69 \pm .42$	μ ,e tags
MAC[M4]	29.0	-7.4 ± 9.2	-19.0	$-.30 \pm .37$	μ tag
MARK II[MK2]	29.0			$-.75 \pm .35$	μ ,e tags
CELLO[C5]	34.0	-40 ± 17	-24.6	$-.81 \pm .69$	μ ,e tag
JADE[J6]	34.4	$-22.8 \pm 6.0 \pm 2.5$	-25.2	$-.45 \pm .12 \pm .05$	μ tag
MARK J[MJ4]	34.6	-21 ± 19	-25.5	$-.41 \pm .37$	μ tag
TASSO[T5]	34.5	-30 ± 18	-25.4	$-.59 \pm .35$	μ ,e tag

4. CONCLUSIONS.

o The relative precision in the Electro-Weak asymmetry measurements, for the leptonic sector, is now better than 10%, and, except as noted below, are in very good agreement with the Standard Model predictions.

o The combined results on μ-pair asymmetry from PETRA are about two std. devs. off from the expected values. Their τ-pair results look normal, although with larger error bars.

o The weak axial coupling constant of c and b quarks have been measured with combined relative precisions of around 20%, and agree with the Standard Model values.

6. AKNOWLEDGMENTS

I gratefully aknowledge the help of my colleagues in the MAC Collaboration, particularly from E. Von Goeler and Horst Blume. Special thanks to B. Naroska who filled me in on the latest from PETRA.

7. REFERENCES

[1] S.L.Glashow, A.Salam, S.Weinberg,Rev.Mod.Phys. **52** ,515 (1980)

[2] M.Böhm and Hollik, Phys.Lett. **139B**,213 (1984)

[3] A.Berends and R.Kleiss, Nucl.Phys. **B177**,237 (1981),
    A.Berends, R.Kleiss and S.Jadach, Nucl.Phys. **B202**,63 (1982)

[4] W.Marciano and A.Sirlin, Phys.Rev. **D29**,945 (1984)

[5] see for example
    R.W.Brown,R.Decker, and E.A. Paschos, Phys.Rev.Lett. ,**52**,1192 (1984);
    B.W. Lynn and R.G.Stuart, Nucl.Phys.,**B253**,216 (1985)

[6] G.J.Feldman et al., Phys.Rev.Lett. **38**,1313 (1977)

[C1] CELLO Coll. H.J.Behrend, et al. Z.Phys. **C14**,283 (1982)

[C2] CELLO Coll. Contributed Paper to the Bari Conf. July,1985

[C3] CELLO Coll. H.J.Behrend, et al. Phys.Lett. **114B**,282 (1982)

[C4] CELLO Coll. H.J.Behrend, et al. Z.Phys. **C16**,301 (1983)

[C5] CELLO Coll. H.J.Behrend, et al. Z.Phys. **C19**,291 (1984)

[HR1] HRS Coll., M.Derrick, et al. Phys. Rev. **D31**,2352 (1985)

[HR2] HRS Coll. K.K.Gan, et al. Phy.Lett. **153B**,116 (1985)

[HR3] HRS Coll. Contributed to this meeting. Preprint ANL-HEP-CP-85-80.

[HR4] HRS Coll. S.Abachi et al. Contributed Paper to the Bari Conf. July,1985

[J1] JADE Coll. W.Bartel, et al. Z. Phys. **C26**,507 (1985)

[J2] JADE Coll. W.Bartel, et al. Phys.Lett. **123B**,353 (1983)

[J3] JADE Coll. W.Bartel, et al. DESY 85-065 (July, 1985)

[J4] JADE Coll. W.Bartel, et al. Z.Phys. **C19**,177 (1983)

[J5] JADE Coll. W.Bartel, et al. Phys.Lett. **146B**,121 (1984)

[J6] JADE Coll. W.Bartel, et al. Phys.Lett. **146B**,437 (1984)

[M1] MAC Coll., W. W. Ash, et al. SLAC-Pub-3741 (1985). To be publ. in Phys.Rev.Lett.

[M2] MAC Coll. E.Fernandez, et al. Phy.Rev.Lett. **54**,1620 (1985)

[M3] MAC Coll. R.Prepost, Proc. 1984 HEP Conf. Leipzig, Vol.I,227 (1984)

[M4] MAC Coll. H.Stephen Kaye, Ph.D. Thesis. SLAC-262 (1983)

[MJ1] MARK J Coll. B.Adeva, et al. Phy.Rev.Lett. **48**,1701 (1982)

[MJ2] MARK J Coll. B.Adeva, et al. Phy.Rev.Lett. **55**,665 (1985)

[MJ3] MARK J Coll. B.Adeva, et al. Phy.Rep. **109**,131 (1984)

[MJ4] MARK J Coll. B.Adeva, et al. Phy.Rev.Lett. **51**,443 (1983)

[MK1] MARK II Coll. M.E.Levi, et al. Phys.Rev.Lett. **51**,1941 (1983)

[MK2] MARK II Coll. M.Nelson, et al. Phys.Rev.Lett. **50**,1542 (1983)

[P1] PLUTO Coll. Ch.Berger, et al. Z. Phys. **C21**,53 (1983)

[T1] TASSO Coll. M.Althoff, et al. Z. Phys. **C22**,13 (1984)

[T2] TASSO Coll. Contributed Paper to the Bari Conf. July,1985)

[T3] TASSO Coll. M.Althoff, et al. Z. Phys. **C26**,521 (1985)

[T4] TASSO Coll. M.Althoff, DESY 84-061 (1984)

[T5] TASSO Coll. M.Althoff, et al. Z. Phys. **C22**,219 (1984)

[TP1] TPC Coll., H. Aihara, et al. Phys. Rev. **D31**,2719 (1985)

RECENT RESULTS FROM DEEP INELASTIC νN EXPERIMENTS

Frank S. Merritt

Enrico Fermi Institute and
Department of Physics
University of Chicago
Chicago, Illinois 60637

ABSTRACT

Recent results from deep inelastic neutrino scattering
experiments at CERN and FERMILAB are reviewed. New
measurements include total cross-sections, structure
functions, neutral currents, opposite-sign dimuons,
and same-sign dimuons. There is reasonably good
agreement between experiments in all these areas.

1. INTRODUCTION

This has been a productive year in neutrino physics. Several
significant results have been published or presented in the last six
months, and some long-standing discrepancies between the results of
different groups have now been resolved.

I will review four areas of deep-inelastic neutrino measurements:
1) total cross-sections, 2) structure functions, 3) neutral currents
and $\sin^2\theta_w$, and 4) multi-muon events. The last of these has already
been covered in other reviews[1] at this conference, so the treatment
here will be comparatively superficial.

2. TOTAL CROSS-SECTION MEASUREMENTS

For the last several years there has been some disagreement among
the total cross-sections measured by different experiments with both
ν and $\bar{\nu}$ beams. The two experiments with highest statistics are CDHS
and CCFRR; the CDHS results[2] have been lower than those of CCFRR[3]
by 11% for ν and 16% for $\bar{\nu}$. The largest errors in both experiments
have been due to flux uncertainties.

The CDHS group has recently presented[4] new results from a high-
statistics run completed in the summer of 1984. This run, undertaken
to improve measurements of σ_{tot}^{ν} and $\sin^2\theta_w$, used a new beam designed
to provide a high-intensity narrow-band neutrino flux and to reduce

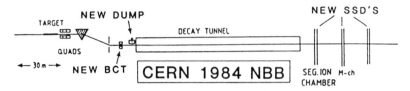

Figure 1: The 1984 CERN Narrow Band Neutrino Beam,
with new beam dump, BCT's, and SSD's.

wide-band backgrounds from decays upstream of the decay pipe. A new
moveable beam dump was installed at the upstream end of the decay pipe
(see figure 1) so that background neutrinos from upstream decays could
be measured by moving this into the beam.

The most important change affecting $\sigma_{tot}^{\nu,\bar{\nu}}$ was the installation of
two new beam monitors: 1) a new beam current transformer (BCT)
immediately upstream of the pipe to measure the hadron flux entering
the pipe, and   2) new solid state detectors (SSD) downstream of the
pipe to measure muons from π and K decays in the pipe. These devices
provided two independent ways of determining the neutrino flux, giving
a cross-check on flux measurements. The two methods agreed to within
the expected 5% errors throughout the run.

The new CDHSW measurements of $\sigma_{tot}^{\nu,\bar{\nu}}/E_{\nu}$ as a function of E_{ν} are
shown in figure 2. The new cross-sections are about 10-15% higher
than the old CDHS measurements of .60 ± .03 for ν and .29 ± .04 for
$\bar{\nu}$, and agree very well with the old CCFRR measurements.

What has changed in the CDHS measurements? In checking possible

σ^{ν}/E_{ν} $(10^{-38}$ $cm^2/GeV)$

CDHSW:	.696 ± .024
CCFRR:	.669 ± .024

$\sigma^{\bar{\nu}}/E_{\bar{\nu}}$ $(10^{-38}$ $cm^2/GeV)$

CDHSW:	.334 ± .011
CCFRR:	.34 ± .02

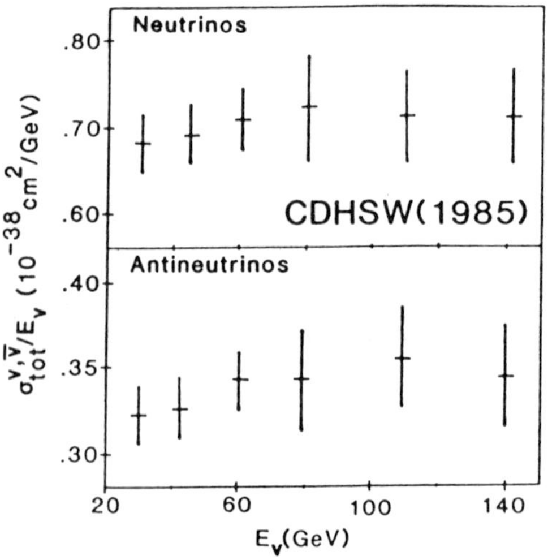

Figure 2: The new CDHSW total cross-section measurements,
showing good agreement with the old CCFRR values.

systematic errors affecting the old data, the CDHS group found[5] an
ohmic connection between the shield and the inner wall of the old BCT
which caused a shift of 6-7% in the calibration. They have also
measured the contribution of δ-rays to the BCT signal (formerly this
was only calculated), and correcting for this effect gave another shift
of 3-3.5% in the same direction. These two effects appear to satisfactorily account for the change in the CDHS measurement. Presumably
the changes in flux calibration affect other CERN measurements of
σ_{tot} in a similar way.

3. STRUCTURE FUNCTIONS

The CDHS group has also extracted new structure function measurements[6] from a wide-band run carried out in 1983, which provided
650,000 ν events and 550,000 ν̄ events. The new data give differential
cross-sections which are systematically higher than older CDHS measurements in the low-x region (see figure 3). The difference is about
30-40% at x=.045 and about 8% at x=.35. The new measurements agree
very well with the old CCFRR measurements, and comparing[7] both of
these experiments with the EMC structure functions gives the expected
mean square quark charge to within 10% (see figure 4).

Two factors have contributed to the change in the measured
structure functions: 1) the 12% change in σ_{tot} discussed earlier,
and 2) improvements in the CDHS analysis program. The most important

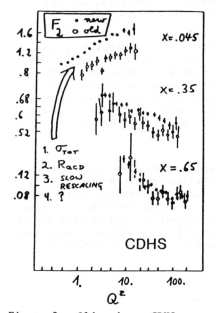

Figure 3: Old and new CDHS
structure functions,
showing 30% change at low x.

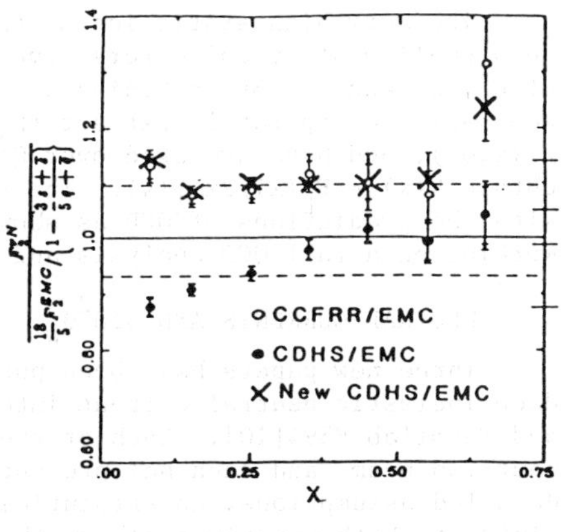

Figure 4: Comparison of structure
functions measured with
ν and μ beams. Ratio of
1 corresponds to the
standard quark model.

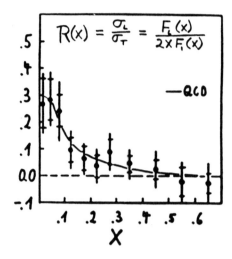

$$R(x) = \frac{\sigma_L}{\sigma_T} = \frac{F_L(x)}{2xF_1(x)}$$

—QCD

Figure 5: New CDHS measurement of R(x), compared to predictions of QCD.

of these improvements are the inclusion of slow rescaling in charmed quark production and an improved parameterization of R(x) = σ_L/σ_t based on measurements from the present data (described below). As a check of the new data, the group has used the new analysis program and the new σ_{tot} measurements to reanalyze the old data, and has found reasonably good agreement with the new results.

The very high statistics of this wide-band run has resulted in very small statistical errors, even after cuts for specific regions of Q^2, x, and y. By combining ν and $\bar{\nu}$ data, the experimenters have been able to separately extract the longitudinal structure function $F_L(x,Q^2)$, and have averaged over Q^2 (the Q^2 dependence is small) to obtain R(x) = $F_L(x)/2xF_1(x)$. The results are in quite good agreement with the predictions of QCD as shown in figure 5. The group is now working on a full QCD analysis and a new determination of $\Lambda_{\overline{MS}}$.

4. NEUTRAL CURRENTS AND $SIN^2\theta_W$

Three new papers have been published in the last six months on deep inelastic neutral current interactions, from CCFRR[8], CDHS[9], and Fermilab E594[10]. Each of these used a high-energy narrow band neutrino beam, and each has presented a complete analysis, including detailed assumptions, uncertainties in parameters, and sensitivity of $sin^2\theta_W$ to both experimental and theoretical uncertainties. This makes it possible to make a precise comparison between the experiments.

Each experiment has used the formulas

$$R_\nu = \sigma^\nu_{NC}/\sigma^\nu_{CC} = \frac{1}{2} - x + \frac{5}{9} x^2 (1 + r) \tag{1}$$

$$R_{\bar{\nu}} = \sigma^{\bar{\nu}}_{NC}/\sigma^{\bar{\nu}}_{CC} = \frac{1}{2} - x + \frac{5}{9} x^2 (1 + \frac{1}{r}) \tag{2}$$

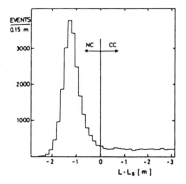

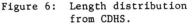

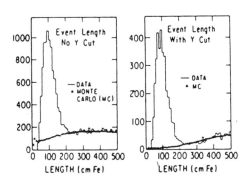

Figure 6: Length distribution
           from CDHS.

Figure 7: Length from CCFRR, with
           and without y cut.

where $x = \sin^2\theta_w$ and $r = \sigma_{CC}^{\bar{\nu}}/\sigma_{CC}^{\nu}$. These are valid in the standard
model for isoscalar targets even when cuts are made on y, so long as
r is defined over the same region of y as R_ν and $R_{\bar\nu}$. Small corrections
must be made for radiative effects, non-isoscalar targets, charmed
quark production, etc. Since $\sin^2\theta_w$ is quite insensitive to $R_{\bar\nu}$, most
of the running has been with neutrinos.

Both CDHSW and CCFRR separate NC and CC events on the basis of
"length", defined as the distance of steel traversed by the most penetrating charged particle produced in the interaction. In a histogram
of all events, NC events appear as a peak in the short length region
(L < 210 cm Fe), as shown in figures 6 and 7 for CDHSW and CCFRR. The
CC background in the short-length region must be extrapolated from
the long-length CC events. This background is lower for CDHS due to
the larger detector size and the focussing effect of the magnetized
target modules. The CCFRR group, in order to reduce this background,
has made a cut to eliminate all events with large y (large E_{had});
since this class of CC events contributes most of the background, this
cut substantially reduces the background as shown in figure 7. The
E594 group has made a similar cut to eliminate events with y > .70.

The published best values of the three experiments are given in
the first column of table I. However, several adjustments need to be
made before a comparison is meaningful:

1) For all three experiments the dominant theoretical uncertainty
   is in the mass of the charmed quark. This mass affects the
   CC cross-sections due to slow rescaling in charmed quark
   production, and thus affects R_ν. CCFRR and E594 have assumed
   $m_c = 1.5$ GeV, while CDHSW uses a Q^2-dependent mass, equivalent
   to a fixed mass of 1.1 GeV, based on a QCD calculation. All
   groups give the sensitivity of $\sin^2\theta_w$ to m_c. I have chosen to
   use an intermediate value of $m_c = 1.3$ GeV, which changes $\sin^2\theta_w$
   by +.002 for CDHSW, -.003 for CCFRR, and -.0025 for E594.

2) Each group quotes its result in a different renormalization
   scheme. I have converted each to the MS scheme used by CDHSW,

which reduces CCFRR by .002 and E594 by .001.

3) CDHSW and CCFRR have included corrections for non-isoscalar targets; this correction is smaller for E594 and has been neglected. I have included it, reducing E594 by .0024.

4) In the CCFRR and E594 results, the first error is statistical and the second is systematic (experimental and theoretical). For CDHSW, the first error is experimental (statistical and systematic) and the second is theoretical (the statistical error for CDHSW is ±.007). I have used the CDHS convention for the errors of all three experiments.

5) CCFRR assumed an error in the charmed quark mass of ±.2 GeV, while CDHSW and E594 assume ±.4. I have increased the CCFRR error assumption to agree with the others.

Adjusting for all of these effects gives the corrected values shown in the second column of Table I. Note that the only one of these adjustments which has any uncertainty associated with it is the first, and this introduces a common (and dominant) theoretical error of ±.005-.006 in all values.

At the recent EPS meeting at Bari, CDHSW and CHARM each presented new preliminary values of $\sin^2\theta_w$ from the 1984 run. The narrow band beam was improved as described earlier, and uncertainties from statistics and wide-band background (the dominant sources of error in the earlier results) were reduced. The CDHSW value was obtained using the same analysis as for their recently published result. The CHARM value is very preliminary and does not contain many of the corrections of the other experiments (this is reflected in the larger CHARM errors).

The weighted average of the corrected values is given in Table I. It is slightly higher than the values obtained from the W and Z masses and from ν_μ-e events, but is within the larger errors of those measurements. The dominant error in the ν_μ-N average now appears to be due to the uncertainty in the effective mass of the charmed quark. This can be experimentally addressed in the analysis of dimuon events using the large data samples from recent CERN and Fermilab runs; perhaps in another year we will have a measurement of this parameter.

TABLE I.   RECENT MEASUREMENTS OF $\sin^2\theta_w$ FROM νN.

EXPERIMENT	QUOTED BEST VALUE	ADJUSTED VALUE
CCFRR (1985)	.242 ± .011 ± .005	.237 ± .012 ± .006
CDHS (1985)	.226 ± .012 ± .006	.228 ± .012 ± .006
E-594 (1985)	.246 ± .012 ± .013	.240 ± .017 ± .005
CDHS (Bari)*	.217 ± .007 ± .006	.219 ± .007 ± .006
CHARM (Bari)*	.215 ± .010	.215 ± .010 ± .006

*=Preliminary

Mean = .224 ± .005 ± .006
χ^2 = 3.5 for 4 D.F.

5. OPPOSITE-SIGN AND SAME-SIGN DIMUONS

Opposite sign dimuon events come primarily from weak charmed quark production ($\nu d \to \mu^- c$ or $\nu s \to \mu^- c$) followed by semileptonic charm decay. The rates and distributions of kinematic variables can be used to extract the size of the strange sea, the semileptonic branching ratio of charm decay, and the energy dependence of charm production.

Both CDHS and CCFRR have presented analyses of these events. The CDHS sample, taken with a wide band beam, contains 14,000 events while the CCFRR sample contains 467 (taken with a narrow band beam). CDHS has extracted the ratio $\kappa = .52 \pm .09$ for the strange-quark fraction of the sea, and a semileptonic charm branching ratio of 7.1% (consistent with e^+e^- measurements). The errors are dominated by systematic uncertainties. CCFRR has extracted similar numbers but the errors are larger and dominated by statistics.

I will leave a discussion of these results to other reviews[1]. However there is one aspect of the data that perhaps deserves attention. After correcting for threshold effects, slow rescaling, and acceptance, one expects the ratio of $\mu^+\mu^-/\mu^-$ to be roughly constant above charm threshold. The CCFRR group has tried to obtain an estimate of the effective charm quark mass by plotting this ratio for various masses, as shown in figure 8. The data seem to favor higher values of m_c. This should <u>not</u> be interpreted as a measurement of m_c, since systematic as well as statistical errors are large. However, it does indicate that the model of slow rescaling with low m_c (< 1.3 GeV) needs more experimental investigation. This subject is interesting in itself and may be relevant to the NC analysis discussed above; it deserves more study with the large $\mu^+\mu^-$ data samples from the CERN runs and from the recent Fermilab quad-triplet run.

New data on same-sign dimuons have been presented by CDHS, CHARM, and CCFRR. Figure 9 shows a summary of current data. Both CDHS and CCFRR see a similar rate, and most of the data of different groups are at least qualitatively consistent. CDHS has obtained 423 same-sign events, and sees a signal 2.5 standard deviations above background; CCFRR sees a similar rate at 1.5 standard deviations. Both groups measure rates about 30 times greater than theoretical predictions. Based on the kinematical distributions of these events, the CDHS group

Figure 8:   The corrected ratio of $\mu^+\mu^-/\mu^-$ for different charm quark masses. The data appear to favor high m_c, although systematic errors may be large.

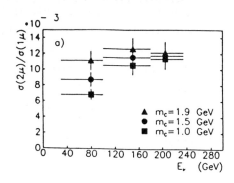

342

has concluded that the second muon is not produced by decays of particles heavier than charm, perhaps by associated $c\bar{c}$ production. However
the origin of same-sign dimuons is still a mystery. The reader is
referred to references 1 for a more detailed review.

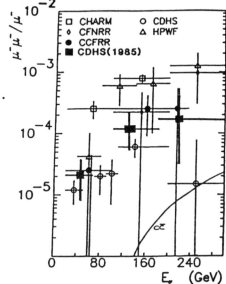

Figure 9:   Same-sign dimuon rates.

REFERENCES

1. K. Lang and M. Murtaugh, presented at this conference.

2. H. Abramowicz et al., Zeit. fur Physik C17, 283 (1983).

3. R. Blair et al., Phys. Rev. Lett. 51, 343 (1983).

4. M. Krasn, presented at International Europhysics Conference on
    High Energy Physics, Bari, Italy (1985).

5. A. Para, private communication.

6. P, Buchholz, presented at International Europhysics Conference
    on High Energy Physics, Bari, Italy (1985).

7. M. Purohit, Ph.D. thesis, Caltech, 1984.

8. P. Reutens et al., Physics Letters 152B, 404 (1985).

9. H. Abramowicz et al., Zeit. fur Physik C28, 51 (1985).

10. D. Bogert et al., to be published in Phys. Rev. Lett., Nov 1985.

11. A. Blondel, presented at International Europhysics Conference on
    High Energy Physics, Bari, Italy (1985).

12. F. Bergsma et al., contributed to International Europhysics
    Conference on High Energy Physics, Bari, Italy (1985).

A STUDY OF THE WEAK NEUTRAL-CURRENT
IN
DEEP-INELASTIC NEUTRINO-NUCLEON SCATTERING

D.Bogert, R.Burnstein, R.Fisk, S.Fuess, J.Morfin,
T.Ohska, L.Stutte, J.K.Walker,and,
J.Bofill, W.Busza, T.Eldridge, J.I.Friedman, M.C.Goodman,
H.W.Kendall, I.G.Kostoulas, T.Lyons, R.Magahiz, T.Mattison,
A.Mukherjee, L.Osborne, R.Pitt, L.Rosenson, A.Sandacz,
M.Tartaglia, F.E.Taylor, R.Verdier, S.Whitaker, G.P.Yeh,and,
M.Abolins, R.Brock, A.Cohen, J.Ernwein,
D.Owen, J.Slate, H.Weerts

Fermi National Accelerator Laboratory
Massachusetts Institute of Technology
Michigan State University

Presented by F.E.Taylor
October 1985

ABSTRACT

A study of the weak neutral-current has been performed on deep-inelastic neutrino-nucleon scattering data taken with a fine-grained calorimeter at FNAL. The neutral-current nucleon structure functions have been compared with those of the charged current to check the predictions of the standard model. A determination of $\sin^2\theta_w$ and ρ have been made. Details of the measurement technique and experimental and theoretical corrections are discussed.[1]

1. INTRODUCTION

The Z^0 boson is a unique probe of the quark structure of the nucleon. In the standard model the Z^0 - quark coupling depends on the weak isospin, the quark electric charge, and $\sin^2\theta_w$. The basic neutral-current (NC) interaction is believed to be an elastic scattering process which is flavor conserving and has no thresholds associated with the mass of the participating quark. The Z^0-nucleon coupling is given by:[2]

$$\frac{g}{\cos\theta_w} \sum_\alpha \langle q_\alpha | T_3^\alpha - Q^\alpha \sin^2\theta_w | q_\alpha \rangle \qquad (1)$$

where T_3^α is the weak isospin of the quark of flavor α, and Q^α is the quark electric charge. The mixing parameter $\sin^2\theta_w$ is free in the theory. The charged-current (CC) interaction is flavor mixing and has energy thresholds associated with light-to-heavy quark mass transitions. The $W^\pm$ - nucleon coupling is given by: $\dfrac{g}{\sqrt{2}} \displaystyle\sum_{\alpha\beta} <q_\beta | U^{\alpha\beta} | q_\alpha>$ (2)

where $U^{\alpha\beta}$ is the Kobayashi-Maskawa[3] mixing matrix. The light-to-heavy quark kinematic factors are accounted for in the CC structure functions by changing $x=Q^2/2M_p E_\nu y$ to the slow rescaling variable: $\xi = x + m_\beta^2/2M_p E_\nu y$, where m_β is the final state quark mass, M_p is the nucleon mass, E_ν is the incident neutrino energy, $y=(E_h-M_p)/E_\nu$, and E_h is the hadron energy. An additional factor $(1-y+xy/\xi)$ is necessary to specify the kinematic suppression of heavy quark production near threshold.[4] The action of these different couplings is of interest in the NC/CC comparison of this experiment.

2. EVENT KINEMATICS RECONSTRUCTION

The kinematics of NC deep inelastic events are reconstructed by measuring both the energy and the angle of the recoil hadron shower. The incident neutrino beam energy is inferred by the energy-versus-radius correlation of the narrow band neutrino beam. The Bjorken scaling variable x is reconstructed from the measured quantities in the small angle approximation by the following:

$$x \overset{\sim}{=} E_h \theta_h^2/2M_p(1-y)$$ (3)

and the inelasticity is given by:

$$y = (E_h-M_p)/E_\nu(r)$$ (4)

where $E_\nu(r)$ is the energy of the incident neutrino beam inferred from the energy-versus-radius correlation of the narrow-band neutrino beam.

The x resolution becomes poor as y approaches 1, thus a y<0.7 cut is imposed. Hadron energies greater than 50 GeV for most of the data are eliminated by this cut. This cut has the additional benefit of reducing the CC to NC signature confusion. A lower energy cut of $(E_h-M_p) > 10$ GeV is required to be well above trigger threshold. The contribution of the incident neutrino energy to the x resolution smearing is minimized by making a radius cut of 1 meter about the incident neutrino beam central axis. This cut reduces the contribution of $K_{\mu2}$ neutrinos to an 11% background of the $\pi_{\mu2}$ events and the three body kaon decays to less than about 1.5%. A typical value of the x resolution when all of the above affects are taken into account is $\sigma_x=x$, but it varies over the kinematic range of the data.

CC and NC events were distinguished by the presence or absence of an outgoing muon track from the neutrino-nucleon interaction vertex. The cuts on the muon track and the hadron energy were chosen so that only a small correction for misclassified events had to be made. The size of the correction was computed by a Monte Carlo simulation which generated realistic events in the detector. We determined that the average identification efficiencies for NC and for CC events were 0.96 and 0.99 respectively.

3. THE NEUTRAL-CURRENT STRUCTURE FUNCTIONS

Data were taken at narrow-band beam secondary momenta of 165, 200, and 250 GeV/c for neutrino production, and -165 GeV/c for antineutrino production. The fiducial mass was 55 metric tons. The number of accepted events before and after the event classification corrections are given in Table I. The mean incident neutrino energy and Q^2 of the accepted events are indicated.

Table I

P_0(GeV/c)	E_ν(GeV)	Q^2(GeV/c)^2	raw NC	CC	corrected NC	CC	R_ν
165	61.3	11.0	950	3235	966	3219	0.300 ± 0.011
200	74.7	12.2	638	2184	647	2175	0.298 ± 0.013
250	92.4	13.8	656	2093	677	2072	0.327 ± 0.014
-165	61.5	8.6	723	1945	740	1928	0.384 ± 0.017

A reduction in possible sources of systematic errors is achieved by comparing the x-dependences of the two interactions in ratio only. Thus we determined the NC structure functions relative to those of the CC. Figure 1 shows the NC/CC ratio as a function of x where all three neutrino beam settings have been combined. We observe that the ratios appear to be approximately flat within statistical errors. The line through the data is the result of Fits 1 and 2 of the NC structure functions to be described.

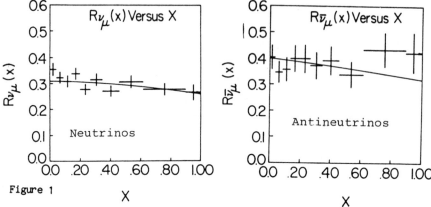

Figure 1

X   X

To extract the NC structure functions we fitted the x dependence for all of the incident neutrino beam settings simultaneously. A simple parameterization of the structure functions was chosen for both the NC and the CC interaction which gave an adequate representation of the world's CC data at our mean $Q^2 = 11 (GeV/c)^2$. The Q^2 evolution of the structure functions described by QCD has been ignored. The forms for the valence and sea structure functions are given by:

$$xV(x) = Ax^{\alpha}(1-x)^{\beta} \tag{5}$$

$$2x\overline{q}(x) = C(1-x)^{\gamma} \tag{6}$$

The charm-quark sea was neglected and the strange-quark sea was assumed to be 20% of the total quark sea $2x\overline{q}(x)$. The Kobayashi-Maskawa matrix and the slow rescaling terms with the charm-quark mass taken to be 1.5 (GeV/c^2) were included in the expression for the CC cross section. Radiative corrections associated with the outgoing muon and the box diagram were performed for the CC events.[5] We fixed the value of $\sin^2\theta_w = 0.246$, however the sensitivity of our result to different values of this parameter were included in our estimate of the systematic errors.

The results of the fits are given in Table II. Two assumptions were made: In Fit 1 the values of A, β, and C have been determined under the constraint that $\alpha = 1/2$ suggested by Regge theory,[6] Fit 2 assumes the Gross Llewellyn-Smith sum rule[7] and the parameters α, β, and C have been determined. The strength of the valence quark distribution is determined by the parameters α and β. For both fits we have assumed that $\gamma_{NC} = \gamma_{CC} = 7$.

We see that there is no systematic difference between the NC and the CC structure functions. The analysis of the NC structure function parameters given in Table II closely parallels the CHARM analysis[8] and there is good agreement between the two experiments.

The systematic errors indicated in Table II are from estimated sensitivities to changes in $\sin^2\theta_w$, the event classification correction, the hadron energy scale, and the upper y cut.

Table II

CC Parameters		Fit 1 NC Parameters	Fit 2 NC Parameters
Valence			
A	3.28	$3.59 \pm 0.63 \pm 0.62$	$A = 3\Gamma(\alpha+\beta+1)/\Gamma(\alpha)\Gamma(\beta+1)$
α	0.50	0.50	$0.48 \pm 0.10 \pm 0.10$
β	3.0	$3.54 \pm 0.40 \pm 0.41$	$3.38 \pm 0.62 \pm 0.54$
Sea			
C	1.0	$1.16 \pm 0.18 \pm 0.13$	$1.21 \pm 0.16 \pm 0.13$
Y	7.0	7.0	7.0
χ^2/degrees of freedom	32.0/37		32.0/37

4. DETERMINATION OF $\sin^2\theta_w$ AND ρ

The extraction of $\sin^2\theta_w$ and ρ in deep-inelastic scattering is based on the assumption that the quark scaled momentum distributions in the NC structure functions are the same as those in the CC structure functions within the slow rescaling effects discussed above. We have tested this assumption in our comparison above of the two interactions. Thus we will assume that this aspect of the standard model is true. We computed the value of $\sin^2\theta_w$ by taking the quark distributions of Duke and Owens,[9] which include the QCD Q^2 evolution of the quark distributions. We found that our result is not very sensitive to the details of the structure function model as long as the same quark distributions are assumed to participate in the two interactions.

We have used the integral NC to CC ratios for both neutrino and antineutrino data to determine $\sin^2\theta_w$ and ρ.

These ratios are defined by: $R_\nu^{(-)} = I(\overset{(-)}{\nu}+N\rightarrow\overset{(-)}{\nu}+X)/I(\overset{(-)}{\nu}+N\rightarrow\mu^{\pm}+X)$ where $I(\overset{(-)}{\nu}+N\rightarrow\overset{(-)}{\nu}+X)$ is the number of NC events which satisfy the acceptance cuts described above, and $I(\overset{(-)}{\nu}+N\rightarrow\mu^{\pm}+X)$ is the corresponding number of accepted CC events. The ratios were calculated separately for all four secondary momentum settings shown in Table I. The neutrino ratios have the greater sensitivity to $\sin^2\theta_w$, while the additional consideration of the antineutrino ratio allows $\sin^2\theta_w$ and ρ to be decoupled partially and thus to be determined simultaneously. Use of these ratios avoids neutrino flux normalization uncertainties.

The values of $\sin^2\theta_w$ and ρ were extracted by a minimum χ^2 fitting procedure which matched a Monte Carlo simulation to the data. The Monte Carlo simulation included all known details of the incident neutrino beam, the experimental resolutions, the backgrounds, and the experimental cuts. In our fit for $\sin^2\theta_w$ alone we found: $\sin^2\theta_w = 0.246 \pm 0.012 \pm 0.013$ where the first error is statistical and the second is systematic. The mean Q^2 of these data was about $11(GeV/c)^2$. The value of ρ was fixed to 0.9915 according to Sirlin and Marciano.[5] The χ^2/degrees of freedom of the fit was 3.9/3. The two dimensional fit with the two radiative corrections discussed above gave: $\sin^2\theta_w = 0.279 \pm 0.027 \pm 0.019$, $\rho = 1.027 \pm 0.023 \pm 0.026$. The χ^2/degrees of freedom for the fit was 2.5/2. The parameters $\sin^2\theta_w$ and ρ are strongly correlated in the two dimensional fit.

The systematic errors of this calculation arise from both experimental and theoretical sources. They are listed in Table III.

To compare our results with other experiments and with theory, we have computed $\sin^2\theta_w$ in two renormalization conventions. In the convention where $\sin^2\theta_w$ is defined in terms of the physical boson masses, we found: $\sin^2\theta_w = 0.247 \pm 0.012 \pm 0.013$. In the minimal subtraction $\overline{MS}$ scheme which is appropriate for the SU(5) theories, we computed: $\sin^2\theta_w = 0.245 \pm 0.012 \pm 0.013$. The errors quoted above are the statistical and systematic errors respectively.

Table III

Experimental sources	$\Delta\sin^2\theta_W/\sin^2\theta_W$		$\Delta\rho/\rho$
Neutral current/Charged current separation	$(\pm 4.2\%)$	$\pm 3.9\%$	$\pm 1.3\%$
Muon elimination in Charged current hadron energy determination	$(\pm 2.0\%)$	$\pm 5.4\%$	$\pm 1.8\%$
Experimental Error	$(\pm 4.7\%)$	$\pm 6.7\%$	$\pm 2.2\%$
Theoretical sources	$\Delta\sin^2\theta_W/\sin^2\theta_W$		$\Delta\rho/\rho$
Strange sea magnitude $(xs(s)=(0.50\pm 0.20)x\bar{u}(x))$	$(\pm 0.4\%)$	$\pm 0.4\%$	$\pm 0.1\%$
QCD corrections $(\Lambda =200\ ^{+200}_{-100}\ \text{MeV/c})$	$(\pm 0.1\%)$	$\pm 0.05\%$	$\pm 0.05\%$
Slow Rescaling $(Mc=1.5\pm 0.4\ \text{GeV/c}^2)$	$(\pm 2.0\%)$	$\pm 1.8\%$	$\pm 1.1\%$
Theoretical Error	$(\pm 2.0\%)$	$\pm 1.8\%$	$\pm 1.1\%$
Total Systematic Error	$(\pm 5.1\%)$	$\pm 6.9\%$	$\pm 2.5\%$

Several aspects of our result have been investigated by considering only the neutrino data, or by changing various elements of the theory. When we use only the neutrino data in our fit, thereby reducing the sensitivity to the antiquark sea, we found $\sin^2\theta_W = 0.246 \pm 0.015$ (statistical error) with a χ^2/degrees of freedom = 2.5/2.

We have relaxed the slow rescaling correction by setting the charm quark mass to zero resulting in a value of $\sin^2\theta_W = 0.228 \pm 0.012$(statistical error) with a poor χ^2/degrees of freedom = 8.0/3. The value of $\sin^2\theta_W$ with no radiative corrections was found to be 0.257 ± 0.012 (statistical error only).

5. SUMMARY

In summary, we have compared the x distributions of NC deep-inelastic neutrino-nucleon scattering with that of the CC interaction. We found no significant difference between these two interactions within the cuts of this analysis. This result tested one of the major assumptions in the determination of $\sin^2\theta_w$ in deep-inelastic neutrino-nucleon scattering. Under the assumption that the same quark distributions participate in the two interactions, we have determined $\sin^2\theta_w$ and ρ. We found for the single parameter fit $\sin^2\theta_w = 0.246 \pm 0.012 \pm 0.013$.

FOOTNOTES:

[1] This work has recently been published. See: Bogert,D. et al., Phys. Rev.Lett. 55, 574 (1985) and Bogert,D. et al. "A determination of $\sin^2\theta_w$ and ρ in deep-inelastic neutrino-nucleon scattering" to be published in Phys.Rev.Lett.
[2] Glashow,S.L. Nucl.Phys. 22, 579 (1961); Weinberg,S., Phys.Rev.Lett. 19, 1264, (1967) Salam,A. in Elementary Particle Theory, ed. N.S.Vartholm, P.367 Almquist and Wiksell,Stockholm (1968).
[3] Kobayashi,M. and Maskawa,K., Prog. Theor. Phys. 49, 652 (1973)
[4] Barnett,R.M., Phys.Rev. D14, 70, (1976).
[5] DeRujula,A. et al., Nucl.Phys. B154, 394 (1979); Sirlin,A. and Marciano,W.J., Nucl.Phys. B189, 442 (1981).
[6] Harari,H., Phys.Rev.Lett. 24, 286 (1970); Kuti,J. and Weisshopf,V., Phys.Rev. D4, 3418 (1971).
[7] Buras,A.; Rev.Mod.Phys. 52, 199 (1980).
[8] Jonker, M. et al. Phys.Lett. 128B, 117, (1983).
[9] Duke,D.W. and Owens,J.F., Phys.Rev. D30, 49, (1984).

II

EXPERIMENT AND PHENOMENOLOGY
BEYOND THE STANDARD MODEL

Session Organizers: C. Baltay and G. Kane

STRONG NEW LIMITS ON GLUINO AND SCALAR QUARK MASSES*

R. Michael Barnett

Lawrence Berkeley Laboratory, University of California
Berkeley, California 94720

Abstract

I conclude from a comprehensive study of the 1984 CERN UA1 monojet and dijet events with missing transverse energy that any excess number of monojets above background is highly unlikely to come from gluino and/or scalar quark production. The new data with the new UA1 cuts and triggers lead to the very restrictive limits: $M(\tilde{g})$ and $M(\tilde{q}) > 60\text{-}70$ GeV. The two intriguing dijet events with $E_T(\text{missing}) > 55$ GeV are not inconsistent with an 80 GeV gluino or scalar quark source.

In this talk I will summarize without details the results of a comprehensive study by H. Haber, G. Kane and myself to determine whether there is evidence from the UA1 experiment[1] at CERN for the production of gluinos and/or scalar quarks. The details of our technique, calculations, results, uncertainties and alternatives may be found in Ref. 2, and other related papers are Refs. 3-7.

The fundamental signature for the production of supersymmetric particles is missing energy. In this work we have usually assumed that the photino is the lightest supersymmetric particle, that R-parity is unbroken and that the photino escapes the detectors unnoticed (giving the appearance of missing energy). In our calculations we chose $M(\tilde{\gamma}) = 0$ but a 10 GeV mass for the photino would in most cases have little affect on our results. We considered the following possible subprocesses:

(a) gluino pair production:

$$gg \rightarrow \tilde{g}\tilde{g} \quad \text{and} \quad q\bar{q} \rightarrow \tilde{g}\tilde{g} \tag{1}$$

(b) scalar-quark pair production

$$gg \rightarrow \tilde{q}\bar{\tilde{q}}, \quad q\bar{q} \rightarrow \tilde{q}\bar{\tilde{q}}, \quad \text{and} \quad qq \rightarrow \tilde{q}\tilde{q} \tag{2}$$

$$q\bar{q} \rightarrow W \rightarrow \tilde{q}\bar{\tilde{q}} \quad \text{and} \quad q\bar{q} \rightarrow Z \rightarrow \tilde{q}\bar{\tilde{q}} \tag{3}$$

*This work was supported by the Director, Office of Energy Research, Office of High Energy and Nuclear Physics, Division of High Energy Physics of the U.S. Department of Energy under Contract Nos. DE-AC03-76SF00098, and by the U.S. National Science Foundation under Agreement No. PHY83-18358

(c) Associated production of gluinos and scalar-quarks

$$qg \to \tilde{q}\tilde{g} \qquad (4)$$

(d) Associated production of photinos with gluinos or scalar-quarks

$$q\bar{q} \to \tilde{g}\tilde{\gamma} \quad \text{and} \quad q\bar{g} \to \tilde{q}\tilde{\gamma} \qquad (5)$$

We have computed the squared matrix elements for all the processes listed above[2].

Other production processes such as scalar-quark production from a gluino component inside the proton (which may be relevant if gluinos are light) and 2-to-3 processes such as $gg \to \tilde{g}\tilde{g}g$ will be discussed later.

We now turn to the decay process. Here, we must consider two separate cases which lead to quite different types of signatures.

Case 1: $M_{\tilde{g}} > M_{\tilde{q}}$. In this case the gluino decays nearly 100% of the time via:

$$\tilde{g} \to q\bar{\tilde{q}} \quad \text{or} \quad \tilde{\bar{q}}q \qquad (6)$$

where the sum is taken over all possible quark flavors which are kinematically accessible. The scalar-quark decays nearly 100% of the time via:

$$\tilde{q} \to q\tilde{\gamma} \qquad (7)$$

Case 2: $M_{\tilde{g}} < M_{\tilde{q}}$. In this case the gluino decays via processes (6) and (7) where the scalar-quark is virtual:

$$\tilde{g} \to q\bar{q}\tilde{\gamma} \qquad (8)$$

Again, one must sum over all quark flavors which are kinematically accessible. A sum over $\tilde{q}_L$ and $\tilde{q}_R$ intermediate states is assumed. In the case of the scalar quark, two decay modes are allowed (making assumptions similar to those above) with branching ratios (B) as indicated:

$$\tilde{q} \to \begin{cases} q\tilde{g} & B = \dfrac{r}{1+r} \\[2ex] q\tilde{\gamma} & B = \dfrac{1}{1+r} \end{cases} \qquad (9)$$

where $r \equiv \alpha_s/(e_q^2\alpha)$. Note that the dominant scalar-quark decay is $\tilde{q} \to q\tilde{g}$; the gluino then decays via Eq. (8). This is the main reason for the difference in signatures between Cases 1 and 2. In Case 1, scalar-quark decay leads to substantial missing energy via the process given by Eq. (7). In Case 2, the dominant process is $\tilde{q} \to q\tilde{g}$, $\tilde{g} \to q\bar{q}\tilde{\gamma}$ leading to far less missing energy as compared with Case 1.

The production processes listed in Eq. (1-5) represent contributions to the lowest order approximation to the inclusive production cross-section of super-symmetric processes. Let us focus here on gluino pair production: $p\bar{p} \to \tilde{g}\tilde{g} + X$. To $O(\alpha_s^2)$, the processes which contribute have been given in Eq. (1). If we consider $O(\alpha_s^3)$ we must include loop corrections to the processes given by Eq. (1), and in addition we must introduce new $2 \to 3$ processes:

$$gg \to \widetilde{g}\widetilde{g}g, \quad q\overline{q} \to \widetilde{g}\widetilde{g}g \quad \text{and} \quad gq \to \widetilde{g}\widetilde{g}q \qquad (10)$$

If the perturbative series is trustworthy, then we should find that the $O(\alpha_s^3)$ contributions are smaller than the $O(\alpha_s^2)$ Born terms. However, Herzog and Kunszt[8] realized that when various triggers and cuts are applied to the total cross section, it is possible that the $O(\alpha_s^3)$ contributions are enhanced significantly. In fact, we find that this indeed occurs for certain ranges of values of the gluino mass when the UA1 triggers and cuts are imposed.

Consider the effect of an E_T^{miss} cut on the processes given by Eq. (10). If the final state gluon (or quark) is hard, then a possible configuration is one where the $\widetilde{g}\widetilde{g}$ pair is emitted in the same hemisphere recoiling against the gluon (or quark). In this configuration, the photinos resulting from the gluino decays are often emitted in the same hemisphere so that it is much easier to have $E_T^{miss} = |p_{T1}\widetilde{\gamma} + p_{T2}\widetilde{\gamma}| \geq E_o$. When the E_T^{miss} cut is applied, the $2 \to 3$ processes result in *more* events passing the cuts as compared with the $2 \to 2$ processes. This has very important consequences for whether light gluinos are excluded by the data.

At present, a full $O(\alpha_s^3)$ calculation does not exist. The Born terms for the $2 \to 3$ processes (Eq. (10)) have been obtained by Herzog and Kunszt.[8] Technical problems involved in the use of their results are discussed in Ref. 2.

Suppose that the gluino mass is on the order of 5 GeV, i.e., slightly heavier than the minimum mass allowed by the beam dump experiments. Such a gluino has properties very similar to that of a b-quark with one important exception: the gluino is a color octet fermion compared with the color triplet b-quark. The resulting gluino distribution function in the proton is small (less than 1% of the gluon distribution) yet not vanishingly small. One reason for this is connected with the color octet nature of the gluino. Comparing the color factors that arise, it follows that $f_{\widetilde{g}}(x,Q^2) \approx 6 f_b(x,Q^2)$ assuming that $M_{\widetilde{g}} = m_b$.

The perturbatively generated gluinos in the proton may be very important[7] in determining the missing energy events expected in the light gluino case. The reason is that new hard scattering processes must now be considered such as $\widetilde{g} + q \to \widetilde{q}$ followed by the subsequent decay of the final state supersymmetric particle. An important issue regarding the gluino distribution function has been raised by Barger et al.[5] in a recent paper. There is not space here to discuss this issue, and I refer the reader to Refs. 2 and 5 for further details. We have chosen to be conservative and to neglect the gluino content of the proton.

Ironically, the UA1 collaboration has introduced a new trigger in their 1984 run[1] which significantly enhances the $\widetilde{g}\widetilde{g}$ processes (Eq. (1)). As a result, the uncertainty of the gluino structure function has in fact become a moot point regarding the analysis of the 1984 data.

In our first paper[6] and other early papers on this subject, fragmentation effects were ignored; however, these effects can play an important role in determining the missing E_T spectrum of a given event and were included in the present analysis. For example, a gluino which is produced by some hard process must first "fragment" into a supersymmetric hadron (say a $g\tilde{g}$ or $\tilde{g}q\bar{q}$ bound state) which then decays weakly, emitting a photino which escapes the detector. If the momentum of the gluino inside the supersymmetric hadron is less than that of the original gluino, then the photino spectrum will be degraded compared to the spectrum which would have resulted had fragmentation been ignored. Let us define z to be the momentum fraction of the gluino (or scalar-quark) inside the supersymmetric hadron. For large gluino or scalar-quark masses the z-distribution is sharply peaked at $z = 1$, and fragmentation effects are not relevant in determining the missing energy spectrum (due to the outgoing photinos). However, for gluino or scalar-quark masses less than about 20 GeV, fragmentation becomes increasingly important and the result is a missing energy spectrum which is softer than it would have been had fragmentation been ignored.

An important quantity to consider is the total scalar transverse energy of an event denoted by E_T. This is defined experimentally by adding in a scalar fashion the transverse energy deposited in all calorimeter cells. This quantity is used by the UA1 collaboration in defining one of the triggers and one of the missing energy cuts, so we must consider it here in detail. It is clear that such a quantity is incalculable in the framework of perturbative QCD; in particular, the sum of the scalar transverse energies of the final-state jets is a severe underestimate of the value of E_T. To get an idea of the magnitude of such an underestimation, consider $p\bar{p}$ collisions with *no* large p_T jets in the final state (the so-called "minimum bias" events). The UA1 collaboration has obtained the E_T distribution of such events,[9]. If one studies the UA1 and UA2 samples[10-11] of large-p_T two jet events, and subtracts out the two jets in each event, the E_T distribution of the "remainder" is substantially harder, roughly twice that of minimum bias.

Given this current state of ignorance, we have rescaled the "minimum-bias" distribution by a factor of 1.67 such that the mean is 40 GeV. (In fact, the median of our rescaled distribution is 32 GeV.) The precise choice for E_r (the remainder energy) can have a tremendous impact on the number of events which pass the UA1 missing energy trigger. This is unfortunate given that E_r is not so well understood theoretically.

Although the details of our simulation of UA1 experimental conditions[1] may not make exciting reading, they are critical to the validity of our results (especially for the question of a light gluino). Furthermore, an understanding of these experimental conditions, which include triggers, cuts, resolution and efficiencies, is essential if one is to attempt an extrapolation to the Tevatron Collider or the Superconducting Super Collider (SSC).

On the basis of extensive consultations with members of the UA1 Collaboration, we have arrived at a procedure for simulating the UA1 conditions. Our approach is not at all the same as those used by others who have analyzed the UA1 data and is described in detail in Ref. 2.

Let us first summarize the newly reported 1984 data from the UA1 Collaboration.[1] They found 23 monojets with at least 15 GeV of missing transverse energy in their 1984 run. They identified 9 as having the characteristics of the $W \rightarrow \tau\nu$ $(\tau \rightarrow \nu + \text{hadrons})$ source. Of the remaining 14 events, UA1 estimates that 6–8 events are due to background sources. We will take 13 events as the 90% confidence limit for the number of events which *could* represent new physics. Since their integrated luminosity in 1984 was about 270 nb^{-1}, the 90% confidence upper limit for monojet production due to new physics is then 4.8 events/100 nb^{-1}. This is actually quite conservative since (as discussed below) our distributions show that most of these events are unlikely to be from supersymmetric sources so that the real limit might be as low as perhaps 2 events/100 nb^{-1}.

Theoretical calculations of monojet rates are subject to the fine distinction between monojets and dijets. We advocate use of the combined monojet + dijet + multijet rate. The UA1 Collaboration reports that after the back-to-back cuts, 2 dijet and no multijet events remain from the 1984 run where they estimate backgrounds at 2 events. The total number of missing-energy events (after τ subtraction) is then 16 with backgrounds at 8–10 events. We again take 13 events to be the 90% confidence level limit giving the limit for the rate for missing-energy events with $E_T^{\text{miss}} > 15$ GeV to be 4.8 events/100 nb^{-1}, while the *dijet* rate's limit is 2 events/100 nb^{-1}.

Our results are summarized in the contour plots, Figs. 1-2. They should only be compared with the 1984 UA1 data. Let us momentarily ignore the region at very low gluino masses where rates are low due to fragmentation effects. From Fig. 1 showing the monojet rate for $E_T^{\text{miss}} > 15$ GeV and the data described above, we set the limits $M_{\tilde{q}} > 50$–60 GeV depending on $M_{\tilde{g}}$ and $M_{\tilde{g}} > 45$–55 GeV depending on $M_{\tilde{q}}$.

We advocate use of all missing-energy events as in Fig. 2. From this plot and the above data, we find $M_{\tilde{q}} > 60$–70 GeV depending on $M_{\tilde{g}}$ and $M_{\tilde{g}} > 50$–70 GeV depending on $M_{\tilde{q}}$. This significant improvement in the limits occurs because, for these large masses, supersymmetry predicts that dijet production should dominate over monojet production even with the back-to-back cuts. So these results combining all missing-energy events are both more limiting and more reliable (since they need make no distinction among numbers of jets).

If one wishes instead to assume that we can accurately separate monojets and dijets, then the above results suggest that it will be useful to examine the dijet rate separately. This is, in effect, done by subtracting Fig. 1 from Fig. 2. Using the above data we then find

$$M_{\tilde{q}} > \begin{cases} 65 & M_{\tilde{g}} \approx 150 \text{ GeV} \\ 75 & M_{\tilde{g}} \approx 80 \text{ GeV} \end{cases} \tag{11}$$

$$M_{\widetilde{g}} > \begin{cases} 60 & M_{\widetilde{q}} \approx 100 \text{ GeV} \\ 70 & M_{\widetilde{q}} \approx 80 \text{ GeV} \end{cases} .$$

The limits quoted above would change by $\Delta M \approx 5$ GeV if our predictions were off by 50%.

How precise should we treat the numbers we have obtained (shown on our contour plots, Fig. 1-2)? There are a number of uncertainties which enter into our calculation, both from theoretical sources and experimental sources (discussed in Ref. 2). In general, we expect the uncertainty in our numbers is less than a factor of two. This implies an uncertainty in our mass limits of

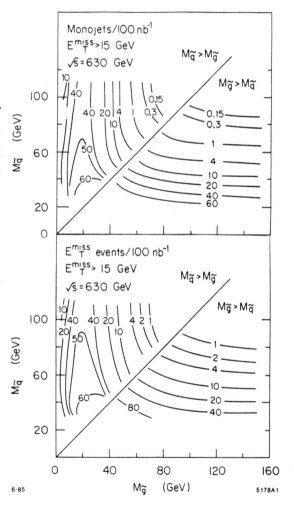

Fig. 1. The number of monojets per 100 nb^{-1} passing the new 1984 UA1 cuts and triggers[1] shown as a contour plot as a function of $M(\widetilde{g})$ and $M(\widetilde{q})$. The 1984 UA1 data have an integrated luminosity of about 270 nb^{-1}.

Fig. 2. Same as Fig. 1 but for monojets, dijets plus multijets.

roughly 5 GeV. However, in the particular case where the gluino is light (say $M_{\tilde{g}} \leq 20$ GeV), much more care must be given to the estimation of uncertainties. In fact, the numbers shown in our contour plots are much less certain in this regime.

As is evident in Figs. 1 and 2, the predicted event rates drop off as $M_{\tilde{g}}$ becomes very small. Very light gluinos lose much of their energy due to fragmentation and gluon bremsstrahlung, and therefore they lead to very little missing energy. As a result very few pass the E_T^{miss} cuts.

The calculations for very low mass gluinos are subject to much larger uncertainties due to fragmentation and to the surviving events being on the tails of the E_T^{miss} and E_T^{jet} distributions. We would predict for $M_{\tilde{q}} \approx 100$ GeV 26 events/100 nb^{-1} for a 5 GeV gluino and 13 events/100 nb^{-1} for a 3 GeV gluino (these are all monojets; dijet production is negligible). While these numbers are much larger than the 4.8 events limit, one cannot neglect the uncertainties intrinsic to theoretical calculations for light gluinos. Our tests convince us that these uncertainties could be as much as a factor of 4 or 5 if added linearly. In spite of this large uncertainty, our predicted event rate is large enough to conclude that $M_{\tilde{g}} = 5$ GeV is ruled out and that $M_{\tilde{g}} = 3$ GeV is very marginal. If the photino mass is non-zero the photino would carry off even more energy, and our results would be strengthened.

This conclusion could not be reached in the papers (including ours) analyzing the 1983 data for two primary reasons. The new missing-energy trigger in the 1984 run is extremely important for $\tilde{g}\tilde{g}$ production when $M_{\tilde{g}}$ 10 GeV. It raises our predictions in this case by an order-of-magnitude, while experimentally this trigger does not dramatically change the observed rate. The calculation of the $\tilde{g}\tilde{g}g$ process by Herzog and Kunszt[8] was not available for the earlier analyses, and we find it increases our predictions for 5 GeV gluinos by a factor of 3. While other recent refinements bring down the rate a little, the end result is that because of the higher rates predicted, it is now possible (or almost possible) to rule out light gluinos.

Could some of the observed monojet events be due to the production of gluinos or scalar quarks? The UA1 Collaboration[1] cannot rule out the possibility that 6–8 of the monojets come from new physics. There are, however, two factors which argue against the monojets coming from gluino or scalar quark production. Both are consequences of the fact that the appropriate event rate (2–3 monojets/100 nb^{-1}) only occurs for large $M_{\tilde{q}}$ or $M_{\tilde{g}}$ (60 GeV). For such masses we would predict 4–6 dijets/100 nb^{-1}, and these certainly have not been observed. Furthermore, at these masses one would expect significant numbers of monojets with $E_T^{miss} > 45$ GeV, and only one was observed in the 1984 run.

The two observed dijet events (surviving the back-to-back cuts) in the 1984 run have $E_T^{miss} \geq 55$ GeV. Although there is a roughly equal background expected, these backgrounds are unlikely to have so much E_T^{miss}. A 70–90 GeV scalar quark could give dijets with such characteristics and with this rate, and would produce very few monojets. Clearly, however, such speculation must await considerably more statistics.

Let us summarize the implications of an alternative scenario where the Higgsino is the lightest supersymmetric particle. With the assumptions discussed in Ref. 2 we would obtain the following allowed regions for $M_{\widetilde{g}}$ and $M_{\widetilde{q}}$:

$$M_{\widetilde{g}} \lesssim 5 \text{ GeV} \quad or \quad M_{\widetilde{g}} \gtrsim 40 \text{ GeV} \tag{12}$$

$$M_{\widetilde{q}} \gtrsim 45\text{--}60 \text{ GeV} \tag{13}$$

What are the implications of our analyses for higher-energy colliders such as the Tevatron and the SSC? Experiments searching for supersymmetry or for other new physics will have to make choices on cuts and triggers, and will have somewhat different resolutions and efficiencies. The choices will be based in part on how backgrounds scale, but also on how quantities such as E_r scale (where E_r is the remaining transverse energy in an event after the jets are removed). Of course, E_r also depends on how the experiment chooses (or needs) to define the jets; both the jet algorithm and the required minimum E_T^{jet} enter this definition. The theoretical calculation of E_r involves knowledge of initial-state radiation and other aspects which theorists are just now learning to include in QCD Monte Carlo programs.

As a result of our ignorance of backgrounds, of future experimental conditions and of the scaling of E_r, it will be difficult to make precise predictions for the Tevatron and the SSC. We do intend, however, to study these questions and to try to find some qualitative answers.

I wish to acknowledge useful conversations with many theoretical colleagues at LBL and CERN, and with experimental colleagues in the UA1 and UA2 collaborations.

REFERENCES

1. J. Rohlf, invited talk at the 1985 Division of Particles and Fields Conference, Eugene, Oregon, August 1985; C. Rubbia, invited talk at the 1985 Lepton-Photon Conference, Kyoto, Japan, August 1985.

2. R.M. Barnett, H.E. Haber and G.L. Kane, LBL report no. LBL-20102 (August 1985), submitted to Nuclear Physics B.

3. J. Ellis and H. Kowalski, Nucl. Phys. 142B, 441 (1984); Nucl. Phys. B246, 189 (1984); Phys. Lett. 157B, 437 (1985); Nucl. Phys. B259, 109 (1985).

4. E. Reya and D.P. Roy, Phys. Rev. Lett. 51, 867 (1983) (E: 51, 1307 (1983)); Phys. Rev. Lett. 53, 881 (1984); Phys. Lett. 141B, 442 (1984); Phys. Rev. D32, 645 (1985); Dortmund preprint DO-TH 85/23 (1985); V. Barger, K. Hagiwara and J. Woodside, Phys. Rev. Lett. 53, 641 (1984); V. Barger, K. Hagiwara and W.-Y. Keung, Phys. Lett. 145B, 147 (1984); V. Barger, K. Hagiwara, W.-Y. Keung and J. Woodside, Phys. Rev. D31, 528 (1985); D32, 806 (1985); A.R. Allan, E.W.N. Glover and A.D. Martin, Phys. Lett. 146B, 247 (1984); A.R. Allan, E.W.N. Glover, and S.L. Grayson, Nucl. Phys. B259, 77 (1985); N.D. Tracas and S.D.P. Vlassopulos, Phys. Lett. 149B, 253 (1984); X.N. Maintas and S.D.P. Vlassopulos, Phys. Rev. D32, 604 (1985); F. Delduc, H. Navelet, R. Peschanski and C. Savoy, Phys. Lett. 155B, 173 (1985); A. De Rujula and R. Petronzio, CERN-TH 4070/84 (1984); G. Altarelli, B. Mele and S. Petrarca, Univ. La Spirenza (Rome), 1985.

5. V. Barger, S. Jacobs, J. Woodside and K. Hagiwara, Wisconsin preprint MAD/PH/232 (1985).

6. R.M. Barnett, H.E. Haber, and G.L. Kane, Phys. Rev. Lett. 54, 1983 (1985).

7. M.J. Herrero, L.E. Ibanez, C. Lopez and F.J. Yndurain, Phys. Lett. 132B, 199 (1983); 145B, 430 (1984).

8. F. Herzog and Z. Kunszt, Phys. Lett. 157B, 430 (1985).

9. G. Arnison et al., CERN-EP/82-122 (1982).

10. G. Arnison et al., Phys. Lett. 132B, 214 (1983); J. Sass, in "Antiproton Proton Physics and the W Discovery," Proceedings of the International Colloquium of the CNRS, Third Moriond Workshop, March 1983, ed. by J. Tran Thanh Van (Editions Frontieres, France, 1983) p.295.

11. P. Darriulat, invited lectures at the 13th SLAC Summer Institute on Particle Physics, July 1985.

SEARCH FOR ANOMALOUS SINGLE PHOTON PRODUCTION AT PEP

ROBERT J. WILSON[*]

Stanford Linear Accelerator Center,

Stanford, California 94305

ABSTRACT

This talk reports a search for the production by e^+e^- annihilation of a single photon accompanied by particles that interact only weakly in matter. The search was performed at PEP ($\sqrt{s}$ = 29 GeV) with a new detector, ASP. No unexpected signal was observed. The limit $N_\nu < 14$ (90% CL) is placed on the number of light neutrino species, and the mass of scalar electrons predicted by theories of supersymmetry is constrained to $m_{\tilde{e}} > 51$ GeV/c^2 (90% CL) for $m_{\tilde{\gamma}} = 0$ and degenerate $\tilde{e}$ mass states.

1. INTRODUCTION

This talk reports the results of a search for the production by e^+e^- annihilation of new particles that interact only weakly in matter and therefore are not directly observable. This is done by searching for production of these new particles accompanied by the radiation of a single photon,

$$e^+e^- \rightarrow \gamma + \text{weakly interacting particles.} \tag{1}$$

The radiative production of neutrino pairs by the standard weak charged and neutral currents contributes to (1).[2] At $\sqrt{s} \ll m_W$,

$$\frac{d\sigma(e^+e^- \rightarrow \gamma\nu\bar{\nu})}{dE_\gamma d\cos\theta_\gamma} \sim \frac{\alpha^3}{\sin^4\theta_W} \cdot \frac{(1 + N_\nu/4)}{p_T^\gamma \cdot \sin\theta_\gamma} \cdot \frac{s}{m_W^4}, \tag{2}$$

---

* For the ASP Collaboration.[1]

where N_ν is the number of light neutrino species, and p_T^γ and θ_γ are the momentum transverse to the beam axis and the polar angle of the radiated photon. For $N_\nu = 3$ the exact cross section[3] corresponds to 1.0 observed event in the acceptance of the search reported in this talk.

Theories of supersymmetry (SUSY) predict the existence of particles that also contribute to (1).[4,5] Many SUSY models contain a photino $\tilde\gamma$ that is stable, interacts weakly in matter, and is produced in pairs by the exchange of a scalar electron $\tilde e$. At $\sqrt{s} \ll m_{\tilde e}$,

$$\frac{d\sigma(e^+ e^- \to \gamma\tilde\gamma\tilde\gamma)}{dE_\gamma \, d\cos\theta_\gamma} \sim \alpha^3 \cdot \frac{1}{p_T^\gamma \cdot \sin\theta_\gamma} \cdot \frac{s}{m_{\tilde e}^4} \cdot \tag{3}$$

There are, in general, two $\tilde e$ mass eigenstates that contribute to the $\gamma\tilde\gamma\tilde\gamma$ final state. If these are degenerate, the effective cross section will be twice that for the case in which one mass is much larger than the other. SUSY also predicts the existence of scalar neutrinos $\tilde\nu$ which can be produced through the normal weak neutral current and through supersymmetric charged currents.[5]

From (2) and (3) it is clear that good photon detection efficiency down to small values of p_T^γ and θ_γ is desirable. Isolation of process (1) from backgrounds requires nearly complete solid angle coverage with electromagnetic calorimetry and charged particle tracking. The most difficult background to eliminate comes from the process $e^+ e^- \to e^+ e^- \gamma$. The kinematics of this process require complete veto capability at $\theta > \theta_{veto} = p_{Tmin}^\gamma / 2E_{beam}$. The acceptance of the experiment reported in this talk is $p_{Tmin}^\gamma > 1$ GeV with $E_{beam} = 14.5$ GeV, i.e. $\theta_{veto} = 34$ mr. The ASP detector described below has no gaps above $\theta = 21$ mr.

2.   THE ASP DETECTOR

The ASP detector is shown in Fig. 1. Photons produced at central polar angles are detected in five-layer stacks of lead-glass bars interleaved with proportional wire chambers (PWCs). These stacks completely surround the interaction point (I. P.) in azimuth. The glass bars are 6 cm $\times$ 6 cm $\times$ 75 cm, and each is read out by a single phototube. The bars are staggered along the beam direction Z (Fig. 1b) to eliminate cracks and provide optimal resolution of the origin of electromagnetic showers. The energy resolution of the calorimeter is measured with radiative Bhabha events (see below) to be 15% at 1 GeV when averaged over all angles with $\theta > 20°$. The PWCs are made from 1.2 cm $\times$ 2.4 cm closed-cell aluminum extrusions and provide position information in the plane perpendicular to the beam.

Charged particles are tracked between the beam pipe and central calorimeter by planes of proportional tubes with resistive sense wires parallel to the beam line. These wires are read out on both ends to provide redundancy and, by charge division, the Z coordinate of charged tracks. The central tracker is surrounded by 2 cm thick veto scintillators with phototubes on both ends. The pair conversion probability of a photon in the material between the I. P. and the veto scintillators is 4.5% at $\theta_\gamma = 90°$.

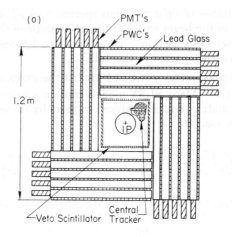

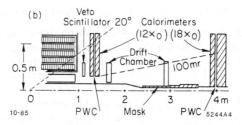

Fig. 1. (a) Cross sectional view of the ASP central calorimeter and tracking system. Only a section of the central tracker is shown; it completely surrounds the I. P. (b) Side view of one quarter of the ASP detector. Only the horizontal lead glass bars are shown in this view; there are 632 bars in the detector.

Calorimeter modules made from lead and scintillator surround the beam pipe in the forward angle region (Fig. 1b). Proportional tubes measure the spatial positions of showers at a depth of 6 X_0 in these modules. The forward calorimeter modules overlap each other and the central calorimeter so that no gaps occur in the detector above $\theta = 21$ mr. Four drift chamber planes track charged particles at polar angles between 21 and 100 mr.

The detector is located 20 m underground. This filters all primary hadrons from the cosmic ray flux and reduces the overall intensity by a factor of ≈ 2.7. Scintillation counters above the detector provide an additional order of magnitude rejection of cosmic rays.

Signals from the lead-glass bars are electronically summed and discriminated to generate hardware triggers. The detector is triggered at a threshold of 1.5 GeV on the total pulse height from all phototubes. It is also triggered above

0.6 GeV by requiring the lead-glass signal be localized to a single quadrant, or adjacent quadrants, and that it be seen in more than one layer. Signals from the forward shower modules are used to trigger on small angle Bhabha scattered electrons. A special trigger for radiative Bhabha $(ee\gamma)$ events is formed by placing the small-angle Bhabha trigger in coincidence with a very low (≈ 0.2 GeV) threshold on the total lead-glass pulse height.

The $ee\gamma$ triggers are kinematically fitted off-line using the forward calorimeters and the charged particle tracking (but not the lead glass) to produce a sample (45,000 events) of electrons and photons with known energies and production angles. This sample is used to study the performance of the central calorimeter. The efficiency of the trigger is measured to be $> 99\%$ for single photons produced at the I. P. with $p_T^\gamma \geq 1$ GeV and $\theta_\gamma > 20°$. This is taken to be the fiducial region for our signal.

3.   EVENT SELECTION

Photon candidates are selected off-line by requiring a cluster of blocks in the central calorimeter with a pattern consistent with an electromagnetic shower. The time of the lead-glass pulse is required to be within $\pm 3\,\sigma_T$ of the known beam crossing time; $\sigma_T = 2.4$ ns at 1 GeV and slightly less at higher energies. The candidate photon is fitted to extract R, the signed projected distance of closest approach to the known I. P. in the XZ or YZ plane. The photon candidate is kept if there is a good fit with $|R| < 30$ cm; $\sigma_R = 3$ cm. (See Fig. 3.) The efficiency of all photon selection criteria is measured as a function of photon energy and angles with the fitted $ee\gamma$ and $e^+e^- \to \gamma\gamma$ events. The average photon detection efficiency is measured to be 75% in the signal region, with little variation in E_γ and θ_γ. The majority of the reconstruction losses occur at azimuths where showers span two lead glass quadrants.

The ability to isolate the single-photon final state from background depends upon electronic noise levels and occupancies in the components of the detector. Triggers taken on randomly chosen beam crossings are used to determine these levels. Radiative Bhabha and $e^+e^- \to \gamma\gamma$ events are used to study occupancies that are correlated with the presence of the signal photon, such as backsplash into the veto scintillators and central tracker, and leakage into the forward shower modules. Cuts are determined for each element of the detector such that no single cut results in more than a 10% loss of efficiency.

The candidate signal events are taken to be those with a single good photon candidate at $\theta_\gamma > 20°$ and no significant signal seen anywhere else in the detector. Three events with single photons with energies consistent with the beam energy were observed in the data sample. These are interpreted as $e^+e^- \to \gamma\gamma$ events in which one of the photons escapes the detector without showering. A study of the observed $e^+e^- \to \gamma\gamma$ data sample predicts 1.5 single photons from this source. The requirement $E_\gamma < 12$ GeV, determined from an analysis of Bhabha events, eliminates this background with negligible loss of signal acceptance. The resulting sample of events with $p_T^\gamma > 0.5$ GeV is displayed in Fig. 2. There are two events with $p_T^\gamma > 1.0$ GeV/c. With the measured efficiencies for the photon reconstruction and isolation cuts, the overall efficiency for photons produced in our signal region is 45%. This includes losses due to photon conversion in the beam pipe and central tracker and the effects of finite resolutions in photon energy and angle.

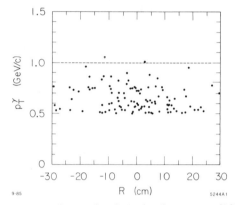

Fig. 2. Final sample of single photon candidates with $p_T^\gamma > 0.5$ GeV/c and $\theta_\gamma > 20°$.

4. ANALYSIS

Signal and background events have different R distributions. The R distribution of signal photons is measured with the $ee\gamma$ sample and is shown in Fig. 3a. The background at low p_T^γ (Fig. 2) is mainly due to interactions between beam particles and residual gas in the beampipe. This background is observed to be flat in R before application of several of the photon pattern cuts that are biased to accept showers with small R. The shape of the final background is measured by relaxing cuts other than these pattern cuts, and is shown in Fig. 3b.

With the measured distributions (Fig. 3) for signal, $P_S(R)$, and background, $P_B(R)$, we maximize the log-likelihood function,

$$ln\left(\frac{e^{-(S+B)} \cdot (S+B)^{N_{ev}}}{N_{ev}!}\right) + \sum_{i=1}^{N_{ev}} ln\left(\frac{S \cdot P_S(R_i) + B \cdot P_B(R_i)}{S+B}\right) \qquad (4)$$

to obtain the best estimate from the experimental data of the number of signal events S and background events B. For a given true number of signal and background events, the confidence level of this experiment is computed by Monte Carlo as the fraction of equivalent experiments that would estimate a number of signal events larger than S. We find that for 2.9 true signal events the confidence level is 90% for the two events that we observe. The 95% CL is given by 3.9 true signal events.

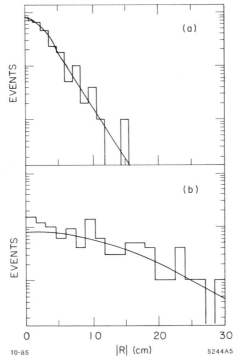

Fig. 3. (a) R-distribution of radiative photons in the signal region. The line is Gaussian with $\sigma = 3$ cm at $|R| < 6$ cm and exponential at $|R| > 6$ cm.
(b) R-distribution of background events with $p_T^\gamma > 0.6$ GeV/c described in the text. The line is a Gaussian with $\sigma = 12$ cm.

5. RESULTS

The integrated luminosity is measured with tracked small-angle Bhabha events to be 68.7 pb^{-1}. With the previously discussed efficiency we obtain the 90% CL limit,

$$\sigma(e^+e^- \rightarrow \gamma + \text{weakly interacting particles}) < 0.094 \text{ pb} \tag{5}$$

for $E_\gamma < 12$ GeV, $p_T^\gamma > 1.0$ GeV/c and $\theta_\gamma > 20°$. Since the detection efficiency is nearly constant with E_γ and θ_γ in the signal region, the limit (5) does not depend greatly on the produced photon spectrum or angular distribution. The cross section for radiative pair production of three neutrino species is[3] 0.032 pb with photons in our acceptance, so (5) places the limit 0.062 pb on any possible anomalous contribution to (1).

From (5) we deduce the limit $N_\nu < 14$ (90% CL) on the total number of light neutrino species. The 95% CL limit is $N_\nu < 20$. The validity of these limits requires no assumptions other than the standard coupling of the Z^0 to neutrinos. From the exact cross section[6] for $e^+e^- \rightarrow \gamma\tilde{\gamma}\tilde{\gamma}$ we deduce the limits shown in Fig. 4 on the $\tilde{e}$ and $\tilde{\gamma}$ masses. From the model of reference 5 for $\tilde{\nu}$ production, we obtain the limit $m_{\tilde{W}} > 48$ GeV/c^2 (90% CL) using the parameter values $m_{\tilde{\nu}} = 0$, $O^+ = 1$, and $m_1 \ll m_2$. These limits are significantly larger than those given by previously published searches[7] and, for $m_{\tilde{\gamma}} = 0$, our limit $m_{\tilde{e}} > 51$ GeV/c^2 already rules out the decay of real Z^0s to degenerate $\tilde{e}$ pairs.

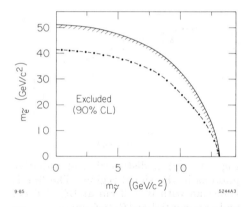

Fig. 4. Limits (90% CL) placed on $\tilde{e}$ and $\tilde{\gamma}$ masses by this experiment. The solid line is the limit for degenerate $\tilde{e}$ masses, and the dashed-dot line is the limit if only one mass eigenstate contributes to the cross section. The 95% CL is $m_{\tilde{e}} > 44$ GeV/c^2 for degenerate mass states and $m_{\tilde{\gamma}} = 0$.

The author wishes to thank M. Peskin for useful discussions. This work was supported in part by the Department of Energy, contract DE-AC03-76SF00515 (SLAC), and by National Science Foundation Grant PHY-8503215.

REFERENCES

1. ASP Collaboration: G. Bartha, D. L. Burke, C. A. Hawkins, R. J. Hollebeek, M. J. Jonker, L. Keller, C. Matteuzzi, N. A. Roe, T. R. Steele, and R. J. Wilson (*Stanford Linear Accelerator Center*), A. S. Johnson and J. S. Whitaker (*Massachusetts Institute of Technology*), C. Hearty, J. E. Rothberg and K. K. Young (*University of Washington*), P. Extermann (*University of Geneva*), P. Garbincius (*Fermi National Laboratory*).
2. E. Ma and J. Okada, Phys. Rev. Lett. $\underline{41}$, 287 (1978).
3. K. J. F. Gaemers, R. Gastmans, and F. M. Renard, Phys. Rev. $\underline{D19}$, 1605 (1979).
4. J. A. Grifols, X. Mor-Mur, and J. Solá, Phys. Lett. $\underline{114B}$, 35 (1982); P. Fayet, Phys. Lett. $\underline{117B}$, 460 (1982); J. Ellis and J. S. Hagelin, Phys. Lett. $\underline{122B}$, 303 (1983).
5. J. S. Hagelin, G. L. Kane, and S. Raby, Nucl. Phys. $\underline{B241}$, 638 (1984).
6. K. Grassie and P. N. Pandita, Phys. Rev. $\underline{D30}$, 22 (1984).
7. A. J. Behrends *et al.*, (CELLO Collaboration), Phys. Lett. $\underline{114B}$, 287 (1982); L. Gladney *et al.*, (Mark II Collaboration), Phys. Rev. Lett. $\underline{51}$, 2253 (1983); W. Bartel *et al.*, (JADE Collaboration), Phys. Lett. $\underline{152B}$, 385 (1985); E. Fernandez *et al.*, (MAC Collaboration), Phys. Rev. Lett. $\underline{54}$, 1118 (1985).

NEUTRINO PRODUCTION OF LIKE-SIGN DIMUONS*

K. Lang, A. Bodek, F. Borcherding, N. Giokaris, I.E. Stockdale;
University of Rochester, Rochester, New York, 14627
P. Auchincloss, R. Blair, C. Haber, S. Mishra, M. Ruiz,
F.J. Sciulli, M. Shaevitz, W.H. Smith, R. Zhu; Columbia
University, New York, NY 10027
Y.K. Chu, D.B. MacFarlane, R.L. Messner, D.B. Novikoff,
M.V. Purohit, California Inst. of Technology, Pasadena, CA 91125
D. Garfinkle, F.S. Merritt, M. Oreglia, P. Reutens;
University of Chicago, Chicago, IL 60637
R. Coleman, H.E. Fisk, Y. Fukushima, Q. Kerns, B. Jin,
D. Levinthal, T. Kondo, W. Marsh, P.A. Rapidis, S. Segler,
R. Stefanski, D. Theriot, H.B. White, D. Yovanovitch;
Fermi National Laboratory, Batavia, IL 60510
O. Fackler, K. Jenkins; Rockefeller Univ., New York, NY 10021

*Presented by K. Lang

ABSTRACT

Neutrino interactions with two muons in the final state were
studied using the Fermilab narrow band beam. A sample of 18
like-sign dimuon events with momentum $P_\mu > 9$ GeV/c yields a prompt
signal of 6.6 ± 4.8 events and a rate of $(1.0 \pm 0.7) \times 10^{-4}$ per
single muon event. The kinematics of these events are compared
with those of the non-prompt sources.

INTRODUCTION

We report on an experimental study of neutrino induced dimuon
events. The dimuon data come from two runs with the same detector
using the Fermilab narrow band neutrino beam as a neutrino source.[1]
The first run (Fermilab experiment E616) took place in 1979 and 1980,
with an integrated proton flux of 5.4×10^{18} on the production
target. The second run took place in 1982 (Fermilab experiment E701)
with a flux of 3.4×10^{18}. Data were taken at six momentum settings
for π and K mesons (+100, ±120, ±140, ±165, ±200, ±250 GeV/c), yielding neutrinos with energies between 40 and 230 GeV. The neutrino
beam[2] was produced by decays of these sign and momentum selected
($\Delta P/P = \pm 11\%$) pions and kaons in a 352 m long evacuated decay pipe.

The neutrino detector is located in Lab E, 1292 m from the
beginning of the decay pipe. The apparatus consists of a target

calorimeter instrumented with scintillation counters and spark chambers followed by an iron toroidal muon spectrometer. The rms hadron energy resolution is 0.89 $\sqrt{E}$ (GeV). The total transverse momentum kick of the toroids is 2.4 GeV/c and the fractional momentum resolution is ±11%.

DIMUON ANALYSIS

Events with two muons were selected from the sample of all charged current triggers (500K events). The computer reconstruction of the dimuon events was examined by physicists and, if necessary, the reconstruction was repeated interactively.

Events are separated into those produced by muon neutrinos from pion decay (ν_π) and from kaon decay (ν_K). The neutrino energy and decay angle (and therefore the event radius at the detector) are kinematically related. Events due to ν_K and ν_π are distinguished by the transverse radial position of the interaction and the measured total energy. Once this separation is made the event transverse vertex position determines the incident neutrino energy with a ±10% error. In order to improve the efficiency for the muon track reconstruction, at least one muon track must have a straight line extrapolation from the target that passes through the first toroid magnet, and intersects the trigger counter placed after it.

All dimuon events must pass the same analysis cuts as the single muon events. Events must have a transverse vertex within a 2.5m × 2.5m square and a longitudinal vertex at least 3.8 m from the downstream end of the target, to ensure containment of the hadron shower. In addition, the vertex for ν_π events is required to have a transverse radius of less than 76 cm.

The like-sign dimuon analysis requires that both muons are momentum analyzed by the toroid magnet with $P_\mu > 9$ GeV/c. Therefore each single muon event used must have a magnetically determined momentum of $P_\mu > 9$ GeV/c. A total of 117,039 charged current events pass these cuts.

The like-sign dimuon sample contains 18 events in which each muon traverses one toroidal magnet and has a momentum greater than 9 GeV/c. Both muons must have angles less than 250 mrad with respect to the beam axis. In addition, both muons must have fitted tracks which originate in a common vertex consistent with counter pulse height and each track must be visible in the spark chambers after the first toroid.

The principal background sources of like-sign dimuons are decays to muons of primary π's or K's at the hadron vertex in a charged current event, and the production of prompt or non-prompt muons from the secondary interactions of the primary hadrons. Calculation of the background uses the inclusive primary hadron spectra obtained with the Lund Monto Carlo Program.[3] The contribution of subsequent interactions of these hadrons is calculated using the measured prompt and non-prompt muon production by hadrons in the Fermilab experiment E379 variable density target.[4,5] This yields the probability for producing a muon, with a momentum greater than a particular cutoff value, as a function of x_{BJ} and the hadron energy E_H.

The model of the background uses the generated E_H and X_{BJ} of charged current Monte Carlo events to produce dimuon events with a particular weight and P_{μ_2}. The produced muon is given a P_T based on transverse momentum fits to hadrons from EMC μ-p data.[6] The Monte Carlo events are reconstructed and required to pass the dimuon analysis cuts. The background is normalized to the number of charged current data events passing the same cuts. The systematic error in the background is $\pm 20\%$.

The non-prompt background for the 18 events is calculated to be 10.7 ± 2.1 events. We have also estimated an additional background due to misclassified trimuon events (originating primarily from hadronic and electromagnetic muon pair production) for which the second muon is hidden in the shower to be 0.6 ± 0.2 events. The background from spatially and temporally coincident charged current neutrino events is calculated to be 0.1 ± 0.1 events. We therefore observe a like-sign dimuon signal above background of 6.6 ± 4.8 events. CDHS reported 91 ± 9 like-sign dimuons with a calculated background of 64 ± 10 events with a momentum out of 10 GeV/c from their 350 GeV wide band run,[8] and 47 events with a background of 30 ± 8 from their 200 GeV narrow band run.[9] The CHARM collaboration reported 74 ± 17 prompt ν_μ induced $\mu^-\mu^-$

and 52±13 prompt $\bar{\nu}_\mu$ induced $\mu^+\mu^+$ with a 4 GeV/c momentum cut from their wide band run.[10] The HPWFOR collaboration reported a prompt signal of 52±31, 65±15, and 37±11 from their quadrupole triplet run[11] with momentum cuts of 5, 10, and 15 GeV/c, respectively. The CFNRR collaboration reported 10 like-sign dimuons with a background of 4.3 events from their quadrupole triplet run.[12]

A calculation of the ratio of prompt like-sign dimuon production to single muon production requires that the numbers of single muon and prompt like-sign dimuon events be corrected for geometrical acceptance. This acceptance must rely on a specific model. In the absence of such a model for like-sign dimuons, we have used a model in which the distribution of prompt dimuon events is the same as that of the non-prompt background (Model 1 $-\pi/K$). The acceptance has also been calculated using a model of gluon bremsstrahlung of charm-anticharm pairs with the anticharm decay producing the second muon[13] (Model 2 $-c\bar{c}$). Although the contribution to like-sign dimuons from D^0-$\bar{D}^0$ mixing is negligible,[14] we have also calculated the acceptance using this model for completeness (Model 3 $-$ D^0-$\bar{D}^0$). The rates calculated with the π and K decay model acceptance are presented with a comparison of acceptance correction for $c\bar{c}$ production and D^0-D^0 mixing in Table 1.

TABLE 1

Energy Bin(GeV)	$\bar{\mu}$ Evnts	$\bar{\mu}\bar{\mu}$ Evnts	Hadron Shower	Other Bkg.	Data $-$Bkg.	Mdl 1 Prompt (π/K)	Mdl 2 Prompt (cc)	Mdl 3 Prompt (D^0-$\bar{D}^0$)	Prompt Rate $\mu^-\mu^-/\mu^-$ ($\times 10^{-4}$)
30-100	108128	2	2.0	0.2	-0.2	-0.9	-1.0	-0.6	-0.1±0.5
100-200	52930	11	5.7	0.4	4.9	12.8	13.7	9.6	2.4±1.7
200-230	16058	5	3.0	0.1	1.9	3.7	3.7	2.9	2.3±2.8
30-230	177116	18	10.7	0.7	6.6	17.1	15.9	13.2	1.0±0.7

A comparison with the prompt like-sign and opposite sign rates from other experiments is shown in Table 2. The energy dependence of the rates of prompt like-sign dimuons to charged current events is shown in Fig. 1. The line in Fig. 1 is the prompt signal calculated from the charm-anticharm production model of Ref. 13, using $\alpha_s = 0.2$ and $M_c = 1.5$ GeV/c^2. The rate scales with α_s^2 and predicts 0.2 events for this experiment.

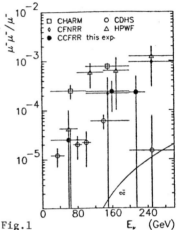

Fig.1

In addition to the overall rates, the event kinematics may be used to determine whether the π and K background can be the sole explanation of the like-sign dimuon signal. Distributions of several kinematic variables for the like-sign dimuons and Monte Carlo events from π and K decay are shown in Fig. 2. The π and K decay shown as the dashed histogram consists of 10.7 events. The distribution of the angle between the two muon tracks projected on a plane perpendicular to the incident neutrino is shown in Fig.2a. The peaking of this distribution near 180° indicates that the second muon is associated with the hadron shower in a large fraction of the events. This is a property expected of π and K decay.

For further comparison with the hypothesis of π and K decay, the second muon is chosen to be the muon which has the smaller momentum in the direction perpendicular to the axis of the hadron shower $(P_{T2\ min})$. The hadron shower direction is determined from the incoming ν beam properties and the chosen first muon. Figure 2b shows the $(P_{T2\ min})^2$ distribution. There is one event with a $(P_{T2\ min})^2$ of 6.0 (GeV/c)^2, which is unlikely to be from π and K decay. Figure 2c displays the momentum of the chosen second muon. Figure 2d shows the distribution of missing energy for all like-sign ν_K dimuon events. The missing energy is the difference between the neutrino energy determined by the transverse vertex radius and the measured energy. The missing energy for the like-sign dimuon data is about 3 standard deviations greater than that expected for the hypothesis of π and K decay. The distribution of $Z_{\mu_2} = P_{\mu_2}/(E_H + P_{\mu_2})$ is shown in Fig. 2e and the distribution of the invariant mass of the $\mu^-\mu^-$ pair is shown in Fig. 2f. There is no evidence of any structure in the mass distribution of the data.

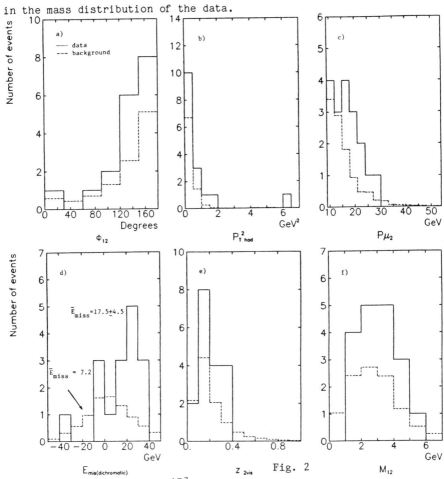

Fig. 2

It has been suggested[15] that enhanced $c\bar{c}$ production could account for the prompt like-sign dimuon signal. Accordingly, in Figure 3 we compare the same kinematic distributions of the data in Fig. 2 with the sum of the π and K decay background (previously displayed) and the rates from the previously described $c\bar{c}$ Monte Carlo multiplied by a factor of 42 (to equal the 6.6 event prompt signal). The kinematics of $c\bar{c}$ do not differ from those of π and K decay sufficiently to attribute the prompt signal to $c\bar{c}$ production any more than to π and K decay. The mean missing energy for $c\bar{c}$ and π/K decay is also about 2 standard deviations away from that of the data.

TABLE 2

Experiment (Ref.)	Beam Type	p_μ cut (GeV/c)	$\bar{\mu}^-\bar{\mu}^-/\bar{\mu}^-$ prompt $\times 10^{-4}$	$\bar{\mu}^-\bar{\mu}^-/\bar{\mu}^-\bar{\mu}^+$ prompt $\times 10^{-2}$
CCFRR (this exp)	NBB	9.	1.0±0.7	4.0±2.9
CDHS (8)	WBB	6.5	0.43±0.23	4.2±2.3
CDHS (9)	NBB	4.5	3±2	5±3
CHARM (10)	WBB	4.	4.5±1.6	14±5
HPWFOR (11)	QTB	10.	3.0±0.8	7±4
CFNRR (12)	QTB	9.	2.0±1.1	–

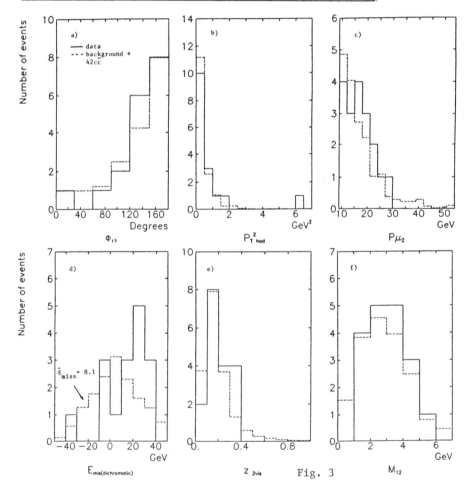

Fig. 3

CONCLUSIONS

The 18 observed like-sign dimuon events with a muon momentum out of 9 GeV/c yield a prompt signal of 6.6±4.8 events. This gives an average rate for prompt like-sign muon production of $(1.0\pm0.7) \times 10^{-4}$ per single muon event. The hadron shower background and/or assoc-iated charm production with antic harm decaying into μ^- cannot explain the measured rates. In addition, although most distributions of the data are similar to those expected from these two sources, the missing energy and p_T kinematic distributions of the data display some features which are not expected from the two models.

REFERENCES

1. R. Blair et al, Phys. Rev. Lett. 51, 3431 (1983).

2. R. Blair et al, Nucl. Instr. & Meth. 226, 281 (1984).

3. T. Sjostrand, Computer Physics Communications, 27, 243 (1982).

4. Fermilab E379/595. (c.f. J. Ritchie, Ph.D. thesis, University of Rochester (1983) and K. W. B. Merritt, Ph.D. thesis, Caltech 91981). Data do not exist for $E_\pi < 40$ GeV, where we use a M.C. calculation.

5. M. Shaevitz, private communications. Data used are in ref. 7.

6. F. W. Brasse, in Proc. of XX Int. Conf., Madison, Wisc., p. 755 (1980).

7. K. Lang, Ph.D. thesis, University of Rochester, UR-908 (1985).

8. J. G. H.deGroot et al, Phys. Lett. 86B, 103 (1970).
    J. Knoblock, in Proc. Neutrino '81, ed. R. J. Cence, E. Ma, A. Roberts, I:421 (1981).

9. M. Holder et al, Phys. Lett. 70B, 396 (1977).

10. M. Jonker et al, Phys. Lett. 107B, 241 (1981).

11. T. Trinko et al, Phys. Rev. D23, 1889 (1981).

12. K. Nishikawa et al, Phys. Rev. Lett 46, 1555 (1981); 54, 1336 (1985).

13. B. Young, T. F. Walsh, T. C. Yang, Phys. Lett. 74B, 111 (1978).
    V. Barger, W. Y. Keung, R. J. N. Phillips, Phys. Rev. D25, 1803 (1982).

14. A. Bodek et al, Phys. Lett. 113B, 82 (1982).

15. R. M. Godbole, D. P. Roy, Z. Phys. C22, 39 (1984).

SAME-SIGN DILEPTON PRODUCTION BY NEUTRINOS

Michael J. Murtagh

Physics Department, Brookhaven National Laboratory
Upton, LI, New York 11973

ABSTRACT

Experiment results on same-sign dilepton production by neutrinos are reviewed.

1. INTRODUCTION

The first observation of same-sign dileptons in neutrino
interactions was reported by the Fermilab E1A collaboration[1] in
1975. They observed 7 $\mu^-\mu^-$ events and concluded that it was unlikely
that the $\mu^-\mu^-$ events could be explained as (π^-, K^-) decay background.
Since then, many neutrino experiments have searched for same-sign dilepton events. While most experiments have an excess of candidates
compared to their expected backgrounds few claim that their excess is
statistically significant.

An obvious source for same sign dileptons within the Standard
Model is associated charm production (the $\bar{c}$-quark decays semi-lepton-
ically to an ℓ^-). The rate for $\mu^-\mu^-$ production from this source calculated in 1st order QCD[2] is significantly ($\lesssim$ 30 times) lower than
the reported experimental rates. Furthermore, bubble chamber experiments have no excess of strange particles as would be expected from $\bar{c}c$
production. It is unclear at present how to accommodate either the
lack of strange particles or a high rate for same-sign dileptons within the Standard Model.

Only neutrino results will be discussed. Similar conclusions,
with poorer statistics, can be drawn from anti-neutrino data.

2. EXPERIMENTAL RESULTS

The measured experimental rates for $\nu_\mu N \to \mu^- \ell^-$ production relative to the total charged current rates are shown in Fig. 1. The number of candidates and expected backgrounds are given in Table I. Apart from a recent high statistics bubble chamber measurement only counter experiment results are shown. Of the previously published[3-7] results, the HPWFOR[6] and CHARM[7] results are the most significant.

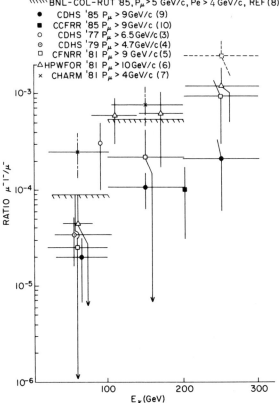

SAME SIGN DILEPTON PRODUCTION BY NEUTRINOS

︴BNL-COL-RUT '85, $P_\mu > 5$ GeV/c, Pe > 4 GeV/c, REF(8)
- ● CDHS '85 $P_\mu > 9$ GeV/c (9)
- ■ CCFRR '85 $P_\mu > 9$ GeV/c (10)
- ○ CDHS '77 $P_\mu > 6.5$ GeV/c (3)
- ⊙ CDHS '79 $P_\mu > 4.7$ GeV/c (4)
- □ CFNRR '81 $P_\mu > 9$ GeV/c (5)
- △ HPWFOR '81 $P_\mu > 10$ GeV/c (6)
- × CHARM '81 $P_\mu > 4$ GeV/c (7)

RATIO μ^{-1^-}/μ^-

E_ν (GeV)

Fig. 1. Measured rates for same-sign dilepton production by neutrinos relative to the charged current rate.

Two recent changes are reflected in Fig. 1. A new analysis of the CFNRR[5] experiment gave a much less significant result and a reduction in the measured rate. The CHARM data points now include systematic errors. This does not affect the result quoted in their paper which is for the rate of same-sign dileptons relative to opposite-sign dileptons.[7]

The major difficulty with $\mu^-\ell^-$ counter experiments is that the dominant background (π/K decays) is comparable to the total observed rate so the estimate of this background and the uncertainty in this estimate are very important. Most experiments use a Monte Carlo calculation where the hadron production details are constrained by available bubble chamber data and the probability that a given hadron is identified as a μ^- is determined by a shower calculation coupled with test beam π/K measurements. The uncertainties in these procedures are such that the large errors quoted on the background estimates are not unexpected. It is also difficult to foresee how this procedure can be significantly improved in the future.

The HPWFOR[6] detector had three distinct target sections with quite different effective collision lengths so, in principle, one could measure the dilepton rate for each target region and extrapolate to infinite density to obtain the true prompt signal. This method was not used directly for the $\mu^-\mu^-$ events because the statistics were limited. It was used for the more abundant opposite-sign ($\mu^-\mu^+$ events) and in this case served as a check on the calculation of the π/K background for the various targets. The straight lines shown in Fig. 2(a-c) for the $\mu^-\mu^-$ data are single parameter fits with the slopes fixed by the same background calculation used for the $\mu^-\mu^+$ events applied to the production and decay of negative hadrons. Given this calculated slope, the intercepts are systematically non-zero and for $P_\mu > 10$ GeV the probability is $< 10^{-4}$ that all the $\mu^-\mu^-$ events are due to π/K background.

A number of new results on same-sign dilepton production are now available.[8-11] The BCR[8] bubble chamber experiment has no candidates with $P_{\mu^-} > 5$ GeV and $P_{e^-} > 4$ GeV. This experiment is a

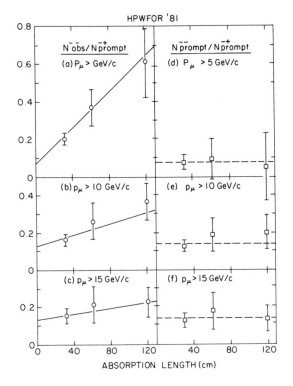

(a) $P_\mu >$ GeV/c — $N^-obs/N^{-+}prompt$

(b) $p_\mu > 10$ GeV/c

(c) $p_\mu > 15$ GeV/c

(d) $P_\mu > 5$ GeV/c — $N^{--}prompt/N^{-+}prompt$

(e) $p_\mu > 10$ GeV/c

(f) $p_\mu > 15$ GeV/c

ABSORPTION LENGTH (cm)

Fig. 2. (a)-(c) Ratio of $N^{obs}(\mu^-\mu^-)/N^{pr}\mu^-\mu^+)$ as a function of absorption length. The solid lines are fits to the data with the slope fixed by the decay calculation. (d)-(f) Ratio of $N^{pr}(\mu^-\mu^-)/N^{pr}(\mu^-\mu^+)$ as a function of absorption length. The errors on the points include the uncertainty in the background calculation. The dashed line represents the weighted average.

Fig. 3. Missing energy for $\mu^-\mu^-$ data and π/K decays in μ^- events.

Fig. 4. Angle ($\Delta\theta$) between μ_1 and μ_2 in the plane normal to $\nu_\mu-$ direction.

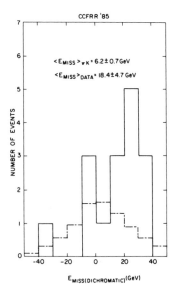

CCFRR '85

$<E_{MISS}>_{\pi K} = 6.2 \pm 0.7$ GeV

$<E_{MISS}>_{DATA} = 18.4 \pm 4.7$ GeV

NUMBER OF EVENTS

$E_{MISS(DICHROMATIC)}$ (GeV)

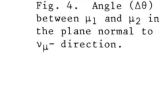

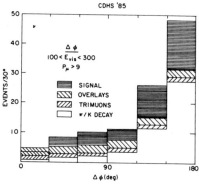

CDHS '85

$\Delta\phi$
$100 < E_{vis} < 300$
$P_\mu > 9$

SIGNAL
OVERLAYS
TRIMUONS
π/K DECAY

EVENTS/30°

$\Delta\phi$ (deg)

high statistics ($\approx$50,000 $\nu_\mu \to \mu^-$ events) experiment in the Fermilab
15' Bubble Chamber filled with a heavy Ne/H_2 mixture. The External
Muon Identifier (EMI) was used to select muon candidates. A total of
91 events with a leaving negative track of momentum $P_L- > 5$ GeV/c
and an electron of momentum $P_e-> 1$ GeV/c were observed. Of these
events 25 were rejected as being δ-rays, Dalitz pairs or K^- decays.
One event was rejected as a trilepton event. Only 12 of the remaining 65 events had an identified muon in the EMI. The expected background due to asymmetric Dalitz pairs or Compton electrons was (9.5 $\pm$
1.7) events so there is no evidence for a signal. None of the 12
events had $P_e- > 4$ GeV/c. The resultant 90% CL upper limits for
$P_\mu- > 5$ GeV/c and $P_e- > 4$ GeV/c are shown in Fig. 1, Table II.

A CERN SPS experiment[12] (WA59) using BEBC filled with a Ne/H_2
mixture has recently reported on a search for μ^-e^- events. No excess
of μ^-e^- events over the expected background was observed in a sample
of $\approx$25,000 charged current events and the resultant limits are also
given in Table II.

The CDHS collaboration[9] has completed the analysis of new narrow and wide band beam exposures at the CERN SPS. The neutrino wide
band beam data set contains $\approx$ 1.5 million $\nu_\mu \to \mu^-$ events with $P_\mu-$
> 9 GeV, a significant increase over previous experiments. The major
concern in this effort was to improve the systematics of the background subtraction.

The π/K background was determined by a Monte Carlo calculation.
The dominant uncertainty comes from the hadron shower fragmentation
and the overall uncertainty in the π/K decay subtractions were estimated to be $\leq$ 15%. Smaller and less uncertain corrections arise from
trimuon events and overlap events where two independent charged current events overlap both in space and time to the extent that the
reconstruction program cannot resolve them. The results (Table I,II)
from this analysis are in agreement with earlier work by the same collaboration. There is an indication of a prompt likesign signal. However, even in the optimum region it is less than 3 standard deviations.

There is also a recent result from the CCFRR collaboration using a narrow band beam at Fermilab.[10] Since the incident neutrino energy is reasonably well known for a narrow band beam one can, as an alternative to a π/K background subtraction, look for kinematic differences between the $\mu^-\mu^-$ candidates and π/K decay events. The missing energy distributions for the two event classes (Fig. 3) are different and indicate a possible real signal.

3. DISCUSSION OF RESULTS

The measured $\nu_\mu N \rightarrow \mu^- \ell^-$ rates are summarized in Table II. The experiments listed in Table IIa used wideband beams with similar energy spectra so the overall rates can be directly compared. The BCR[8] upper limit is significantly below the rate for the CHARM experiment. From the CHARM rate one would expect 7.6 events in the BCR exposure and no events are observed. Taking into account all the appropriate errors the probability for this is $< 10^{-3}$.

The experiments listed in Table IIb have different energy spectra so the energy dependence is relevant. Here the BCR limit of $< 0.76 \times 10^{-4}$ is much less than the overall HPWFOR rate of $(3.4 \pm 0.9) \times 10^{-4}$. It is not possible to directly compare the rates as a function of energy in a meaningful way since the HPWFOR energy dependence is given only for a date set with much less statistical significance than that used for the overall rate.

The more recent experiments do not yield significant new evidence for same-sign dileptons. However, most of them still do have an excess of events over the expected background. It is also worth noting that these new measurements give lower rates than those of earlier measurements but even these are still significantly higher than the rates calculated in 1st order QCD.

There is general agreement that the evidence for a real signal increases as the neutrino energy increases ($E_\nu > 100$ GeV) and as the muon momentum cut increases. Consequently, the present Tevatron experiments at Fermilab could yield new insights into this problem.

4. ORIGIN OF SAME-SIGN DILEPTON EVENTS

There is good agreement among the various experiment on the charactristics of the same-sign dilepton candidates. The angle between the two muons in the plane normal to the neutrino direction peaks at $\approx 180°$ as one would expect if the second lepton (μ_2) is associated with the hadron vertex (Fig. 4). The second muon has limited transverse momentum with respect to the hadron direction $\langle P_T \rangle_\mu \simeq \langle P_T \rangle_{\pi K}$ (Fig. 5). Likewise the momentum component out of the ν-μ_1 production plane is, on average, very similar for μ_2 and for π/K's. The energy dependence of the same-sign events is less clear. The rate relative to the total charged current rate is complicated by the requirement of a second high momentum track in the numerator which introduces an artificial threshold. A better measure, perhaps, is the rate relative to same-sign dileptons. In Fig. 6 the ratio $(\nu_\mu \rightarrow \mu^-\mu^-)/(\nu_\mu \rightarrow \mu^-\mu^+)$ for the HPWFOR[6] and the new CDHS[9] experiments is shown. The data areconsistent with similar energy dependences for both event classes although the HPWFOR data could be consistent with a high energy threshold effect.

Overall the characteristics of same-sign dilepton events are not inconsistent with the production and semi-leptonic decay of a relatively light hadron such as one would expect from $c\bar{c}$ production.

5. CONCLUSION

There has been considerable experimental work on same-sign dilepton production in neutrino interactions over the past 10 years. However, the situation is still unclear. Most experiments do have evidence for an excess of candidates but there is a lack of independent statistically significant results. Present experiments at the Tevatron may provide important new information. However the recent revision of older results and the lower rates from recent experiments indicate that real progress will be difficult. Furthermore, the problems encountered with π/K background subtractions argue for a multi-target experiment. Also, given the conflict with present $c\bar{c}$

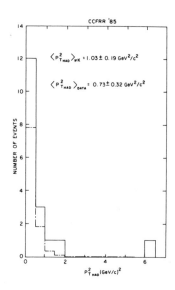

Fig. 5. Transverse momentum of μ_2 in $\mu^-\mu^-$ events and of μ from π/K decay relative to the hadron direction.

Fig. 6. Relative rates of $\mu^-\mu^-$ and $\mu^-\mu^+$ events as a function of neutrino energy.

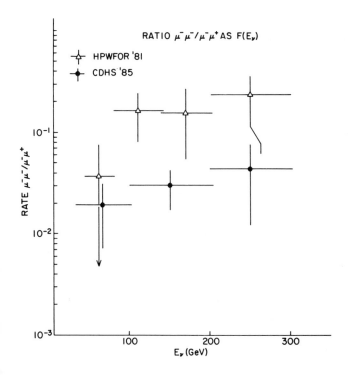

calculations, a high statistics bubble chamber experiment would be useful to establish the presence/absence of $c\bar{c}$ events at the presently measured production rates.

The study of same-sign dilepton production in neutrino interactions is likely to remain an interesting and important topic for some time to come.

This research has been supported by the U.S. Department of Energy under Contract DE-AC02-76CH00016.

6. REFERENCES

1. A. Benvenuti et al. Phys. Rev. Lett. 35, 1199 (1975).
2. H. Goldberg, Phys. Rev. Lett. 39, 1598 (1977);
    B.L. Young et al. Phys. Lett. 74B, 111 (1978);
    V. Barger et al. Phys. Rev. D18, 2308 (1978).
3. M. Holder et al., Phys. Lett. 70B, 396 (1977).
4. J.G.H. deGroot et al., Phys. Lett. 86B, 103 (1979).
5. N. Nishikawa et al., Phys. Rev. Lett. 46, 1555 (1981), Errata Phys. Rev. Lett. 54, 1336 (1985).
6. T. Trinko et al., Phys. Rev. D23, 1889 (1981).
7. M. Jonker et al., Phys. Lett. 107B, 241 (1981);
    K. Winter, private communication.
8. C. Baltay et al., submitted to Phys. Rev. Lett.
9. H. Burkhardt et al., submitted to Zeitschrift für Physik, Particles and Fields.
10. K. Lang et al., Proc. XXth Recontre de Moriond on Electroweak Interactions, Les Arcs, France 1985.
11. A. Haatuft et al., Nucl. Phys. B222, 365 (1983).
12. G. Gerbier et al., Rutherford Appleton Lab. preprint RAL-85-046, May 1985.
13. H.C. Ballagh et al., Phys. Rev. D24, 7 (1981).

Table I

Observed $\mu^-\ell^-$ Event Rates

Exp.	Ref.	P_μ cut(GeV)	Energy	Candidates	Background	Signal
CDHS'77	3	4.5	all	47	30±8	17±11
CDHS'79	4	6.5	all	290	220±30	67±37
CFNRR'81	5	9	all	10	4.3±0.9	5.7±3.3
HPWFOR'81	6	10	all	–	–	65±15
CHARM'81	7	4	all	174	–	74±17
BCR'85	8	5,Pe>4	all	0	1.6	0
CCFRR'85	10	9	all	18	11.1±2.1	6.9±4.7
CDHS'85	9	10	30–100	85	–	20.5±13.5
			100–200	93	–	33±13.5
			200–300	13	–	7±4.9

TABLE IIa. Experiments in Wideband Neutrino Beams with similar E_ν spectra

EXPERIMENT	CDHS[4]	CHARM[7]	BCS[11]	BCR[8]	WA59[12]
P_μ cut(GeV/c)	6.5	4	4.5	5	4.0
P_ℓ cut(GeV/c)	6.5	4	0.8	4	0.8
$\dfrac{\mu^-\ell^-}{\mu^-}$ x10^4 all E_ν	0.34±.18	–	5.3±2.9	≤ 0.76	< 10
$\dfrac{\mu^-\ell^-}{\mu^-\ell^+}$ x10^2 all E_ν	4.1±2.2	14+4±3	22±12	≤ 5.3	< 22

TABLE IIb. Experiments with Narrow Band (NB) or Quad Triplet (QT) Beam.

EXPERIMENT BEAM	BCR[8] WB	CDHS[3] NB	NCFRR[5] QT	HPWFOR[6]* QT	CDHS[9] NB&WB	CCFRR[10] NB	BFHWW[13] QT
P_μ (GeV/c)	5	4.7	9	10	9	9	4
P_ℓ (GeV/c)	4	4.7	9	10	9	9	0.8
$\frac{\mu^-\ell^-}{\mu^-}$ ×10^4 all E_ν	$\leq$0.76	3±2	2.0±1.1	3.4±.9	–	1.4±0.8	–
E_ν<100	$\leq$0.88	–	0.25±0.64	0.45±0.45	.20±.13	0.1±0.5	–
100≤E_ν<200	$\leq$5.4	–	2.2±2.4	6.0±3.0	1.05±.43	3.6±2.1	–
200≤E_ν<300	–	–	9.5±6.5	10.0±7.0	2.1±1.5	2.5±3.1	–
$\frac{\mu^-\ell^-}{\mu^-\ell^+}$ ×10^2 all E_ν	$\leq$5.3	5±3	–	–	1.9±1.2	–	<7
E_ν<100	$\leq$7.2	–	–	–	2.9±1.2	–	<15
100≤E_ν<200	$\leq$23.0	–	–	–	4.2±3.0	–	<12

*In this experiment the division between the two lower energy bins is at 80 GeV. The overall rate is for events originating in the iron target while the energy dependence is taken from Fig. 4(a) of Ref. 15 which is for events originating in the liquid scintillator and iron calorimeters.

UNDERGROUND MUONS FROM CYGNUS X-3

L. E. Price
Argonne National Laboratory*
Argonne, IL   60439

ABSTRACT

Underground detectors, intended for searches for nucleon decay and other rare processes, have recently begun searching for evidence of astrophysical sources, particularly Cygnus X-3, in the cosmic ray muons they record. Some evidence for signals from Cygnus X-3 has been reported. The underground observations are reported here in the context of previous (surface) observations of the source at high energies.

INTRODUCTION

Since its discovery in x-ray emissions in 1966[1], Cygnus X-3 has been observed across the electromagnetic spectrum from radio up to 10^{16} eV. The higher energy ranges (> 0.1 TeV) are observed in air showers where so far there is no direct identification of the primary particle. That these signals are a continuation of the electromagnetic radiation observed at lower energies is inferred from their spatial and temporal coherence, which, given the distance to Cyg X-3 (> 12 kpc) and the intervening magnetic fields, could only be provided by the photon among the known particles.

Some doubt was cast on the identification of the primaries from Cyg X-3 as photons by the Kiel experiment[2], which observed a muon content in extensive air showers from Cyg X-3 almost equal to that from background cosmic ray showers, assumed to be initiated by proton primaries. Photon-initiated showers would be expected to have a muon content lower by about a factor of 10.

It has recently become possible to extend the searches for muons associated with Cyg X-3 (and potentially from other point sources) to much higher energies by exploiting the underground tracking experiments which have begun operations in the last few years, designed primarily to search for nucleon decay. Initial results from these experiments show surprisingly high muon fluxes. If confirmed, these results appear to require either a new type of particle as the cosmic

---

*Work supported by the U. S. Department of Energy, Division of High Energy Physics, under Contract W-31-109-ENG-38.

ray primary or new interactions for photons (or possibly neutrinos) in
the TeV energy range. This paper briefly reviews previous observations of Cyg X-3, then summarizes the observations of underground
experiments, and finally mentions the efforts in progress to account
for the underground muon fluxes.

HISTORY OF CYGNUS X-3 OBSERVATIONS

Cyg X-3 was first observed as a point source of x-rays in a
rocket-borne detector[1]. The x-ray intensity was later found to be
modulated with a regular period of approximately 4.8 hours[3]. TeV air
showers were observed by the Cerenkov technique as early as 1972[4]. A
giant radio outburst in September, 1972,[5] permitted a precise determination of the direction of the source, which led to detection in the
infrared[6]. There followed in short order observations at 30 MeV[7]
and 100 MeV[8]. The first observations in the PeV $(10^{15}$ eV) range were
made by the Kiel extensive air shower array group[9] and confirmation
was provided by the Haverah Park EAS experiment[10].

All Cyg X-3 signals except for radio are correlated with the
x-ray modulation of ca. 4.8 hours, which is generally interpreted as
the orbital period of a close binary system, involving a compact star
such as a neutron star and a normal companion star. Several precise
fits to the x-ray modulation have been made. The one in general use
now is due to van der Klis and Bonnet-Bidaud[11]. Although the x-ray
signals from Cyg X-3 show a fairly continuous modulation with close to
a sinusoidal shape, signals with energies above 100 GeV occur in a
much shorter part of the 4.8 hour period, typically during only 0.1 of
the total period. Observations cluster about two regions of phase at
about 0.2 - 0.3 and 0.6 - 0.7, where the phase of the 4.8 hour period
is defined to run from 0 to 1 and 0 (or 1) corresponds to the x-ray
minimum. Most observations show signals in only one of these two
phase groups.

The energy source for Cygnus X-3 emissions is understood to be
accretion of matter from the companion star onto the compact star.
The x-ray emission is then due to local heating[12], while gamma rays
may arise from acceleration of protons to perhaps 10^{17} eV which then
interact at grazing incidence with the material of the companion star
to produce neutral pions,[13,14] which in turn decay to produce the
observed photons.

Radio absorption measurements indicate that Cyg X-3 is at least
12 kpc (39,000 light years) from the earth. The estimated total
energy emission is 10^{39} ergs/sec. It seems possible that Cygnus X-3
and perhaps a few other sources like it can account for all the high
energy cosmic rays in the galaxy.[14] Observed fluxes from Cyg X-3
are shown in Fig. 1 for energies above 100 GeV. The integral spectrum
is fit fairly well by a simple E^{-1} curve.

MUONS AND UNDERGROUND EXPERIMENTS

Several sophisticated detectors are now situated in underground locations at various depths, intended for searches for nucleon decays, magnetic monopoles, and other phenomena. Some of them have been used to search for muons from Cygnus X-3 with threshold energies set by the overburden ranging from 0.65 TeV to 3 TeV.

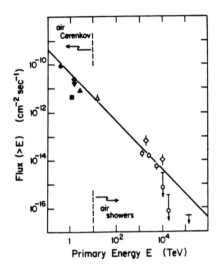

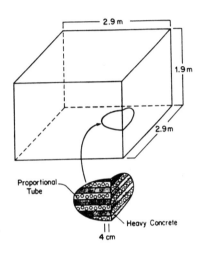

Fig. 1.   Integral spectrum of air showers associated with Cygnus X-3. The solid line is a fit to an E^{-1} shape.

Fig. 2.   The Soudan 1 detector.

As an example, the Soudan 1 detector[5] is shown in Fig. 2. It is constructed of horizontal layers of proportional tubes embedded in heavy concrete. Each layer is 2.9 m square and contains 72 tubes of diameter 2.8 cm. Alternate layers are rotated by 90°, so that two orthogonal views are generated of tracks in the detector. There are a total of 48 layers, making a total detector height of 1.9 m. The detector is placed in the Soudan iron mine in northern Minnesota at a depth of 1800 meters of water equivalent (mwe) at a location 48° N. latitude, 92° W. longitude. Data was collected between September 1981 and November 1983 for a live time of 0.96 year. Cosmic ray muons are observed in the detector as straight tracks. Typical angular resolution is estimated to be $\pm 1.4^{\circ}$ and the uncertainty in the absolute orientation of the detector is estimated to be $\pm 1.5^{\circ}$. The calculated rms multiple coulomb scattering angle for passage through the overburden is 0.8°.

Muons were selected coming from a 3° half angle cone about the nominal direction of Cygnus X-3 ($\delta = 40.8^{\circ}$, $\alpha = 307.6^{\circ}$). Using the ephemeris of Ref. 11 to determine the Cyg X-3 phase of each event yields the histogram of Fig. 3a. Excess events above the background are seen in the phase range 0.65-0.90, amounting to 60 events. The background is determined from off-source α's and in the selected phase range has an rms uncertainty of 17 events, so that the significance of the peak is 3.5σ. Fig. 3b shows the phase plot for nearby off-source directions at the same declination. If the on-source phase peak were due to systematic effects such as a thin spot in the overburden or uneven distribution of live time, similar peaks should appear in the off-source distributions.

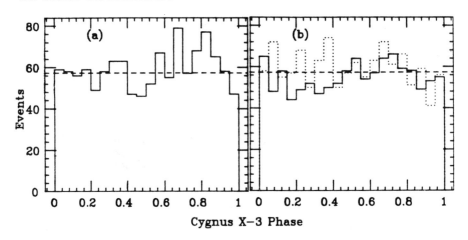

Fig. 3. Cygnus X-3 phase plots from Soudan 1 for (a) on-source muons and (b) off-source, with α less by 10° (dotted) and greater by 10° (solid). For both plots, the average background is shown dashed.

Since there is evidence of considerable variation in the intensity of Cyg X-3 in the air Cerenkov data [16], an attempt has been made to pick out high rate periods in the Soudan 1 data, by selecting pairs of muons within the 3° cone about Cyg X-3 that come within 0.5 hour of each other. The result is Fig. 4, where the average phase of each pair has been plotted. Again a peak appears within the phase region 0.65 - 0.90 containing 29 events with a background uncertainty of 6 events, giving a significance of 4.5 σ.

The NUSEX experiment has also reported[17] results of a search for underground muons from Cygnus X-3. This experiment is located in a road tunnel under Mt. Blanc (45.8° N. latitude, 6.8° E. longitude) at a minimum depth of 4600 mwe, giving a threshold muon energy of 3 TeV. It is also a tracking detector with an area of $(3.5 \text{ m})^2$. It has reported on data with a live time of 2.4 years, taken between June

1982 and February 1985, and has included muons within $\pm5^{\circ}$ of Cyg X-3 in both δ and α. The resulting phase plot is shown in Fig. 5. It shows an excess above background in the phase bin from 0.7 - 0.8, containing 19 $\pm$ 3.6 events (5σ).

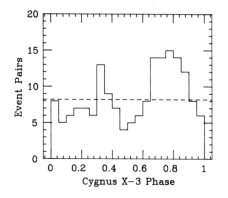

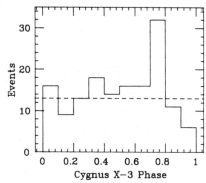

Fig. 4. Average phase of pairs of muons arriving within 0.5 hour in Soudan 1.

Fig. 5. Cygnus X-3 phase plot from NUSEX.

The Frejus experiment, located in the Frejus tunnel, has been taking data while completing the construction of their detector. Their minimum depth is 4400 mwe, corresponding to a threshold muon energy of 2.5 TeV, at a location 45.1° N latitude, 6.7° E. longitude. Completed, it has an area of 72 m^{2}. Early results from the Frejus experiment were reported this summer[8], and are shown in Fig. 6. The angular cuts used are the same as for the NUSEX experiment. Because the detector was growing while data was being taken, this data is predominantly from the first half of 1985. It shows an excess above background for the phase bin 0.6 - 0.7 amounting to 11 $\pm$ 4.3 events (2.5σ). The Frejus collaboration does not consider that at this level their data indicate clear evidence for the presence of a signal from Cyg X-3.

I summarize the fluxes implied by the three underground experiments in Fig. 7, where I have attempted to correct the data from each detector to reflect the rate that would be observed with Cygnus X-3 directly overhead. Fluxes are average over the entire period of observation. The solid line on Fig. 7 is the absolutely normalized E^{-1} line that fits the air shower experiments, indicating that the underground experiments see about the same flux of muons as the air shower experiments see of showers attributed to photons. Four other detectors (IMB, HPW, Homestake, and Kamioka) have reported upper limits consistent with these fluxes.

The magnitude of the problem raised by the underground muon rates is indicated by the calculated points on Fig. 7 of predicted muon

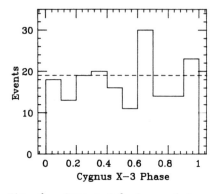

Fig. 6. Cygnus X-3 phase plot
         from Frejus.

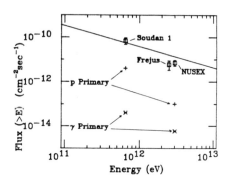

Fig. 7. Integrated flux rates
         for underground muons
         from Soudan 1, NUSEX, and
         Frejus. The solid line
         is the E^{-1} fit from
         Fig. 1. Predicted rates
         are shown if underground
         muons are associated with
         air showers on two
         assumptions about
         primaries.

rates if the air shower primaries are photons or even protons.[19] If
correct, the underground muon rates are very hard to reconcile with
the air shower rates. There is the possibility of a very high flux of
neutrinos, although this would not be predicted from the model of the
source outlined above. This possibility can be ruled out by the zenith
angle subdivisions of the data made by the Soudan 1 and NUSEX
groups. The neutrino hypothesis would suggest a rate independent of
depth. The results from the two experiments are plotted in Fig. 8
with zenith angle translated into the equivalent depth. The result is
clearly inconsistent with the neutrino hypothesis.

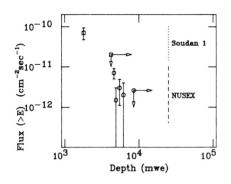

Fig. 8. Depth dependence of
         underground muon flux.

POSSIBLE EXPLANATIONS

Current attempts at understanding the underground data fall into two groups: either known particles have new interactions at energies above 1 TeV or the underground muons (and perhaps the air showers) from Cyg X-3 are produced by previously unknown primary species. In any case, the observed space and time coherence of the signals require that the primary particle be neutral, long lived (> 10^6 sec) and light (m < few GeV).[20]

In the first category are suggestions from Ochs and Stodolsky[21] of a new threshold in the photo-nuclear cross section and from Mohapatra et al.[22] and Ralston[23] of enhanced neutrino cross-sections. None of these authors appear to consider the offered explanations as particularly compelling. In the second category is a suggestion from G. Baym et al.,[24] building on previous suggestions of quark matter in cosmic rays. They suggest that the compact object of Cyg X-3 may be a quark star rather than a neutron star. Chunks of the quark matter may then be knocked off the surface and accelerated to the normal star, where at grazing incidence as above some of them are made electrically neutral for the journey to Earth. A specific particle that fits into this picture is the H (doubly strange dibaryon) proposed by Jaffe.[25]

CONCLUSIONS

While the three underground experiments give rather consistent results, I believe that the overall statistical significance is still rather marginal, and the apparent observations of underground muons from Cyg X-3 should be regarded as interesting but still to be finally proven. The statistical worries are underlined by the fact that each experiment has made some arbitrary choices as to what data is chosen as signal, either in the angular range, in the phase range, or in both, thus decreasing the statistical significance of the observations. A particular point of concern is the approximately 3 times greater solid angle cut used for the two deeper detectors, which is not justified by the inherent detector resolutions or by expected multiple Coulomb scattering in the overburden. Additional data will become available in the next year or two that should clarify matters. In particular, the Frejus and Homestake detectors have large areas. The apparent rate fluctuations may allow time correlations to be made between the various experiments, with a consequent reduction of background. It is also important for the underground detectors to search for other sources known to give TeV air showers, such as Hercules X-1.

REFERENCES

1. Giacconi et al., Ap. J. 148, L119 (1967).
2. M. Samorski and W. Stamm, Proc. 18th Int. Cosmic Ray Conf. (Bangalore), 11, 244 (1983).

3. D. R. Parsignault et al., Ap. J. 209, L73 (1976).
4. Y. I. Neshpor et al., Ap. Sp. Sci. 61, 349 (1979).
5. P. C. Gregory et al., Nature Phys. Sci. 239, 114 (1972).
6. E. E. Becklin et al., Nature Phys. Sci. 245, 302 (1973).
7. A. M. Galper et al., Sov. Astron. Lett. 2, 206 (1976).
8. R. C. Lamb et al., Ap. J. 212, L63 (1977).
9. M. Samorski and W. Stamm, Ap. J. 268, L17 (1983).
10. J. Lloyd-Evans et al., Nature 305, 784 (1983).
11. M. van der Klis and J. M. Bonnet-Bidaud, Astron. Ap. 95, L5 (1981).
12. N. E. White and S. S. Holt, Ap. J. 257, 318 (1982).
13. W. T. Vestrand and D. Eichler, Ap. J. 261, 251 (1982).
14. A. M. Hillas, Nature 312, 50 (1984).
15. M. Marshak et al., Phys. Rev. Lett. 54, 2079 (1985) and Phys. Rev. Lett., to be published.
16. M. F. Cawley, et al., submitted to Ap. J.
17. G. Battistoni et al., Phys. Lett. 155B, 465 (1985).
18. Ch. Berger, talk at Int. Europhysics Conf. on HEP, Bari, Italy, July 18-24, 1985.
19. F. Halzen, Univ. Wisconsin preprint MAD/PH/260 (1985).
20. M. V. Barnhill et al., Univ. Wisconsin preprint MAD/PH/252 (1985).
21. W. Ochs and L. Stodolsky, Max-Planck-Institut preprint MPI-PAE/PTh 54/85 (1985).
22. R. N. Mohapatra et al., Univ. Maryland preprint (1985).
23. J. Ralston, talk at this meeting.
24. Gordon Baym et al., Phys. Lett. 160B 181 (1985); L. McLerran, invited talk at this meeting.
25. R. L. Jaffe, Phys. Rev. Lett. 38, 195 (1977).

NEUTRINO MASS AND RELATED PROBLEMS

Boris Kayser
Division of Physics, National Science Foundation
Washington, D.C. 20550

ABSTRACT

We review some of the physics behind the searches for
neutrino mass, and discuss recent theoretical results
concerning the theory of neutrino mass and the experimental searches for it. These results deal with the
possibility of naturally light Dirac neutrinos, the
possible existence of numerous heavy neutral leptons,
novel ways to search for neutrino mass and for the heavy
leptons, the implications of neutrinoless double beta
decay for neutrino mass, the possible bearing of neutrino oscillation in matter on the solar neutrino problem, and the consistency of a 30 eV electron neutrino
with all existing information.

1. THEORETICAL PREJUDICES

Why is anyone looking for neutrino mass? Why does anybody
think that neutrinos might have mass? One reason is that from the
standpoint of grand unified theories, it is more natural for neutrinos to be massive than massless. This is so for a trivial reason. In any grand unified theory, a given neutrino is put into a
multiplet together with a charged lepton and with several quarks.
Now, apart from the neutrino, all the members of this multiplet are
known to have non-zero masses. Thus, it is natural to expect that

the remaining member of the family, the neutrino, has a non-zero mass too.

Of course, even if neutrinos are massive, we know that they still differ significantly in mass from the quarks and charged leptons to which grand unified theories relate them. Why are they so much lighter than these other fermions? The most popular answer to this question, the "see-saw mechanism,"[1] suggests that neutrinos differ in a fundamental way from the quarks and charged leptons. Namely, neutrinos are their own antiparticles. That is, they are so-called Majorana particles, which transform into themselves under CPT. In this distinguishing characteristic may lie the origin of their relative lightness.

A Majorana particle has just two states: spin up and spin down. By contrast, a Dirac particle, a fermion which is not its own antiparticle, obviously has four states: two for the particle and two for the antiparticle. In the see-saw mechanism, in each generation a four-state Dirac neutrino splits into two Majorana pairs; a light one, ν , identified as the known neutrino, and a heavy one, N . The masses of these Majorana particles are related by $M_\nu M_N \approx M_{\ell \text{ or } q}^2$, where $M_{\ell \text{ or } q}$ is a typical charged lepton or quark mass. In this scheme, it is natural to have $M_N \gg M_{\ell \text{ or } q}$. The fact we have been trying to explain, $M_\nu \ll M_{\ell \text{ or } q}$, then follows.

Attempts have also been made to explain the lightness of neutrinos in a natural way assuming them to be Dirac, rather than Majorana, particles. Though interesting, these models appear to be more complex than the see-saw mechanism. A recent model of this type[2] is a left-right symmetric one containing two distinct neutral Higgs fields ϕ and ϕ' , with ϕ having the correct left- and right-handed weak isospin quantum numbers to generate charged lepton masses, and ϕ' having the correct ones to generate neutral lepton masses. The authors of this model argue that the vacuum expectation value of ϕ' can in a natural way be much smaller than that of ϕ , so that the neutrino masses are, correspondingly, much smaller than the charged lepton masses.

Returning to the Majorana option, we note for later reference that to the extent that CP is a good quantum number in the weak interactions, a Majorana neutrino ν goes into itself not only under CPT but also under CP: $CP|\nu\rangle = \tilde{\eta}_{CP}(\nu)|\nu\rangle$. Here $\tilde{\eta}_{CP}(\nu)$, the intrinsic CP parity of ν , may be +1 or -1 .[3]

We have been talking about the neutrinos that we know to exist. However, as the see-saw model illustrates, many models predict additional, heavier neutral leptons that we do not yet know to exist. An example is the usual form of the left-right symmetric model, in which the W(83 GeV) couples via a left-handed current to the familiar leptons e and ν , while a much heavier W couples via a right-handed current to e and a hypothetical heavy neutral lepton N. The N and ν are Majorana particles, with masses related by the see-saw relation $M_\nu M_N \approx M_e^2$. Thus, since $M_\nu < 50$ eV, $M_N > 5$ GeV. A second example is the O(18) family-unifying model,[4] which unifies the three known generations with each other and with several additional families yet to be discovered. This model contains four neutrinos (ν_e, ν_μ, ν_τ and one more) which couple left-handedly to W(83 GeV), plus four "mirror" neutrinos which couple right-handedly to this same W. All eight neutrinos are light enough to be pair-produced in Z^0 decays. A final example is the very popular superstring theory, which suggests that at energies well below the Planck mass phenomenology is described by an E_6 which contains five two-component neutrinos per generation. A recent analysis[5] of this E_6 emphasizes that it has a problem with respect to neutrino masses. Namely, it has no natural way of keeping the ordinary neutrinos light, and in particular no see-saw mechanism.

2. SEARCHING FOR NEUTRINO MASS

Let us now turn to the question of _how_ people are looking for neutrino mass. In discussing the experimental searches, we assume as usual that in the lepton current $\sum_{\ell=e,\mu,\tau} \bar{\ell}_L \gamma_\mu \nu_{\ell L}$ to which W(83

GeV) couples, the "flavor" eigenstate neutrino ν_ℓ is a linear combination

$$\nu_\ell = \sum_m U_{\ell m} \, \nu_m \tag{1}$$

of the mass eigenstate neutrinos ν_m . The coefficients $U_{\ell m}$ are elements of a unitary mixing matrix.

2.1 Kinematics of $X \to Y + \nu$

Perhaps the most obvious way to look for neutrino mass is to study decays in which a neutrino is emitted, trying to infer the mass of the unseen neutrino from information on the momenta of the other outgoing particles. It should be noted that because of neutrino mixing, a decay $X \to Y + "\nu"$ is really the sum $\sum_m X \to Y + \nu_m$ of the decays into all the neutrino mass eigenstates whose masses M_m permit them to be emitted.[6]

From the beta spectrum in tritium decay, $^3H \to {}^3He + e^- + "\bar{\nu}_e"$, the ITEP group in Moscow has obtained controversial evidence that the dominant ν_m in $"\nu_e"$ has M_m = (20-45) eV .[7] (From this same reaction, evidence had also been reported that one ν_m in $"\nu_e"$ has $M_m \simeq 17$ keV and $|U_{em}|^2 \simeq 0.03$.[8] However, several subsequent experiments on ^{35}S decay saw no sign of this neutrino, and reported that if a neutrino with this mass exists, it must have $|U_{em}|^2 <$ 0.004 or smaller.[9]) From analysis of the decay $\pi \to \mu + \nu_\mu$, an upper limit of 250 keV has been placed on the mass of the dominant ν_m in $"\nu_\mu"$.[10] Finally, from the ν_τ energy spectrum in the decay $\tau \to (3\pi)\nu_\tau$ of taus produced in e^-e^+ collisions, an upper limit of 70 MeV has been placed on the mass of the dominant ν_m in $"\nu_\tau"$.[11]

It has been pointed out[12] that one might be sensitive to somewhat smaller values than 70 MeV for the mass of the heaviest neutrino mass eigenstate (which is presumably the dominant ν_m in ν_τ) if one studies, not τ decay, but the pion spectrum in the rare decay $K^+ \to \pi^+ + \nu + \bar{\nu}$ ($\equiv \sum_m K^+ \to \pi^+ + \nu_m + \bar{\nu}_m$) . Interestingly, as a result of the unitarity of the U matrix, this spectrum does not depend on the $U_{\ell m}$, but only on the neutrino masses. This spectrum also provides a nice illustration of the "Majorana-Dirac confusion theorem": In the absence of right-handed currents, the distinction between a Majorana neutrino and a Dirac one gradually disappears as the neutrino mass goes to zero.[13] In $K \to \pi \nu_m \bar{\nu}_m$, we expect that this distinction will be small when $M_m \ll (M_K/3)$, so that $M_m \ll E_m$. Indeed, it is found[14] that if M_m = 100 MeV, the pion spectrum has an observable dependence on whether ν_m is of Majorana or Dirac character, but if M_m = 50 MeV, this dependence is nearly invisible.

2.2 Neutrinoless Double Beta Decay

A second way to look for evidence of neutrino mass is to search for the nuclear reaction $(A,Z) \to (A,Z+2) + e^- + e^-$, known as neutrinoless double beta decay, or $\beta\beta_{0\nu}$. This reaction would be engendered by the generic diagram in Fig. 1, in which two neutrons in the parent nucleus emit a pair of W bosons W_a and W_b (each of which may or may not be the 83 GeV W boson already discovered), and then W_a and W_b exchange all the neutrino mass eigenstates ν_m . For a given W_a and W_b, the handedness H_a at the W_a ev vertex may be left (H_a = L) or right (H_a = R) or both, so we must sum over all the possible combinations $H_a H_b$ = LL, RR, RL, and LR. As is well-known, the observation of $\beta\beta_{0\nu}$ would imply that the contributing neutrinos ν_m are Majorana particles, and that either (1) these neutrinos have non-zero masses, or (2) a term with $H_a H_b$ = RL or LR is present (so that right-handed currents must exist). Strictly speaking, in case (2) neutrino mass is not required, so a search for $\beta\beta_{0\nu}$ is not exactly a search for neutrino mass. However, it has recently been shown[15] that if the weak interactions are described by a gauge theory, then even an $H_a H_b$ = RL or LR term cannot lead to $\beta\beta_{0\nu}$ unless at least some neutrinos have non-zero masses. Thus, the observation of $\beta\beta_{0\nu}$ would imply neutrino mass, whether or not right-handed currents exist.

402

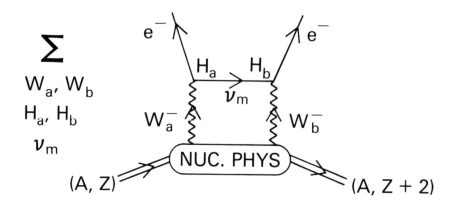

Fig. 1. Neutrinoless double beta decay. The nuclear physics of the W_a and W_b emission is contained in the blob labelled Nuc. Phys.

To see why $\beta\beta_{0\nu}$ requires neutrino mass even when right-handed currents are present, recall that one may introduce gauge theories to a neophyte by asking him to consider the reaction $\nu\bar{\nu} \rightarrow W^-W^+$, where W is the 83 GeV boson with left-handed couplings. Before the advent of gauge theories, the only diagram for this reaction was the charged lepton exchange pictured in Fig. 2a, the amplitude for which is

$$A(\nu\bar{\nu}) = \bar{v}\, \not{\epsilon}_-^*\, \frac{\not{q}}{q^2 + M_e^2}\, \not{\epsilon}_+^*\, (1 + \gamma_5)\, u \; . \tag{2}$$

Here $\epsilon_\pm$ are polarization vectors for the outgoing bosons, and u and v are spinors for the incoming fermions. When the energy $\sqrt{s} \rightarrow \infty$, the cross section corresponding to the amplitude of Eq. (2) violates unitarity. However, it is a hallmark of gauge theories that the complete lowest-order amplitude for any process like $\nu\bar{\nu} \rightarrow W^-W^+$ will have no such unacceptable behavior. In the standard model, the good high-energy behavior results from the presence of an additional diagram, the Z^0 pole pictured in Fig. 2b. This diagram has bad high-energy behavior which exactly cancels that from the diagram in Fig. 2a.

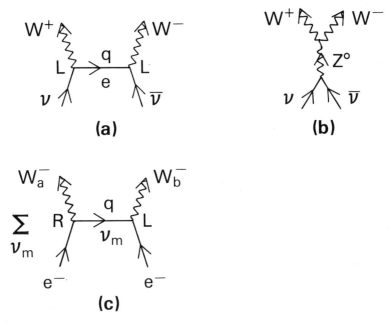

Fig. 2. Diagrams for $\bar{v}v \to W^-W^+$ and $e^-e^- \to W_a^-W_b^-$. The symbols L and R denote the assumed handedness of the current at the vertex, and q is the momentum transfer.

Consider now the diagram in Fig. 2c, which (apart from a time reversal) is just the particle-physics part of the $\beta\beta_{0\nu}$ diagram of Fig. 1 for a specific W_a and W_b, and for $H_aH_b = RL$. The amplitude for the diagram of Fig. 2c is

$$A(e^-e^-) = \sum_m U_{em}^{(a)^*} U_{em}^{(b)^*} \bar{v} \not{\epsilon}_a^* \frac{\not{q}}{q^2 + M_m^2} \not{\epsilon}_b^* (1 + \gamma_5) u , \qquad (3)$$

where M_m is the mass of ν_m, and $U^{(a)}$ and $U^{(b)}$ are the analogues for W_a and W_b of the mixing matrix U for W(83 GeV). As in Eq. (2), $\varepsilon_{a,b}$ are polarization vectors for the outgoing bosons, and u and v are spinors for the incoming fermions. By inspection, we see that the contribution of any one ν_m exchange to the amplitude of Eq. (3) has exactly the same structure as the classic charged lepton exchange amplitude of Eq. (2). Thus, each ν_m exchange must have the same unitarity-violating high-energy behavior as the charged lepton exchange. Now, if the weak interactions are described by a gauge theory, then, as in $\nu\bar{\nu} \rightarrow W^- W^+$, so in $e^- e^- \rightarrow W_a^- W_b^-$ (with $H_a H_b = RL$, in particular) the <u>complete</u> lowest-order amplitude will not violate unitarity. However, in $e^- e^- \rightarrow W_a^- W_b^-$, no Z^0 pole like the one in Fig. 2b can contribute, because the incoming $e^- e^-$ state is not electrically neutral. Thus, in $e^- e^- \rightarrow W_a^- W_b^-$ the good high-energy behavior can only result from a cancellation among the contributions of the various ν_m exchanges themselves. From Eq. (3), we see that when the energy and q^2 become large, the various ν_m contributions differ only in the coefficients $U_{em}^{(a)^*} U_{em}^{(b)^*}$, so that a high-energy cancellation among them requires that

$$\sum_m U_{em}^{(a)} U_{em}^{(b)} = 0 . \tag{4}$$

Due to this constraint, the amplitude of Eq. (3) obviously vanishes even at the low energies relevant to $\beta\beta_{0\nu}$ unless at least one of the masses M_m is non-zero. If all M_m vanish, the $H_a H_b = RL$ contribution to $\beta\beta_{0\nu}$ does too.

The vanishing with the neutrino masses of the $H_a H_b = RL$ contribution to $\beta\beta_{0\nu}$ was noted earlier in the case of left-right symmetric gauge models.[16] The argument just presented shows that such vanishing will occur in <u>any</u> gauge model.

If there are no W bosons besides the one already found, and no right-handed currents, then the diagram of Fig. 1 leads to a $\beta\beta_{0\nu}$ amplitude of the form $M_{eff} N$, where N is essentially a nuclear matrix element, and[17]

$$M_{eff} = \sum_m \frac{\tilde{\eta}_{CP}(\nu_m)}{i} |U_{em}|^2 M_m \tag{5}$$

is an effective neutrino mass. Due to the factors $\tilde{\eta}_{CP}(\nu_m)/i = \pm 1$ in M_{eff}, this quantity can be smaller than all of the actual masses M_m of the mass eigenstates ν_m , even though U is unitary. Calculations of N are uncertain by about a factor of three. The experimental limits on the $\beta\beta_{o\nu}$ decay rate of ^{76}Ge and a popular value of N imply that $M_{eff} < (2-3)$ eV.[18]

2.3 Neutrino Oscillation

A third way to look for evidence of neutrino mass is to search for neutrino oscillation in vacuum, whose observation would imply non-degenerate (hence non-zero) neutrino masses, and non-trivial neutrino mixing. It is interesting that any oscillation in vacuum will be modified by the passage of neutrinos through matter in a way that may have an important bearing on the solar neutrino problem. This problem, a solar ν_e flux on earth which is substantially below predictions, cannot be solved by vacuum neutrino oscillation if neutrino mixing in vacuum is small. However, it was found long ago that if two neutrino types mix with a vacuum mixing angle θ_v and a vacuum oscillation length ℓ_v , then in matter they will have a mixing angle θ_M given by[19]

$$\sin^2 2\theta_M = \frac{\sin^2 2\theta_v}{1 - 2\frac{\ell_v}{\ell_o}\cos 2\theta_v + \left(\frac{\ell_v}{\ell_o}\right)^2} \; . \tag{6}$$

Here ℓ_o is a length characteristic of the matter, depending inversely on its density. Recently, it has been pointed out that the denominator in Eq. (6) has a resonant character.[20] In particular, if the length ℓ_o characteristic of the matter and the length ℓ_v characteristic of the vacuum match in the sense that $\ell_o = \ell_v/\cos 2\theta_v$, then $\sin^2 2\theta_M = 1$, even if $\sin^2 2\theta_v$ is very small. It has been argued that the neutrinos born in the solar core may very well encounter the density that leads to $\ell_o = \ell_v/\cos 2\theta_v$ somewhere during their journey out of the sun.[20] Then they could oscillate appreciably out of their original flavor, ν_e , into another one, and

escape detection in chlorine or gallium solar ν_e detectors on earth.
It is found that the ν_e flux reaching the earth in the energy band
to which a chlorine or a gallium detector is sensitive would be
reduced by a factor of three if $10^{-2} < \sin^2 2\theta_v < 10^{-1}$ and 10^{-7}
$< \delta M^2 < 10^{-4}$ eV^2. Here $\delta M^2 = \left| M_1^2 - M_2^2 \right|$, M_1 and M_2 being the masses
of the two most dominant mass eigenstates in ν_e.

2.4 Heavy Neutral Leptons

Let us now turn from the searches for evidence that the light,
known neutrinos have mass to the searches for _heavy_ neutral leptons.-
Such additional neutral leptons are predicted by numerous models, as
we have discussed, and they are being sought in particle decays,
beam dump experiments, e^+e^- collisions, and elsewhere.[21] It has
been suggested[22] that some of the possibly anomalous CERN "monojet"
events of the form $\bar{p} + p \to$ Jet + (Missing transverse momentum) might
be due to the process

$$
\begin{aligned}
\bar{p} + p \to\ & Z^0\ + \cdots \\
& \downarrow N\ +\ \bar{N} \\
& \qquad\quad \downarrow \text{Unseen particles} \\
& \qquad \downarrow \text{Jet}
\end{aligned}
\tag{7}
$$

Here, a Z^0 (or some heavier $Z^{0'}$) is produced and decays into a pair
of new heavy neutral leptons, one of which subsequently decays into
a final state which is recognized as a jet, while the other decays
into an undetectable final state (such as 3ν), thereby accounting
for the missing transverse momentum. Suppose some of the CERN monojets are due to this process, and that the neutral weak boson in the
process is the Z^0 (94 GeV). Then, given some reasonable assumptions, this same mechanism should also produce monojets in e^+e^-
collisions at a rate which would make them visible at existing e^+e^-
colliders.[23] Thus, monojets have been sought in e^+e^- collisions,

but have not been observed.[24]

Heavy neutral leptons which, like the N of the left-right symmetric model, do not couple appreciably to W (83 GeV) or Z^0 (94 GeV), but would be produced via much heavier weak bosons, may require the SSC for their discovery.[25] Interestingly, once a heavy neutral lepton is found, it would soon be obvious whether it is a Majorana or a Dirac particle. If it is a Majorana particle, it would make this fact strikingly apparent by decaying as often to e^+ + X as to e^- + X .[25] How different this situation would be from the one confronting us in the case of the light neutrinos! Although the latter particles were found long ago, we still have no idea as to whether they are of Majorana or Dirac character.

2.5 Consistency of Experimental Results

Since the ITEP result on the mass of ν_e is controversial, it is of interest to ask how consistent this result is with everything else that we know.[26] The ITEP group found that the dominant ν_m in "ν_e" has M_m = (20-45) eV. Now, suppose we adopt the prejudice that neutrinos are Majorana particles, so that $\beta\beta_{0\nu}$ is relevant, and assume that there are no extra W bosons or right-handed currents. Then we must understand why M_{eff}, Eq. (5), is observed to be less than (2-3) eV. That is, we must understand why $M_{eff} \approx 0$, compared to the mass found at ITEP. To be sure, such vanishing of M_{eff} can occur as a result of cancellation among terms of opposite CP-parity in Eq. (5). If we assume that only two neutrino mass eigenstates are significant pieces of ν_e , the requirement that this cancellation occur becomes

$$M_{eff} = M_1 \cos^2\theta - M_2 S(M_2,A) \sin^2\theta = 0 . \tag{8}$$

Here, θ is the mixing angle that describes the 2 × 2 U matrix, and $S(M_2,A)$ is a suppression factor which may be disregarded for the moment. We are calling the neutrino mass eigenstate which the ITEP group has reported, the dominant one in ν_e, ν_1 , so that $M_1 \approx 30$ eV and $\cos^2\theta \approx 1$, and M_2 must be bigger than M_1 if Eq. (8) is to be satisfied. If $M_2 \gtrsim 10$ MeV, the two participating neutrons in a $\beta\beta$-decaying nucleus A must be close together before the ν_2 exchange contribution to the amplitude of Fig. 1 can be appreciable.[27] This leads to the suppression factor $S(M_2,A)$, which depends on the nucleus A and on M_2 in a more-or-less known way.

With M_1 taken as 30 eV and $S(M_2,A)$ known for given M_2, Eq. (8) may be solved for θ in terms of the one remaining unknown parameter: the mass M_2 of the heavier neutrino. One can then ask what values of M_2 (and the corresponding θ) are allowed by existing laboratory data and cosmological constraints. It is found[26] that, except for a gap between approximately 700 eV and 40 MeV, and another, relatively small gap around 2 GeV, all values of M_2 are excluded by the laboratory data. Furthermore, unless ν_2 has some exotic decay or annihilation mechanism, most of the first of these two gaps is closed by cosmological constraints. Thus, very nearly all M_2 values are excluded, and this is true whether ν_2 has normal weak interactions or is a weak isospin singlet. (Very briefly, the cosmological argument is the following: As M_2 increases, so does the ν_2 decay rate. Now, if M_2 is more than 25 eV, but not too much more, then ν_2 particles which were made in the early universe would still be present now and would over-dominate the present mass density of the universe. If M_2 is somewhat bigger, these ν_2 particles would have decayed before the present. However, depending on their mass and their weak interactions, either their relativistic decay products would have interfered with galaxy formation and distorted the 3° cosmic background photon spectrum, or these neutrinos would have upset the presently observed abundances of the light elements through effects exerted at the time of nucleosynthesis. If M_2 is bigger still, ν_2 is cosmologically "safe".[28])

Since the $\beta\beta_{0\nu}$ nuclear matrix element is somewhat uncertain, and the ITEP group has quoted a range of possible values for M_1, one might ask what happens if we do not require $M_{eff} = 0$, but only M_{eff}

$= \lambda M_1$, with $\lambda \lesssim 1/2$. I have considered this question very briefly, and I believe that almost all values of M_2 are still excluded.

The mass of the electron neutrino is, of course, a quantity of fundamental importance, and the ITEP result is being checked by a number of experiments. The apparent lack of consistency between this result and other existing information adds to the motivation for performing them.

3. CONCLUSION

There are very good theoretical reasons to look for neutrino mass. To be sure, the values of the expected neutrino masses cannot be predicted. However, numerous experiments are probing a variety of mass ranges, and, hopefully, neutrino mass will indeed be found.

REFERENCES

1. M. Gell-Mann, P. Ramond, and R. Slansky, in Supergravity, edited by D. Freedman and P. van Nieuwenhuizen (North Holland, Amsterdam, 1979), p. 315; T. Yanagida, in Proceedings of the Workshop on Unified Theory and Baryon Number in the Universe, edited by O. Sawada and A. Sugamoto (KEK, Tsukuba, Japan, 1979); R. Mohapatra and G. Senjanovic, Phys. Rev. Lett. 44, 912 (1980), and Phys. Rev. D23, 165 (1981).
2. J. Oliensis and C. Albright, Phys. Lett. 160B, 121 (1985).
3. E. Majorana, Nuovo Cimento 14, 171 (1937); G. Racah, ibid 14, 322 (1937). For a discussion, see B. Kayser, Phys. Rev. D30, 1023 (1984).
4. J. Bagger, S. Dimopoulos, E. Masso, and M. Reno, Nucl. Phys. B258, 565 (1985).
5. J. Rosner, University of Chicago preprint EFI 85-34.
6. R. Shrock, Phys. Lett. 96B, 159 (1980).
7. V. Lubimov et al., in Proceedings of the XXII Int. Conf. on High Energy Physics, edited by A. Meyer and E. Wieczorek (Akademie der Wissenschaften der DDR, Zeuthen, DDR, 1984), p. 259.
8. J. Simpson, Phys. Rev. Lett. 54, 1891 (1985).
9. T. Altzitzoglou et al., Phys. Rev. Lett. 55, 799 (1985); T. Ohi et al., Phys. Lett. 160B, 322 (1985).
10. R. Abela et al., Phys. Lett. 146B, 431 (1984).
11. H. Albrecht et al., (ARGUS collaboration), DESY preprint DESY 85-054. Bounds have also been set on this mass by the DELCO,

HRS, MAC, and MARK-II studies of τ decay.

12. N. Deshpande and G. Eilam, Phys. Rev. Lett. 53, 2289 (1984).

13. B. Kayser, Phys. Rev. D26, 1662 (1982).

14. J. Nieves and P. Pal, Phys. Rev. D32, 1849 (1985).

15. B. Kayser, S. Petcov, and S.P. Rosen, to be published.

16. T. Kotani, in Proceedings of the 1984 Moriond Workshop on Massive Neutrinos in Astrophysics and in Particle Physics, edited by J. Tran Thanh Van (Editions Frontieres, Gif sur Yvette, 1984), p. 397.

17. L. Wolfenstein, Phys. Lett. 107B, 77 (1981); B. Kayser and A.S. Goldhaber, Phys. Rev. D28, 2341 (1983).

18. This bound is taken from F. Avignone, III et al., University of South Carolina preprint, in which the data of the Caltech, Guelph-Aptec-Queens, Milano, Pacific Northwest Laboratory -USC, and UCSB-LBL experiments are taken into consideration.

19. L. Wolfenstein, Phys. Rev. D17, 2369 (1978).

20. S. Mikheyev and A. Smirnov, Institute for Nuclear Research, Moscow, preprint.

21. These searches are discussed by M. Gronau, C.N. Leung, and J. Rosner [Phys. Rev. D29, 2539 (1984)], by M. Perl [Proceedings of the Santa Fe Meeting of the Division of Particles and Fields of the American Physical Society, edited by T. Goldman and M. Nieto (World Scientific, Philadelphia, 1985), p. 159], by F. Gilman and S. Rhie [Phys. Rev. D32, 324 (1985)], and by G. Feldman [Stanford Linear Accelerator Center preprint SLAC-PUB-3684; talk presented at the XX Rencontre de Moriond: QCD and Beyond, Les Arcs, 1985].

22. L. Krauss, Phys. Lett. 143B, 248 (1984); M. Gronau and J. Rosner, Phys. Lett. 147B, 217 (1984); J. Rosner, ibid 154B, 86 (1985).

23. S. Glashow and A. Manohar, Phys. Rev. Lett. 54, 526 (1985); J. Rosner, Ref. 22.

24. W. Ash et al., Phys. Rev. Lett. 54, 2477 (1985); G. Feldman et al., Phys. Rev. Lett. 54, 2289 (1985); C. Akerlof et al., Phys. Lett. 156B, 271 (1985).

25. J. Gunion and B. Kayser, in Proceedings of the 1984 Summer Study on the Design and Utilization of the Superconducting Super Collider, edited by R. Donaldson and J. Morfin (Division of Particles and Fields of the American Physical Society, 1985), p. 153; B. Kayser, N. Deshpande, and J. Gunion, to appear in the Proceedings of the Third Telemark Miniconference on Neutrino Mass and Low Energy Weak Interactions.

26. P. Langacker, B. Sathiapalan, and G. Steigman, University of Pennsylvania and University of Delaware preprint.

27. A. Halprin, S. Petcov, and S.P. Rosen, Phys. Lett. 125B, 335 (1983).

28. G. Kane and I. Kani, in University of Michigan preprint UM TH 85-20, point out that cosmological arguments of this kind do not exclude a few isolated, special values of M_2 because of the possibility of resonant $\bar{\nu}_2 \nu_2$ annihilation.

SEARCH FOR HEAVY NEUTRINO
PRODUCTION AT PEP

Gary J. Feldman[*]

Stanford Linear Accelerator Center

Stanford University, Stanford, California 94305, U.S.A.

Abstract

We report a search for long-lived heavy neutrinos produced by the neutral weak current in e^+e^- annihilation at 29 GeV at PEP. Data from the Mark II detector are examined for evidence of events with one or two separated vertices in the radial range of 2 mm to 10 cm. No events were found that were consistent with the hypothesis of heavy neutrino production, eliminating the possibility of heavy neutrinos with decay lengths of 1 to 20 cm in mass range 1 to 13 Gev/c^2.

In e^+e^- annihilation at PEP energies the dominant reaction is the production of a pair of fundamental particles from a single virtual photon. All fundamental particles can be produced copiously in this way, provided enough energy is available to form their mass and provided that they couple to the photon, that is, that they have electric charge. One of the compelling reasons for studying e^+e^- annihilation at the Z pole at SLC is that the dominant reaction will be the formation of pair of fundamental particles from a Z, rather than a

---

[*] This work was supported in part by the Department of Energy, contract DE-AC03-76SF00515.

photon. Thus all fundamental particles which have weak charge will be copiously produced, including, for the first time, electrically neutral particles.[1]

The point of this talk is that one does not necessarily have to wait for SLC turn-on to search for new neutral particles, because even at PEP energies there is a significant coupling to a virtual Z. One example of such a particle could be a heavy neutrino, either from a fourth generation, or from a more exotic source. The cross section for producing a pair of neutrinos is[2]

$$\sigma = \frac{G_F^2 E^2}{192\pi} \left(\frac{(1 - 4\sin^2\theta_W)^2 + 1}{\left(1 - \frac{E^2}{m_Z^2}\right)^2 + \frac{\Gamma_Z^2}{m_Z^2}} \right) \beta(3 + \beta^2), \tag{1}$$

where E is the center of mass energy. At the PEP energy of 29 GeV, this cross section is only 0.34 pb,[3] but the accumulated Mark II data of 208 pb^{-1} yields 71 produced events and thus allows a reasonable search.

If we assume that the GIM mechanism[4] is valid for a heavy neutrino, then it will only be able to decay into one of the known charged leptons (e, μ, or τ) and a virtual W via a (small) mixing angle ϵ.[5] This decay is illustrated in Fig. 1.

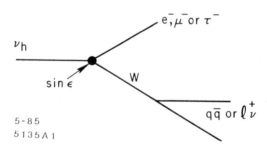

5-85
5135A1

Figure 1. General diagram for heavy neutrino decay.

In this (standard) model, the lifetime of a heavy neutrino is completely calculable given the mixing angle ϵ. It can be expressed in terms of the muon lifetime as

$$\tau(\nu_h \to \ell^- X^+) = \left(\frac{m_\mu}{m_{\nu_h}}\right)^5 \frac{\tau(\mu \to e\nu\bar{\nu}) B(\nu_h \to \ell^- e^+ \nu)}{f(m_{\nu_h}, \ell) \sin^2 \epsilon}, \tag{2}$$

where ℓ represents the lepton to which ν_h primarily couples, and f is a phase space correction which is significant for our application only when $\ell = \tau$ and m_{ν_h} is at most a few times m_τ. The branching fraction B can be calculated in much the same way as in τ decay.[6] Depending on m_{ν_h} and $\sin^2 \epsilon$, the decay lengths of a heavy neutrino can be appreciable. Figure 2 shows the contours of constant decay length as a function of these two variables.

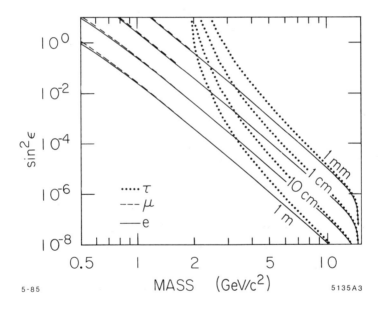

5-85   5135A3

Figure 2. Contours of constant ν_h decay length.

The search was conducted with the Mark II detector at PEP.[7,8] The basic strategy was to look for events with two vertices that are separated from the interaction point and with no tracks coming from the interaction point. Even the observation of a single such event could be a spectacular signal. The main event requirements were

1. Four or more charged tracks. (From Fig. 1, it is clear that each ν_h must decay into at least two charged particles.)

2. One vertex with 2 mm $< r_1 <$ 10 cm, where r_1 is the radial distance between the first vertex and the interaction point. If $r_1 <$ 3 mm or there were only four charged tracks in the event, then there must have been another vertex with $r_2 >$ 2 mm. Otherwise, a second vertex was not required.

3. No vertex within 1 mm of the interaction point.

4. The stability of the interaction point was monitored with beam position monitors. For each run that was used in this analysis, the rms beam position had to be less than 250 μm horizontally and 150 μm vertically.

5. Tracks from identified K_S^0's and Λ's were removed from consideration in finding vertices.

6. Events were rejected if 7.4 cm $< r_1 <$ 8.0 cm, since this was the region of the vacuum pipe.

After applying these cuts, only three events remained. (A Monte Carlo simulation predicted that we would see two events from known sources of background at this point in the analysis.) On further examination of these events, we found that they were all incompatible with the hypothesis of ν_h pair production. In one event the position of the interaction point had moved 3 mm from its assumed position. This was determined by examining the vertex of the events immediately preceding and following the candidate event. A second event had only three charged particles present. The remaining tracks were from two independent photon conversions in the chamber. The final event was kinematically incompatible with the ν_h pair hypothesis because it had a backward-going 8 Gev/c track.

Figure 3 shows the contour of excluded region at the 90% confidence level in the space of decay length and m_{ν_h}. The decay length region between 1 and 20 cm is excluded for $1 < m_{\nu_h} < 13$ GeV/c^2. Figure 4 shows the same contour as a function of $\sin^2 \epsilon$ and m_{ν_h}.

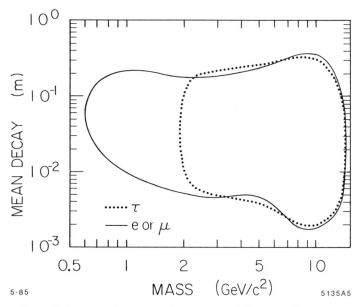

5-85 5135A5

Figure 3. Excluded region for ν_h at the 90% confidence region as a function of decay length and m_{ν_h}.

In conclusion, there is presently no evidence for the existence of long-lived heavy neutrinos and large regions of decay length and mixing angles have been eliminated. We look forward to experiments at the SLC and LEP, where these searches can be conducted with much more sensitivity and generality.

416

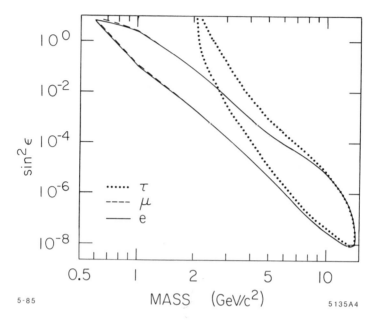

5-85   MASS   (GeV/c^2)   5135A4

Figure 4. Excluded region for ν_h at the 90% confidence region as a function of $\sin^2 \epsilon$ and m_{ν_h}.

References

1. One exception is the production of pairs of identical self-conjugate spinless bosons, which is forbidden by symmetry considerations.

2. See, for example, F. M. Renard, *Basics of Electron Positron Collisions*, (Editions Frontières, Dreux, 1981).

3. This value includes a -5% radiative correction.

4. S. L. Glashow, J. Iliopoulos, and L. Maiani, *Phys. Rev. D* **2**, 1285 (1970).

5. For simplicity, we have assumed here that ν_h mixes primarily with a single lepton species; however, the extension to the general case is obvious.

6. H. B. Thacker and J. J. Sakurai, *Phys. Lett.* **36B**, 103 (1971); Y. S. Tsai, *Phys. Rev.* D **4**, 2821 (1971) and *Phys. Rev.* D **19**, 2809 (1979); J. D. Bjorken and C. H. Llewellyn-Smith, *Phys. Rev.* D **7**, 887 (1973); F. J. Gilman and D. H. Miller, *Phys. Rev.* D **17**, 1846 (1978).

7. The individual who did most of the work on this analysis is Chris Wendt. The other members of the Mark II collaboration at PEP are T. Barklow, A. M. Boyarski, M. Breidenbach, P. Burchat, D. L. Burke, J. M. Dorfan, G. J. Feldman, L. Gladney, G. Hanson, K. Hayes, R. J. Hollebeek,

W. R. Innes, J. A. Jaros, D. Karlen, A. J. Lankford, R. R. Larsen, B. W. LeClaire, N. S. Lockyer, V. Lüth, C. Matteuzzi, R. A. Ong, M. L. Perl, B. Richter, K. Riles, M. C. Ross, D. Schlatter, A. R. Baden, J. Boyer, F. Butler, G. Gidal, M. S. Gold, G. Goldhaber, L. Golding, J. Haggerty, D. Herrup, I. Juricic, J. A. Kadyk, M. E. Nelson, P. C. Rowson, H. Schellman, W. B. Schmidke, P. D. Sheldon, G. H. Trilling, C. de la Vaissiere, and D. Wood from LBL, and M. Levi and T. Schaad from Harvard.

8. The Mark II detector has been previously described. See, for example, R. H. Schindler *el al.*, *Phys. Rev. D* **24**, 78 (1981).

REVIEW OF NEUTRINO MASS MEASUREMENTS

Thomas J. Bowles

Physics Division
Los Alamos National Laboratory
Los Alamos, New Mexico 87545

ABSTRACT

A review of experiments which search for a finite mass of the ν_e, $\bar{\nu}_e$, ν_μ, and ν_τ is presented. At present, the only evidence for a nonzero neutrino mass comes from the ITEP group which claims to observe a 35 eV mass for the ν_e. Many experiments are underway to check this claim and the techniques and status of these experiments are presented.

Experiments to search for a nonzero mass of the known neutrinos (electron, muon, tau and their corresponding antineutrinos) are being carried out with ever increasing sensitivity using a wide range of both nuclear and particle physics techniques. The motivation for these experiments stems from two different, but closely allied, areas of physics: grand unification and cosmology. Until the development of Grand Unified Field Theories (GUT's), there were no theoretical grounds to expect a nonzero neutrino mass; the standard Weinberg-Salam-Glashow theory has zero neutrino mass. Only when one goes to GUT's models such as SO(10) does one have nonzero neutrino masses generated in a natural way. However, the mass scale of neutrinos in GUT's is not determined. Thus, an experimental observation of a nonzero neutrino mass would be tremendously exciting in determining which GUT is correct. At the other extreme of the distance scale, a nonzero neutrino mass has tremendous implications in cosmology. The missing mass problem is well established in that only a small fraction (<10%)

of the mass on the universe is contained in luminous matter[1]. This has been shown at the largest distance scales by measurements of the deceleration parameter of the expansion of the universe. Also it is observed that the velocity of stars in the halos of galaxies continues to increase with distance from the galactic centers, indicating the existence of dark matter. It has been suggested that massive neutrinos could account for this dark matter (although it is difficult to form galaxies with relativistic neutrinos). Since the number density of neutrinos is 10^9 times the number density of baryons in the universe according to the standard model of the Big Bang, we see that a 35 eV neutrino could account for all of the missing mass, and in fact would be sufficient to close the universe. Thus, it is important to search for neutrino masses down to the few eV range in terms of cosmological interest.

The only experiments to probe for neutrino masses in the few eV range are those measuring the beta decay of tritium in which an electron antineutrino is emitted. Thus, I will concentrate on these experiments, but first I will briefly discuss the present limits on the masses of the tau, muon, and electron neutrinos.

The best limits on the mass of the tau neutrino come studies of $e^+e^- \longrightarrow \tau^+\tau^-$ and then observing the decay of $\tau \longrightarrow 3\pi^{\pm} \nu_{\tau}$. The invariant mass distribution for this decay mode is fit with the neutrino mass as a free parameter. The best limit presently comes from the ARGUS collaboration at the DORIS II storage ring. They report[2] observation of 1536 events with $3\pi^{\pm}$ and no π^0's in the decay of one of the τ's with the other τ decaying by 1 charged particle plus neutrals. From analysis of this data they set a limit of

$$m_{\nu_{\tau}} < 70 \text{ MeV} \qquad (95\% \text{ confidence level})$$

where this limit includes all systematic effects. Improved limits on $m_{\nu_{\tau}}$ should be available with higher luminosities and more data to improve the statistical accuracy and with more data to study systematic effects.

The best limits on the mass of the muon neutrino come from observing the decay of $\pi^+ \longrightarrow \mu^+ \nu_{\mu}$ at rest. The masses of the pion and muon are known accurately, so that an accurate measurement of the momentum of the muon provides a test for any missing mass associated with the muon neutrino. The most precise measurement to date gives[3]

$$p_\mu{}^+ = (29.79139 \pm .00083) \text{ MeV/c}$$

which, from the measured masses of the pion and muon gives

$$m_{\nu_\mu} < 250 \text{ keV/c}^2 \qquad \text{(90\% confidence level)}.$$

Future experiments observing the decay of π^+ (and possibly π^-) inflight, in which both the pion and muon momenta are measured, may provide a means of improving these limits[4], but this will also require improved measurements of the masses of the pion and muon.

The best limits on the mass of ν_e comes from measurements of internal bremmstrahlumg emitted during electron capture, as suggested by de Rugjula[5] and Bennett [6]. The Q values in some cases are very small, enhancing the effect of a neutrino mass, and the continuous photon spectrum can be analyzed for a neutrino mass in much the same way as a beta spectrum would be. The best case at present is the measurement of ^{193}Pt by the ISOLDE collaboration[7] who have set a limit of

$$m_{\nu_e} < 500 \text{ eV} \quad \text{(90\% confidence level)}$$

Other promising cases have been investigated, in particular ^{163}Ho[8], but it appears that it will be very difficult to improve on the above limit. Since the CPT theorem requires the mass of particles and antiparticles to be the same, the best (indirect) limits on m_{ν_e} now come from limits on $m_{\bar\nu_e}$ in tritium beta decay.

Thus, at present the only means of searching for neutrino masses in the eV range seems to be in measurements of tritium beta decay. The endpoint energy of this beta decay is the lowest known (18.6 keV) and the complications due to atomic effects are well understood. The beta spectrum for the decay ^{3}H $\longrightarrow$ ^{3}He$^+$ + e$^-$ + $\bar\nu_e$ is given by

$$\frac{dN(E_\beta)}{dE_\beta} = G_f{}^2 \cos^2\theta_c \frac{m_e{}^5 c^4}{2\pi^3 \hbar^7} M_{if} F(Z_f, E_\beta) p_\beta E_\beta (E_o - E_\beta)^2 \times \left\{1 - \frac{m_e{}^2 c^4}{(E_o - E_\beta)^2}\right\}^{\frac{1}{2}}$$

where G_f is the Fermi coupling constant, θ_c is the Cabibbo angle, m_e is the electron rest mass, M_{if} is the nuclear matrix element, $F(Z_f, E_\beta)$ is the Fermi Coulomb factor, p_β and E_β are the beta momenta and energy respectively and E_o is the maximum beta energy for the decay. The effect of a nonzero mass of the antineutrino (hereafter referred to simply as neutrino mass) can be most easily shown in a Kurie plot, in which the phase space and Coulomb effects are removed so that the quantity $\{1/[F(Z_f, E_\beta)\ p_\beta E_\beta]\ dN(E_\beta)/dE_\beta\}^{1/2}$ is plotted. Both the beta spectrum and the Kurie plot of the tritium beta decay spectrum is plotted in figure 1.a. The effect of a 35 eV neutrino mass is shown in the blowup of the endpoint region of beta spectrum in figure 1.b. One sees that the effect of a 35 eV neutrino mass extends only over about the last 100 eV of the spectrum. A measurement of a neutrino mass consists of first measuring the shape of the beta spectrum a few keV below the endpoint where a nonzero neutrino mass has no measurable effect. Then one extrapolates this measurement out to the endpoint assuming a zero neutrino mass. Finally, one measures the beta spectrum in the region of the endpoint and compares this measurement with the extrapolation. A difference between the two can then be interpreted as evidence for a nonzero neutrino mass. In practice of course, the situation is rather more complicated. A very intense source and low background is required since the last 100 eV of the beta spectrum contains only about 3×10^{-7} of the total beta spectrum. The energy resolution of the spectrometer and energy loss in the source distort the beta spectrum and must be very accurately taken into account. And perhaps the most severe complication arises from the fact that in the beta decay, atomic and molecular excitations occur due to the sudden change in nuclear charge. Thus, the spectrum actually consists of many spectra with different endpoints, each with its own probability, corresponding to leaving the atom or molecular in various excited states. Since the first excited state in $^3He^+$ is at 41 eV and is excited with 25% probability in the decay of a free tritium atom, these effects must be very accurately taken into account.

The first evidence for a finite neutrino mass was reported in 1980[9] by a group from the Institute for Theoretical and Experimental Physics (ITEP) in Moscow. This group measured the beta decay of tritium atoms in a valine {NH_2 CT COOH CH $(CH_3)_2$} molecule using a toroidal beta spectrometer. They claim to have observed evidence for a 35 eV neutrino mass with a possible range between 14 and 46 eV at the 99% confidence level depending on what assumptions were made about the final state distributions. The limit of 14 eV was claimed to be a model independent lower limit, where the most extreme assumption possible was made about the final states; 100% of the decays go only to the ground state of $^3He^+$. However, it was pointed out by J. Simpson in 1983[10] that the ITEP measurement had not taken into account the 10 eV line width of the ^{169}Yb source

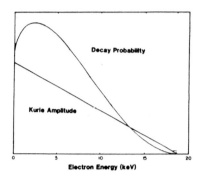

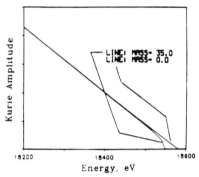

Fig 1.a. Beta spectrum and Kurie plot for tritium beta decay.

Fig 1.b. Endpoint region showing effect of 35 eV neutrino mass with 100% decay to ground state of $^3He^+$.

used to measure the spectrometer resolution. When this was included in the analysis, the data were then consistent with a zero neutrino mass. But, specific calculations of the final state effects in the valine molecule were carried out by Kaplan et al[11]. The ITEP group claimed that these final state effects raised the result for the neutrino mass (which in fact was not true) and that taking into account both the resolution and final state effects, the data were best described by a 35 eV neutrino mass. Also, in 1983 the ITEP group published new data[12] taken with better resolution (20 eV) and substantially lower background (a factor of 20 reduction was obtained by accelerating betas from the source by 4 keV to raise them above betas arising from tritium contamination in the spectrometer). This data required two neutrino masses of 35 eV and 125 eV to produce a reasonable fit. This can occur if neutrino oscillations exist between neutrinos with these masses. However, that the line shape of the conversion electrons used to measure the spectrometer resolution is asymmetric and the ITEP group had used a symmetric Lorentzian shape. This effect was later taken into account and new data published in 1984[13] gave a best fit of 34.8 $\pm$ 1.9 eV and a model independent lower limit of 9 eV at the 90% confidence level. The ITEP data from three runs is shown in figure 2. The statistical accuracy is high, but the primary concerns still are due to possible systematic effects.

The claims of the ITEP group have spurred tremendous experimental activity. At present more than a dozen groups are working on tritium beta decay measurements using a variety of sources and measurement techniques[14]. The sources used are in general solid sources with either ion implanted tritium or chemical compounds containing tritium. These sources have systematic problems associated with final state effects. However, several groups plan to use pure tritium sources, either frozen solid molecular tritium, free molecular tritium gas, or free atomic tritium gas. Only with a free atomic tritium gas source can the final state excitations be calculated to arbitrary accuracy, and so do not present a systematic uncertainty in the results for a neutrino mass. The spectrum of states and excitation probabilities for the beta decay of one of the atoms in a tritium molecule into a $^3He^3H^+$ molecule are considerably different than that of the free tritium atom, but can be calculated reliably[15].

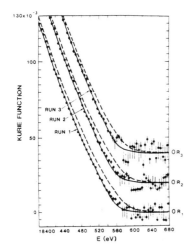

Fig 2. 1984 ITEP data. Three runs with different source thicknesses Dashed line = fit for zero mass. Solid line = fit for 35 eV mass.

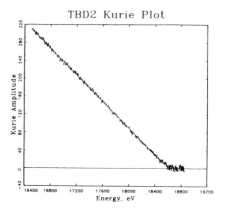

Fig 3. Preliminary data from Los Alamos group. Data taken in three day period with free molecular tritium tritium gaseous source.

The spectrometers used fall into two basic categories: electrostatic or magnetic. The electrostatic spectrometers are generally retarding field spectrometers which transmit all betas above some energy to the focal plane detector. While these spectrometers have very good energy resolution (typically a few eV), they decelerate the electrons near the endpoint to very low energy and have no means of discriminating against backgrounds which produce low energy electrons in this region of the spectrometer (such as electrons from field emission, photoelectrons, tritium contamination, etc.). The magnetic spectrometers have medium energy resolution (typically 20-40 eV), but by accelerating betas from the source by several keV, raise betas near the endpoint above backgrounds due to electrons originating elsewhere in the spectrometer. Thus, the magnetic spectrometers have somewhat worse energy resolution than electrostatic spectrometers in general, but do not have as potentially high backgrounds to deal with. The question of energy resolution on the accuracy of a measurement has been dealt with by Simpson[10] who showed that for an energy resolution R measured incorrectly by a fractional amount , that the difference between the actual neutrino mass m_ν and the neutrino mass M_ν determined using the incorrect resolution is given by $M_\nu^2 - m_\nu^2 = 2\epsilon R$. If the neutrino mass is actually zero, a spurious neutrino mass can be found of $M_\nu = \sqrt{2\epsilon R}$ if the resolution is incorrect. Thus, with a resolution of 30 eV an error of 20% in the resolution generates a 3.5 eV neutrino mass. Thus, the somewhat poorer energy resolution of the magnetic spectrometers requires more care in measuring the resolution, but does not present a limit in measuring neutrino masses down to a few eV.

Taking the above considerations into account, the Los Alamos group is carrying out an experiment using free atomic and molecular tritium as the source. The betas from the source are accelerated by up to 10 KV to minimize backgrounds from tritium contamination in the spectrometer. The spectrometer is a toroidal magnetic spectrometer with very large acceptance (10% of 4π of betas from the gaseous source) with a moderate energy resolution of about 30 eV for 25 keV betas. The system has been described in detail elsewhere[16]. Data acquisition has begun and the quality of data in a three day run with molecular tritium is shown in figure 3. Studies of systematic effects and additional data acquisition are now underway.

The group at Tokyo[17] is also presently taking data with a source containing tritium in an organic compound, but has no results yet. The group at the University of Zurich[18] has completed the first set of data acquisition. They have shown fits of their data to zero and thirty-five eV neutrino masses, which appear to fit a zero neutrino mass reasonably well assuming free tritium atomic final states in their source (which is actually tritium ion implanted in graphite). However, fits to the data indicate a source thickness 3.5 times thicker than the source was made and due to this systematic problem, they are not able to set a limit on the neutrino mass from their data. The groups at Livermore[19] and IBM (20) should also begin data acquisition soon. Thus, there should be several experiments with results on the $\bar{\nu}_e$ mass by the time of the DPF meeting next year.

REFERENCES

1. Faber, S. M. and J. S. Gallagher, 1979 Ann. Rev. Astron. Astrophys. 17:135. (1979).

2. ARGUS collaboration, reported at 1985 Division of Particles and Fields meeting, Eugene, Oregon, August, 1985.

3. Abela, R., M. Daum, G. H. Eaton, R. Frosch, B. Jost, P. R. Kettle, and E. Steiner, Phys. Lett. 146B, 431 (1984).

4. P. Nemethy, private communication.

5. de Rujula, A. Nucl. Phys. B188, 414 (1981).

6. Bennett, C. L., Phys. Lett. 107B, 19 (1981).

7. Jonson, B., Nucl. Phys. A396, 479 (1983).

8. Bennett, C. L., Bull. Am. Phys. Soc. 29, 666 (1984).

9. Lubimov. V. A., E. G. Novikov, V. Z. Nozik, E. F. Tretyakov, and V. S. Kosik, Phys. Lett. 94B, 266 (1980).

10. Simpson, J. J. Proc. Intern. Conf. ICOMAN '83 (Frascati, Italy 1983).

11. Kaplan, I. G., V. N. Smutnyi, and G. V. Smelov, Zh. Eksp. Teor. Fiz. 84, 833 (1983).

12. Boris, S., A. Golutvin, V. Lubimov, V. Nagovizin, E. Novikov, V. Nozik, V. Soloshenko, I. Tichomirov, E. Tret'yakov, Proc. Intern. Europhys. Conf. on High Energy Physics, Brighton, England, p. 386 (1983).

13. V. A. Lyubimov, Proc. XXII Intern. Conf. on High Energy Physics, Leipzig, Germany (1984).

14. R. G. H. Robertson, Los Alamos National Laboratory Preprint LAUR-85-209, Proc. 1985 Aspen Winter Physics Conf., Aspen, Colorado (1985).

15. Martin, R. L. and J. S. Cohen, Phys. Lett. 110A, 95 (1985).

16. Knapp, D. A., T. J. Bowles, J. C. Browne, T. H. Burritt, J. S. Cohen, J. A. Helffrich, M. P. Maley, R. L. Martin, R. G. H. Robertson, and J. F. Wilkerson, Los Alamos National Laboratory preprint LAUR-84-3980. Proceedings of the 1984 Neutrino Mass Miniconference, Telemark, Wisconsin (1984).

17. Ohshima, T., Institute for Nuclear Study, Univ. of Tokyo, private communication.

18. Kundig, W., Univ. of Zurich, Switzerland, private communication.

19. Fackler, O., Lawrence Livermore National Laboratory, priv. comm.

20. Clark, G., IBM, Yorktown Heights, N. Y., private communication.

NEUTRINO OSCILLATIONS: AN EXPERIMENTAL REVIEW

Michael H. Shaevitz

Columbia University Nevis Labs, Irvington, NY 10533

The neutrino has been the main probe in weak inter-action studies over the past two decades. In the standard electro-weak model, the neutrino is a weakly interacting, spin 1/2 particle and is assumed to have zero mass and definite lepton number. In general, there is no inherent reason for neutrinos to be massless and, further, if they are massive, the mass matrix need not be diagonal. If the mass matrix is non-diagonal and at least one neutrino flavor has a finite mass, then neutrino oscillation can occur. Neutrino oscillations provide the only possible tool for studying muon and tau neutrino masses in the eV and below mass range. In addition, oscillation studies can also show the effects of a massive fourth generation neutrino or a wrong-handed neutrino.

For the case of two component mixing, the initial weak eigenstate is given by $|\nu_\alpha\rangle = \cos\theta|\nu_1\rangle + \sin\theta|\nu_2\rangle$, where $|\nu_1\rangle$ and $|\nu_2\rangle$ are the mass eigenstates with $\Delta m^2 = m_1^2 - m_2^2 \neq 0$ and $\sin\theta = U_{\alpha 2} = U_{\beta 1}$ describes the two component mixing. Evolving this state to later times leads to the following probability distributions for the transitions of a weak eigenstate $|\nu_\alpha\rangle$:

i) $\text{Prob}(\nu_\alpha \rightarrow \nu_\alpha) = 1 - \sin^2(2\theta) \sin^2(\frac{1.27 \, \Delta m^2 L}{E})$, and

ii) $\text{Prob}(\nu_\alpha \rightarrow \nu_\beta) = \sin^2(2\theta) \sin^2(\frac{1.27 \, \Delta m^2 L}{E})$;

where L is the distance from the ν source to the detector in km, E is the energy of the ν in GeV, and Δm^2 is in eV^2. An extension to more than one active oscillation channel can easily be made but the number of parameters to be experimentally determined increases. For this reason, most experimental papers present results only for the one channel assumption.

EXCLUSIVE OSCILLATION LIMITS

Two types of oscillation experiments, inclusive (or disappearance) and exclusive (appearance) experiments, are possible corresponding to Eq. i) or ii), respectively. In the exclusive (or appearance) search, an experiment must isolate the anomalous appearance of some neutrino type, ν_β, in a relatively pure beam of another type, ν_α. If no statistically significant signal of ν_β's are observed, the experiment can then set limits on the probability for ν_α to change into ν_β, $P(\nu_\alpha \rightarrow \nu_\beta)$. The limit of $P(\nu_\alpha \rightarrow \nu_\beta)$ can then be turned into a correlated limit on $\sin^2(2\theta)$ and Δm^2 using Eq. ii). The region covered by a given experiment is bracketed by the following limits:

Large $\sin^2(2\theta) \approx 1$ Region	Large $\Delta m^2 (L \gg L_{osc} = \frac{2.5E_\nu}{\Delta m^2})$ Region
$\Delta m^2 > [P(\nu_\alpha \rightarrow \nu_\beta)]^{\frac{1}{2}}/1.27(\frac{L}{E_\nu})$	$\sin^2(2\theta) > 2P(\nu_\alpha \rightarrow \nu_\beta)$

The sensitivity of an experiment is determined by the background rate of ν_β's, the measured rate of ν_α's, and the value of L/E_ν. Large L/E_ν values give the smallest Δm^2 limits but may decrease the $\sin^2(2\theta)$ senstivity because the observed ν_α event rate falls with increasing L or decreasing E_ν.

The current best exclusive limits are given in Fig. 1 and Table I. (A more complete discussion of exclusive limits is given in Ref. 1.) All experiments assume two component mixing in the analysis and the limits are at the 90% C.L. For several of the channels, there exist more sensitive inclusive limits. (Exclusive limits on $\nu \to \bar{\nu}$, i.e. $\nu_\mu^{L.H.} \to \bar{\nu}_e^{L.H.}$, are not presented since they depend in detail on the interactions of wrong-handed neutrinos.)

Table I: Present Exclusive Limits for Neutrino Oscillations

	Δm^2 at $\sin^2(2\theta)=1$	Large Δm^2 $\sin^2(2\theta)$ Limit	Experiment
$\nu_\mu \to \nu_e$	0.6 eV^2	6 x 10^{-3}	Col-BNL[3]
	0.16eV^2	0.038	BEBC/PS[4]
	0.43eV^2	3.4x10^{-3}	BNL-E734[8]
	Positive Result	$\sin^2(2\theta)=0.03^{+}_{-}$ .01 @ $\Delta m^2=5$eV^2	PS-191[9]
$\bar{\nu}_\mu \to \bar{\nu}_e$	1.7 eV^2	8 x 10^{-3}	FNAL/15' [6]
	0.49eV^2	0.028	LAMPF-UCI[7]
$\nu_\mu \to \nu_\tau$	3 eV^2	0.013	FNAL-E531[5]
$\bar{\nu}_\mu \to \bar{\nu}_\tau$	2.2 eV^2	0.044	FNAL/15' [6]
	7.4 eV^2	0.088	FNAL/15' [7]
$\nu_e \to \nu_\tau$	8.0 eV^2	0.6	Col-BNL [3]
$\bar{\nu}_e \to \bar{\nu}_\tau$	None	None	None

The results for the $\nu_\mu \to \nu_e$ channel include three new measurements. The BEBC-PS experiment[4] uses a new high intensity, low energy ($<E_\nu> \approx 1.5$ GeV) horn neutrino beam from the CERN PS which was designed to have a small ν_e contamination. The decay tunnel is 45m long and the BEBC bubble chamber is located 825m from the source. Four electron neutrino events have been found for a sample of 418 muon neutrino events leading to the limits shown above.

For the BNL 734 experiment,[8] the detector was located 96m from the neutrino source, a wide-band horn beam with $<E_\nu> \approx 1.2$ GeV. The experiment detected 418 ν_e

quasi-elastic and 1370 ν_μ quasi-elastic events. After correcting for the difference in the number of protons on target in these samples (a factor of 6.29) and the difference in acceptance due to the high $Q^2 > 0.35$ GeV^2 cut for the ν_μ events (a factor of ~ 10), the measured ν_e/ν_μ flux ratio is given as $\sim 0.5 \times 10^{-2}$. Subtracting the calculated ν_e flux from kaon and muon decay thus gives the limits shown in Fig. 1 and Table I.

The PS 191 experiment[9] uses an electromagnetic calorimeter detector at 130m from the neutrino source. The source is the bare target beam at CERN with an average neutrino energy of 600 MeV. The detector is a fine sampling (.7 χ_0 samples) calorimeter using flash tube chambers and 3mm iron sheets. The fine sampling allows the experiment to study the backgrounds to ν_e events from π^0 production by muon neutrinos. The experiment detects 27 ν_e candidates in a sample of ~ 600 ν_μ events. The expected number is 8.2 ± 1.6 from ν_e's in the beam and π^0 background and leads to an excess of 18.8 ± 5.4 events or $2.9 \pm .9\%$ of the ν_μ flux. If this excess is interpreted as neutrino oscillation, the positive signal results in the allowed region shown in Fig. 1. This result is in conflict with the BNL 734 measurement when beam ν_e's are subtracted but is compatible with the BEBC PS result if $\sin^2(2\theta) < 0.025$.

Several experiments will have results shortly covering an even larger sensitive region. The PS 191 detector is being brought to Brookhaven and will take data in Spring/Summer of 1986. In addition, two experiments are presently underway at Brookhaven (E 776 - Columbia/ Illinois/Johns Hopkins) and LAMPF (E 645 - Ohio St/Argonne/ LAMPF). These experiments should have good sensitivity (E 776 should cover $\sin^2 2\theta > 10^{-3}$ for $\Delta m^2 > 0.7$ eV^2) in the region of the PS 191 result and results are expected in the next year.

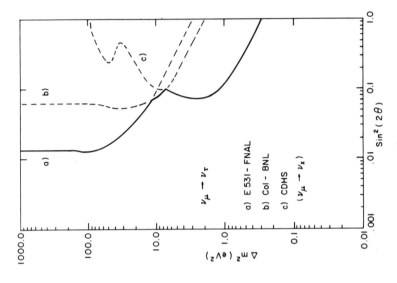

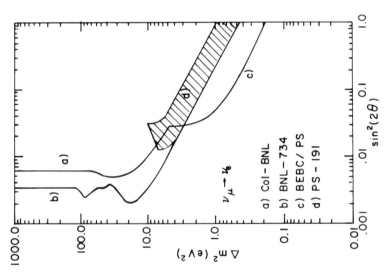

Fig. 1. Exclusive oscillation limits for a) $\nu_\mu \rightarrow \nu_e$ and b) $\nu_\mu \rightarrow \nu_\tau$ (90% C.L.).

INCLUSIVE OSCILLATION LIMITS

An inclusive (or disappearance) oscillation study is made by measuring the change in the number of a given type of neutrino, ν_α, with distance or energy. These experiments are sensitive to all the oscillation channels probed by exclusive experiments and, in addition, to oscillation in which the final state neutrino does not interact with the usual strength. Examples of the latter are: 1) a ν_x associated with a new heavy lepton with mass too heavy to be produced in present neutrino experiments; and 2) wrong-handed neutrinos (i.e. $\bar{\nu}_\mu$ left-handed) with suppressed weak interaction couplings.

The small Δm^2 sensitivity of an inclusive experiment is dependent on the statistical and systematic errors present for the comparison to Eq. i). Measurements of the ν interaction rate at two distances minimizes the dependence of the measurement on knowledge of the energy spectrum and absolute flux at the source. Further, if the data at the two distances are taken simultaneously by using two detectors, any systematic time dependence errors are also eliminated. For an oscillation measurement using data at two distances, there is a high Δm^2 limit to the sensitivity of the experiment. This limit corresponds to the point at which both detectors are many oscillation lengths from the source. At this point, the finite neutrino source size and detector energy resolution reduce the oscillatory behavior of Eq. i). For good sensitivity, the following inequalities should be satisfied:

$$L_{source} < L_{osc} = \frac{2.5\, E_\nu}{\Delta m^2} \quad , \text{ and}$$

$$\frac{\text{Energy}}{\text{Resolution}} = \frac{\Delta E_\nu}{E_\nu} < \frac{0.2}{\text{\# of oscillation lengths } (L_{osc}) \text{ between source and detector}}$$

The high Δm^2 limit is absent if the neutrino intensity as a function of E_ν at the source is known. This knowledge is typically model dependent and therefore introduces some additional systematic uncertainties.

The current situation is that except for the Bugey reactor experiment, all other experiments observe no positive signal and exclude regions in $\Delta m^2/\sin^2(2\theta)$ parameter plane. Table II and Fig. 2 give a summary of the current limits again assuming only two component mixing. The CCFR[11] and CDHS[10] experiments are simultaneous two detector measurements using a high energy ($<E_\nu> \simeq 80$ GeV) and a low energy ($<E_\nu> = 1.2$ GeV) muon neutrino beam respectively. The combination of the two measurements excludes $\nu_\mu \to \nu_x$ oscillations in the region between $1 < \Delta m^2 \leq 800$ eV^2 for mixings with $\sin^2(2\theta) > .1$. Other studies using the up/down asymmetry of cosmic ray neutrinos and the rate of solar neutrinos have been made but the conclusions are model dependent.

Table II: Present Inclusive Limits

	Lower Limit Δm^2 for $\sin^2(2\theta)=1$	Best Limit $\sin^2(2\theta)$	Experiment
$\nu_\mu \to \nu_x$	10^{-3}ev^2	$.6/\Delta m^2$ large	IBM[12] cosmic ray ν's
	$.3$ ev^2	$.06/\Delta m^2=3$ev^2	CDHS-PS[10]
	15 ev^2	$.02/\Delta m^2=120$ev^2	CCFR(NBB)[11]
$\bar\nu_\mu \to \bar\nu_x$	15 ev^2	$.02/\Delta m^2=200$ev^2	CCFR(NBB)[11]
$\nu_e \to \nu_x$	10 ev^2	$.34/\Delta m^2$ large	BEBC(WBB)[13]
$\bar\nu_e \to \bar\nu_x$	$.015$ ev^2	$.17/\Delta m^2$ large	Goesgen(reactor)[14] (spectrum analysis)
	$.12$ ev^2	$.2/\Delta m^2=1$ev^2	Bugey (reactor)[15] (two detector analysis)

positive signal (3.2σ)
$[\Delta m^2=.2$ev^2 $\sin^2 2\theta=.25$]

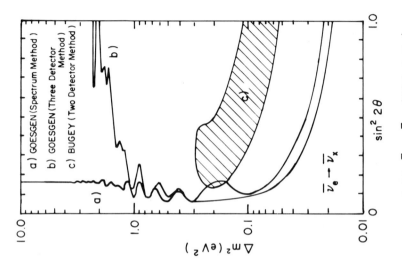

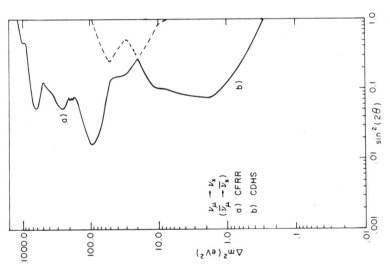

Fig. 2. Inclusive oscillation limits for a) $\nu_\mu \to \nu_x$ and b) $\bar{\nu}_e \to \bar{\nu}_x$ (90% C.L.).

NUCLEAR REACTOR OSCILLATION EXPERIMENTS

Nuclear reactors provide a high flux of $\bar{\nu}_e$ at low
energy (1-5 MeV) that can be used to search for $\bar{\nu}_e \rightarrow \bar{\nu}_x$
oscillations. In these experiments, the flux of $\bar{\nu}_e$ is
either measured at several distances from the reactor
core or compared to a calculated spectrum. The experiments measure the $\bar{\nu}_e$ flux by detecting the reaction:
$\bar{\nu}_e + p \rightarrow e^+ + n$, using liquid scintillation counter target
cells to detect the e^+ and ^3H multiwire chambers for the n.
A time correlated e^+, n event constitutes a valid signature. Pulse shape discrimination in the scintillation
counters is used to eliminate correlated neutron background events.

Two methods have been used by these experiments to
search for neutrino oscillations. For the first method
(multi-position method), event rates at several positions
are compared versus neutrino energy. This would be best
done by using two detectors run simultaneously but the
measurements so far use one detector moved to the different
locations. Since the data at the different distances are
taken at different times, there could be systematic changes
in the reactor flux. The multi-position method, though,
has the advantage of being independent of the knowledge of
the reactor $\bar{\nu}_e$ flux spectrum.

The second method (spectrum method) involves comparing
the observed number of events with a calculated number from
a reactor $\bar{\nu}_e$ flux model. The calculated spectrum is
determined by taking the measured e^+ spectrum obtained from
the BILL spectrometer[16] for ^{235}U and ^{239}Pu, the composition of the reactor, and calculations   for ^{238}U and ^{241}Pu
and unfolding the neutrino spectrum. The measured ^{235}U and
^{239}Pu components contribute 89% of the $\bar{\nu}_e$ flux with an
error of $\pm 3\%$ and the ^{238}U and ^{241}Pu give 11% with an error
of ± 10-20%. The total error on the calculated spectrum

is 5.3%. There are also uncertainties in the absolute
normalization from: 1) the $\bar{\nu}_e p$ cross section, 2) the
number of free protons in the target, 3) neutron detection
efficiency, and 4) reactor core composition and time variation; these uncertainties contribute an error of 5-10% for
the present measurements.

Results have recently been presented from two high
statistics studies. The first is by a Caltech/SIN/TH
Munich group[14] using the Goesgen power reactor (2806 MW).
Data have been recorded and analyzed with one detector
placed at three positions, 37.9, 45.9 and 64.7 m. No
change in the neutrino flux with distance (other than
geometrical solid angle coverage) is observed indicating
an absence of neutrino oscillation. The region excluded
by this measurement is shown in Fig. 2 for the multi-
position and spectrum method.

A second group, Grenoble/Annecy[15], using the Bugey
reactor (2800 MW) has also done a two detector measurement
with one detector placed at 13.6 or 18.3 m. The ratio of
observed events at the two positions, corrected for solid
angle differences and fuel burn-up, lies systematically
above a ratio one (1.102+.014+.028) for the energy range
between 1.5 and 6.5 MeV. The data are, therefore, inconsistent with the no-oscillation hypothesis, which is ruled
out by 3.2 σ or 99.8% C.L. The 95% C.L. allowed region is
shown in Fig. 2; this region is somewhat inconsistent with
the Goesgen multi-position limit and ruled out by the
Goesgen spectrum method limit.

CONCLUSIONS

The search for neutrino oscillations has covered most
of the possible channels but certainly not most of the
allowed parameter space. The majority of experiments
has obtained negative results setting limits on the allowed
oscillation parameters. Two experiments have reported

positive signals. These experiments along with others will continue running in this region with better statistics and control of systematic uncertainties, and should either confirm or refute the positive result.

At present, there is no strong theoretical argument why neutrinos should be massless; in fact, it is natural to assume that they are massive with mixing between generations of weak eigenstates. Setting the scale of these masses and mixings is another problem with little theoretical guidance. For these reasons, experimental measurements should continue probing ever further into the unexplored regions.

REFERENCES

1) C. Baltay, Neutrino 81, Maui, Hawaii (1981);
   A. Bodek, ICOBAN 84, Park City, Utah (1984);
   M. Shaevitz, Proc. Lepton-Photon Symposium,
   Cornell Univ., Ithaca NY (1983).
   M. Shaevitz, NATO ASI on Techniques and Concepts in
   High Energy Physics II (1984).
2) G.N. Taylor et al, Phys. Rev. D$\underline{28}$, 2705 (1984).
3) N.J. Baker et al, Phys. Rev. Lett. $\underline{47}$, 1576 (1981).
4) See R.J. Loveless paper in this Proceedings.
5) N. Ushida et al, Phys.Rev.Lett. $\underline{47}$, 1694 (1981).
6) A.E. Asratyan et al, Phys.Lett. $\underline{105B}$, 301 (1981).
7) H. Chen et al, Irvine-Los Alamos-Maryland experiment
   E225 at LAMPF. See J. Wotschack review talk at Neutrino
   84, Dortmund, Germany.
8) L.A. Ahrens et al, Phys. Rev. D$\underline{31}$, 2732 (1985).
9) F. Vannucci, Proc. EPS-HEP 85 Conf., Bari, Italy (1985).
10) F. Dydak et al, Phys.Lett. $\underline{134B}$, 34 (1984).
11) I.E.Stockdale et al, Phys.Rev.Lett.$\underline{52}$, 1384 (1984);
    C. Haber et al, Fourth Moriond Workshop, France (1984).
    I.E.Stockdale et al, Z. Phys. C$\underline{27}$, 53 (1985).
12) J. Learned, Neutrino 84, Dortmund, Germany.
13) O. Erriquez et al, Phys.Lett. $\underline{102B}$, 73 (1981).
14) K. Gabathuler et al, Phys.Lett. $\underline{138B}$, 449 (1984).
    V. Zacek et al, submitted to Phys.Lett.B.
    (Caltech preprint CALT-63-447).
15) J.-F. Cavaignac et al, Phys.Lett. $\underline{148B}$, 387 (1984).
16) K. Schreckenbach et al, Phys.Lett. $\underline{99B}$, 251 (1981);
    W. Mampe et al, NIM $\underline{154}$, 127 (1978).

DOUBLE BETA DECAY: RECENT EXPERIMENTAL RESULTS

Michael S. Witherell

Department of Physics, University of California
Santa Barbara, California 93106

ABSTRACT

The limit on the lifetime for neutrinoless double beta decay in ^{76}Ge has been raised to 2.5×10^{23} years. This decay is allowed only if the electron neutrino has a Majorana mass term. The present limit corresponds to a limit on the mass of $m_\nu \lesssim 2eV$, using existing calculations of the nuclear matrix element. There is an additional systematic uncertainty in the matrix elements, however, which is demonstrated by the disagreement between theory and experiment in two-neutrino double beta decay.

1.   INTRODUCTION

Double beta decay experiments bear on some of the issues that are most fundamental in our present picture of particle physics.[1] These issues are lepton number conservation, neutrino masses and mixing, and right-handed currents (RHC). Double beta decay can occur in even-even nuclei for which the pairing energy makes ordinary beta decay impossible. Such nuclei can often decay by the process

$$(A,Z) \rightarrow (A,Z+2) + 2e^- + 2\bar{\nu}_e ,$$

which is two-neutrino double beta decay ($\beta\beta_{2\nu}$). This decay is an ordinary second-order weak interaction process and the rate is calculable to the accuracy of the nuclear matrix elements. In addition, the more interesting neutrinoless decay ($\beta\beta_{0\nu}$) may also occur:

$(A,Z) \rightarrow (A,Z+2) + 2e^-$.

Obviously, $\beta\beta_{0\nu}$ is allowed only if lepton number is not conserved. In fact, the neutrinoless decay requires the virtual neutrino emitted at the first weak vertex to be reabsorbed at the second weak vertex. This can only happen if the electron neutrino is a Majorana particle ($\overline{\nu} = \nu$) and if the neutrino helicity is not exactly -1. Boris Kayser,[2] in his paper at this conference, gives a report of recent work by Kayser, Petcov, and Rosen which shows that $\beta\beta_{0\nu}$ implies the existence of at least one massive neutrino.

The recent theoretical interest in double beta decay is due to the predictions of Grand Unified Theories (GUTS), in which baryon number and lepton number are violated. It is natural in these theories for neutrinos to be massive Majorana particles,[2] and it is also natural for the mass to be much less than the associated charged lepton.[3] The only experiment which is sensitive to Majorana neutrino masses as small as a few eV is neutrinoless double beta decay.

2. STATUS OF $\beta\beta_{2\nu}$ DECAY

Since the $\beta\beta_{2\nu}$ rate is calculable, in principle, it can be used to test the nuclear physics involved in calculating the $\beta\beta_{0\nu}$ rate. Unfortunately the situation is confused. Experiments consistently get results for the lifetime which are much longer than the theoretical calculations.

Some of the most sensitive measurements of $\beta\beta_{2\nu}$ come from geochemical experiments. These experiments measure the amounts of noble gas nuclei, which are products of double beta decay, occluded in rocks of known age. To properly calculate the lifetime limit, one needs to measure the age of the rock, and to know that the gas is retained. For the decay $^{82}Se \rightarrow {}^{82}Kr$ the experimental result of $(1.45\pm0.15) \times 10^{20}$ years[4] is much longer than the shell-model calculation of $2.6\times10^{19}y$.[5] Similarly the geochemical result of $(2.6\pm0.3) \times 10^{24}y$[4] for $^{130}Te \rightarrow {}^{130}Xe$ is two orders of magnitude longer than the calculation of $1.7\times10^{19}y$.[5] Recently the Irvine group has reported preliminary results giving a lower limit for ^{82}Se of $T_{1/2} > 7 \times 10^{19}y$,[6] with some candidates at a level consistent with the geochemical result.

We therefore see a discrepancy between experiment and theory of a factor of 6-100 in rates. A recent calculation by Grotz and Klapdor[7] increases the prediction for ^{130}Te, reducing the discrepancy to a factor of 20, but does not change the predictions for ^{82}Se and ^{76}Ge. Because of coherence effects, ^{76}Ge, which is the nucleus in which the best limit for $\beta\beta_{0\nu}$ exists, may not have such severe problems in the matrix element calculation. Unfortunately, the technique which allows a very sensitive search for $\beta\beta_{0\nu}$ in ^{76}Ge is not very sensitive for $\beta\beta_{2\nu}$. The best limit for $\beta\beta_{2\nu}$ in ^{76}Ge is 0.8×10^{20}y,[8] which is still shorter than the calculated lifetime. Until we understand the nature of the $\beta\beta_{2\nu}$ discrepancies, we cannot assume that the $\beta\beta_{0\nu}$ calculations are more accurate than one order of magnitude.

3. STATUS OF $\beta\beta_{0\nu}$ DECAY

The most satisfactory way of comparing limits on $\beta\beta_{0\nu}$ for different nuclei is to compare the limits on the Majorana mass of the neutrino, m_ν. For the reasons discussed above, however, it is prudent to assume a $\times10$ uncertainty in the lifetime calculation. If $\beta\beta_{0\nu}$ is allowed because of a simple Majorana neutrino mass, $\Gamma_{0\nu} \propto m_\nu^2$. Thus the systematic uncertainty in the m_ν scale is a factor of about $\times3$. Of course this factor may be common to all $\beta\beta_{0\nu}$ experiments. If the $\beta\beta_{0\nu}$ decay is due to the existence of right-handed currents as well as neutrino mass, the calculation of parameters is yet more complex.[1),2]

The geochemical experiments cannot distinguish between $\beta\beta_{0\nu}$ and $\beta\beta_{2\nu}$. The most sensitive test on $\beta\beta_{0\nu}$ comes from the result $T_{1/2}(^{128}\text{Te}) > 8\times10^{24}$y.[4] This corresponds to a limit $m_\nu < 1\text{-}5$eV, but the limit is particularly suspect because the Te matrix elements are even more uncertain than those of ^{76}Ge.

The laboratory experiments which give the best limits on $\beta\beta_{0\nu}$ all use the decay ^{76}Ge $\rightarrow$ ^{76}Se. Normal ^{76}Ge is an 8% component of natural germanium, and very pure germanium detectors can be made with energy resolution approaching 0.1%. The energy release is 2040.7 keV, which is large enough for a reasonable sensitivity. A germanium detector serves as both source and detector in such an experiment. One searches for a single narrow line from the summed energy of the two electrons, using the energy resolution to suppress the background by a large factor. The main problem is to reduce backgrounds from cosmic rays and natural radioactivity. This technique for studying $\beta\beta_{0\nu}$ in ^{76}Ge was first used by the Milan group[9] in 1973, giving a result $T_{1/2} > 5 \times 10^{21}$y. The recent experiments have been designed to push that limit to 5×10^{23}y, which makes severe experimental demands. In a single 150 cm^3 Ge counter a lifetime of 5×10^{23}y corresponds to 1 count in 18 months; therefore multiple counter arrays are needed. The background level in a standard Ge counter above ground is greater than 10^3 counts/year in a 1 keV bin for a 100 cm^3 detector. (I will use this unit of c/keV·y·10^2cm^3 consistently throughout the paper.) Thus the background needs to be suppressed by about three orders of magnitude. The solution is to put a multiple-counter array below ground to reduce cosmic ray backgrounds. Very clean materials must be used in the cryostat assembly and in the surrounding shields, usually of copper and lead. With these techniques a number of groups have achieved background levels of about 1 c/keV·y·10^2 cm^3. This is the lowest background level achieved in any low-level counting environment for the 1 MeV energy region.

The sensitivity of an experiment depends on four quantities: (1) the detector resolution, typically 3-4 keV FWHM; (2) the germanium volume; (3) integrated live time; and (4) background level. The experiments which now contribute the best limits are shown in Table I.

Table I: Summary of Recent $\beta\beta_{0\nu}$ Results on ^{76}Ge

Group	Detector	Background	Time	$T_{1/2}$ Limit
	cm^3	c/keV·y·10^2cm^3	y	10^{23}y (68% C.L.)
Milan	115 138	12.5 2.7	2.4 0.8	1.2 0.7 } 1.4
PNL/SC	125	1.0	0.4	1.2
UCSB/LBL	658	1.2	0.4	2.5

The Milan group was the first to begin counting underground, in the Mont Blanc tunnel, and their present limit comes after a very long running time. Because of different backgrounds and live times, their two detectors are listed separately, but together they give a limit of 1.4×10^{23}y.[10] The Battelle-Pacific Northwest laboratory/South Carolina group ran in the Homestake mine with a single detector at very low background, for a limit of 1.2×10^{23}y.[11] The UCSB/LBL group ran with 4 detectors in a powerhouse built inside a mountain at Oroville, California. This apparatus also has a NaI shield 15 cm thick surrounding the germanium, inside the passive shield of borated polyethylene and lead. The 4-detector experiment was completed after 0.4 years of data for a limit of 2.5×10^{23}y[8] for the $\beta\beta_{0\nu}$ lifetime. Using the shell-model calculations of Haxton, et al.[5] or Grotz and Klapdor[7] and the best limit of 2.5×10^{23}y, the limits on the Majorana mass turn out to be about 1-2eV.

The limits on neutrinoless double beta decay also constrain the parameters for neutrino mixing, if the neutrinos are Majorana. Assuming 2 neutrinos, with masses m_1 and m_2, which couple to the electron the limit can be expressed as $|m_1 U_{e1}^2 \pm m_2 U_{e2}^2| \lesssim 2$eV, where the sign depends on the relative CP eigenvalue of the two neutrinos. This leads to a limit on the coupling of a heavy neutrino of mass m_2 to the electron of $|U_{e1}|^2 \lesssim 2\text{eV}/m_2$.[12] This is about a factor of 30 below the limits from π decay[12] in the mass range 10-50 MeV. It is also at least two orders of magnitude below the limits from nuclear beta decay in the mass range 10 keV-10 MeV. Finally, the $\beta\beta_{0\nu}$ limits are more sensitive than neutrino oscillation experiments for the mass range 100eV-10 keV. Of course all of these $\beta\beta_{0\nu}$ limits assume the heavy neutrino is a Majorana particle, and that there is not a cancellation of large masses leading to no $\beta\beta_{0\nu}$ decay.

Although the primary effort in ^{76}Ge has been to search for the ground state transition $0^+ \to 0^+$, there is also an allowed decay to the first excited state, $0^+ \to 2^+$, with an energy of 1481.6 keV. The best limits for this decay are achieved by two groups that have NaI shields which detect the 0.56 MeV γ ray in coincidence with the Germanium signal. The Osaka group,[13] in 0.3 years of counting in the Kamioka mine, set a limit of 4×10^{22}y on this decay. The UCSB-LBL group, on the basis of 0.35 years, achieved a limit of 9×10^{22}y.[8]

4.   FUTURE EXPERIMENTS

In the near future, the main improvement in double beta decay will come in the ^{76}Ge experiments. The UCSB/LBL group has recently installed two more detectors into their apparatus, for a total of about 880 cm^3, and should reach 1300 cm^3 soon. The PNL/SC group is installing a 1200 cm^3 array in the Homestake mine. Finally, a Cal Tech group is installing an array of similar size in the Gotthard tunnel in Switzerland, where they have already obtained low background levels. In about one year, one or more of these experiments will have limits in the range of about 6x10^{23} years, and will slowly approach 10^{24} years. To go beyond this limit, a new idea is needed.

One new approach being pursued is to use purified ^{100}Mo as a source, with silicon detectors. The energy release is 3.0 MeV, which increases the rate by about an order of magnitude for given m$_\nu$, since the decay rate if proportional to E$_0$^{5}. Requiring a coincidence between two separate electron signals helps compensate for the loss of energy resolution. The Osaka group[13] will use 10 layers of ^{100}Mo foils with a total mass of 11g. An LBL-Mt. Holyoke-New Mexico group[14] plans to have 500 foils about 10μm thick, for a total of 170 grams, and to use large area silicon detectors with very thin dead layers. They have already achieved fairly low background levels.

^{136}Xe is an interesting source for double beta decays, because it can be used as a gas or liquid in wire chambers. The Milan group has turned off their Ge experiment to prepare an MWPC using ^{136}Xe at 10 atmospheres. The Cal Tech group is planning a 5 atm. time projection chamber (TPC). An Irvine group has plans for a liquid Xe TPC, and the Moscow group is designing a TPC to operate at atmospheric pressure. The critical test for all of these new detectors for double beta decay will be how much they will be able to suppress background from radioactivity in the detector materials.

5.   CONCLUSIONS

The best present limit on $\beta\beta_{0\nu}$ comes from ^{76}Ge experiments. The recent result of the UCSB/LBL group is $T_{1/2} > 2.5\text{x}10^{23}$ years, which corresponds to a limit on the neutrino mas of m$_\nu$ $\lesssim$ 1-2eV, if the ν_e is a simple Majorana neutrino. Uncertainties in the calculation of the nuclear matrix element raise this limit to about 6eV. The discrepancy between theory and experiment in the 2ν mode must be understood before this uncertainty can be reduced. The limit on the $0^+ \to 2^+$ transition in ^{76}Ge is now 9×10^{22} years.

Because of the importance of the physics issues involved, the nature and the mass of the neutrino, it is worth a large effort to increase the sensitivity for neutrinoless double beta decay. No other experiment has the possibility of seeing neutrino masses in the range 1eV and below in the near future. A new breakthrough in experimental techniques is needed to go much beyond existing limits.

REFERENCES

1) The following are excellent reviews of the theory of double beta decay:

   (a) W.C. Haxton and G.J. Stephenson, Jr., Progress in Particle and Nuclear Physics 12, 409 (1984).

   (b) H. Primakoff and S.P. Rosen, Ann. Rev. Nucl. Part. Sci. 31, 145 (1981).

2) B. Kayser, invited paper at this meeting.

3) M. Gell-Mann, P. Ramond, R. Slansky, Rev. Mod. Phys. 50, 721 (1978).

4) T. Kirsten, H. Richter, and E. Jessberger, Phys. Rev. Lett. 50, 474 (1983) and Z. Physik C16, 189 (1983); T. Kirsten, in Science Underground, AIP Conf. Proc. 96, 396 (1983).

5) W.C. Haxton, G.J. Stephenson, Jr., and D. Strottman, Phys. Rev. Lett. 47, 153 (1981) and Phys. Rev. D25, 2360 (1982).

6) M.K. Moe, A.A. Hahn, and S.R. Elliott, UCI-Neutrino Preprint #133 (1984, unpublished).

7) K. Grotz and H.V. Klapdor, Phys. Lett. 142B, 323 (1984); and Phys. Lett. 157B, 242 (1985).

8) D.O. Caldwell, R.M. Eisberg, D.M. Grumm, D.L. Hale, M.S. Witherell, F.S. Goulding, D.A. Landis, N.W. Madden, D.F. Malone, R.H. Pehl, and A.R. Smith, UCSB preprint 1985, submitted to Phys. Rev. Lett.

9) E. Fiorini, A. Pullia, G. Bertolini, F. Cappelani, and G. Restelli, Phys. Lett. 25B, 602 (1967) and Nuovo Cimento 13A, 747 (1973).

10) E. Bellotti, O. Cremonesi, E. Fiorini, G. Liguori, A. Pullia, P. Sverzellati, and L. Zanotti, Phys. Lett. 146B, 450 (1984).

11) F.T. Avignone, R.L. Brodzinksi, D.P. Brown, J.C. Evans, W.K. Hensley, H.S. Miley, J.H. Reeves, and N.A. Wogman, Phys. Rev. Lett. 54, 2309 (1985).

12) Michael H. Shaevitz, in Proceedings of the 1983 International Symposium on Lepton and Photon Interactions at High Energies, Cornell (1983).

13) H. Ejiri, N. Kamikubota, Y. Nagai, K. Okada, T. Shibata, N. Takahashi and T. Watanabe, in Proceedings of the Santa Fe Meeting, Santa Fe (1984).

14) M. Alston-Garnjost, B. Dougherty, R.W. Kenney, J.M. Krivicich, R.A. Muller, R.D. Tripp, M. Deady, H.W. Nicholson, and B.D. Dieterle, Lawrence Berkeley Laboratory internal document (1985, unpublished).

UCD-85-10

DETECTING THE HIGGS[†]

J.F. Gunion
Department of Physics
University of California, Davis
Davis, California 95616

ABSTRACT

Signals and backgrounds for Higgs particle production are reviewed for both e^+e^- and hadron-hadron colliding beams. The mass range above $m_H = 80$ GeV/c is focused upon. Both the standard model and a simple two doublet supersymmetric model extension thereof are considered. It is found that e^+e^- collisions provide a sufficiently background free environment that, even if $m_H \sim m_Z$, a standard model Higgs will not be particularly difficult to detect. However, two of the three neutral Higgs of a two-doublet SUSY model could be quite hard to detect due to the absence or reduced level of tree graph couplings to W's and Z's. In hadron collisions backgrounds become a very significant problem, especially if m_H lies below $2m_W$. Our conclusion is that in some regions of m_H only purely leptonic decay modes are sufficiently background free that detection is possible. Unfortunately the associated event rates are very low even at $\sqrt{s} = 40$ TeV so that a luminosity of $L = 10^{33}$/cm^2/sec may prove insufficient. A simple SUSY two-doublet model predicts even greater difficulty in discovering two of the three neutral Higgs. We conclude with the suggestion that a high-luminosity, lepton-intensive interaction region should be given serious consideration at the SSC.

I. INTRODUCTION

The mechanism responsible for spontaneous symmetry breaking remains the most essential unresolved problem in physics below the 1

---

†Supported by the Department of Energy.

TeV scale. In the standard SU(2) × U(1) model of electroweak interactions a single neutral Higgs particle is the only remnant of this
symmetry breaking. While some generalizations of the standard model
(SM) replace the Higgs by composite objects or other more complicated
structure, supersymmetric (SUSY) models solve the heirarchy and natural-
ness problems while retaining the elementarity of the Higgs particles.
However at least three neutral and one charged pair of physical Higgs
are required. While it is not known whether either approach is
correct, it is clearly important to assess the sensitivity of future
and present high energy accelerators to the two possible elementary
Higgs particle models.

We consider both e^+e^- and hadron-hadron collisions, focusing, in
the latter case, on the Superconducting Super Collider (SSC) with
assumed energy of $\sqrt{s}$ = 40 TeV and luminosity L = 10^{33}/cm^2/sec. The
rates for Higgs production in the SM have been studied thoroughly in
the literature.[1] More recently predictions of the simplest SUSY two-
doublet model have been explored and result in considerable reassess-
ment. Finally predictions for important background processes are now
available. Thus it is possible to give a more complete survey of
detection possibilities than previously. Not surprisingly e^+e^-
colliders emerge as highly favored over hadron-hadron colliders,
assuming equal Higgs production rates, due to the lack of serious backgrounds. However energies available at e^+e^- facilities in the near
future will place a severe limit on the range of accessible Higgs
masses. Thus hadron-hadron collision must be carefully studied with
regard to their ability to probe the Higgs sector.

II. e^+e^- COLLISIONS

We began by reviewing the production of a SM Higgs in e^+e^-
collisions. Three processes have been proposed as most useful:[4,5]

$$\text{Onium} \rightarrow H + \gamma \tag{1a}$$

$$Z \rightarrow H + Z^* \atop \quad\; \hookrightarrow e^+e^- \tag{1b}$$

$$Z^* \rightarrow H + Z \quad . \tag{1c}$$

Process (1a) produces an adequate event rate up to some fraction of the
onium mass. If toponium were at a mass of 80 GeV/c^2, m_H as large as 30
GeV/c could be accessible. This mode also appears to be quite free of
background. As usual good resolution in the γ energy is required.

Process (1b) is also rate limited. For instance[6] at m_H = 20 GeV/c^2

$$B(Z \rightarrow H\, \ell^+\ell^-) = 3 \times 10^{-5} \tag{2}$$

yielding 30 events for 10^6 Z's. By m_H = 40 GeV/c^2 the event rate is
too low baring higher than expected event rates for Z production at a Z
factory. This is illustrated in Fig. 1.[7]

Process (1c) requires a higher energy e^+e^- machine. Roughly speaking Higgs masses up to

$$m_H \sim \frac{\sqrt{s} - m_Z}{\sqrt{2}} \qquad (3)$$

are accessible. For instance, LEP II with $\sqrt{s}$ = 200 GeV can probe only up to $m_H \sim 80$ GeV.

In exploring backgrounds we have chosen to focus on the most general purpose process for moderate m_H values, (1c). Energies that will be available within a decade imply that only $m_H < 2m_W$ is of any immediate interest. We also choose

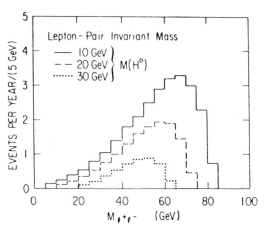

Fig. 1 Dilepton mass spectra for $Z^0 \to H^0 \ell^+ \ell^-$ for several choices of the Higgs mass. The normalization corresponds to 2000 hours of beam time at an average luminosity of 10^{30} cm^{-2} sec^{-2}.

to consider only $m_H \geq 80$ GeV/c^2 since in this region $H \to t\bar{t}$ decay dominates (assuming $m_t \leq 40$ GeV/c^2). Thus the reaction being considered can be depicted as in Fig. 2.

The background derives from continuum production processes of the type

$$e^+e^- \to q\bar{q}Z \qquad (4)$$

and has been studied in Ref. 8. In the region $m_H \sim m_Z$ the graphs that dominate (4) involve nearly on-shell $Z^* \to q\bar{q}$ decay, Fig. 3.

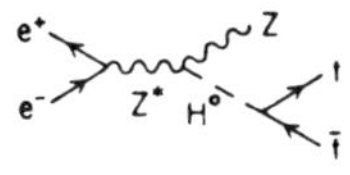

Fig. 2 Feynman graph for process (1c).

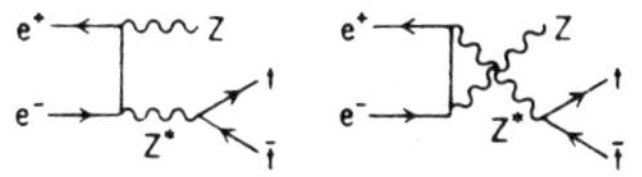

Fig. 3 Feynman graphs for process (4) dominant at $M_{t\bar{t}} \sim m_Z$.

At higher m_H values a variety of other graphs enter, see Ref. 8. To explore the background we imagine first that H is detected as a peak in the missing mass spectrum

$$s_H = (p_{e^+} + p_{e^-} - p_Z)^2 \qquad (5)$$

at $s_H \sim m_H^2$. The best resolution is obtainable if the Z is detected in its leptonic decay modes. In this case a resolution of order

448

$\Delta s_H = .02\, s_0$ is achievable at $\langle s_H \rangle = s_0$. Thus we define an integrated cross section

$$\Sigma(s_0, \Delta s_H) = \int_{s_0 - \frac{\Delta s_H}{2}}^{s_0 + \frac{\Delta s_H}{2}} ds_H \; \sigma_{e^+e^- \to ZX}(s_H). \qquad (6)$$

A typical plot of $\Sigma(s_0, \Delta s_H)$ would show a superposition of a sharp Higgs peak at $\sqrt{s_0} \sim m_H$ and a broader peak from the Z^0 pole peaking at $\sqrt{s_0} \sim m_Z$ but spread out several GeV (of order Γ_Z) on either side. In Fig. 4 we plot the peak signal, $\Sigma_{SIG}(s_0 = m_H^2, \Delta s_H)$, as a function of $\sqrt{s_0}$ in comparison to background process (4), $\Sigma_{BG}(s_0, \Delta s_H)$. The size of Σ_{BG} depends upon the number of $q\bar{q}$ flavors that are summed over in (4). From Fig. 4 we see that only if

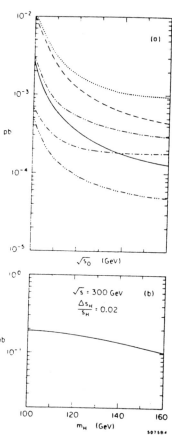

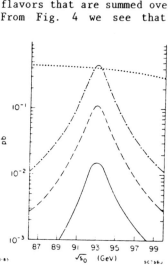

Fig. 4 The curve ····· is $\Sigma_{SIG}(s_0, \Delta s_H)$ $(m_H^2 = s_0)$. The curves ——, ---, and -··- are $\Sigma_{BG}^{Z\text{-pole}}(s_0, \Delta s_H)$ when the final state $q\bar{q}$ pair is $t\bar{t}$; $t\bar{t}$ or $b\bar{b}$; $t\bar{t}$, $b\bar{b}$, $c\bar{c}$, $s\bar{s}$, $u\bar{u}$, or $d\bar{d}$ respectively. $\Delta s_H = .02 s_0$, $\sqrt{s} = 300$ GeV.

Fig. 5 (a) The curves -··-, -·····-, and ···· are $\Sigma_{BG}(s_0, \Delta s_H)$ when the final state $q\bar{q}$ pair is $t\bar{t}$; $t\bar{t}$ or $b\bar{b}$; $t\bar{t}$, $b\bar{b}$, $c\bar{c}$, $s\bar{s}$, $u\bar{u}$ or $d\bar{d}$ respectively. For comparison the curves -·····-··, ——, and --- are the same but using the Z pole approximation to $\Sigma_{BG}(s_0, \Delta s_H)$. $\Delta s_H = .02\, s_0$ throughout, $\sqrt{s} = 300$ GeV. (b) $\Sigma_{SIG}(s_0, \Delta s_H)$ evaluated at $(m_H^2 = s_0)$ for $\Delta s_H = .02 s_0$, $\sqrt{s} = 300$ GeV.

there is no ability to distinguish $t\bar{t}$ from other $q\bar{q}$ pairs will background be a problem and then only if m_H is very near m_Z. Plots at higher values of $s_0 = m_H^2$ are given in Fig. 5. Here many diagrams contribute to Σ_{BG} but clearly the signal peak is far larger than background for a standard model Higgs.

The minimal SUSY two-doublet model extension of the Higgs sector presents special problems.[3] In this model there are three neutral Higgs and two charged Higgs particles, H_1^0, H_2^0, H_3^0 and $H^{\pm}$. Neither H_3 nor H^- have tree level couplings to Z; thus reaction (1c) will not be relevant and their production cross sections will be much smaller than for H_1 and H_2. In this minimal two Higgs doublet model one of these last two neutral Higgs is heavier than the Z and one lighter. We take

$$r_1 = \frac{m_{H_1}}{m_Z} \geq 1, \quad r_2 = \frac{m_{H_2}}{m_Z} \leq 1. \tag{7}$$

(We also remark that the $H^{\pm}$ are heavier than the W and must be pair produced.) The couplings $g(ZZH_1)$ and $g(ZZH_2)$ are functions of r_1 and r_2. As an example we plot in Fig. 6, at fixed $r_2 = .5$, the ratio $[g(ZZH_1)_{SUSY}/g(ZZH)_{SM}]^2$ as a function of r_1. This ratio gives the cross section suppression of the heavier H_1 relative to the standard model neutral Higgs. Though this suppression takes hold very quickly the background process dies away even more quickly as m_H increases (see Fig. 5b) so that even at $m_{H_1} \sim 2m_Z$ the background summed over all $q\bar{q}$ is still not bigger than the signal cross section. On the other hand the event rate is miniscule. At m_{H_1} just below $2m_W$, $\Sigma_{SIG}(m_H^2) \times$ SUSY suppression $\sim 10^{-3}$pb, yielding 1 event per year for a yearly luminosity

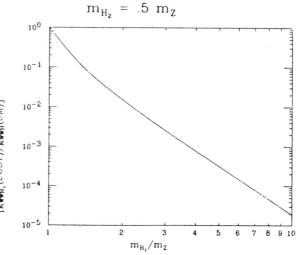

$m_{H_2} = .5\, m_Z$

Fig. 6 The ratio $[g_{WWH_1}(SUSY)/g_{WWH}(SM)]^2$ as a function of m_{H_1}/m_Z for fixed $m_{H_2} = .5m_Z$. The result is independent of the ε sector discussed later in paper.

of $L = 10^{39}/cm^2$. In contrast the $g(ZZH_2)_{SUSY}$ coupling is essentially full strength, $g(ZZH_2)_{SUSY} \sim g(ZZH)_{SM}$, so that H_2 should be easy to find. Thus there is some chance that of the 5 physical Higgs, of the SUSY two-doublet model, only one neutral one will be easily found. How then will we distinguish an SM Higgs from the SUSY H_2? The most feasible approach relies on the fact that couplings of H_2 to $b\bar{b}$ and $t\bar{t}$ quarks will not be the same as SM couplings. The ratio

$$R_{b/t} = \frac{H_2 \rightarrow b\bar{b}}{H_2 \rightarrow t\bar{t}} = \frac{m_b^2}{m_t^2} \tan^2 \alpha \ \tan^2 \beta \qquad (8)$$

is a function of angles α and β that are mixing angles of the SUSY model. For instance

$$\tan \beta = \frac{v_2}{v_1} \qquad (9)$$

where v_2 and v_1 are vacuum expectation values. These angles can take on extreme values in some popular models (generally $v_2/v_1 \gg 1$) resulting in unexpectedly large $b\bar{b}$ decay widths. Thus $R_{b/t}$ could indicate the presence of a non-standard Higgs sector. In any case, both the SM and SUSY should allow observation of at least one neutral Higgs.

III. pp COLLISIONS

In hadron collisions we will find that it is more difficult to detect a neutral Higgs boson. Thus hadron colliders will primarily be of interest when m_H is above the reach of e^+e^- machines. Current plans for a very high energy pp super collider (the SSC at $\sqrt{s} = 40$ TeV, $L = 10^4$ pb^{-1}/yr) coupled with limitations on the e^+e^- center of mass energy that will be available in this century make the exploration of Higgs detection at a pp machine of considerable importance.

There are two distinct regions of m_H that must be analyzed.

A) $100 \text{ GeV} \leq m_H \leq 2m_W$

$$(10)$$

B) $2m_W \leq m_H$.

Region A), studied in Refs. 8 and 9, is sometimes referred to as the "intermediate mass" range, being above planned e^+e^- machine accessibility but below the two W decay threshold. In region A) we presume that

$$H \rightarrow t\bar{t} \qquad (11)$$

and will take $m_t = 40$ GeV. As we shall see, if $m_H < 2m_t$ then detection in the intermediate mass region could become substantially easier. In any case, the problems associated with Higgs detection are very different in the two regions. The discussion will focus on new results.

A) Intermediate Mass

First a brief review of well established facts may be helpful.

1) Inclusive Higgs production (via $gg \rightarrow$ fermion loop $\rightarrow$ Higgs)

$$pp \rightarrow H^o_{} + X$$
$${\scriptstyle \llcorner} t\bar{t} (12)$$

has an impossibly large QCD background from (see e.g. Ref. 1)

$$gg \rightarrow t\bar{t}. (13)$$

2) There is some hope[8,9] for the mechanism

$$pp \rightarrow W^{\pm} + H^o_{}.$$
$${\scriptstyle \llcorner t\bar{t}} (14)$$

To see what is required compute the integrated cross section analogous that considered in eq. (6),

$$\Sigma(s_H, \Delta s_H) = \int_{s_H - \frac{\Delta s_H}{2}}^{s_H + \frac{\Delta s_H}{2}} ds_H \int \text{phase space } d\sigma_{pp \rightarrow W^+ + t\bar{t}} . (15)$$

Let us temporarily adopt a mass resolution

$$\Delta s_H = .1 \; m_H^2. (16)$$

Along with the $W^+ t\bar{t}$ final state we shall also discuss the $W^+ b\bar{t}$ final state since it may be difficult to discriminate between the two. Contributions to these final states come from the reaction of interest, (14), but also from a variety of QCD processes, see fig. 7. The results for Σ are shown in fig. 8. To see the signal we observe that:

a) b/t discrimination must be possible at a level of 1/100;
b) the resolution of eq. (16) must be achievable.

Preliminary investigations are pessimistic regarding this latter point. Cox and Gilman[10] find that the unobserved neutrinos from t decay and the probable inability to classify soft jet fragments will make the

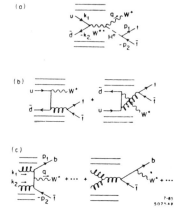

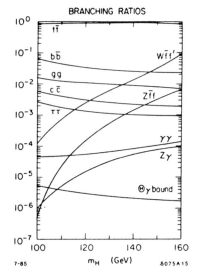

Fig. 7 Processes for pp → W⁺+tt̄ or W⁺+bt̄:
a) The Higgs signal;
b) the QCD background to W⁺tt̄; c) the QCD background to W⁺bt̄.

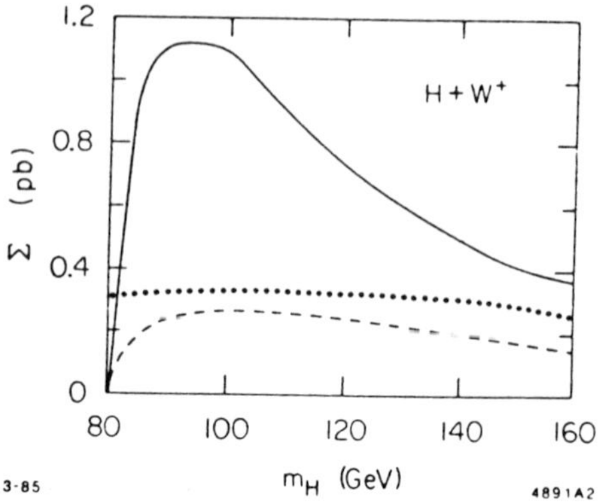

Fig. 8 The cross sections (15) computed for resolution (16) and for processes a), b) and c) of Fig. 7. a) = solid curve; b) = dashed curve; c) = dotted curve × 100. The plot is at $\sqrt{s}$ = 40 TeV with -2<Y_W, Y_H<2, p_T^W>40 GeV. The factor of 1/100 used in plotting c) can be thought of as a 1/100 b/t discrimination factor.

resolution of (16) unachievable. Additional cuts will have to be found to enhance signal over background in order for the mode (14) to be viable for Higgs detection.

If possibilities 1) and 2) are not feasible it is necessary to turn to rare decay modes for the Higgs. A summary of the branching ratios for these rare decays appear in Fig. 9 (m_t = 40 GeV). Certainly at L = 10^4 pb^{-1} yr^{-1} some modes have an event rate that, if background free, would allow H detection. Investigation reveals[8] that, of the various modes, two are of most interest:

Fig. 9 Branching ratios for H⁰ into the indicated modes as a function of m_H. We take m_t = 40 GeV.

$$pp \to H \underset{\substack{\,\,\,\, \searrow W^* \,\, + \,\, W \\ \quad\quad \searrow \bar{f}f'}}{} + X \quad \text{or} \quad H \underset{\substack{\,\,\,\, \searrow Z^* \,\, + \,\, Z \\ \quad\quad \searrow \bar{f}f}}{} + X \qquad (17)$$

and

$$pp \to W^{\pm}, \; Z + H \underset{\substack{\,\,\,\, \searrow \tau^+\tau^-}}{} + X. \qquad (18)$$

In reaction (17) the H is produced via the standard mechanism

$$gg \to \text{fermion triangle} \to H \qquad (19)$$

while reaction (18) proceeds according to the diagram given in Fig. 7a.

Focusing first on reaction (17) we exhibit in Fig. 10 the cross sections, summed over all fermion decay modes for both the W^* and W or Z^* and Z. Obviously significant event rates are possible over much of the intermediate mass range. However purely hadronic decay modes (involving only quarks and anti-quarks) have impossibly large QCD backgrounds.

Mixed hadronic/leptonic decay modes require a bit of discussion. There are several possibilities:

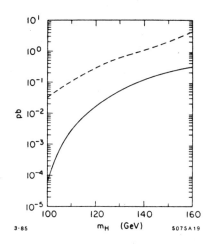

Fig. 10 The $\sqrt{s}=40\,\text{TeV}$ cross section for $pp \to H^\circ{\to}W\bar{f}f'$ (dashed curve) and $pp{\to}H^\circ{\to}Z\bar{f}f$ (solid curve) summed over W^+, W^- and over all light fermion states.

a) $W^* \to$ jets, $\qquad$ $W \to e\upsilon$ or $\mu\upsilon$

b) $W^* \to e\upsilon$ or $\mu\upsilon$ $\qquad$ $W \to$ jets

c) $Z^* \to$ jets, $\qquad$ $Z \to e^+e^-$ or $\mu^+\mu^-$ $\qquad\qquad (20)$

d) $Z^* \to e^+e^-$ or $\mu^+\mu^-$, $\qquad$ $Z \to$ jets

a) In this case m_H can be reconstructed (up to a 2-fold ambiguity) assuming that the missing υ came from on-shell W decay, but the lack of mass restriction in the $q\bar{q}$ jet channel of the W^* decay implies that backgrounds from mixed QCD/EW processes will be enormous (EW = electro-weak).

b) In this case the $q\bar{q}$ channel mass may be restricted but m_H cannot be reconstructed and again QCD/EW backgrounds will be overwhelming. In particular background processes can produce jets with mass m_W and an e/μ of any specific momenta via double on-shell WW production.

c) This case suffers the same problems as in a).

d) Unlike b) m_H can now be reconstructed, but the lower branching ratio for H→Z*Z compared to H→W*W (see Fig. 10) implies that backgrounds are still difficult. In fact, as we shall see in our discussion of the $m_H > 2 m_W$ region, QCD/EW backgrounds are a significant problem for even the on-shell H→ZZ mode when one Z decays hadronically and the other leptonically. Thus we must eliminate mixed hadronic/leptonic decay channels.

The final possibility is, of course, the purely leptonic type final state. The missing neutrinos in the W*W decay case, imply that m_H cannot be reconstructed and that standard EW backgrounds from $q\bar{q}$→W*W will become a problem. However the purely leptonic mode

$$pp \to H \qquad\qquad\qquad + X \qquad\qquad (21)$$
$$ \hookrightarrow Z^* \quad + Z$$
$$ \hookrightarrow \ell\bar{\ell} \quad \hookrightarrow \ell'\bar{\ell}'$$

will allow m_H reconstruction sufficient to suppress the $q\bar{q} \to Z^*Z$ EW background to a manageable level. Unfortunately the sacrifice in event rate is considerable:

$$BR(Z \to \mu^+\mu^- + e^+e^-) \ BR(Z^* \to \mu^+\mu^- + e^+e^-)$$
$$\sim 3.6 \times 10^{-3} \qquad\qquad\qquad\qquad (22)$$

yielding ~ 10 events/year (combine (22) with Fig. 10) at the most favorable large m_H portion of the intermediate mass region. Only ability to use a larger L would salvage reaction (21).

Thus reactions (17) are apparently not viable. We turn therefore to the reactions of eq. (18). Raw event rates are adequate. But not all decay modes of the W's and Z's in (18) allow reconstruction of m_H. We must be able to determine the transverse momentum, $\vec{p}_T^H$, of the Higgs. This implies that the W must decay to jets. The Z can decay to either jets or lepton pairs. When $\vec{p}_T^H$ can be measured we can require

$$|\vec{p}_T^H| > <|\vec{k}_T|> \qquad\qquad (21)$$

where $\vec{k}_T$ is the transverse momentum of the $q\bar{q}$ pair initiating (18). For large m_H and substantial $\vec{p}_T^H$ the decay products of the τ and τ will be approximately colinear but $\vec{p}_{\tau^+}$ will not be parallel to $\vec{p}_{\tau^-}$. This allows m_H reconstruction,[11] see Ref. 18. However if the hadronic decay modes of the W or Z are employed there will be a background from the mixed QCD/EW process initiated by gg fusion

$$pp \rightarrow q\bar{q} + Z_{\rightarrow \tau^+\tau^-} + X. \tag{22}$$

This background needs further investigation but may be controllable. If it is not possible to overcome then we must turn to the purely leptonic channel

$$pp \rightarrow Z_{\rightarrow e^+e^-} + H_{\rightarrow \tau^+\tau^-} + X. \tag{23}$$

These processes have a low event rate at $L = 10^4$ pb^{-1} yr^{-1}. Including both branching ratios and using the cross sections of Refs. 1,8 for pp $\rightarrow$ Z + H + X we find

$$\text{Number of Events} = \begin{cases} \sim 10 & m_H = 110 \text{ GeV} \\ \sim 2 & m_H = 160 \text{ GeV} \end{cases} \tag{24}$$

As for reaction (21) use of (23) can only be comtemplated for a higher L value.

This section on the intermediate mass range can be easily summarized. No mechanism or channel for Higgs detection has been found in which backgrounds will not present a severe problem with the exception of the purely leptonic channels (21) and (23). These complement each other in that (21) probes higher m_H and (23) lower m_H values. However both have low event rates and would require $L \geq 10^5$ pb^{-1} yr^{-1} for viability. Note, however, that if the top quark turns out to be much heavier than $m_t = 40$ GeV, the $\tau^-\tau$ decay modes and Z*Z decay modes of the Higgs will both be enhanced by more than a factor of 10 for $m_H < 2m_t$. The reactions (21) and (23) would then become quite feasible even at the planned SSC luminosity. Of the mixed hadronic/lepton final states reaction (18) might prove feasible should the background (22) be smaller than anticipated.

B) High Mass, $m_H > 2m_W$

In this region the primary decay mode of a standard model Higgs is to W or Z pairs. The simplest Higgs production mechanisms then yield the reaction (see Ref. 1 for a thorough discussion and references)

$$pp \rightarrow H_{\rightarrow W^+W^- \text{ or } ZZ} + X. \tag{25}$$

This process can proceed via both the

$$gg \rightarrow H \tag{26}$$

and

$$W^+W^- \text{ or } ZZ \to H \qquad\qquad (27)$$

subprocesses (Refs. 12 and 13, respectively). For a standard model Higgs (27) dominates (26) above $m_H \sim 350$ GeV. In this review we restrict ourselves to $m_H \lesssim 1$ TeV. Near the upper end of this region the Higgs width becomes enormous and the Higgs sector becomes strongly interacting.

It has been anticipated[1] in the literature that discovery of the Higgs in the mode (25) would be straight forward. The most obvious background is that coming from W^+W^-/ZZ continuum pair production. The complete set of graphs contributing to the WW final state (for instance) is illustrated in Fig. 11. The impact of the Higgs can be illustrated by a sample graph from Ref. 14, Fig. 12a, in which the invariant cross section $d\sigma/dM_{WW}$ exhibits various peaks or enhancements for different values of m_H. These structures can be further enhanced by focusing on longitudinally polarized W's. Recall that the W^+W^-/ZZ continuum contains primarily transversely polarized W's/Z's whereas H couples only to

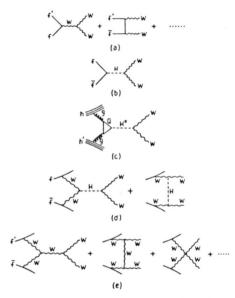

Fig. 11 Feynman graphs for the production of W pairs.

the longitudinal modes. Assuming that the W decay products can be observed and their angle θ^* (defined in the W rest frame) relative to $(\vec{p}_W)$ computed, one can project out the longitudinal decay modes via

$$\sigma_L \propto \int \frac{d\sigma}{d\cos\theta^*} \, (5 \cos^2 \theta^* - 2) \, d\cos\theta^* \qquad\qquad (28)$$

for each W decay. In Fig. 12b we show the results of Ref. 14 for

$$f_L = [d\sigma_{LL}/dM_{WW}]/[d\sigma/dM_{WW}]. \qquad\qquad (29)$$

Clearly Higgs structures are nicely enhanced.

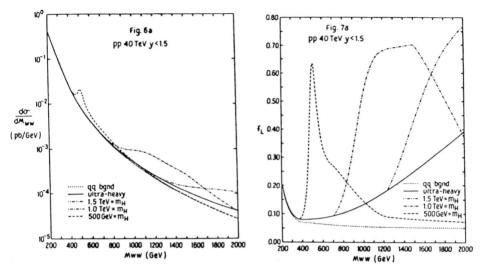

Fig. 12 a) $d\sigma/dM_{WW}$ for pp collisions at $\sqrt{s}$ = 40 TeV - the $q\bar{q}$ continuum background contribution, Fig. 11a, is compared to the predictions of the full graph set of Fig. 11 for various m_H values. b) the fraction f_L of longitudinal W's for the cases plotted in a).

Unfortunately direct continuum production of W or Z pairs is not the primary background except for purely leptonic decay modes. This is certainly obvious for purely hadronic decay channels for both W's or Z's but perhaps not so obvious for the mixed hadronic/leptonic decays. However the relevant backgrounds for the latter case have now been computed[15,16,17] and are very significant. The principle backgrounds derive from the processes (for the W pair case with W^+ decaying to $\bar{\ell}\,\upsilon_\ell$)

$$gg \to q\bar{q}'\ \bar{\ell}\upsilon_\ell \qquad (30)$$

and crossed versions thereof

$$gq \to gq'\ \bar{\ell}\upsilon_\ell$$

$$g\bar{q} \to g\bar{q}'\ \bar{\ell}\upsilon_\ell \qquad (31a)$$

$$q\bar{q}' \to gg\ \bar{\ell}\upsilon_\ell\ ;$$

a generally smaller background derives from reactions such as

$$q_1 q_2 \to q_1' q_2' \, \bar{\ell} \upsilon_\ell . \tag{31b}$$

Reaction (30) has exactly the same final state as the signal reaction decay mode

$$H \to W^- \quad \text{\textsf{¦}} \quad W^+ \quad . \tag{32}$$
$$\ \raisebox{0pt}{$\llcorner$}\ q\bar{q}' \quad \raisebox{0pt}{$\llcorner$}\ \bar{\ell}\upsilon_\ell$$

Reactions (31a) are a background to (32) when a gluon jet cannot be discriminated from a quark jet. Except for b and t jets such discrimination is likely to be very difficult. In what follows it is important to note that since the υ always comes from on-shell W decay its full momentum can be reconstructed (up to a 2-fold ambiguity) from its transverse momentum; the latter can in principle be determined from the $q\bar{q}'$ and ℓ transverse momenta. Note, however, that this reconstruction will prove difficult for $q\bar{q}' = b\bar{t}$ due to "lost" neutrinos and soft particles produced in the t and b decays.

To illustrate the magnitude of the problem focus first on the irreducible background of (30) with the same final state as (32). In Fig. 13a we plot the cross sections[16] for $q\bar{q}'$ W^+ production followed by W^+ decay,

$$\sigma(pp \to q\bar{q}' + \bar{\ell}\upsilon_\ell) \tag{33}$$

coming from Higgs production and decay and from this background process. We give separately the Higgs contributions from gg fusion and from WW/ZZ fusion. We have computed σ with the following restrictions.

1) We take $\Delta M_{q\bar{q}'} = .05 \, m_W$ with $M_{q\bar{q}'}$ centered at m_W.

2) We take $\Delta M_{\bar{\ell}\upsilon_\ell} = .05 \, m_W$ with $M_{\bar{\ell}\upsilon_\ell}$ centered at m_W - note that the υ_ℓ momentum can be computed up to the usual 2-fold ambiguity for this final state.

3) We take ΔM_{WW} as follows:

$$\Delta M_{WW} = \text{Max}(.05 m_H, \Gamma_H); \tag{34}$$

recall that the Higgs width becomes very large at higher values.

4) We impose cuts designed to limit the size of the background:

$$|\vec{P}_T^{\,q,\bar{q}'}| > 5 \text{ GeV}, \quad |Y^{q,\bar{q}'}| < 4, \quad |Y^{\bar{\ell},\upsilon_\ell}| < 4 . \qquad (35)$$

5) We sum over light flavor channels, $q\bar{q}' = \bar{u}d+\bar{c}s$, and consider a single $\bar{\ell}\upsilon_\ell$ flavor channel in the W decay and sum over $q\bar{q}'$ colors.

The result is obvious from Fig. 13; the background is comparable to the signal.

Can this background be reduced relative to the signal by appropriate cuts and projections? We made an extensive effort (but certainly not exhaustive) to explore this question. The most effective cuts come from:

1) increasing the severity of the $|\vec{P}_T^{\,q,\bar{q}'}|$ cuts, which causes a

   much sharper reduction in background than signal;

2) projecting or cutting in $\cos\theta^*$ of the $q\bar{q}'$ and $\bar{\ell}\upsilon_\ell$ "decay" channels, so as to enhance longitudinal W's.

For example the dN/N distributions of signal and background in $\cos\theta^*$ for the leptonic $\bar{\ell}\upsilon_\ell$ channel show that the background has a distribution in $\cos\theta^*_{\bar{\ell}}$ more typical of a transversely polarized W as compared to the signal distribution which peaks, according to $\sin^2\theta^*$, in the $\cos\theta^* \sim 0$ region. In Fig. 13b we plot, as an example, the signal and background curves after performing fairly restrictive cuts:

$$|\vec{P}_T^{\text{jet 1}}| > .3\ m_H, \quad |\vec{P}_T^{\text{jet 2}}| > .1\ m_H, \quad |Y^{\text{jet 1, jet 2}}| < 2,$$

$$\qquad (36)$$

$$|Y^{\bar{\ell},\upsilon_\ell}| < 4, \quad |\cos\theta^*_{\bar{\ell}}| < .2.$$

where jet 1 is defined to have the larger transverse momentum of the two jets. The background is decreased and Higgs observation would be straightforward.

Unfortunately reaction (30) is only a small part of the background if it is experimentally not possible to discriminate between q and g jets. The processes of type (30) and (31) combine to yield the much larger $j_1 j_2\ \ell\bar{\ell}'$ cross sections of Figs. 13a and 13b. (The cuts and mass resolution discussed previously are imposed on the two jets j_1 and j_2 which may be either q or g depending on the particular subprocess.) Nonetheless the signal/background ratio after performing the "standard cuts", Fig. 13b, is not too discouraging.

460

higgs production and background

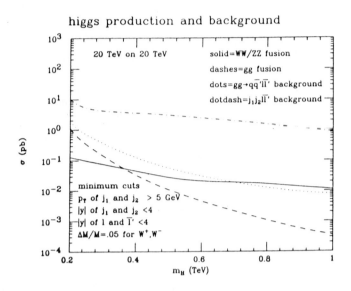

higgs production and background

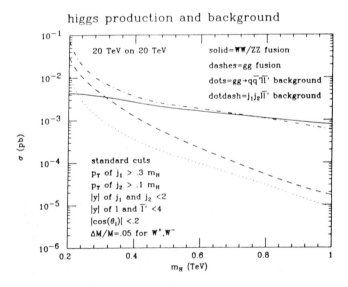

Fig. 13 Plots at $\sqrt{s}$ = 40 TeV of Higgs and background contributions to the $q\bar{q}'$ $\ell\upsilon_\ell$ final state. The cuts are detailed in the text: a) minimal cuts, eq. (35); and b) restrictive cuts designed to enhance signal/background, eq. (36). The background from process (30) is indicated by dots; the total j_1j_2 $\ell\bar{\ell}'$ is indicated by the dot-dash curve.

Consider m_H = .2TeV. We sum over e and μ W decay channels and include the W^+->$q\bar{q}$ and W^-->$\ell\bar{\ell}'$ H decay mode by multiplying the cross sections of fig. 13b by a factor of 4. For a yearly luminosity of 10^4 pb^{-1}, this implies 2800 background events in the W-pair mass bin centered at m_H while in this same bin the Higgs signal would produce 1200 events. This would yield a statistically significant fluctuation.

At m_H = 1 TeV, Γ_H = 450 GeV. In this case signal and background are both integrated over a W-pair mass range of order Γ_H. Even though the ratio of cross sections is 1:1, no clear peak will be evident. In addition the number of events is small, roughly 40 after including the factor of 4 mentioned above. Systematic uncertainties in the exact magnitude of the background might preclude a conclusive decision as to whether or not a broad Higgs enhancement is present.

However we recall that the major contribution to this background derives from final states in which either j_1 or j_2 is a gluon. Were it possible to discriminate gluon jets from quark jets then the primary background (for standard cuts) would derive from the gg->$q\bar{q}'\ell\bar{\ell}'$ subprocess. The cross section for this subprocess is also plotted in the figures. A substantial signal/background ratio is evident at high m_H values in the presence of q/g discrimination. However, this ratio is critically dependent on the mass resolutions in the W and W-pair channels used in computing the cross sections, see eq. (34). Unfortunately q/g discrimination may only be possible in the heavy quark final state, $q\bar{q}'$ = $b\bar{t}$. (We assume the t is light enough for W->bt decay to occur.) The mass resolution in such a W decay channel will be very substantially worse than that assumed due to lost neutrinos, soft particles, and detector resolution. (See the analysis of ref. 10 for a related situation.) In addition, the lost neutrinos will degrade our ability to reconstruct the momentum of the neutrino coming from the direct W->$\ell\bar{\ell}'$ decay. Thus our ability to isolate the W resonance in this channel and to reconstruct the decay angle used to perform the standard cuts will be impacted. The resolution in the W-pair mass will also be poor, but this does not have significant impact at high m_H where Γ_H is very large. Clearly a detailed analysis is required to assess the extent to which a heavy flavor "tag" could be used to enhance signal/background. It is most likely to be useful at high m_H where degradation in the W-pair mass resolution is of less importance and where some enhancement of signal/background would be very helpful.

At m_H values intermediate between these two extremes a transition to the low mass scenario occurs as the Higgs width decreases with decreasing m_H. For example, at m_H=.5 TeV, Γ_H=60 GeV and the cross sections are such that a clear "peak" would probably still be observable, without flavor tagging. Optimistically a combination of good resolution at low m_H using light flavor jets, and heavy flavor tagging at high m_H, could allow for Higgs observation above the two-jet W background throughout the range of m_H considered. Realistically this background may be very difficult to overcome at large m_H.

A second collaboration, Ref. 17, has also examined these issues. Their initial paper is somewhat more pessimistic in its conclusions as was our original letter, Ref. 16. However, we discovered that the programs of the two collaborations each had a single (different) error in them. Ours led to the larger numerical error. The new results quoted here have been obtained by running an adapted and corrected version of their program as well as a corrected version of our own. We thank the authors of Ref. 17 for sending us their program, so that this could be done.

We conclude this section by noting that the mass resolutions we have imposed may be too optimistic and that the cuts we have proposed may be difficult to perform sufficiently well in a real detector, especially given the two fold ambiguity in reconstructing the neutrino momentum. In this case the signal/background ratio could easily decrease to the level at which it might not be possible to observe the Higgs boson in mixed hadronic/leptonic decays of its W^+W^- decay products. In addition an entirely parallel discussion applies to the ZZ Higgs decay modes. This would leave only the purely leptonic final states as a probe of Higgs production. There are two possibilities.

$$H \rightarrow Z + Z \atop \mathrel{\hbox{$\llcorner$}}\ell^+\ell^- \quad \mathrel{\hbox{$\llcorner$}}\ell^+\ell^- \tag{37}$$

$$H \rightarrow W^+ + W^- \atop \mathrel{\hbox{$\llcorner$}}\ell^+\nu_\ell \quad \mathrel{\hbox{$\llcorner$}}\ell'^-\bar\nu_{\ell'} \; . \tag{38}$$

In (37) we suffer the branching ratio loss given in (22), leaving (for $L = 10^4$ pb^{-1}yr^{-1})

$$\text{Number of Events} \atop \text{in } H \rightarrow ZZ \rightarrow \text{leptons} \quad \sim \begin{cases} 100 & m_H = 200 \text{ GeV} \\ 10 & m_H = 1 \text{ TeV.} \end{cases} \tag{39}$$

The longitudinal projection (28) could make such a channel free of ZZ continuum background. However, these are clearly very marginal event rates especially at high m_H.

In (38) the m_H value cannot be reconstructed from the observed charged lepton decay products. A plot of $d\sigma/dM_{\ell\bar\ell}$ for (38) (where $M_{\ell\bar\ell}$ is the invariant mass of the two charged leptons) as compared to the background from WW continuum production reveals that the signal will not be easy to see. We[18] have so far found no cuts that can significantly enhance the signal/background ratio with acceptable loss of event rate. In particular the longitudinal projection (28) cannot be performed since the ν and $\bar\nu$ are not observed.

Thus in both the intermediate mass A) and high mass B) m_H regions Higgs detection at a hadron collider will be very tricky. The only decay modes in which the backgrounds considered here are small compared to the signal are those with purely leptonic final states (21), (23) and (37). The latter are characterized by low event rates at L = 10^4 pb^{-1}yr^{-1}. There is one possibly significant background to such final states that we have not considered. This is

$$pp \rightarrow \text{heavy quarks} \rightarrow \text{leptons and very slow light quarks.} \qquad (40)$$

If all the leptons are fast while the light quarks happen to be slow moving, the final state hadrons could fall below some minimum bias cut and (40) could mimic a purely leptonic final state. While the QCD cross sections for heavy quark (c,b,t) production are very substantial the region of decay phase space that falls in the category (40) is probably miniscule. Nonetheless a careful evaluation of this back-ground should be made. If it proves sufficiently small then consideration should be given to a higher luminosity, lepton-intensive interaction region at which L $\geq 10^5$ pb^{-1}yr^{-1}.

Supersymmetric Scenarios[3]

As a final chapter in our discussion of neutral Higgs observation at a hadron collider we consider the modifications that result in a simple two-doublet super symmetry (SUSY) model for Higgs couplings.[2,3] In this model there are three neutral Higgs. As discussed in the e^+e^- section, one of these, the H_3 will be difficult to make due to an absence of tree level couplings to W's and Z's. We will not discuss it further here. The remaining two, H_1 and H_2, have masses specified by the ratios (7). In fact r_1, r_2 and a sector parameter $\varepsilon = \pm 1$ (which specifies whether v_2/v_1 is smaller or larger than 1, see eq. (9)), are the only parameters of this model. Of course, more complex models (e.g. introduction of a singlet field) are possible. However the simplest model should give some idea of the possibilities. Since we wish to focus on heavy Higgs our discussion explores the scenario for H_1, $m_{H_1} > m_Z$. The lighter particle H_2 with $m_{H_2} < m_Z$ behaves much like a standard model Higgs as far as production is concerned (no strong suppressions emerge), though its decays may be more complex.

The first modifications to the standard model scenario for H_1 that we discuss are those for the production mechanism. These are as follows.

First there is a new mechanism, namely squark fusion

$$\tilde{q}\tilde{q} \rightarrow H_1. \qquad (41)$$

Naively this could be quite important since the coupling involved is the same order as the WW H_1 coupling, e.g.

$$g(\tilde{q}\tilde{q}H_1) \sim gm_Z \cos(\alpha+\beta) \qquad (42)$$

compared to

$$g(WWH_1) = gm_W \cos(\alpha-\beta). \qquad (43)$$

Recall that α and β are certain mixing angles that are functions of r_1, r_2 and the sector parameter ε. In fact in Fig. 6 we found that the ε independent ratio $[g(WWH_1)_{SUSY}/g(WWH)_{SM}]^2$ is strongly suppressed as m_{H_1} increases. In contrast the coupling (42) to squarks is a slowly varying function of m_{H_1}. The importance of (42) will also depend on the masses of squarks (the fusing objects) as well as gluinos, $\tilde{g}$, that are left behind as final state spectators to the fusion process (41). Typical results are illustrated in Fig. 14 which shows that $\widetilde{qq}$ fusion dominates WW/ZZ fusion (for squarks and gluino masses below 100 GeV) once $m_{H_1} > 2m_Z$, the only region where the WW/ZZ fusion process is likely to be of interest. Even if $m_{\tilde{q}}$ and/or $m_{\tilde{g}}$ is extremely large, and $\widetilde{qq}$ fusion unimportant, the WW/ZZ fusion mechanism is itself greatly suppressed compared to the standard model.

squark+squark vs. W+W/Z+Z->Heavy Higgs

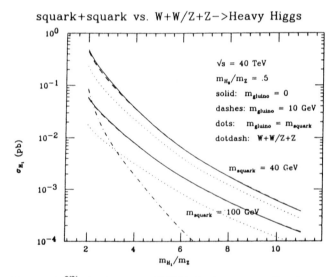

Fig. 14 Plot of $\widetilde{qq}$ fusion vs. WW/ZZ fusion contributions to H_1 production as a function of m_{H_1}. Two different squark mass scales are examined along with a variety of gluino masses. No cuts are performed.

The second production mechanism of interest is gluon fusion,

$$gg \to H_1. \qquad (44)$$

In the SM it is dominated by mediating quark triangle graphs. In SUSY the squark triangle graphs could also contribute. Indeed we have seen that $\widetilde{qq}H_1$ couplings are much larger than qqH_1 couplings, see (42). Unfortunately, the couplings (42), which have various weak isospin factors not written explicitly here, sum to 0:

$$\sum_q g(\widetilde{qq}H_1) = 0. \tag{45}$$

To the extent that squark masses are degenerate this implies that the larger coupling (42) is not effective. In fact squark mass degeneracy is expected to be broken by terms proportional to the associated quark masses squared. This leaves some residual effect but it will turn out to be small. The more important effect derives from alterations to the qqH_1 coupling:

$$g(qqH_1)_{SUSY} \sim g(qqH)_{SM} \begin{cases} \dfrac{\sin\alpha}{\sin\beta} & \text{up flavors} \\[2mm] \dfrac{\cos\alpha}{\cos\beta} & \text{down flavors} \end{cases} \tag{46}$$

Since α and β are functions of r_1, r_2 and ε the alterations to the quark triangle graph contributions to (44) may be easily computed. The results for both triangle graph processes are given in Fig. 15, compared to SM H° production. We have taken $m_t = 40$ GeV.

In Fig. 15a we plot for $\varepsilon = -1$, the ratio of coupling absolute values

$$\frac{|ggH_1 \ (SUSY)|}{|ggH \ (SM)|} \tag{47}$$

as a function of the ratio r_1. The squark loop and quark loop contributions are indicated separately as well as the total. As stated earlier squark loops are at best a small increment to the quark loops. However, the contribution of quark loops (primarily the t quark) is decreased in SUSY (for $\varepsilon = -1$) over the SM. In Fig. 15b we give the resulting uncut SUSY cross section from gg fusion as a function of r_1 (including squark and quark loops) for both $\varepsilon = +1$ and $\varepsilon = -1$ as well as the SM result. It should be noted that most SUSY models prefer the $\varepsilon = -1$ sector which results in a factor 2-3 decrease in the net gg fusion production cross section. Even so, a comparison with Fig. 14 shows that the gg fusion mechanism dominates both $\widetilde{qq}$ fusion and WW/ZZ fusion.

Thus the net effect of SUSY is to decrease the production cross section for the heavier Higgs H_1, while leaving most backgrounds unaffected except for the fact that ΔM_{WW} can be decreased as M_H decreases until the $.05 \ m_H$ resolution limit is reached. The detection of H_1 will become extremely difficult. A thorough analysis is in progress.[3] One must first consider alterations in the branching ratios

466

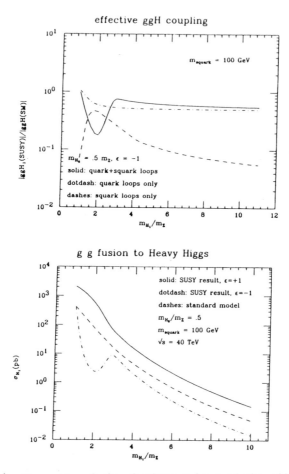

effective ggH coupling

m_{squark} = 100 GeV

$m_{H_6'}$ = .5 m_2, ϵ = -1
solid: quark+squark loops
dotdash: quark loops only
dashes: squark loops only

g g fusion to Heavy Higgs

solid: SUSY result, ϵ=+1
dotdash: SUSY result, ϵ=−1
dashes: standard model
m_{H_6}/m_2 = .5
m_{squark} = 100 GeV
$\sqrt{s}$ = 40 TeV

Fig. 15 a) The ratio of the absolute values of the effective
$ggH_1(SUSY)$ and $ggH(SM)$ couplings, plotted at m_{H_2} = $.5m_Z$ for ϵ = -1
sector. Contributions from squark loops and quark loops as well as the
net sum including interference are shown.
     b) gg fusion cross sections for H_1 production as a function of
H_1. No cuts are incorporated.
     Both plots take the mass scale of squarks at 100 GeV. Individual
squark masses deviate from this value according to the degeneracy
breaking scheme given in Ref. 3.

to the various channels already considered. Secondly there is the
possibility of new decay modes for H_1 to SUSY partners of standard
model particles (Higgsinos ....). Clearly the possibilities are
numerous and results will be very sensitive to SUSY mass scales.
Production of H_2 in SUSY will be close to SM predictions in the (W, Z)
+ H channels dominated by the couplings $g(WWH_2)$ or $g(ZZH_2)$ that are not

greatly altered, except for m_{H_2} very near m_Z where substantial suppression relative to the SM occurs. The gg fusion inclusive H_2 production cross section will also not be drastically altered. However the decay phenomenology of H_2 could be substantially modified. Again a thorough analysis is under way.[3]

IV. CONCLUSIONS

Detection of a standard model Higgs at an e^+e^- collider is relatively straightforward, but limited by planned accelerator energies to $m_H \lesssim 80$ GeV. Supersymmetric models can cause additional subleties and difficulties. In particular the heavier of the two scalar neutral Higgs, H_1, will have suppressed couplings to W's and Z's and its production rate via

$$e^+e^- \to W^{\pm}, Z + H^0$$

will become increasingly small, as m_{H_1}/m_Z increases. The lighter H_2 (with $m_{H_2} < m_Z$ in our simple model) will be produced at a normal rate in the above reaction but may have anomalous decay modes and/or branching ratios.

In pp collisions detection of any heavy Higgs is far from easy. In the standard model only certain reactions yielding purely leptonic final states appear to be clearly background free. At intermediate mass these are

$$pp \to H \qquad\qquad\quad + X$$
$$ \llcorner Z^* \quad + Z$$
$$ \llcorner \ell\bar\ell \quad \llcorner \ell'\bar\ell'$$

and

$$pp \to Z \quad\; + H^0 \qquad\quad + X,$$
$$ \llcorner \ell\bar\ell \quad\; \llcorner \tau^+\tau^-$$

while at high mass the only mode with small background is

$$pp \to H^0 \qquad\qquad\quad + X.$$
$$ \llcorner Z \quad\; + Z$$
$$ \llcorner \ell\bar\ell \quad\; \llcorner \ell'\bar\ell'$$

All suffer from low event rates at $L > 10^4$ pb^{-1} yr^{-1}. The mixed hadronic/leptonic Higgs decay modes may be accessible for $m_H > 2m_W$ and $m_H \lesssim .5$ TeV if sufficiently strong cuts and resolutions can be imposed. At higher m_H values these mixed modes become very problematical. A highly detailed study is required. Finally supersymmetric models complicate the search for the heavier H_1 by a very considerable factor. The complete analysis is still in progress.

Acknowledgements

I would like to thank the Oregon Workshop on Super High Energy Physics for support during the course of much of this work. In addition I would like to acknowledge the important contribution to the efforts summarized herein by my collaborators H.E. Haber and M. Soldate.

References

1. E. Eichten, I. Hinchliffe, K. Lane and C. Quigg, RMP 56 (1984) 579.

2. J.F. Gunion and H.E. Haber, SLAC-PUB-3404 (1984).

3. J.F. Gunion and H.E. Haber, in preparation.

4. See, for example, ECFA/LEP - Specialized Study Group 9 Report, 'Production and Detection of Higgs Particles at LEP' and references therein.

5. B.W. Lee, C. Quigg, H. Thacker, P.R., D16 (1977) 1519; B.L. Ioffe and V.A. Khoze, Leningrad Report No. LINP-274, 1976; J. Ellis, M.K. Gaillard and D.V. Nanopoulos, NP B106 (1976) 292.

6. Wilczek, F., PRL 39 (1977) 1304.

7. Taken from J.M. Dorfan, 'Ann Arbor Theoretical Advanced Study Institute on Elementary Particle Physics' (1984) (QCD 161:145:1984).

8. J.F. Gunion, P. Kalyniak, M. Soldate and P. Galison, 'Searching for the Intermediate Mass Higgs Boson', SLAC-PUB-3604 (1985).

9. J.F. Gunion, P. Kalyniak, M. Soldate and P. Galison, PRL 54 (1985) 1226 and 'Proceedings of the 1984 Summer Study on Design and Utilization of the SSC', Snowmass (1984).

10. B. Cox et al. 'Proceedings of the 1984 Summer Study on Design and Utilization of the SSC', Snowmass (1984).

11. We wish to acknowledge a helpful discussion with F. Paige on this point.

12. H.M. Georgi, S.L. Glashow, M.E. Machacek, and D.V. Nanopoulous, PRL 40 (1978) 692. We use m_t = 40 GeV in the quark loop calculation.

13. R.N. Cahn and S. Dawson, PL B136 (1984) 196; M.S. Chanowitz and M.K. Gaillard, PL B142 (1984) 85; G.L. Kane, W.W. Repko and W.B. Rolnick, PL B148 (1984) 367.

14. M.J. Duncan, G.L. Kane and W.W. Repko, Preprint UMTH-85-18 (1985).

15. The necessary subprocess was computed in J.F. Gunion and Z. Kunszt, PL 161B, 333 (1985); and R. Kleiss and W.J. Stirling, CERN-TH-4186/85.

16. J.F. Gunion, Z. Kunszt and M. Soldate, OITS-298, UCD-85-4, SLAC-PUB-3709, PL to be published.

17. W.J. Stirling, R. Kleiss, and S.D. Ellis, CERN-TH-4209/85.

18. J.F. Gunion and M. Soldate in preparation.

SEARCH FOR SUPERSYMMETRIC PARTICLES AT PEP AND PETRA

Sau Lan Wu

Department of Physics

University of Wisconsin, Madison, Wisconin, USA[*]

and

Deutsches Elektronen-Synchrotron, DESY, Hamburg, Germany

CONTENTS

1. Introduction

2. Pair Production of Unstable Photinos $e^+e^- \rightarrow \tilde{\gamma}\tilde{\gamma}$

3. Search for Scalar Electrons $\tilde{e}^\pm$ and Stable Photinos $\tilde{\gamma}$

4. Search for Scalar Muons $\tilde{\mu}^\pm$ and Scalar Taus $\tilde{\tau}^\pm$

5. Search for Zinos $\tilde{Z}^0$

6. Search for Charginos $\tilde{\chi}^\pm$

7. Conclusion

8. Acknowledgements

9. References

1. INTRODUCTION

Supersymmetry refers to the symmetry between bosons and fermions[1]. The study of supersymmetry has theoretical, but not experimental, motivation, and it is not possible to judge at present whether it will eventually be a useful concept in particle physics[2].

[*] Supported by the US Department of Energy Contract number DE-AC02-76ER00881 and the US National Science Foundation grant number INT-8313994 for travel.

Nevertheless, it is interesting because it introduces a very large number of new particles. In this talk, we give a summary of the recent results from a heroic effort at PEP and PETRA to search for supersymmetry particles.

The Standard Model[3], a gauge theory based on SU(3) x SU(2) x U(1), describes successfully the strong, weak, and electromagnetic interactions. Only the spin-zero Higgs boson, which is needed for spontaneous symmetry breaking in the Standard Model, has not been found so far. As a framework of discussion, we shall therefore use supersymmetric version of the Standard Model[4,5]. More precisely, we shall use N=1 supersymmetry, which means that there is only one set of supersymmetry generators. Thus in N=1 supersymmetry, there is for example just one supersymmetric partner, the wino, $\widetilde{W}^{\pm}$, for the $W^{\pm}$ boson of weak interactions. However, in order for the supersymmetric version to be consistent, the Standard Model has to be modified to contain at least two Higgs doublets, leading to five physical Higgs, three neutral and two charged ones (H^{+} and H^{-}). Their supersymmetric partners are called neutral and charged higgsinos. The resulting list of particles is shown in Table 1.

While the supersymmetric version of the Standard Model is not unique, especially concerning supersymmetry breaking, the particle listed in Table 1 are all present. The only possible exception is the Goldstino $\widetilde{G}$: in supergravity, the Goldstino is eaten by the spin $\frac{3}{2}$ gravitino (the supersymmetric partner of the graviton) to provide its mass, much in the same way as the charged Higgs in the one – doublet Standard Model is eaten by the W to provide a massive W. The net result is that, in supergravity, the gravitino is produced instead of the Goldstino. The partner of the Goldstino $\widetilde{G}$ has properties that depend on the details of the theory and will not be discussed here. In most supersymmetry theories, there is an operator R such that all the usual particles are even under R while the supersymmetric partners are odd, with the consequence that the supersymmetric partner must be produced in pairs. We shall use an

Table 1: List of Particles in N=1 Supersymmetric Standard Model

Spin 0	Spin $\frac{1}{2}$	Spin 1
	Goldstino $\tilde{G}$	
	photino $\tilde{\gamma}$	photon γ
scalar neutrino $\tilde{\nu}$	neutrino ν	
	gluino $\tilde{g}$	gluon g
scalar leptons $\tilde{\ell}_R, \tilde{\ell}_L$	lepton ℓ	
scalar quarks $\tilde{q}_R, \tilde{q}_L$	quark q	
	wino $\tilde{w}^{\pm}$	charged intermediate boson $W^{\pm}$
	zino $\tilde{Z}^{O}$	neutral intermediate boson Z^{O}
neutral Higgs H^{O}	neutral higgsino $\tilde{H}^{O}$	
charged Higgs $H^{\pm}$	charged higgsino $\tilde{H}^{\pm}$	

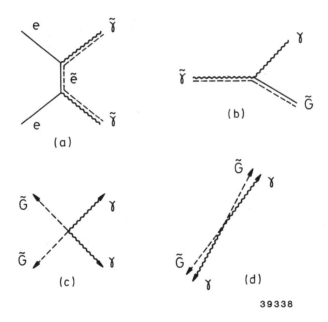

39338

Fig. 1: Search for unstable photinos: this figure gives the Feynman
diagrams for production (a) and decay into $\tilde{G}\gamma$ (b), and the
schematic diagrams for the event when the photino mass is
large (c) or small (d).

additional dotted line to indicate R=-1 particles. If this R- parity is exact, then the lightest particle with R=-1 is stable. It is not known which one of these supersymmetric particles is the lightest; some of the likely candidates are the Goldstino, the photino, and the scalar neutrino.

2.   PAIR PRODUCTION OF UNSTABLE PHOTINOS $e^+e^- \rightarrow \tilde{\gamma}\tilde{\gamma}$

As shown in Table 1 the supersymmetric partner of the photon is the photino ($\tilde{\gamma}$) of spin $\frac{1}{2}$. A pair of photinos could be produced in e^+e^- annihilation by the exchange of a scalar electron ($\tilde{e}$) as shown in Fig. 1(a). If the photino is stable, then this process by itself is almost impossible to detect. If the photino is unstable as permitted in some models, then a possible decay mode[6] for this unstable massive photino is $\tilde{\gamma} \rightarrow \tilde{G}\gamma$, as shown in Fig. 1(b).

For the experimental detection of this production of the unstable photino pairs an important quantity is the photino lifetime. If it is too long, then the photino decays occur outside of the detector. The lifetime is given by[6]

$$\tau_{\tilde{\gamma}} = \frac{8\pi \, d^2}{M_{\tilde{\gamma}}^5}$$

where d is an order parameter. Under the assumption that the photinos decay within the detector, the signature is as follows. If the photino is relatively heavy (a few GeV/c^2 or greater) the signature of the event is a pair of acoplanar photons with missing energy as shown in Fig. 1(c). If the photino is light, the signature of the event is a nearly collinear photon pair with relatively low energies (Fig.1(d)).

Fig. 2 gives the excluded region[7] with 95% confidence level in the photino mass and scalar electron mass plane measured by CELLO[8], JADE[9], MARK J[10], and TASSO[11]. It is assumed that the masses of

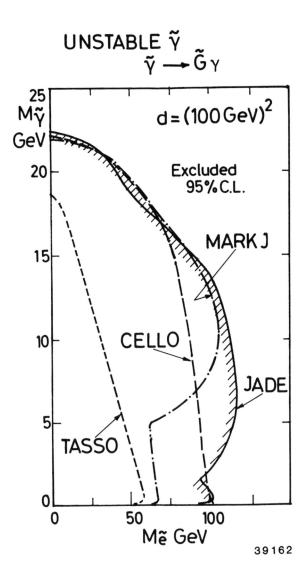

Fig. 2: Excluded region, with 95% confidence level, in $M_{\tilde{\gamma}}$-$M_{\tilde{e}}$ plane for unstable photinos.

the left-handed and right-handed scalar electrons are the same and $d = (100 \text{ GeV})^2$. The results are not sensitive to the assumed value of d. Changing d only alters the lower limit on the photino mass, a part of the curve which is barely visible in Fig. 2.

3. SEARCH FOR SCALAR ELECTRONS $\tilde{e}^\pm$ AND STABLE PHOTINOS $\tilde{\gamma}$

The search[12] for scalar electrons $\tilde{e}$ and stable photinos $\tilde{\gamma}$ has been carried out by ASP[13], MAC[14] and MARK II[15] of PEP and CELLO[16], JADE[17], MARK J[18] and TASSO[19] of PETRA. The following four processes have been used.

A. Pair production of unstable $\tilde{e}^\pm$

$$e^+e^- \rightarrow \tilde{e}^+ + \tilde{e}^-$$
$$\quad \quad \quad \downarrow \quad \downarrow \tilde{\gamma}e^-$$
$$\quad \quad \downarrow \tilde{\gamma}e^+$$

The diagrams for the production and decay are shown in Fig. 3(a) and (b). The signature of such events is acoplanar e^+e^- pair with missing energies and momenta.

B. Single scalar electron production

$$e^+e^- \rightarrow e^\pm \tilde{e}^\mp \tilde{\gamma}$$
$$\quad \quad \quad \downarrow e^\mp \tilde{\gamma}$$

In this process, $\tilde{e}$ is produced by the scattering of an initially radiated photon and an electron as shown in Fig. 3(c). The electron which radiates the photon goes down along the beam pipe and is undetected. The $\tilde{e}$ then decays into a photino and an electron. The signature of this event is a single electron and no other detected particles.

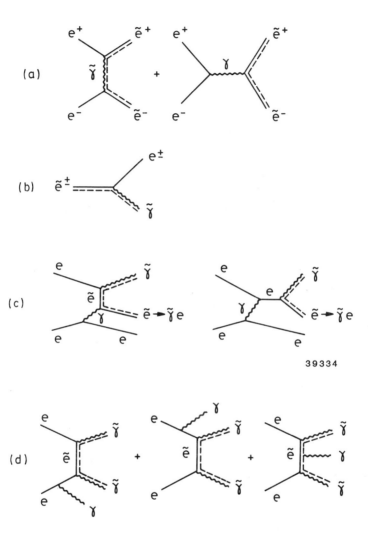

Fig. 3: Feynman diagrams for the production of the scalar electron (a and c) and stable photino (c and d) together with that for the decay of the scalar electron (b).

C. Radiative photino pair production

$$e^+e^- \rightarrow \gamma \, \tilde{\gamma} \, \tilde{\gamma}$$

The diagrams for this process are given in Fig. 3(d). For stable photinos, the signature for this event is a single photon with nothing else.

D. Pair production of stable $\tilde{e}^{\pm}$

$$e^+e^- \rightarrow \tilde{e}^+ \, \tilde{e}^-$$

The signature is a collinear heavy muon pair like event.

The experimentally excluded region[12] in the $M_{\tilde{e}}$ - $M_{\tilde{\gamma}}$ plane is shown in Fig. 4. If the photino mass is assumed to be small, the mass limits can be read off from Fig. 4 by the intercept with the $M_{\tilde{e}}$-axis, and the results are shown in Table 2. The best result, 51.1 GeV at 90% confidence level, is obtained by ASP Collaboration using process C.

4. SEARCH FOR SCALAR MUONS $\tilde{\mu}^{\pm}$ AND SCALAR TAUS $\tilde{\tau}^{\pm}$

As shown in Fig. 5, scalar muons are pair produced in e^+e^- annihilation. The differential cross section is

$$\frac{d\sigma}{d\Omega} \, (e^+e^- \rightarrow \tilde{\mu}^+ \, \tilde{\mu}^-) = \frac{\alpha^2}{8s} \, \beta_{\tilde{\mu}}^3 \, \sin^2 \theta_{\tilde{\mu}}$$

for $M_{\tilde{\mu}_R}$ and $M_{\tilde{\mu}_L}$ not close to each other. Here s = center of mass energy squared and $\beta_{\tilde{\mu}}$ is the velocity of $\tilde{\mu}$ divided by the velocity of light. For $M_{\tilde{\mu}_R} = M_{\tilde{\mu}_L}$ the above cross-section should be multiplied by 2.

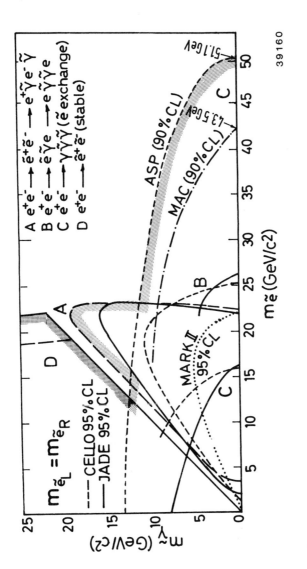

Fig. 4: Excluded region in the $M_{\tilde{\gamma}}$–$M_{\tilde{e}}$ plane for stable photino.

39160

Table 2: Excluded mass regions in GeV/c^2 for scalar leptons. The photino mass is assumed to be zero.

	$M_{\tilde{e}}$ (GeV/c^2)	$M_{\tilde{\mu}}$ (GeV/c^2)	$M_{\tilde{\tau}}$ (GeV/c^2)
ASP (90% C.L.)	<51 (ref.13)		
CELLO (95% C.L.)	<25 (ref.16)	<3.3 to 16 (ref. 21)	M_τ to 3.8 6 to 15.3 } (ref.21)
JADE (95% C.L.)	<25 (ref.17)	<20.3 (ref.20)	M_τ to 18 (ref.22)
MAC (90% C.L.)	<43.5(ref.14)		
MARK II (95% C.L.)	<22 (ref.15)		M_τ to 9.9 (ref.23)
MARK J (95% C.L.)		<20 (ref.18)	M_τ to 17 (ref.18)
TASSO (95% C.L.)	<16.6(ref.19)	<16.4 (ref.19)	

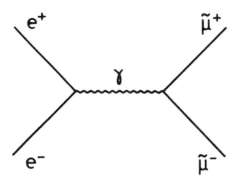

Fig. 5: Feynman diagram for the pair production of scalar muons $\tilde{\mu}$.

JADE[20] has considered the following cases:

A. $\tilde{\mu} \to \mu \tilde{\gamma}$ and $\tilde{\gamma}$ is stable

Here the signature of the event is a pair of acoplanar muons with missing energies and momenta. The excluded region in the $M_{\tilde{\gamma}} - M_{\tilde{\mu}}$ plane with 95% confidence level is given in Fig. 6(a).

B. $\tilde{\mu} \to \mu \tilde{\gamma}$ and $\tilde{\gamma} \to \gamma \tilde{G}$

where $\tilde{G}$ is the massless and non-interacting Goldstino. The signature of the event is $\mu\mu\gamma\gamma$. The excluded region in the $M_{\tilde{\gamma}} - M_{\tilde{\mu}}$ plane with 95% confidence level is given in Fig. 6(b).

Assuming that the photino mass is small, the limits from CELLO[21], JADE[20], MARK J[18], and TASSO[19] are given in Table 2.

Similar searches for the scalar tau have been carried out via $e^+e^- \to \tilde{\tau}^+\tilde{\tau}^-$, with the $\tilde{\tau}$ assumed to decay by $\tilde{\tau} \to \tau\tilde{\gamma}$. Thus the $\tilde{\tau}$ mass is taken to be larger than that of τ. The signature of such an event is an isolated e or μ and a hadronic jet. The mass limits are also shown in Table 2.

5. SEARCH FOR ZINOS $\tilde{Z}^0$

Zino $\tilde{Z}^0$, the supersymmetric partner of the Z^0, is produced through the process

$$e^+e^- \to \tilde{Z}^0 \tilde{\gamma}$$

by an exchange of a scalar electron $\tilde{e}$ as shown in Fig. 7. The following decay modes of $\tilde{Z}^0$ are explored:

A. $\tilde{Z}^0 \to e^+e^-\tilde{\gamma}, \mu^+\mu^-\tilde{\gamma}$

as shown in Fig. 8(a). The signature of the event is a pair of acoplanar leptons with missing energies and momenta (Fig. 8(b)).

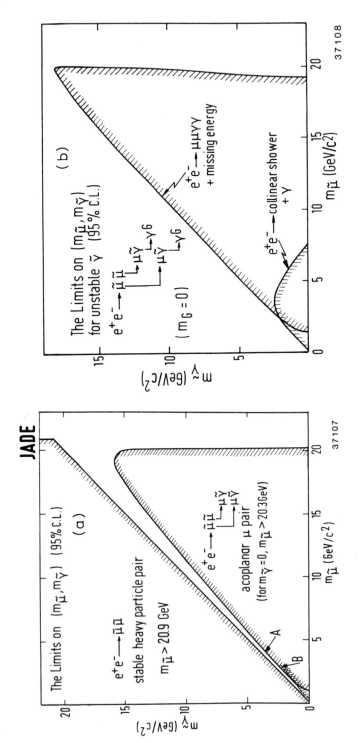

Fig. 6: Excluded region in the $M_{\tilde{\gamma}}$–$M_{\tilde{\mu}}$ plane for (a) stable and (b) unstable photinos. These results are from JADE.

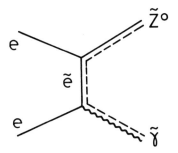

Fig. 7: Feynman diagram for the production of zino $\tilde{Z}^o$.

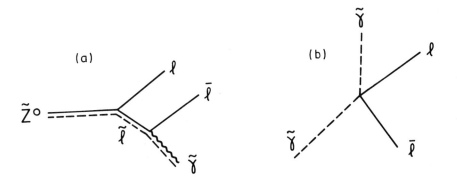

Fig. 8: Decay of $\tilde{Z}^o$ into $\ell\bar{\ell}\tilde{\gamma}$ and schematic diagram of the event.

B. $\tilde{Z}^O \to \bar{q}q\, \tilde{\gamma}$

as shown in Fig. 9(a). The $\tilde{q}$ in this figure can be real or virtual; thus under B we include (i) $\tilde{Z}^O \to q\bar{q}\tilde{\gamma}$ ($\tilde{q}$ virtual), (ii) $\tilde{Z}^O \to q\tilde{\bar{q}}$ with $\tilde{\bar{q}} \to \bar{q}\tilde{\gamma}$, and (iii) $\tilde{Z}^O \to \bar{q}\tilde{q}$ with $\tilde{q} \to q\tilde{\gamma}$. In all three cases, the detected particles are the same, and the signature is a pair of acoplanar jets (for heavy $\tilde{Z}^O$) or one jet (for light $\tilde{Z}^O$) with missing energies and momenta (Fig. 9(b)).

C. $\tilde{Z}^O \to q\bar{q}\tilde{g}$

as shown in Fig. 10. Similar to case B, the $\tilde{q}$ in this figure can be real or virtual. For light $\tilde{q}$ of mass less than 3 GeV, the branching ratio for this $q\bar{q}\tilde{g}$ is dominant (~99%) due to the strong coupling constant α_s.

Excluded regions in $M_{\tilde{Z}^O} - M_{\tilde{e}}$ plane are shown in Fig. 11 and Fig. 12 from results by CELLO[24], JADE[25], and MARK J[26].

It may be appropriate at this point to add the following remark. It is seen from Table 1 that $\tilde{G}$, $\tilde{\gamma}$, $\tilde{Z}^O$ and $\tilde{H}^O$ all have the same quantum numbers (charge 0, spin $\frac{1}{2}$, color singlet, and R=-1). Thus they can mix, i.e., the physical states of definite masses are a mixture of them. These physical states are referred to as neutralinos. The amount of mixing depends on the nature of supersymmetry breaking, and at present there is no way of choosing among the many possibilities. Strictly speaking, this section should be entitled the search for neutralinos through the production of $\tilde{Z}^O$. Since at least one neutralino must have a large $\tilde{Z}^O$ component, this mixing has only relatively minor effects on the results of Fig. 11 and Fig. 12, provided that the assumptions are fulfilled.

6. SEARCH FOR CHARGINOS $\tilde{\chi}^{\pm}$

Similar to the mixing of the neutral supersymmetric particles just discussed, the wino $\tilde{W}^{\pm}$ and the charged higgsino $\tilde{H}^{\pm}$ of Table 1

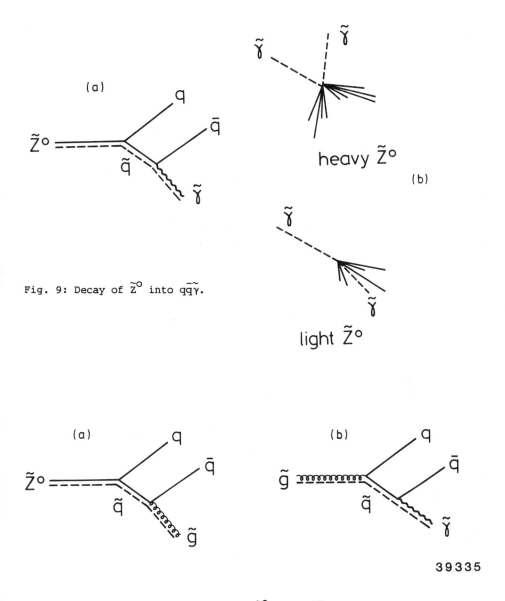

(a)

Fig. 9: Decay of $\tilde{Z}^0$ into $q\bar{q}\tilde{\gamma}$.

heavy $\tilde{Z}^0$

(b)

light $\tilde{Z}^0$

(a)

(b)

39335

Fig. 10: Decay of $\tilde{Z}^0$ into $q\bar{q}\tilde{g}$.

486

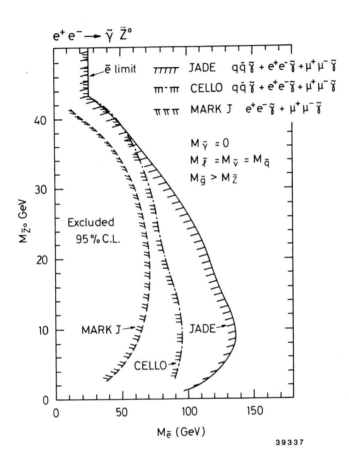

Fig. 11: Excluded region in the $M_{\tilde{Z}^0}$-$M_{\tilde{e}}$ plane from $\tilde{Z}^0 \rightarrow \ell\bar{\ell}\tilde{\gamma}$ and $q\bar{q}\tilde{\gamma}$.

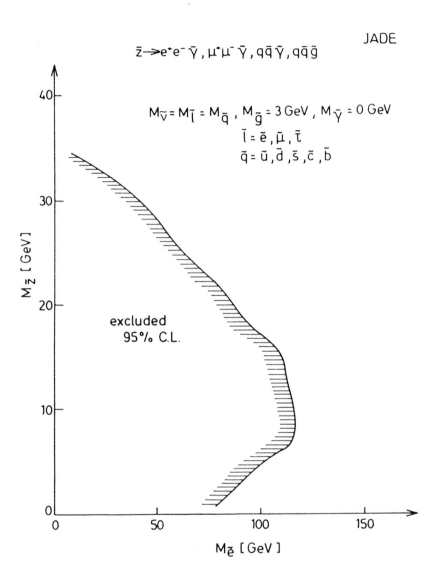

Fig. 12: Excluded region in the $M_{\widetilde{Z}^0}$–$M_{\widetilde{e}}$ plane under the assumption that $M_{\widetilde{g}} = 3$ GeV.

also have the same quantum numbers and hence can mix to produce the mass eigenstates called charginos $\tilde{\chi}^{\pm}$. The pair production of $\tilde{\chi}^{\pm}$

$$e^+ e^- \rightarrow \tilde{\chi}^+ \tilde{\chi}^-$$

can proceed via a virtual photon or a heavy scalar neutrino $\tilde{\nu}_e$ exchange as shown in Fig. 13. Depending on the masses of the supersymmetric particles the charginos can have different dominant decay modes. At PETRA the charginos have been searched through the following modes: (We write down only the ones for χ^-; those for χ^+ are given by charge conjugation.)

A. $\tilde{\chi}^- \rightarrow \ell^- \tilde{\nu}_\ell$

as shown in Fig. 14(a), where ℓ = e, μ, or τ. If both $\tilde{\chi}^+$ and $\tilde{\chi}^-$ decay this way, the signature of the event is a pair of acoplanar leptons with missing energies and missing momenta.

B. $\tilde{\chi}^- \rightarrow q_1 \bar{q}_2 \tilde{g}$ (Fig. 14(b))
      $\hookrightarrow q\bar{q} \tilde{\gamma}$

C. $\tilde{\chi}^- \rightarrow q_1 \bar{q}_2 \tilde{\gamma}$ (Fig. 14(c))

D. $\tilde{\chi}^- \rightarrow \ell^- \bar{\nu}_\ell \tilde{\gamma}$

as shown in Fig. 14(d). The signature is the same as A.

E. Stable $\tilde{\chi}^-$. Such event will look like collinear heavy muon pairs.

CELLO[24], JADE[27], and MARK J[26] have searched for the chargino through A, and the excluded region in the $M_{\tilde{\chi}^\pm} - M_{\tilde{\nu}}$ plane is shown in Fig. 15. In addition, JADE[27] has searched for the chargino through the processes B, C, D, and E, and also through the precise measurement of $R = \dfrac{\sigma(e^+ e^- \rightarrow \text{hadrons})}{\sigma(e^+ e^- \rightarrow \mu^+ \mu^-)}$. Their result from B and R is that $M_{\tilde{\chi}^\pm}$ cannot be below 22.4 GeV, and from E not below 21.1 GeV. The results from C and D, parametrized by the branching ratio $Br(\ell\tilde{\nu}\tilde{\gamma})$ and $Br(q_1\bar{q}_2\tilde{\gamma})$, are shown in Fig. 16.

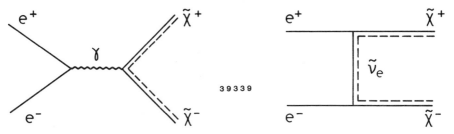

39339

Fig. 13: Feynman diagrams for the production of charginos $\tilde{\chi}^{\pm}$.

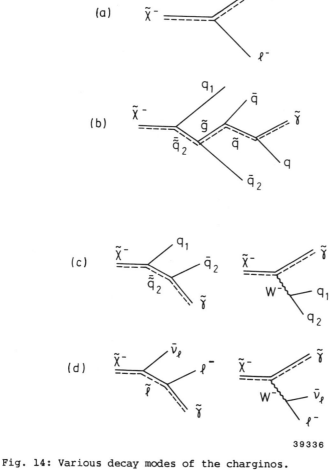

39336

Fig. 14: Various decay modes of the charginos.

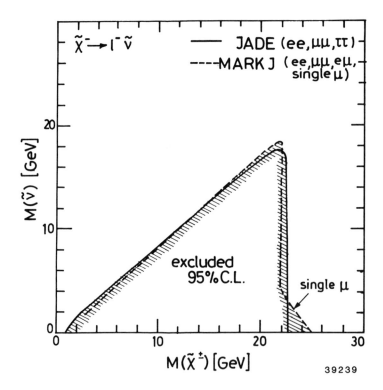

Fig. 15: Excluded region in the $M_{\tilde{\nu}}$-$M_{\tilde{\chi}}$ plane from the decay $\tilde{\chi}^{\pm} \to \ell^{\pm}\tilde{\nu}_{\ell}$.

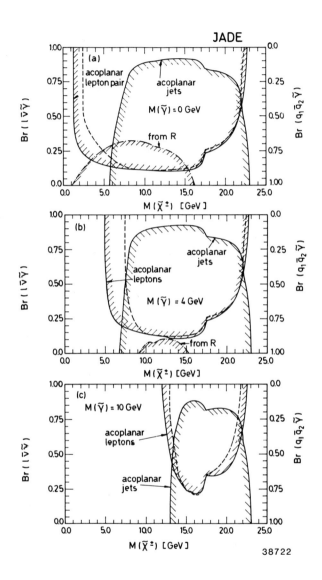

Fig. 16: The limits for chargino masses with 95% C.L. as a function
of the leptonic branching fraction for chargino decay, for
the case of $\tilde{\chi}^{\pm} \rightarrow \ell \nu \tilde{\gamma}$ or $\tilde{\chi}^{\pm} \rightarrow q_1 \bar{q}_2 \tilde{\gamma}$.

7.   CONCLUSION

Due to the efforts of many physicists working at PEP and PETRA,
we have now a great deal of information on the bound on masses of
particles expected on the basis of supersymmetry. No such particle,
however, has been found. Since a possible reason is that the energy
is not high enough, the searches will be continued at other
accelerators, including SLC, LEP, HERA, AND $p\bar{p}$ colliders. We eagerly
wait for the first discovery of such particles; if and when this
happens, it will be a major event for particle physics.

8.   ACKNOWLEDGMMENTS

I would like to thank the conference organizer Professor R. Hwa
and the session organizers, Professors D. Baltay and G. Kane, for
their kind hospitality and for a stimulating meeting. Many thanks to
Dr. S. Komamiya of JADE and Dr. H. Krüger of CELLO for providing me
with the updated data and plots. I would also like to thank the DESY
directorate for their hospitality extended to me while working at
DESY.

9. REFERENCES

1) Gol'fand, Yu.A. and Likhtman, E.P., JETP Lett. 13, 323 (1971);
    Volkov, D.V. and Akulov, V.P., Phys. Lett. 46B, 109 (1973);
    Wess, J. and Zumino, B., Nucl. Phys. B70, 39 (1974);
    Salam, A. and Strathdee, J., Nucl. Phys. B76, 477 (1974).

2) Fayet, P., Phys. Lett. 69B, 489 (1977);
    Nanopoulos, D.V., Savoy-Navarro, A. and Tao, Ch. (organizers),
    Proc. Workshop on Supersymmetry versus Experiment, CERN, Geneva,
    Switzerland, April 1983.

3) Glashow, S.L., Nucl. Phys. 22, 579 (1961);
    Weinberg, S., Phys. Rev. Lett. 19, 1264 (1967);
    Salam, A., Proceedings of the Eighth Nobel Symposium, May 1968,
    ed. Svartholm, N. p. 367 (Wiley, 1968);
    Glashow, S.L., Iliopoulos, J., and Maiani, L., Phys. Rev. D2,
    1285 (1970).

4) Fayet, P., Unification of the Fundamental Particle
    Interactions, ed. Ferrara, S., Ellis, J., and
    van Nieuwenhuizen, P. p. 587 (Plenum Press, New York, 1980).

5) Haber, H.E. and Kane, G.L., Physics Reports 117, 75 (1985).

6) Cabibbo, N., Farrar, G.R., and Maiani, L., Phys. Lett. 105B,
    155 (1981).

7) Komamiya, S., Proceedings of the Topical Conference of the 1985
    SLAC Summer Institute, Stanford, California, July 29-August 9,
    1985.

8) CELLO Collab., Behrend. H. et al., Phys. Lett. 123B, 127 (1983);
    and contributed paper to the International Europhysics
    Conference on High Energy Physics, Bari, Italy, July 18-24 1985.

9) JADE Collab., Bartel, W. et al., Phys. Lett. 139B, 327 (1984);
    and Komamiya, S., private communication.

10) MARK J Collab., Adeva, B. et al., MIT-LNS Reports 139 (1984).

11) TASSO Collab., Althoff, M. et al., Z. Phys. C26, 337 (1984).

12) Komamiya, S., Proceedings of the International Symposium on
    Lepton and Photon Interactions at High Energies, Kyoto, Japan,
    August 19-24, 1985.

13) ASP Collab., Hollebeek, R., Proceedings of the Topical Conference of the 1985 SLAC Summer Institute, Stanford, California, July 29-August 9, 1985.

14) MAC Collab., Fernandez, E. et al., Phys. Rev. Lett. $\underline{52}$, 22 (1984); and Levine, T. private communication.

15) MARK II Collab., Gladney, L. et al., Phys. Rev. Lett. $\underline{51}$, 2253 (1983); and LeClaire, B., private communication.

16) CELLO Collab., Behrend, H.J. et al., Phys. Lett. $\underline{114B}$, 287 (1982); and paper No. 352 (Search for Scalar Electrons and Photinos in e^+e^- Interactions) submitted to the International Symposium on Lepton and Photon Interaction at High Energies, Kyoto, Japan, August 19-24, 1985.

17) JADE Collab., Bartel, W. et al., Phys. Lett. $\underline{152B}$, 385 (1985).

18) MARK J Collab., Adeva, B. et al., Phys. Lett. $\underline{152B}$, 439 (1985).

19) TASSO Collab., Brandelik et al., Phys. Lett. $\underline{117B}$, 365 (1982).

20) JADE Collab., Bartel, W. et al., Phys. Lett. $\underline{152B}$, 392 (1985).

21) CELLO Collab., Behrend, H.J. et al., Phys. Lett. $\underline{114B}$, 287 (1982).

22) JADE Collab., Schneekloth, U., private communication.

23) MARK II Collab., Blocker, C.A. et al., Phys. Rev. Lett. $\underline{49}$, 517 (1982).

24) CELLO Collab., Behrend, H.J. et al., contributed paper to the International Europhysics Conference on High Energy Physics, Bari, Italy, July 18-24, 1985.

25) JADE Collab., Bartel, W. et al., Phys. Lett. $\underline{146B}$, 126 (1984), and Komamiya, S., private communication.

26) MARK J Collab., Adeva, B. et al., Phys. Rev. Lett. $\underline{53}$, 1806 (1984).

27) JADE Collab., Bartel, W. et al., DESY 85-60 (1985).

REVIEW OF RECENT RARE K DECAY EXPERIMENTS

Michael E. Zeller

J.W. Gibbs Lab., Yale University, New Haven, CT. 06520

ABSTRACT

Experiments searching for the decays $K^+ \to \pi^+ \mu^+ e^-$, $K^\circ \to \mu e$, and $K^+ \to \pi^+ \nu \bar{\nu}$ are desribed.

1.   INTRODUCTION

Stimulated by theoretical interest in physics beyond the standard model and the availability of adequate beams and technology, several groups have undertaken searches for very rare decay modes of K mesons. These experiments fall into two categories: Searches for lepton flavor violation, i.e. $K^\circ \to \mu e$ and $K^+ \to \pi^+ \mu^+ e^-$, and searches for decays with missing particles, i.e. $K^+ \to \pi^+ X^\circ$ where X° does not interact. In this paper we summarize and review the current status of the four experiments involved in this effort.

2.   LEPTON FLAVOR VIOLATING SEARCHES

Here we will be discussing three experiments, one of which is looking for $K^+ \to \pi^+ \mu^+ e^-$, (E777 at the AGS), and two which are searching for $K^\circ \to \mu e$, (E780 and 791 at the AGS).

Before beginning let us recall the motivations behind these efforts. The first came with the realization that significant improvement in the current limits can be made with available beams and detector capabilities. All of these experiments are proposing to improve their respective limits by factors of 500 to 1000. Improvements in measurements of any particle physics quantity by factor of these factors is adequate reason to make the effort, but these experiments are going to improve limits on, or perhaps find, lepton flavor violating currents. Because lepton flavor conservation is not understood as a fundamental principle, the limits of this rule have been of great interest since its first observation.

Finally, motivation has recently come from theorists who are attempting to understand the Higgs mechanism and the gauge hierarchy problem. Results of their theories led to predictions that these phenomona could show up at the $10^{-10} - 10^{-11}$ level, within striking distance of our capabilities.

2.1 E777: $K^+ \to \pi^+ \mu^+ e^-$

We now turn to the experimental program. E777 was approved in 1982 and has just completed its first run at the AGS.[1] While the primary purpose of the experiment is as stated above, measurements will also be made of the Dalitz plot distribution for $K^+ \to \pi^+ e^+ e^-$, a search will be made for $e^+ e^-$ states with mass between 140 and 340 MeV, and measurements of the rate of $\pi^0 \to e^+ e^-$ will be made.

In the search for $K^+ \to \pi^+ \mu^+ e^-$ statistics is not a problem. From calculation, and now from preliminary measurements, 3×10^7 K's can be obtained in a beam with 5×10^{11} primary protons on target. They are unfortunately accompanied by 5×10^8 pions.

Triggering is not a real problem either since the first decay mode of K^+ mesons that yields an e^- is $K^+ \to \pi^+ \pi^0$ with the subsequent Dalitz decay of the π^0, $\pi^0 \to e^+ e^- \gamma$. This still does not result in the correct final state particles, and must be accompanied by particle type misidentification.

The difficult part of the experiment is particle identification. Decays such as $K^+ \to \pi^+ \pi^+ \pi^-$ with a subsequent decay of $\pi^+ \to \mu^+ \upsilon$ in the apparatus, and a misidentification of the π^- as an electron can simulate the desired reaction. To keep this background below 10^{-12}, required to make a search senisitive to 10^{-11}, it is necessary to achieve a π/e rejection of 10^{-7}.

We show in Figure 1 a plan view of the apparatus. A $K^+ \to \pi^+ \mu^+ e^-$ event is depicted originating from the 5 meter, evacuated decay region. The system is built around two spectrometer magnets: a 48D48, 48 inch wide by 48 inch long pole tip, with a 24 inch gap, and a 72D18 with a 30 inch gap. The purpose of the first magnet is to direct the decay particles out of the beam and to separate them to either side of the apparatus by charge. The second magnet, in conjunction with proportional chambers P1-P4 serves to analyze the momenta of the decay products.

On the right side of the apparatus, looking upstream, are two hydrogen filled Cerenkov counters, and a lead-scintillator shower calorimeter. On the left side are two Cerenkov counters filled with nitrogen and carbon dioxide, the shower calorimeter, and a muon identifier consisting of seven steel plates sandwiching X-Y arrays of proportional tubes. The choice of gasses in the Cerenkov counters is made to optimize π/e rejection on the right side, while having very high electron detection efficiency on the left. Each election detector has a nominal π/e rejection of 10^{-3}, both by calculation and test beam results, so that the desired 10^{-7} seems achievable.

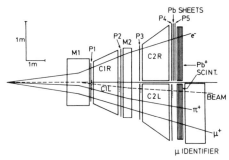

Figure 1. Plan view of the E777 apparatus.

The group has their first run at the AGS, and has tuned the system on the copius $K^+ \to \pi^+ \pi^+ \pi^-$ decay. From these studies they have determined that the flux of K^+ mesons is as expected to within a factor of two, (the results are still being analyzed), and the signal to noise ratio is as expected. Data acquisition for $K^+ \to \pi^+ \mu^+ e^-$ will take place during the next slow extracted beam, (SEB) run at the AGS in March of next year.

2.2 $K^\circ \to \mu e$

Comparing the experimental aspects of the two decays, one sees advantages and disadvantages to each. The neutral beam, while not containing the large charged particle fluxes which can disable detectors in the K^+ beam, does not have a well defined momentum. The beam momentum at the AGS ranges from 4 to 12 GeV/c, thus removing one of the kinematic constraints that exists with a charged beam. The two body final state allows better kinematic constraint at the trigger level, but provides less constraint in vertex determination than that of a three body final state. Finally, particle identification is not as critical for $K^\circ \to \mu e$, but this is true mainly because the correct particle species can be generated from the decay $K^\circ \to \pi e \nu$, with subsequent decay $\pi \to \mu \nu$, at the 5×10^{-4} level. Thus kinematic resolution is of greater importance in this decay mode.

2.2.1 E780. This experiment, approved in 1983, has also just completed its first engineering run at the AGS.[2] In addition to lowering the branching ratio limit for $K^\circ \to \mu e$ to $<10^{-10}$, a search for the decay $K^\circ \to e^+ e^-$ to a comparable level will be made. (We recall that this decay is expected at the $\sim 2 \times 10^{-13}$ level from standard model predictions.)

For this experiment also, statistics is not a real issue. The primary beam of 10^{12} protons per pulse into the 45 μstr beam line will yield $\sim 2 \times 10^7$ K° is per pulse.

A plan view of the apparatus is shown in Figure 2. The spectrometer system is centered about a 72D18 magnet with a 24" gap. The chambers used for momentum measurements, A–D in the figure, are "Mini-Drift" chambers in which sense and field wires are alternately

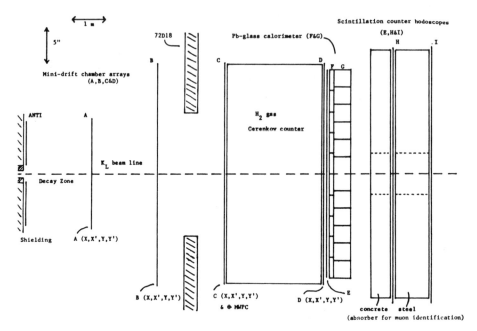

Figure 2. Plan view of E780 apparatus.

placed with a 3mm spacing, and which each have two horizontal and two vertical views. The B plane also has a θ view to remove ambiguities.

Following the Pb-Glass, but preceeded by a concrete shield is a scintillation counter array. This array, along with the electron identifiers, is used at the trigger level to identify muons. Following it is an Iron muon range stack interspersed with scintillation counters.

With special backleg windings, the spectrometer magnet will be capable of a transverse momentum impulse of ~220 MeV/c. This is very close to the 236 MeV of transverse momentum given to the decay products of $K^\circ \to \mu e$ when the decay occurs at 90° from the beam direction. Thus this spectrometer arrangement makes the most probable trajectories of the decay products parallel to the beam; a feature which is employed at the trigger level.

The background events from $K^\circ \to \pi e \nu$ with the pion decaying to a muon and neutrino can be discriminated against using only the drift chambers if the decay occurs outside of the magnetic field region. If it happens in the magnet, the consequence is a mismeasurement of the "muon" momentum. By means of the range stack the range of the muon can be compared with a prediction from the momentum measurement, and thus a large fraction of these events can be rejected.

The group had an engineering run in the spring. During this run prototype chambers, electronics, and part of the trigger were tested and found to work as planned. The number of neutrons in the beam, one per 500 protons on target, was essentially as expected; the rates in the Pb-Glass, $\sim 3 \times 10^7/m^2/10^{12}$ protons on target, were about ten times the expected rates from K decays and neutron interactions. The group is completing their detector construction and plans to begin data acquisition in the next SEB running period at the AGS.

2.2.2 E791. A collaboration between UCLA, LANL, Pennsylvania, Stanford, Temple, and William and Mary, this experiment is a more sensitive exploration for rare K° decays.[3] The goals are to search for $K^\circ \to \mu e$ and $K^\circ \to ee$ with a branching ratio sensitivity of 10^{-12}, to measure the longitudinal muon polarization in $K^\circ \to \mu^+ \mu^-$ to an accuracy of $\sim 14\%$, to search for the CP violating decay $K^\circ \to \pi^\circ e^+ e^-$ to an accuracy of $\sim 10^{-10} - 10^{-11}$, and to search for all other rare K° decays involving electrons, photons and muons in the final state, e.g. $\pi^\circ e \mu$, $\gamma e \mu$, $\gamma e e$, $\gamma \mu \mu$.

The differences between this experiment and E780 which permit this greater sensitivity are an increase in beam on target to 5×10^{12}, an increased acceptance of the beam to 100 μstr, an increase in the length of the decay volume to 8m, and an approximate tripling of data acquisition time to 2000 hours. The K° rate is expected to be 6×10^8/ pulse.

A plan view of the apparatus is shown in Figure 3. The spectrometer system consists of two magnets, a 48D48 with a 30 inch gap and a 96D40 with a 40 inch gap, which allows two measurements of momentum of the final state particles. This double measurement is the method employed to reject the prominent background from $K^\circ \to \pi e \nu$.

A system of five packages of drift chambers, each containing two X and two Y views with a 1 cm cell size, forms the detectors for the spectrometer. Electron identification is determined by a multi-cell gas Cerenkov counter containing an argon-neon mixture, and an array of Pb-Glass detectors. The muons are identified with a range stack consisting of 340 modules, each with 1 1/2 inches of aluminium absorber and a drift chamber with 2 cm cell size. This highly instrumented range stack will also serve as the polarimeter for the $K^\circ \to \mu^+ \mu^-$ polarization measurement.

In order to minimize the number of neutron interactions, (there are $\sim 2 \times 10^{10}$ neutrons per machine pulse in the beam), the beam vacuum pipe is continued through the entire apparatus.

The increased sensitivity of this experiment requires high rate data acquisition capability, and a high level of intelligence in the trigger. Thus, much of the prototype work that has been done is in the area of fast, buffered TDC and ADC units. Along with this, proto-type drift chambers and scintillation counter have been constructed.

The beam is presently under construction and the group hopes for a "warm up" run in the spring of 1986. Their ideal goal is to be

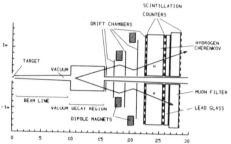

Figure 3. Plan view of the E791 apparatus.

capable of running for two weeks with a beam of 10^{12} and achieve a sensitivity of 10^{-10} in this first run.

3. $K^+ \rightarrow \pi^+ \nu \nu$

We turn now to another type of rare K decay, one which should occur in the standard model. The present limit on the branching ratio for $K^+ \rightarrow \pi^+ \nu \bar{\nu}$ is 1.4 x 10^{-7}, and the group performing this experiment, from BNL, Princeton, and TRIUMF,[4] propose to lower the limit to 2 x 10^{-10}. Using limits on the KM matrix elements derived from the experiment, Gillman and Hagelin[5] have predicted the rate of the decay as a function of the t quark mass. For a 45 GeV t quark their results range from 3×10^{-11} to 1.2×10^{-10} depending on the b quark lifetime. While this experiment cannot quite reach this level, the goal is to improve the existing limit by ~10^3. Thus the window to explore possible new physics outside of the standard model with 6 quarks is quite significant. That physics includes more generations of quarks, and/or neutrinos, some supersymmetric models, technicolor models, and the like. In addition the detector is capable of searching with comparable sensitivity for light, weakly interacting neutral particles such as axions and familons.

A valid $K^+ \rightarrow \pi^+ \nu \bar{\nu}$ event must have a correctly identified π^+ and no other interacting particles. To reject a radiative $K_{\mu 2}$ one must veto the γ or discriminate the muon from a pion; to reject a $K_{\pi 2}$ event one must veto the γ's from the π° decay; and so forth for the other potential backgrounds. The detector thus is designed to measure the π^+ momentum, kinetic energy, and range when stopping. To further insure that the decay product is a pion, it will also observe the $\pi \rightarrow \mu \rightarrow e$ decay chain. The detector also has maximum γ veto coverage over as broad a photon energy range as possible.

In Figure 4 we show a section view of the apparatus; the detector is cylindrically disposed about the beam line. The entire detector is housed in a solenoid magnet which provides a field of 1.0T with a cylindrical volume 2 meters long and 2.9 meters in diameter. The 750-800 MeV/c beam consisting of ~$10^6 K^+$ per machine pulse, accompanied by ~$5 \times 10^6 \pi^+$, is degraded before the target, with ~3×10^5 kaons stopping. The stopping target is a matrix of scintillating fibers, whose number will be between 500 and 1600, which are in a close packed hexagonal array with their length along the beam. The function of this live target is to observe the stopping K^+, to see that only one particle emerges from the decay, and to measure the energy loss of the decay product as it traverses the target.

Travelling radially out from the target, the charged decay product is momentum analyzed by a drift chamber to approximately 2% accuracy. Its energy and range is then measured in a range stack of 21 scintillation counters divided into 3 segments with drift chambers. The pions will come to rest in the range stack, and transient digitizers will record the time and amplitude evolution of pulses in the subsequent $\pi \to \mu \to e$ decay.

Photon veto coverage is divided into a barrel counter consisting of 15 radiation lengths of lead-scintillator, two arrays of BaF_2 crystals of 7.5 radiation lengths viewed by reduced pressure proportional chambers containing TMAE as the photosensitive agent, and two lead-scintillator detectors backing the BaF_2 to increase the depth in the end caps to 19 radiation length. The BaF_2 is employed because of its fast response in the beam region, and because of its greater sensitivity to low energy photons.

The momentum acceptance of the experiment is limited to the region above $K_{\pi 2}$. This is done because the momentum of the π^+ from this decay can be degraded due to nuclear interactions, and then the γ veto efficiency would have to be significantly greater than that presently anticipated. This spectrum cut, the solid angle acceptance,

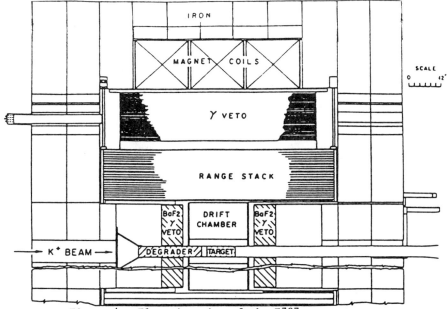

Figure 4.   Elevation view of the E787 apparatus.

502

pion absorption before reaching the end of its range, and various other small inefficiencies, yield an integrated detection efficiency of 1.5%.

It is anticipated that the experiment will take data for 2000 hours with 5×10^{12} protons per pulse on target. It will stop 7.5×10^{11} kaons and have an acceptance of 10^{10} decays.

At the writing of this paper the magnet is under construction, the drift chamber, BaF_2 detector, and scintillating fiber target have been successfully prototyped, and incident beam rates have been verified in test beam runs. It is anticipated that testing will continue in the spring run at the AGS and actual data acquisition will begin next year.

4. CONCLUSION AND ACKNOWLEDGMENTS

The program to study rare K decay physics is well under way. In about two to three years the results of these experiments will be known. Of course the future beyond that point depends on the outcome, but at the least the information gained from this effort will permit the limits to be pushed even further in the next generation of experiments.

Discussions with R. Cousins, L. Littenberg and M. Schmidt are greatfully acknowledged. This work was supported in part by the DOE under contract DE-AC02-76ER03075.

5. REFERENCES

1. AGS Proposal 777, "Search for the Rare Decay Mode $K^+ \to \pi^+ \mu^+ e^-$"; BNL, University of Washington, Yale University, SIN Collaboration; 1982 (unpublished).
2. AGS Proposal 780, "A Search for the Flavor Changing Neutral Currents $K^\circ \to \mu e$ and $K^\circ \to ee$"; BNL, Yale University Collaboration; 1983 (unpublished).
3. AGS Proposal 791, "Study of Very Rare K Decays"; UCLA, LANL, Pennsylvania, Stanford, Temple, William and Mary Collaboration; 1984 (unpublished).
4. AGS Proposal 787, "A Study of the Decay $K^+ \to \pi^+ \nu\bar{\nu}$", BNL, Princeton, TRIUMF Collaboration; 1984 (unpublished).
5. F.J. Gilman and J.S. Hagelin, Phys. Lett. 133B, 443 (1983).

III

QUANTUM CHROMODYNAMICS

Session Organizers: J. Dorfan and A.H. Mueller

THE STATUS OF PERTURBATIVE QCD

John C. Collins*

Institute for Advanced Study,
Princeton, NJ 08540.
U.S.A.

ABSTRACT

Progress in QCD in the past year in reviewed.

1. INTRODUCTION

Over the past year, work on perturbative QCD has shown much vitality. A big impetus has come from the studies that have been made for the SSC, at Snowmass and Oregon, and from the results coming in from the $Sp\bar{p}S$. There is an interesting statistic: At the conference concluding the workshop here, on Super-High Energy Physics[1], 40% of the talks were on perturbative QCD.

Since new accelerators are supposed to look for new physics, this appears paradoxical until one reminds oneself of some of the important properties of QCD. Suppose one produces a Higgs particle that decays to a $t\bar{t}$ pair, which ultimately decays to several light quarks. As these emerge from the scattering they radiate gluons readily, for the QCD coupling is not very small. The gluons themselves radiate more gluons, and as the process gets to longer distances, the effective coupling gets bigger. Eventually they reach the confinement scale, when the large number of partons then turns into a collection of jets of ordinary hadrons.

We all know how cleanly jets appear at the $Sp\bar{p}S$, and they obviously correspond to partons coming out of the the elementary processes inside a collision. But they are not nearly so clean when one wishes to examine them in detail. QCD effects persist to the highest energies in distorting signatures for new physics. As Gunion, Kunszt. and Soldate[2] recently showed, the most serious background to finding the Higgs particle is just ordinary QCD jet production.

The problems in making improved predictions for high energy scattering are all to do with soft partons (especially gluons). Much of the work now is on the small-x problem. which consists of studying the soft partons directly, rather than indirectly through their recoil effects on the hard partons. Not only are these problems of great intrinsic interest, but the answers are vitally needed by experimentalists. In solving the problems, we are going away from the places where one makes the easiest tests of QCD to where the cross-sections are biggest.

---

* Permanent address: Department of Physics, Illinois Institute of Technology, Chicago, Illinois 60616, U.S.A.

Specific areas in which there has been progress are as follows: At last we have a proof[3] of factorization for hadron-hadron collisions – for the Drell-Yan process etc. There are greatly improved techniques for the fast calculation of higher order graphs[4]. Perturbative Monte-Carlo calculations for hadron-hadron scattering have come considerably closer to real QCD[5],[6]. We now understand the applicability of perturbative methods to heavy quark production[7].

Finally, there has been progress in understanding the small-x problem[8],[9], and this subject is making a take-over bid for Regge theory.

2. PROOF OF FACTORIZATION

Proofs of the classical factorization theorem for hadron-hadron collisions have been given[3] by Bodwin, and by Soper, Sterman and myself. These lay to rest the controversy that was started by Bodwin, Brodsky and Lepage[10] as to whether factorization is true. Almost all QCD calculations of cross-sections for high-energy hadron-hadron collisions use factorization, so without the theorem, the predictive power of the theory is lost. The original proofs[11] were known to be incomplete shortly after their publication. When Sterman and myself[12] completed the proof for e^+e^- annihilation, we explicitly did not cover the hadron-hadron case (i.e., the Drell-Yan process etc).

2.1 Factorization

The factorization theorem that we discuss here asserts that the cross-section for a process like Drell-Yan is given as a convolution of a hard scattering cross-section, σ_{hard}, with parton distribution functions, $f(x, Q)$.

The original proofs[11] did not properly treat the cancellation of the effects of soft gluons. At first sight, one appears to have a particular case of the Kinoshita-Lee-Nauenberg and Bloch-Nordsieck cancellations, as has been stated in the literature. However, the simple $O(\alpha_s)$ calculations with incoming partons are misleading in this respect, for soft gluons can be emitted off internal as well as external lines. A simple graph demonstrating this fact is given in Fig. 1. Cancellations of the Bloch-Nordsieck type, such as we are all used to in QED, involve only emission off external lines.

2.2 The proof

The physics of the proof is roughly as follows. First, we know that final-state interactions cancel. The basic ideas come from Ref. 13], and embody the fact that we sum over the possible happenings in the final-state.

Then comes the fact that soft gluons cannot resolve the details of jets. This permits a coherent sum over the emission of soft gluons from different lines in a jet. At the level of Feynman graphs, this is seen as the use of a "soft approximation", after which a certain kind of Ward identity can be used.

After using the Ward identities, we find that the soft gluons factorize: Their emission is effectively off the holes left in the incoming hadrons by the partons

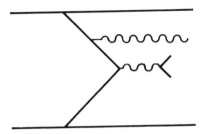

Fig. 1. Emission of soft gluon from internal line.

that went into the hard scattering. The proof of cancellation of soft-gluon effects for jet production in e^+e^- annihilation now applies[12],[14].

The difficulty raised by Bodwin, Brodsky and Lepage is that the soft approximation is not valid in certain non-negligible momentum regions. Because both initial- and final-state interactions are present, one cannot avoid this region by the trivial contour deformation one uses in e^+e^- annihilation[12]. One must first invoke the cancellation of final-state interactions. Unfortunately, the treatment of the final-state cancellations is most easily made in time-ordered perturbation theory, while the Ward identities are only conveniently treated in Feynman perturbation theory. What makes a proof so hard to construct is to combine these ideas in the right order to give both a robust and correct proof, which is what we believe we now have.

2.3 Implications

The techniques used in constructing a full proof of factorization are of general applicability to soft-gluon physics, especially to cases where the effects of the soft gluons do not cancel. The methods apply to all orders, and not merely to the leading logarithm approximation. Already there have been important applications to transverse momentum distributions. The techniques are also crucial in treating the small-x problem.

The proofs are now given in Feynman gauge, rather than in a non-covariant gauge.

3. CALCULATING HIGH ORDER GRAPHS

There were great improvements in techniques to compute high order tree graphs in QCD[2]. The basic ideas have been known for a little while.

First, instead of calculating cut graphs for cross-sections, which is what one normally does, one should calculate amplitudes first. This is because there are many fewer terms to handle. Then one chooses bases for the gluon polarizations cunningly, thus eliminating most of the terms in the couplings. Finally one uses a helicity basis for the fermions; combined with the choice of gluon-polarization basis, this avoids the γ-matrix sprawl that happens so readily when using the traditional methods. The progressive improvements are listed in Ref. 15].

There is hope of extending the methods to treat graphs with massive quarks and rather importantly with loops. To do this would be very valuable, for otherwise we will be deluged with calculations of multi-jet production and the like, from tree-graphs, but with the calculations being incomplete without the virtual corrections. Almost all results for new physics involve heavy fields, of course.

4. MONTE-CARLO PROGRAMS FOR HADRON-HADRON COLLISIONS

There are two methods of making perturbative QCD calculations – the "analytic" methods and the Monte-Carlo methods.

Analytic calculations result in closed formulae for cross-sections, and the numerical work is confined to solving simple differential equations and performing low-dimensional integrals. The basic quantities in this approach are non-logarithmic terms obtained from Feynman graphs; they are capable of systematic improvement by calculating higher orders. But the cross-sections are typically given only for certain inclusive processes, and there is no information on the rest of the event.

The Monte-Carlo approach turns the Feynman graphs, suitably approximated, into a probabilistic algorithm, which can then be used to generate whole events. This approach is therefore very helpful for experimentalists, because the results are easy to compare with their data and fit in well with Monte-Carlo simulations of the apparatus. However, the approximations have often been not very good. Also, it has been difficult to incorporate the full results of perturbative QCD into the algorithms, especially the new results on soft gluons.

The workshops at Snowmass last summer and at Oregon this spring have stimulated much work on improvement in the Monte-Carlos. The aim is to include much more of the known soft gluon physics, and to do this correctly. Initial-state gluon radiation was incorporated in the Monte-Carlos for hadron-hadron scattering. Paige[6] summarized this work in his talk here.

5. PRODUCTION OF HEAVY QUARKS ETC

Another issue that was more-or-less resolved was the applicability of perturbative methods to the production of heavy strongly interacting particles (anything from the charmed quark to possible super-symmetric partners of the usual quarks and gluons). Unfortunately, no one has written down a proof of factorization for these processes. Furthermore, the predictions for total charm cross-sections are notoriously below the quoted experimental values[16]. (At the ISR one talks of experimental cross-sections of between 100 μb and 1 mb, but of theoretical predictions of tens of μb.)

Therefore, many people have proposed that the standard factorization fails for heavy quark production, and have suggested alternative mechanisms for enhancing the charm cross-section. (Examples: flavor excitation, intrinsic charm, pre-binding distortion à la Bethe-Heitler, diffractive excitation.) These proposals have considerable impact on the design of experiments for looking for new heavy flavors. The cross-sections are much larger than from the conventional gluon-fusion mechanism and are typically large contributions in the forward direction.

At the Oregon workshop these issues were investigated. Ultimately, agreement was reached that the standard perturbative mechanism gives the correct QCD prediction. This conclusion[7] was based on an analysis of low-order Feynman graphs and of the coordinate-space physics in the light of what is necessary to make a proof of factorization[3].

Some of the proposed alternative mechanisms are higher twist – like intrinsic charm or pre-binding distortion. These mechanisms might be important for charm production, but not for top production. Others of the proposed mechanisms are actually included in the gluon fusion mechanism, and sometimes, as in the case of flavor excitation, omit relevant graphs.

If there is a substantial fraction of a heavy quark cross-section that is diffractive, as has been suggested from the ISR data[16], then it is a part of the gluon-fusion contribution. This needs further investigation. At the time of writing, the subject of diffractive hard scattering[17] appeared poised for take-off.

The problem with the charm data apparently still remained. Possible explanations are the following: 1. QCD is wrong. 2. The data are wrong. 3. The higher twist corrections are unusually large. 4. The charmed hadron distributions do not follow the charmed quark distributions very well. 5. Higher order corrections bring in larger K-factor than we are used to.

At this meeting, data[18] was presented by the LEBC collaboration. Their cross-section is in agreement with the gluon-fusion predictions, but with a significant excess in the forward region. This is modeled correctly by a Lund string Monte-Carlo calculation for the final-state interactions. The excess forward production is a higher twist effect, according to our new QCD results[7]. But for partons of a couple of GeV, it is not unreasonable to expect such distortions in the final-state interactions.

There are also indications[19],[20] that the higher-order corrections to the gluon-fusion process are unusually large. A full calculation of the $O(\alpha_s^3)$ corrections to heavy quark production would be especially useful. The tree graphs are known, but the virtual corrections are not.

The conclusion then is that the standard perturbative methods for heavy flavor production are reliable, and can be used with confidence to predict cross-sections for new heavy flavors.

6. SMALL X

I will briefly discuss here the progress that has been made in the study of hard collisions whose typical transverse momenta, Q, are much less than the overall center-of-mass energy, $\sqrt{s}$. The quantity x is the ratio $Q/\sqrt{s}$. The motivations for this study I explained in the introduction. Mueller gave a fuller discussion in his talk[8] at this conference, as I did in the proceedings of the workshop conference[9].

The fundamental theoretical issue comes from the logarithms of x that occur in higher order corrections, and from the question of whether factorization is actually true. The logarithms must be resummed in some fashion, for otherwise they ruin the convergence of the perturbation series. The Russian work summarized in the paper of Gribov, Levin and Ryskin[21] led the way. Factorization appears to hold, and there are ways of controlling the logarithms. The techniques that are used are those of the soft gluon type, but mostly so far in leading logarithm approximation. Considerable resummations are need to get final results.

The main phenomenological issues stem from the fact that the Altarelli-Parisi evolution generates a large number of gluons at small x and even not very large Q. The result is that the cross-section for making jets with transverse momenta of a few GeV is of the order of the total hadronic cross-section, at current collider energies. Thus there is a change in the character of minimum bias events: they include perturbative hard scatterings typically. Moreover multiple hard scattering ("parallel") may be common[22], and will have its effect on violations of KNO scaling and the like.

More results can be expected in this field. There is an obvious overlap with Regge theory, which will be explored. Further exploration of the region where partons become overcrowded and recombine within a hadron is continuing; this region sets the ultimate limit that we have at present for the applicability of factorization. This and the study of multiple hard scattering will bring in the need to understand the two-parton correlation functions better. We hope to gain more control over the approximations used. More work on the phenomenology of multiple hard scattering is needed.

7. CONCLUSIONS

1. Perturbative QCD is on solid ground, now that we have a proof of factorization.
2. The range of its applicability is growing. The small-x work is most significant here.
3. The particularly interesting problems that need more work are:
   a. The small-x problem.
   b. More QCD input to the Monte-Carlos.
   c. Understanding the sizes of higher-order graphs. I did not discuss this, but there are many next-to-lowest order corrections that are comparable to the lowest order graphs. Frequently we have plausible excuses as to why this does not wreck the applicability of the perturbation expansion, but it would be nice to work the excuses into a systematic treatment. I am speaking here of the large constant terms, rather than of the large logarithms, which are often well understood.

ACKNOWLEDGMENTS

This work was supported in part by the U.S. Department of Energy under contracts DE-AC02-80ER-10712 and DE-AC02-76ER-02220. I would like to thank the organizers for running such a superb conference.

REFERENCES

1] "Proceedings of the Workshop on Super-High Energy Physics, Eugene OR", Gunion, J.F. and Soper, D.E. (eds.), (World Scientific, Singapore, to appear).

2] Gunion, J.F., Kunszt, Z., and Soldate, M., SLAC-PUB-3709.

3] Collins, J.C., Soper, D.E., and Sterman, G., Nucl. Phys. B261, 104 (1985); G. Bodwin, Phys. Rev D31, 2616 (1985).

4] Gunion, J.F. and Kunszt, Z., Oregon preprint OITS-196.

5] Gottschalk, T., Caltech preprint CALT-68-1241. and in Ref. 1; Sjostrand, T., Fermilab preprint FERMILAB-PUB-85/23-T.

6] Paige, F., these proceedings.

7] Collins, J.C., Soper, D.E., and Sterman, G., Oregon preprint OITS 292, Nucl. Phys. B (to appear).

8] Mueller, A.H., these proceedings; Mueller, A.H. and Qiu, J., Columbia preprint CU-TP-322.

9] Collins, J.C., in Ref. 1.

10] Bodwin, G., Brodsky, S.J., and Lepage, G.P., Phys. Rev. Lett. 47, 1799 (1981).

11] Amati, D., Petronzio, R., and Veneziano, G., Nucl. Phys. B146, 29 (1978); Libby, S.B., and Sterman, G., Phys. Rev. D18, 3252 (1978); Mueller, A.H., Phys. Rev. D18, 3705 (1978); Gupta, S. and Mueller, A.H., Phys. Rev. D20, 118 (1979); Ellis, R.K., Georgi, H., Machacek, M., Politzer, H.D., and Ross, G.G., Nucl. Phys. B152, 285 (1979).

12] Collins, J.C., and Sterman, G., Nucl. Phys. B185, 172 (1981).

13] DeTar, C., Ellis, S.D., and Landshoff, P.V., Nucl. Phys. B87, 176 (1975), and Cardy, J.L., and Winbow, G., Phys. Lett. 52B, 95 (1974).

14] Sterman, G., Phys. Rev. D17, 2773 & 2789 (1977).

15] Berends, F.A., de Causmaeker, P., Gastmans, R., and Wu T.T., Nucl. Phys. B206, 61 (1982); Xu, Z., Zhang, D.H., and Chang, Z., Tsinghu Univ. preprint TUTP-84/3/84; Farrar, G. and Neri, F., Rutgers preprint RU-83-20; Kleiss, R., Nucl. Phys. B241, 61 (1983); Gunion, J. and Kunszt, Z., Oregon preprint OITS-296; Ref. 2].

16] Kernan, A. and VanDalen, G., Phys. Rep. 106, 297 (1985).

17] Ingelman, G. and Schlein, P.E., Phys. Lett. 152B, 256 (1985).

18] LEBC Collaboration, Aguilar-Benitez, M. et al., CERN preprints CERN/EP/ 85-103 and 118.

19] Kunszt, Z. and Pietarinen, E., Zeit. Phys. C2, 355 (1979).

20] Halzen, F. and Hoyer, P., Phys. Lett. 154B, 324 (1985).

21] Gribov, L.V., Levin, E.M., and Ryskin, M.G., Phys. Rep. 100, 1 (1983).

22] Ametller, L., Paver, N., and Treleani, D., Trieste preprint IC/85/118; Humpert, B. and Odorico, R., Phys. Lett. 154B, 211 (1985); Sjostrand, T., Fermilab preprint FERMILAB-PUB-85/119-T.

TESTING QCD IN DEEP INELASTIC LEPTON SCATTERING

John Carr

University of Colorado

Campus Box 390

Boulder, CO   80309

ABSTRACT

We discuss tests of QCD possible with structure
function data available from charged lepton and
neutrino deep inelastic scattering experiments.
Recent data on the "EMC effect" from the BCDMS
collaboration shows that there is no Q^2 dependence
in the ratio F_2^N/F_2^D and so restores confidence that
QCD fits to heavy target data are valid. Measurements of the QCD scaling parameter Λ extracted
from all high Q^2 experiments are shown to be consistent with a value $\Lambda_{LO} = 200 \pm 100$ MeV. Fits
of EMC and SLAC hydrogen F_2 with higher twist
terms included, demonstrate that QCD is consistent
with the data over a wide Q^2 range. An analysis
of the CDHS collaboration shows that QCD fits the
data better than any other gauge theory.

A compilation of high statistics deep inelastic scattering
experiments is given in Table 1. The event totals are approximate and
for the CDHS and BCDMS experiments they show the high statistics

available in structure functions which as yet have only been shown at conferences.

In muon or electron scattering the structure function F_2 is measured from the differential cross-section according to the following relation:

$$\frac{d^2\sigma}{dxdy} \approx \frac{8\alpha^2\pi Em}{Q^4} \left[1 - y + \frac{y^2}{2} \frac{1}{1 + R(x,Q^2)}\right] F_2(x,Q^2)$$

where

$$R(x,Q^2) = \sigma_L/\sigma_T = \frac{F_2}{2xF_1}\left(1 - \frac{4m^2x^2}{Q^2}\right) - 1.$$

For the high energy data, $R(x,Q^2)$ cannot be extracted separately for every data point and F_2 is quoted for a fixed value of R, generally R = 0, or for the variation of $R(x,Q^2)$ expected from QCD. For neutrino data the cross section contains a further structure function F_3 which appears in the cross section with a different sign for neutrino and antineutrino scattering:

$$\frac{d^2\sigma^{\nu(\bar{\nu})}}{dxdy} \approx \frac{G^2 Em}{\pi(1 + Q^2/M_W^2)} \left\{ \left[1 - y + \frac{y^2}{2} \frac{1}{1 + R(x,Q^2)}\right] F_2(x,Q^2) \right.$$

$$\left. \pm (y - \frac{y^2}{2})xF_3(x,Q^2) \right\}$$

In the Quark-Parton Model the neutrino structure function F_2^ν is related to the charged lepton structure function F_2^μ by

$$F_2^\nu = \frac{18}{5} F_2^\mu + \frac{3}{5} x \left[(s + \bar{s}) - (c + \bar{c})\right]$$

Figure 1 shows a comparison of structure functions reproduced from the 1984 Particle Data Group Review. The normalization factors shown are necessary to obtain good agreement between the data sets. New data from the CDHS collaboration shown at the Bari Conference[1] show much better agreement with the CCFRR data, without any artificial normalization shifts.

The discovery of the "EMC effect"[4], showing an x dependence in ratio F_2^{Fe}/F_2^D, casts doubts on the validity of QCD fits to structure functions measured on heavy nuclei targets. From the early EMC and SLAC data it was not possible to clearly exclude a Q^2 dependence in the ratio F_2^{Fe}/F_2^D, so that measurements of the QCD scaling parameter Λ on iron targets might have been confused by unexpected nuclear physics effects. However, the new high statistics results from the BCDMS group[6] shown in Figure 2, exclude a large Q^2 dependence in the ratio F_2^N/F_2^D and so restore confidence in Λ measured on iron targets. The existence of the EMC effect does preclude fits between the high statistics SLAC deuterium data at low Q^2 and the high statistics iron data at high Q^2 from EMC and other experiments, which might otherwise provide precise tests of QCD over a wide Q^2 range. Only hydrogen data may be used for this, as is discussed later.

In Figure 3 we compare measurements of Λ, in leading order QCD, from the high Q^2 experiments. An attempt has been made to simplify the detailed analyses made by each experimental collaboration and choose from each set of data a single value of Λ which represents the best estimate. The error bars plotted are statistics plus systematics added quadratically. From this it can be seen that a value of $\Lambda_{LO} = 200 \pm 100$ MeV is consistant with all data sets. Values of the next-to-leading order parameter $\Lambda_{\overline{MS}}$ for similar fits to the same data do not show a completely consistent trend. For some of the singlet fits to neutrino data[3] the difference $\Lambda_{\overline{MS}} - \Lambda_{LO}$ is about -20 MeV while for the non-singlet fits[7] to EMC data the difference is more consistently around $+20$ MeV.

The EMC group has performed a fit[7] to their hydrogen data and to hydrogen data from the SLAC-MIT group in a range of Q^2 from $3 - 200$ GeV^2. This fit contains higher twist terms parameterized as:

$$F_2 = F_2^{LT} \left(1 + \frac{\mu^2 x^\alpha}{(1 - x)Q^2} \right)$$

The resulting fitted value of $\Lambda_{\overline{MS}} = 115 \pm \begin{smallmatrix} 70 \\ 55 \end{smallmatrix}$ MeV is completely consistent with the value $\Lambda_{\overline{MS}} = 105 \pm \begin{smallmatrix} 55 \\ 45 \end{smallmatrix} \pm \begin{smallmatrix} 85 \\ 45 \end{smallmatrix}$ MeV obtained from fits to the EMC data alone. Figure 4 shows this fit, indicated by the solid curve, compared with the data. The fit allows for a

normalization shift between the data sets and a +7.5% shift of the
SLAC-MIT data results. The dotted line is the fit to the EMC data
alone and the difference between the dotted and solid curves indicates
the magnitude of the higher twist terms in the SLAC data.

A detailed analysis has been performed by CDHS[1] to investigate if
other gauge theories fit the data as well as QCD. Their conclusions
are that QCD clearly fits the data better than the theories they
considered with abelian vector gluons and non-abelian scalar gluons.
This is demonstrated in Figure 5 by comparing the magnitude of the
scaling violations, $dF_2/d \ln Q^2$ for fixed ranges of x, measured in the
data with the best fit predictions from the three theories
considered. This same analysis tried to see evidence for the triple
gluon vertex terms unique to QCD by removing these terms from the
fitted functions. No significant difference in χ^2 of the fits with or
without the triple gluon vertex was observed. Attempts have also been
made, by CDHS and other groups, to see evidence for the asymptotic
freedom variation of the strong coupling constant, α_s, with Q^2 which
is predicted by QCD. However, the statistics on existing data are not
good enough to make a reasonable test.

The new high statistics data coming from CDHS will be matched
with similar new data from CCFRR and will enable much more precise
tests of QCD to be made soon. As an example of the power of the new
data, Figure 6 shows measurements of R from this CDHS data, together
with old CHARM points, compared with QCD predictions. Analysis of
this data will allow detailed tests of QCD predictions of the x
dependence of R rather than only tests of the purely Q^2 dependence
predictions of structure functions possible now.

In conclusion, QCD consistantly fits all existing deep inelastic
data much better than any other gauge theory. The unique features of
QCD associated with its non-abelian nature remain untested as yet, but
new data becoming available will provide higher statistics and should
allow different and more detailed tests of QCD to be performed in the
near future.

REFERENCES

1. CDHS   Z. Phys. C. 17, 283 (1983), Z. Phys C 13, 199 (1982),
             P. Buchholz, Proc. EPS Conference, Bari (1985).

2. CHARM   Phys. Lett. 123B, 269 (1983).

3. CCFRR   Z. Phys. C 26, 1 (1984).

4. BEBC   Phys. Lett. 141B, 133 (1984).

5. FNAL 15'   Fermilab-Pub-85/101-E (1985).

6. BCDMS   Phys. Lett. 104B, 403 (1981), J. Feltesse, Proc. EPS
             Conference, Bari (1985).

7. EMC   Phys. Lett. 105B, 322 (1981), Phys. Lett. 123B, 275
             (1983); Nucl. Phys. B259, 189 (1985).

8. BFP   Phys. Rev. Lett. 51, 1826 (1983), LBL-17108.

9. CHIO   Phys. Rev. D20, 2645 (1979).

10. SLAC   Phys. Rev. D20, 1471 (1979); Phys. Rev. Lett. 52, 727
             (1984).

EXPT	LAB	TARGET	BEAM TYPE	BEAM ENERGY (GeV)	#EVENTS
CDHS	CERN	Fe H_2	ν	30-285	$1.5 \ 10^6$ $9 \ 10^3$
CHARM	CERN	$CaCO_3$	ν	10-160	$2 \ 10^5$
CCFRR	FNAL	Fe	ν	30-250	$2 \ 10^5$
BEBC	CERN	D_2, H + Ne	ν	~10-160	$4 \ 10^4$
FNAL, 15'	FNAL	D_2, N + Ne	ν	~10-150	$2 \ 10^4$
BCDMS	CERN	C, Fe, N_2, H_2, D_2	μ	120-280	$8 \ 10^6$
EMC	CERN	Fe H_2, D_2	μ	120-280	$8 \ 10^5$ $4 \ 10^5$
BFP	FNAL	Fe	μ	93-215	$9 \ 10^5$
CHIO	FNAL	H_2, D_2	μ	97-219	$4 \ 10^4$
SLAC	SLAC	H_2, D_2, Heavy Nuclei	e	< 20	$> 10^8$

Table 1: High statistics deep inelastic scattering experiments.

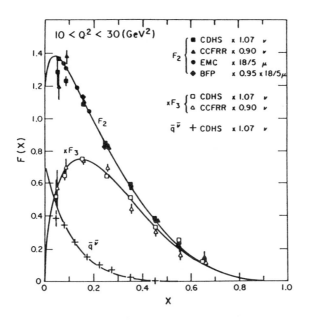

Figure 1:   Comparison of structure functions.

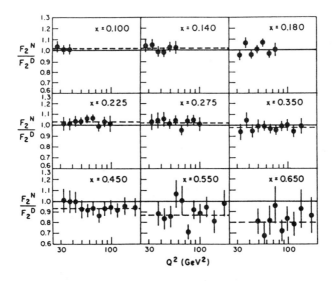

Figure 2:   BCDMS data measuring Q^2 dependence of "EMC effect"

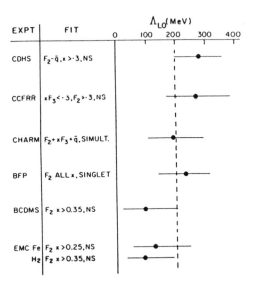

Figure 3: Comparison of Λ_{LO} from different experiments.

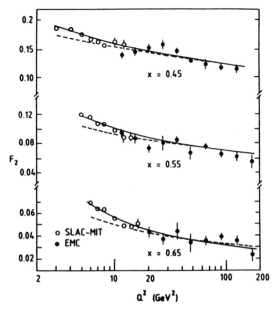

Figure 4: QCD + Higher Twist fits to EMC and SLAC-MIT data.

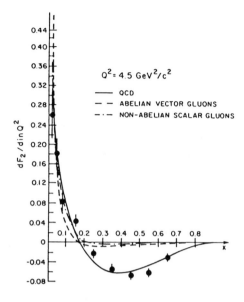

Figure 5: CDHS analysis comparing QCD with other gauge theories.

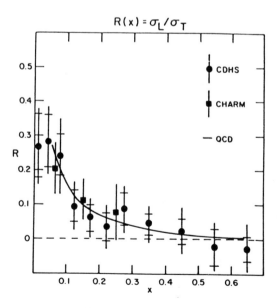

Figure 6: New measurements of R

QCD AT THE CERN p̄p COLLIDER

W.G. Scott (UA1 Collaboration)

Rutherford Appleton Laboratory, UK and
CERN, Geneva, Switzerland

ABSTRACT

This paper reviews briefly some of what we have learned about
QCD related topics from work at the CERN SPS p̄p Collider.

1. INTRODUCTION

It is four years since the first operation of the CERN SPS p̄p Collider[1]. Since that time
the two large counter experiments UA1 and UA2[2,3], installed at the Collider have been
accumulating data in parallel throughout a series of physics runs. The data discussed in this
report come mostly from the 1983 and 1984 runs. In round numbers the integrated luminosity
accumulated in the 1983 run was $\simeq 100$ nb^{-1} per experiment at an energy of $\simeq 273$ GeV per
beam (c.m.s. energy $\sqrt{s} = 546$ GeV) and in the 1984 run $\simeq 300$ nb^{-1} per experiment at an
energy of $\simeq 315$ GeV per beam ($\sqrt{s} = 630$ GeV). In this review emphasis is given to the
presentation of data from the UA1 experiment.

2. INCLUSIVE JET CROSS-SECTION

The inclusive jet cross-section corresponding to the reaction

$$p\bar{p} \rightarrow jet + X \tag{1}$$

is shown in fig. 1a, based on the UA1 data for two different p̄p c.m.s. energies ($\sqrt{s} =$
546 GeV, $\sqrt{s} = 630$ GeV). In UA1 a jet is defined by a jet algorithm[4] which combines
calorimeter hits within a cone specified by a radius of 1 unit in η, ϕ space around the highest

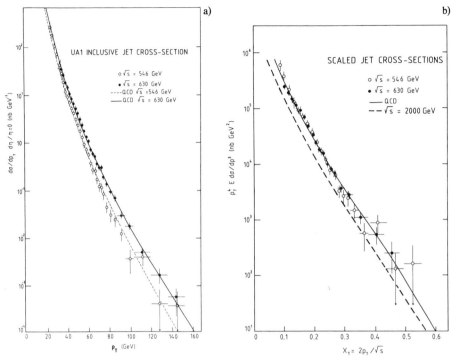

Fig. 1 a) The inclusive jet cross-section; b) the scaled jet cross-section measured in UA1.

transverse energy hits (η is the pseudorapidity and ϕ is the azimuthal angle in radians). The energies and momenta of the jets are computed by taking the scalar and vector sums respectively, over the associated calorimeter hits. The quantity actually plotted is $\langle d^2\sigma/dp_Td\eta\rangle$ averaged over the pseudorapidity interval $|\eta| < 0.7$. Very similar data has been published previously by the UA2 Collaboration[5].

The measured p_T dependence of these cross-sections is clearly in very good agreement with the QCD predictions. Note however that the QCD predictions themselves are incomplete to the extent that the higher order corrections for these processes have not yet been calculated[6]. We can expect that the theoretical predictions will be modified by a K-factor affecting the absolute normalization and, probably to a lesser extent, the p_T-dependence. The usefulness of these measurements as precise QCD tests is probably ultimately limited by the systematic uncertainty on the jet energy scale. The present systematic error on the jet energy scale, believed to be $\simeq \pm 10\%$, is sufficient to make possible higher-order effects essentially unobservable.

The increase in the cross-section between the lower and the higher c.m.s. energies (a factor $\simeq 2$) is comparatively well measured. Figure 1b shows the dimensionless quantity $p_T^4 E\, d^3\sigma/dp^3$ plotted versus $x_T = 2p_T/\sqrt{s}$ for the two different beam energies. On this plot the two sets of data overlap, demonstrating that the observed increase in the cross-section with

522

c.m.s. energy is entirely consistent with perfect scaling. A much larger lever-arm in energy, e.g. $\sqrt{s}$ = 2000 GeV (broken curve) would be needed to be sensitive to non-scaling QCD effects although such effects have been observed previously in ISR experiments[7].

The UA2 collaboration[8] have measured the inclusive (isolated) direct photon cross-section corresponding to the reaction

$$p\bar{p} \rightarrow \gamma X \qquad (2)$$

at $\sqrt{s}$ = 546 GeV and $\sqrt{s}$ = 630 GeV, using a $1.5X_0$ tungsten converter to separate the π^0 and γ contributions statistically. The (preliminary) results for $\sqrt{s}$ = 630 GeV are shown in fig. 2. The data are fully consistent with QCD predictions including the effects of higher order corrections which have been completely calculated for this process[9]. Naturally the systematics on the energy scale for photons are very good ($\simeq \pm 1\%$) and this type of measurement may be regarded as a comparatively precise QCD test. The main disadvantage of direct photons relative to the jets is the very much smaller cross-section which for present luminosities effectively limits the p_T-range over which the theory can be tested.

Before leaving the subject of inclusive cross-sections it is perhaps worth mentionning that measurements of the inclusive single particle cross-sections $\pi^0 \pi^\pm$, etc. may also turn out to be useful QCD tests. Charged pion momenta for example can generally be measured rather accurately in a magnetic spectrometer as in UA1. Furthermore the effects of higher order corrections on the single particle cross-sections seem to be better understood[6] theoretically than is the case for the jets.

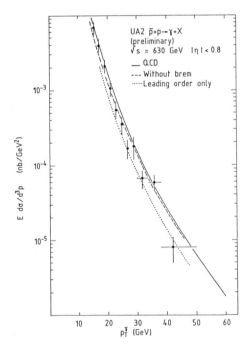

Fig. 2 The direct (isolated) photon cross-section measured by the UA2 Collaboration.

3. TWO-JET CROSS-SECTIONS

It is by now well known that the two-jet cross-section in hadron–hadron collisions to a very good approximation may be expressed in terms of a universal[10] parton–parton cross-section $d\sigma/d \cos \theta \simeq \alpha_s^2/\hat{s} (1 - \cos \theta)^{-2}$ (θ is the c.m.s. scattering angle) multiplied by a product of effective structure functions $F(x_1)/x_1 \cdot F(x_2)/x_2$ where

$$F(x) = G(x) + 4/9 [Q(x) + \bar{Q}(x)] \qquad (3)$$

In eq. (3), G(x), Q(x) and $\bar{Q}(x)$ represent the appropriate momentum weighted parton densities in the proton.

Both UA1 and UA2[11,12] have analysed two-jet data in this context, extracting the effective structure function with results which are largely consistent with each other and with QCD expectations (fig. 3a). Although the conceptual importance of the effective structure function cannot be denied, in practice the usefulness of this measurement as a precise QCD test is limited by the systematic errors on the jet energy scale, as in the case of the inclusive cross-section. For this reason this type of analysis will probably not be persued very much further at the CERN collider. We note only that the data rather naturally suggest an exponential form for the effective structure function and may be parametrized by[12]

$$F(x) = 6.2 \, e^{-8.3x} \qquad (4)$$

over most of the range in x.

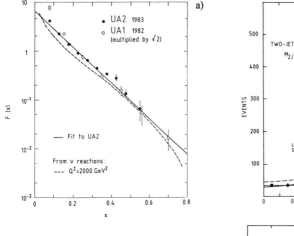

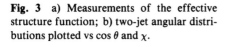

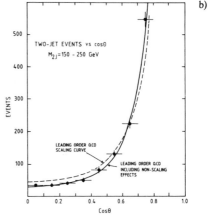

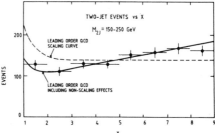

Fig. 3 a) Measurements of the effective structure function; b) two-jet angular distributions plotted vs $\cos \theta$ and χ.

524

Of more practical importance is the angular distribution for the jet pairs shown in fig. 3b as a function of $\cos\theta$ and as a function of $x = (1 + \cos\theta)/(1 - \cos\theta)$. The data are plotted for two-jet masses $m_{2J} = 150$–250 GeV and are based on the UA1 data from the 1983 running[13]. Note that the shape of the angular distribution can be measured independent of the systematic error on the jet energy scale and that remaining systematic errors, coming for example from uncertainties in the energy response as a function of angle, are believed to be at the level of a few percent only. In fig. 3b the broken curves represent the scaling predictions for the angular distributions while the solid curves incorporate various non-scaling effects, i.e. variation of the effective structure function and α_s with Q^2. The data show clear evidence for non-scaling effects in the angular distribution and we can expect that a quantitative analysis including the effects of higher-order corrections could be refined into a sensitive QCD test. It is interesting to remark that non-scaling effects which one might reasonably have assumed to be rather unimportant at these energies turn out to be large and measurable. This is to be contrasted with the case of the leading order differences between the various subprocesses $gg \rightarrow gg$, $q\bar{q} \rightarrow q\bar{q}$, etc. which seemed at one time to be an all important complication in hadron–hadron physics and which turn out to be almost completely unobservable.

Preliminary UA1 data[14] on the two-jet angular distribution at the highest subprocess c.m.s. energies ($m_{2J} = 250$–300 GeV) based on the 1984 running are shown in fig. 4. The data are consistent with the QCD prediction represented by the solid curve. The broken curves show the effect of contact interactions which are expected to be present if the partons (in this case quarks) are themselves composite. Contact interactions lead to more spherically symmetric angular distributions and hence to an excess of events in the region $\theta \simeq 90°$, $\chi \simeq 1$. No evidence for contact interactions is found in the data and a (preliminary) limit on the mass scale associated with the contact interaction is obtained $\Lambda_{contact} \gtrsim 400$ GeV. This result is representative of the potential sensitivity of the high mass two-jet angular distribution to new phenomena which typically populate preferentially the $90°$ region relative to the QCD distribution, e.g. new resonances $W' \rightarrow$ 2-jets, $Z' \rightarrow$ 2-jets, etc. Very naively formulae of ref. 6 would suggest that the existence of extra space-dimensions would have a similar effect

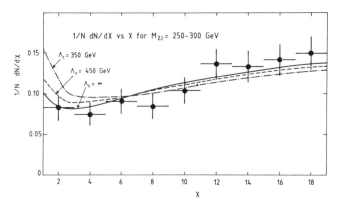

Fig. 4 The high mass two-jet angular distribution from UA1.

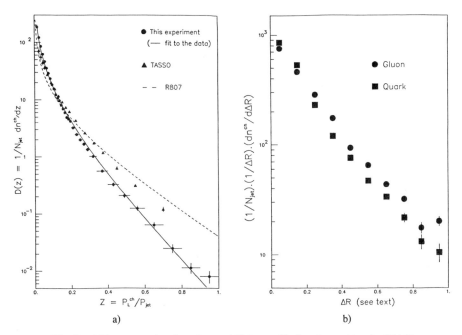

Fig. 5 a) Fragmentation function and b) jet profile for charged tracks (UA1).

on the angular distribution at sufficiently high-energy, although clearly the limits that could be obtained in practice would hardly be relevant to superstring theories!

Finally we show the measured fragmentation function for UA1 [15] jets in two-jet events compared to lower energy e^+e^- [16] and ISR (pp) [17] data (fig. 5a). The fragmentation function has large systematic errors due to the uncertainties in the jet energy scale which enters in the determination of z. Nonetheless the measured fragmentation function seems to be significantly softer than the fragmentation functions measured at PETRA and at the ISR. This is due mainly to the effects of the scaling deviations relative to the lower energy data rather than to intrinsic quark/gluon differences at fixed Q^2 which seem to be rather small. It is interesting to note that the empirical parametrization (solid curve) which fits the UA1 data very well

$$zD(z) = 3.4 \, e^{-7.0z} \tag{5}$$

mirrors closely the empirical fit to the effective structure function discussed above.

To an extent quark and gluon jets have been separated statistically in the UA1 experiment, exploiting the difference in shape of the quark and gluon structure functions in the proton. The most significant difference between quark and gluon jets appears to be in the distribution of the charged tracks around the jet axis as a function of the separation ΔR in η, ϕ space (fig. 5b). On the basis of this data one may conclude that the gluon jets have a tendency to be broader than quark jets consistent, at least qualitatively, with QCD expectations.

4. THREE-JET CROSS-SECTIONS

The UA1 three-jet analysis based on the 1983 data has been published previously[13]. That analysis is outlined briefly here and new (preliminary) data based on the 1984 running are overlayed on the published data where appropriate.

The UA1 three-jet, analysis exploits directly the (approximate) scale invariance property of QCD. At fixed subprocess c.m.s. energy the three-jet and two-jet cross-sections are expressed in terms of energy independent dimensionless variables, specifically angles and energy ratios. Under these circumstances the total two-jet and three-jet cross-sections, at the parton level, have identical energy dependence

$$\sigma_{2J} = C_{2J}\alpha_s^2/\hat{s} \qquad \sigma_{3J} = C_{3J}\alpha_s^3/\hat{s} \qquad (6)$$

where C_{2J} and C_{3J} are numerical coefficients, calculable from the theory, which depend on the exact cuts and on the subprocess considered. Although C_{2J} and C_{3J} themselves depend strongly on the subprocess, it turns out that the ratio C_{3J}/C_{2J} is remarkably similar for all incoming parton combinations. Thus the three-jet to two-jet ratio is directly related to α_s in the leading order:

$$\sigma_{3J}/\sigma_{2J} = \langle C_{3J}/C_{2J}\rangle \alpha_s \qquad (7)$$

and rather independent of the parton densities which largely cancel in the ratio.

At fixed subprocess c.m.s. energy the three-jet matrix element depends on four independent variables. The three-jet Dalitz plot x_3 vs x_4 is shown in fig. 6a and the three-jet angular distributions ϕ vs $\cos\theta_3$ are shown in fig. 6b. In fig. 6 the closed circles correspond to

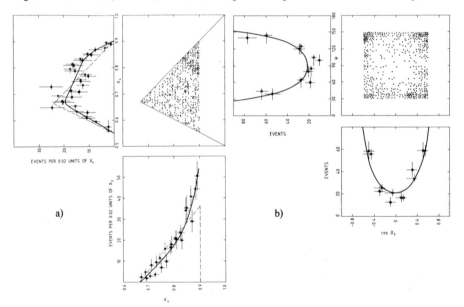

Fig. 6 a) Dalitz-plot x_2 vs x_4; b) angular distributions $\cos\theta_3$ vs 4 for three-jet events in UA1.

the published data ($\sqrt{s}$ = 546 GeV, $\sqrt{\hat{s}}$ = 150–250 GeV) while the crosses represent the preliminary 1984 data ($\sqrt{s}$ = 630 GeV, $\sqrt{\hat{s}}$ = 180–350 GeV). Since as has been emphasized the variables plotted are dimensionless data sets at different beam energies and different subprocess c.m.s. energies may readily be superposed. The variables x_3, x_4, cos θ_3 and ψ are phase-space variables so that the deviations from uniformity apparent in fig. 6a and fig. 6b truly reflect the variation of the matrix element. It is interesting to note that the agreement with QCD predictions (solid curves) is only qualitative. In particular the angular distributions have a tendency to be more strongly peaked than is predicted by the leading order (scaling) QCD matrix element. It is perhaps not unreasonable to suppose that as in the two-jet case the agreement between theory and experiment will be improved by the inclusion of non-scaling effects and that a full quantitative analysis will yield information on the form of the higher-order corrections.

Finally fig. 7 shows the three-jet to two-jet ratio plotted as a function of the subprocess c.m.s. energy. The closed circles correspond to the published 1983 data ($\sqrt{s}$ = 546 GeV) while the crosses represent the preliminary 1984 data ($\sqrt{s}$ = 630 GeV). The solid curve represents the leading order QCD prediction (taking Λ = 0.2 GeV and six quark flavours) computed on the assumption that the Q^2-scales for three-jet and two-jet events are identical. The broken curve which fits the data better represents the expected modification of the prediction if the Q^2-scale for the three-jet sample is intrisically lower than for the two-jet sample (Q_{3J} = 2/3 Q_{2J}). No significant dependence on subprocess c.m.s. energy is observed as would be expected from eq. (7).

Unfortunately the determination of α_s from this data is confounded by the theoretical uncertainty related to the higher-order corrections to the two-jet and three-jet cross-sections which have not yet been calculated. As is well known this problem is closely related to the uncertainty regarding the Q^2-scale definition for three-jet and two-jet events. If the Q^2-scales for two-jet and three-jet production are not identical at fixed subprocess c.m.s. energy then eq. (7) acquires a correction due to the non-cancellation of the factor α_s^2 and due to the non-cancellation of the structure function due to the scaling deviations. Since this uncertainty cannot be eliminated until (if ever) the appropriate theoretical calculations have been done both UA1 and UA2 quote results for $\alpha_s \cdot (K_3/K_2)$ where the factor K_3/K_2 represents the effect of the unknown higher order correction. For comparable angular acceptance the UA1 and UA2 results are fully consistent and yield values for $\alpha_s \cdot (K_3/K_2) \approx 0.23$.

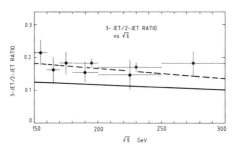

Fig. 7   The three-jet/two-jet ratio plotted subprocess c.m.s. energy (UA1).

528

The UA1 Collaboration have taken the further step of comparing the three-jet sample with a subset of the two-jet events with smaller scattering angles obtaining a significantly smaller result:

$$\alpha_s \cdot (K_3/K_2) = 0.16 \pm 0.02 \pm 0.03 \ . \tag{8}$$

On the assumption that Q^2-scale for three-jet events is indeed lower than the Q^2-scale for two-jet events at the same subprocess c.m.s. energy, for comparable angular acceptance one might reasonably suppose that K_3/K_2 is closer to unity for this comparison. At any rate it is clear that the value of $\alpha_s \cdot (K_3/K_2)$ for whatever reason can depend strongly on the angular acceptance and due care will have to be taken in comparing results from different experiments. Naturally the ideal solution would be that the angular dependence of K_3/K_2 is calculated theoretically and compared with the existing experimental results.

References

1) The staff of the CERN proton–antiproton project, Phys. Lett. **107B,** 306 (1981).

2) G. Arnison et al. (UA1 Collaboration), Phys. Lett. **107B,** 320 (1981).

3) M. Banner et al. (UA2 Collaboration), Phys. Letters **115B,** 59 (1982).

4) G. Arnison et al. (UA1 Collaboration), Phys. Letters **132B,** 214 (1983).

5) P. Bagnaia et al. (UA2 Collaboration), Phys. Letters **160B,** 349 (1985).

6) R.K. Ellis et al., Fermilab–Pub–85/152–T (1985).

7) T. Åkesson et al. (AFS Collaboration), Phys. Lett. **123B,** 133 (1983).

8) P. Hansen (UA2 Collaboration), Proc. XVI Symposium on Multiparticle Dynamics, Kiryat-Anavim (Israel) 1985.

9) P. Aurenche et al., Nucl. Phys. **B168,** 296 (1980).

10) F. Halzen, Contribution to this meeting.

11) G. Arnison et al. (UA1 Collaboration), Phys. Lett. **136B,** 294 (1984).

12) P. Bagnaia et al. (UA2 Collaboration), Phys. Lett. **144B,** 283 (1984).

13) G. Arnison et al. (UA1 Collaboration), Phys. Lett. **158B,** 494 (1985).

14) A.K. Nandi (UA1 Collaboration), to appear in Proc. Topical European Meeting on the Quark Structure of Matter, Strasbourg-Karlsruhe, 1985.

15) G. Arnison et al. (UA1 Collaboration), in preparation.

16) K. Althoff et al. (TASSO Collaboration), Z. Phys. **C22,** 357 (1984).

17) T. Åkesson et al. (AFS Collaboration), preprint CERN–EP/85–164 (1985).

THE STATUS OF PERTURBATIVE QCD (An Update)

FRANCIS HALZEN

Physics Department, University of Wisconsin
Madison, Wisconsin 53706
U.S.A.

ABSTRACT

This is an update on the status of testing perturbative QCD in hadronic interactions. We concentrate on recent $p\bar{p}$ collider results. Here the highest Q^2 can be reached and signatures for photons (real, virtual or weak) and jets are clean. We discuss: (i) jets and direct photons, (ii) the importance of tests directly comparing W with Z cross sections, and (iii) heavy-quark production which is the last frontier of perturbative QCD.

1. JETS AND PHOTONS

It is now more than ten years ago that Bjorken dreamed up the ultimate QCD experiment in which the differential cross section $d\sigma_{ij}/d\theta$ of partons i,j in two colliding hadrons is directly measured by observing two wide-angle jets in the final state. Schematically

$$\frac{d\sigma^{\bar{p}p}}{dx_{\bar{p}}dx_p d\theta} = \sum_{ij} f_i(x_{\bar{p}})\, f_j(x_p) \frac{d\sigma_{ij}}{d\theta} \ . \tag{1}$$

Here f_i, f_j are the probabilities for finding partons i,j in the interacting $\bar{p},p$ with momentum fractions $x_{\bar{p}}$, x_p. Even neglecting heavy quarks, the sum in Eq. (1) runs over 49 terms as $i,j = u,d,s,\bar{u},\bar{d},\bar{s},g$ and it is therefore doubtful that one can ever extract $d\sigma_{ij}/d\theta$ from experimental data. This problem is

530

that one can ever extract $d\sigma_{ij}/d\theta$ from experimental data. This problem is finessed by factorization,[1] indeed to a very good approximation

$$\frac{d\sigma_{ij}}{d\theta} \cong \frac{d\sigma}{d\theta} \tag{2}$$

and therefore Eq. (1) collapses to a single term

$$\frac{d\sigma^{\bar{p}p}}{dx_{\bar{p}}dx_p d\theta} \cong \left[\sum_i f_i(x_{\bar{p}})\right]\left[\sum_j f_j(x_p)\right]\frac{d\sigma}{d\theta} . \tag{3}$$

Equation (2) states that all two-body partonic cross sections (or at least those contributing significantly to Eq. (1)) have a universal shape. That $q\bar{q} \to q\bar{q}$, $qg \to qg$ and $gg \to gg$ have to a very good approximation the same angular distribution is shown in Fig. 1. The hint in the early UA1 analysis that the agreement with Eqs. (2) and (3) is not perfect has been confirmed.[2] The origin of the discrepancy is not approximation (2); differentiating subprocesses will indeed be a heroic task, as can be judged from Fig. 1. The deviation between theory and experiment is the effect of scaling violations which are expected[1] to distort the partonic relation (3). Observing $\ell n Q^2$ QCD effects has become a relatively easy task with the collider. This can be exploited[2] to directly study the scaling violations of the generic factorized parton distributions in Eq. (3).

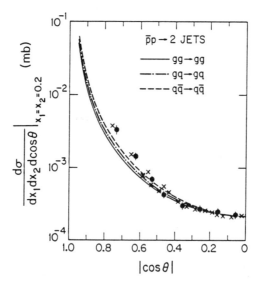

Fig. 1. Almost universal angular distribution of $2 \to 2$ parton cross sections is compared with two-jet data.[2]

As a test of QCD, hadronic jet physics has however reached an impasse. The reason is best illustrated by the three-jet analysis discussed in the accompanying experimental review.[2] Let us symbolically write Eq. (3) in the form

$$\frac{d\sigma^{2j}}{d(\cos\theta)} = \frac{\alpha_s^2(?)}{\hat{s}} \left[\begin{array}{c} \text{angular} \\ \text{distribution} \end{array} \right] \left[\begin{array}{c} \text{generic} \\ \text{parton distribution} \end{array} \right]. \qquad (3')$$

Here $\hat{s}$ is the parton c.m. energy and the angular distribution is basically the Rutherford cross section $(1 - \cos\theta)^{-2}$. The three-jet $2 \to 3$ cross section σ^{3j} can be cast in the same form[3] except for the normalization which is now $\alpha_s^3(?)$. So naively the ratio $\sigma^{3j}/\sigma^{2j} = \alpha_s$; actually

$$\frac{\sigma^{3j}}{\sigma^{2j}} = \alpha_s \frac{K(3j)}{K(2j)}. \qquad (4)$$

The strong coupling α_s is not determined as the K factors in (4) are not known. The problem is related to the choice of scale (?) in Eq. (3) in which α_s (and the parton distributions) run. Any choice is acceptable "?" $= p_T^2, \hat{s}, \hat{s}/4.8 \ldots$; they just represent different perturbative expansions of the all-order result which is of course independent of the choice made. This is however not the case for the leading-order calculation and in the absence of higher-order calculations the choices cannot be distinguished. The "?" can furthermore represent a different scale in the $2 \to 2$ and the $2 \to 3$ processes leading to the two unknown corrections in Eq. (4).

This problem has been at least partially tackled[4] for the closely related process $\bar{p}p \to \gamma X$. Prompt photon physics has a distinguished record as a test of QCD: the cross section and many other features of this process were true predictions of the theory anticipating the experiments and accommodating the results in a quantitative sense. The recent UA2 data is in agreement with the leading-order prediction[5] ($O(\alpha\alpha_s)$ diagrams in Fig. 2) made in 1978; see Fig. 3.

The discrepancy between theory and experiment for $p_T \lesssim 20$ GeV was also anticipated;[6] its origin is the neglect of the $O(\alpha\alpha_s^2)$ diagrams in Fig. 2, where quarks produced with very large cross sections radiate a prompt photon. More important, however, is the fact that now Aurenche *et al.* have completed a full higher-order $\alpha\alpha_s^2$ calculation including loop diagrams.[4] So here we can tackle the problem illustrated (but unsolved) in the jet case, see Eq. (4). Schematically

532

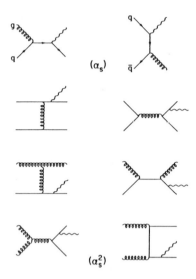

Fig. 2. $O(\alpha_s)$ and $O(\alpha_s^2)$ diagrams for direct photon production.

the full calculation has the form

$$E \frac{d^3\sigma}{d^3p} = \frac{\alpha_s(\mu)}{\pi}$$

$$\times \left\{ \text{BORN} \right.$$

$$+ \frac{\alpha_s(\mu)}{\pi} \left[\frac{b}{2} \ln \frac{\mu^2}{\Lambda^2} - 2\ln \frac{M^2}{\Lambda^2} P_{qq} \otimes \right] \text{BORN}$$

$$+ \left. \frac{\alpha_s(\mu)}{\pi} K \otimes \right\} f_i(M^2) \otimes f_j(M^2) ,$$

(5)

where b is related to the number of flavors $b = \frac{11}{2} - \frac{N_f}{3}$. The terms in curly brackets respectively represent the Born diagrams in Fig. 2(a), the higher-order corrections related to the coupling α_s, the emission of photons by quarks, and finally constant terms. The renormalization mass μ and factorization scale M in (5) are again undetermined. Here, however, we can attempt to get further information on μ, M from general renormalization group arguments.[7] As the result to all orders in perturbation theory is independent of μ, M, one can attempt to optimize the truncated result (5) by requiring that it have a minimal

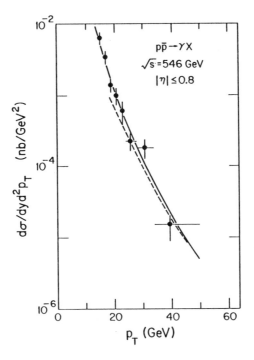

Fig. 3. Direct photon cross section compared with the leading-order prediction[5] (dashed line) and a calculation including the next-order contribution.[4]

dependence on these scales, *i.e.*

$$\mu \frac{\partial \sigma}{\partial \mu}\bigg|_{\mu=\mu_{\text{opt}}} = 0 \,, \qquad M \frac{\partial \sigma}{\partial M}\bigg|_{M=M_{\text{opt}}} = 0 \,. \qquad (6)$$

Here $\sigma(\sqrt{s}, p_T)$ is the cross section given by (5); its dependence on μ, M is illustrated in Fig. 4 for c.m. energy $\sqrt{s} = 63$ GeV and photon transverse momentum $p_T = 10$ GeV. There clearly is a stability (saddle) point. Notice that the result is remarkably stable in M^2 once we fix $\mu = \mu_{\text{opt}}$. At this point there is no large K factor, so requiring rapid convergence would yield the same result within 10%. On the contrary, for a conventional choice such as $\mu = M = p_T$ the higher-order calculation leads to a K factor as shown in Fig. 4. Although this procedure can strictly speaking only be applied to the non-singlet cross section $\sigma(p\bar{p} \to \gamma X) - \sigma(pp \to \gamma X)$ for which the structure function $f_{i,j}$ in (5)

534

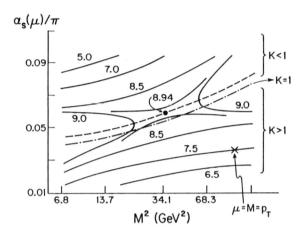

Fig. 4. $E d^3\sigma/d^3p$ in units 10^{-37} cm^2/GeV^2 at $\sqrt{s} = 63$ GeV, $p_T = 10$ GeV as a function of the renormalization mass μ and the factorization scale M.

can be defined beyond leading order, this is a first illustration that improved perturbation theory works.[8] A phenomenological analysis[8] along these lines yields the result shown in Fig. 3.

As was already the case for lower-energy experiments, direct γ's might again be a tool superior to jets for testing QCD. In this context we remind the reader of the fact that the process $qq \to qq\gamma$ provides us with the ultimate test[9] of color gauge theory. It displays radiation zeroes and tests color gauge theory just like $W \to q\bar{q}\gamma$ tests the radiative corrections to the electroweak theory. The accumulated luminosity of the data is approaching values which make this process accessible for scrutiny.

2.   COMPARING W AND Z RATES: A "HIGH" PRECISION TEST OF QCD

It is a little appreciated fact that QCD makes a prediction[10] readily computable beyond $O(\alpha_s^2)$, symbolically

$$\frac{\sigma(q\bar{q} \to W)}{\sigma(q\bar{q} \to Z)} = \text{"1"} + O(\alpha_s^2) \left[\frac{m_t^2 - m_b^2}{m_Z^2}\right], \qquad (7)$$

i.e. up to $O(\alpha_s^2)$ the diagrams producing W and Z are identical and the cross sections are the same up to known standard model couplings and phase space factors. In this order the neutral Z can be produced via the additional diagrams shown in Fig. 5. Even these would vanish as a result of the anomaly cancellation of the fermions in the triangle. Only the (t, b) generation contributes because of their large mass difference. The partonic relation (7) translates into a prediction[11] for W, Z production by $\bar{p}, p$ beams

$$\frac{\sigma(p\bar{p} \to W^{\pm})}{\sigma(p\bar{p} \to Z)} = 3.3 \pm 0.2 . \tag{8}$$

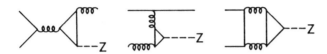

Fig. 5. Diagrams producing Z, not W in $q\bar{q}$ interactions.

The error in (8) is dominated by imprecise knowledge of the structure functions, more specifically the difference between $u(x)$, $d(x)$ which is poorly known. Low-energy experiments can really make a contribution here; see Ref. 11 for a discussion. In the standard model this prediction is tested in the form[11]

$$R = 9 \frac{\Gamma_Z(\text{stand}) + (N_\nu - 3)\Gamma_{Z \to \nu\bar{\nu}}}{\Gamma_W(\text{stand})} . \tag{9}$$

This test of QCD is well-known and usually advertised as a bound on the number of light neutrinos N_ν.

The fact that many ambiguities (e.g. K factors) cancel in the ratio of W, Z cross sections can also be exploited in their production with large p_T. For $\sqrt{s} = 540$ GeV the large p_T cross sections are calculable from leading-order perturbation theory for $p_T \gtrsim 25$ GeV. I.e. here multigluon emission[12] does not significantly modify the result computed from the leading-order $q\bar{q} \to (W, Z)+g$ and $qg \to (W, Z)+q$ diagrams.[13] They would cancel in the W-to-Z ratio anyway. The cross sections are shown in Fig. 6 in the form of the number $W \to e\nu$ and $Z \to \nu\bar{\nu}$ monojet events as a function of p_T for $\sqrt{s} = 630$ GeV. The $W \to e\nu$ large p_T cross section agrees well with the observations[15] of both UA1, UA2; see Fig. 7. More interesting, however, is that the equality of the large p_T yields

$$B(W \to e\nu)\frac{d\sigma}{dp_T}(W^{\pm}) = B(Z \to \nu\bar{\nu})\frac{d\sigma}{dp_T}(Z) , \tag{10}$$

536

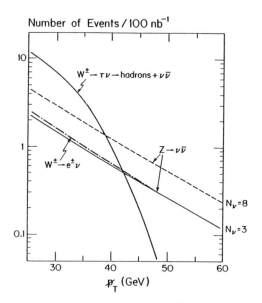

INTEGRAL DISTRIBUTION, $\sqrt{s}=630$ GeV

Fig. 6. Integral missing p_T distribution for $W \to \tau\nu$ and $Z \to \nu\bar{\nu}$ (solid: $N_\nu = 3$; dashed: $N_\nu = 8$) at $\sqrt{s} = 630$ GeV. The p_T scale for $W \to e\nu$ shows not the missing p_T, but the p_T of the W.

shown in Fig. 6. What is the origin of this equality? In the limit $M_W \simeq M_Z$ and $u(x) \simeq d(x)$ we derive

$$\frac{B(Z \to \nu\bar{\nu})\, d\sigma(Z)/dp_T}{B(W \to e\nu)\, d\sigma(W)/dp_T} = \frac{3}{4(1 - x_W)} \frac{1}{\cos^2 \theta_c} \left[1 + O(x_W^2)\right] . \qquad (11)$$

Here $x_W = \sin^2 \theta_W$ and (10) follows from the fact that $x_W \simeq \frac{1}{4}$ and the Cabibbo angle $\cos^2 \theta_c \simeq 1$. Corrections for $M_W \neq M_Z$ and $u(x) \neq d(x)$ have opposite signs and cancel approximately. Again relation (10) depends on $N_\nu = 3$. Tests of (10) involve measurement of the monojet cross section $Z(\to \nu\bar{\nu})$+gluon by the missing p_T technique. A large background from $W \to \tau\nu_\tau$ events has to be subtracted; see Fig. 6. Nevertheless, experiments are reaching the sensitivity to test (10). In the meantime, it provides us with a way to estimate the standard model monojet rate.

This particular type of phenomenology transcends perturbative QCD. Observation of deviations from Eqs. (8), (9) or (10) signals new physics:[11] more neutrinos, new quarks or leptons, Higgs, supersymmetric matter.... In this sense they play the same role as R measurements in e^+e^- collisions: breakdown of the QCD prediction indicates the discovery of a new threshold.

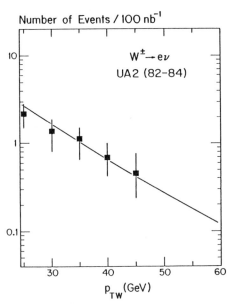

Fig. 7. Integral p_T distribution for $W \to e\nu$ events [13] compared with UA2 data. [15]

3. HEAVY QUARKS

Heavy-quark production is the last frontier of perturbative QCD with the leading $O(\alpha_s^2)$ diagrams $gg \to Q\bar{Q}$ and $q\bar{q} \to Q\bar{Q}$ unable to explain the rate and especially the Feynman x dependence of the data. [16] Here $Q = c, b$. It is not clear yet whether these problems will persist for truly heavy quarks, with $\sqrt{\hat{s}} \simeq m_Q$. Over the past year evidence has been accumulating that the higher-order diagrams play an important role, and might be the key to the problem. $O(\alpha_s^3)$ diagrams like $gg \to gQ\bar{Q}$ (see Fig. 8(a)) diverge however when the final-state gluon's momentum becomes soft. We therefore start the discussion by computing the $O(\alpha_s^3)$ cross section for $p_{Tj} \geq p_{T\,\min}$, where p_{Tj} is the transverse momentum of the jet opposite the Q (or $\bar{Q}$). We can compare this cross section

538

to the $O(\alpha_s^2)$ yield subjected to the same cut. Here the jet opposite Q (or $\bar{Q}$) is of course a heavy quark $\bar{Q}$ (or Q); see Fig. 8(b). Illustrative calculations[17] at $\sqrt{s} = 540$ and 43 GeV are shown in Fig. 9. Two important results emerge: (i) the higher-order $2 \to 3$ process dominates at higher energies, and (ii) the $2 \to 3$ process has a flatter rapidity distribution with heavy quarks produced in the $y = 2 \sim 3$ region where none are expected on the basis of the "leading" fusion diagrams.

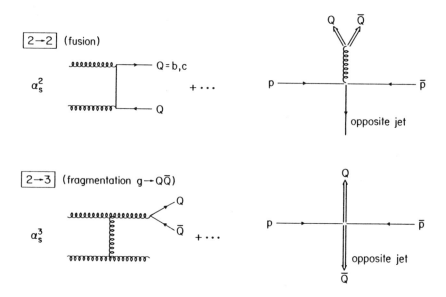

Fig. 8. (a) QCD graphs for heavy-quark production. (b) Collider view of the diagrams in (a).

There is really no surprise here—the result should be familiar from QED. When high-energy γ's interact with matter $O(\alpha^3)$ pair production $\gamma Z \to \gamma e^+ e^- Z$ dominates over $O(\alpha^2)$ Compton scattering $\gamma e \to e\gamma$. The origin of the dominance of $2 \to 3$ in Fig. 8(a) can be further clarified by visualizing the process as a two-jet interaction followed by the fragmentation of a final-state gluon jet into $Q\bar{Q}$. To leading log

$$\sigma(gg \to Q\bar{Q}g) \simeq \sigma(gg \to gg)\,P(g \to Q\bar{Q}) \tag{12}$$

$$P(g \to Q\bar{Q}) \simeq \int \frac{ds_{Q\bar{Q}}}{s_{Q\bar{Q}}}\,dx\,\frac{\alpha_s}{2\pi}\,P_{g \to Q\bar{Q}}(z)\,. \tag{13}$$

Here $P_{g \to Q\bar{Q}}(z)$ is the Altarelli-Parisi gluon fragmentation function and $s_{Q\bar{Q}}$ the heavy pair's invariant mass. We can use (12), (13) to estimate the ratio of the $(2 \to 3)/(2 \to 2)$ processes at $\theta = 90°$

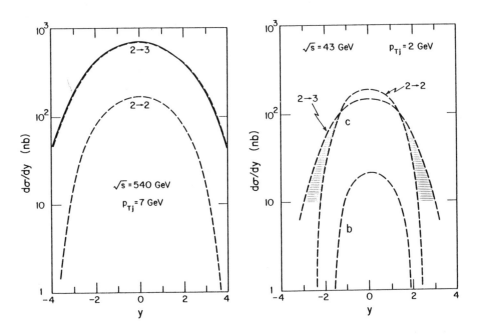

Fig. 9. (a) Rapidity distribution for heavy-quark production $(c, b$ added) by $2 \to 2$ and $2 \to 3$ diagrams in Fig. 8. (b) Same as (a) at $\sqrt{s} = 43$ GeV and $p_{Tj} = 2$ GeV.

$$\frac{\sigma(gg \to Q\bar{Q}g)}{\sigma(gg \to Q\bar{Q})} = \left[\frac{\sigma(gg \to gg)}{\sigma(gg \to Q\bar{Q})} \right] \left[\frac{\alpha_s}{3\pi} \ln\left(\frac{\hat{s}}{4m_Q^2} \right) \right] \qquad (14)$$

Here

$$\frac{\sigma(gg \to gg)}{\sigma(gg \to Q\bar{Q})} = \frac{27N^3}{N^2 - 2} = 104 \qquad (15)$$

is just a color factor which happens to be 104! It compensates for the fact that the second factor in (14) which represents the $g \to Q\bar{Q}$ fragmentation rate is $O(10^{-2} - 10^{-1})$ (more about this later). Therefore, we conclude that heavy-quark production by fragmentation of the gluon jet can be the dominant process. This dominance is further enhanced by the fact that the underlying process $gg \to gg$ proceeds via vector exchange compared with fermion exchange for the leading fusion mechanisms.

We next comment on the new source of heavy quarks in the $y = 2 \sim 3$ range from the $g \to Q\bar{Q}$ mechanism. It is a simple kinematic fact that for charm quarks this mechanism populates the x range associated with "diffractive" production; see Table 1. Whether it is the origin of diffractive production is at present a speculation. It depends on computation of the regularized cross section when $p_{Tj} \to 0$. We here face the problem of the convergence of the perturbative series, a problem solved[18] in principle. Its phenomenological implications depend however on the computation of all $O(\alpha_s^3)$ diagrams including loops. This has not been done.

Table 1. Regions in Feynman x where $2 \to 3$ processes are expected to dominate the fusion mechanism.

$\sqrt{s}$(GeV)	$y = 2 - 3$ corresponds to
20	$x = 0.7 \sim 1$
43	$x = 0.3 \sim 1$
62	$x = 0.2 \sim 0.7$
540	$x = 0.03 \sim 0.07$

A final comment about the charm fragmentation function $P(g \to c\bar{c})$ which has been measured[19] by direct observation of D^*'s in jets. The perturbative QCD prediction is given by (13)

$$\frac{dP(g \to c\bar{c})}{dz} \cong \frac{\alpha_s}{\pi} \ell n \frac{\hat{s}}{4m_c^2} P_{g \to c\bar{c}}(z) \tag{16}$$

$$\cong 0.1\left[z^2 + (1 - z)^2\right], \tag{17}$$

which is of order 10% whatever the value of z, as previously mentioned. Data[19] (in a biased subsample of jets and with large error bars) indicated a rate at least ten times larger, possibly suggesting non-perturbative effects. The fragmentation function (16) is however one of the few observables where power corrections to the leading-log estimates are computable. They are small.[20] At this conference, however, a new result of the UA1 group fortunately falls in line with the perturbative calculations.[19] The status of perturbative QCD is good.

REFERENCES

1. Halzen, F. and Hoyer, P., Phys. Lett. B130, 326 (1983); Combridge, B. and Maxwell, C., Nucl. Phys. 239, 429 (1984); Cohen-Tannoudji, G. et al., Phys. Rev. D28, 1626 (1983).

2. Scott, W., this conference.

3. Combridge, B. and Maxwell, C., Nucl. Phys. B151, 299 (1985).

4. Aurenche, P. et al., Phys. Lett. B140, 87 (1984).

5. Halzen, F. and Scott, D. M., Phys. Lett. B78, 318 (1978).

6. Contogouris, A. P. et al., Phys. Rev. D25, 990 (1982); Dechantsreiter, M. et al., Phys. Rev. D24, 2856 (1981); Gandhi, R. et al., Phys. Lett. B152, 261 (1985).

7. Stevenson, P. M., Phys. Rev. D23, 2916 (1981).

8. Baier, R., private communication.

9. Hagiwara, K. et al., Phys. Lett. B135, 324 (1984).

10. Hikasa, K., Phys. Rev. D29, 1939 (1984).

11. For a recent discussion, see Deshpande, N. G. et al., Phys. Rev. Lett. 54, 1757 (1985).

12. Multigluon emission has been extensively reviewed elsewhere, e.g., Halzen, F., in Proc. XV Symp. on Multiparticle Dynamics, Lund, Sweden (1984), G. Gustafson and C. Peterson, eds. (World Scientific); Altarelli, G., in Proc. Inter. Europhysics Conf. on High Energy Phys., Bari, Italy (1985).

13. Calculation from Cudell, J. R. et al., Phys. Lett. B157, 447 (1985). The calculation uses the structure functions of M. Glück et al., Z. Phys. C13 119 (1982), takes p_T^2 for the renormalization and factorization scale and includes higher-order corrections following Ellis R. K. et al., Phys. Lett. B104, 45 (1981).

14. Glover, E.W.N. and Martin, A. D., Durham preprint DTP/85/10 (1985).

15. Darriulat, P., this conference.

16. Halzen, F., in Proc. of the 21st Inter. Conf. on High Energy Physics, J. de Phys., Colloque C-3, supplément au n° 12, p. 401 (1982).

17. Calculations from Halzen, F. and Hoyer, P., Phys. Lett. B154, 324 (1985).

18. Collins, J. C. et al., Oregon Workshop on Super High Energy Physics, OITS 292 (1985); Collins, J. C., this conference.

19. Frey, R., this conference.

20. Mueller, A. H. and Nason, P., Columbia preprint CU-TP-303 (1985).

TESTING QCD WITH TWO PHOTONS

David O. Caldwell

Physics Department, University of California
Santa Barbara, California 93106

ABSTRACT

With the settling of some theoretical problems, measurements of the photon structure functions again emerge as sensitive tests of QCD and as an excellent way to determine the QCD parameter, $\Lambda_{\overline{MS}}$. Total $\gamma\gamma$ cross sections demonstrate point-like behavior but do not yet require the details of QCD. While $\gamma\gamma$ production of baryon-antibaryon pairs in the kinematic region explored so far does not show agreement with QCD predictions, the similar production of meson pairs is in accord with no-parameter calculations. Just as 2γ widths of mesons give information on their quark content, so this parameter also provides constraints on gluon content. In particular, a new limit on $\iota(1450) \to \gamma\gamma$ gives additional evidence that this particle is a glueball, the existence of which is a consequence of the non-Abelian nature of QCD.

1.   INTRODUCTION

Two-photon physics has long been touted as a great hope for incisive tests of QCD. By discussing four areas which are probably the best explored so far, we shall see how well these prophecies have been fulfilled. These topics are photon structure functions, two-photon total cross sections, the two-photon-to-hadron-antihadron continuum, and the two-photon widths of mesons, with their implications for gluon content.

Before discussing these specific sub-fields of two-photon physics, it is worthwhile to see in a general way why there is hope that a good testing ground for QCD can be provided by the interaction of two photons. For example, consider inclusive hadron production by two photons versus that by one photon (e^+e^- annihilation). The quark propagator occurs only as an $O(\alpha_s)$ correction to the e^+e^- annihilation process, which then gives three-jet production, whereas it appears in the lowest order diagram in the two-photon process.

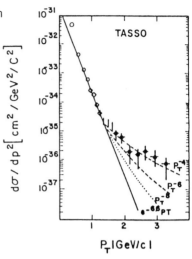

Fig. 1. Inclusive hadron production by two photons.

The sensitivity to the point-like character of the photon coupling to hadronic matter is illustrated in Fig. 1, where the inclusive hadron production by two photons is shown as a function of p_T.[1] A point-like p_T^{-4} dependence (two-jet creation) appears at a remarkably small value of p_T.

The total cross section for the production of two quarks by two photons is dependent on the mass scale involved. Thus $\sigma \propto \ln(W/m_q)^2$, if quarks are treated as massive free particles, whereas in lowest order QCD the scale is set by the renormalization parameter, Λ_{LO}, and $\sigma \propto \ln(Q^2/\Lambda_{LO}^2)$. If higher order corrections are included, the sensitivity to the scale $\Lambda_{\overline{MS}}$ is preserved.

2. PHOTON STRUCTURE FUNCTIONS[2]

Deep inelastic photon-photon scattering has given special hope of a decisive QCD test because QCD predicts not only the Q^2 evolution of the photon structure functions, as it does for many processes, but also its shape and absolute normalization, up to second order in the strong coupling constant, $\alpha_s(Q^2)$, in terms only of the QCD scale parameter, $\Lambda_{\overline{MS}}$, to which it appeared to be remarkably sensitive.

However, after the initial optimism, two problems raised serious doubts. The first difficulty is that in addition to the calculable point-like part of the structure function, there is a hadronic or VDM (vector meson dominance model) component which cannot be calculated in perturbative QCD. The second and completely unexpected difficulty was that the next-to-leading-order corrections gave a negative structure function at small x, even at large Q^2. Here Q^2 is the square of the four-momentum of the more virtual photon, $x \approx Q^2/(Q^2+W^2)$, and W is the center of mass energy in the collision of the virtual photon with the nearly real photon.

This second and much more serious issue seems now to have been re-solved. The solution to the problem in second order, as proposed by Bardeen and worked out by Antoniadis and Grunberg[3] and by Rossi,[4] is that there is a cancellation between matching singularities in the point-like and hadronic components. In other words, the problem arises only because there is an artificial separation in the calculation of these two parts. Because the hadronic part cannot be perturbatively calculated for one nearly real photon, an extra parameter, t, has to be introduced which gives a hadronic contribution to the structure function in addition to some assumed form of the hadronic part taken from the VDM. Thus the higher order structure function, F_2^{HO}, is given by the now regularized F_2^R, which is calculated in perturbative QCD, plus these hadronic contributions:

$$F_2^{HO}(x,Q^2) = F_2^R(x,Q^2,\Lambda_{\overline{MS}}) + \Delta(x,t) + F^{VDM}(x,Q^2). \qquad (1)$$

Fortunately the value of t may be obtained from experiment for small values of x, and the t correction has very little effect on the result for $x>0.3$. It is in this low x, and especially low Q^2, region that the hadronic contribution is large, and hence even F^{VDM} can be determined in one kinematic domain and $\Lambda_{\overline{MS}}$ in another (high x and Q^2). Fig. 2 shows a comparison of PEP4/9 data, which is at the lowest Q^2 yet measured, with forms of the photon structure function using F^{VDM} obtained in the VDM from two different parameterizations[5] of the pion structure function. The correct form of F^{VDM} appears to lie somewhere between the two commonly used forms shown, but notice that even at this low Q^2,

F_2^{HO} at higher x is little affected by the choice of F^{VDM}. Nevertheless, more work needs to be done to determine better Δ and F^{VDM} experimentally, and PEP4/9 has found that a better separation of the hadronic and point-like parts can be obtained from the topology of the events.

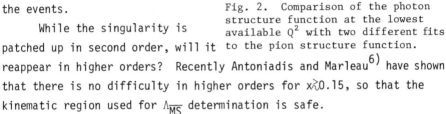

Fig. 2. Comparison of the photon structure function at the lowest available Q^2 with two different fits to the pion structure function.

While the singularity is patched up in second order, will it reappear in higher orders? Recently Antoniadis and Marleau[6] have shown that there is no difficulty in higher orders for $x \gtrsim 0.15$, so that the kinematic region used for $\Lambda_{\overline{MS}}$ determination is safe.

Given restored faith in the QCD calculation, how well do the measurements agree with QCD? First, the Q^2 evolution of the structure function is as predicted, as one can see from Fig. 3. The data follow the predicted slope in the entire Q^2 range from 0.5 to 200 GeV². In addition, the absolute normalization is correct using VDM plus higher-order QCD with $\Lambda_{\overline{MS}}$=200 MeV. Second, the predicted x-shape of the structure function is confirmed by experiment, as displayed in Fig. 4. Note that this shape is rather unusual because of the rise with x.

While the Q^2 and x dependences are consistent with QCD and make excellent negative tests, Fig. 4 also shows a problem with these as positive tests, namely that VDM plus the quark parton model (QPM) give a fairly good fit to the data as well. This raises the question as to whether it is just the point-like nature of the interaction which is being seen and not gluonic effects at all. The use of the QPM here is somewhat deceptive in that if current quark masses are used instead of effective masses, the QPM predictions can be ruled out by experiment. In other words, gluon effects are seen in the sense that the bare quarks must acquire masses $\approx$ 300 MeV by gluonic interactions in order to get agreement with the data.

Even with adjustment of the quark mass, the data give a somewhat better fit to QCD than to the QPM. An important feature of that figure is the difference displayed between low-order (LO) and high-order (HO) QCD. Already the measurements favor the latter, but a 30% change in the VDM correction could alter this result. Thus the measurements are not yet precise enough to test unambiguously higher order effects.

In several of these comparisons between theory and experiment $\Lambda_{\overline{MS}}$ = 200 MeV has been used to get the correct normalization. Fig. 5 gives some idea of how sensitive the structure function F_2 is to this parameter. $\Lambda_{\overline{MS}}$ determined in this way is comparable in accuracy to the best determinations made in other ways, as is seen in Fig. 6, taken from Wagner's review.[2] Using the PLUTO, JADE, and TASSO two-photon data, he obtains[2] a value $\Lambda_{\overline{MS}}$ = (230±70) MeV, including systematic errors. This result is insensitive to the hadronic component, since if F^{VDM} is even set to zero, Λ changes by only 70 MeV. With the singularity problem solved and better determinations of F^{VDM}, and hence of $\Delta(t,x)$, which are now possible with the low Q^2 and x data, plus improved statistics, this will be a very powerful way to measure the QCD scale parameter.

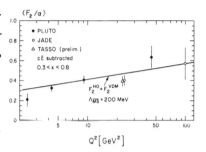

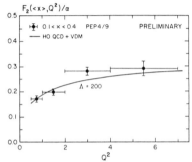

Fig. 3. Q^2 evolution of the photon structure function in two x regions, compared with QCD plus VDM.

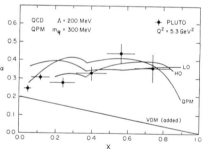

Fig. 4. Comparison of the QPM and QCD in leading and higher order.

3. PHOTON-PHOTON TOTAL CROSS SECTION

In principle, photon-photon total cross sections can also provide information on QCD. At the present state of measurement accuracy, however, really only point-like behavior is demonstrated. In the energy range covered the hadronic contribution is still large, contributing to the uncertainty, and the precision is not sufficient to distinguish the QPM from QCD. This is shown in Fig. 7 for single-tagged PLUTO data.[1] Double-tagged PEP9 data[7] goes to higher energy and may have smaller systematic errors, but the conclusion is the same. In short, QCD is checked, but no new information on QCD is obtained. Although, except for Fig. 1, we shall not discuss two-photon jet analysis here, it is sufficient to say that exactly the same conclusion can be reached from the accuracy of that existing data.

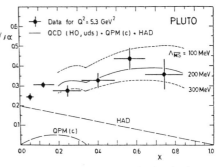

Fig. 5. Sensitivity of the hadronic structure function to $\Lambda_{\overline{MS}}$.

4. CONTINUUM HADRON PAIRS

The production by two photons of particle-antiparticle pairs of hadrons at non-resonant invariant masses can provide quite a different sort of check on QCD. So far there is disagreement for baryons and good agreement for mesons.

Data on $\gamma\gamma \to p\bar{p}$ by the TASSO collaboration[8] in the mass range $2.0 < W_{\gamma\gamma} < 3.1$ GeV agree with the predicted $W_{\gamma\gamma}^{-10}$ dependence given by the QCD derived scaling law. However, the data are a factor ~ 5 above the calculated[9] rate and have the wrong angular dependence. PEP4/9 observations are in approximate agreement with the TASSO result. Explanations of the discrepancy are

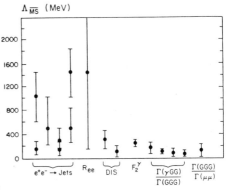

Fig. 6. Comparison of $\Lambda_{\overline{MS}}$ values determined in different experiments.

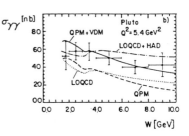

Fig. 7. Total $\gamma\gamma$ cross section as a function of center of mass energy.

that either the data are not at a sufficiently large value of momentum transfer squared for the theory to be applicable, or that there is resonant production of $p\bar{p}$.

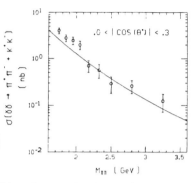

Fig. 8. Mark II measurement of $\gamma\gamma\to\pi^+\pi^-+K^+K^-$.

Similar problems beset the reaction $\gamma\gamma\to\Delta\bar{\Delta}$ observed by PEP4/9. Although the cross section for $\gamma\gamma\to\Delta^{++}\bar{\Delta}^{--}$ is predicted[9] to be an order of magnitude larger than that for $p\bar{p}$, the observed cross sections are also in this ratio, making the $\Delta^{++}\bar{\Delta}^{--}$ production also ~ 5 times the QCD prediction. However, the observed $p\bar{p}\pi^+\pi^-$ events in the $\Delta\bar{\Delta}$ mass region are also compatible with being just phase space, so that it is not clear that any $\Delta\bar{\Delta}$ are produced. Assuming these are $\Delta\bar{\Delta}$ events, the angular distribution is again isotropic, instead of the expected steep rise. In this case instead of the predicted $W_{\gamma\gamma}^{-10}$ dependence, the data follows $W_{\gamma\gamma}^{-5}$.

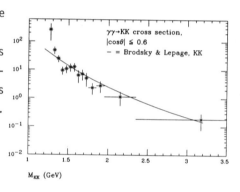

Fig. 9. PEP4/9 measurement of $\gamma\gamma\to K^+K^-$ compared with QCD calculation.

In contrast, the production of pairs of pseudoscalar mesons is in fair agreement with a no-parameter QCD calculation. Brodsky and Le-Page[10] obtained a scaling law, $d\sigma/d\cos\theta_{cm} = c\alpha^2\alpha_s^2 W_{\gamma\gamma}^{-6}\sin^{-4}\theta_{cm}$, for $\gamma\gamma\to\pi\pi$, where c is a known absolute normalization constant proportional to the $\pi\to\mu$ decay constant, f_π. Furthermore, $d\sigma(\gamma\gamma\to KK)=(f_k/f_\pi)^4 d\sigma(\gamma\gamma\to\pi\pi)$. The problem in testing this prediction is that there are a small number of $\pi\pi$ pairs in a sea of $\mu\mu$ and ee pairs.

Two groups have been able to make such comparisons. Mark II has a larger quantity of data but cannot separate π's from K's, so they use the Brodsky-LePage ansatz, $d\sigma/dt(\gamma\gamma\to K^+K^-)=2d\sigma/dt(\gamma\gamma\to\pi^+\pi^-)$, and one of their results[11] is shown in Fig. 8. While there is good agreement above 2 GeV, there is a discrepancy at lower masses, which is presumably still in the resonance region. A similar conclusion can be reached for the PEP4/9 $\pi\pi$ data, which do not have as good statistics but which have separated π's and K's. The PEP4/9 KK data is shown in Fig. 9 and is in better agreement with the calculations, presumably because resonances are not such a problem.

5.   TWO-PHOTON WIDTHS OF MESONS AND GLUEBALLS

While numerous tests of QCD have been discussed here, these have not been very sensitive to the gluonic aspect of the theory. The non-Abelian nature of QCD would be manifested most directly if the attraction between the color charges of gluons resulted in glueballs. A favorite way to search for glueballs is in J/ψ radiative decay, since in perturbation theory this process is dominated by $J/\psi\rightarrow\gamma+2$ gluons. For a particle copiously produced in this way, complementary information can be provided by its two-photon width. As a clue to gluon content, these measurements may be combined to give the property "stickiness", defined by Chanowitz[12] as $S\propto(m_x/k^*_{\psi\rightarrow\gamma x})^{2L+1}\Gamma(\psi\rightarrow\gamma x)/\Gamma(x\rightarrow\gamma\gamma)$, which eliminates phase space factors ($k^*_{\psi\rightarrow\gamma x}$ being the photon's momentum in the ψ center of mass).

Two-photon widths have been measured for the pseudoscalar and tensor mesons, which are of interest in the glueball hunt. For the $J^{PC}=2^{++}$ particles, if the s-wave dominates in the decays, L=0 in the formula for S, and if the normalization is chosen so that for the f meson, $S_f=1$, then the stickiness ratios are $S_f:S_\zeta:S_{f'}:S_\theta=1:(>1):14:(>20)$. The inequality signs result from upper limit measurements for the $\gamma\gamma$ widths. The results indicate that the f' has a considerable gluonic content and the glueball candidate, the $\theta(1690)$, to have even more. A better determination of $\Gamma(\theta\rightarrow\gamma\gamma)$ could be quite important.

The $\gamma\gamma$ widths of the 0^{-+} mesons have been especially well determined, and much improved limits have been obtained for the glueball candidate, the $\iota(1450)$. The only published result[13] is by Mark II of $\Gamma(\iota\rightarrow\gamma\gamma)\cdot B(K\bar{K}\pi)<8$ keV, although a preliminary result from TASSO was <7 keV and more recently is reportedly <2.2 keV (95% CL). Mark II also has a better limit[11] of 2.0 keV (90% CL). The best limit is from PEP4/9 of <1.3 keV (95% CL), and their result is shown in Fig. 10, where all three events are assumed to be $\imath$'s. If $B(K\overline{K}\pi) \approx 0.7$ as indicated by Mark III results,[14] then $\Gamma(\imath \rightarrow \gamma\gamma) < 1.9$ keV. Using this result, L=1 in S, and normalizing to $S_\eta = 1$, we have $S_{\pi^0} : S_\eta : S_{\eta'} : S_\imath = 0.02 : 1 : 2.5 : (>50)$. Both from this high value of stickiness and from the fact that the $\Gamma(\imath \rightarrow \gamma\gamma)$ limit is even less than in any of the many calculations of that quantity from η-η'-$\imath$ mixing, where $\imath$ is assumed to be mainly a glueball, it can be concluded that the $\imath$ is likely to be a glueball with very little $q\overline{q}$ mixing. Once more, two photons prove to be a useful tool in probing QCD.

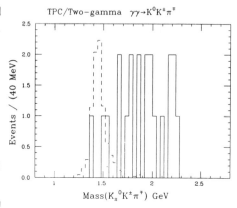

TPC/Two-gamma $\gamma\gamma \rightarrow K^0 K^\pm \pi^\mp$

Mass($K_s^0 K^\pm \pi^\mp$) GeV

Fig. 10. $K_s^0 K^\pm \pi^\mp$ mass spectrum. The dashed histogram is the $\imath$ signal which should have been seen (with no background) if its $\gamma\gamma$ width were as large as the limit given by this experiment. Three events are assumed to be in the $\imath$ region.

REFERENCES

1. These results were taken from a review by U. Maor, Proceedings of XX Recontre de Moriond (1985, to be published).

2. W. Wagner, Proceedings of the VI International Workshop on Photon-Photon Collisions (World Scientific, ed. R. Lander, 1984), p. 241 gives an excellent review of this topic and should be consulted for references not given here.

3. I. Antoniadis and G. Grunberg, Nucl. Phys. B213, 445 (1983).

4. G. Rossi, Phys. Letters 130B, 105 (1983); Phys. Rev. 29D, 852 (1984).

5. C.B. Newman, et al., Phys. Rev. Lett. 42, 951 (1979).

6. I. Antoniadis and L. Marleau, SLAC-PUB-3691 (1985, to be published in Phys. Lett.).

7. D. Bintinger, et al., Phys. Rev. Lett. 54, 763 (1985).

8. M. Althof, et al., Phys. Lett. 130B, 449 (1983).

9. G.R. Farrar, E. Maina, and F. Neri, Rutgers University preprint RU-85-08 (1985, unpublished).

10. S.J. Brodsky and G.P. LePage, Phys. Rev. <u>D24</u>, 1808 (1981).

11. G. Gidal, Proceedings of the XVI Symposium on Multiparticle Dynamics (1985, to be published).

12. M. Chanowitz, Proceedings of the VI International Workshop on Photon-Photon Collisions (World Scientific, ed. R. Lander, 1984), p.95.

13. P. Jenni, et al., Phys. Rev. <u>D27</u>, 1031 (1983).

14. J.D. Richman, Ph.D. Thesis, CALT-68-1231 (1985); N. Wermes, SLAC-PUB-3730 (1985).

STATUS OF α_s DETERMINATION IN e^+e^- ANNIHILATION

Ren-yuan Zhu

Lauritsen Laboratory
California Institute of Technology
Pasadena, California 91125 USA

ABSTRACT

I review the experimental measurements of the strong coupling constant α_s in e^+e^- annihilation. The problem of the large spread of α_s values reported in 1983 has been resolved, and is explained by approximations in the $O(\alpha_s^2)$ QCD calculations, and inappropriate choices of jet variables in some cases. Summarizing all results using the exact QCD calculation the strong coupling constant at $\sqrt{s}$ = 34.6 GeV is: α_s = 0.13 ± 0.02, corresponding to $\Lambda_{\overline{MS}}^{(5)}$ = 150 ± 100 MeV. This value of Λ is in good agreement with the measurements of deep inelastic scattering and Y decay.

1. INTRODUCTION

This report reviews the status of α_s determinations in e^+e^- annihilation and is a summary of a rapporteur talk given at the 1985 DPF Conference, Oregon. In 1983, the spread of α_s values[1] presented at the EPS conference, Brighton, and at the Lepton-Photon Conference, Cornell (Fig. 1) raised strong doubts as to whether the strong coupling α_s could be extracted from e^+e^- data. Since then many studies and comments[2,3,4,5,6] have addressed this topic. According to these studies, almost all α_s values determined by PETRA/PEP groups by using some approximate $O(\alpha_s^2)$ calculations[7] have been systematically too large. The only exception is the MARK J group analysis[10a] which uses its own calculation,[4] based on the matrix elements of Refs. 13 and 17.

* Supported in part by U.S. Department of Energy Contract No. DE-AC03-81-ER40050.

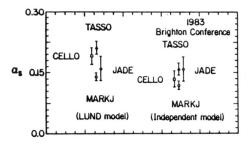

Fig. 1.   The α_s values obtained by 4 PETRA groups at 1983 Brighton conference.

Section 2 of this report comments on the existing $O(\alpha_s^2)$ QCD calculations. Section 3 covers the experimental α_s measurements in e^+e^- annihilation. The results from MARK J,[10a,b] PLUTO[11] and TASSO[5] which use EECA and the ERT calculation[13] agree quite well. A review of α_s obtained from e^+e^- resonances shows the overall consistency of QCD tests in e^+e^- annihilation.

2. $O(\alpha_s^2)$ QCD CALCULATION FOR $e^+e^- \rightarrow$ PARTONS

In order to avoid confusion due to the renormalization scheme dependence of QCD,[12] only the modified minimum subtraction scheme $(\overline{MS})$ is used throughout this report. In this scheme the Q^2 dependence of the strong coupling $\alpha_s(Q^2)$ can be parametrized as:

$$\alpha_s = \frac{2\pi}{\dfrac{(33 - 2N_f)}{6} \ln(Q^2/\Lambda^2) + \dfrac{(153 - 19N_f)}{(33 - 2N_f)} \ln(\ln(Q^2/\Lambda^2))} \quad , \tag{1}$$

where N_f is number of quark flavors with mass less than $Q/2$, and Λ is the flavor dependent scale parameter of the theory.

In most cases perturbative QCD calculations are not directly comparable with the experimental data, however, since the experimentally observed particles are hadrons which are the products of the so-called "fragmentation" of partons. Since the fragmentation process is not calculable in perturbative QCD, Monte Carlo models are used. The most widely used models in the determinations of α_s are the independent jet fragmentation model (IF)[15] and the color string jet fragmentation model (SF).[16,8]

In these models "jet resolution" cuts are used to decide whether a final state is composed of 3 versus 4 partons, or 2 versus 3 partons, etc. There are two commonly used jet resolution criteria:

(1) ε–δ cuts, which require a minimum parton energy as well as a minimum open angle between each pair of partons; and

(2) y cuts, which require a minimum invariant mass for each pair of partons.

2.1 Existing $O(\alpha_s^2)$ Calculation for $e^+e^- \rightarrow$ Partons

Three groups of physicists have calculated perturbative QCD to complete second order. While the ERT[13] and VGO[14] calculations are exact and for bare partons only, the FKSS[7] calculation is approximate, with jet resolution already imposed. The FKSS calculation[7] is widely used by most groups in PETRA and PEP, and in the standard SF model.[8]

The MARK J group[4] factorizes the $O(\alpha_s^2)$ corrections to 3-parton final states as a 2-dimensional Ratio function in the differential Dalitz distribution:

$$\frac{1}{\sigma_0} \frac{d\sigma_{3-\text{parton}}}{dx_g d(x_q - x_{\bar{q}})} = A_0 \frac{\alpha_s}{\pi} \left\{ 1 + \text{Ratio}(x_g, x_q - x_{\bar{q}}) \frac{\alpha_s}{\pi} \right\} \qquad (2)$$

where x's are normalized momenta of partons, $x_i = 2P_i/\sqrt{s}$, and the subscripts g, q and $\bar{q}$ refer to gluon, quark and antiquark. The matrix elements used to evaluate the Ratio function are from ERT,[13] and are written according to a procedure developed by Kunszt and Ali,[17] to improve the numerical convergence of the computation.

2.2 Comparison of $O(\alpha_s^2)$ Calculations

Precise numerical comparison of the different theoretical $O(\alpha_s^2)$ calculations is difficult, since the kinematics of the 3-parton final states of these calculations is different, i.e., the difference changes for different kinematic variables.

Reference 4 compared MARK J's ERT + jet resolution calculation with the FKSS calculation,[7]. Tables 1 and 2 show the comparison of the integrated thrust distribution, where the differences can be summarized as follows:

(1) (5-10)% when y cuts are used, and

(2) (20-30)% when ε–δ cuts are used.

A comparison[4] of the integrated EECA cross-section from $\cos(\chi) = 0$ to 0.75 shows a linear dependence in ε for the FKSS calculation, in contrast to cut independent results obtained from the ERT+jet resolution calculation, as shown in Fig. 2.

In Reference 5 a study by the TASSO group together with Gutbrod and Schierholz reveals three problems in the FKSS calculation that causes the higher α_s values:

(1) discarding partons when imposing the ε–δ cuts, 15%;

(2) neglecting terms of order ε and δ^2 (or y), 5%; and

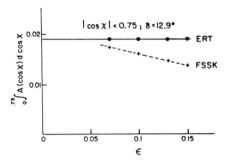

Fig. 2.   The integrated EECA cross-section at the parton level as a function of ε, shown for two calculations: ERT as solid dots and FKSS as crosses.[2a,4]

(3)   missing some 4-parton final states, 6%.

With those missing states in (3) inserted the so-called "extended FKSS" calculation recently used by TASSO would still give 20% higher α_s values than those obtained using the exact ERT calculation.

The above conclusions are further confirmed by an independent calculation of Gottschalk and Shatz.[6] Starting from the ERT matrix element[13] they made an analytical integration to evaluate the $O(\alpha_s^2)$ 3-parton final state cross-section with y cuts imposed. A comparison is made and shows that almost all the α_s values determined according to Refs. 7 and 8 are systematically too large.

3. EXPERIMENTAL MEASUREMENTS

In order to limit the systematic errors in the measurement of α_s associated with our relatively poor knowledge of the fragmentation process and contributions from higher order QCD processes (beyond second order) it is necessary to choose a kinematic variable with the following characteristics:

1)   Experimentally can be well measured (with good resolution);

2)   Has small high order corrections;

3)   Is not sensitive to the jet resolution parameters imposed in models;

4)   Has small fragmentation effects:

   a)   Should be a linear sum of final state momenta, and therefore insensitive to the final state details depending on fragmentation;

   b)   Should have a relatively flat s dependence, since fragmentation effects decrease as a power of s, while the QCD contribution varies logarithmically with s.

3.1 α_s Determinations from the Shape Variables

As discussed above, the shape variable of choice is the energy-energy correlation asymmetry.[91] Studies of the EECA[4,211] show that it has small second order corrections compared to other kinematic variables, and is the only variable which is cut independent. Table 3 lists the K values for four different kinematic variables, where K is a measure of the relative size of the second order corrections. Figure 3 shows the integrated cross-sections of the thrust, oblateness, and EECA as function of the cuts. The energy dependence of the integrated EECA cross-section from the MARK J data is shown in Fig. 4 together with calculated $q\bar{q}$ contribution and two Monte Carlo model predictions for IF and SF with $\Lambda_{\overline{MS}} = 100$ MeV.

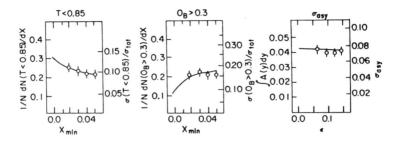

Fig. 3. Monte Carlo results of fraction of events with: (1) $T < 0.85$, (2) $O > 0.3$, and (3) $|\cos(\chi)| < 0.75$ as function of cut-off y and ε. The curves, corresponding to the right-hand scales, are the parton level calculation. The points, corresponding to the left-hand scale, are data.[2a,4]

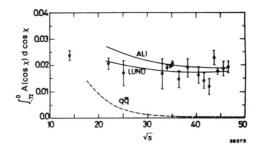

Fig. 4. The integrated EECA data (points) of MARK J together with the $q\bar{q}$ contribution (dashed curve) are shown as function of $\sqrt{s}$. The Monte Carlo predictions of two models for $\Lambda_{\overline{MS}} = 100$ MeV are shown as solid curves.[10b]

Two points are clear from Fig. 4:

(1) The EECA has a flat s dependence; and

(2) Fragmentation effects, which are significant for $\sqrt{s}$ < 22 GeV, gradually die out at higher energies.

All 5 groups at PETRA measured α_s at $\sqrt{s}$ = 34.6 GeV by using shape variables, especially the EECA. The first $O(\alpha_s^2)$ α_s determination was done by the JADE group[24a] in 1982. By using a 3 jet cluster analysis the data was fitted to the $O(\alpha_s^2)$ FKSS (y cuts) calculation. The α_s values they obtained are: 0.16 for both SF and IF with no model dependence observed.

The MARK J group[10a] made the first $O(\alpha_s^2)$ α_s determination by using the EECA in 1983. As discussed in Section 2.2, the ERT + jet resolution (ε–δ cuts) scheme was used in this determination. By fitting to the perturbative region ($|\cos\chi|$ < 0.72) of the EECA distribution the MARK J obtained α_s values 0.12 for IF and 0.14 for SF with a statistical error of ± 0.01. MARK J has summarized this result as α_s = 0.13 ± 0.01 (statistical) ± 0.02 (systematic) for $\sqrt{s}$ = 34.6 GeV (Fig. 5). The result is stable under variations of the cuts. Although the observed fragmentation model dependence is moderate it was pointed out that this is the major source of systematic error.

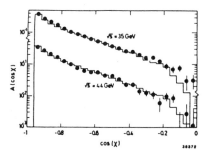

Fig. 5.   EECA data of MARK J at $\sqrt{s}$ = 35 and 44 GeV are shown as a function of $\cos(\chi)$.   The histograms are full Monte Carlo simulations using the Ali fragmentation model with $\Lambda_{\overline{MS}}$ = 100 MeV.[10b]

The CELLO[23] group performed a similar analysis with considerably less statistics than the MARK J data. The α_s values they obtained were higher and somewhat more model dependent: 0.12-0.15 for IF and 0.19 for SF. Since they used the FKSS calculation (y cuts)[8] in their $O(\alpha_s^2)$ QCD corrections it is difficult to directly compare their results with the MARK J result. The larger model dependence has finally been understood as being due to the energy-momentum conservation scheme used in their IF models.[3]

The JADE[24b] group have also determined α_s by using the FKSS calculation (y cuts)[8] and EECA. With similar statistics to the MARK J, their α_s values are only slightly higher than MARK J: 0.165 for *SF* and 0.11-0.14 for *IF*. A comparison between JADE and CELLO shows that CELLO's data is systematically higher than JADE's data, but the difference is not statistically significant, due to the large error bars resulting from the small size of CELLO's data sample.

By using both the extended FKSS (ε–δ cuts) calculation and a calculation similar to MARK J's method, namely ERT + jet resolution (ε–δ cuts), the TASSO[5] group made two determinations of α_s from their EECA data. The α_s values they obtained are 0.19 for *SF* and 0.14-0.16 for *IF* by using the former, and 0.16 for *SF* and 0.117-0.127 for *IF* by using the latter method. The numerical difference between extended FKSS (ε–δ cuts) and ERT + jet resolution (ε–δ cuts) is 20%. As discussed earlier, these results can not be compared directly with CELLO and JADE, but the consistency of the ERT results from TASSO and MARK J is evident. Figure 6 shows TASSO's data, compared with QCD Monte Carlo predictions from the *SF* and *IF* models, where the QCD calculations are based on the ERT matrix elements.

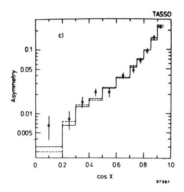

Fig. 6.   The corrected EECA measured by TASSO at 34.6 GeV. The solid (dashed) line is the prediction of the independent (string) jet model, as obtained from fits to the region $|\cos(\chi)| < 0.7$.[5]

The TASSO group[5,22] also performed an analysis in which they choose groups of kinematic variables as "shape parameters". Due to the choice of variables this analysis depends more on the details of fragmentation and on the unknown higher order QCD corrections. In summary, the major source of the systematic error in TASSO's results, α_s between 0.12 and 0.23, comes from the approximations in the FKSS calculation.

In a fashion similar to the MARK J and TASSO groups, the PLUTO group[11] analyzed their EECA data by use of the ERT + jet resolution (ε–δ cuts) calculation. They obtained 0.145 for *SF* and 0.136 for *IF* which is summarized as $\Lambda_{\overline{MS}}^{(5)}$ between 100 and 300 MeV. This result is in good agreement with the corresponding results of MARK J and TASSO. Figure 7 shows their data compared to QCD predictions.

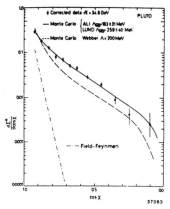

Fig. 7.   The corrected EECA measured by PLUTO at 34.6 GeV. The solid line is
the QCD Monte Carlo expectation. The dashed (dash-dotted) line represents
the expectations from the Webber $(q\bar{q})$ model.[11]

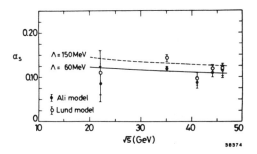

Fig. 8.   The MARK J strong coupling constant α_s measurements, from the EECA,
are shown as a function of $\sqrt{s}$. Solid points are obtained from the Ali model
and open circles from the LUND model. Curves for $\Lambda_{\overline{MS}}^{(5)} = 60$ and 150
MeV are also shown.[10]

The latest MARK J analysis[10b] compared their data with QCD predictions over the
energy region 14 to 46.78 GeV. Figure 5 shows the MARK J asymmetry data at 35 GeV,
and at higher energies (39.7-46.78 GeV), compared with the Ali Monte Carlo predictions with
$\Lambda = 100$ MeV. The α_s values as a function of $\sqrt{s}$, obtained from fitting the EECA data in 3
GeV bins are shown in Fig. 8. A fit to the energy dependence of the data gives
$\Lambda_{\overline{MS}}^{(5)} = 60 \pm 12_{-20}^{+25}$ MeV for IF and $150 \pm 30_{-20}^{+50}$ MeV for SF. This result is summarized as
$\Lambda_{\overline{MS}}^{(5)} = 100 \pm 30_{-45}^{+60}$ MeV, or $\alpha_s = 0.12 \pm 0.02$ at $\sqrt{s} = 44$ GeV.

3.2 α_s Determination from e^+e^- Quarkonium Resonances

According to QCD, the decay of a quarkonium resonance would proceed through gluon exchange governed by α_s at the scale of the quark mass. The measurements of the decay width of quarkonia therefore provide a testing ground to determine α_s. By using the data from the CLEO and CUSB groups[26] the α_s values are determined by Brodsky, Lepage and Mackenzie[25] as: $\alpha_s(0.157M_\Upsilon) = 0.226^{+0.067}_{-0.042}$ from the radiative decay of Υ, $\alpha_s(0.157M_{\Upsilon'}) = 0.197^{+0.123}_{-0.055}$ from the radiative decay of Υ', $\alpha_s(0.48M_\Upsilon) = 0.165^{+0.006}_{-0.003}$ from Υ leptonic decay, and $\alpha_s(0.48M_\Upsilon) = 0.151^{+0.026}_{-0.018}$ from Υ' leptonic decay.

These α_s values can be transferred to a 4 flavor $\Lambda^{(4)}_{\overline{MS}}$ value 80 to 140 MeV according to (1).

3.3 Summary

Table 4 lists all $O(\alpha_s^2)$ α_s determinations obtained in $e^+e^- \rightarrow$ hadrons[5,10,11,23,24] at $\sqrt{s} = 34.6$ GeV. The following observations are in order:

(1) All measurements using the approximate FKSS calculation gave higher α_s values;

(2) Summarizing all results we therefore include only those which are based on the exact ERT calculation and obtain a mean value of strong coupling α_s at 34.6 GeV:

$$\alpha_s = 0.13 \pm 0.02$$

The error quoted includes both statistical and systematic contributions.

(3) The dominant systematic error of this determination is from the fragmentation model dependence, which is $(20 \pm 10)\%$ for $\alpha_s(SF)/\alpha_s(IF)$.

(4) Comparing with the first order determinations, α_s from the second order measurements is lower by 25-35%.

By using equation (1) above, the α_s value can be transferred to the QCD scale parameter for 5 flavors:

$$\Lambda^{(5)}_{\overline{MS}} = 150 \pm 100 \text{ MeV} .$$

This Λ value is in good agreement with the values obtained from e^+e^- resonances (Section 3.3) and with those obtained from deep inelastic scattering measurements.[27] Figure 9 shows the world's α_s values together with the QCD predictions for $\Lambda^{(5)}_{\overline{MS}}$ as determined from $e^+e^- \rightarrow$ hadron data. As the figure shows, QCD remains a viable candidate theory for the strong interaction.

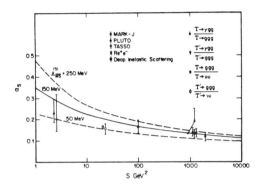

Fig. 9.   A compilation of α_s obtained from $e^+e^- \rightarrow$ hadron compared with α_s obtained from deep inelastic scattering and Y decay. Curves for $\Lambda_{\overline{MS}}^{(5)} = 50$, 150 and 250 MeV are also shown.

4. ACKNOWLEDGEMENTS

I wish to thank Drs. A. Ali, G. Altarelli, F. Barreiro, J. Branson, M. Chen, J. Dorfan, T. Gottschalk, G. Kramer, H. Newman, G. Schierholz, T. Sjöstrand, G. Swider and S.C.C. Ting for many useful discussions.

REFERENCES

1.   P. Soeding, *Proceedings of the International Europhysics Conference on High Energy Physics*, Brighton (U.K.), 20-27 July 1983, p. 567;
     J. Dorfan, *Proceedings of the Int. Symp. on Lepton and Photon Interactions at High Energies*, Cornell, 686 (1983).

2.   (a) B. Adeva *et al.* (MARK J), Phys. Rep. <u>109,</u> 133 (1984);
     (b) M. Chen, MIT-LNS Report 137 (1983), 147 (1985) and Symposium of Subnuclear Science, Erice, Aug. 1983;
     L. Garrido, invited talk at Europhysics Conference on High Energy Physics, Bari (Italy), July 1985.

3.   T. Sjöstrand, Zeit. f. Phys. <u>C26,</u> 93 (1984).

4.   R. Y. Zhu, MIT-LNS Report No. RX-1033 (1983), Ph.D. Thesis, and *Proceedings of the 1984 DPF Conference*, Santa Fe, Oct. 31-Nov. 3, 1984, p. 229.

5.   M. Althoff *et al.* (TASSO), Zeit. f. Phys. <u>C26,</u> 157 (1984).

6. T. D. Gottschalk and M. P. Shatz, Phys. Lett. 150B, 451 (1985); Caltech Preprints CALT-68-1172 and CALT-68-1173.

7. K. Fabricius, I. Schmitt, G. Schierholz and G. Kramer, Phys. Lett. 97B, 431 (1980); F. Gutbrod, G. Kramer and G. Schierholz, Zeit. f. Phys. C21, 235 (1984).

8. T. Sjöstrand, Comp. Phys. Comm. 28, 229 (1983).

9. C. Basham, L. Brown, S. Ellis and S. Love, Phys. Rev. Lett. 41, 1585 (1979) and Phys. Rev. D19, 2018 (1979).

10. B. Adeva et al. (MARK J), (a) Phys. Rev. Lett. 50, 2051 (1983); and (b) Phys. Rev. Lett. 54, 1750 (1985).

11. Ch. Berger et al. (PLUTO), DESY Report 85/039 (1985).

12. G. t'Hooft, Nucl. Phys. B61, 455 (1973);
T. Applequist and A. deRujula, Phys. Rev. Lett. 34, 43 (1975);
T. Applequist and H. D. Politzer, Phys. Rev. Lett. 34, 365 (1975);
W. A. Bardeen et al., Phys. Rev. D18, 3998 (1978);
M. Dine and S. Sapirstein, Phys. Rev. Lett. 43 688 (1979);
A. Buras, Rev. Mod. Phys. 52, 199 (1980);
W. J. Marciano, Phys. Rev. D29, 580 (1984).

13. R. K. Ellis, D. A. Ross and A. G. Terrano, Phys. Rev. Lett. 45, 1226 (1980); and Nucl. Phys. B178, 421 (1981).

14. J. A. M. Vermaseren, K. J. F. Gaemers and S. J. Oldham, Nucl. Phys. B187, 301 (1981).

15. A. Ali et al., Phys. Lett. 93B, 155 (1980); and Nucl. Phys. B168, 490 (1980);
P. Hoyer, P. Osland, H. G. Sander, T. F. Walsh and P. M. Zerwas, Nucl. Phys. B161, 349 (1979).

16. B. Andersson, G. Gustafson and T. Sjöstrand, Z. Physik C6, 235 (1980); and Nucl. Phys. B197, 45 (1982).

17. Z. Kunszt, Phys. Lett. 99B, 429 (1981); and Phys. Lett. 107B, 123 (1981);
A. Ali, Phys. Lett. 110B, 67 (1982).

18. W. A. Bardeen et al., Phys. Rev. D18, 3998 (1978);
M. Dine and S. Sapirstein, Phys. Rev. Lett. 43, 668 (1979);
K. G. Chetyrkin, A. L. Kataev, F. V. Tkachov, Phys. Rev. Lett. 85B, 277 (1979);
W. Celmaster and R. J. Gonsalves, Phys. Rev. Lett. 44, 560 (1979) and Phys. Rev. D21, 3112 (1980).

19. See D. W. Duke and R. G. Roberts, Phys. Rep. 120, 275 (1985) and references therein.

20. T. F. Walsh, *Proceedings of the International Europhysics Conference on High Energy Physics*, Brighton (U.K.), 20-27 July 1983, p. 545.

21. D. G. Richards, W. J. Stirling and S. D. Ellis, Nucl. Phys. B229, 317 (1983);
    A. Ali and F. Barreiro, Phys. Lett. 118B, 155 (1982) and Nucl. Phys. B236, 269 (1984).

22. TASSO Collaboration, data reported by G. Rudolph at International Europhysics Conference on High Energy Physics, Brighton (U.K.), 20-27 July 1983;
    G. Wolf, DESY Report 83/096 (1983).

23. H.-J. Behrend *et al.* (CELLO), Phys. Lett. 138B, 311 (1984).

24. W. Bartel *et al.* (JADE), (a) Phys. Lett. 119B, 239 (1982); and (b) Zeit. f. Phys. C25, 231 (1984).

25. P. B. Mackenzie and G. P. Lepage, Phys. Rev. Lett. 47, 1244 (1981);
    S. J. Brodsky, P. B. Mackenzie and G. P. Lepage, Phys. Rev. D28, 228 (1983).

26. P. Avery *et al.* (CLEO), Phys. Rev. Lett. 50, 807 (1983);
    C. Klopfenstein *et al.* (CUSB), CUSB 83-07.

27. G. Altarelli, review talk at International Europhysics Conference on High Energy Physics, Bari (Italy), July 1985. The $\Lambda_{\overline{MS}}^{(4)}$ value from deep inelastic scattering is 100-500 MeV.

TABLE 1. Comparison of $O(\alpha_s^2)$ integrated 3-parton final state fractions from ERT matrix element and FKSS calculation with y cuts.[4]

y Cut	Integration Range	ERT	FKSS	ERT/FKSS
0.01	$T < 0.85$	0.034 ± 0.001	0.0323	1.04
0.01	$T < 0.90$	0.081 ± 0.001	0.0752	1.08
0.01	$T < 0.95$	0.223 ± 0.003	0.2019	1.11
0.02	$T < 0.85$	0.049 ± 0.001	0.0486	1.01
0.02	$T < 0.90$	0.111 ± 0.001	0.1054	1.05
0.02	$T < 0.95$	0.278 ± 0.003	0.2560	1.09
0.04	$T < 0.85$	0.059 ± 0.001	0.0589	1.00
0.04	$T < 0.90$	0.129 ± 0.001	0.1233	1.05
0.04	$T < 0.95$	0.308 ± 0.003	0.2809	1.10

TABLE 2. Comparison of $O(\alpha_s^2)$ integrated 3-parton final state fractions from ERT matrix element and FKSS calculation[7] with $\varepsilon-\delta$ cuts.[4]

ε	δ	Integration Range	ERT	FKSS	ERT/FKSS
0.05	15	$T < 0.85$	0.027 ± 0.001	0.0260	1.04
0.05	15	$T < 0.90$	0.067 ± 0.002	0.0591	1.13
0.05	15	$T < 0.95$	0.200 ± 0.003	0.1568	1.29
0.07	12.9	$T < 0.85$	0.031 ± 0.001	0.0293	1.04
0.07	12.9	$T < 0.90$	0.074 ± 0.002	0.0652	1.13
0.07	12.9	$T < 0.95$	0.222 ± 0.003	0.1669	1.33
0.10	12.9	$T < 0.85$	0.039 ± 0.001	0.0364	1.06
0.10	12.9	$T < 0.90$	0.093 ± 0.002	0.0787	1.18
0.10	12.9	$T < 0.95$	0.257 ± 0.003	0.1911	1.35

TABLE 3.   A calculation of K values, $O(\alpha_s^2)/O(\alpha_s)$, from a Monte Carlo integration of ERT matrix elements.[4]

Variable	Integration Range	K
Thrust	$T < 0.85$	18.9
Oblateness	$O > 0.30$	3.5
EEC	$|\cos(\chi)| < 0.75$	i2.3
EECA	$|\cos(\chi)| < 0.75$	3.1*

* The small K value here is in agreement with Ref. 21.

TABLE 4.   A summary of α_s values from e^+e^- at $\sqrt{s} = 34.6$ GeV.[5,10,11,23,24]

Group	Date	Variable	QCD	α_s (SF)	α_s (IF)	SF/IF
JADE	Sep. '82	Cluster	FKSS (y)	0.16	0.16	1.0
MARK J	May '83	EECA	ERT (ε,δ)	0.14	0.12	1.17
CELLO	Dec. '83	EECA	FKSS (y)	0.19	0.15-0.12	1.26-1.58
CELLO	Dec. '83	Cluster	FKSS (y)	0.18	0.15-0.12	1.38-1.50
JADE	Jun. '84	EECA	FKSS (y)	0.165	0.14-0.11	1.18-1.47
TASSO	Jun. '84	EECA	ERT (ε,δ)	0.159	0.127-0.117	1.24-1.34
TASSO	Jun. '84	EECA	EXT-FKSS (ε,δ)	0.19	0.16-0.14	1.21-1.36
TASSO	Jun. '84	Shape	EXT-FKSS (ε,δ)	0.20	0.15	1.33
PLUTO	Jun. '85	EECA	ERT (ε,δ)	0.145	0.136	1.07

Simulation of Hadronic Reactions

Frank E. Paige

Physics Department
Brookhaven National Laboratory
Upton, NY 11973, USA

Monte Carlo event simulation is a basic tool for connecting perturbative QCD with experimental data. Recent improvements in models for hadronic reactions are reported.

Perturbative QCD provides a very good description of the data on interactions at large momentum transfer. But perturbative QCD is formulated in terms of quarks and gluons, not the observed hadrons. Hadrons are formed by nonperturbative aspects of QCD characterized by small momentum transfers, implying the creation of a jet of hadrons with limited p_T from each quark or gluon. This nonperturbative hadronization is not understood in any fundamental way, but it can be described by a variety of models. Since the pioneering work of Field and Feynman,[1] Monte Carlo programs based on a combination of perturbative QCD and nonperturbative models have played an increasingly important role in particle physics. They are now widely used to extract results on jets from experimental data, to study signatures and backgrounds for various processes, and to correct data for detector effects.

Any useful Monte Carlo program must describe the complete range of Q^2 from the initial hard scattering down to the formation and decay of hadrons at $Q^2 \ll 1 \, \mathrm{GeV}^2$. The largest Q^2 values are generally described by the appropriate perturbative cross section calculated to leading order in α_s. The smallest values are in the confinement region, so phenomenological models must be used. At intermediate values the effects of QCD radiative corrections are important. Partons (quarks and gluons) of a given Q^2 radiate additional partons having $Q'^2 \sim \alpha_s Q^2$. This radiation is most important when the radiated partons are colliner; in this limit the probability for each additional radiation is given to leading order by a simple factor. Thus perturbative QCD radiation is equivalent in leading order to a classical cascade, known as the branching approximation, which can naturally be implemented in a Monte Carlo approach.[2]

Ever since the branching approximation was first introduced for final state radiation, the need has been recognized to extend it to include initial state radiation in hadronic reactions. The first programs to do this[3,4] clearly demonstrated its importance, but they either were slow[3] or had to pretabulate shower configurations.[4] Recently, Gottschalk[5] and Sjostrand[6] have suggested a new approach in which the cascade is generated backwards from the hard scattering to the incoming hadron. This algorithm is relatively straightforward and efficient, and it has been implemented both in ISAJET[7] and in PYTHIA.[6]

With the inclusion of QCD radiation from the initial state partons, Monte Carlo programs for hadronic reactions now correctly describe perturbative QCD to leading-log accuracy. Only limited attempts have been made so far to incorporate higher order corrections. Because of factorization, the nonperturbative hadronization of jets should be independent of the process, and a phenomenological description of it can be extracted from e^+e^- data. The largest uncertainty in the description of hadronic interactions, therefore, is that associated with the beam jets arising from the spectator partons. This involves completely nonperturbative aspects of QCD, so progress will probably be made only by more experimental studies.

Hard Scattering

The first step in simulating a hadronic interaction is to generate a primary hard scattering according to a cross section $\hat{\sigma}$ calculated in QCD perturbation theory and convoluted with structure functions incorporating QCD scaling violations. (An alternative[6] is to generate the scaling violations from the QCD radiative corrections.) That is, the hard scattering cross section has the standard form of the QCD-improved parton model,

$$\sigma = \hat{\sigma} F(x_1, Q^2) F(x_2, Q^2),$$

where x_1 and x_2 are the usual momentum fractions with $x_1 x_2 = \hat{s}/s$ and Q^2 is the characteristic momentum transfer scale. This is basically straightforward, but because the cross sections vary rapidly some care is needed to make the program efficient. As a result, the number of processes available is limited. ISAJET now includes:[7]

Minimum Bias: No hard scattering at all, so that the event consists only of beam jets.

QCD Jets: All of the $O(\alpha_s^2)$ processes for two-body QCD scattering, including the production of heavy quarks.

Drell-Yan: Production and decay of a γ, W^+, W^-, or Z^0 either alone or at high p_T in conjunction with a jet.

W Pairs: Production of W^+W^-, Z^0Z^0, $W^\pm Z^0$, or $W^\pm \gamma$, including only the contribution from $\bar{q}q$ annihilation.

Supersymmetry: Production of pairs of supersymmetric particles in the simplest model with global supersymmetry.

Other programs generally include even fewer examples of new physics.

EUROJET[8] has taken a somewhat different approach by explicitly including the QCD cross sections for the production of two and three jets, using a cutoff to separate them. Then to avoid double counting it cannot use the branching approximation to produce multijet events. This approach is good at moderate energies for which multi-parton configurations are not important.

QCD Radiative Corrections

Consider the radiation of one extra gluon from a quark line,

$$q(p) \rightarrow q(p') + g(p'').$$

This radiation is most important in the collinear limit, $p^2 \rightarrow 0$, since it is this region which produces the leading-log scaling violations. In this region the cross section is given by σ_0, the cross section without the extra gluon, times a factor,

$$\sigma = \sigma_0 \left[\frac{\alpha_s(p^2)}{2\pi p^2} P(z) \right],$$

where $P(z)$ is an Altarelli-Parisi[10] function and

$$z = \frac{p'_+}{p_+} = \frac{p'_0}{p_0} = \frac{p'_0 + |\vec{p}'|}{p_0 + |\vec{p}|} = \dots .$$

The various choices for z are equivalent for collinear radiation but give different results when continued to large angles. Thus the most important part of the QCD radiation can be expressed in terms of probabilities rather than amplitudes. By using exact noncollinear kinematics multiple jet states can be included, at least approximately. This provides a natural basis for a Monte Carlo algorithm called the branching approximation.[2]

The branching approximation incorporates the leading-log scaling violations for the structure functions and for jet fragmentation. It also gives correctly the structure of jets in QCD perturbation theory, since the typical mass of a jet is small, $M^2 \sim \alpha_s p_T^2 \ll p_T^2$, so that nearly collinear radiation dominates. Finally, the branching approximation turns out to reproduce the leading order three-jet cross section within a factor of about two over all of phase space.[11]

One defect[12] of the branching approximation is that it overestimates the multiplicity of a jet at large Q^2. While it correctly sums the leading series in $\ln Q^2$, it does not sum the $\ln(1/x)$ terms, which are comparable for the multiplicity. More physically, the leading series in $\ln Q^2$ comes from independent radiation of gluons from each parton. Many of the gluons which contribute to the multiplicity have low momentum and are not independent. The correct expression has the same form but with the restriction that the emission angles as well as the masses are ordered, so it can be implemented in a Monte Carlo approach.[13]

Final State Radiation: Consider for simplicity only gluon radiation from a quark line. The cross section for the emission of n gluons is

$$\frac{d\sigma}{dp_1^2 dz_1 \dots dp_n^2 dz_n} = \sigma_0 \frac{1}{n!} \prod_i \left[\frac{\alpha_s(p_i^2)}{2\pi p_i^2} P(z_i) \right],$$
$$Q^2 > p_1^2 > p_2^2 > \dots > p_n^2.$$

Note that this contains both collinear singularities, the explicit $1/p_i^2$, and infrared singularities, the $1/(1 - z_i)$ in the $P(z_i)$.

The infrared singularities cancel in the usual way with those from the virtual graphs. This can be implemented simply by treating $P(z)$ as a distribution.[10] The collinear singularities do not cancel; they are needed to build up the leading-log QCD scaling violations. They are handled by introducing a cutoff $p_i^2 = t_i > t_c$ and assuming that QCD below the cutoff is described by the nonperturbative hadronization model. If t_c is relatively large, then the model must explicitly produce jets, but if $t_c \sim 1\,\mathrm{GeV}^2$, then it can be as simple as cluster decay with two body phase space. The choice made in ISAJET is to take $t_c = (6\,\mathrm{GeV})^2$ and to use independent fragmentation for the nonperturbative model.

The basic quantity needed to set up the Monte Carlo algorithm is the probability $\Pi(t_0, t_1)$ for evolving from an initial mass t_0 to a final mass t_1 emitting no gluon radiation greater than the cutoff. The formula is simple if the cutoff is taken to be not $t > t_c$ but rather $z_c < z < 1 - z_c$. Then the t and z integrations separate, and the result is[2]

$$\Pi(t_0, t_1) = \left[\frac{\alpha_s(t_0)}{\alpha_s(t_1)}\right]^{2\gamma(z_c)/b_0}, \qquad \gamma(z_c) = \int_{z_c}^{1-z_c} dz P(z).$$

For a fixed z cutoff the mass t_1 at which the first resolvable branching occurs is determined from $\Pi(t_0, t_1)$, a value of z is for the branching is generated according to the appropriate $P(z)$, and the proceedure is iterated until all partons reach the cutoff mass t_c. To obtain instead the more physical cutoff at fixed t, the same proceedure is followed using the minimum value of z_c obtained from the initial mass; each branching is rejected if it falls outside the z limits for its mass.

Initial State Radiation: There are two apparent problems in extending the branching approximation to initial state gluon radiation. The virtual masses are spacelike, so that kinematics alone does not order them, and the QCD evolution starting at the proton almost never gives the correct momentum. As has been recently suggested by Gottschalk[5] and by Sjostrand,[6] the solution to both problems is to do the evolution backwards from the desired hard scattering, forcing the ordering of the virtual masses t_i by hand. In Gottschalk's approach, the QCD radiation is used to produce the scaling violations for the structure functions. In Sjostrand's approach, the nonscaling structure functions are assumed to be known and are used to calculate the probabilities for radiation. This allows the initial hard scattering to be generated according to the cross section with nonscaling structure functions, so it has been adopted in ISAJET.

Consider the emission of one extra gluon from a process producing a state X with mass s_{12}. As for final state radiation, all definitions of the momentum fraction z are equivalent if the gluon is collinear, but a choice must be made for noncollinear gluons. The choice for which the branching approximation reproduces first order QCD is $z = s_{12}/s_{13}$. Then the cutoff $|t| > t_c$ plus two-body kinematics for the process $2 + 3 \to X + g$ implies an upper limit on z, while the lower limit is set just by the available beam momentum.

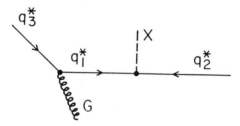

Fig. 1: Radiation of one extra initial state gluon.

In Sjostrand's approach structure functions are used which satisfy the Altarelli-Parisi equations[10] From these equations it follows that the probability for a parton b not to disappear during a backwards evolution from $|t_0|$ to $|t_1|$ is

$$S_b = \exp\left\{ -\int_{|t_1|}^{|t_0|} dt' \frac{\alpha_s(t')}{2\pi t'} \sum_a \int_{z_{\min}}^{z_{\max}} dz \left[\frac{(x/z)f_a(x/z,t)}{x f_b(x,t)} \right] P_{a\to bc}(z) \right\}.$$

This formula, the basis for the initial state Monte Carlo algorithm, is similar to the final state formula for $\Pi(t_0, t_1)$ except for the ratio of structure functions in the square brackets. This ratio implies that a branching can occur only if the structure function $(x/z)f_a(x/z,t)$ for the new initial parton a is large enough so that finding it in the incoming hadron is not improbable.

In generating events according to S_b it is necessary to have a bound for the ratio of structure functions; this is derived by assuming the usual Regge behavior. Otherwise, the generation of the backwards evolution from S_b is done by standard Monte Carlo techniques[6,7] generally similar to those used for the final state.

For a heavy quark the $f_b(x,t)$ vanishes at the threshold $t_q^* = 4m_q^2 x/(1-x)$, making the ratio of structure functions infinite and so forcing the branching $g \to q\bar{q}$ to occur. Hence a heavy quark in the initial state will be accompanied by an associated antiquark.

Jet Fragmentation

The fragmentation of partons into jets involves physics at the confinement scale and hence must rely on models. The factorization of collinear singularities in QCD perturbation theory, however, strongly suggests that jet fragmentation should be independent of the process, so that e^+e^- results can be used.

The simplest model is independent fragmentation.[1,14,15] To fragment a light quark q of momentum p, a $q'\bar{q}'$ pair is generated with small transverse momenta $\pm \vec{k}_T$. The $\bar{q}'$ is combined with the q to form a 0^- or a 1^- meson carrying a momentum $p'_+ = zp_+$, where z is choosen according to some distribution $f(z)$ such as

$$F(z) = 1 - a + a(b+1)(1-z)^b.$$

Hadrons with values of z so small that p_L is negative are discarded. The proceedure is then iterated for the new q' until all the momentum is used. The algorithm can be generalized to include heavy quarks, gluons, and baryon production through diquarks.

Independent fragmentation is very simple and incorporates the most important features of jet fragmentation, but it has a number of obvious defects. Since a massless parton is fragmented into massive hadrons, energy and momentum cannot be conserved except by arbitrary proceedures. Similarly, flavor is not conserved, since hadrons with $p_L < 0$ and the final quark are discarded. Finally, a nearly collinear branching leads to a larger multiplicity even if the mass is so small that the new partons could not possibly be resolved. Such branchings must be included in the QCD jet evolution to get the correct scaling violations, but then the structure of the events depends on the cutoff.

The Lund model[16] is more complicated, but it avoids many of the problems of independent fragmentation by treating hadronization as the breakup of a relatavistic string between the quark and the antiquark. Since the whole system is fragmented together, energy, momentum, and flavor can all be conserved. A gluon is treated as a kink on the string, so a $q\bar{q}g$ event smoothly merges into a $q\bar{q}$ one as the gluon becomes parallel to the quark or as its energy becomes small.

Cluster models[2,17,18] attempt to derive all the properties of jets by carrying the QCD cascade down to masses of order 1 GeV and then hadronizing low-mass clusters using a simple model such as phase space. Existing models use both the incoherent, leading-log branching approximation discussed above and a branching including coherent effects which reproduce the correct asymptotic multiplicity growth. The use of perturbation theory down to small masses gives such models a lot of predictive power, but it need not be correct.

For e^+e^- interactions the Ali et al.[14] and Hoyer et al.[15] independent fragmentation models, the Lund string model,[16] the Gottschalk incoherent cluster model,[17] and the Webber coherent cluster model[18] all fit most of the data from PEP and PETRA very well. However, the Lund and Webber models, which include coherent effects between jets, do a better job of describing the slow particles in the events.[19] Since there are other sources of soft particles in hadronic reactions, independent fragmentation has been chosen for ISAJET on the basis of simplicity.

Beam Jet Fragmentation

After a hard scattering event has been generated, something must be done with the remaining constituents of the proton. While factorization in QCD perturbation theory implies that high p_T jets must be treated like jets in e^+e^- reactions, there is very little theoretical guidance on what to do with the beam jets. The assumption made in ISAJET is that they are identical to a minimum bias event at the reduced energy, but this is surely too simplistic.

While the standard pictures of multiparticle production give short-range rapidity correlations and essentially a Poisson multiplicity distribution, the minimum bias data show long-rang correlations and a broad multiplicity distribution. One solution[20] to this conflict is that the large fluctuations arise from unitarity. The basic amplitude is a single chain, or cut Pomeron, having only short range correlations and giving an average multiplicity $\bar{n}$ with Poisson fluctuations.

But unitarity requires that this amplitude be iterated, leading to graphs with K cut Pomerons giving multiplicity $n \sim K\bar{n}$. A different discontinuity of the two Pomeron graph gives the elastic cross section, so the probability for $2\bar{n}$ should be of order $\sigma_{el}/\sigma_{tot} \approx .20$. Furthermore, events with high multiplicity in one region generally have several cut Pomerons and hence high multiplicity everywhere.

A version of this scheme has been implemented in ISAJET. A number K of cut Pomerons is selected, the energy is divided among them, and each cut Pomeron is hadronized in its own center of mass using a modified independent fragmentation model. Since the single particle distribution is determined by the single chain graph, the observed increase in dn/dy is incorporated by making the splitting function is made energy dependent. The probabilities P_K for K cut Pomerons are taken to be independent of energy and are adjusted to fit the experimental data.

An alternative scheme, which is used in the latest version of PYTHIA,[6,21] is to treat the hadronization of beam jets like that of e^+e^- jets. Then it is necessary to introduce multiple scattering to reproduce the minimum bias data. The number of scatterings is taken to be proportional to the QCD jet cross section,

$$\frac{dN}{dp_T} = \frac{1}{\sigma_0} \frac{d\sigma}{dp_T}$$

where σ_0 should be of order the total cross section. At very high energies something like this approach is necessary, since events with jets comprise a large fraction of the total cross section. At $S\bar{p}pS$ energies it offers the possibility of relating the beam jets to e^+e^- physics, but any obvious connection with unitarity is lost.

Results and Discussion

ISAJET jet events with $p_T > 35\,\text{GeV}$ were generated at $\sqrt{s} = 540\,\text{GeV}$. Jets were found by the UA1 algorithm, and for those with $p_T > 20\,\text{GeV}$ the scalar transverse energy flow dE_T/dy was plotted versus $y - y_{\text{jet}}$. The results and the data from the UA1 Collaboration[21] are shown in Fig. 2. Compared to the data, ISAJET gives slightly too narrow jets, but the transverse energy flow away from the jets is about right. This is in contrast with previous versions which did not include initial state QCD radiation and which gave too little dE_T/dy by a factor of two away from the jets.

Results from PYTHIA with only single scatterings[6] are also shown in Fig. 2. This model gives too low a value for dE_T/dy away from the jets, even though it treats initial state QCD radiation in the same way as ISAJET. Part of the difference occurs because PYTHIA uses Lund string fragmentation, which tends to merge two nearly collinear jets and so gives fewer hadrons from the same initial state gluons. The larger difference, however, is that PYTHIA treats the beam jet as a string system, giving a minimum bias multiplicity significantly below the data at $S\bar{p}pS$ energies.

PYTHIA can obtain agreement[21] with the data by including multiple hard scatterings with $p_T > 1.6\,\text{GeV}$ with a normalization cross section $\sigma_0 = 10\,\text{mb}$. The value of σ_0 seems rather small, and the fit to the minimum bias data requires

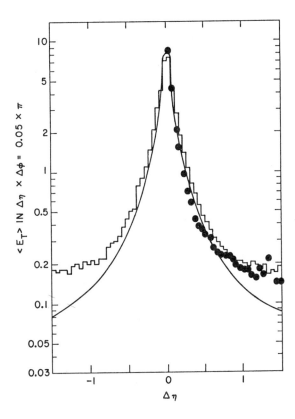

Fig. 2: Transverse energy flow dE_T/dy in $\Delta\phi < 90°$ vs. $y - y_{\text{jet}}$.
The histogram is data from the UA1 Collaboration,[22] the circles
are from ISAJET,[7] and the smooth curve is from PYTHIA.[6]

$\sigma_0 = 40\,\text{mb}$, so that the jet model does not extrapolate smoothly to low p_T. This
may be a detail, or it may indicate that the physics of beam jets is really different.

There is no reason to believe that the underlying beam jet must be the same in
a hard scattering event as in a minimum bias event. Indeed, one might expect that
a hard scattering would disrupt the leading proton and so change the beam jet.
Since string fragmentation seems to work better than independent fragmentation
for e^+e^- interactions, it may be that less of the dE_T/dy should be attributed to
gluon radiation and more to soft physics than is the case in ISAJET.

Initial state gluon radiation also reproduces the transverse momentum distribution for the $W^{\pm}$ quite well.[7,22]

At higher mass scales the perturbative QCD effects should dominate the nonperturbative hadronization. For the SSC energy and p_T range ISAJET predicts[25]

that dE_T/dy away from the jet increases almost in proportion to p_T, as it must in a theory with no scale. Such an increase will have important consequences for experiments at very high energies.

I wish to thank S. Protopopescu, T. Gottschalk, and T. Sjostrand for many valuable discussions. This work is supported in part by the United States Department of Energy under Contract No. DE-AC02-76-CH00016.

References

1. R.D. Field and R.P. Feynman, Nucl. Phys. <u>B136</u>, 1 (1978).
2. G.C. Fox and S. Wolfram, Nucl. Phys. <u>B168</u>, 285 (1980).
3. R.D. Field, in *1984 Summer Study on the Design and Utilization of the Superconducting Super Collider* (Snowmass, CO, 1984) p. 713.
4. R. Odorico, Computer Phys. Comm. <u>32</u>, 139 (1984).
5. T.D. Gottschalk, CALT-68-1241 (1985).
6. T. Sjostrand, Phys. Letters <u>157B</u>, 321 (1985).
7. F.E. Paige and S.D. Protopopescu, BNL-37066 (1985).
8. A. Ali, private communication.
9. R.K. Ellis, D.A. Ross, and A.E. Terrano, Phys. Rev. Letters <u>45</u>, 1226 (1980);
   J.A.M. Vermasseren, K.J.F. Gaemers, and S.J. Oldham, Nucl. Phys. <u>B187</u>, 301 (1981);
   K. Fabricius, G. Kramer, G. Schierholz, and I. Schmidt, Phys. Letters <u>97B</u>, 431 (1980).
10. G. Altarelli and G. Parisi, Nucl. Phys. <u>B126</u>, 298 (1977).
11. T.D. Gottschalk, E. Mondsay and D. Sivers, Phys. Rev. <u>D21</u>, 1799 (1980).
12. A.H. Mueller, Phys. Lett. <u>104B</u>, 161 (1981);
    A. Bassetto, M. Ciafaloni, G. Marchesini and A.H. Mueller, Nucl. Phys. <u>B207</u>, 189 (1982).
13. G. Marchesini and B.R. Webber, Nucl. Phys. <u>B238</u>, 1 (1984).
14. A. Ali, E. Pietarinen, G. Kramer and J. Willrodt, Phys. Lett. <u>93B</u>, 155 (1980).
15. P. Hoyer, P. Osland, H.G. Sander, T.F. Walsh and P.M. Zerwas, Nucl. Phys. <u>B161</u>, 349 (1979).
16. B. Andersson, G. Gustafson, G. Ingelman and T. Sjostrand, Comp. Phys. Comm. <u>27</u>, 243 (1982); ibid., <u>28</u>, 229 (1983).
17. T.D. Gottschalk, Nucl. Phys. <u>B207</u>, 201 (1983).
18. B.R. Webber, Nucl. Phys. <u>B238</u>, 492 (1984).
19. JADE Collaboration, DESY 85-036 (1985).
20. V.A. Abramovskii, O.V. Kanchelli and V.N. Gribov, in *XVI International Conf. on High Energy Physics* (Batavia, IL, 1973) Vol. <u>1</u>, p. 389.
21. T. Sjostrand, Fermilab-Pub-85/19 (1985).
22. UA1 Collaboration, CERN-EP/84-160 (1984).
23. V. Barger, T.D. Gottschalk, J. Ohnemus, and R.J.N. Phillips, Mad/Ph/247 (1985).
24. F.E. Paige and S.D. Protopopescu, to appear in *Supercollider Physics Topical Conference*, (Eugene, OR, 1985).

APPLICATION OF MONTE CARLO RENORMALIZATION GROUP METHODS IN NON-ABELIAN LATTICE GAUGE THEORIES

Frithjof Karsch

Department of Physics,
University of Illinois at Urbana-Champaign,
1110 W. Green Street, Urbana, IL 61801

ABSTRACT

We discuss results for the SU(3) β-function obtained with Monte Carlo Renormalization Group techniques at intermediate coupling values. We find evidence for a pronounced dip in the β-function below $\beta = 6.0$ which implies that the asymptotic value is approached from below. Deviations from this asymptotic value seem to be small above $\beta = 6.05$.

I) INTRODUCTION

During the recent years we have seen that Monte Carlo simulations of lattice gauge theories provide a powerful tool to obtain interesting results for non-perturbative observables of Quantum Chromodynamics (QCD) like the string tension, σ, or the deconfinement temperature, T_d. A detailed analysis of these observables on large lattices over a wide range of couplings, however, has shown that they did not scale as expected from the perturbative β-function of QCD. The approach to the continuum limit thus seems to be more complicated then expected and a more detailed study of the non- perturbative β-function seems to be neccessary in order to to get a better understanding of the observed deviations from asymptotic scaling at intermediate couplings. Monte Carlo Renormalization Group (MCRG) techniques have been developed to deal with such problems.[1,2]

On a discreet Euclidean space-time lattice the action of a SU(3) gauge theory is defined as

$$S_G = \beta \sum_{plaq.} (1 - \frac{1}{3} \text{ReTr} U_{x,\mu} U_{x+\mu,\nu} U^\dagger_{x+\nu,\mu} U^\dagger_{x,\nu}) \tag{1}$$

with $\beta \equiv 6/g^2$. In order to recover continuum physics the lattice cutoff has to be removed. This requires a renormalization of the bare parameter g^2 in order to leave physical observables invariant. However, it is only in the continuum limit that a unique scaling relation $g(a)$ exists which ensures that all physical observables remain unchanged during the renormalization process $(a,g) \rightarrow (a',g')$. This regime of coupling constant values is generally referred to as scaling regime, lattice artefacts are small in this regime and physical observables can be extracted from lattice simulations once the β-function

$$B(g) = -a \frac{d}{da} g(a) \tag{2}$$

is known. However, in general we know only the regularization scheme independent perturbative result for the SU(N) β-function

$$B(g) = -b_0 g^3 - b_1 g^5 + O(g^7) \tag{3}$$

with

$$b_0 = \frac{11N}{48\pi^2} \quad ; \quad b_1 = \frac{34}{3} (\frac{N}{16\pi^2})^2 \tag{4}$$

In the asymptotic ($g^2 \rightarrow 0$) scaling regime, where eq.(3) is valid, we obtain an analytic relation between the lattice cutoff a and the bare coupling g^2

$$a \Lambda_L = (b_0 g^2)^{-b_1/2b_0^2} \exp\{-1/(2b_0 g^2)\} \tag{5}$$

We thus can remove the lattice cutoff a from measured dimensionless observables like σa^2 or $T_d a$ in favor of the Λ_L-parameter and check whether $\sqrt{\sigma}/\Lambda_L$ or T_d/Λ_L are independent of the bare coupling g^2. ine fact that this is not the case for $\beta \simeq 6.0$ shows that the application of the asymptotic β-function was unjustified in this regime of coupling constants. It may, however, still be true that physical observables scale according to a more complicated non- perturbative β-function in this regime and indeed reflect continuum physics already at these intermediate coupling values.

II) THE MCRG APPROACH

In the MCRG approach one does not determine the β-function itself but a related quantity $\Delta\beta(\beta)$, which gives the change in β neccessary to compensate for a change in lattice spacing by a factor b (in the following we will assume b=2),

$$\Delta\beta(\beta) = \beta(a) - \beta'(2a). \tag{6}$$

$\Delta\beta(\beta)$ is related to the β-function through the integral equation

$$\int_{\beta-\Delta\beta}^{\beta} \frac{dx}{x^{3/2}B(\sqrt{6/x})} = - \frac{2 \ln 2}{\sqrt{6}}. \tag{7}$$

The RG-ideas underlying the operative procedure to determine the $\Delta\beta(\beta)$ -function are illustrated in fig.1)

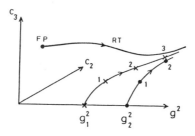

FIG. 1: Schematic diagram of the anticipated RG flow in a three dimensional coupling constant space. FP denotes the fixpoint in the $g^2 = 0$ hyperplane while RT denotes the renormalized trajectory.

Starting with a one parameter action, i.e. the Wilson action for pure SU(3) gauge theory with one bare coupling g^2, all kind of new interactions will be generated even after the first RG-transformation step. The values of the corresponding couplings describe a flowline in a multidimensional coupling constant space. QCD has a ultraviolet attractive fixpoint in the $g^2 = 0$ hyperplane which has one direction of instability, the renormalized trajectory (RT). Starting with an action close to the fixpoint the flowlines get attracted by the RT. Successive points on the flowlines describe identical long range physics but the corresponding lattice scale changes by a factor b=2 after each blocking step. Thus by arranging the couplings g_1^2, g_2^2 of the starting configurations such that after a few blocking steps both sequences coincide on the RT but differ by one blocking step the couplings β, β' can be determined which correspond to starting lattices with a factor b=2 different lattice scales a, a'=2a.

III) RESULTS FOR THE SU(3) β-FUNCTION

The application of the above described RG-ideas in a MC-simulation require some additional care as these simulations have to be performed on finite lattices. In order to minimize finite size effects blocking sequences from L^4 lattices after n-blocking steps are compared with sequences obtained from $(L/2)^4$ lattices after (n-1) blocking steps. As the finite size of the starting lattices allow only a small number of blocking steps it is also desirable to use a RG-transformation which allows to reach the RT as soon as possible.[3,4] Using a RG-transformation with one or more parameters the RG-transformation can be systematically optimized by demanding good (consistent) matching results for as many observables as possible already after the first blocking step. A possible choice of a RG-transformation with one free parameter has been suggested by R.H.Swendsen[3]: The new block link is selected with the probability

$$\text{prob.}(V_{AB}) \sim \exp\{\frac{p}{6} \text{ Tr } (V_{AB}^\dagger X + \text{h.c.})\}, \qquad (8)$$

were X is taken to be the sum of the matrix products along 7 different paths connecting the sites A and B as indicated in fig.2 (For possible other choices of X see ref 4.))

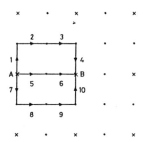

FIG. 2: Construction of the block link using 7 paths of length 2 and 4.

In practice the matching results of a large set of Wilson loop expectation values are used to fix P and to determined $\Delta\beta(\beta)$. Table I illustrates how the optimization of the parameter P can be achieved.

P	Blocking step	□	▭	⌐	⌐⌐
	1	0.592(6)	0.600(6)	0.534(5)	0.503(5)
	2	0.487(9)	0.475(10)	0.474(9)	0.467(10)
25	3	0.44(2)	$0.45\binom{+2}{-3}$	$0.44\binom{+2}{-3}$	$0.45\binom{+2}{-3}$
	$\frac{1}{3}(4\Delta\beta(n-2)-\Delta\beta(n-1))$	0.451(14)	0.433(15)	0.454(13)	0.456(14)
	$\frac{1}{3}(4\Delta\beta(n-3)-\Delta\beta(n-2))$	0.43(3)	$0.44\binom{+3}{-4}$	$0.43\binom{+3}{-4}$	$0.44\binom{+3}{-4}$
	~1	0.356(3)	0.439(5)	0.347(3)	0.348(4)
	2	0.421(9)	0.426(11)	0.419(10)	0.419(12)
40	3	$0.428\binom{+14}{-15}$	0.43(2)	0.43(2)	$0.43\binom{+2}{-3}$
	$\frac{1}{3}(4\Delta\beta(n-2)-\Delta\beta(n-1))$	0.442(13)	0.422(17)	0.443(14)	0.443(17)
	$\frac{1}{3}(4\Delta\beta(n-3)-\Delta\beta(n-2))$	0.43(2)	0.43(3)	0.43(3)	$0.44\binom{+3}{-4}$

TABLE I: The matching predictions $\Delta\beta(\beta)$ for $\beta = 6.3$ and P=25 and 40 together with estimates for $(\Delta\beta)^{n=\infty}$ for an infinite number of blocking steps.[4]

The matcing predictions after 1,2 and 3 blocking steps obtained for
two different values of P are given. It can be seen that while the
lower value of P leads to decreasing values for $\Delta\beta(\beta)$ with increasing
number of blocking steps the larger value of P gives increasing
values. There is obviously an optimal choice for P which gives more
or less constant values for $\Delta\beta$ already after the first blocking step.
Of course all values of P should lead finally to the same value for
$\Delta\beta(\beta)$, suggesting that the P-dependence should become weaker with
increasing number of blocking steps. That this is indeed the case is
shown in fig.3 where the P-dependence of the matching predictions
after the first and second blocking step for all observables measured
is shown.

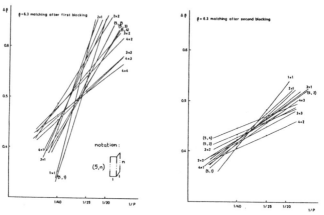

Fig.3: The matching predictions $\Delta\beta(\beta)$ obtained from various block
loops after the first (a) and second (b) blocking step at $\beta=6.3$ as a
function of the free parameter P appearing in the RG-transformation
defined in eq.(8). Planar Wilson loops of size RxT have been measured
as well as a few non-planar loops as indicated in fig.3a.

The above RG-transformation has been used to study the β-function at
three values of β, β = 6.0, 6.3 and 6.6.[4] The resulting values for
$\Delta\beta(\beta)$ are given in table II along with results obtained with the 1-
loop improved ratio method.[5,6]

	$\Delta\beta(\beta)$	
β	blocking	ratio method
5.8		.41 ± .09
6.0	.35 ± .02	.356 ± .03
6.2		.45 ± .04
6.3	.43 ± .03	.449 ± .03
6.4		.50 ± .05
6.6	.56 ± .06	.56 ± .05

Table II: Results for $\Delta\beta(\beta)$ obtained with the 1-loop improved ratio method[5,6] and optimized block transformations.[4]

IV) DISCUSSION AND CONCLUSIONS

The data obtained from optimized block transformations and 1-loop improved ratio tests together with the asymptotic scaling prediction are shown in fig.4.

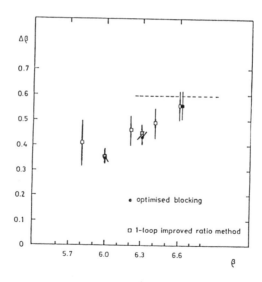

Fig.4: The shift $\Delta\beta(\beta)$ as a function of β. The data given in table II are plotted together with asymptotic scaling result from eq.(3).

Obviously there are large deviations from the asymptotic scaling result below $\beta = 6.0$. However, the result for $\Delta\beta(\beta)$ at $\beta = 6.6$, $\Delta\beta$ (6.6) $= 0.56 \pm .05$ indicates that the scaling violations are small in the whole intervall (6.04,6.6). Similar conclusions have been drawn from the analysis of the β-function with a $\sqrt{3}$ RG-transformation.[7] The same qualitative behavior has been observed in measurements of the string tension[6] and deconfinement temperature.[8] This would suggest that there is a early onset of scaling already in the region $5.5 < \beta < 6.0$, where different observables follow a unique scaling behavior. However, it appears that the SU(3) mass gap has a qualitative different behavior in this regime[9] and is more consistent with asymptotic scaling. A more detailed analysis is certainly neccessary to pindown the starting point of the scaling regime where a unique β-function exists which describes the scaling behavior of all observables.

A similar structure found for the SU(3) β-function has also been observed in string tension[10] and MCRG studies[11] of the SU(2) gauge theory. The dip, however, is not as pronounced as in the SU(3) case. This is in agreeement with the intuitive picture that the dip is related to a nearby second order phase trasnition in the fundamental-adjoint coupling constant plane.[4,12] This critical point is closer to the fundamental axis in the case of SU(3) than for SU(2) and thus suppresses $\Delta\beta(\beta)$ stronger in the vicinity of the transition point. This would suggest that the approach to the continuum limit is smoother along lines $\beta_f/\beta_a = -c$ [c>0] in the fundamental-adjoint plane. An analysis of $\Delta\beta$ along the line $\beta_f = -\frac{1}{6}\beta_a$, however, shows similar deviations from asymptotic scaling as on the fundamental axis.[13]

In conclusion we have seen that MCRG studies gave a quantitative understanding of the non-perturbative β-function at intermediate couplings. However, more work is neccessary to clarify whether a unique scaling regime below the onset of asymptotic scaling exists. In addition the analysis of the β-function in a larger coupling space would be interesting to understand the origin of the dip in the β-function and its relation to the singularities in the fundamental-adjoint coupling plane.

References

1) Ma, S. K., Phys. Rev. Lett. 37 461 (1976); Swendsen, R. H., Rev. Lett. 42 859 (1979); K. G. Wilson in "Recent Developments in Gauge Theories", ed. G. t'Hooft et al.,(Plenum Press 1980).

2. For a recent review see: Hasenfratz, P., CERN-TH.3999/84, Lectures given at the International School of Subnuclear Physics, Erice 1984.

3) Swendsen, R. H., Phys. Rev. Lett. 47 (1981) 1775 and in "Statistical & Particles Physics", Proc. Scottish Universities Summer School in Physics, ed. K. C. Bowler and A. J. McKane (SUSSP 1984).

4) Bowler, K. C., Hasenfratz, A., Hasenfratz, P., Heller, U., Karsch, F., Kenway, R. D., Meyer-Ortmanns, H., Montvay, I., Pawley, G. S. and Wallace, D. J., Nucl. Phys. B257 [FS14] (1985).

5) Hasenfratz, A. Hasenfratz, P. Heller, U. and Karsch, F., Ph.. Lett. 140B 76 (1984).

6) Bowler, K. C., Gutbrod, F., Hasenfratz, P., Heller, U., Karsch, F., Kenway, R. D., Montvay, I., Pawley, G. S., Smit, J., and Wallace, D. J., "The β-function and potential at β=6.0 and 6.3 in SU(3) Gauge Theory", ITFA-85-07 and references therein.

7) Gupta R., Guralnik, G., Patel, A., Warnock, T. and Zemach, C., Phys. Rev. Lett. 53 1721 (1984).

8) Kennedy, A. D., Kuti, J., Meyer, S., and Pendleton, B. J., Phys. Rev. Lett. 54, 87 (1985).

9) de Forcrand, P., Schierholz, G., Schneider, H., and Teper, M., DESY preprint 84-107 (1984).

10) Gutbrod F., and Montvay, I., Phys. Lett. 136B 411 (1984); Karsch F., and Lang, C. B., Phys. Lett. 138B 176 (1984).

11) Mackenzie P. B., in "Gauge Theory on a Lattice: 1984", Proc. Argonne National Laboratory Workshop (ANL 1984); Heller U., and Karsch, F., Phys. Rev. Lett. 54 1765 (1985); Patel, A., Otto S., and Gupta, R., Phys. Lett. 159B 143 (1985)..

12) Makeenko Yu., and Polikarpov, M. I., Phys. Lett. 135B 133; (1984) Martinelli G., and Polikarpov M. I., ITEP-145 (1984).

13) Chalmers, D. L., Phys. Lett. 160B 133 (1985).

LATTICE GAUGE THEORY WITH FERMIONS: A PROGRESS REPORT

John B. KOGUT

Department of Physics, University of Illinois at Urbana-
Champaign, 1110 West Green Street, Urbana, IL 61801

We present a progress report in lattice gauge theory computer
simulations which include the effects of light, dynamical
fermions. Microcanonical and hybrid microcanonical-Langevin
alogrithms are presented and discussed. Physics applications
such as the thermodynamics of Quantum Chromodynamics, hierarchal
energy scales in unified gauge theories, and the phase diagram
of theories with many fermion species are discussed. Prospects
for future research are assessed.

I.   LATTICE GAUGE THEORY WITH FERMIONS

The four dimensional Euclidean Action density S for lattice
gauge theory with fermions reads generically,

$$S = \sum_{ij} \bar{\psi}_i [\not{D}(U)+m]_{ij} \, \psi_j + S_o(U) \tag{1.1}$$

where ψ_i is a Grassman field at site i, $A_{ij} = [\not{D}(U)+m]_{ij}$ is the gauge
covariant Dirac operator and $S_o(U)$ is the pure gauge field Action on
the lattice.[1] The precise form of the gauge covariant discrete
difference operator $\not{D}(U)$ depends on the lattice fermion method
employed. We will be considering staggered fermions[2] in this article
so ψ_i will be one component objects and the fermion contribution to
Eq. (1.1) reads,

$$\sum_n \bar{\psi}(n)\{\frac{1}{2} \sum_{\mu=1}^{4} \eta_\mu(n)[U_\mu(n)\psi(n+\mu)-U_\mu^+(n-\mu)\psi(n-\mu)]+ m\psi(n)\} \tag{1.2}$$

where $\eta_\mu(n)$ are phase factors that carry the spin-1/2 character of the
continuum Dirac field and $U_\mu(n)$ is the SU(3) rotation matrice residing
on the link between sites n and n + μ. For the purposes of this
discussion all these details are not essential. Suffice it to say
that Eq. (1.2) has the good feature of describing four species of
Dirac fermions which become massless when m $\to$ 0 in a natural fashion.
$\langle\bar{\psi}\psi\rangle$ is a good order parameter for chiral symmetry, one of the two
basic quantities (confinement is the other) of interest here.

Since the subject of this talk is the status of computer simulations of lattice gauge theory with fermions, our interest focuses on the partition function,

$$Z = \int \prod_i d\psi_i \prod_j d\bar{\psi}_j \prod_{n,\mu} dU_\mu(n)\exp(-S) \qquad (1.3)$$

Since the ψ_i are anti-commuting numbers a direct simulation of Eq. (1.3) is not practical. Instead the fermions can be integrated out of Eq. (1.3) since Eq. (1.1) is a quadratic form in ψ,

$$Z = \int \prod dU_\mu(n) \det[\not{D}(U)+m] \exp(-S_0(U))$$

$$= \int \prod dU_\mu(n) \exp(-S_0(U) + tr \ln[\not{D}(U)+m]) \qquad (1.4)$$

It is not so clear, however, that this step represents real progress since tr $\ln[\not{D}(U)+m]$ is an effective, non-local interaction among the U-variables. Such Actions are not well studied and classified in the context of traditional statistical mechanics approaches to critical phenomena. At least the determinant in Eq. (1.4) is positive semi-definite for staggered fermions.

We all recognize the physical origin for the determinant here. It represents closed fermion loops, virtual quark-antiquark pairs, and the plus sign, $+tr \ln[\not{D}(U)+m]$, in Eq. (1.4) is responsible for the perturbation theory rule: -1 for each closed fermion loop.

Various numerical approaches to evaluating Eq. (1.4) and physically relevant matrix elements have been proposed. Monte Carlo methods, the so-called pseudo-fermion algorithms[3], are being studied as well as microcanonical[4,5] and Langevin equations.[5] I will concentrate on the latter two methods in this review.

II. THE MICROCANONICAL ENSEMBLE AND MOLECULAR DYNAMICS

We begin by reviewing the molecular dynamics approach[4] to problems in equilibrium statistical mechanics. Consider a boson field ϕ which might be defined on a lattice. The theory has an action $S(\phi)$ which determines its Path Integral and equilibrium statistical mechanics properties. This system has no natural dynamics which would govern its approach to equilibrium. However, it can be given dynamics in several ways -- the molecular dynamics and the Langevin equations are two alternatives. In the molecular dynamics approach we associate $S(\phi)$ with a potential $V(\phi) \equiv \beta^{-1} S(\phi)$ and construct a fictitious Hamiltonian,

$$H = T + V = \sum_i \frac{1}{2} p_i^2 + V(\phi) \qquad (2.1)$$

where i labels lattice sites and p_i will soon be interpreted as the momentum conjugate to ϕ_i. Using Eq. (2.1) we could consider the classical statistical mechanics based on the invariant phase space

$\Pi\ dp_i d\phi_i$ and the Boltzmann factor $\exp(-\beta H)$. Since the p_i-integrals are trivial, ,this formulation reduces to the original Path Integral formulation of the boson field theory.

To give this approach some meat, we identify p_i with the momentum conjugate to ϕ_i by introducing a 5th dimension τ into the problem,

$$p_i = d\phi_i/d\tau \qquad (2.2)$$

Then the ensemble given by the phase space measure $\Pi\ dp_i d\phi_i$ and the Boltzmann factor $\exp(-\beta H)$ defines the usual canonical ensemble of classical statistical mechanics. There is still no advantage in all this until one passes to the microcanonical ensemble. Now the energy is fixed $H = E$ and the measure in phase space is $\Pi\ dp_i d\phi_i \delta(H-E)$. Observables in the system $\theta(p,\phi)$ have expectationivalues,

$$\langle\theta\rangle = \frac{1}{Z} \int \Pi_i\ dp_i d\phi_i\ \delta(H-E)\ \theta(p,\phi) \qquad (2.3)$$

If θ is just a function of ϕ, then standard arguments apply to show that $\langle\theta\rangle$ calculated in the microcanonical ensemble is the same as $\langle\theta\rangle$ calculated in the canonical ensemble in the large volume $V \to \infty$ limit.[6]

But $\langle\theta\rangle$ can also be calculated from the time evolution of the classical system. This is the molecular dynamics approach to the problem. Let $(\phi(\tau),p(\tau))$ describe the phase space point of the physical system. Then a time-average of θ can be calculated,

$$\langle\theta\rangle = \lim_{T\to\infty} \frac{1}{T} \int_0^T \theta(p(\tau),\phi(\tau))d\tau \qquad (2.4)$$

This time average reproduces the expectation value Eq. (2.3) if the Ergodic Hypothesis works for this physical system. Roughly speaking, one must assume that the Hamiltonian dynamics of the system carries the phase space point $(\phi(\tau),p(\tau))$ uniformly over the energy shell $H = E$.

The final ingredient in this molecular dynamics approach is the computation of the coupling β given the system's fixed energy. The necessary correspondence follows from the equi-partition theorem,

$$\langle T\rangle = \frac{1}{2} \beta^{-1} N \qquad (2.5)$$

where N is the number of independent, excited degrees of freedom in the system.

Eq. (2.4) and (2.5) coupled with the Hamilton equations of motion following from Eq. (2.1) and (2.2) represent a clear alternative to Monte Carlo simulation procedures of pure bose systems. This formulation has several interesting points: (1) It is fully deterministic, (2) It involves ordinary coupled differential equations and (3) It generalizes to a practical method for fermions. Let's review the fermion method before discussing its strengths and weaknesses further.

Now we wish to invent a classical system in 4+1 dimensions involving only complex numbers whose molecular dynamics generates the Path Integral Eq. (1.4) with the infamous fermion determinant. Consider the Lagrangian[5],

$$L = -S_0(U) + \frac{1}{2} \sum_{n,\mu} \dot{U}_\mu^\dagger(n) \hat{P} \dot{U}_\mu(n) + \sum_{ij} \dot{\phi}_i^\dagger [A^\dagger A]_{ij} \dot{\phi}_j - \omega^2 \sum_i \phi_i^\dagger \phi_i \quad (2.6)$$

where A is the lattice Dirac operator defined earlier and $\hat{P}$ is a projection operator. Thus L consists of kinetic energy terms for the gauge fields and the pseudo-fermions, and potential terms for both. $A^\dagger A$ appears in L rather than A itself to insure positivity. This unusual form for the pseudo-fermion kinetic energy will generate the fermion determinant with the correct sign. Note that this L is local because A couples only nearest neighbors.

It is straight-forward to identify the canonical momenta of this physical system,

$$p_\mu(n) = \dot{U}_\mu(n) \qquad P_i = [\dot{\phi}^\dagger A^\dagger A]_i \quad (2.7)$$

and construct the Hamiltonian

$$H = \frac{1}{2} \sum p^2 + \sum P^\dagger (A^\dagger A)^{-1} P + S_0(u) + \omega^2 \sum \phi^\dagger \phi \quad (2.8)$$

and consider the Hamiltonian equations of motion,

$$\dot{P}^\dagger = \frac{d}{d\tau} (A^\dagger(U)A(U)\dot{\phi}) = -\omega^2 \phi$$

$$\dot{p} = \ddot{U} = -\frac{\partial}{\partial U^\dagger} S_0(U) + \dot{\phi}^\dagger \frac{\partial}{\partial U^\dagger} (A^\dagger(U)A(U))\dot{\phi} \quad (2.9)$$

These equations are generic in character. The real equations which are simulated choose a convenient parametrization for the $U_\mu(n)$ matrices and incorporate constraints appropriately.[7] But the point to be stressed here is simply that Eq. (2.9) is a tractable set of coupled ordinary differential equations. The fermions introduce the complication of requiring the solution of a sparse set of linear equations for $\dot{\phi}$ of the form $A^\dagger A \dot{\phi} = \ldots$ for each time step. This is done very efficiently with good control of errors by standard methods such as the conjugate-gradient algorithm. As the bare quark mass approaches zero, these iterative sparse matrix algorithms require more computer time, but they prove to be quite practical.[8]

Our last task is to check that L really gives the original Path Integral. The canonical ensemble based on Eq. (2.8) reads,

$$Z = \int Du\,Dp\,D\phi\,D\phi^\dagger\,DP\,DP^\dagger \exp(-H/T) \quad (2.10a)$$

All the variables except U enter H quadratically, so the integrals can be done,

$$Z = \text{const.} \int DU \; \det{}^2 A(U) \exp(-S_o(U)/T) \qquad (2.10b)$$

which is the required answer except for the second power of the determinant. However, since $A^\dagger A$ in the staggered fermion method does not couple nearest neighbor pseudo-fermion fields, ϕ can be set to zero on every other lattice site.[8] In this final scheme $\det{}^2 A$ is replaced by $\det A$.

Now we see clearly the character of the tricks in Eq. (2.6) and (2.8). The pseudo-fermion kinetic energy in L is "$\frac{1}{2} mv^2$" with $m \sim A^\dagger A$. When the H is constructed we have "$p^2/2m$" and the $(A^\dagger A)^{-1}$ here was responsible for the positive power of $\det A^\dagger A$ in Eq. (2.10b). The nice feature of this scheme is that the full non-local character of the determinant is avoided by the algorithm. In each time step $A^\dagger A \phi = \dots$ is solved for ϕ -- this is a local operation since $A^\dagger A$ only couples nearby degrees of freedom.

The last ingredient in the algorithm is the calculation of the coupling constant β. If we identify the number of active, independent degrees of freedom N^* of the system, this can be done using the equipartition theorem,

$$\frac{1}{2} \beta^{-1} N^* = \langle T \rangle = \langle \dot{\phi} A^\dagger A \dot{\phi} + \frac{1}{2} \sum \dot{U}^2 \rangle \qquad (2.11)$$

The calculation of N^* for particular parametrizations of the U matrices is discussed in ref. 7.

III.   QUANTUM CHROMODYNAMICS SIMULATIONS

Now let's discuss the status of large scale simulations of SU(2) and SU(3) gauge theories with four light, dynamical Dirac fermions -- simulations close to the real theory QCD. Various projects are in progress.

First is the thermodynamics of the continuum field theory. Here one wants to understand QCD at finite temperature and study the transition from hadronic matter to a quark-gluon plasma. One wants to know if there are true non-analyticities in the thermodynamic quantities of interest such as the entropy and internal energy densities. The SU(2) and SU(3) theories without fermion feedback showed such non-analyticities and their behavior is well-understood in the context of traditional statistical mechanics. The situation is relatively unclear when fermion feedback is accounted for and the subject is quite controversial. A chiral restoring transition is certainly present, but fermion screening may be qualitatively similar for all temperatures rendering the thermodynamics of the "transition" smooth.

The SU(2) theory has been studied on a 6×12^3 lattice at 6 β values and three fermion mass values (0.10, 0.075 and 0.050) with the microcanonical algorithm for each point.

In Fig. 1 I show the scaling regions of the pure SU(2) theory and the theory with $N_f = 4$ species of fermions. The agreement with asymptotic freedom for $\langle \bar\psi\psi \rangle$ is quite nice. Note that fermion feedback shifts the $\langle \bar\psi\psi \rangle$ curve toward stronger coupling as N_f is increased <u>and the slope of $\ln\langle \bar\psi\psi \rangle$ vs. β changes appropriately.</u>

In Fig. 2 I show $\langle \bar\psi\psi \rangle$ extrapolated to zero mass and the Wilson line for a 6×12^3 lattice in the SU(2), $N_f = 4$ theory. It appears that the transition from hadron to quark-gluon matter is abrupt.

IV. HIERARCHY PROBLEMS IN UNIFIED GAUGE THEORIES

I want to illustrate that lattice methods can be applied to theories "beyond QCD" which might have interesting mass scales at arbitrarily high energies. Unfortunately, the most interesting schemes involve chiral fermions and these cannot be attacked by lattice methods because we cannot place a

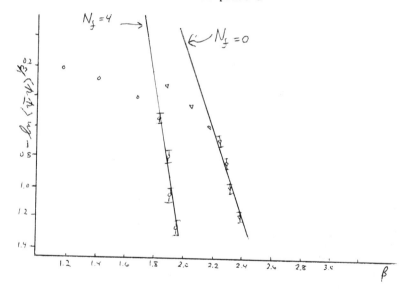

Fig. 1 $\langle \bar\psi\psi \rangle$ vs. β for $N_f = 0$ and 4.

590

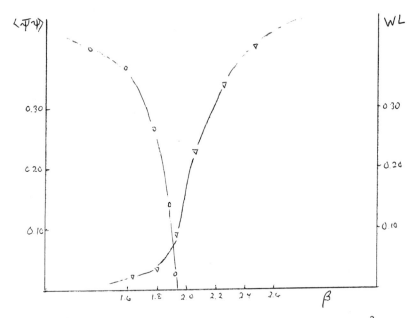

Fig. 2 $\langle\bar{\psi}\psi\rangle$ and the Wilson line for SU(2), $N_f = 4$, 6×12^3 lattice.

single neutrino on the lattice with a conventional Action. Anyway, in the realm of vector theories we can ask whether a theory can support disparate mass scales without the need to fine tune a fundamental parameter. Chiral symmetry breaking and asymptotic freedom can conspire to do this, as suggested by Raby, Dimopolous and Susskind.[9] By considering single gluon exchange they suggest that when $C_f g^2 \sim O(1)$ massless fermions of color charge C_f will condense into a chiral condensate. By asymptotic freedom, this criterion leads to an exponential sensitively of the characteristic energy scale of the condensate to the fermion's color charge. Changes of scale of 10^{5-10} are possible in such "technicolor" schemes although realistic models do not exist.

The validity of the underlying feature of the scenerio, that $C_f g^2 \sim O(1)$ leads to condensation, can be tested by lattice methods. In Fig. 3 I show data for the pure SU(2) theory ($N_f = 0$) in which fundamental and adjoint condensates have been measured. The $\ell = 1$ condensation occurs at much weaker coupling (shorter physical distances) in general support of the scenerio.

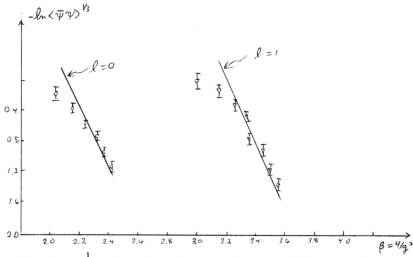

Fig. 3 $\ell = \dfrac{1}{2}$ and $\ell = 1$ condensates in the SU(2) $N_f = 0$ theory.

Does this hierarchal structure survive the inclusion of fermion feedback? Consider SU(2) with $N_f = 4$ Majorana quarks.[10] Two mass scales can be searched for by simulating the theory of finite temperature and measuring $\langle \bar{\psi}\psi \rangle$ for the $\ell = 1$ quarks and the string tension for $\ell = 1/2$ static quarks. In Fig. 4 I show data from a 4×8^3 simulation depicting $\langle \bar{\psi}\psi \rangle$ and the Wilson line. Clearly the deconfinement and the chiral symmetry restoration temperatures are distinct.

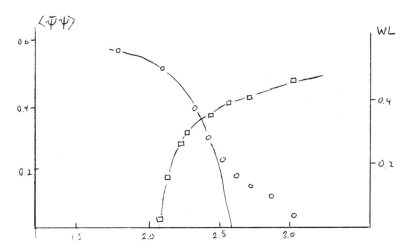

Fig. 4 $\langle \bar{\psi}\psi \rangle$ and the Wilson line for SU(2) theory with adjoint quarks.

592

ACKNOWLEDGEMENT

This work is partially supported by the National Science Foundation under grant number NSF-PHY82-01948.

REFERENCES

1) Wilson, K. G., Phys. Rev. D14, 2455 (1974).

2) Kogut, J.B. and Susskind, L., Phys. Rev. D9, 3501 (1974); Phys. Rev. D11, 395 (1975). Susskind, L., Phys. Rev. D16, 3031 (1977).

3) Weingarten, D. and Petcher, D., Phys. Lett. B99, 33 (1981). F. Fucito, E. Marinari, G. Parisi and G. Rebbi, Nucl. Phys. B180[FS2], 369 (1981).

4) Callaway, D. and Rahman, A., Phys. Rev. Lett. 49, 613 (1982).

5) Polonyi, J. and Wyld, H.W., Phys. Rev. Lett. 51, 2257 (1983). Polonyi, J., Wyld, H.W., Kogut, J.B., Shigemitsu, J. and Sinclair, D.K., Phys. Rev. Lett. 53, 644 (1984).

6) Guha, A. and Lee, S.-C., Phys. Rev. D27, 2412 (1983).

7) Kogut, J., Polonyi, J., Shigemitsu, J, Sinclair, D.K. and Wyld, H.W., Nucl. Phys. B251[FS13], 311 (1985).

8) See the second reference in item 5 above, for example.

9) Raby, S., Dimopoulos, S. and Susskind, L., Nucl. Phys. B169, 373 (1980).

10) Kogut, J., Polonyi, J., Sinclair, D.K. and Wyld, H.W., Phys. Rev. Lett. 54, 1980 (1985).

Lattice Gauge Theory With Special Processors and Super Computers

Norman H. Christ

Department of Physics, Columbia University
New York, New York 10027

In this talk I would like to address the computer requirements of lattice gauge theory, examining the computer resources currently being applied to the problem and those expected to be available in the next two or three years. Although I will attempt to be general, my experience lies in the direction of specially built computers and I will tell you about the progress of our project at Columbia and discuss the results we have obtained so far.

Let us ask what is required for a satisfactory numerical calculation. Recall that the lattice version of non-Abelian gauge theory approximates space-time by a 4-dimensional mesh of points with lattice spacing "a" [1]. The gauge variables are associated with the links joining neighboring sites: an SU(3) matrix is introduced for each link, so that an observable $O[U]$ can be computed from the path integral

$$\langle O \rangle = \frac{\prod_\ell \int d[U_\ell] \exp\{\frac{2}{g_0^2} \sum_p \text{re tr} [\prod_{\ell \in p} U_\ell]\} O[U]}{\prod_\ell \int d[U_\ell] \exp\{\frac{2}{g_0^2} \sum_p \text{re tr}[\prod_{\ell \in p} U_\ell]\}}. \tag{1}$$

Here the sum in the exponent is over all elementary plaquettes p in the lattice. One can deduce from Eq.(1) that as g_0 approaches zero, the product of the U matrices around a plaquette approaches I and it becomes possible to represent

$$U_\ell \simeq I - i g_0 \ell^\mu A_\mu^i \lambda^i \tag{2}$$

where A_μ^i is the usual vector potential so that

$$\frac{2}{g_0^2} \text{tr} [\prod_{\ell \in p} U_\ell] \rightarrow - \frac{1}{2} F_{\mu\nu}^i F_{\mu\nu}^i \qquad (\mu,\nu) = p \tag{3}$$

which is the usual Yang-Mills action. Thus, for sufficiently small

g_o, we have the continuum Yang-Mills theory with "a" representing a Lorentz non-invariant short-distance cutoff. The asymptotic freedom of the theory implies that as $g_o \to 0$, "a" measured in physical units should also vanish.

Thus a good test that we are in fact simulating QCD is to see that the scaling of physical quantities $Q_i(g_o,a)$ predicted by asymptotic freedom is actually observed:

$$Q_i = \left[\frac{f(g_o)}{a} \right]^{d_i} c_i \tag{4}$$

where

$$f(g_o) = \left(\frac{1}{g_o^2} \right)^{\frac{51}{121}} e^{-\frac{24\pi^2}{33g_o^2}} \left(1 + 0(g_o^2) \right) \tag{5}$$

and d_i is the physical mass dimension of the quantity Q_i. As I will discuss later, this may be true if $\beta = 6/g_o^2 \geqslant 6.3$.

How large should the lattice be? Recall that the Euclidean path integral in Eq.(1) for a finite space-time volume, performed with periodic boundary conditions in the time direction is actually a calculation of

$$\frac{\text{tr} \{ e^{-H/kT} \, 0 \, \}}{\text{tr} \{ e^{-H/kT} \}} \tag{6}$$

where the temperature T is related to the number of sites N_t in the time dimension of the lattice by $kTa = 1/N_t$. For $\beta = 6.3$ the critical temperature for color deconfinement corresponds to $N_t = 12$. Thus, the zero temperature properties of QCD at $\beta = 6.3$ should be computed on a lattice with linear dimension $N \gg 12$. (Note the value $\beta = 6.0$ used in recent calculations corresponds to $N=8$. Perhaps $N=32$ which would make $N/N_t \sim 3$ is sufficiently large that finite temperature and size effects could be measured and corrected for.

Accumulating 50,000 Monte Carlo sweeps on such a lattice would require $(5 \times 10^4 \times 4 \times 32^4$ links) $\times$ 80 μsec/link, or 5000 CRAY1 hours. If we wanted three different values of β, two different lattice sizes and an equal amount of time for computing fermion propagators in this no-quark-loop approximation, we would need ~10 CRAY1 years.

All of this omits the effects of quark loops. The very promising techniques being experimented with for including quark loops are not yet well enough understood to be certain of the additional time they will require. If one assumes that the above lattice size is adequate for Wilson fermions and that the Langevin method for treating fermion loops requires an additional factor of 10 in time, as concluded by Ukawa and Fukugita [2], then the requirement becomes 100 CRAY1 years!

How is this to be achieved? A variety of powerful computers are becoming increasing available for this type of work. Let me review them briefly. The various supercomputer-class commercial machines are described in Table I. All of these are "vector" machines, capable of executing strings of identical floating point operations very efficiently. They are all multi-million dollar mainframes except for the last one, the ST100, which is an array processor available for hundreds of thousands of dollars. Two ST100's are currently being used for lattice gauge theory, one at Santa Barbara and one at Argonne.

Table 1. Commercial Supercomputers shown with their peak speed in millions of floating point operations and their availability for lattice gauge theory calculations. The integer n represents the number of processors in the particular system.

type	speed	availability
CRAY1	160 Mflops	now
CRAY XMP	n x 200 Mflops	now
CRAY2	1,000 Mflops	'85
CYBER 205	n x 100 Mflops	now
ETA-10	10,000 Mflops	'87
ST100	100 Mflops	now

Table 2.   Special purpose machines being constructed for lattice gauge theory.

Columbia		
16-node	256 Mflops	now
64-node	1,000 Mflops	Fall '85
256-node	8,000 Mflops	Spring '87
Rome-CERN (APE)		
4-node	256 Mflops	Spring '86
16-node	1,000 Mflops	
IBM (GF-11)		
576-node	11,000 Mflops	

The second table lists the three large scale special purpose machines presently under construction for lattice gauge theory. All of these machines will actually be quite flexible computers able to be programmed to execute a variety of algorithms for lattice gauge theory or other homogenous physical problems. They all achieve significant speed and economy from a high degree of parallelism. The Columbia machine [3] is interconnected as a two-dimensional mesh. Each node is controlled by an Intel 80286 microprocessor and simple microcode on each node. The Italian computer [4] is arranged as a 1-dimensional mesh controlled by microcode generated by a single 3081/E. The nodes of the IBM project [5] are linked by a highly configurable switch capable of realizing a number of interconnection schemes. They are centrally controlled by microcoded instructions.

Finally, I would like to discuss the status of our project at Columbia and describe our present results [6] for the scaling behavior of the color deconfinement phase transition for SU(3) gauge theory without quarks. Our results come from the 16-node machine which has been performing pure gauge theory Monte Carlo calculations since last April. The machine cost ~$100,000 to build and presently runs at 40% of the speed of a CRAY1. Program improvements and hardware changes made to a single node have more than doubled this speed. We expect the entire 16-node machine to exceed the speed of a CRAY1 for lattice gauge theory by the end of the month.

The quantity that we have been studying is the temperature of the color deconfinement phase transition. We adjust the parameter $\beta = 6/g_0^2$ until the critical temperature, which should behave as in Eq.(4),

$$kT_c = \frac{t(\beta)}{a} , \tag{7}$$

equals the temperature of the lattice, $kT = 1/N_t a$. If we call that value β_c, then

$$\frac{1}{N_T} = t(\beta_c) \tag{8}$$

and varying N_t allows us to determine $t(\beta)$ and compare with the scaling prediction of Eq.(5).

The critical value of β can be recognized by measuring the expectation value of the Wilson line operator

$$O_L(U) = \text{tr} \left\{ \prod_{\ell \in L} U_\ell \right\} . \tag{9}$$

This operator is constructed as the trace of the product of the link matrices lying along a line L in the time direction. It is the large mass limit of the propagator of a single quark fixed at the spacial position of the line L. We can increase our statistics by averaging over all possible lines. Thus we expect

$$< O_L(U) >_{T < T_c} = 0 . \tag{10a}$$

$$< O_L(U) >_{T > T_c} \neq 0. \tag{10b}$$

Unfortunately, this hypothesis has two difficulties. Even if $<O(U)> \neq 0$ for $T > T_c$, it is still very small $\propto \exp\{-E_q \cdot N_t \cdot a\}$ Where E_q is the linearly diverging self energy of a massive point charge. Second, for a finite volume with periodic boundary conditions, we must worry about the inconsistency of Gauss's law if only a single charge is present in the volume. This is realized by the presence of a zero mode, the integration over which makes the expectation value in Eq.(10) vanish for any value of T.

For the non-Abelian theory, this zero mode is the symmetry $U_\ell \rightarrow (1)^{\frac{1}{3}} U_\ell$ for all time-like links ℓ in a particular time plane. Since this Z_3 symmetry multiplies $<O(U)>$ by $(1)^{\frac{1}{3}}$, integration over the three transforms of any gauge-field configuration makes $<O(U)>$ vanish. As the volume becomes infinite such global changes of phase become impossible. Thus for a finite volume calculation we must limit the length of Monte Carlo time over which we can average to avoid these jumps in Z_3 phase. This in turn limits the precision with which Eq.(10a) can be measured.

Our results to date for a lattice of spacial volume 16^3 are shown in Fig.1. For each of the three temporal dimensions, the magnitude of the Wilson line is plotted as a function of β. This quantity is obtained from our Monte Carlo calculation by first averaging the trace in Eq.(9) over all 16^3 lines in our spacial volume and then averaging over blocks of 100 Monte Carlo sweeps. The magnitudes of the complex numbers obtained from each block are then averaged and the fluctuations among them (including the effects of correlations) used to compute the errors. We interpret the rise in $|<O(U)>|$ over the narrow range of β shown in Fig.1 as arising from the deconfining phase transition. Assuming that the critical value of beta lies in the region of greatest increase, we deduce that β_c take values 6.05±.05, 6.275±.025, 6.35±.05 for lattices with N_t = 10,12 and 14 respectively.{1}

{1} These preliminary conclusions are confirmed and strengthen by additional Monte Carlo data and a more detailed analysis as presented in Ref. [6]

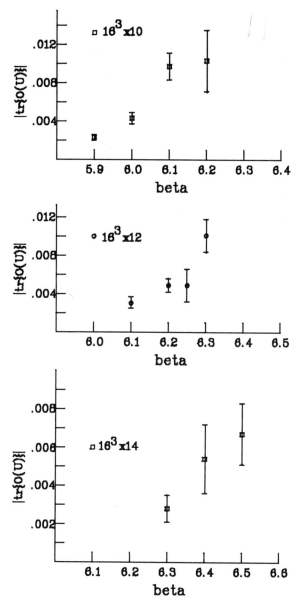

Fig.1. The magnitude of the expectation value of the Wilson line operator, averaged over blocks of 100 Monte Carlo sweeps for lattices with a spacial volume of 16^3 and 10, 12 and 14 sites in the time direction.

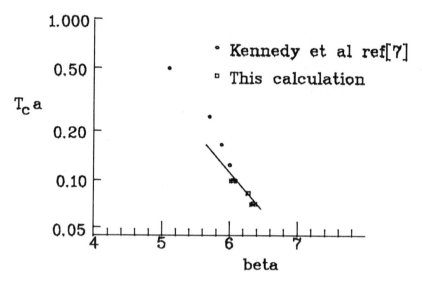

Fig.2. The color-deconfinement critical temperature as a function β.
The straight line has the slope predicted by the continuum
renormalization group.

These results are compared to values of β obtained earlier [7]
on smaller lattices and to the predictions of scaling in Fig.2. The
essentially straight line shown in the figure is the prediction of
Eq.(5) with the multiplicative constant adjusted to make the curve
pass through our three points. As can be seen, the region β > 6.05
shows a different β dependence than that seen for lower β values,
a behavior consistent with the predictions of continuum perturbation
theory and the renormalization group.

If β > 6.3 is in fact sufficiently large to correspond to
continuum physics, then one might expect the 32^4 lattice calculation
discussed at the beginning of the talk to be adequate for studying
the spectrum of hadrons and bulk properties of QCD. In which case,
the most powerful special purpose computer projects or the
commercial supercomputers available in two years (if dedicated to such

calculations) might be realistically expected to give physical results
including the effects of fermion loops. We are very pleased to be
able to report results from the first phase of our project and plan
over the next few months to extend these confinement studies to
spacial volumes of 20^3.

References

[1] For a review of lattice Quantum Chromodynamics see J. Kogut,
     Rev. Mod. Phys., 55, p775 (1983).

[2] A. Ukawa and M. Fukugita, Phys. Rev. Lett. 55, 1854 (1985)

[3] N. Christ and A. Terrano, IEEE Transactions on Computers,
     C-33, No.4, p344 (1984).

[4] P. Bacilieri, N. Cabibbo, E. Marinari, G. Parisi, F. Costantini,
     G. Fiorentini, S. Galeotti, D. Passuello, R. Tripiccione,
     A. Fucci, R. Petronzio, F. Rapuano, F. Marzano, S. Petraca,
     G. Salina, D. Pascoli, E. Remiddi, Universita di Roma preprint
     IFUP – TH 84/40 (1984).

[5] J. Beetem, M. Denneau, D. Weingarten, IBM preprint (1984).

[6] N. Christ and A. Terrano, Phys. Rev. Lett., to be published.

[7] A. Kennedy, J. Kuti, S. Meyer, and B.J. Pendleton, Phys. Rev.
     Lett. 54, 87 (1985).

RECENT RESULTS FROM ARGUS [†]

Robin Davis
University of Kansas,
Lawrence, Kansas. 66045. U.S.A.

(representing the ARGUS collaboration[*])

Abstract

Results are reported from data accumulated by ARGUS, operating at center of mass energies in the region of 10 Gev in the DORIS II e^+e^- storage ring at DESY. In radiative $\Upsilon(2S)$ decays three clearly resolved photon lines are observed corresponding to transitions to the ^3P_J states with J=0,1 and 2 of the $b\bar{b}$ system. The observation of the decay $D^0 \rightarrow K_s^0 \phi$ with a branching ratio BR$(D^0 \rightarrow \overline{K}^0 \phi)$ = $(1.41\pm0.43)\%$ represents the first direct evidence for flavor annihilation by W exchange in charmed particle decays. The inclusive decay B $\rightarrow$ J/Ψ + X was observed and this branching ratio was measured to be $(1.37\pm0.6)\%$. A new charmed meson, the $D^*(2420)$, was observed through its decay into $D^*(2010)\pi$.

Introduction

The ARGUS detector [1] has been collecting data for 2 years. During 1983 and 1984 a total luminosity of 82.2/pb was taken comprising 21.6/pb, 36.2/pb, 11.4/pb and 13.0/pb respectively on the $\Upsilon(1S)$, $\Upsilon(2S)$, $\Upsilon(4S)$ and in the continuum or during scanning. The event sample comprises about 300000 multihadronic events from the continuum, 215000 $\Upsilon(1S)$ decays, 130000 $\Upsilon(2S)$ decays, 9000 $\Upsilon(4S)$ decays. This permits studies of various topics in $b\bar{b}$ spectroscopy, as well as in the production and in the decays of quarks. In the work reported here, charged particles were identified using information from dE/dx, from time of flight and shower counter measurements, as well as from the muon chambers. The details have been given earlier[2].

Radiative decays of the $\Upsilon(2S)$

The lowest ^3P states of the Υ system should be accessible by E1 transitions from the $\Upsilon(2S)$. According to the quarkonium model the center of gravity of the three ^3P states should lie only about 130 MeV/c^2 below the mass of the $\Upsilon(2S)$ with a fine splitting of about 20 MeV/c^2. Excellent energy resolution for low momentum photons is therefore needed to detect and resolve these states. This is achieved by using the ARGUS detector as a pair spectrometer: those photons converting before the chamber are measured by detecting the conversion electrons in the drift chamber. Using the full information on the e^+ and e^- tracks a 5C fit is performed by requiring 1. the e^+e^- mass be small, 2. the conversion point of the photon be at

[†]Invited talk given at the meeting of the Division of Particles and Fields of the American Physical Society held at the University of Oregon, Eugene, Oregon. August 12 - 15, 1985.

the beam tube or the inner wall of the drift chamber and 3. that the reconstructed photon point to the main vertex of the event. All e^+e^- combinations giving an acceptable fit were accepted as photons. This procedure was carefully checked by reconstructing π^0's which produced two converted photons. The resulting π^0 mass of (134.2 ± 0.7) MeV/c^2 is in good agreement with the accepted value and confirms the energy calibration of the converted photons. In Fig.1 we show the raw single photon spectrum without background subtraction for $\Upsilon(2S)$ decays. As expected it shows clearly three narrow lines on a smooth continuum background, which results mainly from photons from π^0 decays. By fitting the lines with three gaussians with fixed width $\sigma_\gamma = 1.1$ MeV (determined from a Monte Carlo analysis) and the background by a 3rd order polynomial the resulting photon energies and branching ratios for the three lines are determined and are displayed in the table. The significance for all lines is well above four standard deviations.

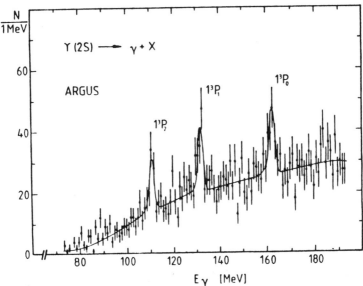

Figure 1

If these transitions had occurred between the P states and the $\Upsilon(1S)$, then Doppler broadening from the recoil motion of the P states would have produced line widths in excess of those observed. Our excellent resolution allows us to confirm that these transitions occur between the 2S and the P states.

The photon energies and branching ratios are in good agreement with other experiments[4]. They also agree well with the spin assignments J=2,1 and 0 as seen from a comparison of the measured value R_{BR} = BR($\Upsilon(2S)$ $\rightarrow$

$\gamma^3 P_J)/BR(\Upsilon(2S) \rightarrow \gamma^3 P_1)$ and the theoretical one R_{TH}, given by the $BR \sim (2J+1) \times E_\gamma^3$ normalized to the line with J=1. The center of gravity of these ^3P states turns out to be $M_{c.o.g.}(^3P) = 9899.0 \pm 1.2$ MeV/c^2. Also the relative fine splitting

$$R_{hs} = \frac{M(^3P_2) - M(^3P_1)}{M(^3P_1) - M(^3P_0)} = 0.69 \pm 0.04 \tag{1}$$

is measured with high accuracy. This number can be used to discriminate between various potential models [5], in particular to conclude that the vector part of the confining force is less than 30%.

TABLE 1. Results on the $\Upsilon(2S) \rightarrow \gamma^3 P_J$ transitions

J	E_γ(MeV)	N_{Event}	BR(%)	R_{TH}	R_{BR}
2	110.6±0.3±0.9	50±11	9.8±2.1±2.4	0.99	1.07
1	131.7±0.3±1.1	67±13	9.1±1.8±2.2	1.00	1.00
0	162.1±0.5±1.4	69±15	6.4±1.4±1.6	0.62	0.71

First evidence for W exchange processes in charmed meson decays

The difference between the lifetimes of the neutral and charged D mesons is not fully understood. A possible mechanism which can enhance D^0 decays over D^+ decays is flavor annihilation by W exchange which is Cabibbo allowed for the former but forbidden for the latter. The decay $D^0 \rightarrow K_s^0 \phi$ could proceed by a spectator diagram: this would, however, be OZI forbidden and has a calculated[6] branching ratio below 10^{-5}. This decay is therefore expected to proceed mainly through a W exchange process and its observation provides a crucial test for the presence of such a process.

The decay channel $D^0 \rightarrow K_s^0 \phi$ was observed as a signal in the $K_s^0 K^+ K^-$ invariant mass distribution with the $K^+ K^-$ invariant mass lying in the ϕ region. Since this work has recently been published[7], only a brief summary is presented here. The invariant $K_s^0 K^+ K^-$ mass spectrum, shown in Fig.2, has a pronounced D^0 signal at a mass of (1863.9±1.9)MeV/c^2. After making a background subtraction 36.7±8.0 entries were found in the signal.

Branching ratios for the observed D^0 decays were determined by comparing the results given above with the decay $D^0 \rightarrow K_s^0 \pi^+ \pi^-$ observed in the same data sample using the same cuts. This channel is well established, with a branching ratio recently determined[8] to be $(7.7 \pm 1.6 \pm 1.2)\%$. The result finally obtained was:

$$Br(D^0 \to \overline{K}^0\phi) = (1.41 \pm 0.43)\% \qquad (2a)$$

$$Br(D^0 \to \overline{K}^0K^+K^-) = (1.40 \pm 0.45)\% \qquad (2b)$$

The result for the branching ratio for $D^0 \to \overline{K}^0K^+K^-$ is in good agreement with that derived from preliminary MARK III results, which gave $(1.5 \pm 0.5 \pm 0.3)\%$, while the observed signal for $D^0 \to \overline{K}^0\phi$ is well within their quoted limit of 2.4% at the 95% confidence level.

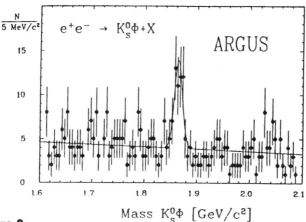

Figure 2

The measured rate for the decay $D^0 \to \overline{K}^0\phi$ is 3 orders of magnitude above that predicted by the spectator model where only an OZI violating process can contribute. We therefore conclude that in a simple quark picture the decay $D^0 \to \overline{K}^0\phi$ can only proceed via a W exchange diagram. Our observation represents the first direct evidence for W exchange in heavy quark decays.

The decay $B \to J/\Psi + X$

In data collected on the $\Upsilon(4S)$ resonance, which decays into a pair of B mesons, we have observed the inclusive decay of the resulting B mesons into the J/Ψ[9]. This process is believed to occur through the color suppressed quark spectator diagram shown in Fig. 3a. The J/Ψ mesons were detected through their decay into e^+e^- or into $\mu^+\mu^-$. Multihadron events with more than two charged tracks were used: in addition events with three and four charged tracks were required to have more than four and two photons respectively. These latter cuts were imposed to suppress $\tau^+\tau^-$ events feeding into our sample. Muons were identified by requiring a hit in the muon chambers, together with an energy deposition of $\leq$ 0.6 GeV in the shower counters. Electrons were identified by utilising information from dE/dx, time of flight and shower counter measurements[2]. The resulting efficiences for observing J/Ψ decays into e^+e^- and into $\mu^+\mu^-$ were estimated to be 30% and 50% respectively.

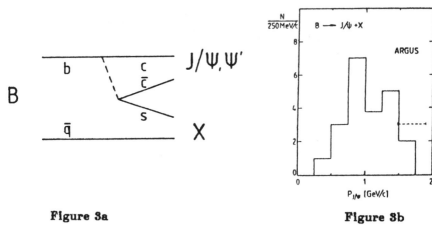

Figure 3a   **Figure 3b**

The combined lepton pair invariant mass spectrum is presented in Fig. 4, where the muon pair events are shown cross hatched. Extensive investigations of various contributions to the background have been made, including studies of nearby continuum data, of Monte Carlo data and studies employing event mixing techniques. The continuum data was used to estimate background contributions from the continuum under the 4S. The contributions from $b \rightarrow c \rightarrow s$ sequential decays were estimated using Monte Carlo techniques. Finally the event mixing procedures were used to study background contributions from events where the leptons came one each from the decay of opposite daughter B mesons from an $\Upsilon(4S)$. Together these gave a satisfactory description of the background shape. There are 22 events in the J/Ψ region, above an estimated background of 4.5±0.8 events, giving a 4.5 σ effect. This yields a branching ratio for the process

$$\mathrm{Br}(B \rightarrow J/\Psi + X) = (1.37 \pm 0.6)\% \tag{3}$$

The recoil momentum spectrum of the J/Ψ is shown in Fig. 3b. It is soft, indicating that high masses dominate in the system that is recoiling against the J/Ψ. The K and K* region is shown by the dashed line and contains only two events. Similar results were also reported at this conference[10].

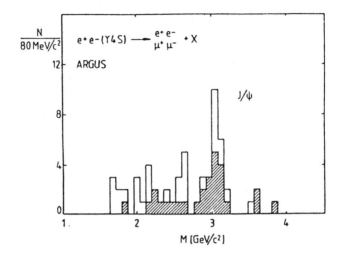

Figure 4

Observation of a new charmed meson, the $D^*(2420)$

Several potential models have predicted the existence of a D^* state in the mass region around 2.4 GeV[11]. We have accordingly searched for a D^* state decaying into $D^{*0}(2010)\pi^+$. (It is to be understood, in this discussion, that the charge conjugate states are always included in the data presented here). The sample of $D^{*+}(2010)$ events was obtained by exploiting the Q value of 5.8 MeV in the decay $D^{*+} \to D^0\pi^+$, with the D^0 decaying into either $K^-\pi^+$, $K^-\pi^+\pi^+\pi^-$ or $K^-\pi^+\pi^0$. In the latter case the π^0 was undetected: together these channels represent 37% of all D^0 decays. A clean sample of $D^{*+}(2010)$ was obtained by requiring the mass difference, $M(D^0\pi^+) - M(D^0)$ to lie between 144 and 147 MeV/c^2, and the quantity $x_p = p/p_{max} > 0.45$. The resulting $D^{*+}(2010)$ was then combined with all other π^- candidates and the resulting mass difference $M(D^{*+}\pi^-) - M(D^{*+})$ was studied. Additional cuts were made on the scaled momentum of the $D^{*+}\pi^-$ system, requiring $x_p > 0.6$ for the D^{*0}. A cut was also made on θ, the angle between the $D^{*+}\pi^-$ direction and the $D^{*+}(2010)$ momentum vector defined in the $D^{*+}(2010)\pi^-$ rest frame, requiring $\cos\theta < 0$. The first cut is motivated by the nature of charm quark fragmentation, which results in a hard momentum spectrum for the produced heavy meson[2], while light hadronic background is concentrated at lower x_p. The second cut reduces background which peaks at forward angles, due to the combination of the $D^{*+}(2010)$ with random low momentum pions.

The mass difference spectrum, $M(D^{*+}\pi^-) - M(D^{*+})$, for combinations passing these cuts, is shown in Fig. 5. A prominent peak is seen around 410 MeV. A

Breit-Wigner for the signal, plus a threshold factor times a second order polynomial for the background, were fitted to the mass difference distribution, yielding the results shown in Table 2. All sources of systematic error, including that introduced by the assumed mass dependence of the background, are negligible in comparison with the statistical uncertainties. Monte Carlo study shows the detector resolution to be 15 MeV/c^2 in this mass region, while the observed width is much larger, indicating that this new state decays strongly. The statistical significance of the enhancement is 3.9 standard deviations. In the following we refer to this state as the $D^{*0}(2420)$. This observation offers an explanantion for the observed enhancement near 2.44 GeV/c^2 in the recoil spectrum for $D^0 \to K^- \pi^+$ reported by MARK II at SPEAR[12].

TABLE 2. Properties of $D^{*0}(2420)$ determined from fits to the distribution of mass difference, $M(D^{*+}\pi^-) - M(D^{*+})$.

Channel	$D^0 \to K^-\pi^+$ $D^0 \to K^-(3\pi)^+$	$D^0 \to K^-\pi^+\pi^0$	Combined
Mass difference [MeV/c^2]	411 ± 7	410 ± 11	410 ± 6
$M(D^{*+}(2420))$ [MeV/c^2]	2421 ± 7	2420 ± 11	2420 ± 6
Full width Γ [MeV/c^2]	64 ± 26	75 ± 36	70 ± 21
Number of events	82^{+28}_{-21}	52^{+24}_{-18}	135^{+34}_{-29}

Supporting evidence for the observation was obtained using the D^0 decay channel, $D^0 \to K^-\pi^+\pi^0$, which results in a satellite peak in the $K^-\pi^+$ mass distributions, shifted to lower masses by the missing π^0 [13]. A cut of $x_p > 0.45$ was required for the $D^{*+}(2010)$. Candidate events containing D^{*+} decays in this channel were selected by requiring $M(D^0\pi^+) - M(D^0) < 152$ MeV/c^2. The momentum and decay angle cuts described above were then applied to the $D^{*+}\pi^-$ combinations and the resulting mass difference plot for $M(D^{*+}\pi^-) - M(D^{*+})$ is shown in Fig. 6. Results obtained from a fit with a Breit-Wigner plus background are also shown in Table 2. The effect of the missing π^0 increases the detector resolution to 25 MeV/c^2, but this is still smaller than the natural width of the state. Monte Carlo studies showed that there was negligible shift in the mass difference due to the missing π^0. Masses and widths for the three channels are consistent: the combined significance of the effect is 4.9 standard deviations and is shown in Fig. 7.

Two different studies have been made to confirm that the kinematic cuts used in this analysis do not produce this enhancement. These were made using (a) a sideband of the D*+, and (b) wrong charge combinations, that is D*+π+. No significant enhancement was found in either approach.

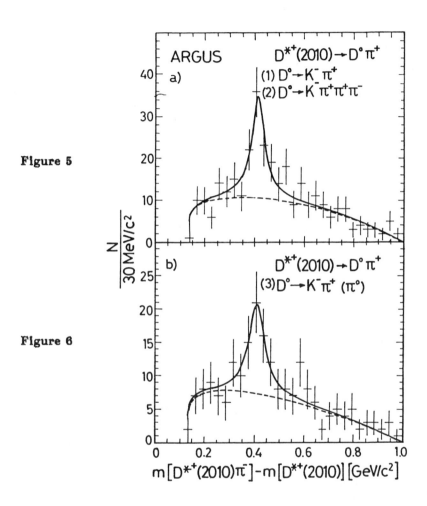

Figure 5

Figure 6

After folding in the acceptance of our detector, obtained from Monte Carlo studies, and the effects of our cuts on cos(θ) and on x_p for the D*⁰(2420), we conclude that $(24^{+8}_{-6} \pm 8)\%$ of observed D*+(2010) result from the decay of the D*⁰(2420). Using results quoted in references 2,8 and 14, we obtain σ[D*+] = (940 ± 150 ± 270) pb at $\sqrt{s} \approx 10$ GeV. Using this value, and isospin coefficients to correct for the unseen neutral decay channel D*⁰(2420) → D*⁰(2010)π⁰, we finally obtain, $\sigma[D^{*0}(2420)] \cdot BR[D^{*0}(2420) \to D^*\pi] = (340^{+190}_{-180})$ pb.

This state is probably one of the 1P states of c and ū quarks[11]. These states are expected to be the lowest lying of the orbitally excited charmed states and to be more easily produced than higher angular momentum states in e^+e^- annihilations. The possible assignments are $^3P_J(0^+, 1^+, 2^+)$ and $^1P_1(1^+)$: however the observation of a strong decay to a vector plus a pseudoscalar rules out the 0^+ assignment. Mixing of the $^3P_1$ and $^1P_1$ states would complicate predictions of their masses and widths. All model calculations[11] have predicted P states lying within 100 MeV/c^2 of the value reported here. The mass splittings of some of these states are less than their natural widths, so it is possible that this signal is produced by more than one resonance.

Figure 7

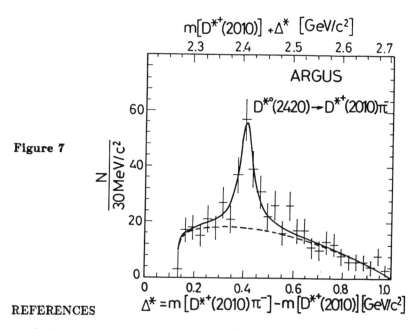

REFERENCES

* Current members of the ARGUS collaboration are: H.Albrecht, U.Binder, G.Harder, I.Lembke-Koppitz, A.Philipp, W.Schmidt-Parzefall, H.Schröder, H.D.Schulz, R.Wurth (DESY), A.Drescher, J.P.Donker, U.Matthiesen, H.Scheck, B.Spaan, J.Spengler, D.Wegener (Dortmund), J.C.Gabriel, K.R.Schubert, J.Stiewe, R.Waldi, S.Weseler (Heidelberg), K.W.Edwards, W.R.Frisken, Ch.Fukunaga, D.J.Gilkinson, D.M.Gingrich, M.Goddard, H.Kapitza, P.C.H.Kim, R.Kutschke, D.B.MacFarlane, J.A.McKenna, K.W.McLean, A.W.Nilsson, R.S.Orr, P.Padley, P.M.Patel, J.D.Prentice, H.C.J.Seywerd, B.J.Stacey, T.-S.Yoon, J.C.Yun (IPP Canada), R.Ammar, D.Coppage, R.Davis, S.Kanekal, N.Kwak (Kansas),

G.Kernel, M.Pleško (Ljubljana), L.Jönsson, Y.Oku (Lund), A.Babaev, M.Danilov, A.Golutvin, V.Lubimov, V.Matveev, V.Nagovitsin, V.Ryltsov, A.Semenov, V.Shevchenko, V.Soloshenko, V.Sopov, I.Tichomirov, Yu.Zaitsev (ITEP-Moscow), R.Childers, C.W.Darden, and H.Gennow (S.Carolina).

1) H.Albrecht et al. (ARGUS collaboration), Phys.Lett. **134B** (1984), 137.

2) H.Albrecht et al. (ARGUS collaboration), Phys.Lett. **150B** (1984), 235.

3) H.Albrecht et al. (ARGUS collaboration), Phys.Lett. **160B** (1985), 331.

4) C.Klopfenstein et al. (CUSB collaboration), Phys.Rev.Lett. **51** (1983), 160.
   P.Haas et al. (CLEO collaboration), Phys.Rev.Lett. **52** (1984), 799.
   R.Nernst et al. (Crystal Ball collaboration), Phys.Rev.Lett. **54** (1985), 2195.

5) P.M.Tuts, Proc. of the 1983 Intl. Lepton/Photon Symposium, Cornell (1983), 244.
   The author wishes to thank J.Rosner for a useful discussion on this point.

6) I.Bigi and M.Fukugita, Phys.Lett. **91B** (1980) 121.

7) H.Albrecht et al. (ARGUS collaboration), Phys.Lett. **158B** (1985), 525.

8) J.Hauser (MARK III collaboration), Ph.D. Thesis, Caltech (1985).

9) H.Albrecht et al. (ARGUS collaboration), Phys.Lett. **162B** (1985), 395.

10) A.Jawahery (CLEO collaboration), To appear in these proceedings.

11) A.De Rújula et al., Phys.Rev.Lett. **37** (1976), 398;
    E.Eichten et al., Phys.Rev. **D21** (1980), 203;
    S.Godfrey and N.Isgur, Phys.Rev. **D32** (1985), 189;
    B.Klima and U.Maor, DESY preprint 84-029 (1984);
    S.Jena, Phys.Rev. **D28** (1983), 2326.

12) G.Goldhaber et al. (MARK II collaboration), Phys.Rev.Lett. **69B** (1977), 503.

13) G.Goldhaber, Proc. of the XVIIIth Rencontre de Moriond, La Plagne, 1983.

14) Particle Data Group, Rev.Mod.Phys. **56** (1984), S203 and Erratum.

IV

HADRON PHYSICS

Session Organizers: J. Donoghue and F. Halzen

Study of Minimum-Bias-Trigger events at $\sqrt{s}$=0.2-0.9 TeV with magnetic and calorimetric analysis at the CERN proton-antiproton Collider

UA1 Collaboration
Aachen-Amsterdam-Annecy-Birmingham-CERN-College de France-
Harvard-Helsinki-Imperial College-Kiel-Padova-Queen Mary College-
Riverside-Roma-Rutherford-Saclay-Vienna-Wisconsin

presented by
G. Piano Mortari
Dipartimento di Fisica, Universita' "La Sapienza", Roma
Istituto Nazionale di Fisica Nucleare, Sezione di Roma, Italy

at the DPF Meeting of American Physical Society
Universtity of Oregon, USA, August 1985.

Abstract
The general features of Minimum Bias proton-antiproton collisions at the CERN SPS Collider are studied, with magnetic and calorimetric analysis in the UA1 detector, from 0.2 to 0.9 TeV centre of mass energy. We present evidence that several new features observed by the Collider experiments can be related to the emergence of the low-x jets signal.

---

1. Introduction

Since the first operation of the CERN proton-antiproton Collider [1], the study of the general characteristics of hadron-hadron collisions in this new energy domain has shown new effects with respect to lower energy data, the most spectacular being the emergence of the jet signal and the jet dominance in large transverse energy events [2].

From the analysis of Minimum-Bias-Trigger (MB) events UA1 has reported a remarkable dependence of the average transverse momentum of the charged particles, $\langle p_T \rangle$, on the event multiplicity, n_{ch} [3], and UA5 has reported violations of the KNO scaling in the charged particle multiplicity distribution due to an excess of large multiplicity events [4].

We now present new preliminary results based on a sample of MB events collected during the 1985 run of the Pulsed Collider from $\sqrt{s}$=200 GeV to 900 GeV. We present the charged particle p_T and multiplicity

distribution and the correlation between $\langle p_T \rangle$ and n_{ch}. We show evidence for the production of a non negiglible fraction of events containing low
E_T jets and we will correlate the large multiplicities and large transverse
momenta with the emergence of the jet signal [5].

2. The Pulsed Collider and the detector

During March '85 the CERN SPS Collider attained its maximum center
of mass energy of 0.9 TeV. The motivation for boosting the Collider operation up to maximum beam momentum was mainly the search for Centauro's
or other unusual events observed in cosmic ray experiments [6] in the region of 10^3 TeV laboratory momentum.

The maximum energy of the SPS operated in Collider mode is limited
by the power consumption in the main ring magnets, 47 MW at a beam momentum of 315 GeV/c. To reach the maximum momentum of 450 GeV/c, it
was proposed [7] to cycle the coast beams so that the large power consumption at 450 GeV/c is compensated by a lower one at lower beam momentum. The Collider was run in pulsed mode [8], between 100 and 450 GeV/c
beam momentum, in a cycle of 21.6 seconds: 4.0 sec at flat top, 8.2 sec at
flat bottom and 9.4 sec for the ramps at a rate of 80 GeV/sec. To limit the
power consumption in quadrupoles, the Collider was operated with normal
β, $\beta_v = \beta_h = 50$ m, and the β values were measured to be independent of the
beam momentum, ensuring that the luminosity is proportional to the beam
energy to an accuracy better than 1%. With typically $6*10^{10}$ protons and
$7*10^9$ antiprotons per bunch, with two p-bunches against one $\bar{p}$-bunch, the
beam lifetime was a few hours and the average luminosity 10^{26} cm^{-2}s^{-1}.

The UA1 detector was taking data through all the SPS cycle, the information on beam energy being recorded on tape. It is worth noticing that,
when comparing data taken at different energies, most of the systematic
effects due to the detector response and to the beam parameters cancel
out since the data were recorded at the same time and with the same circulating beams.

The UA1 detector covers the full solid angle with magnetic and calorimetric analysis. The main parts relevant to the present analysis are
the Central Detector (CD), which measures the momentum of charged par-
ticles in the pseudorapidity interval $|\eta| < 3.5$, and the calorimeter, which
samples the energy deposited in $|\eta| < 5.5$ over the complete azimuth angle.
The field of the dipole magnet, usually set at 0.70 T during the fixed ener-

gy runs, was lowered to 0.21 T in order to reduce the beams displacement in the center of the detector at low beam momenta.

The trigger requires at least two charged particles emitted in opposite rapidity hemispheres in the pseudorapidity range $1.5 < |\eta| < 5.5$ in coincidence with the beam crossing; it accepts almost 100% of the inelastic non single diffractive cross section. For the normalization we assume that the trigger cross section is equal to 2/3 of the total cross section obtained interpolating ISR and UA4 data [9].

3. Data sample and selection criteria

The present analysis is based on a sample of 110000 events collected during the '85 Pulsed Collider run subdivided as follows:

$\sqrt{s}$ (GeV)	200	250-450	580-680	900
no. of events	40000	6000	15000	50000

Two samples of 40000 MB events at $\sqrt{s}$=546 GeV and of 70000 MB events at $\sqrt{s}$=630 GeV, collected during the '83 and '84 Collider runs respectively, are also used for comparison with the Pulsed Collider one and with the data collected in the same two periods at fixed energy with a high-E_T-jet trigger.

An event is retained for analysis if it fulfils requirements on the timing information of the trigger hodoscopes, on the vertex position along the beam direction, on the ratio of vertex-associated to non-associated tracks and on the total energy deposited in the calorimeter. These cuts reduce the contamination from beam-gas interactions and halo particles to a negligible fraction with very high efficiency.

Charged particles, reconstructed in the Central Detector, are accepted if they fulfill requirements on track length, fit χ^2 and vertex association; to reduce the contamination from photon conversion and secondary interactions in the beam pipe, only tracks with $p_T > 0.15$ GeV/c are considered for the analysis. The reconstruction efficiency is about 96% independent of the event multiplicity. The CD acceptance has been evaluated assuming a flat distribution in η and ϕ; regions with acceptance lower than 20% have been removed and the data have been corrected for the acceptance and for the low p_T cut. Only tracks in the pseudorapidity interval $|\eta| < 2.5$ are considered for the analysis. Corrections due to strange particle decays, photon conversions and secondary interactions in the beam pipe and the uncertainties in the acceptance evaluation contribute to a systematic error of $\pm 3\%$ to the quoted charged particle multiplicity.

Jets are defined with the UA1 jet finding algorithm based on transverse energy clusters in $\eta - \phi$ space. To define a jet only the calorimeter data are used: an energy vector $\vec{E}_c$ is associated to each calorimeter cell, pointing to the interaction vertex. Cells with E_T larger than 1.5 GeV may initiate the clustering of a jet; cells within a distance $\Delta R = 1$, $\Delta R \equiv (\Delta \eta^2 + \Delta \phi^2)^{1/2}$, are associated to the jet initiator. The transverse energy of the jet, $E_T{}^J$, and the jet axis are defined by the sum of all the transverse energy of the cells in the cone of radius $R = 1$.

Studies have been done to check the stability of the jet definition when changing the algorithm parameters; in particular, when dealing with low E_T jets, the initiator threshold could introduce a strong bias in the jet search. On the other hand, lowering the initiator threshold results in splitting clusters and in ambiguities in the definition of the jet axis and energy. In the present analysis we have used only jets with the axis in the pseudorapidity interval $|\eta| < 1.5$ and azimuthal angle $> 30°$ from the vertical, where the two halves of the calorimeter join.

4. Transverse momentum dependence on multiplicity

The charged particle inclusive invariant cross section, $E * d^3\sigma/dp^3$, measured at two extreme energies for the Pulsed Collider run (200 and 900 GeV), is shown in Fig.1 as a function of transverse momentum p_T. The error bars include the systematic error due to acceptance correction and the smearing effect due to the p_T measurement error has been unfolded assuming a gaussian resolution function with $\sigma(1/p_T) = 0.02$, where the units are in GeV/c. Curves on the data points are fits of the form

$$E \frac{d^3\sigma}{dp^3} = \frac{A (p_T)^\alpha}{(p_T + p_{TO})^\alpha} \tag{1}$$

which reproduces rather well the data over the complete p_T range. The distributions tend to flatten at higher energy indicating an increasing of the average transverse momentum with $\sqrt{s}$; $\langle p_T \rangle$ is obtained from the parameters of the fit, extrapolating the p_T distribution to $p_T = 0$:

$$\langle p_T \rangle = 2 * p_{T0}/(\alpha - 3) \qquad (2)$$

In Fig. 2 $\langle p_T \rangle$ vs. $\sqrt{s}$ is plotted; also the point obtained from 1983 data at 546 GeV is shown. The average p_T is increasing from 0.391 GeV/c at $\sqrt{s} = 200$ GeV to 0.447 GeV/c at $\sqrt{s} = 900$ GeV.

Fig. 3 shows the p_T invariant distribution at $\sqrt{s} = 900$ GeV for three bands of charged multiplicity: again, as already observed after the first Collider run in 1981 [3], a clear flattening off of the distributions with increasing multiplicity is observed.

Another way to point out the same effect is to plot $\langle p_T \rangle$ as a funtion of the event charged multiplicity: in fig. 4a the data at 200 and 900 GeV from the Pulsed Collider run are shown, in fig. 4b the data obtained at the highest ISR energy[10], where we can observe a first indication of this effect. The effect of the 0.15 GeV/c p_T cut has been corrected for using the parameters (α and p_{T0}) of the p_T distribution fits for the three multiplicity intervals. The average transverse momentum increases from ~ 0.370 to ~ 0.525 Gev/c at the highest Collider energy.

5. Production of low transverse energy jets

A possible interpretation of the new effetcs observed at the Collider is the presence, in the MB sample, of events containing low E_T jets produced in hard scattering of low-x partons [11]. Events containing high E_T jets are characterized by higher multiplicities and higher transverse momenta in the "underlying event", indipendently of the jet E_T[12]. The tranverse energy flow around the jet axis, $d^2E_T/d\Delta\eta d\Delta\phi$, shows that both the width of the jet and the transverse energy density outside the jet, the "underlying event", are independent of the jet E_T. Fig. 5a shows the E_T flow around the jet axis, integrated over ϕ in the jet hemisphere, for events collected at 630 GeV with a 2-Jet-Trigger (plotted for $30 < E_T^J < 40$ GeV), and for MB events with at least one jet with $E_T > 5$ GeV at the same energy: the transverse energy is clustering around the jet axis, while the E_T density away from the jet axis is the same for both samples. For all the other MB events, when an axis is randomly chosen in the jet acceptance, the E_T density is $\sim 1/2$ of the plateau value of the Jet sample. In Fig. 5b the E_T flow around the jet axis is plotted for events collected during the Ramping Collider run at 900 GeV, toghether with the E_T flow of events without jet.

The experimental definition of jet becomes meaningless at low E_T, when the low E_T jets mix with high p_T particles. Low E_T jets can be generated by the algorithm: the search for a jet is started by a single high p_T particle and the jet could be built up by fluctuations in the E_T density in the cone of radius R=1 in the $\eta - \phi$ space around the initiator. In this case the jet profile is narrow, the width of the of the jet being in practice the size of a calorimeter cell. This situation is illustrated in fig. 5c, where the E_T flow, at $\sqrt{s}$=200 GeV, is plotted for jets of $E_T \gtrsim 5$ GeV and for "jets" of $E_T < 3$ GeV (but $\gtrsim 1.5$ GeV, the initiator value); we can also obsverve how the "plateau" value away from jet axis in the latter case is closer to that of events without jets, for which a random axis has been chosen.

To find out the minimum E_T to define a good jet, we have used the ratio between the transverse energy contained in a cone of radius $R_0 < 1$ and that contained in the cone of radius R=1. This ratio,

$$F(R_0) = \frac{\Sigma E_T(R=R_0)}{\Sigma E_T(R=1)} \tag{3}$$

measures the shape of the jet and its value is close to 1 for an isolated high p_T particle. Fig. 6 shows the distribution of the average value of $F(R_0)$, for R_0=0.2, as a function of the jet E_T for all jets with $E_T^J > 1.5$ GeV (initiator threshold) for two energies of the Ramping Collider run and for the MB data collected at 630 GeV in 1984. From the shape of $<F>$ vs. E_T^J we can see as the jet profile doesn't change for $E_T^J \gtrsim 5$ GeV independently of the c.m. energy; the same conclusion is obtained for different values of R_0 (0.3,0.4).

We can now define a sub-sample of events (Jet) containing all the events with at least one jet of $E_T \geq 5$ GeV and axis in the region, $|\eta|<1.5$ and $|\Delta\phi|<30°$ from the vertical. The fraction of Jet-events increases roughly like $\ln(s)$ going from 5.9% at 200 GeV to 17.2% at 900 GeV (fig. 7); about 30% of these events contain two jets in the acceptance, which exhibit the tipical collinear behaviour in the transverse plane (figs. 8a,8b).

6. General features of Jet and No-Jet events

Jets have been searched for and defined using only the calorimeter information. Complementary information on the charged particle content of the events is independently obtained analyzing the data coming from the Central Detector. For the two sub-samples of the events, Jet and No-Jet, the average charged multiplicity, the multiplicity distribution, the transverse momentum distribution and the correlation between $\langle p_T \rangle$ and multiplicity are completely different.

figs. 9a and 9b show the multiplicity distribution in terms of the KNO variable $z=n_{ch}/\langle n_{ch}\rangle$ for the two samples at $\sqrt{s}=200, 350, 630$ and 900 GeV. The distributions do not depend on the c.m. energy, $\langle n_{ch}\rangle$ for the Jet sample is twice as large as for the No-Jet one and the distribution is much narrower.

Average charged multiplicity in $|\eta|<2.5$

$\sqrt{s}$ (GeV)	200	350	630	900
$\langle n_{ch}\rangle_{No-Jet}$	13.81±0.07	14.55±0.14	15.06±0.09	15.93±0.07
$\langle n_{ch}\rangle_{Jet}$	26.49±0.23	30.24±0.49	32.21±0.26	32.89±0.13

Fig. 10 shows the invariant differential cross section as a function of p_T for the Jet and No-Jet samples at $\sqrt{s}=900$ GeV; while the No-Jet distribution is normalized to its cross section, the Jet one is simply normalized to the first point of the No-Jet one so to show immediately that the average p_T is larger.

The average transverse momentum dependence on multiplicity is shown in figs. 11a (200 GeV) and 11b (900 GeV) for the Jet and No-Jet events; the distribution is almost flat for the Jet events, while a smaller dependence of $\langle p_T \rangle$ on multiplicity is still present in the No-Jet sample, probably due to the presence of events with jets outside the acceptance. The average transverse momentum for the Jet events increases with the c.m. energy faster than for No-Jet ones (fig. 12).

<u>Average transverse momentum in $|\eta|\leq 2.5$</u>

$\sqrt{s}$ (GeV)		200	350	630	900
$\langle p_T \rangle_{No-Jet}$	(GeV/c)	.382±.005	.394±.006	.407±.006	.411±.005
$\langle p_T \rangle_{Jet}$	(GeV/c)	.474±.007	.476±.008	.502±.007	.516±.006

7. Jet cross sections

The number of observed jets has to be corrected for the background due to fluctuations in the transverse energy density and for the efficiency of the jet finding algorithm. Both corrections are evaluated with ISAJET[13] MonteCarlo generated events which are then processed and analyzed exactly as the real data. For the evaluation of background the hard scattering process is switched off in the MonteCarlo and the p_T distribution has been tuned to reproduce the No-Jet event data at 200 GeV, where we expect a small contribution from undetected jets. The resulting background correction is 35% for $E_T^J = 5$ GeV and less than 8% for $E_T^J = 10$ GeV with little dependence on c.m. energy.

The jet algorithm efficiency has been evaluated with the ISAJET jet event generator. Results have been checked with the (unbiased) sample of jets, with $E_T^J > 5$ GeV, recoiling against large transverse momentum $W^\pm$ produced in the UA1 detector [14]; the algorithm efficiency accounts fairly well for the number of jets assuming $E_T^J = E_T^W$. The jet finding efficiency is 45% for $E_T^J = 5$ GeV and is larger than 90% for $E_T^J = 10$ GeV.

After applying both corrections, the inclusive jet differential cross section, $d^2\sigma/dE_T d\eta$, is obtained correcting the number of events for the azimuthal angle acceptance and normalizing to the trigger cross section as outlined in sec. 2.

Fig. 13 shows the jet cross section at $\sqrt{s}=546$ GeV vs. E_T^J (MB data); also results from the Jet-Trigger 1983 data[15] are plotted. The cross section extends over nine orders of magnitude for E_T^J in the range 5-120 GeV. Fig. 14 shows the jet cross sections at $\sqrt{s}=200, 350, 630$ and 900 GeV. Fig. 15 shows the integrated jet cross section for $E_T^J > 5$ GeV and $|\eta| < 1.5$ as a function of c.m. energy; the errors are statistical only, while the estimated systematic error due two background and efficiency evaluation is $\sim 20\%$. The jet cross section is increasing like $\sim \ln(s)$ and gives a considerable contribution to the inelastic non-single diffractive cross section, σ_{NSD}, when approaching values of $\sqrt{s} \approx 1$ TeV.

If we now subtract the jet integrated cross section, σ_{Jet}, from the trigger cross section, which is compared in fig. 16 with the inelastic non-single diffractive cross section measured by the UA4 and UA5 Collaborations [16], we observe that the resulting cross section for soft processes is almost costant with $\sqrt{s}$; the emergence of the jets signal accounts for the increase of the inelastic $p\bar{p}$ cross section in the energy range explored by the CERN Pulsed Collider.

8. Conclusions

We have presented the general features of MB events at c.m. energy from 0.2 to 0.9 TeV at the CERN Pulsed Collider, with the UA1 detector with magnetic and calorimetric analysis.

We have shown evidence for the production of low transverse energy jets with a non negligible cross section. While jets are defined by the calorimeter, complementary information from the Central Detector shows that the characteristics of Jet and No-Jet events are completely different, supporting the interpretation of a naive two-component model.

The No-Jet (soft) component exhibits low multiplicity with large dispersion and low transverse momenta, the Jet (hard) component is characterized by high multiplicity with small dispersion and high transverse momenta.

The fraction of soft processes stays almost constant in the Pulsed Collider energy range, while the fraction of hard processes increases roughly like $\ln(s)$.

Combining the results of minimum bias and large transverse energy triggers the jet inclusive cross section is measured over nine orders of magnitude down to $x_T \approx 0.01$.

The contribution of hard processes to the inelastic non-single diffractive cross section becomes large when approaching 1 TeV c.m. energy and is mainly responsible for the increase of the $p\bar{p}$ inelastic rate.

I would like to thank Prof. C. Rubbia who encouraged me to give this presentation, Prof. A. Kernan who invited me to this Meeting and all my friends in the UA1 Collaboration. Special thanks are due to C. Albajar,

V. Cecconi, F. Ceradini, G. Ciapetti, A. DiCiaccio, F. Ghio, F. Lacava,
L. Nisati,C. Zaccardelli and L.Zanello for their help in preparing this talk.

References

1) The Staff of the CERN pp Project: Physics Letters 107B, 306 (1981).
2) M. Banner et al. : Physics Letters 118B, 203 (1982).
    G. Arnison et al. : Physics Letters 123B, 115 (1983).
3) G.Arnison et al. : Physics Letters 118B, 167 (1982).
4) K. Alpgard et al : Physics Letters 121B, 209 (1982).
5) G. Ciapetti:Proceedings of the 5th Topical Workshop on Proton-Anti-
        proton Collider Physics, Saint Vincent, Italy, 488 (1985).
6) For a review see : C.M.G. Lattes, Y. Fujimoto and S. Hasegawa,
        Physics Reports 65C, 151 (1980).
7) J.G. Rushbrooke : "Proposal for achieving pp collisions at up to 1 TeV
        c.m. energy by means of a cycle variation of stored
        beam energy in the SPS Collider", CERN-EP/82-6.
8) R. Laukner : "Pulsed operation of the CERN SPS Collider",
        CERN-SPS/85-21, presented at the 1985 Particle Acce-
        lerator Conference, Vancouver, Canada.
9) M. Bozzo et al. : Physics Letters 147B, 392 (1984).
10) A. Breakstone et al. : Physics Letters 132B, 463 (1983).
11) G. Pancheri and C. Rubbia : Nuclear Physics A418, 117 (1984).
    G. Pancheri:Proceedings of the 5th Topical Workshop on Proton-Anti-
        proton Collider Physics, Saint Vincent, Italy, 505 (1985).
    G. Pancheri and Y. Srivastava : Physics Letters 159B, 69 (1985).
12) G. Arnison : Physics Letters 132B, 214 (1985).
13) F. Paige and S. Protopopescu : BNL report 31987 (1981).
14) G. Arnison et al. : Lettere al Nuovo Cimento 44, 1 (1985).
15) E. Buckley: Proceedings of the 5th Topical Workshop on Proton-Anti-
        proton Collider Physics, Saint Vincent, Italy, 96 (1985).
    K. Sumorok : Proceedings of the XX Rencontres de Moriond, La Plagne,
        France, (1985).
16) V. Palladino : CERN-EP/85-154, pres. at the Workshop on Elastic and
        Diffractive Scattering,Chateau de Blois,France, (1985).
    J.G. Rushbrooke : CERN-EP/85-124, pres. at the XVI Symposium on
        Multiparticle Dynamics,Kiryat-Anavim,Israel,(1985).

Figure captions

Fig. 1 : Charged particles inclusive invariant cross section vs. transverse momentum at $\sqrt{s}$=200 and 900 GeV.

Fig. 2 : Average transverse momentum as a function of $\sqrt{s}$.

Fig. 3 : Charged particles inclusive invariant cross section vs. transverse momentum at $\sqrt{s}$=900 GeV for three different bands of charged multiplicity.

Fig. 4 : Average transverse momentum as a function of the event charged multiplicity : a) Collider at 200 and 900 GeV, b) ISR at 63 GeV.

Fig. 5 : Transverse energy flow around the jet axis, $dE_T/d\eta$, integrated over the jet ϕ hemisphere : a) for jet-trigger events ($30 < E_T^J < 40$ GeV) and MB events at 630 GeV ($E_T^J > 5$ GeV), b) for MB events at 900 GeV ($E_T^J > 5$ GeV) , c) for MB events at 200 GeV ($E_T^J > 5$ GeV and $1.5 < E_T^J < 3$ GeV). For the 3 plots also the "plateau" level for No-Jet events is shown.

Fig. 6 : $\langle F(R_0) \rangle$ as a function of jet transverse energy at $\sqrt{s}$=200,900 and 630 GeV (R_0=0.2).

Fig. 7 : Fraction of Jet-events vs. $\sqrt{s}$.

Fig. 8 : $\Delta\phi$ for the 2-jets events at a) 200 GeV and b) 900 GeV.

Fig. 9 : Multiplicity distribution in terms of the KNO variable $z=n_{ch}/\langle n_{ch}\rangle$ for a) Jet and b) No-Jet sample at $\sqrt{s}$=200,350,630 and 900 GeV.

Fig. 10: Charged particles inclusive invariant cross section vs. p_T for Jet and No-Jet samples at 900 GeV.

Fig. 11: Average transverse momentum vs. the event charged multiplicity for the Jet and No-Jet samples at a) 200 and b) 900 GeV.

Fig. 12: Average transverse momentum vs. $\sqrt{s}$ for Jet and No-Jet samples.

Fig. 13: Inclusive jet differential cross section as a function of E_T^J at $\sqrt{s}$=546 GeV for Jet-trigger and MB data.

Fig. 14: Inclusive jet differential cross section vs. E_T^J for $\sqrt{s}$=200,350, 630 and 900 GeV.

Fig. 15: Integrated jet cross section, for $E_T^J > 5$ GeV and $|\eta| < 1.5$, vs. $\sqrt{s}$.

Fig. 16: Inelastic non-single diffractive cross section from FNAL to Pulsed Collider energies.

626

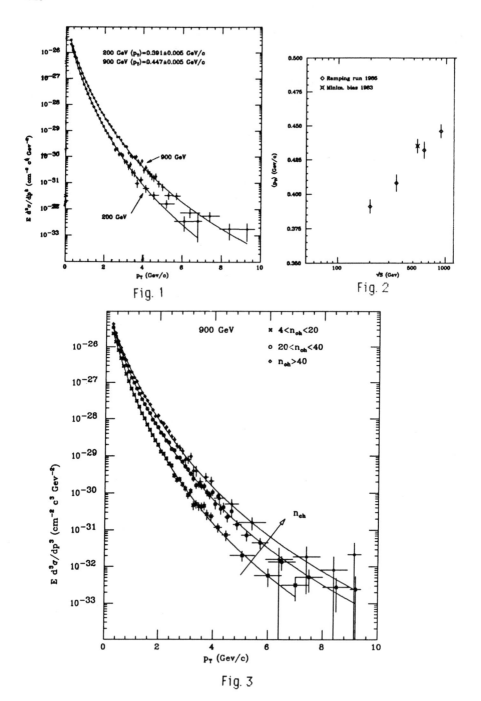

Fig. 1

Fig. 2

Fig. 3

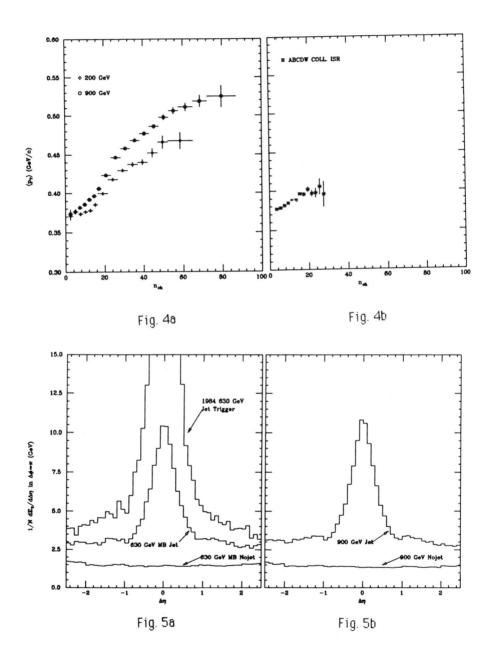

Fig. 4a

Fig. 4b

Fig. 5a

Fig. 5b

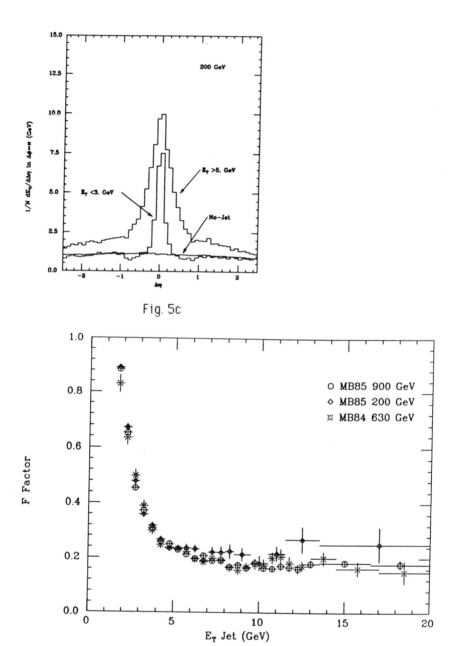

Fig. 5c

Fig. 6

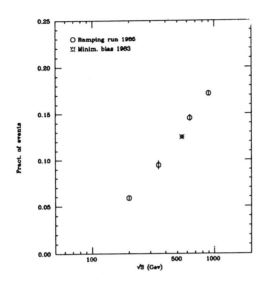

Fig. 7

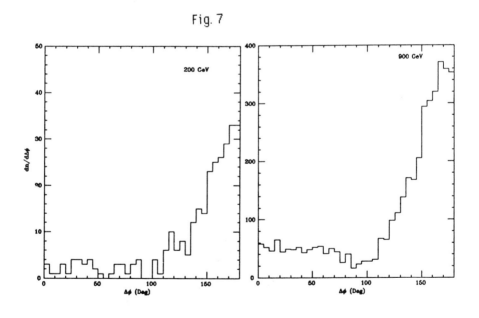

Fig. 8a   Fig. 8b

630

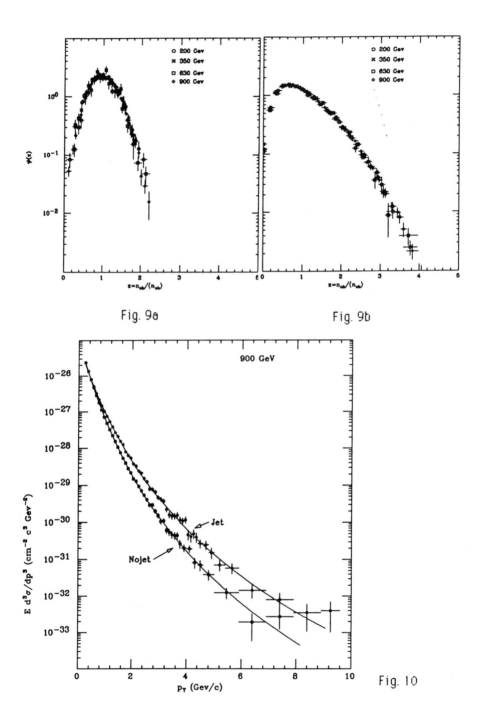

Fig. 9a

Fig. 9b

Fig. 10

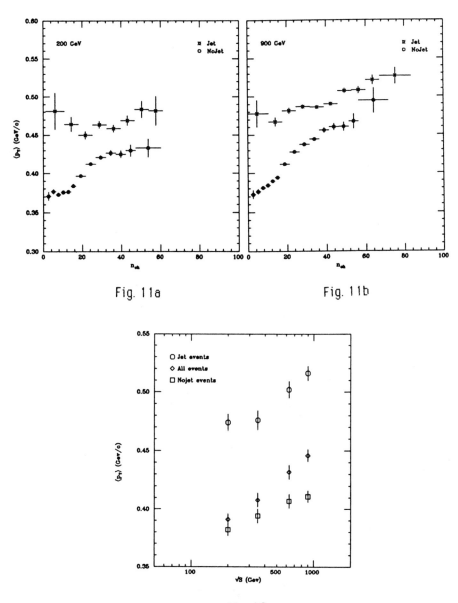

Fig. 11a

Fig. 11b

Fig. 12

632

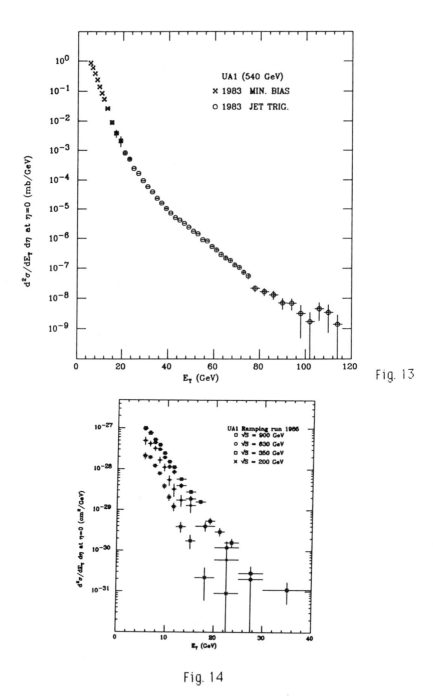

Fig. 13

Fig. 14

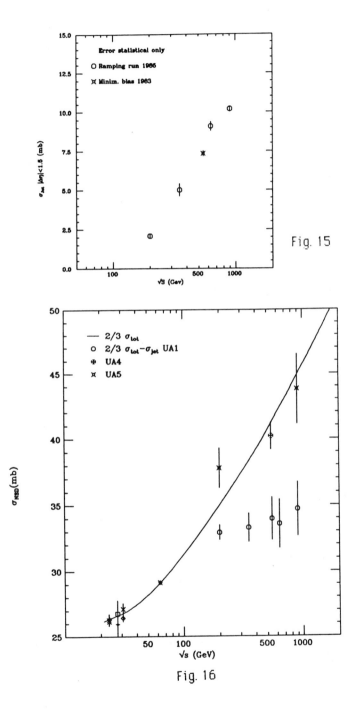

Fig. 15

Fig. 16

SMALL-X BEHAVIOR AND MINIJETS IN QCD

A. H. Mueller

Institute for Advanced Study
Princeton, N.J. 08540

and

Department of Physics
Columbia University
New York, NY 10027*

1. Introduction and Physical Picture of Small-x Behavior[1]

In calculating jet cross sections at Collider energies the gluon density in a proton, $G(x, M)$, enters quadratically. For example, the cross section for producing a pair of jets with a jet $p_\perp > M$ is [2,3]

$$\sigma(M) = \left(\frac{\alpha C_A}{\pi}\right)^2 \frac{\pi^3}{2M^2} \int \frac{dx_1}{x_1} \frac{dx_2}{x_2}$$

$$\theta\left(x_1 x_2 - \frac{4M^2}{s}\right) x_1 G(x_1, M^2) x_2 G(x_2, M^2), \tag{1}$$

so long as $\frac{M^2}{s} \ll 1$, the region where gluon scattering dominates jet production. Now one usually calculates the x and M^2 dependence of G by using the Altarelli-Parisi equation

$$M^2 \frac{\partial}{\partial M^2} xG(x, Q^2) = \frac{\alpha C_A}{\pi} \int_x^1 \frac{dx'}{x'} \frac{x}{x'} \gamma\left(\frac{x}{x'}\right) x' G(x', M^2), \tag{2}$$

which leads to

$$xG(x, Q^2) \sim \exp\left\{2\sqrt{\frac{C_A}{\pi b} \ln\left(\frac{\ln Q^2/\Lambda^2}{\ln Q_0^2/\Lambda^2}\right) \ln\frac{1}{x}}\right\} \mathcal{G}(x, Q_0^2, Q^2) \tag{3}$$

* Permanent Address

This research is supported in part by the U.S. Department of Energy Grant No. DE-AC02-76ERO220.

Talk given at the Division of Particles and Fields Meeting, Eugene, August 12-15, 1985.

where G is slowly varying. Thus $xG(x, M^2)$ grows rapidly for small x and leads to large jet cross sections for M^2 not too large. This growth in xG has an interesting physical interpretation to which we now turn. [1]

Suppose we consider a proton, of momentum $\vec{p}$, in a frame where $\vec{p} = \vec{e}_z p$ and p is large compared to the proton mass. In this frame $xG(x, M^2)$ represents the number of gluons in the proton per unit rapidity interval, with longitudinal momentum centered about xp and having transverse size, $|\Delta b| \le 1/M$. Now when $xG(x, M^2) \ge M^2 R^2$, with R the proton radius, the gluons within this unit of rapidity begin to spatially overlap in the longitudinally thin disc which they occupy. As x is decreased even further we may expect the gluons in this rather dense system to scatter and annihilate with one another, eventually reaching a saturation limit as $x \to 0$. When the gluon density becomes large enough that interactions are no longer negligible, the Altarelli-Parisi equation ceases to be valid [1,4-7] and (3) does not represent the actual gluon density. In the following section I shall describe the equation which replaces the Altarelli-Parisi equation when gluonic interactions are included, but where $xG(x, M^2)$ is still far from saturation. That is, the equation which describes gluon interactions in the low density limit will be given.

2. Modified Altarelli-Parisi Equation.

Although the physical picture presented in the last section is compelling in that large gluon densities must include annihilation and scattering, as well as the usual evolution, it is not straightforward to derive a modification of the standard Altarelli-Parisi equation without a detailed use of Feynman diagrams.[1,7] If the graphs in Fig. 1 represent the usual evolution and if the vertex at the bottom of those graphs represents the measurement of a gluon with longitudinal momentum xp and transverse size $1/M$, then the graphs of Fig. 2 represent the leading higher twist effects. (The graphs of Fig. 2 are a sampling of the graphs included in the three categories schematically represented in that figure.) The different cut graphs of Fig. 2 are related by the relative weights listed below each figure. The negative term comes from graphs one would definitely associate with shadowing corrections in that the gluon $k_1 - \ell$ undergoes an elastic scattering with the gluon k_2. The graphs of Fig. 2a include recombination. The graphs in Fig. 2a have been calculated in a leading logarithmic approximation for longitudinal momentum integrals.[7] Using the Abramovskii, Gribov, Kanchelli[8] weights we find

$$M^2 \frac{\partial}{\partial M^2} \, xG(x, M^2) = \frac{\alpha C_A}{\pi} \int_x^1 \frac{dx^1}{x^1} \frac{x}{x^1} \gamma \left(\frac{x}{x^1} \right) x^1 G(x^1, M^2)$$

$$- \frac{4\pi^3}{N^2 - 1} \left(\frac{\alpha C_A}{\pi} \right)^2 \frac{1}{M^2} \int_x^1 \frac{dx^1}{x^1} (x^1)^2 \, G^{(2)}(x^1, M^2) . \tag{4}$$

as the modified evolution equation. $G^{(2)}$ is the two gluon distribution per unit area of the proton, where the normalization of $G^{(2)}$ will be given in the next section. The second term on the right hand side of (4) is non-linear and has a negative sign corresponding to a depletion of the gluon number as the leading non-linear effect.

3. The Size of the Non-Linear Term.

Equation (4) is an interesting equation in its own right,[1] however we shall be interested here only in determining the region of M^2 and x for which the non-linear term is a small correction to the normal evolution term. Perhaps the easiest way to specify the normalization of $G^{(2)}$ is to say that for a very loosely bound large system, like a nucleus,

$$G^{(2)}(x, M^2) = \frac{[G(x, M^2)]^2}{8/9\pi R^2} \tag{5}$$

with R the radius of the nucleus. The identification of $G^{(2)}$ as a two gluon number density per unit area is now manifest. Consider two simple models to estimate when the non-linear term in (4) is important.

(i) Suppose we imagine the valence quarks in the proton as forming a bound system. Then, in analogy with (5) we take

$$G^{(2)}(x, M^2) = \frac{2}{3} \frac{[G(x, M^2)]^2}{\frac{8}{9}\pi R^2} \tag{6}$$

with R now the radius in which the valence quarks move. (The $\frac{2}{3}$ in (6) comes about because we require the two gluons causing the non-linearity to come from different valence quarks.) Then the two terms on the right hand side of (4) are equal when

$$xG(x, M^2) \approx \frac{25}{\pi\alpha} M^2 R^2 . \tag{7}$$

with R expressed in fm and M in Gev. Taking $R = 1 fm$ and $\alpha \approx 1/3$ we see that (7) is far from being satisfied. The non-linear term is very small effect in (4) so long as $M \geq 1 Gev$.

Rather than trying to saturate the whole area of the proton with gluons of transverse size $1/M$ we might try to saturate a smaller area. In this case it is more natural to have the non-linear term arise from gluons both of which come from the same valence quark whose evolution we imagine starts from $Q_0 \approx 1 Gev$. In this case one finds

$$x^2 G^{(2)}(x, M^2) \approx \frac{1}{\pi(2/Q_0)^2} \left[\frac{x}{x_0} G_q \left(\frac{x}{x_0}, M^2 \right) \right]^2 \tag{8}$$

where the valence quark is taken to have momentum fraction x_0 and G_q is the gluon distribution in an elementary quark; a quantity calculable in perturbative QCD. (8) represents the saturation condition over a transverse radius equal to $\frac{2}{Q_0}$. The two terms on the right hand side of (4) are now equal when

$$\left[\frac{x}{x_0} G_q \left(\frac{x}{x_0}, M^2 \right) \right]^2 \approx \frac{8}{3\pi\alpha} \left(\frac{M}{Q_0} \right)^2 xG . \tag{9}$$

Taking $Q_0 = 1 Gev$, and $M = 3 Gev, x_0 = 0.5, \wedge = 0.2 Gev$ we find the right hand side of (9) is about 100 while the lefthand side of (9) is about 7 at $x = 5 \times 10^{-4}$.

Again non-linear effects are very small. We conclude, then, that the non-linear terms in (4) are not important in calculating evolution for the starting point of the evolution at a Gev or so.

Before leaving this topic, however, let me make two comments. (i) In case the hadron being discussed is a large nucleus the non-linear (shadowing) corrections can be much larger. As an estimate when the non-linear and linear terms of (4) are equal, take $xG = AxG_{Proton}$ in (5), in which case equality occurs when

$$1 = \frac{4\pi^3}{N^2 - 1} \frac{\alpha C_A}{\pi} \frac{A}{\frac{8}{9}\pi R^2} \frac{xG_{Proton}}{M^2} , \tag{10}$$

that is, when

$$xG_{Proton}(x, M^2) \approx 15 \, M^2 A^{-1/3} \tag{11}$$

with M in Gev. At $M \approx 1Gev$ we see that the non-linear term is comparable to the linear term for $x \leq 0.03$ or so. This corresponds to a strong shadowing of the gluon distribution. Unfortunately the shadowing of the quark distribution is not so direct and it is not yet clear whether a computation of shadowing effects for nuclear structure functions can be reliably done within perturbative QCD.

4. Minijets

Having seen that non-linear effects are small we may use (1) with more confidence in calculating the cross sections for low energy jets. As an example take $M = 2Gev, \alpha \approx \frac{1}{3} xG \approx 3$ and integrate x_1 and x_2 over the central 6 units of rapidity for a rough approximation for the production of jet pairs, where the jet transverse energies are greater than $2Gev$ each at the CERN collider. One finds a cross section of about 30 mb in agreement with more precise evaluations.[2,3]

A question which immediately arises is whether one can associate the jet cross with the rise in the total inelastic cross section as one goes from ISR to collider energies. I think it is clearly questionable to associate the complete rise of the total cross section with minijet production. However, rather than trying to give a complete and precise answer to this question I would like, instead, to rephrase the question. Suppose one has an event with some given number of minijets being produced. We should count this as a true increase in the cross section if the event would not have occurred without the minijet production. Since minijets are most often produced in small impact parameter collisions one might guess that soft particle production would likely occur whether or not the hard gluons interacted to produce the minijets. Nevertheless, a part of the increase in the total inelastic cross section may be due to minijet production, but I would guess that the major rise is due to processes not having gluon transverse momenta as large as $2Gev$.

What about multiple minijet production? There are at east two separate sources of multiple minijets. (i) There may be two, or more, independent pairs of minijets due to scatterings of several gluons in each hadron, with the gluons in general at transverse separations on the order of 1 fm.[10] The situation is illustrated in Fig. 3. The cross section for such events is

$$\sigma_4(M) = \left[\left(\frac{\alpha C_A}{\pi} C_A \right)^2 \frac{\pi^3}{2M^2} \right]^2 \int dx_1 dx_2 dx_1' dx_2'$$

$$\Theta\left(x_1 x_2 - \frac{4M^2}{s}\right)\ \Theta\left(x_1' x_2' - \frac{4M^2}{s}\right)\ .$$

$$\cdot \frac{1}{2!}\ G_{(2)}(x_1, x_1', M^2, \underline{\delta}^2)\ G_{(2)}(x_2, x_2', M^2, \underline{\delta}^2)\ \frac{d^2\underline{\delta}}{(2\pi)^2} \tag{12}$$

in the leading logarithmic approximation, and where each of the jets is required to have transverse momentum greater than M. (The reader should be warned, however, that possibly important virtual corrections to (1) and (12) have not yet been included.) δ is a momentum transfer which can be defined by taking the amplitude complex conjugate to that of Fig. 3, and which occurs in calculating the cross section, to have momenta $k_1 - \delta, k_1' + \delta, k_2 - \delta, k_2' + \delta$ corresponding to the momenta $k_1 k_1', k_2, k_2'$ in Fig. 3. $G_{(2)}$ in (12) is related to the $G^{(2)}$ in (4) by

$$\int \frac{d^2\underline{\delta}}{(2\pi)^2}\ G_{(2)}(x, x, M^2, \underline{\delta}^2) = G^{(2)}(x, M^2)\ . \tag{13}$$

In terms of graphs $G_{(2)}(x_1, x_2, M_1^2, M_2^2, \delta^2)$ is given by those graphs indicated in Fig. 4, where p_- is large $p_\perp = 0$ and we suppose $\delta_- << x_1 p_-, x_1' p_-$. The generalization to N semi-hard inclusive scatterings of this type is straightforward.

(ii) Another mechanism for producing multiple minijets is shown in Fig. 5 where we might require $|k_1^\perp - \ell_1^\perp|, |\ell_1^\perp - \ell_2^\perp|, \cdots |\ell_{N-1}^\perp + k_2^\perp| \geq M$ as a jet criterion as in our previous case. The cross section is, in the leading logarithmic limit,

$$\sigma_N = \frac{\pi^3}{2M^2}\left(\frac{\alpha C_A}{\pi}\right)^N \int dx_1 dx_2 G(x_1, M^2)\ G(x_2, M^2)\ \frac{1}{(N-2)!}$$

$$\ln^{N-2}\frac{x_1 x_2 s}{4M^2} \times \int_R \frac{d^2\underline{\ell}_1}{\pi}\ \frac{d^2\underline{\ell}_2}{\pi} \cdots \frac{d^2\underline{\ell}_{N-1}}{\pi}\ \frac{1}{(\underline{\ell}_1)^2(\underline{\ell}_1 - \underline{\ell}_2)^2(\underline{\ell}_2 - \underline{\ell}_3)^2 \cdots (\underline{\ell}_{N-1})^2}$$

$$+\text{virtual corrections}\ , \tag{14}$$

where the R on the transverse momentum integral signifies the constrained intergration discussed above. (The virtual corrections to (14) have not yet been included.) σ_N in (14) is proportional to $\frac{1}{M^2}$ and so dominates the mechanism, described by (12) in the example of four jets, discussed earlier when M is large. However, mechanism (i) has the enhancement due to the growth of the gluon distributions and so, for M^2 fixed, dominates in the high energy limit. The expression in (14) is especially interesting in that the integrand of that expression is exactly the same as occurs in calculations of the bare Pomeron in QCD.[11,12] (Does this mean that there is some contribution of minijets to the rise in σ^9]? Yes, but there is no quantitative relationship yet.)

Minijets area fascinating new subject. They are presumably responsible for the slow falloff of the large-n tail of the multiplicity distribution at Collider energies, and they are also likely responsible for the growth of $< p_\perp >$ in large multiplicity minimum bias events. The question which remains is how much can be done quantitatively in the context of perturbative QCD.

REFERENCES

1] Gribov, L. V., Levin, E. M. and Ryskin, M. G., Phys. Reports 100, 1(1983).

2] Gaisser, T. and Halzen, F., Phys. Rev. Lett. 54, 1754(1985).

3] Sjöstrand, T., Fermilab Preprint.

4] Durand, L., in "Design and Utilization of the SSC, Snowman, 1985" (Fermilab, 1985).

5] Collins, J. C., in "Design and Utilization of the SSC, Snowman, 1985" (Fermilab, 1985).

6] Durand, L. and Putikka, W., Mod/TH/85-3, June 1985.

7] Mueller, A. H. and Qiu, J., CU-TP-322, September 1985.

8] Abramovski, V. A., Gribov, V. N. and Kancheli, O. V., Sov. J. Nucl Phys. 18, 308(1974).

9] Cline, D., Halzen, F. and Luthe, J., Phys. Rev. Lett. 31, 491(1973).

10] Paver, N. and Treleani, D., Z. Phys. C 28, 187((1985).

11] Kuraev, E. A.,Lipatov, L. N. and Fadin, V. S., Sov. Phys. JETP 45, 199(1978).

12] Balitskii, Ya., Ya. and Lipatov, L. N., Sov. J. Nucl. Phys. 28, 822(1979).

640

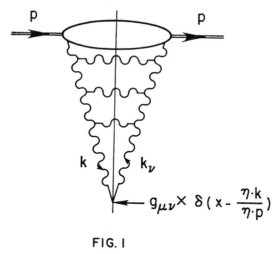

FIG. I

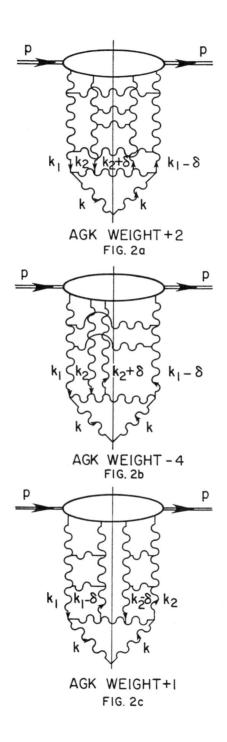

AGK WEIGHT + 2

FIG. 2a

AGK WEIGHT - 4

FIG. 2b

AGK WEIGHT + 1

FIG. 2c

642

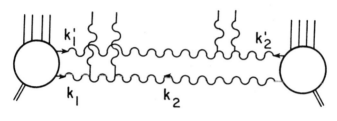

FIG. 3

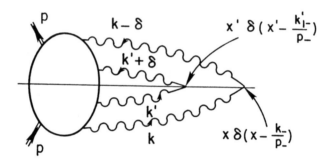

$$G_{(2)}(x, x', M^2, M'^2, \underline{\delta}^2) = 2p_- 2E_p (2\pi)^3 \int \frac{d\delta_+}{2\pi}$$

FIG. 4

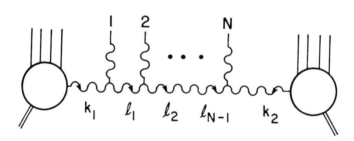

FIG. 5

A TWO-COMPONENT MODEL FOR SOFT HADRONIC INTERACTIONS
AT HIGH ENERGY

T.K. Gaisser

Bartol Research Foundation of the Franklin Institute

University of Delaware

Newark DE 19716

ABSTRACT

We interpret energy-dependence of inclusive cross sections and multiplicity distributions of minimum-bias events in a two-component model motivated by QCD.

1. INTRODUCTION

A prominent feature of minimum-bias events at the $S\bar{p}pS$ collider is the correlation between multiplicity and transverse momentum. The mean p_T per event is larger in events with more secondaries [Arnison et al., 1982, Banner et al., 1982,1983]. The mean p_T increases with energy, and the multiplicity increases with energy at a rate that is faster than logarithmic. The latter feature is related to the fact that the central rapidity density also increases with energy. It is natural to interpret these features in terms of a QCD-motivated picture as being due to the increasing production of soft jets.

It is possible to quantify these ideas [Gaisser & Halzen, 1985] by defining a parameter p_T^{min} by the relation

$$\sigma_{inel} = \sigma_{soft} + \sigma_{jet}(p_T^{min}),$$ (1)

where the jet cross section is calculated in perturbative QCD.

Each interaction proceeds either through the constant background channel (which is assumed to consist only of beam fragments which obey Feynman scaling) or the jet channel in which jet fragmentation is superimposed on beam fragmentation to give higher multiplicity and higher transverse momentum to these events. (See Figure 1). Since the basic hard scattering cross section has the Rutherford form, this component increases with energy as smaller and smaller angles $(p_T^{min}/\sqrt{s})$ become accessible. In addition, as energy increases $x_T^{min}=2p_T/\sqrt{s}$ decreases and more partons (mainly gluons) can contribute to jet production. Fig. 2 shows the rapidity distribution expected in the model.

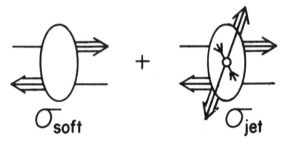

Figure 1. Illustration of the two-component model.

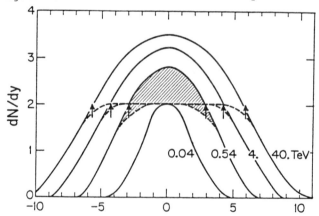

Fig. 2. Sketch of the two-component rapidity distribution. At each energy the rapidity distribution consists of a plateau on which is superimposed the excess of hadrons near y=0 from fragmentation of the scattered jets. Arrows indicate rapidities of scattered jets. The excess is cross-hatched for 540 GeV.

This procedure has several obvious problems: (1) A sharp cutoff between hard and soft processes is clearly artificial. (2) The question of whether the hard component simply adds to the cross section is unclear [A. Meuller, 1985]. (3) At small enough p_T(jet) multiple scattering processes will become important. (4) In order to fit the observed energy dependence of the data it is necessary to have the "parameter" p_T^{min} increase slowly with energy. A bonus, however, is that if Eq. 1 is used to define the parameter, it is possible to extrapolate predictions for inclusive cross sections to SSC energies. This is because there is cosmic ray data on inelastic cross sections up to that energy [Baltrusaitas et al., 1984] as well as extrapolations of fits to total and elastic cross sections at accelerator energies [Block & Cahn, 1985]. Despite these limitations, the basic picture is clearly supported by the recent observation {Ciapetti, 1985] by the UA1 group that high multiplicity events show jet-like structure.

2. MULTIPLICITY DISTRIBUTIONS

Recently we have applied this two-component model to multiplicity distributions [Gaisser, Halzen & Martin, 1985]. In the model the overall multiplicity distribution is given by

$$P_n(\sqrt{s}) = \frac{\sigma_{soft}}{\sigma_{inel}} P_n^{soft}(\sqrt{s}) + \frac{\sigma_{jet}}{\sigma_{inel}} \sum_{\substack{n_1 \\ n_2 = n - n_1}} P_{n_1}^{soft}[(1-\langle x\rangle)\sqrt{s}] P_{n_2}^{jet}[\langle x\rangle\sqrt{s}]. \quad (2)$$

If one uses the same KNO form for P_n^{jet} and P_n^{soft} and the same form $\bar{n} = a + 2\ell n(s)$ for both components, Eq. (2) then gives a prediction for SSC energies. Details of the fit at $\sqrt{s} = 540$ GeV are not good, however. Since the jet distribution is known to be narrower than the low energy KNO form it is more correct to use a different multiplicity distribution for the jet component as well as a different dependence of $\bar{n}$(jet) on energy. This has been done recently [Gaisser, Halzen, Martin & Maxwell, 1985] with the result shown in Fig. 3.

646

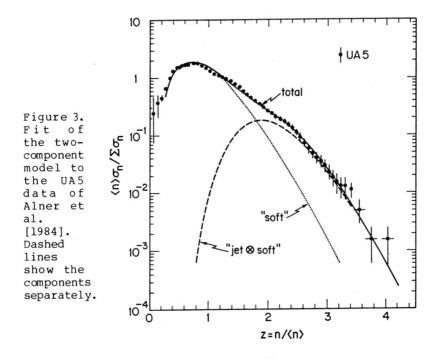

Figure 3. Fit of the two-component model to the UA5 data of Alner et al. [1984]. Dashed lines show the components separately.

The fit in Fig. 3 has $\sigma_{jet}/\sigma_{inel} = 0.1$, smaller than the value of about 0.4 needed to describe the rapidity distribution at 540 GeV (see Fig. 2). A similar fraction of 10% was obtained by Pancheri and Srivastava [1984] in a similar two-component fit to the multiplicity distribution at 540 GeV. Equation (2) contains a substantial approximation that must be removed in order to obtain self-consistent fits to cross section, multiplicity distribution and rapidity densities. Fluctuations in x_1 and x_2 (the momentum fractions carried by the scattering partons) from event to event will contribute significantly to the overall multiplicity distribution. Because the dominant gluon distribution is peaked at small x, the mode of the x distribution is less than the mean. Taking account correctly of this source of fluctuations will therefore broaden P_n^{jet} toward smaller n values and allow a larger ratio of $\sigma_{jet}/\sigma_{inel}$ to fit the data.

ACKNOWLEDGEMENTS

This work was done in collaboration with Francis Halzen and in part with Alan Martin and C.J. Maxwell. Research supported in part by the U.S. Department of Energy under DE-AC02-78ER05007.

REFERENCES

Alner, G.J. et al. (UA5 collaboration), Physics Letters 138B, 304 (1984).

Arnison, G. et al. (UA1 collaboration) Physics Letters 118B, 167 (1982) and 123B, 115 (1983).

Baltrusaitas, R.M. et al., Phys. Rev. Letters 52, 1380 (1984).

Banner, M. et al. (UA2 collaboration) Physics Letters 118B, 203 (1982) and Z. Phys. C 20 117 (1983).

Block, M. and R. Cahn, Revs. Mod. Phys. 57 563 (1985).

Ciapetti, G. Proc. p̄p collider workshop, Aosta, Italy (1985).

Gaisser, T.K. and F. Halzen, Phys. Rev. Letters 54, 1754 (1985).

Gaisser, T.K., F. Halzen and A.D. Martin, preprint MAD/PH/258 (also DPT/85/16) (1985).

Gaisser, T.K., F. Halzen, A.D. Martin and C.J. Maxwell, submitted to Physics Letters (1985).

Mueller, A. talk at this conference (DPF 1985).

Pancheri, G. and Y. Srivastava, Proc. Sante Fe DPF Meeting (World Scientific, ed.T. Goldman and Michael Martin Nieto)p.293 (1984) and earlier references therein.

GLUONS AND THE RISING TOTAL CROSS SECTION

P. L'Heureux, B. Margolis and P. Valin

Physics Department, McGill University
3600 University St., Montreal H3A 2T8
CANADA

ABSTRACT

We discuss here the proposition that rising
total cross sections in hadron-hadron interaction are mainly the result of gluon-gluon
interaction. We present a model of this
effect.

INTRODUCTION

Particles of quark structure similar to that of the proton are
easily produced by fragmentation but those of different composition
are produced mostly centrally. These include mesons and antibaryons.
Figure 1 shows the quantity $M^3\sigma(M,s)/\Gamma$ plotted against s/M^2. Here
$\sigma(M,s)$ is the inclusive cross section for producing mass M in p - p
interaction. This quantity, for central production, scales as [1],[2]

$$\sigma(M,s) \simeq f(M) \, F_{gg} \, (M^2/s) \tag{1}$$

where $\quad F_{gg}(\tau) = (\frac{n+1}{2})^2 \int_\tau^1 \frac{dx}{x} (1-x)^n (1-\tau/x)^n \tag{2}$

is the gluon-gluon structure function with n = 5. This scaling is
independent of quark content. Particles produced by fragmentation
(e.g. $\Delta^{++}(1232)$) behave differently as can be seen in Figure 1. As
suggested by the scaling law of equation (1) we choose as the

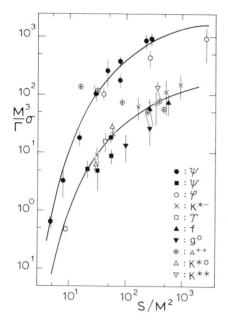

The quantity $M^3\sigma(M,s)/\Gamma$ plotted against s/M^2. M and Γ are the resonance mass and total width respectively, and $\sigma(M,s)$ the production cross section of M in p - p interaction. The curves are proportional to $F_{gg}(M^2/s)$.

Figure 1.

mechanism for central production the dominant QCD parton process gluon-gluon→gluon-gluon, this to be followed by fragmentation of the gluon into particles.

We study elastic diffraction in the pp and p̄p channels at high energies through an eikonal model built in impact parameter space out of QCD and parton model concepts. Following the considerations on particle production above, the eikonal function is separated into two terms, a contribution from valence quarks and a gluon-gluon initiated term. The latter is responsible for the energy dependence of the model at high energies.

The Model

The two colliding protons are considered as a collection of partons, carrying a fraction x_n of the longitudinal momentum of its host proton labelled by n = 1,2. The parton distributions $G_k(x_n, \vec{b}_n)$ for parton of type k depend on x_n and $\vec{b}_n$, the transverse position vector.

We write for the eikonal $\chi(s,b)$

$$\chi(s,b) = \frac{i}{2} \sum_{ij} \int G_i(x_1,\vec{b}_1)G_j(x_2,\vec{b}_2)\delta^{(2)}(\vec{b}_2-\vec{b}_1-\vec{b})\frac{\sigma_{ij}(\hat{s}_{ij}(b))}{\sigma^0_{ij}}$$

$$d^2\vec{b}_1 d^2\vec{b}_2 dx_1 dx_2 \tag{3}$$

Where $\sigma_{ij}(\hat{s}_{ij}(b))$ is the parton-parton cross section at C.M. energy squared $\hat{s}_{ij}(b)$ assumed to depend on b. σ^0_{ij} is a constant, with dimensions of area, characteristic of the pair i,j.
We write

$$G_k(x_n,\vec{b}_n) = f_k(x_n)h_k(\vec{b}_n) \tag{4}$$

where the $f_k(x_n)$ are the x distributions of the parton model with

$$F_{ij}(\tau) = \tau\int_0^1\int_0^1 f_i(x_1)\, f_j(x_2)\delta(x_1 x_2-\tau)dx_1 dx_2 \tag{5}$$

The $h_k(\vec{b}_n)$ are density profiles for each parton convoluting into a structure function in impact parameter space

$$W_{ij}(\vec{b}) = \iint h_i(\vec{b}_1)h_j(\vec{b}_2)\delta^{(2)}(\vec{b}_2-\vec{b}_1-\vec{b})\, d^2\vec{b}_1 d^2\vec{b}_2 \tag{6}$$

with $W_{ij}(0) = 1$. We have then

$$\chi(s,b) = \frac{i}{2} \sum_{i,j} \frac{1}{\sigma^0_{ij}} W_{ij}(b) \int_0^1 F_{ij}(\tau)\sigma_{ij}(\hat{s}_{ij}(b))d\tau/\tau \tag{7}$$

where we take $\hat{s}_{ij}(b) = \tau s W_{ij}(b)$; the subprocess energy is proportional to the fractional overlap of the colliding matter.

Our previous considerations for particle production (see the particle production behaviour in Figure 1) suggest a two component eikonal model.

The Valence Quark Component

This consists of contributions of nine pairs of valence quarks for p - p scattering and corresponding quark-antiquark pairs for $\bar{p}p$. According to formula (7) for X(s,b) these nine contributions are to be added together. This fragmentation component is not very energy dependent. We parametrize the integral in (7) by a constant then and write the contribution to $\chi(s,b)$ as $\chi_V(b)$

$$\chi_V(b) = i/2\ T_{vv}(b)\sigma_{vv} \quad + \text{Regge Corrections} \tag{8}$$

Here $\quad W_{vv}(b) = T_{vv}(b)/T_{vv}(0) \quad , \quad \int T_{vv}(b)d^2\vec{b} = 9 \tag{9}$

It follows that $\sigma_{vv}^0 = 9/T_{vv}(0)$.

The Gluon Component

Gluon-gluon scattering is the dominant process for central production according to QCD perturbation theory. Lowest order perturbation theory cut-off for soft gluons yields a cross section which can be written as [3]

$$\sigma_{gg\to gg}(M^2) = 9\pi\alpha_s^2(M^2)\ [\frac{17}{12(M^2+\delta^2)} + \frac{1}{\delta^2} - \frac{1}{M^2+\delta^2}\ \ell n\,(\frac{M^2+\delta^2}{\delta^2})] \tag{10}$$

with $\quad \alpha_s = \dfrac{12\pi}{25\ell n M^2/\Lambda^2}$. We take Λ = 127 MeV. $\tag{10a}$

Here δ^2 is the gluon cut-off. We have the dominant central contribution to the eikonal then

$$\chi_G(s,b) = \frac{i}{2\sigma_{gg}^0}\ W_{gg}(b) \int_0^1 F_{gg}(\tau)\sigma_{gg}(\hat{s})d\tau/\tau \tag{11}$$

where $\hat{s} = \tau s W_{gg}(b)$. Here $\sigma_{gg}(\hat{s})$ and $F_{gg}(\tau)$ are taken from formulae (10) and (2) respectively. Otherwise one or both might be deduced from the scattering data. σ_{gg}^0 can be considered as an effective gluon number parameter. We will come back to this point below. Our eikonal has basically two terms then plus Regge corrections

$$\chi(s,b) = \chi_V(b) + \chi_G(s,b) + \text{Regge corrections} \qquad (12)$$

The first term is of the Chou-Yang, Durand-Lipes energy independent form. The energy dependence is provided by the second term alone aside from the small transient Regge corrections.

Comparison with Gluon-Gluon Born Approximation

The contribution to the total cross section from gluon-gluon interaction in lowest Born approximation is given by

$$\sigma^{Born} = \int_{m_0/s}^{1} d\tau/\tau F(\tau)\sigma_{gg}(\tau s) \qquad (13)$$

where m_0 is the lowest gluon-gluon energy. Using the approximate form which is good for $s > \delta^2$

$$\sigma_{gg}(s) \simeq \frac{9\pi\alpha_s^2(s)}{\delta^2} \qquad (14)$$

with α_s given by formula (10a) and $F(\tau)$ by (2) leads to

$$\sigma^{Born} \propto \log s \qquad \text{for } s \gg \Lambda^2, m_0^2 \qquad (15)$$

If σ_{gg} instead went as a constant as s gets large, the Born contribution would go as $\log^2 s$. Using our eikonal picture

$$\sigma_{tot} = 4\pi f_{el}(0) \qquad (16)$$

$$f_{el}(t) = \frac{1}{2\pi} \int_0^{\infty} (1-e^{i\chi(s,b)}) J_0(b\sqrt{-t})d^2\vec{b} \qquad (17)$$

with $\chi(s,b)$ given by (7).

One could identify the large s Born approximation to (13) with the contribution of g-g interaction as described in the eikonal model above. Then for the range of s where $\chi_G(s,b)$ is not too large this contribution

$$\sigma_{tot}^{gg} = 2\int [e^{i\chi_v} - e^{i(\chi_v + \chi_G)}]d^2\vec{b} \approx -2i\int \chi_G e^{i\chi_v}d^2\vec{b}$$

$$= \frac{1}{\sigma_{gg}^0}\int W_{gg}(b)e^{i\chi_v}d^2\vec{b}\int F_{gg}(\tau)\sigma[\tau sW(b)]d\tau/\tau \qquad (18)$$

Identifying (18) with (13) for s large enough (but not too large) yields

$$\sigma_{gg}^0 = \int d^2\vec{b}\ W_{gg}(b)\ e^{i\chi_v(b)} \qquad (19)$$

Calculations and Comparison with Experiment

We have taken $T_{vv}(b)$ and $W_{gg}(b)$ to be Fourier transforms of dipole form factors

$$T_{vv}(b) = \frac{3\mu^2}{32\pi}\ (\mu b)^3 K_3(\mu b) \qquad (20)$$

$$W_{gg}(b) = \frac{1}{8}(\mu'b)^3 K_3(\mu'b) \qquad (21)$$

where μ and μ' are the respective inverse ranges of the quark and gluon distributions [3]. Regge behaviour has been introduced first very roughly in Model I by replacing $\chi_v(b)$ by

$$\chi_v(s,b) = \frac{i}{2}\ (1+C/\sqrt{s})\ T_{vv}(b)\sigma_{vv} \qquad (22)$$

The real part of the amplitude for scattering is introduced by replacing s by $se^{-i\pi/2}$ in χ_G.

In Model II the Regge term is written as [3]

$$R(s,t) = -\beta_+ e^{a_+ t}\ (e^{-i\pi\alpha_+} +1)E^{\alpha_+ -1}$$

$$+\xi\beta_- e^{a_- t}(e^{-i\pi\alpha_-} -1)E^{\alpha_- -1} \qquad (23)$$

where E is the laboratory energy and

ξ = +1, -1 for pp and $\bar{p}p$ scattering respectively. Here we take $a_+ = a_-$ and $\alpha_{\pm} = 1/2 + t$.

It is well know that use of the dipole form factor (20) produces a diffraction pattern in elastic scattering with multiple dips. A form factor which removes diffraction zeros at higher values of momentum transfer has been given in the literature[3],[4]. We write this as

$$\tilde{T}_{vv}(t) = \frac{\gamma^2+t}{\gamma^2-t}\left[(1-t/m_1^2)(1-t/m_2^2)\right]^{-2} \tag{24}$$

where $\tilde{T}_{vv}(t)$, the Fourier transform of $T_{vv}(b)$ replaces $(1-t/\mu^2)^{-4}$ of Models I and II. The use of (24) with Regge corrections given by (23) defines Model III.

Figure 2 shows calculation and experimental data for pp elastic scattering using Model I. Figure 3 shows high energy $\bar{p}$-p scattering calculations and measured cross sections using Models I and II. Figure 4 shows σ_{tot}, forward slope B(0) and forward real to imaginary scattering amplitude ratio $\rho(0)$ for models I and II. The open circle data is for $\bar{p}$-p scattering in figure 4. Figure 5 shows p-p elastic scattering using Models II and III for $|t|>1GeV^2$ showing how Model III fills the second dip.

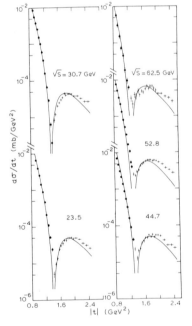

Figure 2.

Elastic differential cross sections for p-p scattering at $\sqrt{s}$ = 23.5 to 62.5 GeV. Data from reference (5). Theoretical curve is Model I with no real part. σ_{vv} = 3.06 mb and C = 5.33 GeV. Other parameters as in text.

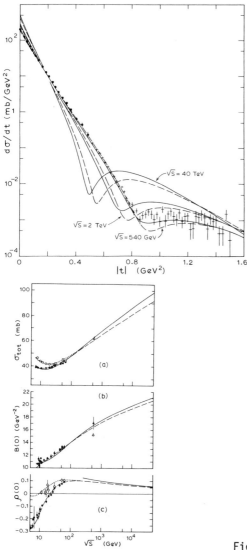

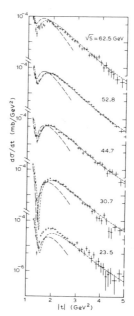

Figure 3.

Elastic differential cross section for p̄p(≅pp, at high energies in our model). Data from reference (6). Predictions are given for the Tevatron and SSC. Full curves are Model I and long dashed curves Model II calculations.

Figure 4. σ_{tot}, forward slope B(0) and Ref(0)/Imf(0)=ρ(0). Full curves are Model I, long dashed curves Model II. Short dotted curves are predictions for p̄p with Model II. For data, see Ref (3).

Figure 5.

Large |t| elastic differential cross section at ISR energies for Models II and III (long and short dashes resp.). Data from reference (5).

Both Models II and III have not had the parameters optimized to fit
the total cross section and so our best prediction for the total cross
section are with Model I.

The form of $\chi_G(s,b)$ is such that the ratio $\sigma_{gg}/\sigma_{gg}^0$ in eq. (11)
depends to good approximation on the product $\sigma_{gg}^0 \delta^2$ because of the form of
eq. (14). From rather rough considerations on particle production[3] we
had estimated $\delta \approx .54 \text{GeV}$ and this was used in the fits of figures 2 to 5.
However the value of σ_{gg}^0 determined from the fits is found to be considerably larger than the estimate given by eq.(19). This indicates one
should be using a considerably larger δ. Further a large gluon cut-off
suggests a larger cut-off $m_0 \sim 2\delta$ rather than $\sim 2m_\pi$ as has been used in the fits
of reference (3) shown here. In fact the fits of Models I and II with $\delta = .54 \text{GeV}$,
$m_0 = 2m_\pi$ yield $(1/\sigma_{gg}^0) = 1.27 \times 10^{-4} \text{GeV}^2$ whereas formula (19) yields
$1/\sigma_{gg}^0 = 1.33 \times 10^{-2} \text{GeV}^2$ using $\mu' = .64$ GeV and $\mu = .94$ GeV.

Using the value of σ_{gg}^0 estimated from eq.(19) and taking $m_0 \sim 2\delta$ we
estimate that $\delta \sim 2.3 \text{GeV}$. This is an interesting result. Our physical
picture of the rising cross section is that it is in good part from fairly
hard processes and this would include jet formation then. Central production has a strong contribution from these harder processes, the
lighter produced particles π, ρ, K, K^* etc. coming as secondaries.

The model[3] features a rising ratio of σ_{el}/σ_{tot} and a forward
slope rising slower than σ_{tot}. The model does not have a factorizable
eikonal nor geometric scaling. Finally we point out that we have a
rising total cross section but we do not saturate the Froissart bound.

References

1) Y. Afek, C. Leroy, B. Margolis and P. Valin, Phys. Rev. Lett. 45,
   85 (1980).
2) T.K. Gaisser, F. Halzen and E.A. Paschos, Phys. Rev. D15, 2572
   (1977); F. Halzen and S. Matsuda, Phys. Rev. D17, 1344 (1978).
3) P. L'Heureux, B. Margolis and P. Valin, Phys. Rev. D Oct. 1 (1985)
4) C. Bourrely, J. Soffer and T.T. Wu, Phys. Rev. D19, 3249 (1979).
5) K.R. Schubert, in "Landolt-Bornstein Numerical Data and Functional
   Relationships in Science and Technology."New Series, Group I,
   Vol. 9, edited by H. Schopper, Springer-Verlag, Berlin, 1980.
6) M. Bozzo et al. ; UA4 Collaboration, Phys. Lett. 147B, 385 (1984).
   See also reference (3) for compilation of other data.

RECENT RESULTS ON J/ψ DECAYS FROM THE MARK III

William J. Wisniewski [‡]

University of Illinois at Urbana-Champaign, Urbana, Illinois 61801

representing the Mark III Collaboration

ABSTRACT

An update of results on radiatively produced $\rho\rho$ and $\omega\omega$ is presented, as well as a coupled channel analysis of $\iota(1460)$ decays. Preliminary results on the hadronic decays of the J/ψ to $\omega\iota$ and $\phi\iota$ are discussed. New results on the ξ are presented.

The Mark III has acquired three blocks of data at the J/ψ. The first two samples, with a total of 2.7×10^6 produced J/ψ, were collected in 1982-83. The third set of $3.1 \times 10^6 J/\psi$ was acquired in Spring, 1985. Most of the analyses discussed in this report were limited to the first two data sets. The entire data sample was used for the analysis of the $\gamma K^+ K^-$ and $\gamma K_s^0 K_s^0$ channels.

1. RADIATIVELY PRODUCED $\rho\rho$ AND $\omega\omega$ FINAL STATES

The $\gamma 4\pi$ final state has been analyzed in both the $\gamma\pi^+\pi^-\pi^+\pi^-$ and the $\gamma\pi^+\pi^0\pi^-\pi^0$ modes.[1] This analysis is complicated by large background contamination due to $J/\psi \to 5\pi$. The 4π mass distribution containing contributions from both modes as well as 5π background is shown in fig. 1. Two peaks are evident at ~ 1.55 and ~ 1.80 GeV/c^2. After background subtraction, branching ratios for $\gamma 4\pi$ with $m_{4\pi} < 2$ GeV/c^2 are found to be

$$B(J/\psi \to \gamma\pi^+\pi^-\pi^+\pi^-) = (3.05 \pm 0.08 \pm 0.45) \times 10^{-3} ,$$
$$B(J/\psi \to \gamma\pi^+\pi^0\pi^-\pi^0) = (8.3 \pm 0.2 \pm 3.1) \times 10^{-3} .$$

The ratio of these ratios is consistent with expectations for isoscalar 4π. A 10 channel spin-parity analysis has been performed. The region below 2 GeV/c^2 is dominated ($51.3\pm8.5\%$) by $\rho\rho$ with $J^P = 0^-$. The $0^-\rho\rho$ spectrum is displayed in fig. 2. The structure at 1.55 GeV/c^2 is primarily

0^-. The 0^- component falls off at the 1.8 GeV/c^2 peak. The product branching ratio to pseudoscalar $\rho\rho$ for the region below 2 GeV/c^2 is

$$B(J\psi \to \gamma X_{0^-}) \cdot B(X_{0^-} \to \rho\rho) = (4.7 \pm 0.3 \pm 0.9) \times 10^{-3}.$$

Upper limits at 90% confidence level (CL) are set for structures in the 2^+ channel:

$$B(J/\psi \to \gamma\theta) \cdot B(\theta \to \rho\rho) < 5.5 \times 10^{-4},$$
$$B(J/\psi \to \gamma g_T) \cdot B(g_T \to \rho\rho) < 6.0 \times 10^{-4}$$

where 2.1 GeV/c$^2 < m_{g_T} <$ 2.4 GeV/c^2.

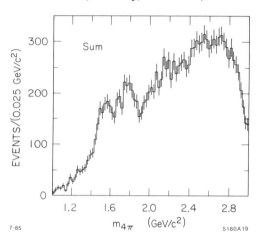

Fig. 1. 4π invariant mass distribution before background subtraction. The plot is the sum of $J/\psi \to \gamma\pi^+\pi^-\pi^+\pi^-$ and $J/\psi \to \gamma\pi^+\pi^0\pi^-\pi^0$.

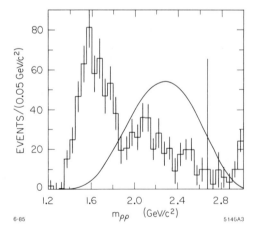

Fig. 2. Mass distribution for the pseudoscalar $\rho\rho$ component. The solid line shows P-wave phase space for $J/\psi \to \gamma\rho\rho$.

The five γ, four charged π final state has been studied in a search for $J/\psi \to \gamma\omega\omega$. [2] Requiring that each opposite $\pi^+\pi^-\pi^0$ triplet have the ω mass yields the mass plot in fig. 3a, where a clear signal is present. The background, which is drawn cross-hatched, is gotten from the ω sidebands. There is no background from $J/\psi \to \omega\omega$ and $\omega\omega\pi^0$ since these processes are forbidden by C-parity. The background is mainly due to $J/\psi \to \omega 4\pi$.

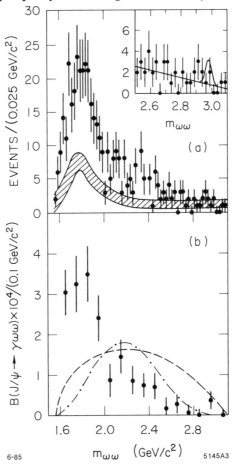

6-85   5145A3

Fig. 3. (a) $\omega\omega$ invariant mass distribution. The band represents the background. The mass region from 2.5 GeV to 3.1 GeV is shown in the insert with the 90%CL curve for the η_c superimposed. (b) $B(J/\psi \to \gamma\omega\omega$ as a function of $m_{\omega\omega}$. S-wave phase space is indicated by the dashed line, P-wave phase space by the dotted-dashed line.

The product branching ratio for the $\omega\omega$ signal is

$$B(J/\psi \to \gamma\omega\omega) = (1.76 \pm 0.09 \pm 0.45) \times 10^{-3},$$

with 70% of this below 2 GeV/c^2. This rate is consistent with expectations for isoscalar $\rho\rho$. The branching ratio of $\eta_c \to \omega\omega$ is found to be less than 3.1×10^{-3} at 90% CL. A 7 channel spin parity analysis indicates that the 0^- channel dominates in the region below 2 GeV/c^2, accounting for (85 $\pm$ 19)% of the signal. Limits are set, at 90% CL, for production of θ and g_T:

$$B(J/\psi \to \gamma\theta) \cdot B(\theta \to \omega\omega) < 5.5 \times 10^{-4},$$
$$B(J/\psi \to \gamma g_T) \cdot B(g_T \to \omega\omega) < 6.0 \times 10^{-4}.$$

2. COUPLED CHANNEL ANALYSIS OF $\iota(1460)$

The Mark III has observed several peaks in different radiatively produced channels in the mass region below 2 GeV/c^2. The $\iota(1460)$ is seen in the $K\overline{K}\pi$ channel with a width of 98 ± 15 MeV/c^2, spin-parity 0^-, and product branching ratio $B(J/\psi \to \gamma\iota) \cdot B(\iota \to K\overline{K}\pi) = 5.0 \pm 0.8 \times 10^{-3}$. The $\eta\pi^+\pi^-$ spectrum shows structure at 1380 MeV/c^2 with a width of about 100 MeV/c^2 and product branching ratio $B(J/\psi \to \gamma X(1380)) \cdot B(X(1380) \to \eta\pi\pi) \approx 2 \times 10^{-3}$. The spin parity of this structure has not yet been determined. The radiatively produced $\gamma\rho$ spectrum exhibits a peak at about 1420 MeV/c^2 with width $\sim$150 MeV/c^2, spin-parity consistent with 0^- and branching ratio $B(J/\psi \to \gamma X(1420) \cdot B(X(1420) \to \gamma\rho) \approx 1 \times 10^{-4}$. In addition there is pseudoscalar structure in both $\rho\rho$ and $\omega\omega$. Since the pseudoscalar ground state multiplet is full, and eight of the nine members of the radially excited pseudoscalar multiplet have been found, the presence of additional 0^- particles may point to the existence of non-$q\bar{q}$ mesons.

A coupled channel analysis has been performed in an effort to determine if these structures have a common origin. The idea is that the observed mass and width of the parent resonance may differ in the separate channels due to threshold effects. Details of this analysis may be found elsewhere.[3] The mass spectra for the four final states included in this analysis, $K\overline{K}\pi, \rho\rho, \omega\omega$ and $\gamma\rho$, are shown in fig. 4, where the spectra have been corrected for relative efficiencies. The fit includes two interfering Breit-Wigners centered at $\sim$1.5 GeV/c^2 and $\sim$1.8 GeV/c^2, where the higher mass structure in $\rho\rho$ has been taken to be pseudoscalar. The lower mass Breit-Wigner couples to all four channels. It is assumed that the higher mass Breit-Wigner does not couple to $K\overline{K}\pi$. The results of the fit are superposed on the spectra in fig. 4. The analysis indicates that the structure in the lower portion of the $\rho\rho$ spectrum can be explained by a resonance below threshold, in this case the ι. The parameters of the ι in the $K\overline{K}\pi$ are left unaffected by the fit. The $\eta\pi\pi$ channel can not be accommodated by the fit. The ratio $g_{\rho\rho}/g_{\omega\omega}$ of the couplings of the parent state to the $\rho\rho$ and $\omega\omega$ channels is expected to be 3 from SU(3) symmetry. The fit yields 5.0 ± 0.7, in reasonable agreement with the expected value. The ratio $g_{\rho\rho}/g_{\gamma\rho}$ of the couplings to the $\rho\rho$ and $\gamma\rho$ channels is expected to be 400 from vector meson dominance. The fit yields 3300 ± 600. The agreement between expectation and measurement is poor in this case.

If the lower mass $\rho\rho$ and $\omega\omega$ structures are associated with the ι, then

$$B(J/\psi \to \gamma\iota) \cdot B(\iota \to \rho\rho) = (1.5 \pm 0.2) \times 10^{-3},$$
$$B(J/\psi \to \gamma\iota) \cdot B(\iota \to \omega\omega) = (0.3 \pm 0.1) \times 10^{-3} \text{ and}$$
$$B(J/\psi \to \gamma\iota) > (6.9 \pm 0.4 \pm 1.0) \times 10^{-3}.$$

This speculative identification gives the ι the largest radiative production rate in J/ψ decays aside from the η_c, as well as associating with it decay modes to non-strange mesons. Both characteristics enhance the glueball interpretation of the ι.

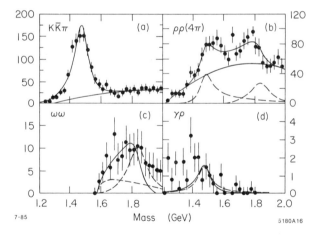

Fig. 4. Invariant mass distributions for (a) $K\overline{K}\pi$, (b) $4\pi(\rho\rho)$, (c) $\omega\omega$, and (d) $\gamma\rho$ corrected for relative efficiencies (arb. units, 0.025 GeV bins). The curves represent the results of the coupled channel analysis.

3. HADRONIC DECAYS OF J/ψ TO $\omega\iota$ AND $\phi\iota$

The Mark III Collaboration has recently completed an analysis [4] that maps out the quark content of the η and η' based on a mixing model parametrization by Rosner.[5] The results suggest that the η' wave function can have substantial contibution, $(35\pm18)\%$, from mixing with a pseudoscalar glueball or radially excited $q\bar{q}$ state. This analysis was extended by Haber and Perrier [6] to include ι. They predict branching ratios $B(J/\psi \rightarrow \omega\iota) = (7.5 \pm 1.2) \times 10^{-4}$ and $B(J/\psi \rightarrow \phi\iota) = (0.9 \pm 0.2) \times 10^{-4}$. The light quark content of the ι is tested by the first decay mode. The strange quark content is tested by the second decay mode.

The $\omega\iota$ hadronic decay of the J/ψ is studied in the $\omega K^{\pm}K_s^0\pi^{\mp}$ and $\omega K^{+}K^{-}\pi^0$ final states. The search uses the $\pi^+\pi^-\pi^0$ decay mode of the ω. The ω candidate is required to have a mass within $15\text{MeV}/c^2$ of the ω mass. This produces the $K^{\pm}K_s^0\pi^{\mp}$ spectrum seen in fig. 5a, which has clear structure at 1450 MeV/c^2. (Structure is also seen in $K^{+}K^{-}\pi^0$). Fitting with a Breit-Wigner yields a mass of 1449 ± 8 MeV/c^2 and a width of 44^{+29}_{-16} MeV/c^2. The product branching ratio, $B(J/\psi \rightarrow \omega X(1450)) \cdot B(X(1450) \rightarrow K\overline{K}\pi) = (7.5 \pm 2.5 \pm 2.0) \times 10^{-4}$, is measured using the $K^{\pm}K_s^0\pi^{\mp}$ channel.

The $\phi\iota$ hadronic decay of the J/ψ is searched for in the $\phi K^{\pm}K_s^0\pi^{\mp}$ and $\phi K^{+}K^{-}\pi^0$ channels. The search uses both the $K^{+}K^{-}$ and $K_s^0K_L^0$ decay modes of the ϕ. The spectrum obtained for $K\overline{K}\pi$ recoiling against the ϕ is shown in fig. 5b. There is no peak in the $1450\text{MeV}/c^2$ region. An upper limit is set, at 90% CL, on the branching ratio $B(J/\psi \rightarrow \phi X(1450)) \cdot B(X(1450) \rightarrow K\overline{K}\pi) < 2.3\text{x}10^{-4}$.

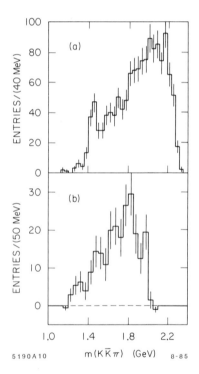

5190A10 $m(K\overline{K}\pi)$ (GeV)   8-85

Fig. 5. Invariant mass spectra for (a) $K^{\pm}K_s\pi^{\mp}$ from $J/\psi \rightarrow \omega K^{\pm}K_s\pi^{\mp}$ and (b) $K\overline{K}\pi$ from $J/\psi \rightarrow \phi K\overline{K}\pi$.

If the peak seen in the $K\overline{K}\pi$ spectrum opposite the ω is taken to be the ι, then this analysis indicates that the ι is not a pure glueball, but may have substantial $q\bar{q}$ mixing. It is also possible to make the identification of this peak with the E(1420), though the interpretation of the results is cloudier since the E is expected to be an $s\bar{s}$ state.

4. $\xi(2.2)$

4.1 Review of the Status of the ξ

The Mark III Collaboration first reported the observation of a narrow state ξ at 2.2 GeV/c^2 during the summer of 1983.[7] The signal was observed in the radiative K^+K^- spectrum with a significance of 4.7 standard deviations(SD). The product branching ratio was measured to be $B(J/\psi \to \gamma\xi) \cdot B(\xi \to K^+K^-) = (3.8 \pm 1.3 \pm 0.9) \times 10^{-5}$. Supporting evidence was seen in radiative $K_s^0 K_s^0$ where a product branching fraction of $(2.8 \pm 1.4 \pm 0.7) \times 10^{-5}$ was measured for a structure with significance 2 SD. These observations were based on a sample of $2.7 \times 10^6 J/\psi$.

The DM2 Collaboration, using a detector with drift chamber and shower counter resolution comparable to that of the Mark III and substantially poorer time-of-flight(TOF) resolution (520ps vs ~200ps), acquired $8 \times 10^6 J/\psi$ soon afterward. They do not observe a signal at 2.2 GeV/c^2 in either the K^+K^- or $K_s^0 K_s^0$ channels and set 95% CL limits on the product branching ratios [8] based on the assumption that the ξ has zero width and spin 0 : $B(J/\psi \to \gamma\xi) \cdot B(\xi \to K^+K^-) < 1.2 \times 10^{-5}$ and $B(J/\psi \to \gamma\xi) \cdot B(\xi \to K_s^0 K_s^0) < 2.0 \times 10^{-5}$.

The Mark III ran again at the J/ψ in Spring 1985 in an attempt to confirm the ξ in a new data sample. A sample of $3.1 \times 10^6 J/\psi$ was acquired. The results of studies of the full $5.8 \times 10^7 J/\psi$ sample are presented here.

4.2 $\gamma K_s^0 K_s^0$

The $K_s^0 K_s^0$ final state is studied in the $\pi^+\pi^-\pi^+\pi^-$ channel. The analysis requires four charged particles and at least one neutral. An event is selected as a $K_s^0 K_s^0$ candidate if there are two opposite $\pi^+\pi^-$ pairs whose invariant mass is within 50 MeV/c^2 of the K_s^0 mass. These events are 4C fit to the hypothesis $\gamma\pi^+\pi^-\pi^+\pi^-$. The momenta used in the kinematic fit are calculated at the distance of closest approach of the π's in the $\pi^+\pi^-$ pairs which form the candidate K_s^0. Any event with a fit χ^2-probability greater than 1% is retained. The parameters of the K_s^0 after the fit, which are consistent with Monte Carlo expectations, are 497 MeV/c^2 for the mass and 4.7 MeV/c^2 for the mass resolution. In addition, the proper decay length found in the data agrees well with the Monte Carlo prediction. The final sample of $K_s^0 K_s^0$ events is selected by requiring $\delta^2 \leq .0004$ GeV/c^2, where $\delta^2 = (m_{\pi_1^+\pi_1^-} - .497)^2 + (m_{\pi_2^+\pi_2^-} - .497)^2$. This corresponds to a radial cut in the $(\pi^+\pi^-)vs(\pi^+\pi^-)$ mass plane at 20 MeV/c^2 from the K_s^0 mass. The resulting $K_s^0 K_s^0$ mass spectrum is seen in fig. 6b. The $f'(1515)$ and $\theta(1720)$ are clearly visible, as is the ξ, perched atop a broad $K_s^0 K_s^0$ structure. The background is estimated by taking a ring in the $(\pi^+\pi^-)vs(\pi^+\pi^-)$ mass plane of area equal to that of the $K_s^0 K_s^0$ signal: $.0008 < \delta^2 < .0012$ (GeV/c^2)^2. These events are shown cross-hatched in fig. 6b. The background is seen to produce no peak in the ξ region. There are ~14 background events in the 1.9-2.6 GeV/c^2 mass region. This region is fit to a Breit-Wigner convoluted with a 10 MeV/c^2 Gaussian resolution function. Both the mass and width are free to vary when the unbinned maximum likelihood fit is performed. The fit is displayed in the inset of fig. 6b. The parameters obtained are:

$$m(\xi) = 2.232 \pm 0.007 \pm 0.007 \text{ GeV/c}^2$$
$$\Gamma(\xi) = 0.018^{+0.023}_{-0.015} \pm 0.010 \text{GeV/c}^2$$

where the first error is statistical and the second systematic. The statistical significance of this fit is 3.6 SD. The significance is sensitive to the choice of the background parameterization as well as the mass region over which the fit is performed. Variations in these lead to significance extremes between 3.0 and 4.7 SD. The fit finds 23 events in the ξ signal. The Monte Carlo generated detection efficiency is 28% for the process $J/\psi \to K_s^0 K_s^0$, where both K_s^0's decay to charged pions. This efficiency is insensitive to various spin hypotheses. The product branching ratio is

$$B(J/\psi \to \gamma\xi) \cdot B(\xi \to K_s^0 K_s^0) = (3.2^{+1.6}_{-1.3} \pm 0.7) \times 10^{-5}.$$

The Dalitz plot is shown in fig. 7b. Diagonal bands due to the f′ and θ are clearly visible at the upper edge of the plot. A band is discernable at the ξ.

4.3 $\gamma K^+ K^-$

The $J/\psi \to \gamma K^+ K^-$ events are selected by requiring two oppositely charged particles and a least one detected photon. These events are kinematically fit to the $\gamma K^+ K^-$ hypothesis. Events with fit proabability

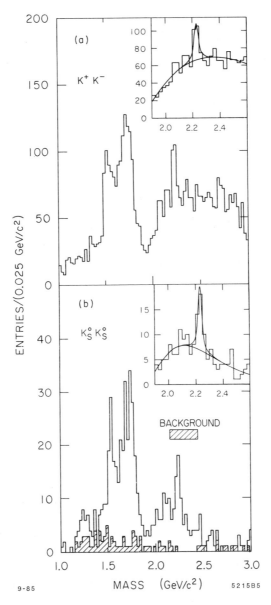

Fig. 6. $K\overline{K}$ invariant mass distributions for the full sample of $5.8 \times 10^6 J/\psi$ are shown for the K^+K^- final state (a) and for the $K_s^0 K_s^0$ final state, where the four pion background is shown cross-hatched(b). Fits to the 1.9–2.6 GeV/c^2 mass region are displayed in the insets.

greater than 1% are retained. The TOF system is then used to identify tracks on an event basis. The quantity Δ is defined as

$$\Delta = t_+^m - \frac{1}{2}(t_{\pi^+} + t_{K^+}) + t_-^m - \frac{1}{2}(t_{\pi^-} + t_{K^-})$$

where $t_\pm^m$ is the measured time and $t_{\pi^\pm}$ and $t_{K^\pm}$ are the predicted times for the the π and K hypotheses. This quantity is required to be greater than 100ps. This cut discriminates against $\pi^+\pi^- + neutrals$ and $e^+e^- + neutrals$. The effects of this TOF requirement are independent of the K^+K^- mass. Radiative e^+e^- events are further reduced by eliminating events where the energy measured in the shower counter and associated with each of the charged tracks is greater than 1 GeV. In an effort to reject backgrounds from $\rho\pi$ and K^*K, events with two photons are fit to the $\gamma\gamma\pi^+\pi^-$ and $\gamma\gamma K^+K^-$ hypotheses. If the resulting $m_{\gamma\gamma}$ is within 50 MeV/c^2 of the π^0 mass, and the fit has probability $> 2\%$, the event is rejected. The resulting K^+K^- mass distribution is shown in fig. 6a where clear f', θ and ξ signals are visible. The Dalitz plot for this final state is shown in fig. 7a. There are clear f' and θ bands, as well as contamination at the edge of the Dalitz plot due to γe^+e^- and horizontal and vertical bands due to highly asymmetric π^0 decays from K^*K. In order to check that the ξ signal is not due to these two backgrounds, the border ($|cos\theta_{\gamma K}| > .99$) and K^* bands ($.7 < m_{\gamma K}^2 < .9$ (Gev/c^2)^2) have been excluded. The ξ signal remains. Each of these additional cuts reduces the signal by 12%. However, the significance varies little. These cuts bias heavily against higher spins and have not been applied when determining the parameters of the ξ.

The cuts applied to the total data sample in the analysis presented here are milder than those imposed in the analysis of the first $2.7 \times 10^6 J\psi$ reported earlier.[7] This loosening of cuts was mandated by problems with the performance of the detector in the new data sample. The loss of the outermost 3 layers of the drift chamber coarsened the momentum resolution by 25%. In addition, the TOF system suffered a drop in resolution from 190 to 220ps. Part of this degradation can be traced to the bunch lengthening of the beams at SPEAR after the installation of mini-β. Since these cuts used in this analysis are less restrictive, the background contamination is increased. However, this set of cuts is found to have consistent efficiency for both the 2.7 and $3.1 \times 10^6 J/\psi$ samples.

The spectrum between 1.9 and 2.6 GeV/c^2 is fit with a Breit-Wigner convoluted with a 10 MeV/c^2 Gaussian resolution function and a background function. The ξ parameters determined by this unbinned maximum likelihood fit are

$$m(\xi) = 2.230 \pm 0.006 \pm 0.014 \text{ GeV}/c^2$$
$$\Gamma(\xi) = 0.026^{+0.020}_{-0.016} \pm 0.017 \text{GeV}/c^2$$

in good agreement with the parameters found in $K_s^0 K_s^0$. The signal is found to contain 93 events and have a significance of 4.5 SD. Depending on the fit window and the background parameterization, this significance can vary between 3.9 and 5.8 SD. The significance is lower than that reported for the 1982-83 data, but the number of events seen in the new analysis is the consistent with the number seen in the earlier analysis. The significance decreases in the new analysis because of the introduction of more background . Fig. 8a shows the 1982-83 sample signal from this new analysis. Fig. 8b shows the 1985 sample signal. The number of events in the ξ peak is comparable in these data sets. The detection efficiency is 38% and is independent of spin hypothesis. It varies slowly between 1 and 3 GeV/c^2 by less than 20%. The product branching ratio is

$$B(J/\psi \rightarrow \gamma\xi) \cdot B(\xi \rightarrow K^+ K^-) = (4.2^{+1.7}_{-1.4} \pm 0.8) \text{x} 10^{-5}.$$

4.4 Interpretation

Searches for additional decay modes of the ξ have been done. They are summarized in table 1. The searches were performed on the $2.7 \times 10^6 J/\psi$ sample for all modes but $\mu^+\mu^-$ where the full data sample was used.

Final State	$B(J/\psi \rightarrow \gamma\xi) \cdot B(\xi \rightarrow X)$
$\xi \rightarrow \mu^+\mu^-$	$< 5 \times 10^{-6}$
$\xi \rightarrow \pi\pi$	$< 2 \times 10^{-5}$
$\xi \rightarrow K^* K$	$< 2.5 \times 10^{-4}$
$\xi \rightarrow K^* \bar{K}^*$	$< 3 \times 10^{-4}$
$\xi \rightarrow \eta\eta$	$< 7 \times 10^{-5}$
$\xi \rightarrow p\bar{p}$	$< 2 \times 10^{-5}$

The ratio of branching ratios of the ξ to $K^+ K^-$ and $K_s^0 K_s^0$ is consistent with the value 2 expected for the decay of an isoscalar meson. The possible quantum numbers of the ξ are $J^{PC} = (even)^{++}$.

668

There has been considerable theoretical speculation on the nature of the ξ. It has been suggested that it might be a low mass Higgs particle.[9] There are difficulties with this interpretation caused by measurements performed in the Υ region at CESR,[10] but they are not insurmountable. Glueball and hybrid models predict state in the 2-2.5 GeV/c^2 region with the possibility that decays to $s\bar{s}$ states dominate.[11] It is also possible to fit this state into the $q\bar{q}$ spectrum as an L=3, S = 2 $s\bar{s}$ bound state.[12].

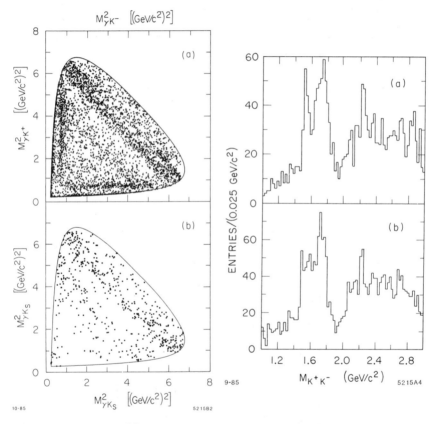

Fig. 7. Dalitz plots for (a) the K^+K^- channel and for (b) the $K_s^0 K_s^0$ channel.

Fig. 8. K^+K^- invariant mass distributions for (a) the $2.7 \times 10^6 J/\psi$ sample and for (b) the 3.1×10^6 sample.

5. SUMMARY

The radiative $\rho\rho$ and $\omega\omega$ spectra are found to be dominated by pseudoscalar structure below 2 GeV/c^2. A coupled channel analysis suggests that part of this structure could be associated with the ι, giving it an even larger production rate in J/ψ radiative decays. The ι is also seen produced opposite the ω suggesting that the ι wavefunction has a non-strange quark component. A narrow state $\xi(2.2)$ has been observed again in doubling the old data sample. It is seen in radiative decays to both $K_s^0 K_s^0$ and $K^+ K^-$. Its nature remains unclear.

‡ This work was supported in part by the Department of Energy under contract DE-AC02-76ER01195.

REFERENCES

1. R. M. Baltrusaitis et al., SLAC–PUB–3682, May 1985.

2. R. M. Baltrusaitis et al., Phys. Rev. Lett. **55**,1723 (1985).

3. N. Wermes, SLAC–PUB–3730, July 1985.

4. R. M. Baltrusaitis et al., SLAC–PUB–3435, Feb. 1985.

5. J. L. Rosner, Phys. Rev. **D27**, 1101 (1985).

6. H. E. Haber and J. Perrier, SCIPP 85/39, July 1985.

7. K. Einsweiler, SLAC–PUB–3202, Sept. 1983,
   W. Toki, SLAC–PUB–3262, Nov. 1983,
   D. Hitlin, Proceedings of the 1983 International Symposium on Lepton and Photon Interactions.

8. J. Augustin et al., contributed paper to 1985 International Symposium on Lepton and Photon Interactions. Preprint: LAL 85/27

9. H. E. Haber and G. L. Kane, Phys. Lett. **135B**, 196 (1984),
   R. S. Willey, Phys. Rev. Lett., **52**, 585 (1984),
   R. M. Barnett et al., Phys. Rev. **D30**, 1529 (1984).

10. S. Behrends *et al.*, Phys. Lett.**137B**, 277 (1984),
    S. Yousseff *et al.*, Phys. Lett.**139B**, 332 (1984).

11. M. Chanowitz and S. R. Sharpe, Phys. Lett. **132B**, 413 (1983),
    B. F. L. Ward, Phys. Rev. **D31**, 2849 (1985).

12. S. Godfrey *et al.*, Phys. Lett. **141B**, 439 (1984).

PARTIAL WAVE ANALYSIS OF KKPI SYSTEM IN D AND E/IOTA REGION

S.U. Chung, R. Fernow, H. Kirk, S.D. Protopopescu, D.P. Weygand
Brookhaven National Laboratory, Upton, New York 11973

D. Boehnlein, J.H. Goldman, V. Hagopian, D. Reeves
Florida State University, Tallahassee, Florida 32306

R. Crittenden, A. Dzierba, T. Marshall, S. Teige, D. Zieminska
Indiana University, Bloomington, Indiana 47401

Z. Bar-Yam, J. Dowd, W. Kern, H. Rudnicka
Southeastern Massachusetts University, N. Dartmouth, Massachusetts

Presented by S.D. Protopopescu

ABSTRACT

A partial wave analysis and a Dalitz plot analysis of high-statistics data from reaction $\pi^- p \to K^+ K_S \pi^- n$ at 8.0 GeV/c show that the D(1285) is a $J^{PG} = 1^{++}$ state and the E(1420) a $J^{PG} = 0^{-+}$ state both with a substantial $\delta\pi$ decay mode. The 1^{++} $K^*\overline{K}$ wave exhibits a rapid rise near threshold but no evidence of a resonance in the E region. The assignment of $J^{PG} = 0^{-+}$ to the E is confirmed from a Dalitz-plot analysis of the reaction $pp \to K^+ K_S \pi^- X^\circ$.

Since the observation that the J/ψ has a substantial radiative decay to the $\iota(1440)$ making it a prime glueball candidate[1] there has been renewed interest in the spin parity of the E(1420). The E(1420) was originally discovered by Armenteros et al.[2] in $\bar{p}p$ annihilations at rest with a preferred J^{PG} assignment of 0^{-+}. However, a subsequent experiment by Dionisi et al.[3] concluded, from a Dalitz plot analysis of the reaction $\pi^- p \to K^+ K_S \pi^- n$ at 4.2 GeV/c, that the E(1420) is a 1^{++} state coupling predominantly to $K^*\overline{K}$. This conclusion was supported by the higher statistics experiment of Armstrong et al.[4] by a similar analysis of central production of the E(1420). Our Dalitz plot analysis of the $K^+ K_S \pi^-$ system in the reaction

$$\pi^- p \rightarrow K^+ K_S \pi^- n \text{ at } 8.0 \text{ GeV/c.} \qquad (1)$$

with more then 10 times the statistics of Dionisi et.al. contradicts the two latter claims and concludes that the E(1420) is a 0^{-+} state[5]. We will present here, in addition to our original results, a partial wave analysis (PWA) of the same data and a Dalitz plot analysis of the reaction

$$\bar{p}p \rightarrow K^+ K_S \pi^- X^0 \text{ at } 6.6 \text{ GeV/c.} \qquad (2)$$

They all support our conclusion that the E(1420) is a $J^{PG} = 0^{-+}$ state.

The data come from two experiments performed with the Brookhaven National Laboratory Multiparticle Spectrometer (MPS). The layout is described in ref. 5 for the experiment with a beam at 8.0 GeV/c. It consists basically of a tagged beam impinging on liquid-hydrogen target located inside the MPS magnet and surrounded on four sides by a lead-scintillator veto box. Downstream of the target but still inside the magnet are seven drift-chamber modules with seven measuring planes each. Interspersed with the drift-chamber modules are three proportional wire chambers (P1, P2, P3) for triggering purposes. Downstream of the magnet there is a large high-pressure Cherenkov counter hodoscope (C1) with γ threshold = 10 and two scintillation counter hodoscope (H1, H2). The only difference in the apparatus between the two runs was the use of a 60-cm target for $\bar{p}$ instead of a 30-cm target for π^-.

The trigger required a positive particle with momentum > 1.5 GeV/c going through C1, H1 and H2 without emitting light in C1. This was achieved using a RAM-trigger, a three dimensional coincidence matrix system using random-access memories (RAM's). There were actually two RAM's in coincidence, one using P2*P3*H2 for momentum selection and another using P2*P3*(H1*$\bar{C}$1) for nonpion identification. For the π^- beam run we required in addition a multiplicity of 2 in P1, 4 in P2 and no signal in the veto box.

For the p̄ beam run the only requirements beyond the RAM's were a multiplicity of at least 2 in P1 and at least 4 in P2. The beam fluxes were 10^6/pulse for π^- and 10^5/pulse for p̄, and the trigger rates 10/pulse and 20/pulse respectively (with about 1500 pulses/hour). The total number of triggers accumulated were $1.5*10^6$ for a 200-hour π^- run and $4.5*10^6$ for a 350-hour p̄ run. The total number of events for reaction (2) is 16,000 and for reaction (1) 15,000 after requiring the missing-mass squared to be between 0.4 and 1.3 (GeV) and -t to be less than 1.0 (GeV).

The $K^+K_S\pi^-$ mass spectra, given in Figs. 1a and 1b, show clearly the D and E states. The background in reaction (2) is much higher than for reaction (1), in part because we do not separate K^+ and p and we expect reaction (2) to be contaminated by p̄pK_S events. The missing-mass squared for reaction (2) is featureless and the t distribution is flat, consistent with production by annihilation in flight. A fit to the spectra with two simple Breit-Wigner functions and a polynomial function give for reaction (1): m_D =1285 $\pm$ 2MeV and Γ_D = 22 $\pm$ 2 MeV; m_E = 1421 $\pm$2 MeV and Γ_E =60 $\pm$ 10 MeV, and for reaction (2): m_D =1277 $\pm$ 3 MeV and Γ_D =32 $\pm$ 8 MeV; m_E =1424 $\pm$ 3 MeV and Γ_E =70 $\pm$ 15 MeV.

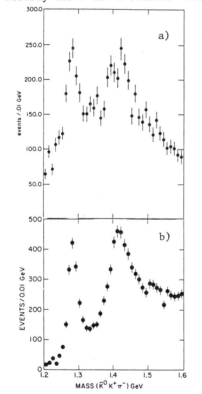

Fig. 1a. Effective mass spectrum for reaction (2).
Fig. 1b. Same as 1a) for reaction (1) requiring 0.4<MM^2<1.3 (GeV)^2, 0.48<M(K_S)<0.52 GeV and -t <1.0 (GeV)^2.

The above values are consistent with those observed in other hardronic reactions.

For reaction (1) we have performed a complete partial wave analysis with amplitudes that depend on a set of five variables, to be denoted τ: three Euler angles (α, β, γ) which rotate from the Gottfried-Jackson frame to the Dalitz system and two Dalitz-plot variables (s_1, s_2). The differential cross section is given by:

$$I(\tau) \equiv \frac{d\sigma}{d\alpha \, d\beta \, d\gamma \, ds_1 \, ds_2} \propto \sum \rho_{ab} \, A_a \, A_b^*$$

(3)

where a,b are the set of quantum numbers needed to describe the production and decay, i.e. $a = (J^{PG}M^\eta \text{ (isobar)})$. We have assumed for the analysis that only two isobar states are needed, K^* and δ. Since the recoil baryon has spin 1/2 the density matrix is constrained to have rank ≤ 2 for a given t bin. This constraint plus the positivity of the density matrix can be imposed by requiring that:[6]

$$\rho_{ab} = \sum_{k=1,2} V_{ak} \, V_{bk}^*$$

(4)

where V_{ak} are complex parameters in the fit.

From threshold to 1.6 GeV we found that only the states $J = 0^{-+}, 1^{++}, 1^{+-}, 1^{-+}$ with all allowed M^η values and incoherent background were needed to fit the data. The decay modes $\delta\pi$ and $K^*\bar{K}$ are allowed for 0^{-+} and 1^{++}, while for 1^{+-} and 1^{-+} only the $K^*\bar{K}$ mode is allowed. The total number of parameters in the partial wave analysis is of the order of 30, compared to ≈ 10 for a Dalitz-plot fit, but the PWA deals with a 5-dimensional space instead of 2-dimensional space for the Dalitz-plot analysis.

The analysis was done using MINUIT[7] and a program developed at BNL[8] to find the maximum likelihood with the log of likelihood given by:

$$L = \sum_{i=1}^{n} \ln \frac{I(\tau_i)}{\int I(\tau) \, A(\tau) d\tau}$$

(5)

where $A(\tau)$ represents the finite acceptance of our apparatus. The
results are displayed in Fig. 2, where the different J^{PG} waves
are displayed as a function of $K\bar{K}\pi$ mass. We chose to show only the
summed partial waves (over M^{η} and decay modes) because we found
that the separation between decay modes depends sensitively on the
δ parameterization while the sum is more stable. The δ parameterizations we used is the coupled channel formula of Flatté, following
previous analyses. We have used alternative parameterizations
(such as an S-wave Breit-Wigner) and found that the results shown
in Fig. 2 did not vary significantly.

As can be seen in Fig. 2 the 1^{++} wave shows a peak in the
D-region and a rapid rise across the $K^*\bar{K}$ threshold in the
E-region. This confirms that the D is 1^{++} state but not the E. On
the other hand the 0^{-+} wave shows a clear and significant peak in
the E-region.

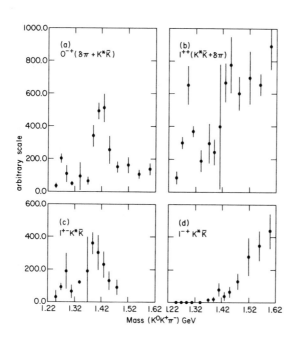

Fig. 2 Acceptance
corrected partial
waves for reaction
(1) obtained by a
PWA.
a) $J^{PG} = 0^{-+}$ wave
adding $\delta\pi$ and $K^*\bar{K}$
contributions
b) $J^{PG} = 1^{++}$
c) $J^{PG} = 1^{+-}$ wave

In Fig. 3 we show the relative phases. The rapid positive in-
creases increase of the $0^{-+}(\delta)$ phase relative to the $1^{++}0^{+}(K^*\bar{K})$
phase is characteristic of a 0^{-+} resonance interfering with a
non-resonant 1^{++} background. The $1^{+-}(K^*\bar{K})$ shows some peaking in
that region, however, there is no noticeable phase motion respect
to the 1^{++} wave (Fig. 5a), so it is probably not resonating. The
other waves in the fit, 1^{-+} and incoherent backgound (not shown),
are negligible below 1.4 GeV and rise slowly up to 1.6 GeV. This
analysis supports the original conclusion from a Dalitz-plot
analysis of the same data that the D is a 1^{++} state and the E a 0^{-+}
state with no evidence for a $1^{++}(K^*\bar{K})$ state in the E-region. For
comparison we show in Fig. 4 the results of the Dalitz-plot

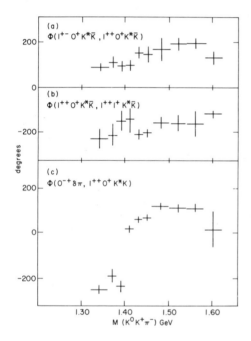

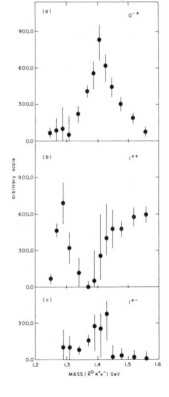

Fig. 3 Relative phases a function
of $\bar{K}K\pi$ mass. The notation is J^{PG}
(isobar).

Fig. 4 Spin-parity content of $\bar{K}K\pi$ system
produced in reaction (1).

analysis using Zemach amplitudes. Although there are some
quantitative differences between the two sets of results (not
surprising given the very different parameterizations), the
conclusions to be drawn are identical.

For reaction (2) a partial wave analysis is not likely to add
any more information than a Dalitz-plot analysis since it is an
inclusive annihilation channel with very little production
information. Therefore only a Dalitz-plot analysis with Zemach
amplitudes was done. The partial waves between 1.36 and 1.52 GeV
are given in Fig. 5. The large amount of background in the
D-region makes it impossible to separate $1^{++}(\delta)$ from $0^{-+}(\delta)$. In
the E-region we found that the only required waves were 0^{-+}, 1^{++},
1^{+-} and a flat phase space background. The analysis supports the
conclusion that the E is a 0^{-+} state.

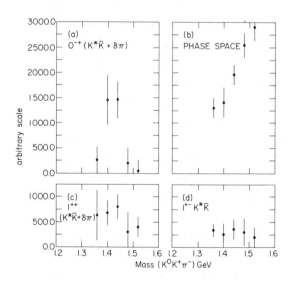

Fig.5
Spin parity content
of $\bar{K}K\pi$ system in
reaction (2).

In Table I we show the difference in likelihood for various
hypothesis (a similar table is given in ref. 5 for reactions (1)).
Note that if we had done the fits one wave at a time with a flat
background we could have erroneously concluded that the data
favored a 1^{++} interpretation for the E.

Table I. Differences in Log of Likelihood from
Best Fit (0^{-+}, 1^{++},1^{+-} and Background)

Waves in fit*	1.38-1.42 GeV	1.42-1.46 GeV
1^{++},1^{+-}	-10	-7.
1^{++}	-20	-25.
0^{-+}	-25	-38.

*All fits include flat phase space background.

In summary, we conclude that the D(1285) is a 1^{++} state and the E(1420) is a 0^{-+} state, and both of them require substantial $\delta\pi$ decay modes. We do not quote at this time a $K^*K/\delta\pi$ branching ratio as its value depends on the precise parameterization of the δ and is subject to a large systematic error. The 1^{++} ($K^*\bar{K}$) wave does not show a resonant behaviour in the E-region rather, in reaction (1), shows a rapid rise near threshold. Our results contradict those of Dionisi, et al.[3] and Armstrong, et al.[4] who find the E is a 1^{++} ($K^*\bar{K}$) state. They are in good agreement with those of Baillon and of the J/ψ radiative decay,[11] so that the ι(1440) and the E(1420) may very well be the same object, weakening the glueball interpretation for the ι(1440). However, the recently measured values for the ι(1440), M= 1458 $\pm$ 7 and = 99 $\pm$ 6 MeV in J/ψ decays, are higher than those of the hadro-produced E(1420), so the question cannot be considered settled.[11]

ACKNOWLEDGEMENTS

We gratefully acknowledge the helpful support of the MPS group and the alternating-gradient synchrotron crew throughout the course of this experiment. We are also indebeted to J. Albright, P. Meehan, H. Neal, and K. Turner for their help in carrying out this experiment. This research was supported by the U.S. Department of Energy under Contracts No. DE-AC02-76-CH00016, No. DE-AS05-76ER03509, and No. DE-AC02-84ER40125, and by the National Science Foundation Grant No. PHY-8120739.

REFERENCES

1. Sharre, D.L., et al., Phys. Lett. 97B, 329 (1980);
    Edwards, C., et al., Phys. Rev. Lett. 49, 259 (1982).

2. Armenteros, R. et al., in Proceedings of the International
    Conference on Elementary Particle Physics, Siena, Italy,
    1963, edited by G. Bernardini and G. Puppi (Societa Italiana
    di Fiscia Bologna, Italy, 1963), p. 287; Baillon, P., et al.,
    Nuovo Cimento 50A, 393 (1967).

3. Dionisi, C. et al., Nucl. Phys. B169, 1 (1980).

4. Armstrong, T.A., et al., Phys. Lett. 146B, 273 (1984).

5. Chung, S.U. et al., Phys. Rev. Lett. 55, No. 8, 779 (1985).

6. James, F. and Roos, M., MINUIT CERN Program Library D516.

7. Chung, S.U., Program PWAVE.

8. Chung, S.U. and Trueman, T.L., Phys. Rev. D11, 633 (1975).

9. Flatté, S., Phys. Lett. 63B, 224 (1976); Gay, J.B. et al.,
    Phys. Lett. 63B, 220 (1976).

10. Baillon, P., in Experimental Meson Spectroscopy, 1983, edited
    by S.J. Lindenbaum, American Institute of Physics Conference
    Proceedings No. 113 (American Institute of Physics, New York,
    1983), p. 78.

11. Perrier, J., presented at the Conference on Physics in
    Collision, University of California, Santa Cruz, California,
    22-24 August 1984 (unpublished).

NEW RESULTS AT $\sqrt{s}$ = 200 AND 900 GeV
FROM THE CERN COLLIDER

UA5 Collaboration

Bonn - Brussels - Cambridge - CERN - Stockholm

Presented by

P. Carlson

University of Stockholm, Sweden

ABSTRACT

Results from the pulsed operation of the CERN $\bar{p}p$ Collider at $\sqrt{s}$ = 200 and 900 GeV are presented. Data on cross-sections, multiplicity distributions, the energy dependence of the average charged multiplicity and scaling properties of the pseudorapidity distribution in the fragmentation region are presented as well as the results from a first search for Centauro type events.

INTRODUCTION

In March-April 1985 the antiproton proton Collider at CERN operated for the first time in a pulsed mode with beams circulating for several hours in a 21.6 s cycle with momenta between 100 and 450 GeV/c. This scheme was suggested in 1982 to achieve a c. m. energy $\sqrt{s}$ = 900 GeV by reducing the average power dissipation [1]. A new run with the UA5 apparatus that had been used earlier in two runs with beam momenta 273 GeV/c [2] ($\sqrt{s}$ = 546 GeV) and one run at the ISR with $\sqrt{s}$ = 53 GeV [3] was proposed in the same year [4]. The main aim was to search for exotic phenomena seen in cosmic ray experiments [4,5] and to make a survey of the energy dependence of normal hadronic interactions.

The duty cycle at $\sqrt{s}$ = 900 (200) GeV was 18%(38%) and the initial luminosity of a typical two hour run was about 10^{26} cm^{-2}s^{-1} with a luminosity lifetime of some two hours. The background conditions were better during this pulsed collider run than during the earlier $\sqrt{s}$ = 546 GeV runs. The integrated luminosity at $\sqrt{s}$ = 900 GeV was 7 μb^{-1}. Presented to this conference are preliminary data on inelastic and total cross-sections, further evidence for non-scaling multiplicity distributions, data on the energy dependence of the average charged multiplicity, a comparison of rapidity distributions in the fragmentation region and a first report on the search for Centauro type events at $\sqrt{s}$ = 900 GeV.

2. DETECTORS, TRIGGERS AND DATA TAKING.

The layout of the UA5 detector is shown in fig. 1. Two large streamer chambers, 6m×1.25m×0.5m are placed above and below the 2 mm thick beryllium vacuum pipe. Each chamber is viewed by three cameras and each frame contains a stereo pair of views. At each end of the chambers there are two trigger hodoscopes, a "forward hodoscope" and a "trigger hodoscope" (fig. 1). Together they cover the pseudorapidity range 2<|η|<5.6 ($\eta = -\ln \tan \theta/2$ where θ is the c. m. production angle).

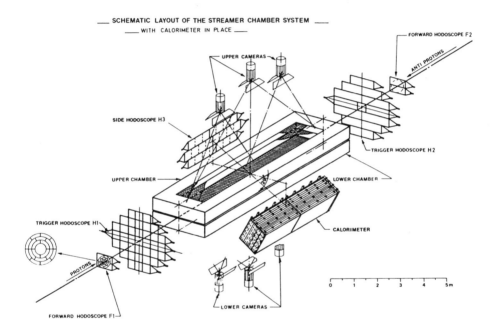

Fig. 1. Layout of the UA5 experiment.

For part of the run a 1.5 m long lead converter plate was inserted between the vacuum pipe and the upper streamer chamber in order to study the production of photons, relevant to the Centauro search. A lead-iron sandwich calorimeter, covering the pseudorapidity range |η|<1 and with 10% azimuthal coverage (fig. 1) was also used in the search for Centauro type events. Details on the apparatus can be found in ref. [6].

The streamer chambers were triggered at the flat bottom ($\sqrt{s}$ = 200 GeV), at the flat top ($\sqrt{s}$ = 900 GeV) and also during the ramps. Two different triggers were used: the 2-arm trigger and the 1-arm trigger. During the 1 s deadtime of the streamer chambers data from the calorimeter and rates for the 1-arm and 2-arm triggers were collected. The data collected during the run are summarized in table 1.

Table 1. Summary of data taken during the 1985 pulsed Collider run.

$\sqrt{s}$ (GeV)	Number of recorded streamer chamber events	Number of recorded electronic events
900	54 500	275 000
800*	17 500	65 000
546*	2 500	10 000
200	41 000	150 000
All	115 000	500 000

* Data taken during ramping of the accelerator. The energies are approximate.

The preliminary results presented here are based on 2000 measured events at 200 and 900 GeV.

3. CROSS-SECTION MEASUREMENTS

The rates N_1 and N_2 for the 1-arm and 2-arm triggers are related to the luminosity L through $N = L\sigma$. The relations between the physical processes elastic scattering, the two parts of the inelastic scattering (single diffraction SD and non-single diffraction NSD) with the different triggers are illustrated in fig. 2. The trigger cross-sections σ_1 and σ_2 are linear functions of the physical cross-sections σ_{SD} and σ_{NSD} with constants, trigger efficienses ε, that are determined by simulations [7].

CROSS SECTIONS   TRIGGERS

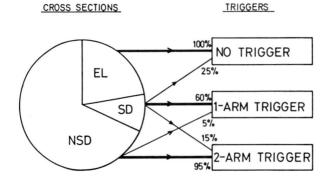

Fig. 2.
Relations between different physical processes and the different triggers. The figures refer to 900 GeV.

The relations are:

$$\begin{pmatrix} \sigma_1 \\ \sigma_2 \end{pmatrix} = \begin{pmatrix} \varepsilon_1^{SD} & \varepsilon_1^{NSD} \\ \varepsilon_2^{SD} & \varepsilon_2^{NSD} \end{pmatrix} \cdot \begin{pmatrix} \sigma_{SD} \\ \sigma_{NSD} \end{pmatrix} \tag{1}$$

Here ε_1^{SD} is the 1-arm triggering efficiency for single diffractive events etc. Given measured rates N_i, the relevant cross-sections can be determined through eq. (1). The absolute value of the machine luminosity is only known to about ± 10%. However, the ratio of the luminosities at 200 and 900 GeV is known to about 1% (0.224 ± 0.002) which, through eq. (1), makes possible an accurate determination of the ratio of the inelastic cross-sections. At each energy we can also estimate the ratio σ_{SD}/σ_{NSD} through eq. (1).

The electronic trigger rates were corrected using the streamer chamber photo-graphs to distinguish between beam-beam and beam-gas interactions. For the 2-arm triggers, the fraction of beam-beam interactions was found to very high, > 90%. For the 1-arm triggers, however, this fraction was much lower being about 8% and 27% for 200 and 900 GeV, respectively. The ratios of the trigger rates at 900 to those at 200 GeV was found to be constant during each run and also from run to run. Table 2 summarizes our measured cross-section ratios and gives the deduced values using σ_{inel} = 42.5 ± 0.7 mb at 200 GeV, obtained by interpolation of σ_{tot} and σ_{el}/σ_{tot} between ISR energies and 546 GeV [8,9].

Table 2. Preliminary values of measured cross-section ratios and deduced values of the cross-sections. First error is statistical, second is systematic.

Quantity	$\sqrt{s}$ = 900 GeV	$\sqrt{s}$ = 200 GeV	Measured ratio 900 GeV/200 GeV
σ_{inel} (mb)	50.9±1.3±1.3	42.5±0.7 (input)	1.20±0.02±0.03
Ratio σ_{SD}/σ_{NSD}	0.161±0.015±0.05	0.125±0.015±0.04	
σ_{SD} (mb)	7.1±0.5±1.7	4.7±0.8±1.3	1.52±0.11(+0.12−0.27)
σ_{NSD} (mb)	43.8±0.7±1.8	37.8±0.8±1.3	1.16±0.02(+0.02−0.03)

By extrapolation we find the ratio σ_{el}/σ_{tot} at 900 GeV to be 0.235±0.008 [9] and using the value of σ_{inel} from table 2 we arrive at the following preliminary value for the total cross-section at 900 GeV:

$$\sigma_{tot}(\bar{p}p, \ 900 \ GeV) = 66.5 \pm 1.8 \pm 1.6 \ mb$$

The first error is statistical and the second systematic.

684

Fig. 3 shows the energy dependence of the total cross-section. The line is a fit using dispersion relations taken from Amos et al. [10]. Our value at 900 GeV falls nicely on the curve.

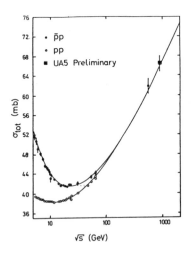

√s̄ (GeV)

<u>Fig. 3.</u>
Energy dependence of the total cross-sections for pp and p̄p. The figure is from ref. [10] and the UA5 point at 900 GeV has been added.

Recently fits to total cross-sections have been made using analytic functions of two forms, giving different asymptotic behaviour. One gives a cross-section that grows like $(\ln s)^2$ when $s \to \infty$ (the Froissart-Martin bound), the other gives a local rise as $(\ln s)^2$ but a constant limiting value as $s \to \infty$. Block and Cahn [11] use data below 63 GeV (ISR energy range and below), whereas Bourrely and Martin [12] use ISR data and the value at $\sqrt{s}$ = 546 GeV from the Collider [8].

In table 3 we compare our measured ratio of the inelastic cross-sections at 900 GeV and 200 GeV, 1.20±0.02, with the predictions for the different forms of the asymptotic behaviour. We use the values σ_{el}/σ_{tot} = 0.235 and 0.187 at 900 GeV and 200 GeV, respectively (see above).

Table 3. The ratio σ_{inel}(900 GeV)/σ_{inel}(200 GeV).

	$\sigma \to$ constant locally $\sigma \propto (\ln s)^2$	$\sigma \to (\ln s)^2$
Block and Cahn [11]	1.22	1.38
Bourrely and Martin [12]	1.21	1.26
Measured UA5	1.20±0.02±0.03	

We conclude that our preliminary measurements suggests a local growth of the cross-sections like $(\ln s)^2$ but with a constant asymptotic value.

4. MULTIPLICITY DISTRIBUTIONS.

The UA5 collaboration has published evidence for non scaling multiplicity distributions in non-single diffractive interactions [2], with an increased probability for high multiplicity events at $\sqrt{s}$ = 546 GeV as compared to ISR energies ($\sqrt{s}$ = 20-60 GeV). In a recent paper [13] we have shown that multiplicity distributions for energies $\sqrt{s}$ > 10 GeV can be well fitted with a negative binomial distribution:

$$P(n;<n>,k) = \begin{bmatrix} n + k - 1 \\ k - 1 \end{bmatrix} \left[\frac{<n>/k}{1+<n>/k}\right]^n \frac{1}{(1+<n>/k)^k}$$

The two parameters of the distribution, the average multiplicity $<n>$ and k, a parameter affecting the shape of the distribution, were found to vary with energy in a regular manner.

Fig. 4 shows our preliminary results on the distributions of observed multiplicities. The solid lines are fits to these data of a negative binomial distribution that has been transformed to an observed distribution using our standard techniques [7]. The fits are excellent and the resulting values of the parameter k^{-1} are shown in fig. 5 together with data at 546 GeV and lower energies [13]. We confirm our earlier conclusion that multiplicity distributions do not obey scaling at high energies and that they are well fitted by the negative binomial distribution with smoothly varying parameters.

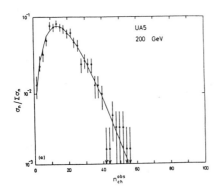
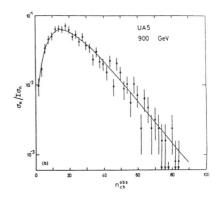

Fig. 4. Distributions of observed multiplicities at 200 GeV (a) and 900 GeV (b). The lines are negative binomial fits discussed in the text.

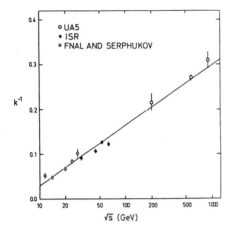

Fig. 5. The energy dependence of the parameter k^{-1} of the negative binomial distribution. From [13] to which has been added the new points at 200 and 900 GeV.

The average charged multiplicity $\langle n \rangle$ for non-single diffractive events is 34.6±0.7 and 22.1±0.7 at 900 and 200 GeV, respectively. In fig. 6 is shown the energy dependence of $\langle n \rangle$ for non-single diffractive interactions. The two curves are fits of the data to the forms $\langle n \rangle = A + B \ln s + C(\ln s)^2$ and $\langle n \rangle = \alpha + \beta s^{\gamma}$. The preliminary values of the parameters are $A = 3.0\pm0.7$, $B = -0.1\pm0.2$, $C = 0.173\pm0.015$, $\alpha = -6.6\pm1.4$, $\beta = 6.9\pm1.1$ and $\gamma = 0.13\pm0.01$.

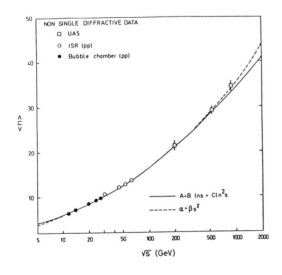

Fig. 6. The energy dependence of the average multiplicity for non-single diffractive events. Data from [13] and new data at 200 and 900 GeV.

5. SCALING IN THE FRAGMENTATION REGION

The pseudorapidity distributions for inelastic data, i.e. including the diffractive part, are shown in fig. 7. Our preliminary data at 200 and 900 GeV are
shown together with our data at 546 GeV [2] and at 53 GeV [3]. In order to
test scaling in the beam fragmentation region the rapidity density is plotted
against $y^{*}_{beam} - \eta$, representing the rapidity in the restframe of the beam
particle. The data suggests scaling within some 20% in about two units of
rapidity from the beam particle.

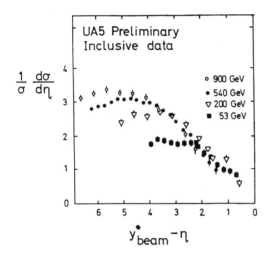

Fig. 7. Pseudorapidity
densities plotted vs. $y^{*} - \eta$,
i. e. approximatively the
rapidity of the beam particle.
Data at 53 GeV from [3],
at 546 GeV from [7].

6. SEARCH FOR CENTAURO LIKE EVENTS.

The search for Centauro events [4,5] is conducted along two main lines. Firstly
the data taken with the photon converter (see section 2) will allow the
identification of individual photons and thus make possible a search for events
with no or a very small number of photons. This investigation is in progress
and no results will be presented here.

Secondly the data from the calorimeter has been compared with our normal
Monte Carlo simulation and with a "Centauro" Monte Carlo, in which the average
transverse momentum was high, 1.5 GeV/c, and in which only nucleons ($\bar{p}p$, $\bar{n}n$)
were produced. The Centauro type events tend to give more energy per particle
in the calorimeter, they are more penetrating. From a first study of 600 events
we do not obtain any significant signal with this method, giving an upper limit
to the production of about 1%.

REFERENCES

1. J. G. Rushbrooke, CERN-EP/82-6 (1982)
2. G. J. Alner et al., Phys. Lett. 138B(1984)304.
3. K. Alpgård at al., Phys. Lett. 112B(1982)183.
4. UA5 Collaboration, Proposal CERN/SPSC/P184(15.10.1982).
5. C. M. G. Lattes et al., Phys. Rep. C65(1980)151
    K. Alpgård et al., Phys. Lett. 115B(1982)71.
6. UA5 Collaboration, Phys. Scr. 23(1981)642.
7. UA5 Collaboration, to be submitted to Phys. Rep.
    D. R. Ward, Proc. of the 3rd Topical Workshop on Proton-Antiproton
    Collider Physics, Rome 1983, in CERN 83-04, p. 75.
    B. Åsman, "From tracks on film to corrected multiplicity distributions",
    University of Stockholm, Department of Physics report USIP 85-17 (1985).
8. M. Bozzo et al., Phys. Lett. 147B(1984)392.
9. See e. g. R. Castaldi and G. Sanguinetti, CERN-EP/85-36, to appear in
    Annu. Rev. Nucl. Part. Science .
10. N. Amos et al., CERN-EP/85-94, submitted to Nucl. Phys.
11. M. M. Block and R. N. Cahn, Rev. Mod. Phys. 57(1985)563.
12. C. Bourrely and A. Martin, CERN-TH 3931 (1984).
13. G. J. Alner et al., Phys. Lett. 160B(1985)199.

IS CYGNUS X-3 STRANGE?

Gordon Baym, Edward W. Kolb, Larry McLerran and
T. P. Walker
Theory Department
Fermi National Accelerator Laboratory
Batavia, IL 60510

R. L. Jaffe
Center for Theoretical Physics
Laboratory for Nuclear Science
Massachusetts Institute for Technology
Cambridge, Mass. 02139

ABSTRACT

We discuss the recently reported measurements of the properties of high energy cosmic rays arriving from the direction of the compact binary X-ray source Cygnus X-3. We argue that the source of these events may be a strange quark star, and that the primary which directly produces them is a low baryon number neutral hadron with multiple strangeness which is stable up to (at least) simultaneous double strangeness changing weak decays.

1. The System

Recently the Soudan 1 underground proton decay detector has reported[1] observations of high energy muons from the direction of the compact binary x-ray source Cygnus X-3 (2030+4047), with a distribution of arrival times apparently modulated with the 4.8 hour orbital period, P, of Cyg X-3. Preliminary results from the Nusex detector[2] as well as previously reported measurements from the Kiel air shower array[3] tend to confirm these observations.

The zenith angle dependence of the Soudan events indicates that the muons are produced in air showers originating in the upper atmosphere or the upper crust of the earth ($1-10^3$ m). Such muons must have an energy at least .6 TeV to penetrate rock and reach the Soudan detector. The datas suggest that the initiating particle might be a neutral hadron, for reasons which we shall soon discuss. Several candidates have been proposed, including strange quark droplets, which either are supposed to arrive in large baryon number globs[4] or are supposed to produce a large neutrino flux at the star which is assumed to be what produces the muons which are observed in the detectors[5] The possibility that the initiating particles are neutrinos seems to be ruled out by the observed of muons arriving in detectors. In this paper, we shall assume that the experimental observations are correct and explore unconventional sources for the muons. We suggest that the muons are produced by metastable neutral low-baryon number strange hadrons originating from the condensed star in Cyg X-3, argue that for this to be the case most probably the condensed star must be entirely composed of matter with a large strangeness fraction, and describe how such a strange star might be produced in a supernova explosion.

As several authors have already noted[4] the observation that the absolute flux of muons at the Soudan detector, 7×10^{-11} $cm^{-2}sec^{-1}$ is comparable to that of air showers from Cyg X-3 themselves extrapolated into this energy range indicates that the initiating particle is unlikely to be either a photon or neutrino, if the photon or neutrino cascade by conventional processes. The flux of muons is two to three orders of magnitude too large to be produced by such air showers if the showers are initiated by high energy gamma rays. On the other hand, since the overall neutrino and gamma ray fluxes from Cyg X-3 should be comparable at high energies, if they are both produced by high energy hadronic interactions, the absolute neutrino flux at the Soudan detector should be below the limit of detectability.[6] If the muons are produced by neutrino interactions, which do not generally produce air showers, the zenith angle distribution of such produced muons would be different from that observed. The inference therefore is that the most likely candidate for the initiating particle is a hadron, and in order for interstellar magnetic fields not to alter its direction relative to that of the source, the hadron must be neutral. We shall refer to this particle here as the "cygnet."

In order that the cygnet may arrive from Cygnus X-3 without a tremendous reduction in flux due to decays, the cygnet must have a minimum lifetime $\geq d/\gamma c$, where $d \sim 12.5$ kpc. (This lower limit might be violated by perhaps as much as a factor of two and still allow for a reasonable flux of cygnets, but much more reduction forces the production rate of cygnets at the source to be unrealistically large.) The energy threshold for muons in the Soudan detector is about .5 Tev, corresponding to hadron primary energy of 5-10 Tev, or to $\gamma \sim 5-10$ Tev/m, with m the mass of the hadron. This value of γ is about a factor of 5-10 higher for the NUSEX data. In the Soudan detector, the pulses appear in an interval of about .3 units of phase, and in order that the dispersion in travel times not smear out the observed orbital modulation of the signal, γ must be $> (d/.6Pc)^{1/2} \sim 10^4$. For NUSEX, the signal is in .1 units of phase, and the corresponding limit on γ is 10^3. The minimum mass for these energies is therefore $m \lesssim 1-2$ Gev. The minimum lifetime is ~ 10 yr. These considerations rule out the possibility that cygnets might be neutrons. A possible cygnet candidate is the doubly strange H dibaryon first proposed in Ref.(7). This hadron has the quantum numbers of two lambdas, and is a tightly bound 6-quark bag. If $m_H < m_p + m_\Lambda$ then the particle can only decay by doubly weak processes, and can have a lifetime in the requisite range.

The observed 4.8 hr period of Cygnus X-3 has been associated with the orbital period of a binary system, thought to consist of a young pulsar and a 4 M_{sun} companion.[8] Vestrand and Eichler[8] have used these components to generate short pulses of E > 1 Tev radiation assuming the magnetosphere of the pulsar to be an efficient accelerator of charged particles[9] which then collide with the atmosphere of the companion star.

The cygnets may either be present in some form in the beam which originates from the neutron star, or arise from conventional nucleon interactions of nucleons from the neutron star with the companion star. In the latter case, the production cross section for cygnets is large, and it is difficult to imagine that such a particle would have escaped detection in accelerator experiments. We shall therefore assume that the cygnets are components of the beam of particles originating from the neutron star. We picture then the cygnets originating tightly bound to charged hadrons and being stripped by the interactions of such complexes with matter in the companion star.

If cygnets or their primaries are already present in the flux leaving the neutron star then they must be present in some form in the neutron star itself. The alternative, that they are created by surface bombardment by counter-accelerated particles in the high energy beam, probably requires too large a production cross section to be consistent with laboratory data. It has long been suggested that neutron stars could be largely composed of strange quark matter.[10] If this is true they provide a copious source of H's. A possibility is that strange quark matter exists in the core of the star and is somehow brought up to the surface.[5] We are unable to find any mechanism which efficiently transports matter from the core to the surface by diffusion, convection, or excavation. The other possibility, which we now pursue, is that exotic matter exists stably up to the stellar surface in sufficient quantity to form a rich component of the emitted cosmic rays[11]

The scenario we describe is based on the possibility, suggested by Witten[11] that high density strange quark matter is absolutely stable with respect to ordinary nuclear matter, i.e., the energy per baryon of strange quark matter, with strangeness per baryon f_s, around a density ρ_s is less than that of ordinary nuclear matter at $\rho_{nm} \sim 0.16$ fm^{-3}. In this scenario, ordinary nuclei cannot decay into strange matter because of the presence of a barrier as a function of strangeness fraction. Conversion to strange quark matter would require a very high order weak interaction[11-12] with a lifetime far in excess of the age of the universe. If strange matter is stable at all it is stable in bulk and for all baryon numbers above some A_{min}[12] A_{min} cannot be too small lest light nuclei decay by first order or second order weak processes into strange matter. A strangelet (a droplet of strange matter of nuclear dimensions) with $A < A_{min}$ would decay by sequential first order weak processes and by α particle and nucleon emission. Strange matter must have positive electric charge on the quarks. Otherwise, without a Coulomb barrier, a single strangelet would rapidly gobble up all ordinary matter with which it came in contact[12] For generic values of model parameters the Coulomb barrier of strange matter, while lower than that of ordinary matter, suffices to prevent strangelets from absorbing ordinary nuclei at ordinary stellar temperatures.

In this scenario the lowest energy configuration of the star would have to contain strange matter out to its edge; the problem is to understand how it forms in the star. Let us first look at the high density core. In the cases discussed$(11-12)$ non-strange quark matter becomes stable relative to nuclear matter at some sufficiently high density, ρ_{crit}. If, in the formation of the neutron star during a supernova, the central density exceeded ρ_{crit}, then the core would form quark matter over strong interaction time scales and later relax to finite strangeness fraction via single weak processes. These processes should be rapid [~ 10^{-10}sec, or longer, for strangness-changing hadronic weak decays in dense matter] compared with deleptonization times for the supernova core, and so the rate at which the initially degenerate neutrinos leave the core determines the timescale for formation of a strange core. (Until the neutrinos have left, the matter is diffuse and not at sufficiently high density to rapidly form quark matter) On the other hand, if the mean baryon densities reached are not high enough to overcome the barrier, a strange quark region can still form through fluctuations in the local density. Once formed they would not only be absolutely stable, they would proceed to expand by converting the neighboring normal matter -- either quark matter or neutrons -- to strange quark matter. For example, if a neighboring neutron crosses the surface of the strange region, it would disassemble into its component quarks; the up quark would be converted to strange either by a direct semi- leptonic process or by purely hadronic weak interactions.

Eventually the entire core would be turned into a strange quark core. The conversion process releases an energy of order tens of MeV per baryon; however, because the core after deleptonization is bound with ~ 100 MeV per baryon, the burning should not lead to explosive disassembly of the core.

After the core has been converted to strange quark matter, the strange matter begins to eat its way out from the core. At the interface between the strange matter and

ordinary nuclear matter, ordinary matter is absorbed
through the interface and is converted to quark matter, if
the matter being eaten through is sufficiently neutron
rich, or if the temperature of the star remains
sufficiently high that there are some particles in the
nuclear matter with kinetic energies sufficient to overcome
the Coulomb barrier of strange matter. As the strange
quark matter burns its way to the surface, it may also
preheat the matter in front of the conversion region and
generate enough particles with kinetic energies above the
Coulomb barrier to maintain burning. Our estimates
indicate that this is indeed possible, and that a time of
about a year is necessary to burn a neutron star into a
quark star.

The matter emitted from the surface of the strange
quark star would consist of strangelets of relatively low
baryon number (including H's) in addition to the expected
mix of nucleons, hyperons, and mesons. Strangelets that
are so light as to be unstable via strong interaction
processes decay rapidly away. Those that are stable and
many which decay weakly survive long enough to be stripped
in the atmosphere of the companion yielding nucleons,
hyperons, H's in addition to high energy photons and
neutrinos (from meson decay). The strangelets which decay
in flight from the surface of the strange quark star might
provide a source of high energy neutrinos with a flux not
constrained by the photon flux, and might be measurable[5]
Stable strangelets which miss the companion seed the galaxy
with stable quark matter. Such matter, in the form of
large baryon number globs may be the source of
Centauros[13-14] and explain a number of cosmic ray
anomalies[15] In addition the stable quark matter produced
in this manner by Cyg X-3 like objects in the past would
have seeded the solar nebula and led to a substantial
abundance of strange matter, which might be detectable in

terrestial searches.

In order to relate the flux of cygnets to the flux of muons, the detailed properties of cygnet-hadron interactions must be understood. To a first approximation, if the cygnet is an H particle, its interactions are similar to those of a proton with an energy equal to that of an H. The properties which might allow for a more sophisticated computation of the showering of the H particle are the following. The cross section for H-p interactions is somewhere midway between that of a proton and that of a deuteron. The fractional energy loss for the H, so long as the H holds together should be roughly half that of a proton, since the mass is twice as large and the energy loss per collision should be about the same. After several interactions, the H particle might fall apart into two Λ's or protons and kaons. The Λ's and kaons are in the projectile fragmentation region, and should quickly decay, generating fast muons. When the Λ decays into πp, the pion is faster than is typical for centrally produced π's. and should increase the fast muon flux. Kaons have smaller cross sections and mean free paths than pion and therefore decay before interacting as much as pions, and therefore produce fast muons.

Estimates of the total flux of hadronic primaries which would produce a flux of muons corresponding to the rate seen at Soudan give a flux which is about a factor of twenty times larger than the rate observed in air showers extrapolated into this energy range. Because the H particle may shower somewhat differently than a proton primary, this discrepancy may not be so severe. Also, since the flux from Cyg X-3 may be variable, and the measurements were not made at the same time, this discrepancy, which deserves more study, does not yet appear to rule out the H particle hypothesis.

There is a puzzling feature of the NUSEX data which is

not explained by our proposal. The NUSEX detector sees the
muon signal from a region many degrees on a side around
Cygnus X-3. We have no mechanism for dispersing H
particles over such a large angular range. This result
seems difficult to explain by any mechanism since the NUSEX
experiment sees a wider angular dispersion than Soudan, and
since NUSEX measures higher energy particles than Soudan,
one would expect that the dispersion at Soudan would be
larger. For example, if there was production of a new
particle high in the atmosphere with high transverse
momentum, the spread at Soudan should be larger by a factor
of about 5-10 compared to NUSEX since the energy of
particles detected is lower by about this amount. Two
plausible, although unattractive explanations for the
increased angular broadening would be either multiple muon
scattering in the rock beyond that computed using multiple
Coulomb scattering Monte-Carlo computations, or an improper
determination of the detector orientation.

Acknowledgements

We all thank C. Alcock, J. D. Bjorken, S. Errede, T.
Gaisser, M. Gell- Mann, F. Halzen, S. Kahana, K.
Ruddick, T. Stanev, M. Turner and G. Yodh.

References

1. M. L. Marshak, J. Bartlet, H. Courant, K. Heller,
T. Joyce, E. A. Petersson, K. Ruddick, M. Shupe, D. S.
Ayres, J. Dawson, T. Fields, E. N. May, L. E. Price and
K. Sivaprasad, U. Phys. Rev. Lett. 54, 2079 (1985).

2. NUSEX Collaboration, Talk presented at the 1'st
Symposium on Underground Physics, Saint Vincent, Italy,
April (1985).

3. M. Samorski and W. Stamm, Astrophys. J. 268, L17,
(1983).

698

4. M. V. Barnhill, T. K. Gaisser, T. Stanev and F. Halzen, U. Wisc. preprint MAD/PH/243 (1985).

5. G. L. Shaw, G. Benford, and D. J. Silverman, U. C. Irvine preprint 85-14, (1985).

6. E. W. Kolb, M. S. Turner, and T. P. Walker, Fermilab preprint, 1985. (1985); T. Gaisser and T. Stanev, Bartol Preprint BA-85-12 (1985),

7. R. L. Jaffe, Phys. Rev. Lett. 38, 195, (1977).

8. W. T. Vestrand and D. Eichler, Ap. J. 261, 251 (1982).

9. M. A. Ruderman and P. G. Sutherland, Ap. J. 196, 51 (1975).

10. G. Baym and S. A. Chin, Nuc. Phys. A262, 527, (1976); G. Chapline and M. Nauenberg, Nature, 264, 23, (1976); B. A. Freedman and L. D. McLerran, Phys. Rev. D16, 1169, 1976 and Phys. Rev. D17, 1109, (1978); K. Brecher Astrophys J. 215, 117 (1977); W. Fechner and P. Joss, Nature, 274, 347, (1978).

11. E. Witten, Phys. Rev. D30, 272 (1984).

12. R. L. Jaffe and E. Farhi, Phys. Rev. D30, 2379, (1984).

13. J. D. Bjorken and L. D. McLerran, Phys. Rev. D20, 2353, (1979); S. A. Chin and A. Kerman, Phys. Rev. Lett. 43,1292 (1979); A. K. Mann and H. Primakoff, Phys. Rev. D22, 1115, (1980); F. Halzen and H. C. Liu, Phys. Rev. Lett. 48, 771, (1982); F. Halzen and H. C. Liu, Phys. Rev. D25, 1842, (1982); J. Elbert and T. Stanev University of Utah Preprint UU/HEP82/4 (1982).

14. C. M. G. Lattes, Y. Fujimoto, and S. Hasegawa, Phys. Rept. 65, 151, (1980).

15. H. C. Liu and G. Shaw, Phys. Rev. D30, 1137, (1985).

FORMATION AND SIGNATURES OF QUARK-GLUON PLASMA

Rudolph C. Hwa

Institute of Theoretical Science and Department of Physics
University of Oregon, Eugene, Oregon   97403
USA

ABSTRACT

The formation of quark-gluon plasma in heavy-
ion collisions is studied in an approach that
does not depend on the details of the strong
interaction dynamics. It is shown that the
initial temperature does not increase
indefinitely with increasing incident energy,
apart from nonscaling effects, and that it
varies with A as 180 $A^{1/6}$ MeV. The
thermalization time is found to be around
0.15 fm/c. Two methods of identifying the
signatures of the quark-gluon plasma are
discussed. One is through lepton-pair
production and the other involves the
dissolution of hadrons in the plasma.

The title of this talk, being made up by the
organizers of this session, implies a broader coverage of
the subject than what I intend to present. I shall
discuss the work that I have done on the subject, mostly
in collaboration with K. Kajantie. The two parts of the
problem, formation and signatures, are not interdependent
in our treatment, although the details about the

signatures are obviously dependent on certain aspects of the formation, such as the temperature and the formation time. Because the formation problem is more difficult and has been treated with less rigor, let me discuss first the signature problem.

The signals usually suggested for a quark-gluon plasma are to be found in lepton-pair, direct photon, and strange-particle productions. I shall discuss some new features about lepton-pairs as a diagnostic tool[1] and make some remarks about a novel scheme which is still in its developmental stage.

The main difference between lepton-pair production in hadron-hadron and nucleus-nucleus collisions is that in the latter case the dilepton can be emitted over an extended region in space-time so that one has to keep track of both energy-momentum and space-time. In the h-h case one usually considers only the Drell-Yan (DY) process. In the A-A case there is also the DY process, which we define to be the creation of dilepton from the annihilation of quark and antiquark originiated from the incident nuclei. Those pairs must be distinguished from the ones emitted in the quark-gluon plasma as well as from the hadron phase if a distinct signature of the quark matter is to be identified. Let us use "thermal emission" to refer to those dileptons emitted from the system in thermal equilibrium, in contrast to the DY process.

In the thermal regime the lepton pairs are still produced by the annihilation of quarks and antiquarks, but the q and $\bar{q}$ distributions bear no resemblance to those in the original nuclei. If they are taken to be proportional to $\exp(-k_1^\mu u_\mu/T)$ in the Maxwellian approximation, where k_1 (k_2) is the momentum of a quark (antiquark) in a space-time cell at $u^\mu = (ch\eta, sh\eta)$, η being the spatial rapidity $\frac{1}{2}\ln(t+z)/(t-z)$, then the product of the two thermal distributions yields another thermal distribution for the dilepton, $\exp(-q^\mu u_\mu/T)$. Since u_μ and T are space-time dependent, the probability of detecting a dilepton of mass M at rapidity y and transverse momentum q_T involves a non-trivial integration over all space-time where thermal dileptons can be produced:

$$\frac{dN}{dM^2 dy dq_T^2} \propto \int d^4x \ \exp\left[-\frac{M_T}{T} \cosh (y - \eta)\right] \tag{1}$$

where $M_T = (M^2 + q_T^2)^{\frac{1}{2}}$. M_T enters the expression because the energy of the dilepton is $E = M_T \cosh y$.

On the basis of (1) it was first concluded that the M_T dependence of the dilepton emission rate is exponential;[2] later it was suggested that the dependence is power-law behaved.[3] As it turns out, both are right in different M_T regions. We show where the thermal pairs dominate over the Drell-Yan pairs and how the normalization depends on pion multiplicity.[1]

The integration of (1) over the transverse coordinates yields a trivial factor πR_A^2, if $R_A \leqslant R_B$, under appropriate impact parameter conditions. The remaining integrations over dt dz are, however, non-trivial; in terms of the proper time τ and η it is $\tau d\tau d\eta$. The η integration can be carried out in the approximation $\cosh (y - \eta) \cong 1 + \frac{1}{2}(y - \eta)^2$, giving rise to a factor $(T/M_T)^{\frac{1}{2}}$, and leaving essentially

$$\int d\tau \ \tau \ \left(\frac{T}{M_T}\right)^{\frac{1}{2}} \exp\left(-\frac{M_T}{T}\right) \tag{2}$$

The limits of integration should be from τ_i when the quark-gluon system is first thermalized to τ_f when the hadron phase ends with freeze-out of the hadrons. Actually, there are other factors that should modify (2) as soon as the system enters into the mixed phase at the critical temperature T_c; they involve hadron form factors, etc., due to hadronic annihilation into virtual photon. That part of the integral from τ_Q (end of quark phase) to τ_f is complicated but is damped by the Boltzmann factor

$\exp(-E/T_c)$ for $M_T \gg T_c$. We therefore expect its contribution to the final total rate to be proportional to $M_T^{-1/2} \exp(-M_T/T_c)$. To focus on the quark phase we shall concentrate on the part of the integral from τ_i to τ_Q and identify the range of M_T for which the thermal emission from quark matter is dominant.

To evaluate (2) we need the τ dependence of T which is known for hydrodynamical expansion.[4] A useful way of expressing that dependence is to write[1]

$$T^3\tau = c(dN_\pi/dy_\pi) \tag{3}$$

where c is some constant that depends on the geometry of the collision. The right-hand of (3) is proportional to the entropy density which in turn can be related to the observable pion multiplicity under the assumption of adiabatic expansion through the mixed and hadron phases. It then follows from (2) that the thermal rate from quark matter can be written as

$$\frac{dN_{QM}}{dM^2 dy\, d^2 q_T} \propto \left(\frac{dN_\pi}{dy_\pi}\right)^2 M_T^{-1/2} \int_{T_c}^{T_i} dT\ T^{-6.5}\ e^{-M_T/T} \tag{4}$$

where $T_i = T(\tau_i)$. To get maximum contribution from this integral M_T should be chosen such that the peak of the integrand (occurring at $M_T/5.5$) is between the limits of integration, i.e.

$$T_c < M_T/5.5 < T_i \tag{5}$$

For such values of M_T, the contributions from the mixed and hadron phases are unimportant by comparison and the dependence of the dilepton rate on M_T is power-behaved, namely M_T^{-6}.[3]

It is (5) that we want to bring to the attention of the experimentalists so that they know where to look for the best signals of the quark-gluon plasma. Unfortunately, T_c and T_i are both theoretical quantities at this stage. If we take $T_c \approx 150$ MeV and $T_i \approx 450$ MeV, then (5) implies $1 \lesssim M_T \lesssim 2.5$ GeV. For $M_T < 1$ GeV there would be significant contamination from hadronic annihilation and resonance decays. For $M_T > 3$ GeV the thermal rate begins to be damped exponentially and eventually at higher M_T the DY dileptons dominate. The best way to be certain that a dilepton signal is of quark-matter origin is to check the multiplicative factor $(dN_\pi/dy_\pi)^2$ in (4). Thus, a dilepton experiment designed to look for such signals should measure pion multiplicity in addition to dileptons and should be triggered by high-multiplicity events.

A new idea in the search for plasma signatures is based on the recognition that the quarks in a hadron become unconfined if the hadron is immersed in a plasma.[5] Consider the production of a hadron h by the annihilation of a q and a $\bar{q}$ in the incident nuclei. If no plasma is formed by the nuclear collision, then h would be detected, most likely through its decay products. On the other hand, if a plasma is formed, then in the rest frame of h it sees an environment of unconfined quarks and gluons which can interact with the quarks in h and destroy the mechanism for confinement, since the potential between those quarks no longer increases when they are separated. If the plasma state persists for long enough time for h to dissolve, then the detector would see no signal for h. So the failure to detect h would be a signature for the existence of the plasma.

In practice there are a number of complications to consider. The mass of h should be high enough so that it cannot be created readily on the surface of the thermal plasma. Yet it should be low enough so that it can be produced by $q\bar{q}$ annihilation in a relativistic heavy-ion collider. J/ψ may not be massive enough because it can be thermally emitted and yet not effectively attenuated in the plasma.[6] Υ is probably the upper limit for the energy of RHIC being considered at BNL. Given a good candidate for h, one should be mindful of the dependence of the dissolution phenomenon on the impact parameter of the collision, the initial temperature and lifetime of the plasma, and the geometrical size of h. These factors should be taken into account in relating the observed cross section of h to the existence or nonexistence of the plasma. Work on the problem is still in progress.

The last topic that I shall discuss is on the formation of the plasma—more specifically, the initial temperature T_i when the quark-gluon system is first thermalized. Usually the estimate is done by extrapolation from the final state, using whatever phenomenological inputs that are relevant, such as the hadron-hadron and hadron-nucleus data.[4,7] Our approach is from the initial state before the collision takes place.[8] Since the thermalization process is a difficult problem to treat at a high level of precision,[9] we shall avoid dynamics insofar as possible and use conservation laws to derive both T_i and τ_i. We pose the question: as s → ∞, does T_i → ∞ also, or does it reach a finite limit, apart from scaling violating effects?

Our dynamical input is nuclear transparency, which we take to mean that a fast quark has a long degradation length (100fm) in propagation through nuclear matter.[10] It does not mean that a proton has equally long degradation length because there are wee quarks in a hadron, which interacts strongly with the target. Thus regarding an A + A collision at relativistic energy as the collision of two super bags of quarks and gluons with momentum distribution as specified by low Q^2 electroproduction, one may ignore the partons in the fragmentation regions and concentrate only on the thermalization process in the central region. The lack of long-range correlation in rapidity is the essential dynamical input that renders the kinematical calculation for T_i feasible.

Basically the scheme is as follows. If we switch off the interaction responsible for thermalization, we can easily calculate the energy density ϵ' of the quark-gluon system at any space-time point using free-particle trajectories on the initial system. When the thermalizing interaction is switched on, those trajectories will bend in complicated ways. But so long as the interaction is short-ranged in rapidity, there is only semi-local rearrangement of the quarks and gluons in the central region independent of what happens in the fragmentation region. By Lorentz invariance within the central region, the semi-local rearrangement cannot change the energy density in that region. Consequently, after thermalization is complete, at τ_i, the energy density ϵ should be identified with the initial ϵ' before the interaction is turned on, but evaluated at τ_i. The initial parton number density as inferred from electroproduction is not too far from what is implied from the rapidity-plateau height of the produced particles in hadron–hadron and hadron–nucleus collisions. Thus it is not a poor approximation if the number density ρ is regarded as conserved, as the interaction is turned on. In essence this statement reflects a judicious choice of the initial parton density because that density is not known a priori without some interaction, including electroproduction. Once ϵ and ρ are known, the determination of the initial temperature T_i is straightforward.

There is, however, one additional point of crucial importance. The initial trajectories of the partons without interaction would all emanate from the tip of the light cone if the nuclei are infinitely contracted as $s \to \infty$. But not all partons are similarly restricted in their spatial extension at t = 0, since they have different momenta. We require that there be distributed contraction in the sense that wee partons with y = 0 have an uncertainty in longitudinal positions of order l, the typical hadronic size, and that partons with rapidity y are restricted to $\Delta z = \mathit{l}/\cosh y$.

With the dual input of nuclear transparency and distributed contraction, we find that the temperature T_i cannot increase indefinitely with increasing s. The reason is that the fast partons are compressed, but they do not interact, while the slow ones which do interact strongly are not compressed. Hence the thermal energy density cannot increase without bound, resulting in a limiting temperature. When the numbers are put in, we found[8]

$$T_i = 180 \ A^{1/6} \ \text{MeV}, \quad \tau_i = 0.27 \ A^{-1/6} \ \text{fm/c} \quad (6)$$

For $16 \leqslant A \leqslant 238$, it implies $300 \leqslant T_i \leqslant 500$ MeV and $0.1 \leqslant \tau_i \leqslant 0.2$ fm/c. The average temperature obtained is quite sufficient for the formation of quark-gluon plasma.

The only scale used in the problem is $\ell \sim m_\pi^{-1}$ and/or average transverse mass m_T of partons before collision. It is interesting that results as specific as (6) can follow from such general considerations.

I thank Francis Halzen for inviting me to give this talk. The collaboration with Keijo Kajantie was essential in arriving at the results reported here. This work was supported in part by the US Department of Energy under contract number DE-AT06-16ER10004.

REFERENCES

1. R.C. Hwa and K. Kajantie, Phys. Rev. D32, 1109 (1985).
2. K. Kajantie and H.I. Miettinen, Z. Phys. C9, 341 (1981); 14, 357 (1982).
3. L.D. McLerran and T. Toimela, Phys. Rev. D31, 545 (1985).
4. J.D. Bjorken, Phys. Rev. D27, 140 (1983).
5. R.C. Hwa and K.F. Liu (unpublished).
6. J. Cleymans and R. Philippe, Z. Phys. C22, 271 (1984).
7. R. Anishetty, P. Koehler and L. McLerran, Phys. Rev. D22, 2793 (1980).
8. R.C. Hwa and K. Kajantie, OITS-293 (1985).
9. G. Baym, Phys. Lett. 138B, 18 (1984); R.C. Hwa, Phys. Rev. D32, 637 (1985).
10. R.C. Hwa and M.S. Zahir, Phys. Rev. D31, 499 (1985).

THE STATUS OF SKYRMIONS

Eric Braaten

Dept. of Physics & Astronomy
Northwestern University
Evanston, IL   60201

and

High Energy Physics Division
Argonne National Laboratory
Argonne, IL   60439

ABSTRACT

The motivation for the Skyrme model from QCD is
discussed, and its phenomenological applications are
reviewed.

1.   BIRTH AND REBIRTH OF THE SKYRME MODEL

The SU(2) Skyrme model[1] is perhaps the oldest realistic model of
low-energy strong interactions. Skyrme suggested in 1960 that the
pions be described by an SU(2)-valued field $U(x) = \exp\left(i\vec{\pi}(x)\cdot\vec{\tau}/F_\pi\right)$.
The resulting field theory can have solitons (skyrmions), i.e.
classical extended objects with a conserved topological charge B (for
baryon number). The quantum states of the skyrmion can be identified
as baryons.

Do these "baryons" satisfy fermionic statistics? Although the
fundamental fields are bosons, Finkelstein and Rubinstein[2] showed that
a topological property of SU(2) allows the solitons to be fermions in
the quantum theory. Aside from this work and a few other notable
exceptions,[3] Skyrme's idea was almost completely ignored for over 20
years. One reason was that the success of the SU(2) Skyrme model
appeared to be a coincidence. The baryons of QCD remain fermions when
the theory is generalized to any number of flavors N_f. In contrast,
if the SU(2) Skyrme model is naively generalized to $SU(N_f)$, the
Finkelstein-Rubinstein mechanism fails for $N_f \geqslant 3$ and the skyrmions
are automatically bosons.

It was the solution of this problem by Witten[4] that launched the
recent revival of the Skyrme model. Witten showed that if $N_f \geqslant 3$, one

must take into account the effects of anomalies in QCD by adding to
the action a "Wess-Zumino term" proportional to the number of colors
N_c. One then finds that the skyrmions are fermions if N_c is odd and
bosons if N_c is even, just like the baryons of QCD. It is therefore
plausible that the Skyrme model could arise as an approximation to
QCD. This can be made more plausible by considering 1) the low-energy
effective lagrangian for pseudoscalar mesons and 2) the $1/N_c$ expansion
for QCD.

2.   MOTIVATION FOR THE SKYRME MODEL FROM QCD

The chiral limit of QCD is the limit in which N_f flavors of
quarks are exactly massless. Its lagrangian then has a $G = SU(N_f)_L \times$
$SU(N_f)_R$ chiral symmetry. On the other hand, its vacuum is understood
to be invariant only under a diagonal subgroup $H = SU(N)_V$.
Goldstone's theorem then implies the existence of $N_f^2 - 1$ massless
Goldstone bosons, which can be identified as pseudoscalar mesons.
They can be described by an "order parameter" field $U(x)$, with values
in $G/H = SU(N_f)$.

This theory has a mass gap separating the massless pseudoscalar
mesons from the next lightest particles. If one is interested in
processes involving only massless mesons with energies and momenta
below the mass gap, the massive particles in the spectrum can be
integrated out to give a low-energy effective lagrangian L_{eff} for the
massless particles only. It should be expressible in terms of the
field $U(x)$ and it should be invariant under chiral symmetry just like
the QCD lagrangian. It must then have an expansion

$$L_{eff} = L_2 + L_4 + L_6 + \cdots + N_c L_{WZ} \quad , \tag{1}$$

where L_n is of nth order in derivatives of U and L_{WZ} is the Wess-
Zumino term, whose importance was emphasized by Witten. It is of 4th
order in derivatives of U but it cannot be expressed in terms of U in
any simple way. Its coefficient must be proportional to N_c to
reproduce the effects of anomalies in QCD. The long distance behavior
of the theory is governed by the unique 2-derivative term

$$L_2 = \frac{1}{4} F_\pi^2 Tr \left(\partial_\mu U \; \partial^\mu U^\dagger \right) \tag{2}$$

where $F_\pi = 93$ MeV is the pion decay constant. All of the other coefficients in the effective lagrangian (1) are unknown, although they could in principal be calculated from QCD using lattice techniques. One can construct a model for QCD by truncating L_{eff}. An example is the original Skyrme model[1] which corresponds to

$$L_4 = \frac{1}{e^2} \, Tr \left(\partial_\mu U (\partial_\mu U^\dagger \partial_\nu U - \partial_\nu U^\dagger \partial_\mu U) \partial_\nu U^\dagger \right) \tag{3}$$

and $L_6 = L_8 = \ldots = 0$. Thus a model of this type arises automatically as a low energy effective lagrangian for QCD in the chiral limit.

The $1/N_c$ expansion for QCD was introduced by 't Hooft,[5] who showed that if QCD, a SU(3) gauge theory with coupling constant g, is generalized to $SU(N_c)$ with $g^2 N_c$ held fixed, it has a well-defined expansion in powers of $1/N_c$. In the limit $N_c \to \infty$, it reduces to a theory of infinitely many free mesons and glueballs. For finite N_c, the theory can be expressed as a $1/N_c$ expansion around this free $N_c = \infty$ theory. A puzzling feature about this expansion is that it appears to describe only mesons and glueballs. Where are the baryons? Witten has argued that they will appear as solitons of the meson and glueball fields because, in the $1/N_c$ expansion, they behave qualitatively like solitons.[6]

One can imagine formulating the $1/N_c$ expansion in terms of a lagrangian with infinitely many fields and infinitely many interaction terms of order $1/\sqrt{N_c}$, $1/N_c$, etc. The pseudoscalar mesons will be among the particles that appear as fundamental fields in this lagrangian. Upon integrating out the heavy fields, one would again obtain an effective lagrangian, which, in the chiral limit, must have the form of L_{eff} in (1). Does L_{eff} retain any memory of the baryons being solitons in the $1/N_c$ expansion? Amazingly enough, the answer is yes! Solitons arise very naturally in a lagrangian of the form (1), and, when quantized semiclassically, they turn out to have exactly the quantum numbers of baryons.

As has been emphasized by Witten[7], if baryons arise as solitons of the meson fields in the $1/N_c$ expansion, then their properties are completely determined by the properties of mesons. The fact that they appear as solitons in L_{eff} suggests that their properties are determined primarily by those of low energy mesons. Donoghue, Golowich, and Holstein[8] have extracted a prediction for the nucleon mass from π-π scattering data: $M_N = 880$ GeV $\pm$ 300 GeV. The large error indicates that very precise measurements of meson properties would be required to obtain accurate predictions for the properties of baryons.

3.   CLASSIFICATION OF MODELS AND METHODS

If one truncates L_{eff} down to the 2-derivative term L_2 in (2), simple scaling arguments indicate that the solitons are classically unstable and will shrink down to a point. The various alternative forms of the Skyrme model can be classified according to what is added to the lagrangian to stabilize the skyrmion.

<u>Type I</u> skyrmion - stabilized by adding a constraint.[9]

<u>Type IIa</u> skyrmion - stabilized by adding higher derivative terms in the pseudoscalar meson field $U(x)$. The simplest example is the original model of Skyrme, but a realistic model might require other terms.[10]

<u>Type IIb</u> skyrmion - stabilized by adding heavy meson fields to the lagrangian, such as the ω.[11]

<u>Heterotic</u> skyrmion - stabilized by replacing the core of the skyrmion by a bag of confined quarks (also known as the hybrid chiral bag model).[12]

We will restrict our attention to Type IIa and Type IIb skyrmions, since they arise most naturally out of the $1/N_c$ expansion for QCD. Note that any model of Type IIb will be equivalent at low energies to some model of Type IIa, since one can integrate out the heavy meson fields to get an effective lagrangian for the pseudoscalar mesons.

The classical skyrmion is described by a static solution of the following form

$$U_s(\vec{r}) = \exp\left(iF(r)\hat{r}\cdot\vec{\tau}\right) , \qquad (4)$$

where $F(r)$ is some model-dependent function that satisfies $F(0) = \pi$, $F(\infty) = 0$. This solution is invariant under simultaneous rotations of the isospin and coordinate axes. Other static solutions degenerate in energy are generated by translations $U_s(\vec{r} - X)$ and isospin rotations $AU_s(\vec{r})A^\dagger$. In models of Type IIb, the skyrmion will act as a source for the heavy meson fields, and the solution will require nontrivial static forms for these fields also.

The quantum skyrmion is usually treated semiclassically. The quantum states of the skyrmion are superpositions of rotated and translated skyrmions $AU_s(\vec{r}-X)A^\dagger$, weighted by a wavefunction $\Psi(A,X)$ corresponding to definite spin and isospin and definite momentum. A more complete description of the quantum state requires the inclusion of pion fluctuations about the skyrmion, which in leading order can be taken to be Gaussians in the normal modes.

Two distinct approaches fall under the label of semiclassical methods. One of them concentrates on deriving model-independent relations[13], which follow from the symmetries of the soliton solution and do not depend on the detailed form of the lagrangian. The other semiclassical approach consists of numerical investigations of specific models. While initially focused on Skyrme's original model, this approach has since been applied to other models of Type IIa and Type IIb.

A severe difficulty of these models is that, although they can be regarded as truncations of renormalizable theories, they are not themselves renormalizable. The semiclassical method, which can be organized into an expansion in $1/N_c$, is therefore not systematic. The only unambiguous predictions that can be made are the form of the leading $1/N_c$ contribution to various quantities. One advantage of models of Type IIa is that one can systematically treat the quantum effects of low-energy pseudoscalar mesons in a simultaneous expansion in $1/N_c$ and in the energy. Such an expansion was advocated for the meson sector by Weinberg[14] and preliminary steps in extending it to the soliton sector have been made by Schnitzer.[15]

Skyrmions have also been studied using nonperturbative approximations, which enable one to address questions which would not be practical in a strict semiclassical approach. Some examples are a study of the apparent rotational instability of the skyrmion[16] and a treatment of the electromagnetic form factor of the skyrmion at large momentum transfer.[17]

4. PHENOMENOLOGY OF THE SKYRME MODEL

The pioneering phenomenological work on the Skyrme model by Adkins, Nappi, and Witten[18] focussed on the static properties of baryons in the original SU(2) Skyrme model. This work has since been extended to other models as well as to other processes.

The most spectacular results have come from model-independent relations in the SU(2) case.[13] Examples are the predictions $(\mu_n - \mu_p)/\mu_{N\Delta} = \sqrt{2}$ and $g_{\pi N\Delta}/g_{\pi NN} = 3/2$ which are correct to within 2%. In contrast, the nonrelativistic quark model with $N_c = 3$ gives predictions which differ by a factor of 4/5 and are therefore only accurate to 20%. In fact, it has been shown that model-independent predictions for static properties in the Skyrme model agree with the $N_c \to \infty$ limit of the quark model.[19]

Numerical calculations of the static properties of baryons in the SU(2) model have not been as successful. The original calculations of

Adkins, Nappi, and Witten gave predictions which were off by about 30%. This was not too bad for a model with only 2 parameters, but calculations with more elaborate models have not brought about a corresponding increase in accuracy. For example, adding a pion mass term[20] made some more properties calculable but did not give significantly better results.

The calculations of static properties were first extended to the SU(3) Skyrme model by Quadagnini.[21] Both the numerical predictions and the model-independent predictions seem to be much less accurate than the SU(2) predictions. The discrepancies seem to be larger than one would expect from symmetry breaking by the strange quark mass, and the reason for this does not seem to be understood.

A very interesting prediction of the SU(3) Skyrme model is the existence of a classically stable dibaryon.[22] Its lowest quantum state is an SU(3) singlet with spin 0 and 2 units of strangeness. A reliable estimate for its binding energy is not available, since this requires the calculation of zero-point energy contributions from pion fluctuations around the dibaryon as well as around the single skyrmion. There is an interesting puzzle concerning the parity of this state. The Skyrme model seems to predict the existence of two degenerate states which form a parity doublet, while a similar state in the bag model is predicted to have even parity. This is the only known case in which the Skyrme model and bag model disagree on the quantum numbers of a particle.

Dynamical properties of baryons are also amenable to treatment within the Skyrme model. Phase shifts for pion-nucleon elastic scattering have been calculated. Model-independent relations between phase shifts seem to be rather well satisfied.[23] Furthermore, baryon resonances appear in these phase shifts and the energies of most of the known resonances are reproduced to within 10%.[24] These calculations break down at threshold, but the threshold behavior should be computable using the soft-pion techniques advocated by Schnitzer.[15] This model can also be applied to many other dynamical properties, including baryon-baryon elastic scattering and photon-baryon elastic scattering.

The Skyrme model has also been applied to nuclear physics. This work has focussed on extracting a nucleon-nucleon potential from the interaction energy between two static skyrmions.[25] The resulting potential has many qualitative features of the phenomenological potentials of nuclear physics. An alternative approach is to identify a nucleus with atomic number A with a quantum state of a static soliton with baryon number A.[26] In the case of the deuteron, this approach gives automatically the correct quantum numbers and predicts static properties with accuracy comparable to the predictions for baryons.

5. CONCLUSIONS

About two years have elapsed since the rebirth of the Skyrme model ended twenty years of oblivion. In those two years, it has shown a surprising ability to describe aspects of strong interaction physics that one would have expected to be beyond its range of validity. With further effort, it may surpass the nonrelativistic quark model and the bag model in providing a conceptual framework in which to understand the strong interactions.

REFERENCES

1. T. H. R. Skyrme, Proc. Roy. Soc. London A260, 127 (1961); Nucl. Phys. 31, 556 (1962).
2. D. Finkelstein and J. Rubinstein, J. Math. Phys. 9, 1762 (1968).
3. J. G. Williams, J. Math. Phys. 11, 2611 (1970); N. K. Pak and H. C. Tze, Ann. Phys. 117, 164 (1979); A. P. Balachandran, V. P. Nair, S. G. Rajeev, and A. Stern, Phys. Rev. Lett. 49, 1124 (1982); Phys. Rev. D27, 1153 (1983).
4. E. Witten, Nucl. Phys. B223, 422 (1983); B223, 433 (1983).
5. G. 'tHooft, Nucl. Phys. B72, 461 (1974); B75, 461 (1974).
6. E. Witten, Nucl. Phys. B160, 57 (1979).
7. E. Witten, in Solitons in Nuclear and Elementary Particle Physics, ed. A. Chodos, E. Hadjimichael, and C. Tze (World Scientific, Singapore, 1984).
8. J. Donoghue, E. Golowich, and B. Holstein, Phys. Rev. Lett. 53, 747 (1984).
9. J. W. Carlson, Nucl. Phys. B253, 149 (1985).
10. M. Lacombe, B. Loiseau, R. Vinh Mau, and W. N. Cottingham, Orsay preprint IPND/TH 85-17; A. Jackson, A. D. Jackson, A. S. Goldhaber, G. E. Brown, and L. C. Castillo, Stony Brook preprint.
11. G. S. Adkins and C. R. Nappi, Phys. Lett. 137B, 251 (1984).
12. See for example P. J. Mulders, Phys. Rev. D30, 1073 (1984).
13. G. S. Adkins, in Solitons in Nuclear and Elementary Particle Physics, ed. A. Chodos, E. Hadjimichael and C. Tze (World Scientific, Singapore, 1984).
14. S. Weinberg, Physica 96A, 327 (1979).
15. H. Schnitzer, Brandeis preprint BRX TH 179.
16. R. Ingermanson, Lawrence Berkeley preprint LBL-19122.
17. J. P. Ralston, Los Alamos preprint Print-85-0787.

18. G. S. Adkins, C. R. Nappi, and E. Witten, Nucl. Phys. B228, 552 (1983).
19. A. V. Manohar, Nucl. Phys. B248, 19 (1984).
20. G. S. Adkins and C. R. Nappi, Nucl. Phys. B233, 109 (1984).
21. E. Quadagnini, Nucl. Phys. B236, 35 (1984).
22. A. P. Balachandran, A. Barducci, F. Lizzi, V. G. J. Rodgers, and A. Stern, Phys. Rev. Lett. 52, 887 (1984).
23. M. P. Mattis and M. P. Peskin, Phys. Rev. D32, 58 (1985).
24. A. Hayashi, G. Eckart, G. Holzworth, and H. Walliser, Phys. Lett. 147B, 5 (1984); M. P. Mattis and M. K. Karliner, Phys. Rev. D31, 2833 (1985).
25. A. Jackson, A. D. Jackson, and V. Pasquier, Nucl. Phys. A432, 567 (1985).
26. E. Braaten and L. Carson, Argonne preprint ANL-HEP-CP-85-67, Northwestern preprint NUTP-85-3.

A DEPENDENT EFFECTS IN
PARTICLE PRODUCTION

Don D. Reeder

Department of Physics
University of Wisconsin,
Madison, Wisconsin 53706

Abstract

Recent results on the A dependence of both the production of strange and charm particles and the production of hadrons at large transverse momentum provide additional support for the model of multiple scattering of constituents in the nucleus.

The study of rare processes often requires the use of nuclear targets, which distort the picture of the fundamental nucleonic interaction. Although the effect of the nuclear environment can be of interest in its own right, it must be removed in order to extract the characteristics of the fundamental interaction.[1] Moreover, the use of photons and leptons to probe the consituents of the nucleon can also be complicated by this effect as, for example, the rather subtle variation with atomic number A of the inelastic form factors deduced from deep inelastic muon scattering by the EMC collaboration and which have been the subject of many recent investigations both experimental and theoretical.[2] The cross section for photoproduction also shows effects which are ascribed to the "shadowing" of target nucleons when the photon becomes hadronic via vector dominance.[1] In this short review I wish to focus on the rather large nuclear effects on the production of particles, particularly of those containing heavy quarks. New data have been obtained in experiment E-605 at Fermilab[3] on the A dependence of particle multiplicity at large transverse momentum and on the production of dihadrons and dileptons of large invariant mass. The E-613 collaboration[4] at Fermilab has reported the first measurement of the A dependence of the production of open charm.

Nuclear effects are different from those resulting from the use of a thick target. In the latter the interactions are independent events – a cascade – in the

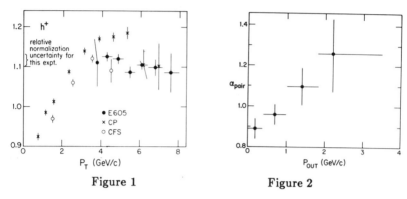

Figure 1   Figure 2

loose collection of nuclei which comprise the target. This is a consequence of the fact that the characteristic length describing the formation of hadrons after an interaction is short compared to the absorption length. When data obtained in beam dump experiments are extrapolated to zero absorption length to remove thick target effects, the extrapolation is done subject to this condition.

When the nucleus – a collection of nucleons – is itself a target, the ratio of the nuclear cross section to that of the nucleon is an "effective" A commonly parameterized as A^α. For example, the absorption cross section is found to vary as $A^{0.7}$, suggestive of a "black disc" of radius $A^{1/3}$. Phenomenologically, this empirical variation is extended to the differential cross section:

$$\left(\frac{E d^3\sigma}{dp^3}\right)_A = A^{\alpha(x_F, P_t)} \left(\frac{E d^3\sigma}{dp^3}\right)_N$$

Over a decade ago a Chicago-Princeton[5] collaboration found that the production of single hadrons at large transverse momentum (p_t) varied as A^α where $\alpha \gtrsim 1$. The detailed values were strongly correlated with the quantum numbers of the hadron. This collective effect (called an anomalous nuclear enhancement) was also observed in the production of hadron pairs, although the anti-shadowing is only observed in asymmetric pairs depending on the magnitude of the difference in p_t. Multiple scattering of the nucleon constituents – quarks and gluons – (the CMS model) has been reasonably successful in describing these data.[6]

The E-605 collaboration[3] at Fermilab submitted to this conference new results. Using targets composed of beryllium, copper and tungsten, they measured α for π^+ and K^+ production at large p_t, which are shown in Figure 1 together with earlier measurements.[7-9] The CP data had suggested that α might decrease near $p_t = 6$ GeV/c^2, however, this is not corroborated by the new results. Within the normalization uncertainties, α increases to ~ 1.1 to 1.15 and remains relatively constant, consistent with the CMS model.

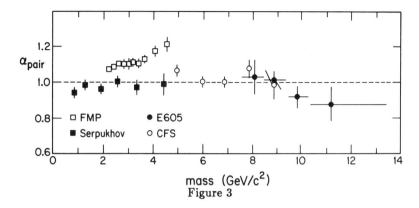

mass (GeV/c^2)

Figure 3

The values of α obtained by E605 for dihadron production, predominately symmetric pairs, are plotted in Figure 3 together with earlier data. In this case (except for the FMP data) there is agreement that $\alpha \simeq 1.0$, which indicates that the symmetric pairs arise from a single hard scattering of the constituents. As the net p_t increases from zero, multiple interactions are increasingly important. The present data of E-605 is limited in the values of net p_t which can be reached. However, the multiple scattering contribution should be more visible in those events for which the plane of the dihadrons does not contain the incident beam momentum. They define p_{out} for a pair as the component of the lower p_t hadron's momentum perpendicular both to the beam and the higher p_t hadron's momentum. As can be seen in Figure 2, the α appears to increase as p_{out} increases, a result expected in the CMS picture.

At small momentum transfer the effect is reversed. Shown in Figure 4 are selected results of Barton et al.[10] on the production of protons at $p_t = 0.3$ Gev/c for two values of x_F.

Note that, as before, the A^α parameterization correctly describes the data; however, the linear extrapolation of the fit to $A = 1$ overestimates the hydrogen cross section by up to a factor of 2! This feature appears to be common to most measurements of the A dependence at small p_t and it means that the inference of A dependent effects from measurements on a single complex nucleus together with those on hydrogen should be done with great caution.

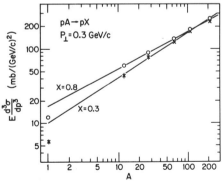

Figure 4

The production of heavy quarks

can be described within the context of QCD by both non-perturbative and perturbative processes, which are distinguished by the A dependence of the partial cross section. In general, non-perturbative or diffractive models favor an $A^{\frac{2}{3}}$ dependence while perturbative QCD models require A^1. Nuclear effects, of course, may alter the specific predictions.

Halzen[11] has pointed out the issues concerning A dependent effects in interpreting the production of charm. Skubic et al.[12] have measured the A^α dependence of the production of the strange quark (Λ, K^0) and found $\alpha < \frac{2}{3}$ for $p_t \lesssim 0.3$ Gev/c and for $x_F > 0.2$. The variation of α as a function of x_F for neutral strange hadrons and mesons is apparent in Figure 5. These values are relatively independent of the type of hadron although there is some suggestion that α varies with the number of strange quarks. A surprising feature of these data in addition to the small value of α is that fact that the ratio of multiplicities is independent of A, despite enormous variations in the spectrum ($d\sigma^\Lambda/dx \sim (1-x)^{0.5}$ and $d\sigma^{\bar\Lambda}/dx \sim (1-x)^{9.0}$)

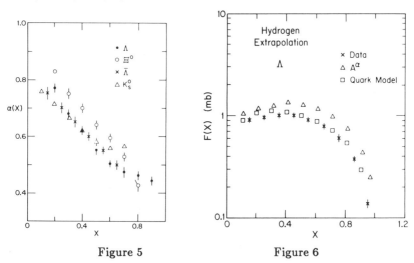

Figure 5   Figure 6

Pondrom in his review of strange particle production[13] obtained a good description of the production cross section and the A dependence using a modification of an empirical Collision Model by Das and Hwa.[14,15] In this model the multiplicity is written:

$$\frac{x\,d\sigma}{\sigma_{inel}dx} = \int F(x_1, x_2) R(x_1, x_2, x) \frac{dx\,dx_2}{x_1\,x_2}$$

where F is the probability of finding quarks with fractions of momentum x_1 and x_2 (x_1, x_2 limited by phase space, $(x_1 + x_2) \leq 1$); and R is the probability of the two quarks to recombine to form a meson with $x = x_1 + x_2$. In extending the model to include A dependence, Pondrom leaves R unchanged, since the formation length is large compared to the nuclear radius, but F is modified to include quark interaction of "straggling" within the nucleus. The nuclear F is found by convoluting the nucleonic F with a collision function for each quark of the form:

$$Q(z) = \lambda z + (1 - \lambda)\, \delta(z - 1)$$

In Figure 6 the data for Λ^0 production on hydrogen are compared with the predictions of the A^α parameterization and the model calculation. Again note that the naive extrapolation overestimates the data.

The E613 collaboration at Fermilab have measured the A dependence of the production of the charm quark in a beam dump experiment[4] using a 400 GeV proton beam which was directed into thick targets of tungsten, copper, and beryllium. The resultant hadronic cascade is absorbed in the target, and the muons produced are either absorbed or deflected in magnetized iron toroids located immediately downstream of the target. The neutrinos escaped and were detected 56 m downstream in a lead-scintillator calorimeter.

The primary source of these neutrinos is the semileptonic decay of charm quarks. Because of their short lifetime most charm particles decay before interacting, producing prompt neutrinos whose rate is independent of the target density. Of course, some pions and kaons also decay into neutrinos, producing a flux of non-prompt neutrinos, but their rate does depend on target density. In beam dump experiments[17−20] the variation of the observed rate with target density is one method which can be used to separate the prompt neutrinos from the non-prompt background.

In the analysis of the tungsten data the $\nu_e + \bar{\nu}_e$ CC events are identified using two independent methods: (1) subtraction of known backgrounds, and (2) use of the longitudinal shower length to directly estimate the $\nu_e + \bar{\nu}_e$ rate. Both methods agree.[21] In the first method the rates are computed for: ν_μ CC events for which the muon is not accepted in the spectrometer, ν_μ NC events, and ν_e NC events, and, finally, the non-prompt $\nu_e + \bar{\nu}_e$ CC events, which are primarily (85%) from K_{e3} decays. These background rates are then subtracted from the raw rate of events with no identified muon to obtain the prompt $\nu_e + \bar{\nu}_e$ CC event rate.

These prompt $\nu_e + \bar{\nu}_e$ rates are shown in Figure 7 for each target. Since they are nearly equal for the three targets, the charm cross section, if parameterized as A^α, has a value of α similar to that which describes proton absorption cross sections[10] $(\sigma_{pp} \propto \sigma_0 A^{.72})$. Assuming the charm production cross section varies as $s^{1.3}$, where s is the square of the energy in the proton-nucleon center of mass, the change in the prompt rate due to finite target length is calculated. The correction to the rates due to secondary production in the cascade varys depending on α from no effect for $\alpha = 0.72$ to -1% for W and +6% in Be for $\alpha = 1.0$.

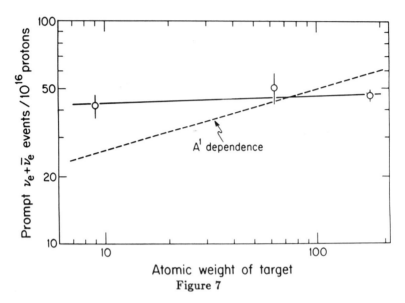

Figure 7

A linear fit to the data determines

$$\alpha = 0.75 \pm 0.05$$

with a $\chi^2 = 0.5$ for 1 degree of freedom, the solid line in Figure 7. The dashed line is a fit assuming an A^1 dependence corrected for finite target length; it has $\chi^2 = 24.6$ for 2 degrees of freedom.

Although limited by the statistical accuracy of the data, the variation of α with neutrino energy was investigated. The analysis was done independently in five energy regions with the results shown in Figure 8. No significant variation with neutrino energy can be inferred, although the lower values at large energy are suggestive that the small values of α seen in the strange quark production at large x_F may prove more universal.

The measurement of the A dependence from the ν_μ CC rates is much less precise because of the substantial non-prompt rates. The ν_μ and $\bar{\nu}_\mu$ CC data are plotted in Figure 9 for the full and partial density Cu and W targets.

In Figure 9 the lines show the non-prompt backgrounds calculated using a parameterization of the hadronic production processes and the detector geometry and normalized to the full density Cu and W targets. The agreement suggests that the non-prompt flux calculation is a good representation of the effects of both target density and atomic number.

Thus for the first time the variation of the charm production cross section with atomic number has been measured. The result is nearly identical to that of

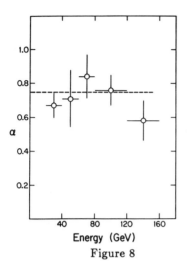

Energy (GeV)

Figure 8

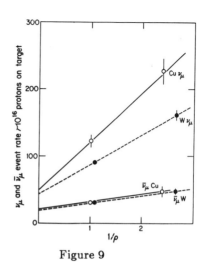

Figure 9

the proton-nucleus absorption cross section, and different from the dependence of Drell-Yan lepton pair production.[22,23] The variation is qualitatively similar to that observed for strange particle production, suggesting that the cause is rescattering effects characteristic of the nuclear environment rather than single hard scattering of the beam quark.

The effect of this result on the values of the charm cross section reported to date are shown in Figure 10, taken from the review of Kernan and VanDalen[24]. The data appear in agreement and are greater than the hydrogen value reported by LEBC collaboration[25] as would be expected from the other hadron production at small p_t.

In one of the few theoretical calculations, Badalian and Zhamkochian[26] calculate the A dependence of open charm using a multiple scattering model. They require as input the measured values of the $d\sigma/dx$ for D production on hydrogen; the energy dependence of $\sigma(pN \to DX)$; and the D nucleon elastic scattering cross section. After making reasonable assumptions for the unmeasured functions, they predict a value of α slightly smaller than the measured value decreasing slightly at large x. Considering the arbitrary nature of the assumptions, the agreement is surprisingly good. In contrast another calculation by Likhoded and Slabospitsky[27] who use only the structure functions as modified by nuclear effects (the EMC effect) fail completely to describe the data.

In conclusion, experiment indicates that nuclear effects play an important role in several aspects of particle production, especially heavy quark production. The indication is that A dependent effects of the charm production cross section are more similar to those of strange quark rather that the Drell Yan production of

722

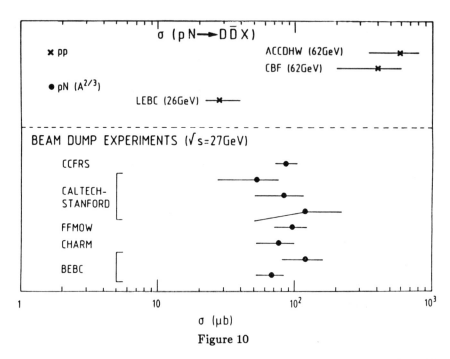

Figure 10

lepton pairs. The CMS model appears to provide a promising framework within which the panoply of phemonema can be understood. Among the remaining issues are: to explain the A dependence in all kinematic regions from first principles and to reconcile constituent scattering with the long formation length for the final state hadrons.

REFERENCES

1. N.N. Nikolaev, *Sov. J. Part. Nucl.* **12(1)**, 63 (81).

2. E.L. Berger, Proceedings of the Topical Seminar on Few and Many Quark Systems, San Miniato, Italy, March 1985 (to be published).

3. (E605) Y.B. Hsiung et al., *Phys. Rev. Lett.* **55**,457 (85).

4. M.E. Duffy et al. **Phys. Rev. Lett.**, to be published

5. (CP) D. Antreasyan et al., *Phys. Rev.* **D19**, 764 (79). This also references previous work.

6. M. Lev and B. Petersson, *Zeit. fur Physik C -Particles and Fields* **21**,155 (83).

7. D.A. Finley et al., *Phys. Rev. Lett.* **42**,1031 (79).

8. (Serpukov) V.V. Abramov et al., *JETP Lett.* **42**,352 (83).

9. (CFS) R.L. McCarthy et al., *Phys Rev Lett.* **40**,213 (78).

10. D. S. Barton et al., *Phys. Rev.* **27D**, 2580 (83).

11. F. Halzen, Proceedings of the XXI International Conference on High Energy Physics, les Editions de Physique, Paris, pg. C3-381, 1982.

12. P. Skubic et al., *Phys. Rev.* **D18**, 3115 (78).

13. L.G. Pondrom et al., *Physics Reports* **122**,57 (85).

14. k.P. Das and R.C. Hwa, *Phys Lett.* **68B**,459 (77).

15. R.C. Hwa, *Phys. Rev Lett.* **52**,492 (84).

16. The apparatus has been described in R.C. Ball et al., *Phys. Rev. Lett.* **51**, 743 (83).

17. P. Fritze et al., *Phys. Lett.* **96B**, 427 (80).

18. M. Jonker et al., *Phys. Lett.* **96B**, 435 (80).

19. H. Abramowicz et al., *Z. Physik* **C13**, 179 (82).

20. A. Bodek et al., *Phys. Lett.* **113B**, 77 (82).

21. M.E. Duffy et al., *Phys. Rev. Lett.* **52**, 1865 (83).

22. S.D. Drell and D.M. Yan, *Phys. Rev. Lett.* **25**, 316 (70).

23. M. Binkley et al., *Phys. Rev. Lett.* **37**, 571 (76).

24. A. Kernan and G. VanDalen, *Physics Reports* **106**, 297 (84).

25. M. Aguilar-Benitez et al., *Phys. Lett.* **123B**, 98 (83).

26. A.R. Badalian and V.M Zhamkochian, Yerevan Physics Institute preprint 755(70)-84, Yerevan, USSR

27. A.K.Likhoded and S.R. Slabospitsky, Serpukov preprint 84-78, Serpukov, USSR

V

THEORY BEYOND THE STANDARD MODEL

Session Organizers: L.S. Brown and B. Ovrut

HIGHER DIMENSIONS, SUPERSYMMETRY, STRINGS

Peter G. O. Freund

Enrico Fermi Institute and Department of Physics
University of Chicago, Chicago, IL 60637 USA

Grand unification,[1] the maximal extension of the unadulterated Yang-Mills idea, fails to address a number of phenomenological and theoretical issues. It provides no theoretical criteria for the choice of the gauge group, of the (spin 0 and 1/2) matter fields and of the Yukawa couplings. Although it involves scales close to the Planck scale it does not include gravity. Over the last few years, it has become increasingly clear that these problems are most cleanly dealt with by going to theories in which the world manifold has one time dimension but more than 3 space dimensions. When combined with supersymmetry this approach singles out two space-time dimensionalities: 10 and 11.

The maximal[2] supergravity theory[3] involves 11-dimensions. This theory provides a natural mechanism[4] for the compactification of 7 space-like dimensions leading to an "effectively" 4-dimensional final world. In spite of this remarkable feature, and of the beautiful compactification geometries[5] that implement it, this theory has its own problems. At the phenomenological level it fails to account for the observed chiral fermion spectrum in any natural way,[6] and it leads to an immense anti-de-Sitter-like cosmological constant. At the theoretical level, while inextricably including gravity, it does not appear to yield a meaningful quantum theory of gravity.

It has been known for some time that the chiral fermion problem is curable in 10-dimensions, where supersymmetric Yang-Mills fields can be coupled into the system.[6] But 10 is also the critical dimension for superstring theory. This has led Green and Schwarz[7] to propose that superstring theory holds the key to the meaningful incorporation of gravity at the quantum level.

At first sight one would worry that an arbitrary gauge group has been reintroduced into the theory. This however is not so. If the superstring theory is to have any gauge group at all (in 10-dimensions) then anomaly cancellation requires[7] this group to be either $Spin(32)/Z_2$ or $E_8 \times E_8$. There are then at least four (by now really five) such theories. Without gauge group in 10 dimensions we have the type IIa and IIb theories of closed superstrings, each exhibiting $N=2$ supersymmetry (in 10 dimensions). Beyond these there are $E_8 \times E_8$ and $Spin(32)/Z_2$ strings (two types of the latter) and they exhibit $N=1$ supersymmetry in 10 dimensions. This however is too much of a good thing, and one may wonder as to how to eliminate some of these candidate theories or how to provide for a unified treatment thereof. All this is reminiscent of the plethora of 4-dimensional supergravity theories with $N=1,2,3,4,8$ supersymmetries, all of which are ultimately, just different compactifications of one and the same 11-dimensional theory.[4],[5] In this spirit I have proposed[8] that the $E_8 \times E_8$ and one of the $Spin(32)/Z_2$ theories are but different compactifications of one and the same 26-dimensional string theory. This is relatively easily understood. Suppose we start with strings in $d=10+N$-dimensional space-time. Were we to compactify N of these dimensions on the maximal torus of some nonabelian Lie group G of rank N, then beyond the obvious N Kaluza-Klein abelian gauge symmetries (connected with the isometries of the torus) there would appear a larger nonabelian gauged G-symmetry. The G-gauge bosons other than those corresponding to the Cartan subalgebra of the Lie algebra of G (i.e.,other than the Kaluza-Klein ones), arise as solitons as noted by Kac, Frenkel, Goddard and Olive.[9] Both $Spin(32)/Z_2$ and $E_8 \times E_8$ have the same rank $N=16$, which implies $d=10+16=26$ for the dimensionality of the primordial string theory. But this is precisely the critical dimension of the bosonic Veneziano-Nambu-Goto string. We are thus led to suggest that superstrings in 10-dimensions are toroidal compactifications of a primordial bosonic string theory in 26 dimensions. There are a number of problems which have to be addressed in order to implement this suggestion.

First of all, if we start from closed strings in 26 dimensions, we would obtain one such gauged G group from the "left movers" and another gauged G group from the "right-movers" on the string. That is twice as much as we need.

Second, the original 26-dimensional theory is purely bosonic, whereas the supersymmetric superstrings obviously involve both fermions and bosons in a carefully balanced way. Where do these fermions come from?

Third, the 26-dimensional theory has a tachyon, whereas the 10-dimensional superstrings do not have this shortcoming.

How is one to get around these problems? The tachyon of the 26-dimensional theory is of course not a flaw of the theory itself, but rather of the vacuum around which one expands it: 26-dimensional flat Minkowski space. The presence of the tachyon only indicates the instabilitiy of this vacuum. Were we to find a supersymmetric vacuum, tachyons would then be automatically ruled out. Solving the fermion problem would then automatically remove the tachyon problem. There are two ways to address the fermion and G-doubling problems.

The first way, considered by Gross, Harvey, Martinec and Rohm[10] consists in treating the left- and right-movers independently. For, say, the right-movers they start in 26 dimensions, proceed with the toroidal compactification described above, and acquire thereby the $G=E_8 \times E_8$ or $Spin(32)/Z_2$ gauge symmetry. Since they do not repeat the construction for the left-movers they do not encounter G-doubling. For the left- movers they start directly in 10-dimensions and this yields $N=1$ supersymmetry and the corresponding fermions. The left-movers then bring the fermions and the supersymmetry whereas the right-movers the gauged rank 16 internal symmetry. Consistency at one loop then forces the weight lattice of G to be even and self-dual thus unambiguously restricting G to $E_8 \times E_8$ and $Spin(32)/Z_2$. This approach is quite asymmetric between left- and right-movers whence its name "heterotic".[10]

A second approach starts from the observation that string theory is a two-dimensional field theory and can therefore present fermionic solitons even when starting from a purely bosonic lagrangian. I have therefore suggested[8] that these fermionic solitons combine with the bosons into a supersymmetric spectrum. But then how do these fermions acquire the 10-dimensional spinorial Lorentz quantum numbers required by the spin-statistics theorem? In a recent paper Casher, Englert, Nicolai and Taormina[11] have suggested that the second "unwanted" G group in closed string compactification may be the key to this problem. They identify the transverse $SO(8)$ part of the 10-dimensional Lorentz group, not with the corresponding rotation subgroup $SO(8)_M$ of 10-dimensional Minkowski space, but rather with the diagonal $SO(8)$ subgroup of an $SO(8)_M \times SO(8)_{int}$ group, where $SO(8)_{int}$ is an $SO(8)$ subgroup of the rank 16 group G' on whose maximal torus the 26 → 10 compactification for half the closed string modes is performed. For this to yield spinors, G' must be $E_8 \times E_8$, but unlike in what was said above, this $E_8 \times E_8$ *does* get broken and ultimately yields "half" the transverse $SO(8)$.

In other words both for right movers and left movers there are two different types of toroidal compactification; the one, described at the beginning, which yields the gauged $G = E_8 \times E_8$ or $Spin(32)/Z_2$ symmetry, and the new compactification which yields fermionic solitons and breaks a $G' = E_8 \times E_8$ down to $SO(8)_{int} \times H$ where H is a group acting only on the vacuum state, and which is involved in the manufacture of topological solitons. These solitons then do provide the fermions required for supersymmetry. Let us call *internal* that type of toroidal compactification which yields a gauged internal symmetry group $E_8 \times E_8$ or $Spin(32)/Z_2$, and *supersymmetric* the type which yields the $N=1$ supersymmetry. Starting from 26 dimensions one then encounters four possibilities. Compactifying left movers supersymmetrically and right movers internally one gets the $N=1$ supersymmetric heterotic strings with gauge group $E_8 \times E_8$ or $Spin(32)/Z_2$. Obviously the same end result is obtained were one to interchange left and right movers in this compactification. Compactifying both left- and right movers supersymmetrically one ends up without any internal gauge symmetry but with more supersymmetry: one recovers the $N=2$ supersymmetric type IIa and IIb superstrings. Finally, compactifying both left- and right-movers internally, no supersymmety results and the corresponding string has a tachyon, so that this compactification is unstable. In short then, there appears to exist a unique 26-dimensional string theory such that on the one hand the Veneziano-Nambu-Goto string with its tachyon is found when expanding this theory around the unstable vacuum provided by 26-dimensional Minkowski space, but the supersymmetric heterotic and type II strings are obtained when expanding around less trivial vacua of this theory. The various superstrings are then expansions of the same string theory around different supersymmetric vacua.

As always, finding the true vacuum becomes then a complicated, but hopefully manageable *dynamical* problem. Four of the five apparently different superstring theories have thus a common origin in a theory in 26-dimensions (the fifth theory which involves open and closed $SO(32)$ superstrings with Chan-Paton rules will not be discussed here). It is important to further study the interacting string in order to fully demonstrate the validity of such a picture.

Similar results were obtained also by Vafa and Witten[12] from a rather different approach. They impose twisted boundary conditions on the 26-dimensional string and notice that the 26-dimensional bosonic string can be viewed as a double cover of a 10-dimensional superstring. They further point out that the 10-dimensional theory could be viewed as the double cover of a theory in 6 dimensions, or the fourfold cover of a theory in 4 dimensions(!) or the eightfold cover of a theory in 3 dimensions. These dimensions 3, 4, 6, 10 are of course precisely those in which $N=1$ supersymmetric Yang-Mills systems arise. These are presumably related to the zero-modes of the corresponding string theories (*assuming* these to be consistent at the quantum level). What kinds of new string theories would these be? Given the higher twistedness of the corresponding boundary conditions, I would expect these theories to be rleated to parastatistics. But in precisely these dimensions 3, 4, 6 and 10 parastring theories have been considered a long time ago by Ardalan and Mansouri.[13] Beyond Lorentz invariance, the absence of Mandelstam analyticity violations at the one-loop level has been checked for these theories.[15] Could these be yet further consistent string theories? Is there a relation[15] here as well to 2-dimensional σ-models with Wess-Zumino term with quantized coefficient N such that $|N|>1$? In view of the fundamental level at which these theories are believed to be operating, these questions deserve answers.

I now wish to mention some further open problems in string theory.

1. First of all, superstrings were revived as a possible cure for the problems of quantum gravity. One loop results are of course very encouraging, but it is essential to show that these theories are finite to all loops. Remarkable progress in this direction is being made by Mandelstam[16] and by Brink.[17]

2. Second, at the phenomenological level, compactification of heterotic $E_8 \times E_8$ strings from 10 to 4 dimensions on Calabi-Yau manifolds are very promising as can be seen in the beautiful work of Candelas, Horowitz, Strominger and

Witten.[18] The Ricci-flat metrics on these manifolds are not explicitly known. It is therefore all the more remarkable that the pattern of Yukawa couplings of the effective E_6 unified theory in 4-dimensions can be determined even without the knowledge of the Calabi-Yau metric. The phenomenological importance of these couplings is obvious.

3. Once one abandoned point particles for extended objects, one may ask why stop at strings. Could membranes or higher dimensional objects[19],[20] lead to further interesting theories?

4. The low energy limit of superstring theory involves the local field theory of a supersymmetric Yang-Mills and supergravity system. As a rule such a theory yields solitons (e.g., magnetic monopoles). These should have counterparts at the string level, but being nonperturbative effects would not be noticed in the usual perturbative approach to string theory. Could it be that the ordinary "electric" string has a "magnetic" counterpart so that there is an infinite wealth of such soliton states involving even an "electric-magnetic" self-duality à la Montonen-Olive.[21] Given that all excited string states are already superheavy, these "magnetic monopoles" should not be ignored simply in view of their high mass.

5. We are dealing here with a gravity theory different from Einstein's. what happens to the singularity theorems of Penrose and Hawking? Could this be the cure?

6. Finally, there is the problem of finding a geometrical description of strings. Gell-Mann has a nice analogy with general relativity for stating the urgency of this problem. There, Einstein using the physical evidence of the equivalence of inertial and gravitational mass and the *gedankenevidenz* of travel in elevators, has come up with the remarkable geometrical principle of general covariance. It took then still quite a number of years before the dynamics which respects this principle was nailed down. In string theory, just the opposite seems to have happened. We know all the dynamics but ignore the geometrical principle which underlies it, the way general covariance underlies general relativity.

One possible avenue is to consider the geometry of loop space,[22]-[28] i.e., the infinite-dimensional space of (closed) string configurations. An alternative advocated by Witten[29] considers the extension of twistorial ideas to string theory. The two approaches may in fact be related. To see this we note that, as pointed out in references 26] and 28], the geometrical constructs in loop space involve so much freedom that constraints must be imposed, in a way similar to the constraints in the superfield formulation of supergravity and supersymmetric Yang-Mills theories. But those constraints can be derived with twistorial methods.[30]–[33] One may therefore be tempted to generalize this derivation of constraints to superstrings. Let me be more specific. First let me recall the ordinary $N=1$ super-Yang-Mills and supergravity cases in 4-dimensions.[30]–[33] In 4-dimensions (- + + + signature) one has Majorana spinors. From any *c-number* Majorana spinor ψ one can construct the 4-vector $v_a(\psi) = \psi^T C \gamma_a \psi$. This vector is null: $v_a(\psi)v^a(\psi) = 0$, as can readily be checked using the equivalence of Majorana and Weyl spinors in 4-dimensions. Now consider the superspace covariant derivatives D_α and D_a (the Dirac index $\alpha = 1, \cdots, 4$; the Lorentz index $a = 0,1,2,3$). These covariant derivatives depend on whether the supersymmetric gauge fields or supergravity are present or not. So does their algebra. Now consider the fermion combination ($\psi_\alpha = Bose$, $D_\alpha = Fermi$)

$$Q(\psi) = \psi^\alpha D_\alpha \tag{1a}$$

and the bosonic combination (v_a and D_a both Bose)

$$D(\psi) = v^a(\psi)D_a \tag{1b}$$

We can require that $Q(\psi)$ and $D(\psi)$ obey the simple quantum mechanical supersymmetry algebra

$$[Q(\psi),Q(\psi)] = 2D(\psi)$$
$$[Q(\psi),D(\psi)] = [D(\psi),D(\psi)] = 0 \tag{2}$$

for every Majorana spinor ψ_α (here the bracket [ ] is a commutator in all cases except one, namely when both bracketed objects are Fermi, in which case it is an anticommutator). Given the basic (tangent superspace) geometric relations

$$[D_A, D_B] = R_{AB} - T^C_{AB}D_C \tag{3}$$

(the bracket again as above), we can translate the requirements (2) into constraints on the curvatures (R_{AB}) and torsions (T^C_{AB}). These constraints are precisely those imposed in the superspace formulation of supersymmetric Yang-Mills and supergravity theories. Here these constraints acquire geometrical meaning as integrability conditions in all the lightlike superplanes defined by ψ^α, $v_a(\psi)$ for all ψ. These considerations can be extended to the case of $N>1$ supersymmetries. For the $N=4$ super Yang-Mills case and for the $N=5,6,78$ supergravities a remarkable thing happens. In these cases the constraints along with the Bianchi identities imply the field equations![31],[33]

Now to the generalization to strings. Instead of D_α and D_a we now have[28] $D_\alpha(\sigma)$ and $D_a(\sigma)$ (actually Bardakci[28] considers the nonsupersymmetric case in which there is only a $D_a(\sigma)$, but for our considerations supersymmetry is important, as the power of twistorial techniques is amplified in the supersymmetric case). So we again define a c-number Majorana-Weyl spinor $\psi_\alpha(\sigma)$ along the string. Again $v_a(\psi(\sigma)) = \psi^T(\sigma)C\gamma_a\psi(\sigma)$ is null as can readily be shown. Define the combinations

$$Q(\psi(\sigma)) = \psi^\alpha(\sigma)D_\alpha(\sigma) \quad D(\psi(\sigma)) = v^a(\psi(\sigma))D_a(\sigma) \ .$$

Even for given $\psi(\sigma)$ there are now infinitely many Q's and D's, one for each value of σ. Just as we required before, that $Q(\psi)$ and $D(\psi)$ obey the superalgebra[2] connected with a point particle, it is now natural to require $Q(\psi(\sigma))$ and $D(\psi(\sigma))$ to obey Virasoro superalgebra connected with a string. This constrains the curvatures and torsions in the infinite dimensional geometry. We might even suspect that these constraints when combined with the Bianchi identities lead to field equations as well. It remains to be seen whether this generalization of Eq. (2) does yield the usual string theories.

To conclude then, as of now we have theories that appear to reconcile general covariance with quantum theory, thus solving a problem dating back half a century. At the phenomenological level we have the promising $E_8 \times E_8$ heterotic string. At this level there are still many problems most notably with proton decay, too much freedom in the realm of Calabi-Yau manifolds, etc...None of these problems looks insuperable, nor does any of them lead to some contradiction. At the formal level, a lot of work remains to be done before these theories acquire the degree of elegance and self-containedness so manifest in general relativity and in the 4-dimensional formulation of Maxwell's electrodynamics.

References

1. See, e.g., H. P. Nilles, Phys. Rep. **110**, 1 (1984).

2. W. Nahm, Nucl. Phys. **B135**, 149 (1978).

3. E. Cremmer, B. Julia and J. Scherk, Phys. Lett. **76B**, 409 (1978).

4. P.G.O. Freund and M. A. Rubin, Phys. Lett. **97B**, 233 (1980).

5. F. Englert, Phys. Lett. **119B**, 339 (1982); M. J. Duff, Nucl. Phys. **B219**, 389 (1983); P. van Nieuwenhuizen in *Relativity, Groups and Topology II*, B. de Wit and R. Stora, eds. (North Holland, Amsterdam, 1984); B. de Wit and H. Nicolai, Phys. Lett. **148B**, 60 (1984); L. Castellani, R. d'Auria and P. Fré, Nucl. Phys. **B239**, 610 (1984).

6. E. Witten in *Shelter Island II*, R. Jackiw et al., editors (MIT Press 1985) p. 227.

7. M. B. Green and J. H. Schwarz, Phys. Lett. **109**, 444 (1982); ibid. **149B**, 117 (1984).

8. P.G.O. Freund, Phys. Lett. **151B**, 387 (1985).

9. I. B. Frenkel and V. G. Kac, Inv. Math. **62**, 23 (1981); I. B. Frenkel, J. Func. An. **44**, 259 (1981); P. Goddard and D. Olive in *Vertex Operators in Mathematics and Physics*, J. Lepowsky et al., editors (Springer, N.Y. 1985) p. 51.

10. D.J. Gross, J. Harvey, E. Martinec and R. Rohm, Phys. Rev. Lett. **54**, 502 (1985) and Princeton preprints.

11. A. Casher, F. Englert, H. Nicolai and T. Taormina, preprint CERN-TH. 4220/85.

12. C. Vafa and E. Witten, Princeton preprint.

13. F. Ardalan and F. Mansouri, Phys. Rev. **D9**, 3341 (1974).

14. F. Mansouri, private communication, 1985.

15. I. Antoniadis and C. Bachas, SLAC preprint.

16. S. Mandelstam, private communication.

17. L. Brink, in *Symposium on Anomalies, Geometry, Topology*, W. A. Bardeen et al., eds. World Sci. Pub. 1985, p. 325.

18. P. Candelas, G. Horowitz, A. Strominger and E. Witten, Nucl. Phys. **B258**, 46 (1985).

19. S. Randjbar-Daemi, A. Salam, E. Sezgin and J. Strathdee, Phys. Lett. **151B**, 351 (1985); P.G.O. Freund and F. Mansouri, Z. f. Physik **C14**, 279 (1982); A. Sugamoto, Nucl. Phys. **B215**, 381 (1983); P.A. Collins and R.W. Tucker, Nucl. Phys. **B112**, 159 (1976); P. S. Howe and R. W. Tucker, J. Math. Phys. **19**, 869, 981 (1978).

20. J. Hoppe, MIT thesis, 1982.

21. C. Montonen and D. Olive, Phys. Lett. **72B**, 117 (1977); H. Osborn, Phys. Lett. **83B**, 321 (1979).

22. W. Siegel, Phys. Lett. **151B**, 396 (1985).

23. T. Banks and M. Peskin, preprint SLAC-PUB 3740 (1985).

24. D. Friedan, preprint EFI 85-27.

25. A. Neveu and P. West preprint CERN-TH. 4200/85.

26. W. Siegel and B. Zwiebach, Berkeley preprint UCB-PTH 85/30.

27. M. Kaku and J. Lykken, in *Symposium on Anomalies, Geometry, Topology,* W. A. Bardeen et al., eds. World Sci. Pub. 1985, p. 360.

28. K. Bardakci, Berkeley preprint UCB-PTH 85/33.

29. E. Witten, Princeton preprint.

30. A. Ferber, Nucl. Phys. **B132**, 55 (1978).

31. E. Witten, Phys. Lett. **77B**, 394 (1978).

32. J. Crispim-Romao, A. Ferber and P.G.O. Freund, Nucl. Phys. **B182**, 45 (1981).

33. L. L. Chau and C. S. Lim, BNL preprint 1985.

SUPERSTRING PHENOMENOLOGY

P. Candelas

Center for Theoretical Physics
University of Texas, Austin TX 78712

Gary T. Horowitz

Department of Physics, University of California
Santa Barbara, California 93106

Andrew Strominger

The Institute for Advanced Study
Princeton, N.J. 08540

Edward Witten

Joseph Henry Laboratories, Princeton University
Princeton, N.J. 08544

ABSTRACT

We discuss some recent work on Kaluza-Klein compactifications of superstring theories and the resulting low
energy phenomenology. The string theory approach to
compactification is emphasized as opposed to the effective
field theory approach. An attempt is made to keep the
discussion non-technical.

A number of remarkable developments have recently taken place[1,2]
in the theory of superstrings.[3] These new developments inspire
optimism that superstrings provide both the first mathematically consistent quantum theory of gravity and the first truly unified theory
of all forces and matter. The ultimate test of any physical theory,
however, is not mathematical consistency, but confrontation with experiment. Historically, this confrontation has been crucial in the
development as well as the testing of new theories. This has made the
development of a quantum theory of gravity exceedingly difficult, since
quantum gravitational effects are presumably only directly observable
at 10^{19} GeV, and are thus beyond the reach of present day experiment.

A remarkable feature of superstring theories is that they should in principle lead to many low energy predictions. An attempt at exploring these predictions was made in ref. [4] and will be reviewed here. The basic reason predictions are possible is that it is very difficult to construct a consistent quantum theory of gravity, and those theories that have been found to survive the many consistency requirements have almost no freedom to adjust coupling constants,[5] alter the gauge group[1] or add or subtract particles.[*3] Thus, if the theory is correct, it must predict all the masses and couplings of the observed elementary particles!

Superstring theories appear to be consistent only if the dimension of spacetime is ten and the gauge group, if any, is $E_8 \times E_8$ or $SO(32)$. On the one hand these constraints are desirable in that they are precisely the type of restrictions one would hope to arise in a truly unified theory. On the other hand, they are at first sight disappointing since we certainly do not observe ten dimensions or $SO(32)$ or $E_8 \times E_8$ gauge symmetries. Clearly, if these theories are to have some connection with nature we must find a way to reduce both the dimension of spacetime and the gauge group. The idea we are going to discuss is based on the modern incarnation[6] of the old Kaluza-Klein program that some of the spacelike dimensions are compactified and small enough to not be directly observable.

One can discuss compactifications either in terms of a low energy effective field theory determined from the string, or directly in terms of the string theory. Here we discuss only the latter approach. The reader is referred to reference [4] for an effective field theory approach which leads to equivalent conclusions. We will try to review the main results in a non-technical manner and ignore certain generalizations. For further details see reference [4].

As discussed by Green and Schwarz,[3] the superstring theory is a field theory with an infinite number of fields. Among these many fields is a massless spin two field. In view of general arguments,[7]

---

*However, some loss of predictive power could occur if the full quantum theory turns out to have degenerate vacua.

this strongly suggests that superstrings must in a suitable approximation give rise to general relativity. In general relativistic field
theories the geometry of spacetime is not fixed, and one can expand
around curved backgrounds. One is thereby led to attempt an expansion
of the superstring theory around backgrounds other than flat ten dimensional Minkowski space. Phenomenologically interesting backgrounds
are of the form $M^4 \times K$, where K is some small six dimensional space
and M^4 is four dimensional Minkowski space.

It is not at all obvious that there exists a consistent expansion
around such backgrounds. Indeed, a considerable effort was required
to demonstrate the consistency of the superstring theory expanded
around flat M^{10}. On a curved background, we begin with a generalization of the superstring action:

$$S = - \frac{1}{4\pi\alpha'} \int d\sigma d\tau \ \sqrt{-q} \ (q^{\alpha\beta} g_{\mu\nu} \partial_\alpha X^\mu \partial_\beta X^\nu + \text{superpartners})$$

$$\alpha,\beta = 0,1$$

$$\mu,\nu = 0,1,\cdots,9 \tag{1}$$

where we simply replace the flat metric $\eta_{\mu\nu}$ on M^{10} with the curved
metric $g_{\mu\nu}(X)$ on $M^4 \times K$. $X^\mu(\sigma,\tau)$ is the location of the string world
sheet in spacetime and $q_{\alpha\beta}$ is the world sheet metric. S contains no
kinetic term for $q_{\alpha\beta}$, so one does not expect to treat $q_{\alpha\beta}$ as a dynamical variable of the theory. Instead, one attempts to solve for $q_{\alpha\beta}$ (up
to gauge freedom) via its equation of motion

$$\frac{\delta S}{\delta q_{\alpha\beta}} = 0 \tag{2}$$

The theory described by equation (1) can be viewed as a two dimen-
sional supersymmetric non-linear σ-model with dynamical fields $X^\mu(\sigma,\tau)$
(plus fermions) and internal space $M^4 \times K$. Equation (2) requires the
stress energy tensor $T_{\alpha\beta}$ of this two dimensional field theory to vanish.
Classically, this can always be achieved. However, quantum mechanically, (2) becomes an operator equation. The trace of this equation
takes the form:

$$T^\alpha_{\ \alpha} = aR + \beta_{\mu\nu} \partial_\alpha X^\mu \partial_\beta X^\nu q^{\alpha\beta} + \text{superpartner} \tag{3}$$

where R is the scalar curvature of the metric $q_{\alpha\beta}$, a is a coefficient depending on the dimension of spacetime and $\beta_{\mu\nu}$ is the β-"function" associated with the spacetime metric. If this trace does not vanish, then the equation of motion (2) for $q_{\alpha\beta}$ cannot be imposed and additional dynamics must be introduced for $q_{\alpha\beta}$.[8] Whether or not this can be done in a consistent fashion remains an open question. Even if it can, however, the resulting "Polyakov" string theory probably does not contain massless particles and is therefore not relevant as a theory of the fundamental interactions.

Fortunately, there are special cases in which $T^{\alpha}{}_{\alpha} = 0$. This condition is related to a local scale invariance of the two dimensional σ model. Classically, S is invariant under the transformations:

$$X^{\mu} \rightarrow X^{\mu}$$
$$q_{\alpha\beta} \rightarrow \Omega^2 q_{\alpha\beta} \tag{4}$$

which implies $T^{\alpha}{}_{\alpha} = 0$. However in the quantum theory, ultraviolet regulators usually destroy this scale invariance. For finite theories, however, this is not the case. Supersymmetric non-linear σ models on a flat two dimensional world sheet are known to be finite (assuming the existence of appropriate regulators) for Ricci flat internal spaces and N = 2 or N = 4 supersymmetry, and are believed to be finite for Ricci flat spaces and N = 1 supersymmetry.[9] Scale invariance on a curved world sheet requires in addition that the dimension of spacetime be the critical dimension 10. Thus Ricci flatness of the internal space $M^4 \times K$ is, in the present framework, essential for consistent superstring compactifications.[10]

Of particular interest is the heterotic superstring,[2] because it probably has the best phenomenology (and because it's hard to believe that nature declined such a good opportunity to base itself on exceptional groups). Recall that in this theory the fields are divided into left moving and right moving modes on the world sheet. In the fermionic representation there is a right moving anticommuting ten dimensional Majorana-Weyl spinor S, and 32 left moving anticommuting two dimensional Majorana-Weyl spinors ψ^m m = 1,$\cdots$,32 which are spacetime scalars. For the SO(32) superstring, ψ^m is in the fundamental representation of the group, while for $E_8 \times E_8$ the situation is more complicated.[2] The presence of gauge interactions in this theory suggests that one should be able to expand about a background configuration that has a non-zero Yang-Mills field as well as non-zero curvature. In this case, the light cone action for the heterotic superstring becomes[2,11]:

$$S_{LC}^{H} = -\frac{1}{4\pi\alpha'} \int d\sigma d\tau (\partial_\alpha X^i \partial^\alpha X^j g_{ij} + i\bar{S}\gamma^- D_+ S$$

$$+ \psi^m D_- \psi^m + F_{ij}^A \bar{S}\gamma^- \gamma^i \gamma^j S T_{mn}^A \psi^m \psi^n) \qquad (5)$$

where $i,j = 1,\cdots,8$ run only over the transverse space $R^2 \times K$ and S obeys the light cone condition $\gamma^+ S = 0$. F_{ij}^A is the SO(32) or $E_8 \times E_8$ background Yang-Mills field, and T_{mn}^A the appropriate generators. $D_\pm$ is a covariant derivative with respect to both the background gauge and spin connections. It acts on the left movers as $D_-\psi = \partial_-\psi + (\partial_- X^j)A_j\psi$ (with $\partial_\pm = \frac{\partial}{\partial\tau} \pm \frac{\partial}{\partial\sigma}$) and on the right movers as $D_+ S = \partial_+ S + (\partial_+ X^j)\omega_j S$. It is related to the spacetime gauge covariant derivative D_j by $D_\pm = (\partial_\pm X_j)D_j$.

The heterotic superstring corresponds to an $N = \frac{1}{2}$ supersymmetric non-linear σ model, as the supersymmetry acts on the right movers only. In general this is insufficient to insure that the σ model trace anomaly vanishes, even on a Ricci flat background. However, if you set the background gauge field A_i equal to the spin connection ω_i then a remarkable thing happens. Since ω_i must take values in a subgroup of SO(8) rather than SO(32) or $E_8 \times E_8$, many of the ψ's decouple from the gauge field and just satisfy a free Dirac equation. Their contribution to the trace anomaly will be cancelled by ghosts just as in the expansion about flat space without background gauge fields. The remaining ψ's now enter symmetrically with the components of the spinor S, and one recovers a left right symmetry in this sector of the theory.

An N = 1 (at least) world sheet supersymmetry then results[*] and the trace anomaly will vanish if the geometry is Ricci flat.

Another consistency requirement is the absence of anomalies in the four dimensional Lorentz algebra. In particular, there are potential anomalies in the commutator $[M^{i-}, M^{j-}]$ associated with the value of the two point function of T_{++} in the σ model. In the critical dimension, for background geometries $M^4 \times K$, this anomaly has been shown not to arise in lowest non-trivial order when K is Ricci-flat,[11] and presumably does not arise at higher orders either.

So far we have found that if one looks for a string state that can be described just by a background geometry, and assumes the relevant conditions are obeyed for all values of the sigma model coupling, then consistency forces the background geometry to be Ricci flat, and the background gauge connection A to equal the background spin connection ω. Have we found all the consistency requirements? Hopefully the answer is no, because these constraints do not uniquely determine a background configuration. Further consistency requirements might be useful in narrowing the range of potential vacuum configurations for the superstring theory.

We have argued that (modulo the above comments) one can construct a consistent first quantized string theory on Ricci flat manifolds with $A = \omega$, but it remains to be shown that such backgrounds provide solutions to the full superstring theory. What is meant by this statement is that the vacuum expectation values of all tadpoles vanish. In a scalar field theory with potential $V(\phi) = \phi$, for example, setting the background scalar field to zero is not the semiclassical approximation to a solution of the theory because the tadpole $\langle 0|\phi|0\rangle$ does not vanish. The superstring theory is a field theory with an infinite number of fields, so one similarly requires that $\langle 0|\Phi|0\rangle = 0$, where Φ is any operator that creates a tadpole.

---

[*] Actually, the N = 1 world sheet supersymmetry will not be obvious in the form of the action (5) unless the geometry is not only Ricci flat but Kähler. Otherwise the supersymmetry is apparent in the Ramond Neveu-Schwarz formulation of this theory. See Reference [11] for details.

There is a simple argument which shows that to all orders in the σ-model, but tree level in the string theory, all the tadpoles vanish for the background configurations we have discussed. Consider, for example, the operator $\Phi_D = q^{\alpha\beta} g_{ij} \partial_\alpha X^i \partial_\beta X^j$, which creates a scalar dilaton field. For backgrounds with no trace anomaly, this operator scales as $\Phi_D \rightarrow \Omega^{-2} \Phi_D$ under the two dimensional scale transformations (4). However, since the world sheet in tree level string calculations can (after stereographic projection) be taken to be a plane which is invariant under these transformations, the expectation value $\langle 0|\Phi_D|0\rangle$ must vanish. Similar arguments can be made for the other tadpoles.

Having shown that Ricci flat manifolds with $A = \omega$ provide solutions of the superstring theory at tree level, one still needs to check their classical stability before they can be called vacuum solutions.[12] A general argument for perturbative stability can be made in the special case where the resulting theory has $N = 1$ spacetime supersymmetry (as distinct from world sheet supersymmetry). This may also be the most phenomenologically interesting case, since without spacetime supersymmetry at the compactification scale there is likely to be a serious hierarchy problem. With $N = 1$ spacetime supersymmetry, the Hamiltonian for perturbations around the supersymmetric background can be written $H = Q^+ Q$, where Q is the supercharge operator for spacetime supersymmetry. In light cone gauge the absence of ghosts is manifest, so H acts on a positive norm Hilbert space and its expectation value in any state is therefore non-negative. This ensures perturbative vacuum stability.

What is the condition for $N = 1$ spacetime supersymmetry? In flat space, the action (5) is invariant under the supersymmetry transformations[2,3]

$$\delta_\varepsilon X^i = (p^+)^{-\frac{1}{2}} \bar\varepsilon \gamma^i S$$

$$\delta_\varepsilon S = i(p^+)^{-\frac{1}{2}} \gamma_- \gamma_\mu (\partial_- X^\mu) \varepsilon \qquad (6)$$

where $\mu = 0, \cdots, 9$ and ε is a right moving ten dimensional Majorana-Weyl spinor.

The commutator of two supersymmetry transformations is a

translation:

$$[\delta_1, \delta_2] X^i = \xi^- \partial_- X^i + a^i$$

$$[\delta_1, \delta_2] S = \xi^- \partial_- S$$

where

$$\xi^- = - \frac{2i}{p^+} \bar{\varepsilon}^{(1)} \gamma_- \varepsilon^{(2)}$$

$$a^i = - 2i \bar{\varepsilon}^{(1)} \gamma_i \varepsilon^{(2)} \tag{7}$$

A curved background admits spacetime supersymmetry if there exists supersymmetries parametrized by $\varepsilon^{(1)}$ and $\varepsilon^{(2)}$ such that a^i is any spatial translation of the transverse $R^2 \times K$. (Four dimensional Lorentz invariance then insures the full four dimensional Poincaré supersymmetry.)

Unlike S, ε need not obey the light cone condition $\gamma^+ S = 0$. There are correspondingly two types of supersymmetry transformations characterized by $\gamma^+ \varepsilon_+ = 0$ and $\gamma^- \varepsilon_- = 0$ of positive and negative "transverse chirality." If $\varepsilon^{(1)}$ and $\varepsilon^{(2)}$ have the same transverse chirality then a^i can be easily seen to vanish. Both types of world sheet supersymmetries are thus necessary to ensure spacetime supersymmetry. The transformation laws for ε_+ are particularly simple:

$$\delta X^i = 0$$

$$\delta S = - 2i \sqrt{p^+} \varepsilon_+ \tag{8}$$

The action (5) will be invariant under this transformation if ε_+ is a covariantly constant spinor on the transverse space.[*] This follows since the first and third term in (5) are trivially invariant, and the second term gives a total derivative when $\mathcal{D}_+ \varepsilon_+ = 0$. The last term also vanishes because when $A = \omega$, the gauge and metric curvatures are also equal and $F^A_{ij} \gamma^i \gamma^j \varepsilon_+ = 0$ is just the integrability condition for $\mathcal{D}_j \varepsilon_+ = 0$. Under these conditions there are always ε_- supersymmetries as well since these are just the usual supersymmetries of the nonlinear σ model.

---

[*] More precisely, ε_+ is the restriction to the world sheet of a covariantly constant spinor on $\mathbb{R}^2 \times K$.

Thus we have learned that N = 1 (or greater) spacetime supersymmetry follows from the existence of a covariantly constant spinor on the internal manifold K. This restriction can be understood in terms of a group known as the holonomy group of K. If a spinor (or vector) is parallel transported around a closed loop in K, it will undergo some O(6) rotation. The group of all rotations that can be generated in this way is the holonomy group. For example, the holonomy group of flat space is trivial while the holonomy group of the standard metric connection on the six sphere is O(6). The existence of a covariantly constant spinor says that the holonomy group of K cannot be the full O(6) since there is at least one spinor that is not rotated under parallel transport. In fact it implies that the holonomy group is at most an SU(3) subgroup of O(6). (We are interested in the case where it is precisely SU(3), otherwise the manifold K must have zero Euler number which we will soon see leads to severe problems with phenomenology.) A manifold with U(3) (or smaller) holonomy is known as a complex Kähler manifold.

The requirement that K be a three complex dimensional Ricci flat Kähler manifold is quite a restrictive one. To better understand the nature of this restriction, we consider the requirements on K individually. Recall that a six dimensional real manifold can be viewed as patches of $\mathbb{R}^6$ which are "glued together" at the edges by identifying points of one patch with those of another in a smooth (i.e. C^∞) manner. Similarly, a three dimensional complex manifold can be viewed as patches of $\mathbb{C}^3$ glued together in a holomorphic i.e. complex analytic manner. Since holomorphic functions are always smooth, it is clear that every complex three manifold can be viewed as a real six manifold. However, the converse is not always the case. For example, although it is not obvious, the familiar manifolds S^6 and $S^2 \times S^4$ cannot be viewed as complex manifolds, although $S^3 \times S^3$ and $S^1 \times S^5$ can.

To understand the Kähler condition we must discuss the metric. The analog of a positive definite metric for a real manifold is a hermitian metric one can define a unique (torsion free, metric compatible) covariant derivative. Now consider a vector V such that for any function f depending only on the complex coordinates $\bar{z}^i$ and not on z^i we have

$$V^i \nabla_i f = 0 \qquad (9)$$

Such a vector is called holomorphic. In most cases, if one starts with a holomorphic vector and parallel transports it along a curve, then the vector will not remain holomorphic. However, there are special metrics for which this difficulty does not occur, namely those with U(3) holonomy. These special metrics are called Kähler. Not every complex manifold admits a Kähler metric. For example, no metric on $S^3 \times S^3$ or $S^1 \times S^5$ can be Kähler, although $S^2 \times S^2 \times S^2$ does admit Kähler metrics. One can view a Kähler manifold as the nicest type of complex manifold in that the metric structure and the complex structure are compatible in the above sense.

Having restricted ourselves from arbitrary Riemannian six manifolds to complex manifolds to Kähler manifolds, there remains one final restriction -- our space must be Ricci flat. In general there is a topological obstruction to finding a Ricci flat metric on a Kähler manifold. This is known as the first Chern class. For example, in two dimensions, the first Chern class is related to the Euler number. In this case it is well known that the only compact two manifold which admits a Ricci flat metric is the one with zero Euler number i.e. the torus. Calabi conjectured that in higher dimensions the first Chern class was the only obstruction and that if it vanishes a Ricci flat metric always exists.[13] Calabi's conjecture was proved twenty years later by Yau.[14] This result is of great significance to us for the following reason. Whereas Ricci flat Kähler metrics are in general quite complicated, Kähler manifolds with vanishing first Chern class can be constructed quite easily -- as we now indicate. We call such manifolds Calbi-Yau manifolds (if their Euler number is non-zero) although it should be noted that the theorems proved by these mathematicians are really more general than we consider here.

Perhaps the simplest example of a Calabi-Yau manifold is the following. Start with $\mathbb{C}^5$ with coordinates $(z_1, \cdots, z_5)$ and consider the subspace satisfying

$$\sum_{i=1}^{5} z_i^5 = 0 \qquad (10)$$

This is a four dimensional complex manifold. Notice that if (z_i) is a solution so is $\lambda(z_i)$. Now identify each non-zero solution (z_i) with $\lambda(z_i)$ for all complex $\lambda \neq 0$. Let Y_0 denote the resulting three complex dimensional manifold. One can show that Y_0 has anishing first Chern class and hence admits a Ricci flat Kähler metric. Although this space is simple to construct, its topology is surprisingly complicated. For example, it turns out that the Euler number χ of Y_0 is -200. Therefore every vector field on this space must vanish in at least 200 points!*

Many other examples of Ricci flat Kähler spaces can be constructed.[15,16] However, it is important to note that although there are an infinite number of complex three manifolds that admit Kähler metrics, it is likely that only a <u>finite</u> number of these admit Ricci flat Kähler metrics.[15]

Having found a vacuum configuration of the superstring theory, we can now extract the low energy effective field theory. Recall that the gauge connection was assumed to acquire an expectation value equal to that of the spin connection, which is valued in SU(3) (because K has SU(3) holonomy). For the heterotic superstring with an $E_8 \times E_8$ gauge field this breaks the gauge group down to $E_8 \times E_6$. The theory contains fermions (gluinos) in the adjoint representation. The adjoint (248) of E_8 decomposes as (1,78) ⊕ (3,27) ⊕ $(\bar{3},\overline{27})$ ⊕ (8,1) under SU(3) × E_6.

Let n_{27}^L and n_{27}^R be the number of left and right handed massless fermion multiplets in four dimensions transforming as a 27 under E_6, and let $N = n_{27}^L - n_{27}^R$. Since the difference between left and right is just a matter of definition, the number of generations is $|N|$. Because the fundamental theory contains ten dimensional Weyl fermions massless chiral fermions on M^4 are associated, in the usual way, with chiral zero modes of the Dirac operator on K. Fermions transforming in the 27 of E_6 also transform in the 3 of SU(3). Hence $N = n_{27}^L - n_{27}^R$

---

*Strictly speaking, this statement is only true for vector fields whose zeros are non-degenerate.

is just the index of the (gauge coupled) Dirac operator on K acting on spinors in the 3 representation of SU(3). Now since the SU(3) Yang-Mills potential is equal to the spin connection, this index can depend only on the geometry of K. But the index is a topological invariant. Hence, it can depend only on the topology of K. In fact it turns out that N is simply one half the Euler number χ of K. So the number of generations is $|N| = \frac{1}{2} |\chi(K)|$.

This is a solution of the long-standing chiral fermion problem in Kaluza-Klein theories.[17] To our knowledge there is no previous case of a supersymmetric Kaluza-Klein compactification that produced a chiral spectrum of fermions. In the present case, not only are chiral fermions naturally obtained, but they are in the correct representation (27) of a realistic grand unified group E_6.[18] It is remarkable that superstring theory, which was developed to solve the problem of quantum gravity, also solves the chiral fermion problem.

Returning now to the manifold Y_0, we see that this particular example would predict 100 families, which is considerably more than we need and in serious conflict with cosmology (among other things). Fortunately, there is a general procedure for constructing a new Calabi-Yau manifold from an old one such that the absolute value of the Euler number is reduced. This procedure is applicable whenever there exists a group of discrete symmetries that leave no point fixed (except, of course, for the identity which leaves every point fixed). A simple example of this is provided by S^2, which has a Z_2 symmetry that maps each point to the antipodal point. If we identify points related by this Z_2 we obtain a new manifold $(S^2/Z_2) = RP^2$. Roughly speaking RP^2 is "half as big" as S^2, and has half the Euler number. This can be easily seen by using the formula for the Euler number expressed as an integral of the curvature. In general, if the discrete group has n elements the new Euler number will be 1/n times the old one.

For the manifold Y_0, consider the group generated by

$$A: (z_1,z_2,z_3,z_4,z_5) = (z_5,z_1,z_2,z_3,z_4)$$
$$B: (z_1,z_2,z_3,z_4,z_5) = (\alpha z_1,\alpha^2 z_2,\alpha^3 z_3,\alpha^4 z_4,z_5) \tag{11}$$

where $\alpha \equiv e^{2\pi i/5}$. It is clear that A and B each take solutions of (10) into solutions and hence are symemtries of Y_0. It is also clear that $A^5 = B^5 =$ identity so that the group of symmetries generated by A and B is $Z_5 \times Z_5$. Somewhat less clear but not hard to verify is that no element of this group has fixed points except for the identity. It is this last requirement which rules out more general symmetries such as non-cyclic permutations and multiplying the z_i with powers of α other than those given by powers of B. By identifying points related by this symmetry group, we obtain a new Ricci flat Kähler manifold $Y = Y_0/Z_5 \times Z_5$ with Euler number

$$\chi(Y) = \chi(Y_0)/25 = -8 \qquad (12)$$

(Unlike RP^2, Y is still orientable.) So the superstring theory formulated on Y predicts four generations. Other Calabi-Yau manifolds have been found that predict one, two, three or four generations.[15,16]

So far we have seen that the superstring theory can produce a reasonable number of generations of quarks and leptons in the required representation of a realistic grand unified group. However there are at least two more ingredients that are essential for a realistic E_6 theory: mechanisms for breaking E_6 down to a baryon number conserving group at the GUT scale and for Weinberg-Salam symemtry breaking.[18] The Higgs fields responsible for the latter can be found in the superpartners of the quarks and leptons. The former occurs in a fashion unique to Kaluza-Klein theories, as we now explain.

Quotient spaces such as $Y_0/Z_5 \times Z_5$ or S^2/Z_2 are not simply connected. Consider for example the line on the two sphere extending from the north to south pole. On S^2/Z_2 this line becomes a non-contractible loop. However if we go twice around this loop on S^2/Z_2, it can be obtained from a closed loop on S^2 which is contractible (i.e. $\pi_1(S^2/Z_2) = Z_2$). Similarly Y has loops which are contractible after transversing five times. Now consider the expectation value:

$$U = \langle \exp(i \oint_\Gamma A_m \, dx^m) \rangle \qquad (13)$$

where A_m lies in an Abelian subgroup of E_6 and Γ is a non-contractible loop. U is then an element of E_6. In general U can acquire an expectation value even if the E_6 field strength vanishes.[19] However if we wrap five times around Γ Stokes theorem can then be used to relate the path ordered exponential to the integral of the curvature over a two surface spanning the loop. Since the E_6 field strength vanishes for vacuum configurations, this implies $U^5 = 1$.

An expectation value for U breaks the E_6 gauge symmetry. Fermion zero modes that are not neutral under U acquire masses, as do gauge fields that do not commute with U. Thus the effects of U resemble in this respect those of a Higgs field in the adjoint representation. This is precisely what is required for GUT symmetry breaking, and allows one to break E_6 down to, for example, $SU(3) \times SU(2) \times U(1)$[3].

Having obtained these initial successes, one should now proceed to compute particle lifetimes, masses and coupling constants. These quantities turn out to depend on more refined topological properties of the internal manifold K.[16,20] For the Y manifold discussed in this paper there is a serious problem with rapid proton decay.[5,21] There might be a Calabi-Yau manifold with appropriate topological properties for which such difficulties are absent. Another problem is supersymmetry breaking.[12] At present no phenomenologically viable mechanism has been found. Finally, assuming a Calabi-Yau manifold is found that has all the right properties, one would like to understand why superstring theory prefers this one over any other. Or, indeed, why is four spacetime dimensions preferred at all?

These are serious problems and certainly worthy of attention. However, it is an unprecedented and exciting state of affairs that we are judging a potential unification of quantum mechanics and gravity on the basis of low energy phenomenology.

Acknowledgments

We are grateful to E. Martinec for useful conversations. This work was supported in part by NSF Grants PHY80-19754, PHY81-07384, and PHY82-05717.

REFERENCES

1. Green, M. B. and Schwarz, H. J., Phys. Lett. 149B, 117 (1984).
2. Gross, D. J., Harvey, J. A., Martinec, E., and Rohm, R., Phys. Rev. Lett. 54, 502 (1985), and Princeton preprints.
3. Green, M. B. and Schwarz, J. H., Nucl. Phys. B181, 502 (1981); B198, 252 (1982); B198, 441 (1982); Phys. Lett. 109B, 444 (1982); Green, M. B., Schwarz, J. H., and Brink, L., Nucl. Phys. B198, 474 (1982).
4. Candelas, P., Horowitz, G. T., Strominger, A., and Witten, E., to appear in Nucl. Phys. B. (1985).
5. Witten, E., Phys. Lett. 149B, 351 (1984).
6. Scherk, J. and Schwarz, J. H., Phys. Lett. 57B, 463 (1975); Cremmer, E. and Scherk, J., Nucl. Phys. B103, 399 (1976).
7. Feynman, R. P., Proc. Chapel Hill Conference (1957); Weinberg, S., Phys. Rev. D138, 988 (1965); Deser, S., Gen. Relativity and Grav. 1, 9 (1970).
8. Polyakov, A. M., Phys. Lett. 103B, 207, 211 (1981).
9. Freedman, D. Z. and Townsend, P. K., Nucl. Phys. B177, 443 (1981); Alvarez-Gaumé, L. and Freedman, D. Z., Phys. Rev. D22, 846 (1980); Commun. Math. Phys. 80, 282 (1981); Morozov, A. Ya., Perelmov, A. M., and Shifman, M. A., ITEP preprint; Hull, C. M., unpublished; Alvarez-Gaumé, L. and Ginsparg, P., in Proceedings of the Argonne Symposium on Anomalies, Geometry and Topology.
10. This point was independently made by Friedan, D. and Shanker, S., unpublished talk given at the Aspen Summer Institute (1984). See also Witten, E., Comm. Math. Phys. 92, 455 (1984) and Lovelace, C., Phys. Lett. 135B, 75 (1984).
11. Callan, C., Friedan, D., Martinec, E., and Perry, M., Princeton preprint (1985).
12. Whether or not they remain solutions at the quantum level is of course an important and non trivial question that we do not address here. See Dine, M., Rohm, R., Seiberg, N., and Witten, E., to appear in Phys. Lett. B (1985); Dine, M. and Seiberg, N., IAS preprint (1985), and Kaplunovsky, V., to appear.

13. Calabi, E., in Algebraic Geometry and Topology: A Symposium in Honor of S. Lefschetz (Princeton University Press, 1957), p. 78.

14. Yau, S.-T., Proc. Natl. Acad. Sci. $\underline{74}$, 1798 (1977).

15. Yau, S.-T., in Proceedings of the Argonne Symposium on Anomalies, Geometry and Topology.

16. Strominger, A. and Witten, E., to appear in Comm. Math. Phys.

17. Witten, E., Nucl. Phys. $\underline{B186}$, 412 (1981). Work on this problem can be found in Chapline, G. and Slansky, R., Nucl. Phys. $\underline{B209}$, 461 (1982); Wetterich, C., Nucl. Phys. $\underline{B223}$, 109 (1983); Randjbar-Daemi, S., Salam, A., and Strathdee, S., Nucl. Phys. $\underline{B214}$, 491 (1983); Witten, E., to appear in the Proceedings of the 1983 Shelter Island II Conference (MIT Press, 1985); Olive, D. and West, P., Nucl. Phys. $\underline{B217}$, 248 (1983); Chapline, G. and Grossman, B., Phys. Lett. $\underline{125B}$, 109 (1984); Frampton, P. H. and Yamamoto, K., Phys. Rev. Lett. $\underline{125B}$, 109 (1984); Weinberg, S., Phys. Lett. $\underline{138B}$, 47 (1984); Frampton, P. H. and Kephardt, T. W., Phys. Rev. Lett. $\underline{53}$, 867 (1984); Koh, I. G. and Nishino, H., Trieste preprint ICTP/84/129.

18. Gürsey, F., Ramond, P., and Sikivie, P., Phys. Lett. $\underline{60B}$, 177 (1976); Gürsey, F. and Sikivie, P., Phys. Rev. Lett. $\underline{36}$, 775 (1976); Ramond, P., Nucl. Phys. $\underline{B110}$, 214 (1976).

19. Hosotani, Y., Phys. Lett. $\underline{129B}$, 193 (1983).

20. Witten, E., to appear in Nucl. Phys. B.

21. Dine, M., Kaplunovsky, V., Nappi, C., Mangano, M., and Seiberg, N., Princeton preprint (1985).

THE HETEROTIC STRING

Jeffrey A. Harvey[*]

Joseph Henry Laboratories
Princeton University
Princeton, New Jersey 08544

ABSTRACT

The structure of the heterotic string is reviewed
with emphasis on the differences between the
heterotic string and other superstring theories.

The last year has seen an enormous revival of interest in superstring theories as unified theories of all fundamental interactions.
This revival was triggered by the discovery of Green and Schwarz that
gauge and gravitational anomalies cancel in ten-dimensions only for
gauge group $G=SO(32)$ or $E_8 \times E_8$.[1] The anomaly cancellation for $SO(32)$
follows from the existence of a one-loop finite and anomaly-free Type I
string theory with $G=SO(32)$.[1,2] The discovery of the $E_8 \times E_8$ heterotic
string[3] and of compactifications of this theory to four dimensions with
realistic grand unified gauge groups[4] have made the goal of a unification of all forces seem tantalizingly close. In this talk I will
review the heterotic string, other aspects of superstring theory are
discussed by other speakers at this meeting.

---

[*]Research supported in part by NSF Grant PHY80-19754.

The raison d'etre of string theories is to provide a consistent finite quantum theory of gravity. It is remarkable that open string theories which do not contain gravitons were found to be inconsistent with unitarity at the one loop level unless additional closed string states (including the graviton) were added to the spectrum[5] and the dimension of space time was chosen to be 26 (for the bosonic string)[6] or 10 (for the super string)[7]. Thus in string theories one is forced to include gravitational interactions. Yang-Mills interactions were included in earlier string theories by assigning Chan-Paton factors to open string diagrams which roughly corresponds to imagining charges on the ends of open strings and assigning group theory factors as in quark diagrams.

The ad-hoc nature of the Chan-Paton rules and the simplicity of closed string theories suggest that a string theory where the charges live "inside" the closed string might give a more satisfactory way,[11] of introducing gauge interactions. In such a theory the motion of the charges on the string would be described by a set of currents j_α^a (σ,τ) where a is a group index, σ and τ are coordinates on the string world sheet and α is a world sheet index. In two dimensions it is convenient to split the current into left and right handed pieces j_L^a $(\tau+\sigma)$, j_R^a $(\tau-\sigma)$. Consider just the left-handed components for the moment. Then in two dimension the currents satisfy an algebra of the form

$$[j_L^a(u), j_L^b(u')] = if^{abc} j_L^c(u) \, \delta(u-u') + \frac{ic}{2\pi} \delta^{ab} \, \delta'(u-u') \qquad (1)$$

where f^{abc} are the structure constants of a group G, $u = \tau+\sigma$ and the coefficient c of the Scwinger term is a constant. If we have closed strings with spacelike parameter $0 < \sigma < \pi$ and periodic boundary then conditions j_L^a $(u+\pi) = j_L^a$ (u). We can expand j_L^a

$$j_L^a (u) = \frac{1}{\sqrt{\pi}} \sum_{n=-\infty}^{+\infty} j_n^a \, e^{-2inu} \qquad (2)$$

and the commutation relations then read

$$[j_n^a, j_m^b] = if^{abc} j_{n+m}^c + c \delta_n^{ab} \delta_{n+m,0} \qquad (3)$$

For the n=m=o components these are just the commutation relations of a Lie algebra. For general n and m this algebra is known in the mathematical literature as an affine Lie algebra and is denoted by $\hat{g}$ if g is the corresponding Lie algebra. Mathematicians in constructing representations of g were naturally led to the vertex operators which occur in string theory. The construction of the heterotic string is in part based on this vertex operator construction of affine Lie algebras due to Lepowsky and Wilson,[7] Frenkel and Kac,[8] and Segal.[9] The papers of Frenkel[10] and Goddard and Olive[11] describe this construction in language more familiar to physicists. Current algebras composed out of bose fields first appeared in ref. 12. The suggestion that this construction be used to generate $E_8 \times E_8$ or $SO(32)$ starting from 26 dimensions was made in ref. 13.

The upshot of this construction is that the left- or right-handed currents alone may be constructed simply out of left or right moving bosonic string coordinates X^I $(\tau \pm \sigma)$ which parametrize the maximal torus of g with I = 1... rank g. Thus in order to obtain $E_8 \times E_8$ or $SO(32)$ with rank 16 we would take say 16 left-handed coordinates $X^I(\tau + \sigma)$ I = 1...16. Since the bosonic string is only consistent in 26 dimensions we must also include 10 coordinates for spacetime X^μ (τ, σ). The spacetime coordinate must contain both left and right moving parts in order to correspond to non-compact dimensions. For consistency there must be additional right-moving degrees of freedom. In order to obtain a supersymmetic spectrum we can add right-moving spinor degrees of freedom S^a $(\tau - \sigma)$. This gives a consistent string theory since the right-moving fields are just those of the right-moving superstring in its critical dimension d=10. The physical (transverse) degrees of freedom of the heterotic string and their normal mode expansions are therefore

$$X^I (\tau+\sigma) = x^I + p^I (\tau+\sigma) + \frac{i}{2} \Sigma_n \frac{\tilde{\alpha}_n^I}{n} e^{-2in(\tau+\sigma)} \qquad I = 1...16$$

$$X^i(\tau+\sigma) = \frac{1}{2} x^i + \frac{1}{2} p^i(\tau+\sigma) + \frac{i}{2} \Sigma_n \frac{\tilde{\alpha}_n^i}{n} e^{-2in(\tau+\sigma)} \qquad i = 1...8$$

$$X^i(\tau-\sigma) = \frac{1}{2} x^i + \frac{1}{2} p^i(\tau-\sigma) + \frac{i}{2} \Sigma_n \frac{\alpha_n^i}{n} e^{-2in(\tau-\sigma)}$$

$$S^a(\tau-\sigma) = \Sigma_n S_n^a e^{-2in(\tau-\sigma)} \tag{4}$$

with commutation relations

$$[x^i, p^j] = i\delta^{ij}, \quad [\alpha_n^i, \alpha_m^j] = [\tilde{\alpha}_n^i, \tilde{\alpha}_m^j] = n \, \delta_{n+m,0} \, \delta^{ij}$$

$$[\alpha_i^n, \tilde{\alpha}_j^m] = 0 \quad \{S_n^a, S_m^b\} = \frac{1}{4}((\gamma^0+\gamma^9)(1+\gamma^{11}))^{ab} \, \delta_{n+m,0}$$

$$[x^I, p^J] = \frac{i}{2} \delta^{IJ} \tag{5}$$

The factors of 1/2 in the commutation relations of x^I and p^J arises from the fact that X^I is subject to the second class constraint $(\partial\sigma-\partial\tau) X^I = 0$.

The left-moving coordinates X^I correspond to a particular "compactification" of the left-moving part of the bosonic string on the maximal torus of $E_8 \times E_8$ or $SO(32)$. To understand what this means and why these groups are singled out, consider the compactification of both left and right moving parts on a torus of radius R. We would write down an expansion

$$X^I = x^I + p^I\tau + 2L^IR\sigma + \text{oscillators}. \tag{6}$$

Here the L^I corresponding to the winding numbers of the closed string around the torus and correspond to points on the lattice defining the torus. The p^I are the allowed momenta and just as in solid state physics must lie on the dual lattice. In the heterotic string we must choose a particular value of R in order to have physical states and since X^I depends only on $\sigma+\tau$ the expansion (5) only makes sense geometrically if the lattice is self dual. The groups $E_8 \times E_8$ and $SO(32)$

(actually Spin(32)/Z_2) are the only groups with even self-dual 16 dimensional weight lattices.

The argument that only these groups are allowed can be made more rigorous by looking at one-loop string diagrams. At one-loop level the string world sheet is a torus. For consistency string theories must be invariant under reparametrizations of the string world sheet. At the one loop level there are global reparametrizations known as modular transformations that cannot be reached continuously from the identity. Invariance under the group of modular transformations uniquely picks out $E_8 \times E_8$ or Spin(32)/Z_2.

Given the normal mode expansions (4) it is easy to construct the spectrum of the heterotic string. The mass operator is given by

$$\frac{\alpha'}{2} (\text{mass})^2 = N + (\tilde{N}-1) + \frac{1}{2} \sum_1 (p^I)^2 \qquad (7)$$

where α' is the Regge slope, N is the number operator of the right-moving superstring and $\tilde{N}$ is the number operator for the left-moving bosonic string. The -1 corresponds to the usual zero point energy of the bosonic string which usually gives rise to a tachyon. However there is also a constraint which follows from demanding that there be no distinguished point on the string which gives

$$N = \tilde{N} - 1 + \frac{1}{2} \sum_1 (p^I)^2 \qquad (8)$$

This projects out the tachyon and gives as massless particles precisely those of N=1 supergravity coupled to N=1 super Yang-Mills in 10 dimensions with gauge group $E_8 \times E_8$ or Spin(32)/Z_2.

To see this we construct the massless states which are just those with N=0 and $\tilde{N}$=1 or $\sum_I (p^I)^2$=2. The states of the heterotic string are direct products of Fock space states $|>_R \times |>_L$ of the right-moving superstring and the left-moving bosonic string. The right-moving ground state is annihilated by α_n^i, S_n^a (n>0) and N. It forms an irreducible representation of the Clifford algebra of the zero mode oscillators S_0^a and consists of 8 bosonic states $|a>_R$. The left-handed ground state is annihilated by $\tilde{\alpha}_{n_j}$, $\tilde{\alpha}_n^I$ $(n>0)$, $\tilde{N}$ and p^I. The physical massless states are of the form $\tilde{\alpha}_{-1}^j|o\rangle_L$, $\tilde{\alpha}_{-1}^I|o\rangle_L$, or $|p^I$, $(p^I)^2 = \rangle_L$. The states $|i$ or $a\rangle_R \times \tilde{\alpha}_{-1}^j|o\rangle_L$ form an irreducible N=1 D=10 supergravity multiplet while the 16.16 states $|i$ or $a\rangle_R \times \tilde{\alpha}_{-1}^I|o\rangle_L$ together with the 16.480 states $|i$ or $a\rangle_R \times |p^I$, $(p^I)^2 = 2\rangle_L$ form an N=1 D=10 super Yang-Mills multiplet with $G = E_8 \times E_8$ or $Spin(32)/Z_2$.

In order to describe the coupling of gauge fields to the heterotic sring it is necessary to express the currents described earlier in terms of the coordinates X^I. The diagonal currents are given by a form familiar from bosonization of fermions in two dimensions

$$H^I \equiv p^I \equiv \frac{\partial X^I}{\partial(\tau+\sigma)} \tag{9}$$

The off diagonal currents can be constructed as

$$E_p = :e^{2ip^I X^I}: C(p) \tag{10}$$

where p is a root vector, the double dots indicate normal ordering and C(p) is essentially a Klein factor which corrects the sign in certain commutation relations. By expanding H^I and E_p in normal modes it is easy to see that they do satisfy the current algebra (3) reexpressed in the Cartan-Weyl basis.

The bosonic form of the current algebra described here bears a close resemblance to known string theories. However there are at least two other equivalent descriptions, one in terms of fermion fields and another in terms of strings moving in group manifolds with Wess-Zumino term. This latter description does not seem very practical for actual calculations however. In either the bosonic or fermionic forms it is straightforward to cosntruct the vertex operators which describe the interactions of various string states and to use these to calculate various simple scattering process at both the tree and one-loop level.[3] The heterotic string is one-loop finite and presumably anomaly free since it is finite and modular invariant. The heterotic string incorporates gauge interactions in a fundamentally different way than open string theories. In its $E_8 \times E_8$ form it also appears very promising from a phenomenological point of view. E_8 is the last in the E-series of grand unified groups $E_4 = SU(5)$, $E_5 = SO(10)$, E_6, E_7, E_8. The structure of the heterotic string is such that it rather naturally gives rise to $N=1$ supersymmetric E_6 grand unified theories upon compactification to four dimensions.[4] Hopefully our understanding of string theory will soon improve to the point that we can determine whether or not the heterotic string actually describes nature.

REFERENCES

1. M.B. Green and J.H. Schwarz, Phys. Lett. 149B (1984) 117.

2. M.B. Green and J.H. Schwarz, Phys. Lett. 151B (1985) 21.

3. D. Gross, J. Harvey, E. Martinec and R. Rohm, Phys. Rev. Lett. 54 (1985) 502, Nucl. Phys. B256 (1985) 253, and Princeton preprint.

4. P. Candelas, G. Horowitz, A. Strominger and E. Witten, Nucl. Phys. B256 (1985) 46.

5. D.J. Gross, A. Neveu, J. Scherk and J.H. Schwarz, Phys. Rev. D2 (1970) 697.

6. C. Lovelace, Phys. Lett. 34B (1971) 500.

7. J. Lepowsky and R.L. Wilson, Comm. Math. Phys. 62 (1978) 43.

8. I.B. Frenkel and V.G. Kac, Invent. Math. 62 (1980) 23.

9. G. Segal, Comm. Math. Phys. 80 (1981) 301.

10. I.B. Frenkel, J. Funct. Analysis 44 (1981) 259.

11. P. Goddard and D. Olive, Proc. of a Conference on Vertex Operators in Mathematics and Physics, ed. J. Lepowsky (Springer-Verlag, 1984).

12. M.B. Halpern, Phys. Rev. D12 (1975) 1684.

DOE/ER/40033B/101

Report Number RU/85/136

THE ONE-LOOP EFFECTIVE LAGRANGIAN OF THE SUPERSTRING *

Burt A. Ovrut[†]

Department of Physics

The Rockefeller University

New York, New York 10021

ABSTRACT

Using the low energy effective Lagrangian of the superstring, we discuss the breaking of supersymmetry and the internal gauge group. We calculate the quadratically divergent part of the one-loop potential energy and examine the stability of the vacuum.

Of paramount importance is the form of the four-dimensional, low energy $N=1$ supergravity allowed by the superstring. This problem has been approached from two points of view:

a) A truncation of the $d=10$, modified Chapline-Manton $N=1$ supergravity theory[1] to four spacetime dimensions.[2]

---

*Invited talk at the Meeting of the Division of Particles and Fields, American Physical Society, Eugene, Oregon, August, 1985.

[†] On leave of absence from the University of Pennsylvania, Philadelphia, Pa. 19104.

b) A new, d=4, "off-shell" irreducible, N=1 supergravity multiplet[3] which incorporates the symmetries and particle spectrum of the superstring[4].

Although these theories do not fully reflect the complexity of compactification on Calabi-Yau manifolds, they provide a reasonable starting point. After elimination of all auxiliary fields the Lagrangian can be written in the form of a minimal supergravity theory[5]. Remarkably, however, the Kahler potential of the theory is fixed by the superstring. The Lagrangian contains in addition to the graviton and gravitino the following fields:

1) A vector superfield transforming as a 248 of E_8 (or one of its subgroups, if E_8 is broken).

2) Vector fields $A_\mu{}^a$ and their fermionic superpartners λ^a that arise from the decomposition of the 78 vector superfield of E_6 under $\mathcal{G}$, the low energy gauge group.

3) Two complex scalar gauge singlet fields S and T and their fermionic partners χ_S and χ_T.

4) Complex scalars C^A and their fermionic partners $\chi_C{}^A$ that arise from the decomposition of the 27 chiral superfields of E_6 under $\mathcal{G}$. They include quarks, leptons, and Higgs bosons.

The Kahler potential and matter superpotential are given by[2,4]

$$K = M^2 \left[- \ln\left(\frac{S + \bar{S}}{M}\right) - 3 \ln \left\{ \left(\frac{T + \bar{T}}{M}\right) - 2 \frac{\bar{C}_A C^A}{M^2} \right\} \right] \qquad (1)$$

$$W_C = d_{ABC} \, C^A C^B C^C \qquad (2)$$

where $M = M_{Planck}/\sqrt{8\pi}$ and d_{ABC} are proportional to the invariant tensors of $\mathcal{G}$. We have suppressed family indices for the matter fields in K and W_C. It was shown in Ref. 2 that the kinetic energy term for the $\mathcal{G} \times E_8$ supergauge fields is of the form

$$\delta_{ab} \, Re \, \frac{S}{M} \, W^{a\alpha} W_\alpha^b \qquad (3)$$

where $W^{a\alpha}$ is the superfield covariant curl of the gauge vector supermultiplet. The E_8 sector of the theory can become strongly interacting for energies below mass scale, Λ_{cond}. For such energies it is possible, by using (3), to integrate out the E_8 supergauge fields and obtain an effective superpotential for S. It is found to be [6,7]

$$W_S = M^3 b \; e^{-3S/2b_o M} \tag{4}$$

where b is an arbitrary constant, and b_o is the coefficient of the E_8 gauge coupling beta founction (or the beta function of some subgroup of E_8, if E_8 is spontaneously broken during compactification). Typically, Λ_{cond} is much larger than 10^{10} GeV. In this paper we want to discuss energy scales of $O(10^{10}$ GeV) or smaller. Therefore, we omit all E_8 supergauge fields from the effective Lagrangian, and add W_s to the superpotential. In addition, Dine et al.[6], point out that the antisymmetric tensor field strength can acquire a v.e.v. This leads to an additional term in the superpotential of the form $M^3 a$, where a is a constant. The effective superpotential is then

$$W = M^3 (a + b \; e^{-3S/2b_o M}) + d_{ABC} \; C^A C^B C^C \tag{5}$$

Given K and W one can calculate the effective low energy Lagrangian consistent with the superstring. More generally, one could take

$$W = W_S(S) + d_{ABC} \; C^A C^B C^C \tag{6}$$

where W_S is not restricted to be of the form in (5). We assume that W_S is such that the associated potential energy has a minimum for finite S, and that W_S is non-vanishing at this minimum (e.g. a $\neq$ 0 in Eqn.(5)). Supersymmetry breaking is then introduced through the W_S sector of the theory. In this paper we examine how the super-symmetry breaking is communicated to the matter fields C^A and whether the vacuum is stable when radiative corrections are included. First, we calculate the tree-level potential energy. If we define

$$\hat{S} = S + \bar{S} \ , \quad \hat{T} = T + \bar{T}$$

$$\hat{Q} = \hat{T} - 2 \ \frac{\bar{C}_A C^A}{M^2} \tag{7}$$

then the Kahler metric is given by

$$g = \begin{pmatrix} \dfrac{M^2}{\hat{S}^2} & 0 & 0 \\[2ex] \hline 0 & \dfrac{3M^2}{\hat{Q}^2} & -6MC^B \\[2ex] 0 & \dfrac{-6M\bar{C}_A}{\hat{Q}^2} & \dfrac{6M}{\hat{Q}} \ (\delta_A^{\ B} + \dfrac{2\bar{C}_A C^B}{M\hat{Q}}) \end{pmatrix} \tag{8}$$

The inverse Kahler metric is

$$g^{-1} = \begin{pmatrix} \dfrac{\hat{S}^2}{M^2} & 0 & 0 \\[2ex] \hline 0 & \dfrac{\hat{Q}\hat{T}}{3M^2} & \dfrac{\hat{Q}C^B}{3M^2} \\[2ex] 0 & \dfrac{\hat{Q}\bar{C}_A}{3M^2} & \dfrac{\hat{Q}}{6M} \delta_A^{\ B} \end{pmatrix} \tag{9}$$

Substituting these results into the bosonic part of the interaction Lagrangian given by Cremmer et al.[5]

$$\mathcal{L}_{B,INT}/e \quad = \quad -M^4 \ e^G \ \{ G_i' \ (G''^{-1})^i_{\ j*} G'^{j*} - 3 \}$$

$$- \ \frac{M^4}{2} \ g^2 \ \mathrm{Re} \ f_{ab}^{-1} \ (G_i'(T^a)^i_{\ j} \ z^j)(G_k'(T^b)^k_{\ \ell} \ z^\ell) \tag{10}$$

where, from (3),

$$f_{ab} = \delta_{ab} \ \frac{S}{M} \tag{11}$$

and T^a are the Lie algebra generators of $\mathcal{G}$, we find that the tree level potential energy is given by

$$V/e = \frac{M^4}{S Q^3} \left\{ \frac{\hat{S}^2}{M^2} \left| D_S W \right|^2 + \frac{\hat{Q}}{6M} \left| \frac{\partial W}{\partial C^A} \right|^2 \right\} + \frac{18 g^2 M^3}{\hat{Q}^2} \, \text{Re} \, \frac{1}{S} \, (\bar{C}_A (T^a)^A{}_B C^B)^2$$

$$(12)$$

Note that V is non-negative. This is due to an exact cancellation of the -3 part of the first term in Eqn. (10). This cancellation is easily traced to two properties of the effective theory induced from the superstring:

1) the $-3 \, \ell n \, (\hat{Q}/M)$ term in the Kahler potential K, and

2) the T-independence of the superpotential W.

The tree-level vacuum state is determined by minimizing V and is given by

$$\langle D_S W \rangle = 0$$

$$\langle C^A \rangle = 0 \tag{13}$$

The v.e.v. $\langle T \rangle$ is undetermined at tree level. Note that V = 0 at the minimum. The gravitino mass is given by

$$m_{3/2} = \frac{1}{(\langle \hat{S} \rangle \, \langle \hat{T} \rangle^3)^{1/2}} \, \langle W_S \rangle \tag{14}$$

Since W is non-vanishing at the minimum, it follows that $m_{3/2}$ is non-zero and that supersymmetry is spontaneously broken. The value of $m_{3/2}$ is not fixed at tree level since $\langle T \rangle$ is undetermined. We conclude that the low energy effective theory has, at tree-level, a "stable" vacuum state ($\langle S \rangle$, $\langle T \rangle$, and $\langle C^A \rangle$ are finite) with a "naturally" vanishing cosmological constant (no fine tuning of parameters). Furthermore, this vacuum spontaneously breaks supersymmetry but does not set the scale of this breaking ($m_{3/2}$ is non-zero but arbitrary). It is important to note that if any one of the terms $D_S W$, $\partial W/\partial C^A$, or $\bar{C}_A (T^a)^A{}_B C^B$ did not vanish then $\langle \hat{S} \rangle \langle \hat{Q} \rangle \to \infty$, at least one of $\langle S \rangle$, $\langle T \rangle$, or $\langle C^A \rangle$ would be infinite, and the vacuum "unstable".

Having found the vacuum state, we expand S, T, and C^A around this vacuum $(z^i = \langle z^i \rangle + z'^i)$ and calculate the tree level Lagrangian up to operators of dimension four. The Lagrangian can then be written as

$$\mathcal{L} = \mathcal{L}_{KE(1)} + \mathcal{L}_{KE(2)} + \mathcal{L}_{SUSY} + \mathcal{L}_{SB} + \mathcal{L}_H \qquad (15)$$

where

$$\mathcal{L}_{KE(1)} = -D^\mu \bar{C}_A D_\mu C^A - \bar{\chi}_C{}^A{}_L \not{D} \chi_{\overline{CAR}}$$

$$- \frac{1}{4} F^a_{\mu\nu} F^{a\mu\nu} - \bar{\lambda}^a_L \not{D} \lambda^a_R \qquad (16)$$

$$\mathcal{L}_{KE(2)} = -\partial^\mu \bar{S} \partial_\mu S - \partial^\mu \bar{T} \partial_\mu T - \bar{\chi}_{SL} \not{D} \chi_{\overline{SR}} \qquad (17)$$

$$\mathcal{L}_{SUSY} = -\left(\left| \frac{\partial W_C}{\partial C^A} \right|^2 + \frac{g^2}{2} (\bar{C}_A (T^a)^A{}_B C^B)^2 \right)$$

$$\qquad (18)$$

$$- \left(\frac{1}{2} \frac{\partial^2 W_C}{\partial C^B \partial C^A} \chi_C{}^A{}_L \chi_C{}^B{}_L + i\sqrt{2}\, g(T^a)^A{}_B C^B \bar{\lambda}^a_R \chi_{\overline{CAR}} + h.c. \right)$$

$$\mathcal{L}_{SB} = -\left(\left(\frac{m_{3/2}}{M} \right) \left[(1+\delta^2)|S|^2 - \delta(S^2 + \bar{S}^2) \right] \bar{C}_A C^A \right.$$

$$- \left(\frac{m_{3/2}}{M} \right) \left[(S - \delta\bar{S}) W_C + h.c. \right]$$

$$\left. - \left(\left(\frac{m_{3/2}}{M} \right) \left[\frac{1}{4} (S - \delta\bar{S}) \bar{\lambda}^a_L \lambda^a_L + h.c. \right] \right) \right. \qquad (19)$$

and

$$\delta = -\frac{\langle \hat{S} \rangle^2}{\langle W \rangle} \langle \frac{\partial^2 W}{\partial S^2} \rangle \qquad (20)$$

$\mathcal{L}_H$ is rather complicated. The physically interesting part of $\mathcal{L}_H$ is

$$\mathcal{L}_H = -m^2_{3/2} \left[(1+\delta^2)|S|^2 - \delta(S^2 + \bar{S}^2) \right] + m_{3/2} \left[\frac{\delta}{2} \bar{\chi}_{SL} \chi_{SL} + h.c. \right] + \ldots \qquad (21)$$

The section of the Lagrangian involving normal matter only, $\mathcal{L}_{KE(1)} + \mathcal{L}_{SUSY}$, can be shown to be globally supersymmetric. That is,

$$\mathcal{L}_{KE(1)} + \mathcal{L}_{SUSY} = \frac{1}{4} \left[\int d^2\theta \; W^{a\alpha} W^a_{\alpha} + h.c. \right]$$
$$+ \int d^4\theta \; C^\dagger_A \; e^{2g(T^a V^a)^A}_{B} C^B + \left[\int d^2\theta \; W_C(C^A) + h.c. \right] \tag{22}$$

where C^A is the chiral superfield associated with $(C^A, \chi_C^{\;A})$, V^a is the vector superfield associated with $(A_\mu^{\;a}, \lambda^a)$ and gauge group $\mathcal{G}$, and $W^{a\alpha}$ is the superfield covariant curl of V^a. The potential energy for this part of the Lagrangian is

$$V = \left| \frac{\partial W_C}{\partial C^A} \right|^2 + \frac{g^2}{2} \; (\bar{C}_A (T^a)^A_{\;B} C^B)^2 \tag{23}$$

which is non-negative and quartic in the fields. All mass terms and cubic couplings vanish. Hence, $\mathcal{G}$ is not spontaneously broken at tree-level. One might hope that radiative corrections would spontaneously break $\mathcal{G}$ through a Coleman-Weinberg type mechanism.[8] However, since $\mathcal{L}_{KE(1)} + \mathcal{L}_{SUSY}$ is globally supersymmetric, all radiative corrections to V from this part of the Lagrangian vanish. The Coleman-Weinberg mechanism might be induced by supersymmetry breaking terms in the Lagrangian. The explicit supersymmetry breaking terms involving normal matter are contained in $\mathcal{L}_{SB}$. They differ from the supersymmetry breaking terms in flat Kahler potential theories[9] in two fundamental ways.

1) The terms in $\mathcal{L}_{SB}$ all couple normal matter to "hidden" sector field S.

2) The terms in $\mathcal{L}_{SB}$ are "hard" operators of dimension four. Note from $\mathcal{L}_H$ that S has mass of $O(m_{3/2})$. It follows that S is not necessarily heavy with respect to the electroweak scale. Therefore S should not be decoupled from the effective Lagrangian. Unfortunately, the inclusion of S in the effective Lagrangian leads to

another problem. Since the operators in $\mathcal{L}_{SB}$ are "hard", they induce quadratically and logarithmically divergent terms in the radiatively corrected potential energy for which there are no counterterms. It follows that the divergent terms must be cut off at a scale Λ, and the result added to the effective potential. Assuming $\Lambda \gg m_{3/2}$, we expect the quadratic terms to dominate over the logarithmic one.

We now calculate, to the one-loop level, the quadratically divergent contributions to the C,T field sector of the potential energy which are proportional to $m_{3/2}$ (there are quadratically divergent terms proportional to V but these are uninteresting). First, we must determine the part of the Lagrangian relevant to this specific calculation. The necessary terms may be obtained from the Lagrangian of Cremmer et al.[5] They are (note we are not limiting ourselves to terms with dimension $d \leq 4$)

$$
\begin{aligned}
\mathcal{L}_{KIN}/e = & - (\tfrac{M}{S})^2 \, \partial_\mu \bar{S} \, \partial^\mu S \\
& - \Big\{ (\tfrac{M}{S})^2 \, \bar{\chi}_{SL} \not{\partial} \chi_{SR} + 3(\tfrac{M}{Q})^2 \, \bar{\chi}_{TL} \not{\partial} \chi_{TR} - \frac{6MC^A}{\hat{Q}^2} \, \chi_{TL} \not{\partial} \bar{\chi}_{CAR} \\
& - \frac{6M\bar{C}_A}{\hat{Q}^2} \, \bar{\chi}_C{}^A{}_L \not{\partial} \chi_{TR} + \frac{12\bar{C}_B C^A}{\hat{Q}^2} \, \bar{\chi}_C{}^B{}_L \not{\partial} \bar{\chi}_{CAR} + h.c. \Big\}
\end{aligned}
\tag{24}
$$

and

$$
\begin{aligned}
\mathcal{L}_{INT}/e = & \frac{-M^2}{\hat{S}\,\hat{Q}^3} \left| W - \hat{S} \frac{\partial W}{\partial S} \right|^2 + \frac{M^2}{(\hat{S}\hat{Q}^3)^{1/2}} \Big\{ - \frac{\partial^2 W}{\partial S^2} \, \bar{\chi}_{SL} \chi_{SL} \\
& - \frac{6W}{\hat{Q}^2} \, \bar{\chi}_{TL} \chi_{TL} + 12 \frac{\bar{C}_A W}{MQ^2} \big[\bar{\chi}_{TL} \chi_C{}^A{}_L + \bar{\chi}_C{}^A{}_L \chi_{TL} \big] \\
& - 24\bar{C}_A \bar{C}_B \frac{W}{M^2 \hat{Q}^2} \, \bar{\chi}_C{}^A{}_L \chi_C{}^B{}_L + h.c. \Big\}
\end{aligned}
\tag{25}
$$

Now let $S = \langle S \rangle + S'$ and expand up to terms quadratic in S' (since we are interested in S' loops).

Define

$$\hat{m}_{3/2} = \frac{1}{(\langle \hat{S} \rangle \, \hat{Q}^3)^{1/2}} \; (\langle W_S \rangle + W_C) \qquad (26)$$

The gravitino mass term is given by

$$\hat{m}_{3/2} [\bar{\psi}_{\mu R} \sigma^{\mu\nu} \psi_{\nu R} - \bar{\psi}_R \cdot \gamma \; \eta_L] + \text{h.c.} \qquad (27)$$

where η_L is the Goldstone fermion defined by

$$\eta_L = \sqrt{3} \; \chi_{TL} \qquad (28)$$

Combining the $\bar{\chi}_T \chi_T$ mass term with Eqn. (27) we find

$$\hat{m}_{3/2} [\bar{\psi}_{\mu R} \sigma^{\mu\nu} \psi_{\nu R} - \bar{\psi}_R \cdot \gamma \; \eta_L - \frac{2}{3} \bar{\eta}_L \eta_L] + \text{h.c.} \qquad (29)$$

The remaining part of the Lagrangian can be written as

$$\mathcal{L} = \mathcal{L}_{KE} + \mathcal{L}_{SUSY} + \mathcal{L}_{SB} \qquad (30)$$

where

$$\mathcal{L}_{KE} = - \partial_\mu \bar{S} \, \partial^\mu S - \bar{\chi}_{SL} \slashed{\partial} \chi_{SR} \qquad (31)$$

$$\mathcal{L}_{SUSY} = - \hat{m}_{3/2}^2 \; \delta^2 |S|^2 + \frac{\hat{m}_{3/2}}{2} \, \delta [\bar{\chi}_{SL} \chi_{SL} + \text{h.c.}] \qquad (32)$$

$$\mathcal{L}_{SB} = - |\hat{m}_{3/2}|^2 \; |S|^2 \qquad (33)$$

and δ is defined in Eqn. (20) The first two terms, $\mathcal{L}_{KE} + \mathcal{L}_{SUSY}$, are globally supersymmetric. That is

$$\mathcal{L}_{KE} + \mathcal{L}_{SUSY} = \int d^4\theta \; S^\dagger S + [\int d^2\theta (\frac{-\hat{m}_{3/2}}{2} \; \delta S^2) + \text{h.c.}] \qquad (34)$$

where S is the chiral superfield associated with (S, χ_S). It follows that the interactions in $\mathcal{L}_{SUSY}$ do not contribute to quadratic divergences. The quadratically divergent contribution to the C,T potential energy due to $\mathcal{L}_{SB}$ is easily computed. It is found to be

$$+ \quad \frac{\hat{m}^2_{3/2} \Lambda^2}{(4\pi)^2} \tag{35}$$

One can choose a gauge in which the $\psi_R \cdot \gamma \eta_L$ term in Eqn.(29) vanishes. In this gauge, the quadratically divergent contribution to the C,T potential energy due to η_L is

$$- \quad 4 \; \frac{\hat{m}^2_{3/2} \Lambda^2}{(4\pi)^2} \tag{36}$$

Finally, we must add the gravitino contribution, which, following Barbieri and Ceccotti[10], is given by

$$+ \quad 2 \; \frac{\hat{m}^2_{3/2} \; \Lambda^2}{(4\pi)^2} \tag{37}$$

Adding these three terms together, we conclude that the quadratically divergent contribution to the C,T sector of the potential energy, at the one-loop level, is

$$V_{1\text{-loop}, \Lambda^2} = - \frac{\hat{m}^2_{3/2} \Lambda^2}{(4\pi)^2} \tag{38}$$

where $\hat{m}_{3/2}$ is a function of C^A and T defined in Eqn. (26). Also we note that not only does the gaugino mass vanish to tree level, but there is also no quadratically divergent contribution.

Let us now examine the stability of the vacuum. Adding together Eqns. (23) and (38), we have

$$V = \left|\frac{\partial W}{\partial C^A}\right|^2 + \frac{g^2}{2}(\bar{C}_A(T^a)^A{}_B C^B)^2$$

$$- \frac{\hat{m}_{3/2}^2 \Lambda^2}{(4\pi)^2} \tag{39}$$

which is minimized for $\langle C \rangle = 0$, and $\langle T \rangle = 0$. This implies that $m_{3/2} \rightarrow \infty$, and that the vacuum is unstable. This instability arises because, at tree level, we had a flat potential for T, but the radiative corrections introduce a potential $\sim -1/(\hat{T})^3$ for which there is no counterterm. We conclude that using the tree level Lagrangian to calculate one-loop corrections to the action is (unlike flat Kahler metric theories) too naive. A more ambitious approach, using the full effect of the superstring at the one-loop level, is apparently necessary and is currently under study.

ACKNOWLEDGEMENTS

All work discussed in this talk was done in collaboration with J. Breit and G. Segre and appeared as a University of Pennsylvania Preprint (1985). Topics of a similar nature have been discussed by M. Mangano in a Princeton Preprint (1985), and by P. Binetruy and M. Gaillard in an LBL Preprint (1985). Models of the type discussed in this talk were first introduced by J. Ellis, A. Lahanas, D. Nanopoulos, and K. Tamvakis in Phys. Lett. 134B(1984) 429, within the context of N=1 supergravity theories.

REFERENCES

1) A.H. Chemseddine, Nucl. Phys. B185 (1981) 403. E. Bergshoff,
   M. De Roo, B. De Wit and P. Van Nieuwenhuizen, Nucl. Phys. B195
   (192) 97.
   G.F. Chapline and N.S. Manton, Phys. Lett. 120B (1983) 105.

2) E. Witten "Dimensional Reduction of Superstring Models"
   Princeton preprint 1985.

3) G. Girardi, R. Grimm, M. Muller and J. Wess, Z. Phys. C26
   (1984) 427; Phys. Lett. 147B (1984) 81.

4) W. Lang, J. Louis and B.A. Ovrut, Pennsylvania preprint
   UPR-0280T (1985) and Karlsruhe preprint KA Thep 85-2 (1985).

5) E. Cremmer, B. Julia, J. Scherk, S. Ferrara, L. Girardello and
   P. van Nieuwenhuizen, Nucl. Phys. B147 (1979) 105.

6) M. Dine, R. Rohm, N. Seiberg and E. Witten, "Gluino Condensation
   in Superstring Models", I.A.S. preprint 1985.

7) Gluino condensation in the hidden sector is also discussed in
   J.P. Derendinger, L.E. Ibanez and H.P. Nilles, CERN preprint
   TH 4123/85 (1985).

8) S. Coleman and E. Weinberg, Phys. Rev. D7 (1973) 1888.

9) L. Hall, J. Lykken, and S. Weinberg, Phys. Rev. D27 (1983) 2369;
   H. Nilles, M. Srednicki, and D. Wyler, Phys. Lett. 120B (1983)
   346.

10) R. Barbieri and S. Cecotti, Z. Phys. C17 (1983) 183.

MAKING SENSE OF ANOMALOUS GAUGE THEORIES*

R. Jackiw

Center for Theoretical Physics
Laboratory for Nuclear Science
and Department of Physics
Massachusetts Institute of Technology
Cambridge, Massachusetts 02139 U.S.A.

1. DESCRIPTION OF AN ANOMALOUS GAUGE THEORY

Because quantum mechanics is not a freestanding, autonomous physical theory, relying as it does on antecedents in classical mechanics, the procedure for building a quantal model with consistent dynamics is to construct first the model so that it possesses consistent and non-trivial dynamics within classical physics, and then quantize. For a theory involving gauge fields $F_{\mu\nu}$ with matter sources giving rise to a current J_ν,

$$D^\mu F_{\mu\nu} = J_\nu \tag{1}$$

we require that the classical dynamics governing the sources be such that the matter current is covariantly conserved,

$$D^\mu J_\mu = 0 \tag{2}$$

since this is implied by the gauge field equation (1).

$$D^\nu J_\nu = D^\nu D^\mu F_{\mu\nu} = 0 \ . \tag{3}$$

Current conservation is assured when the matter dynamics admits a group of continuous symmetry transformations, and J_μ is the associated Noether current.

However, when the sources include chiral fermions, the well-known *anomaly phenomena* typically destroys conservation of the current because of *quantum mechanical violation of the classical symmetry*. While there are many ways of deriving this effect, ranging from the original perturbative calculations to

* This work is supported in part by funds provided by the United States Department of Energy (D.O.E.) under contract number DE-AC02-76ER03069.

topological/cohomological analyses that are popular these days, the *physical* origin of quantum mechanical/anomalous symmetry breaking remains obscure. The most we can say is that chiral symmetry of fermions in interaction with gauge fields is broken when the negative energy Dirac sea is filled to define the second quantized vacuum.

Failure to conserve the matter current also signals loss of gauge invariance in the fermion determinant $D(A)$.

$$D(A) \equiv \det(\partial\!\!\!/ + A\!\!\!/) \quad . \tag{4}$$

Since the expectation value of the current is proportional to the variation of $D(A)$,

$$\langle J^\mu \rangle = \frac{\delta}{\delta A_\mu} i \, ln \, D(A) \quad . \tag{5}$$

The following equations show that $D(A)$ is not gauge invariant when J^μ is not conserved.

$$D(A + D\theta) - D(A) \propto \int tr(D_\mu \theta) \left(\frac{\delta}{\delta A_\mu} D(A) \right) = iD(A) \int tr \, \theta D^\mu \langle J_\mu \rangle \quad . \tag{6}$$

Lack of manifest consistency in the field equation (3), and loss of gauge invariance (6) raises questions about the gauge field theory, which is called *anomalous*, when it is afflicted with these problems. Specifically, one worries about the renormalizability and unitarity of the model.[1]

2. CONVENTIONAL APPROACH TO ANOMALOUS GAUGE THEORIES

One proposal for dealing with an anomalous gauge theory is to adjust fermion content so the anomaly vanishes.[1] For example, in a 4-dimensional theory the anomalous divergence is

$$(D_\mu J^\mu)_a = \mp \frac{i}{24\pi^2} \, tr \, T^a \epsilon^{\mu\alpha\beta\gamma} \partial_\mu \left(A_\alpha \partial_\beta A_\gamma + \frac{1}{2} A_\alpha A_\beta A_\gamma \right) \tag{7}$$

where the matrix T^a represents the action of the Lie algebra on the fermions, and the sign depends on fermion chirality. The trace is proportional to the invariant symmetric tensor $d_{abc} \equiv \frac{1}{2} \, tr \, T^a \{T^b, T^c\}$, which reflects the group factor associated with the basic triangle graph, with carries the essence of the axial anomaly.[2] Hence, if the group or the representation has vanishing d_{abc}, there is no anomaly.

Insisting that anomalies be absent by the above mechanism has been a successful idea. It leads to the prediction that the number of quarks equals that of the leptons. The discovery of new particles has kept pace with this constraint, which is our only theoretical basis for the equality. Also, demanding that unified theories be anomaly-free has placed strong constraints on acceptable models.

For string theories the limitation is so restrictive that the dramatic reduction of possibilities has persuaded some that the ultimate theory describing Nature is to be found among string theories.

Owing to the importance of and current interest in anomaly-free chiral theories, it is interesting to inquire whether the above is the only resolution of the anomaly problem, or whether other, more subtle, mechanisms are available. This question also arises when one appreciates the mathematically coherent frame that has recently been given to the anomaly phenomena, which I shall now describe.

3. MATHEMATICALLY COHERENT FRAME FOR ANOMALIES

The non-invariance of the fermion determinant against infinitesimal gauge transformations, exhibited in (6), may also be presented for finite transformations by the formula

$$D(A^g) = e^{i2\pi\omega_1(A;g)} D(A)$$
$$A^g_\mu \equiv g^{-1} A_\mu g + g^{-1}\partial_\mu g$$

(8)

Performing a second gauge transformation, and using the composition law $g_1 g_2 = g_{12}$ shows that ω_1 satisfies the 1-cocycle condition.

$$\omega_1(A^{g_1}; g_2) - \omega_1(A; g_{12}) + \omega_1(A; g_1) = 0 \;(\text{mod integer})$$

(9)

Moreover, when we consider the unitary operators $U(g)$ which implement static gauge transformations on the quantum field theoretic Hilbert space, we find that in an anomalous gauge theory their composition law does not follow the group composition law, rather an additional phase occurs.[3]

$$U(g_1)U(g_2) = e^{i2\pi\omega_2(A;g_1,g_2)} U(g_{12}) \;.$$

(10)

Again, a condition on the phase emerges when another gauge transformation is performed: ω_2 must satisfy the 2-cocycle condition.

$$\omega_2(\mathbf{A}^{g_1}; g_2, g_3) - \omega_2(\mathbf{A}; g_{12}, g_3) + \omega_2(\mathbf{A}; g_1, g_{23})$$
$$- \omega_2(\mathbf{A}; g_1, g_2) = 0 \;(\text{mod integer})$$

(11)

Equivalent to the above is the statement that the algebra of infinitesimal generators G_a,

$$U(g) = e^{i \int dr \theta^a G_a}$$
$$g = e^{\theta^a T^a}$$

(12)

does not follow the group's Lie algebra, rather there is an extension.

$$i[G_a(\mathbf{r}), \; G_b(\mathbf{r}')] = f_{abc} G_c(\mathbf{r})\delta(\mathbf{r} - \mathbf{r}') + S_{ab}(\mathbf{A}; \mathbf{r}, \mathbf{r}')$$

(13)

Here G_a is composed of two pieces: $(D_i F^{0i})_a$ generates transformations on the gauge field degrees of freedom and J_a^0 does the same job on the matter degrees of freedom, and includes fermion bilinears in an anomalous gauge theory. The extension S_{ab} is the infinitesimal part of ω_2 [with the infinitesimal gauge functions θ_1 and θ_2 "stripped off"].

The cocyles may be determined by *a priori* cohomological reasoning, and the cohomological result agrees with explicit perturbative calculations: The anomalous divergence (7) agrees with the infinitesimal portion of ω_1, and similarly the computed anomaly in the commutator agrees with the infinitesimal portion of the cohomological prediction for ω_2.[4]

4. COPING WITH AN ANOMALOUS GAUGE THEORY

From the above, we know that chiral anomalies arise because the gauge symmetry is realized with a projective representation rather than faithfully. This in itself is not remarkable in a quantum theory – for example, Galilean symmetry is represented projectively in non-relativistic quantum mechanics. However, the difficulty arises when Gauss' law is imposed, which in the classical theory requires setting G_a to zero, since G_a is the time component of the field equation. In the quantum theory, $G_a = 0$ cannot be satisfied "strongly" as an operator identity because G_a does not commute with the basic dynamical variables; indeed, it generates static gauge transformations on them. If the algebra of the constraints closes, as it does in the absence of an anomaly [*i.e.*, in (13) S vanishes], one recognizes them as "first class" and imposes them as a restriction on physical states: G_a should annihilate physical states and $U(g)$ should leave them invariant. This is the usual quantization procedure for non-anomalous gauge theories and insures gauge invariance. The projective presentation (10) and the corresponding extension in the algebra of constraints as in (13) elevates them to "second class" and prevents proceeding in the above manner. But it is not self-evident that no other way of imposing the constraints is open.

Although no progress can be reported on this subject, it has been conjectured that a successful gauge invariant quantization [as yet unknown] might give rise to the Wess-Zumino term [1-cocycle] in the effective action of the fermion-gauge field system.[5] Then the gauge non-invariance of the fermion determinant would be compensated by the gauge non-invariant Wess-Zumino term. [The scalar field occuring in the Wess-Zumino term is now viewed as a (functional) dummy integration variable, needed to give a local expression to a non-local gauge functional. In this way, it is analogous to a Faddeev-Popov ghost.]

There are some indications that something like this could happen.

(i) When the fermion determinant in a 3-dimensional gauge theory is evaluated with a gauge invariant but parity violating regulation [like Pauli-Villars] one finds in the effective action the Chern-Simons term, which is the odd-dimensional analog of the Wess-Zumino term.[6]

(ii) Consider a 4-dimensional gauge field theory with paired chiral fermions so that it is anomaly-free and also with Higgs fields to give the fermions a mass. By sending the parameters of the Higgs Lagrangian to infinity, one may decouple one of the two paired chiral fermions. The resultant theory remains anomaly-free and gauge invariant, because in the limit a Wess-Zumino term survives.[7] However, renormalizability is lost, at least perturbatively.[8]

One may also turn to 2-dimensional models and see whether anything useful can be learned in that simple setting. Of course, issues of renormalizability will not be faced because the models are finite. However, the following analysis of a massless Weyl fermion, interacting with an Abelian gauge field – the chiral Schwinger model – shows that the anomaly need not spoil unitarity of the theory, although gauge invariance is lost.[9]

The Lagrangian is

$$
\begin{aligned}
\mathcal{L} &= -\frac{1}{4} F^{\mu\nu} F_{\mu\nu} + \overline{\psi}\left(i\not{\partial} - e\frac{1 - i\gamma_5}{2}\not{A}\right)\psi \\
&= -\frac{1}{4} F^{\mu\nu} F_{\mu\nu} + \overline{\psi}_+(i\not{\partial} - e\not{A})\psi_+ + \overline{\psi}_-\not{\partial}\psi_-
\end{aligned}
\tag{14}
$$

The fermion determinant may be easily computed; it leads to the action

$$
I_F = -\frac{e^2}{8\pi} \int A_\mu (g^{\mu\alpha} + \epsilon^{\mu\alpha}) \frac{\partial_\alpha \partial_\beta}{\Box} (g^{\beta\nu} - \epsilon^{\beta\nu}) A_\nu + \frac{e^2}{8\pi} a \int A^2
\tag{15}
$$

where a is an undetermined constant governing with arbitrary strength a local polynomial in the potential. Such undetermined polynomials reflect the infinities of fermion determinants and connot be fixed without invoking some additional principle. Usually gauge invariance selects a unique value [this is what happens in the conventional Schwinger model]. Here, however, no value for a makes I_F gauge invariant – the theory is anomalous. But rather than rejecting it outright, let us continue the analysis to see whether sensible physics can emerge, at least with some values for a.

Since $-\frac{1}{4} \int d^2x \, F^{\mu\nu} F_{\mu\nu} + I_F$ is quadratic, the model is non-interacting, and the particle spectrum is readily deduced. One finds a sensible unitary theory with positive energy, provided $a > 1$. Then the particle spectrum consists of massive vector mesons with mass $\infty > m > \frac{e}{\sqrt{\pi}}$. Also, there are massless excitations, which appear to be deconfined fermion anti-fermion pairs.

A similar conclusion can be reached for the non-Abelian, 2-dimensional chiral theory, although the analysis is less explicit since the effective action is no longer quadratic.[10]

5. SUMMARY

In an anomaly-free gauge theory, current conservation and consistency of the field equation is assured by a symmetry of the matter dynamics. In an anomalous gauge theory, the current is not conserved and its divergence is given in terms of gauge fields. But it may be possible to maintain consistency if the gauge field dynamics has the further consequence that the gauge field expression, which gives the anomalous divergence, vanishes. Evidently, this happens in the 2-dimensional examples analyzed by us.[9,10]

We see that not only are these simple 2-dimensional anomalous models consistent and unitary, but also they exhibit a faint image of the physics in the electro-weak sector: chiral fermions interacting with gauge fields giving rise to massive vector mesons and deconfined fermions. Gauge invariance is broken, not explicitly or spontaneously, but quantum mechanically. If these ideas can be realized in a realistic 4-dimensional model, we would find that anomalies resolve yet another outstanding physics puzzle: how to break a gauge symmetry without invoking the unaesthetic Higgs mechanism or the inconclusive dynamical symmetry breaking scenario.

I have been asked to what purpose does one abandom the conventional approach to anomalous theories[1] since, it provides stringent, therefore useful, constraints on unified models and seems experimentally validated. My answer is that in addition to unification we need to understand gauge symmetry breaking and the origin of mass. Neither grand unified field theories nor superstrings have anything conclusive to say. The anomaly phenomenon seems to touch on this subject and it is interesting to explore its ramifications.

6. REFERENCES

1. D. Gross and R. Jackiw, *Phys. Rev. D* **6**, 477 (1972); C. Bouchiat, J. Iliopoulos and Ph. Meyer, *Phys. Lett.* **38B**, 519 (1972).

2. I. Gerstein and R. Jackiw, *Phys. Rev.* **181**, 1955 (1969).

3. L. Faddeev, *Phys. Lett.* **145B**, 81 (1984); L. Faddeev and S. Shatashvili, *Teor. Mat. Fiz.* **60**, 206 (1984) [*Theor. Math. Phys.* **60**, 770 (1984)]; J. Mickelsson, *Comm. Math. Phys.* **97**, 361 (1985); I. Singer, (to be published).

4. The anomaly in commutators of fermion bilinears was given by R. Jackiw and K. Johnson, *Phys. Rev.* **182**, 1459 (1969). The anomaly in the commutator of the full generator was calculated by L. Faddeev, Nuffield workshop, 1985; I. Frenkel and I. Singer (unpublished); S.- G. Jo, *Nucl. Phys. B.* (in press), *Phys. Lett. B* (in press); A. Niemi and G. Semenoff, *Phys. Rev. Lett.* **55**, 927 (1985). The last two analyses are especially interesting since they relate the anomalous commutator to Berry's phase in the quantum adiabatic theorem.

5. Faddeev, ref. 4.

6. N. Redlich, *Phys. Rev. Lett.* **52**, 18 (1984), *Phys. Rev. D* **29**, 2366 (1984); L. Alverez-Gaumé and E. Witten, *Nucl. Phys.* **B234**, 269 (1984).

7. E. D'Hoker and E. Farhi, in *Anomalies, Geometry and Topology*, A. White, ed., World Scientific (in press).

8. E. D'Hoker and E. Farhi (in preparation).

9. R. Jackiw and R. Rajaraman, *Phys. Rev. Lett.* **54**, 1219 (1985), (E) **54**, 2060 (1985); R. Rajaraman, *Phys. Lett.* **154B**, 305 (1985).

10. R. Rajaraman, CERN preprint TH.4227/85 (1985) (unpublished).

TORSION AND GEOMETROSTASIS IN COVARIANT SUPERSTRINGS

Cosmas Zachos

HEP Division, Argonne National Laboratory, Argonne, IL 60439

The covariant action for freely propagating heterotic superstrings consists of a metric and a torsion term with a special relative strength. It is shown that the strength for which torsion flattens the underlying 10-dimensional superspace geometry is precisely that which yields free oscillators on the light cone. This is in complete analogy with the geometrostasis of two-dimensional σ-models with Wess-Zumino interactions.

I am reporting on some observations made in collaboration with T. Curtright and L. Mezincescu[1] concerning the geometrical structure of covariant free superstrings. To this end, I shall exploit the striking analogy of this formulation to the geometrostasis[2] of two-dimensional σ-models with torsion[3], i.e. a Wess-Zumino interaction[4].

In σ-models, when the strength of the torsion term in the action is such that the underlying group geometry is parallelized, the renormalization of this geometry ceases and the system reduces to a free theory[2]. The covariantly formulated heterotic string is a σ-model whose fiber (target) coset is 10-dimensional superspace. It appears like an interacting theory, but it reduces to the light-cone oscillators comprising the free superstring when the strength of the torsion term relative to the metric term is such that the underlying superspace is flattened--and only then. This specific connection to the σ-model should be contradistinguished from treatments which associate effective descriptions of strings propagating in non-trivial backgrounds with generalized σ-models whose cosets are not, in any case, superspace[5]. Recently, however, Witten has formulated the problem super-covariantly in curved superspace[6].

I briefly review some relevant background on plain bosonic chiral σ-models with torsion[2]. The base space is flat two-dimensional space-time, while the group fiber is the coset $G_L \times G_R / G_V$. The projective coordinates of this group manifold are the "pions" ϕ^a, i.e. the Nambu-Goldstone bosons which shift under the nonlinear axial transformations. Exponentiating $i\phi^a T^a$ yields the standard group elements U, and hence $U^{-1} dU$ are elements in the algebra. All group manifold tensor functions of ϕ^a are expressed in terms of flat tangent space indices $(i,j,k,\ldots)$, converted from curved group indices $(a,b,c,\ldots)$ through the vielbein one-form and the counterpart currents defined in the fibration,

$$v^j \equiv \frac{i}{2} \, \text{Tr} \; T^j U^{-1} dU \equiv J_\mu{}^j dx^\mu \, , \qquad\qquad J_\mu{}^j \equiv \frac{i}{2} \, \text{Tr} \; T^j U^{-1} \partial_\mu U \; . \quad (1)$$

The conventional σ-model is then simply the flat contraction of space-time and group indices in the current bilinear, amounting to a metric term, the "Sugawara action":

$$I_1 = \frac{1}{2\lambda^2} \int d^2x \; \eta^{\mu\nu} \delta_{jk} J_\mu{}^j J_\nu{}^k \, , \quad (2)$$

where λ is the dimensionless coupling $(=1/f_\pi)$.

The properties of this model are modified dramatically if a torsion (Wess–Zumino) interaction term I_2 is introduced. The group manifold considered is typified by a fundamental invariant three-form constructed from the structure constants of the algebra:

$$\Omega_3 \equiv \eta \, f_{ijk} v^i v^j v^k = -\frac{\eta}{2} \, \text{Tr} \; U^{-1} dU \; U^{-1} dU \; U^{-1} dU \equiv S_{abc} d\phi^a d\phi^b d\phi^c \, , \quad (3)$$

where η is an arbitrary numerical strength. This form is associated to Cartan's torsion two-form, and the corresponding rank-three anti-symmetric tensor $S^a{}_{bc}$ is the torsion completing the Christoffel symbols $\Gamma^a{}_{bc}$ inside covariant derivatives on the group manifold. Ω_3 is closed, $d\Omega_3 = 0$, but not exact, i.e. it may only be derived from a potential two-form Ω_2 locally on the group manifold: $\Omega_3 = d\Omega_2$.

Using this, the torsion interaction term

$$I_2 = \frac{1}{2\lambda^2} \frac{2\eta}{3} \int d^3x \; \epsilon^{\lambda\mu\nu} f_{ijk} J_\lambda{}^i J_\mu{}^j J_\nu{}^k \quad (4)$$

may be converted to a two-dimensional interaction through integration by parts. However, the coefficient η is constrained by the cohomology of the manifold[4] to equal $N\lambda^2/c$, where N is an arbitrary integer and c is a normalization characteristic of the group; e.g. $c = 2\pi$ for the conventional hyperspherical model based on $SU(2)_L \times SU(2)_R/SU(2)_V$.

At $\eta = \pm 1$, it may be seen by use of the Maurer–Cartan equation $dv^i = -f^i{}_{jk} v^j v^k$ that $\Omega_3 = -\eta v^i dv^i$ reduces to the parallelizing torsion, i.e. the one dictating the vanishing of the generalized curvature two-form $\mathscr{R}^{ij}$. (For $\eta = -1$ this is obvious, since the spin connection vanishes, while the curvature is definable as its covariant curl.) When the geometry which drives the renormalization of these models is flattened, renormalization comes to a standstill, as the infrared geometrostatic fixed point is reached[2,3,4]. This fact was checked to two loops in ref. [2] and and should not be unexpected[4]; it was finally proved to all orders in perturbation theory in ref. [7]. The only bosonic geometries with this property are actually the group manifolds discussed above. At the geometrostatic point, $\eta = \pm 1$, these theories may also be shown to be free[4].

This result persists upon N=1 supersymmetrization, i.e. when Majorana spinor superpartners of the ϕ^a's are introduced, which are not projective coordinates themselves, however[3].

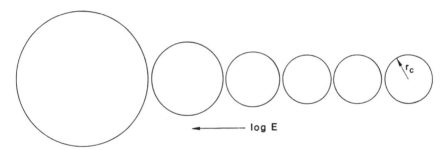

Figure. *The renormalization of the bosonic hypersphere σ-model with torsion: At high energies the coupling λ decreases indefinitely, i.e. the radius of the group hypersphere $r = f_\pi = 1/\lambda$ increases with energy. At low energies however, in lieu of infrared slavery, λ increases to a critical value $1/r_c = \sqrt{2}\,\pi/N$ and stops evolving with decreasing energy, i.e. achieves conformal invariance.*

Formulating free superstrings in a covariant language, so that their spacetime symmetries and supersymmetry are manifest, results in lagrangians which appear at first to describe nonlinear interactions. Covariant superstrings[8] may be regarded as *bosonic* two-dimensional σ-models[9] in curved spacetime (the world-sheet), with constraints which arise from the extra variations of the world-sheet metric, and from the appropriate boundary conditions. I will discuss the 10-dimensional heterotic superstring.[10] The fiber coset of this string is 10-dimensional N=1 superspace. In contrast to ordinary spacetime which may be flattened without torsion, flat superspace needs torsion, and hence, unlike the case of bosonic strings, a torsion term is necessary in formulating strings covariantly.

The projective coordinates corresponding to the nonlinearly realized 10-translation and the 32(16)-component Majorana-Weyl supersymmetry are X^μ and θ^m respectively, where μ and m are flat tangent space vector and (Majorana-Weyl) spinor indices, respectively. In addition, the fermionic representation of the heterotic string also utilizes 32 ancillary Majorana-Weyl world-sheet spinors ψ^j,[10] which are scalars under the action of the above superspace coset. They are thus to be contrasted to the above projective coordinates (X^μ, θ^m) which are world-sheet scalars but, of course, transform nontrivially in 10-dimensional superspace.

The supervielbein one-form and the counterpart currents are

$$\omega^M \equiv (\omega^\mu, \omega^m) = (dX^\mu - i\bar\theta\Gamma^\mu d\theta, \ d\theta^m) \equiv \omega_\alpha{}^M d\xi^\alpha \ , \tag{5}$$

where ξ^α denote the two world-sheet coordinates. The metric term of the action (which $alone$ contains the bosonic string as $\theta \to 0$) is

$$I_1 = \frac{1}{2\pi} \int d^2\xi \ \sqrt{-g} \ g^{\alpha\beta} \eta_{\mu\nu} \omega_\alpha{}^\mu \omega_\beta{}^\nu \ . \tag{6}$$

This is manifestly world-sheet reparameterization invariant, and globally Lorentz and supersymmetry invariant in ten dimensions.

The fundamental invariant three-form constructed from the supersymmetry structure constant is

$$\Omega_3 = -i\eta(C\Gamma_\mu)_{mn} \omega^\mu \omega^m \omega^n = -i\eta dX^\mu (d\bar\theta\Gamma_\mu d\theta) \ . \tag{7}$$

Because of incomplete antisymmetry of the above structure constant, ω^μ appears not thrice but once in the above expression, which will result in an extra factor of 3 in the normalization of the action analog of Eq. (4).

Clearly, Ω_3 is not only closed, but also exact: $\Omega_3 = d\Omega_2$, $\Omega_2 = i\eta \ dX^\mu(\bar\theta\Gamma_\mu d\theta)$.

The resulting torsion term for the heterotic superstring is thus

$$I_2 = -\frac{2i\eta}{2\pi} \int d^3\xi \ \varepsilon^{\alpha\beta\gamma} \partial_\alpha X^\mu (\partial_\beta\bar\theta\Gamma_\mu\partial_\gamma\theta) = \frac{2i\eta}{2\pi} \int d^2\xi \ \varepsilon^{\alpha\beta} \partial_\alpha X^\mu (\bar\theta\Gamma_\mu\partial_\beta\theta) \ , \tag{8}$$

which contains the string tension factor just like the metric term.

This term has no analog for the purely bosonic string in flat space. Note that because of the lack of a nontrivial topology in the superspace considered, there are no cohomological reasons to constrain the coefficient η to be a priori quantized.[11]

The parallelizing torsion in superspace is $(d\omega^\mu, 0)$: for that value the spin connection and hence the curvature vanishes. The fundamental three-form corresponds to this value for $\eta = 1$, since $\Omega_3 = \eta\, \omega_\mu d\omega^\mu$. For both σ-models and superstrings, the torsion and its normalization may be identified from equations of motion of $I_1 + I_2$. The heterotic superstring is not chirally symmetric, and thus the value $\eta = -1$ amounts to reversing the sign of the Γ_μ matrices and reversing the chiralities of the two sectors of the string discussed below to yield and equivalent mirror-image theory—actually this theory conforms to the sense chosen in ref. [10].

The geometrical correspondences between σ-models and the covariant superstring covered so far are collected below:

CHIRAL σ-MODEL WITH TORSION	COVARIANT N=1 HETEROTIC SUPERSTRING
Base space: 2-dim. spacetime	2-dim. curved world-sheet
Coset: $G_L \times G_R/G_V$	N=1 SUSY & Poincaré / SO(9,1)
Projective coordinates: $\phi^i \equiv V_a^{\ i}\phi^a$	(X^μ, θ^m)
Vielbein form: $V_a^{\ i}d\phi^a = J_\mu^{\ i}dX^\mu$	$\omega^\Gamma \equiv (dX^\mu - i\,\bar\theta\Gamma^\mu d\theta, d\theta^m) \equiv \omega_\alpha^{\ \Gamma}d\xi^\alpha$
Maurer-Cartan Eq. $dV^i = -f^i_{jk}V^jV^k$	$d\omega^\Gamma = \left(-i(\Gamma^\mu)_{mn}\omega^m\omega^n,\ 0\right)$
Metric term: $$I_1 = \frac{1}{2\lambda^2}\int d^2x\ \eta^{\mu\nu}\delta_{jk}J_\mu^{\ j}J_\nu^{\ k}$$	$$I_1 = \frac{1}{2\pi}\int d^2\xi\ \sqrt{-g}\ g^{\alpha\beta}\eta_{\mu\nu}\omega_\alpha^{\ \mu}\omega_\beta^{\ \nu}$$
3-form-torsion: $\Omega_3 = \eta\, f_{ijk}V^iV^jV^k$ $d\Omega_3 = 0$ $\Omega_3 \sim d\Omega_2$ locally	$\Omega_3 = -i\eta(C\Gamma_\mu)_{mn}\omega^\mu\omega^m\omega^n$ $d\Omega_3 = 0$ $\Omega_3 = d\Omega_2$
Torsion term: $$I_2 = \frac{1}{2\lambda^2}\frac{2\eta}{3}\int d^3x\ \varepsilon^{\lambda\mu\nu}f_{ijk}J_\lambda^{\ i}J_\mu^{\ j}J_\nu^{\ k}$$	$$I_2 = \frac{-2i\eta}{2\pi}\int d^3\xi\ \varepsilon^{\alpha\beta\gamma}\partial_\alpha X^\mu \partial_\beta\bar\theta\Gamma_\mu\partial_\gamma\theta$$

Eqs. of motion:

$$\partial_\mu J^{\mu i} - \eta f^{i}_{jk} J^{j}_{\mu} J^{k}_{\nu} \epsilon^{\mu\nu} = 0$$

At $\eta = \pm 1$, $\Omega_3 = \mp v^j dv^j$ and $\mathscr{R}^{ij} = 0$
free theory

$$\partial^{\alpha} \omega_{\alpha}{}^{\mu} + i \eta \bar{\omega}_{\alpha} \Gamma^{\mu} \omega_{\beta} \epsilon^{\alpha\beta} = 0$$

At $\eta = 1$, $\Omega_3 = \omega_\mu d\omega^\mu$ and $\mathscr{R}^{MN} = 0$
free string

The central observation in this discussion is that it is only for the special strength $\eta = 1$ selected by the geometric criterion of parallelizability that the string action $I_1 + I_2$ possesses a local fermionic invariance: this invariance is essential for its reduction to free oscillators on the light-cone , i.e. the free superstring, to be detailed below. Naturally, one expects that this is not a mere coincidence, but no complete causal connection is available so far.

The following full covariant formulation of the heterotic superstring, in addition to I_1 and I_2 with $\eta = 1$, also contains the reparameterization invariant kinetic term for the ancillary, non-supersymmetric fermions:

$$I = \frac{1}{2\pi} \int d^2 \xi \{ \sqrt{-g} \, g^{\alpha\beta} \omega_{\mu\alpha} \omega^{\mu}{}_{\beta} + 2i \epsilon^{\alpha\beta} \partial_\alpha X^{\mu} \bar{\theta} \Gamma_\mu \partial_\beta \theta + i \bar{\Psi}^j \gamma_\alpha \sqrt{-g} \, P^{\alpha\beta}_{+} \partial_\beta \Psi^j \} \, , \quad (9)$$

where $\gamma_\alpha = e_\alpha{}^a \gamma_a$ are world-sheet Dirac matrices defined from tangent space ones through the world-sheet zweibein $e_\alpha{}^a$, γ_P is the pseudo-scalar matrix, and $P^{\alpha\beta}_{\pm} \equiv (g^{\alpha\beta} \pm \frac{\epsilon^{\alpha\beta}}{\sqrt{-g}})/2$ is a projection operator which enforces the Weyl chirality on Ψ:

$$P^{\alpha\beta}_{+} \gamma_\alpha \partial_\beta \Psi = \gamma^\alpha \partial_\alpha \frac{(1 + \gamma_P)}{2} \Psi \qquad (10)$$

As a result, the Ψ's describe only 16 degrees of freedom. I will conclude by sketching how this action merely describes a free string.

As already remarked, the action (9) possesses the following local fermionic symmetry[8], characterized by a 10-dimensional spinor, world-sheet vector parameter $\kappa^{\alpha m}$, with $P_- \kappa = \kappa$:

$$\delta\theta = i\,\Gamma_\mu\omega_\alpha{}^\mu\kappa^\alpha \qquad\qquad \delta X^\mu = i\,\bar{\theta}\Gamma^\mu\delta\theta$$

$$\delta e_a^\alpha = -4P_-^{\alpha\beta}\partial_\beta\bar{\theta}e_{a\gamma}P_-^{\gamma\delta}\kappa_\delta \qquad\qquad \delta\Psi^j = 0 \ . \tag{11}$$

This, along with the other symmetries of the string, may be used to prune out the superfluous degrees of freedom in the equations of motion. The light cone is defined by $X^\pm \equiv \dfrac{X^0 \pm X^9}{\sqrt{2}}$ and $\Gamma^\pm = \dfrac{\Gamma^0 \pm \Gamma^9}{\sqrt{2}}$. Recalling that $\Gamma^+\Gamma^+ = 0$, the invariance (11) permits rotating half of the components of θ away (a total of 8, corresponding to $\Gamma^-\chi$), so as to gauge fix on the light cone frame: $\Gamma^+\theta = 0$. Reparameterization invariance is then used to go to the conformal gauge, $\sqrt{-g}\,g^{\alpha\beta} = \eta^{\alpha\beta}$; and conformal invariance to fix to the evolution time, $X^+ = p^+\tau$. Lastly, consistently to the above, a two-dimensional tangent space rotation fixes the zweibein to $(e^0{}_\tau - e^0{}_\sigma) + (e^1{}_\sigma - e^1{}_\tau) = 2$.

The equations of motion of (12) then collapse to free equations derivable from the light cone action

$$I = \frac{1}{2\pi}\int d^2\xi\{-\partial_\alpha X^k\partial^\alpha X^k + \frac{i}{\sqrt{2}}\bar{\theta}\Gamma^-(\partial_\sigma - \partial_\tau)\theta + \frac{i}{\sqrt{2}}\bar{\Psi}^j\gamma^+(\partial_\sigma + \partial_\tau)\Psi^j\} \tag{12}$$

where the transverse coordinates X^k take 8 values. This yields the hybrid spectrum of the heterotic string. The left-moving sector contains 8 bosonic components and 8 fermionic ones (θ) and is supersymmetric; the right-moving sector contains 8 bosonic components and 16 fermionic ones (Ψ) without supersymmetry[12]—upon bosonization of the latter, they all add up to 24, equal to the transverse dimensions of the bosonic string. To recapitulate, for the special value $\eta = 1$ a local supersymmetry emerges so as to reduce the generalized superspace σ-model to the free heterotic string.

A reasonable question to ask at this point is whether properly defined quantum theories with $\eta \neq 1$ are consistent, and if so, whether they are related to interacting strings. Furthermore, in that case, would the renormalization group push them to the $\eta = 1$ limit in the infrared, analogously to σ-models?

Extension of this geometric connection to the Green-Schwarz superstring is not straightforward. This is because the fundamental closed form Ω_3 breaks the $O(2)$ symmetry which connects the two fermionic coordinates with each other, in the 10-dimensional $N = 2$ superspace on which they are defined. On the other hand, the parallelizing torsion does not[1]. Aspects of this feature may be understood in terms of the recent curved superspace formulation of the IIB superstring.[13]

REFERENCES

1. T. Curtright, L. Mezincescu, and C. Zachos, Phys. Lett. 161B, in press.
2. E. Braaten, T. Curtright, and C. Zachos, Nucl. Phys. B260, (1985) 630.
3. T. Curtright and C. Zachos, Phys. Rev. Lett. 53 (1984) 1799; J. Gates, C. Hull, and M. Roček, Nucl. Phys. B248 (1984) 157.
4. E. Witten, Commun. Math. Phys. 92 (1984) 455; A. Polyakov and P. Wiegmann, Phys. Lett. 131B (1983) 121; 141B (1984) 223; O. Alvarez, Commun. Math. Phys. 100 (1985) 279.
5. J. Scherk and J. Schwarz, Phys. Lett. 52B (1974) 347; E. Fradkin and A. Tseytlin, Phys. Lett. 158B (1985) 316; P. Candelas et al., Nucl. Phys. B288 (1985) 46; A. Sen, Fermilab preprints, Pub-85/77-T, May 1985; 81-T, June 1985; C. Callan et al., Princeton preprint, June 1985; S. Jain, R. Shankar, and S. Wadia, Tata preprint TIFR-TH-85-3; D. Nemeschansky and S. Yankielowicz, Phys. Rev. Lett. 54 (1985) 620; E. Bergshoeff et al., Trieste preprint IC/85/51; R. Nepomechie, Washington preprint 40048-20-P5, July 1985.
6. E. Witten, Princeton preprint 85-0458; for connection to parallelism, see P. Yasskin and J. Isenberg, Gen. Rel. Grav. 14 (1982) 621; further see E. Fradkin and A. Tseytlin, Lebedev preprint N150, 1985.
7. S. Mukhi, Tata Institute preprints TIFR-TH-85-12;13, May 1985.
8. M. Green and J. Schwarz, Phys. Lett. 136B (1984) 367; Nucl. Phys. B243 (1984) 285; M. Green, *Symposium on Anomalies, Geometry, Topology*, edited by W. Bardeen and A. White, (World Scientific, 1985).
9. M. Henneaux and L. Mezincescu, Phys. Lett. 152B (1985) 340.
10. D. Gross et al., Nucl. Phys. B256 (1985) 253.
11. For a detailed discussion, see J. Rabin, Chicago-EFI Preprint, September 1985.
12. Nevertheless, the free right-handed fermion lagrangian is a realization of spontaneously broken world-sheet N=32/2 supersymmetry (T. Curtright, to be published). The reason is that, in two dimensions, the interaction terms of the chiral Volkov-Akulov lagrangian for nonlinearly realized supersymmetry vanish.
13. M. Grisaru et al., Cambridge preprint 85-0603.

AMBITWISTORS (AND STRINGS?)

James Isenberg
Department of Mathematics, University of Oregon
Eugene, Oregon 97403
U.S.A.

Philip Yasskin
Department of Mathematics, Texas A & M University
College Station, Texas 77843
U.S.A.

ABSTRACT

The ambitwistor approach to the analysis of field theories is an outgrowth of the twistor programme of Penrose. We review some of the ideas and successes of the approach, and speculate on how ambitwistors might be useful in studying superstring theory.

In a certain sense, the goals of superstring theory and of the ambitwistor program are quite similar. In both cases, one hopes to build a theory of the fundamental interactions of physics using mathematics which involves embeddings of 1-dimensional objects in generalized spacetimes. It may be that this is as far as the relationship goes. On the other hand, it is possible that the tie-in between the two approaches will become a strong one. Anticipating this possibility, we wish to present here a very brief sketch of some of the ideas and methods of the ambitwistor program to date. Towards the end of this sketch, we will discuss tentative ideas for using ambitwistor techniques in string theory.

The ambitwistor approach is an outgrowth of the twistor programme of Penrose.[1] There are two themes of the twistor programme: (1) Focus on conformally invariant geometric structures in spacetime (e.g., null paths and null planes); and (2) Use holomorphic (i.e., complex analytic) analysis techniques. Thus the workhorse of the twistor programme, projective twistor space ($\mathbb{PT}$), corresponds to the set of (anti-self-dual) null planes in complex Minkowski spacetime ($\mathbb{CM}$), and perhaps the most successful application of twistors has been the Atiyah-Ward-Penrose correspondence which relates self-dual Yang-Mills fields on $\mathbb{CM}$ to certain holomorphic vector bundles over $\mathbb{PT}$.

In using ambitwistor spaces[2] rather than twistor spaces, one remains faithful to the two themes--conformally invariant structure and holomorphic analysis. However, the ambitwistor approach allows one to obtain general (non-self-dual) solutions of interesting field equations as well as the self-dual (and anti-self-dual) solutions obtained using the twistor spaces themselves. To see why this happens as well as to illustrate the techniques, we shall now elaborate a bit.

We first discuss the spaces. Twistor space itself is $\mathbb{T} = \mathbb{C}^4$, with a specified action of each of the groups SL(4,$\mathbb{C}$) and SU(2,2) on $\mathbb{T}$. SL(4,$\mathbb{C}$) is the 4-fold cover of the conformal group of (compactified) complex Minkowski spacetime $\mathbb{CM}$, while SU(2,2) is the 4-fold cover of the conformal group of (compactified) real Minkowski spacetime $\mathbb{M}$ (sitting inside $\mathbb{CM}$). Based on $\mathbb{T}$, one defines[t]

Projective Twistor Space:

$$\mathbb{PT} = \{\text{twistors, modulo complex conformal scale factors}\} \quad (1)$$
$$\simeq \mathbb{P}^3(\mathbb{C})$$

Dual Twistor Space:

$$\mathbb{T}^* = \{\text{linear maps } W: \mathbb{T} \to \mathbb{C}: Z \mapsto \langle W|Z\rangle\} \quad (2)$$
$$\simeq \mathbb{C}^4$$

Projective Dual Twistor Space:

$$\mathbb{PT}^* = \{\text{dual twistors, modulo complex conformal scale factors}\} \quad (3)$$
$$\simeq \mathbb{P}^3(\mathbb{C})$$

---

[t]These spaces, as well as others which play an important role in the twistor programme, are all "flag manifolds" of $\mathbb{T} = \mathbb{C}^4$. See Wells, ref [1].

Ambitwistor Space:

$$\mathbf{A} = \{(Z,W) \in \mathbb{T} \times \mathbb{T}^* \text{ such that } \langle W|Z\rangle = 0\}$$
$$\simeq 7 \text{ dimensional quadric in } \mathbb{C}^4 \times \mathbb{C}^4 \tag{4}$$

Projective Ambitwistor Space

$$\mathbb{P}\mathbf{A} = \{([Z],[W]) \in \mathbb{P}\mathbb{T} \times \mathbb{P}\mathbb{T}^* \text{ such that } \langle[W]|[Z]\rangle = 0\}$$
$$\simeq 5 \text{ dimensional quadric in } \mathbb{P}^3(\mathbb{C}) \times \mathbb{P}^3(\mathbb{C}) \tag{5}$$

One of the keys to the utility of these various spaces in describing fields in spacetime is the correspondence between geometric objects in $\mathbb{C}\mathbf{M}$ and geometric objects in these spaces. These results are easily derived algebraically[3] from a certain spinorial equation—it appears in the literature in the form $i x^{AA'} \pi_{A'} = \omega^A$ —relating points $x^{AA'}$ in $\mathbb{C}\mathbf{M}$ and twistors $Z = \begin{matrix} \pi_{A'} \\ \omega^A \end{matrix}$. We sketch the important ones in the following table.

$\mathbb{C}\mathbf{M}$	$\mathbb{P}\mathbb{T}$	$\mathbb{P}\mathbb{T}^*$	$\mathbb{P}\mathbf{A}$
point	projective line $(\mathbb{P}^1(\mathbb{C}))$	projective line $((\mathbb{P}^1(\mathbb{C}))$	projective line $(\mathbb{P}^1(\mathbb{C}) \times \mathbb{P}^1(\mathbb{C}))$
null geodesic line	–	–	point
self-dual null plane ("β-plane")	projective plane $(\mathbb{P}^2(\mathbb{C}))$	point	projective plane $(\mathbb{P}^2(\mathbb{C}))$
anti-self-dual null plane ("α-plane")	point	projective plane $(\mathbb{P}^2(\mathbb{C}))$	projective plane $(\mathbb{P}^2(\mathbb{C}))$

In this table, we find the key result that

$$\mathbb{P}\mathbf{A} \approx \{\text{null geodesics in } \mathbb{C}\mathbf{M}\}$$

[Note that for convenience, here and throughout this paper, we ignore the subtleties regarding points and lines "at ∞."[1] It should be pointed out, however, that for most applications (such as for the correspondences described below) one works not with $PT \simeq P^3(\mathbb{C})$ but rather with $P\tilde{T} \simeq P^3(\mathbb{C}) - P^1(\mathbb{C})$].

Perhaps the most important success thus far of the twistor programme (as well as of its ambitwistor offshoot) is the treatment of Yang-Mills fields. One obtains the following 1-1 correspondences between Yang-Mills fields on (subsets of) $\mathbb{CM}$ and holomorphic structures on (subsets of) PT and PA[4],[2].

$$\left\{ \begin{array}{c} \text{Self-Dual} \\ \text{Yang-Mills Solutions} \\ \text{on } \mathbb{CM} \end{array} \right\} \leftrightarrow \left\{ \begin{array}{c} \text{Vector}^\dagger \text{ Bundles} \\ \text{over } PT \\ \text{(trivial on all } P^1(\mathbb{C}) \text{ subspaces)} \end{array} \right\} \qquad (6)$$

$$\left\{ \begin{array}{c} \text{Anti-Self-Dual} \\ \text{Yang-Mills Solutions} \\ \text{on } \mathbb{CM} \end{array} \right\} \leftrightarrow \left\{ \begin{array}{c} \text{Vector Bundles} \\ \text{over } PT^* \\ \text{(trivial on all } P^1(\mathbb{C}) \text{ subspaces)} \end{array} \right\} \qquad (7)$$

$$\left\{ \begin{array}{c} \text{All} \\ \text{Connections} \\ \text{on } \mathbb{CM} \end{array} \right\} \longleftrightarrow \left\{ \begin{array}{c} \text{Vector Bundles} \\ \text{over } PA \\ \text{(trivial on all } P^1(\mathbb{C}) \times P^1(\mathbb{C}) \\ \text{subspaces)} \end{array} \right\} \qquad (8)$$

$$\left\{ \begin{array}{c} \text{All} \\ \text{Yang-Mills Solutions} \\ \text{on } \mathbb{CM} \end{array} \right\} \leftrightarrow \left\{ \begin{array}{c} \text{Vector Bundles} \\ \text{over } PA \\ \text{(trivial on all } P^1(\mathbb{C}) \times P^1(\mathbb{C}) \\ \text{subspaces)} \\ \text{(3rd order extendible to } PT \times PT^*) \end{array} \right\} \qquad (9)$$

The first of these correspondences (with the addition of a certain reality condition tied to the fibring of $P^3(\mathbb{C})$ over S^4) has led to a fairly complete understanding of the instantons[5] on S^4. The last of the correspondences awaits further study to determine its utility.

To understand why the PT bundles correspond to self-dual solutions and the PA bundles correspond to fields with no such

---

†The dimension of the vector bundles of interest corresponds to that of the Lie algebra in which the Yang-Mills fields take values.

restriction, note that self-dual connections are exactly those which are flat on anti-self-dual planes. Now, in carrying out the constructions implied by these correspondences, one finds that the field which corresponds to a given bundle B over a twistor-type space Q is necessarily flat over the geometric structures in $\mathbb{CM}$ corresponding to points in Q. Hence bundles over $\mathbb{PT}$ determine connections which are flat over anti-self-dual null planes (therefore self-dual) while bundles over $\mathbb{PA}$ determine connections which are flat over null lines (therefore no restriction).

To treat gravitational fields, one does not work directly with $\mathbb{T}$, $\mathbb{PT}$, $\mathbb{A}$, $\mathbb{PA}$, or any of the other spaces defined above. Indeed, $\mathbb{T}$ and all the others have the geometry of Minkowski spacetime built into them, so one must work with modified twistor spaces if one wants to consider spacetimes with curvature. These modified ("twistoroid") spaces are deformations which preserve some selected part of the rigid structure of the original spaces but are otherwise free (apart from holomorphic conditions). The geometric correspondences between the twistoroid spaces and the curved complex spacetimes are to an important extent much like those outlined above. (See table on previous page.)

The "nonlinear graviton" of Penrose[6] is the most successful example of the twistoroid treatment of curved spacetimes. In this case, one considers deformations $\mathbb{PJ}$ of $\mathbb{PT}$ which, like $\mathbb{PT}$, a) are bundles over $\mathbb{P}^1(\mathbb{C})$, and b) have a well-defined holomorphic volume element. One can then prove that the following is a 1-1 correspondence:

$$
\left\{
\begin{array}{c}
\text{Self-Dual} \\
\text{Torsion-Free} \\
\text{Einstein Solutions} \\
\text{on } \mathbb{C}^4
\end{array}
\right\}
\longleftrightarrow
\left\{
\begin{array}{c}
\mathbb{PJ} \\
\text{(Deformations of } \mathbb{PT}) \\
\text{with bundle structure } \mathbb{PJ} \to \mathbb{P}^1(\mathbb{C}) \\
\text{and volume form } w
\end{array}
\right\}
\qquad (10)
$$

A key to the verification of this correspondence is the proof that the sections of any $\mathbb{PJ} \to \mathbb{P}^1(\mathbb{C})$ form a $\mathbb{C}^4$ family of projective lines in $\mathbb{PJ}$. As with $\mathbb{PT}$, these projective lines correspond to points in spacetime. Also important is the observation that the fibres of the bundle endow the spacetime with a family of geodesic anti-self-dual null planes which are in 1-1 correspondence with the points of $\mathbb{PJ}$. This is where the self-duality of the solutions comes from.

In spite of many attempts,[7] no one knows how to code a spacetime with non-self-dual curvature into the structure of a deformation of **PT**. One is thus led to consider deformations of **PA**. It is straightforward to obtain a 1–1 correspondence between general holomorphic spacetimes (on $\mathbb{C}^4$) and certain deformations[8],[9] $\mathcal{PB}$ of **PA**. The more difficult problem is building in Einstein's equations (just as the Yang–Mills equations are built into the correspondences (9)). There are two approaches currently being used for studying this problem: ours,[8] and that of LeBrun.[9] Ours works with spaces $\mathcal{PB}$ which (like **PA**) fibre over $\mathbb{P}^1(\mathbb{C}) \times \mathbb{P}^1(\mathbb{C})$ and involves teleparallel fields as an intermediary, while LeBrun's deformations forego the fibration. Neither approach has yet been carried to completion, However, one presumes that the correspondence will be of the form

$$\left\{ \begin{array}{c} \text{Einstein Solutions} \\ \text{on } \mathbb{C}^4 \end{array} \right\} \quad \overset{?}{\longleftrightarrow} \quad \left\{ \begin{array}{l} \mathcal{PB} \\ \text{(Deformations of } \textbf{PA}) \\ \text{with} \text{_____} \end{array} \right\} \qquad (11)$$

Whatever the final form of this correspondence, two features are clearly expected: 1) that the points in $\mathcal{PB}$ will correspond to null geodesics in the spacetime, and 2) that embedded $\mathbb{P}^1(\mathbb{C}) \times \mathbb{P}^1(\mathbb{C})$'s in $\mathcal{PB}$ will correspond to points in the spacetime.

Most of the attempts in recent years to construct field theories for the fundamental interactions of physics have worked with spacetimes of dimension greater than 4 and have also attempted in some way to incorporate supersymmetry. Are twistor techniques of any use in such studies?

Of course the standard twistor spaces (**T**,**A**, etc.) as well as the twistoroid spaces ($\mathcal{PJ}, \mathcal{PB}$, etc.) are tied to 4 dimensions with no supersymmetry. They, together with many of their remarkable features,[1] must be abandoned. To replace them, there are a variety of possibilities, depending upon which of their features one considers to be essential. For example, for replacing twistor space itself, one might focus on the correspondence with anti–self–dual null two planes, on the action of the conformal group,[10] on the tie–in with spinors,[11] on the "twistor equation,"[12] or on the tie–in with the collection of complex structures,[13] among many other things. None of these approaches has yet been developed or applied very extensively, so it is not yet clear how useful any of them will be.

There has been some interesting work based on generalizations of ambitwistor space, however. Taking the point of view that elements of $\mathbb{P}A_{(d,s)}$ are the null geodesic "paths" of massless particles in a given spacetime of regular dimension d and superdimension s, Witten[14] has constructed the spaces $\mathbb{P}A_{(4,12)}$ and $\mathbb{P}A_{(10,16)}$ corresponding to super Minkowski spacetimes of the indicated dimensions. Using these, he has obtained ambitwistor-type correspondences for super-Yang-Mills fields:

$$
\left\{
\begin{array}{c}
N = 3 \\
\text{Super Yang-Mills Solutions} \\
\text{on } \mathbb{C}M_{(4,12)}
\end{array}
\right\}
\leftrightarrow
\left\{
\begin{array}{c}
\text{Vector Bundles} \\
\text{over } \mathbb{P}A_{(4,12)} \\
\text{(trivial on all } Q_{(4,12)} \\
\text{subspaces)}
\end{array}
\right\}
\qquad (12)
$$

and

$$
\left\{
\begin{array}{c}
N = 1 \\
\text{Super Yang-Mills Solutions} \\
\text{on } \mathbb{C}M_{(10,16)}
\end{array}
\right\}
\leftrightarrow
\left\{
\begin{array}{c}
\text{Vector Bundles} \\
\text{over } \mathbb{P}A_{(10,16)} \\
\text{(trivial on all } Q_{(10,16)} \\
\text{subspaces)}
\end{array}
\right\}
\qquad (13)
$$

where $Q_{(d,s)}$ refers to the space consisting of all null geodesic 'paths' of massless particles which can pass through a point p in $\mathbb{C}M_{(d,s)}$ [for d = 4 and s = 0, one has $Q_{(4,0)} = \mathbb{P}^1(\mathbb{C}) \times \mathbb{P}^1(\mathbb{C})$, as in (8) and (9)]. It is interesting that in these correspondences (12)–(13), even though one has full (super) Yang-Mills solutions, there is no need for bundle extendibility conditions, as appear in (9). This is because null geodesic "paths" of massless particles in superspace are not simply 1-dimensional paths. For example, in $\mathbb{C}M_{(10,16)}$, there are eight dimensions (accomodating eight dimensions worth of supersymmetric gauge invariance of paths.) As usual with the Yang-Mills twistor correspondences, flatness over these paths is automatic, and this leads to the super Yang-Mills field equations.

Witten has also [14] considered the possibility of relating deformations of the generalized ambitwistor spaces $\mathbb{P}A_{(d,s)}$ (especially in the case d = 10, s = 16) to solutions of the supergravity field equations. He reports some progress, but as in the non-super case, much work remains.

The success of the twistor and ambitwistor techniques in reformulating the problem of finding classical solutions of the Yang-Mills and Einstein field equations [as well as solutions of certain other equations, such as the Yang-Mills-Dirac-Higgs field equations[15], and all massless linear field equations, (arbitrary spin)[16]] does not imply that such techniques will be useful in the analysis of superstring theory. However, there is strong motivation for studying the possibility. Firstly, one notes the rather ad hoc way in which gravitational and Yang-Mills fields appear in superstring theory as it is presently understood. There is of course nothing wrong with treating such fields as just low-energy manifestations of a more comprehensive all-energy field. However, in view of the geometric nature of both gravity and Yang-Mills, one would like a more geometric understanding of how these fields are obtained from the string theory. The ambitwistor construction of gravitational fields from the holomorphic structure of the space of null geodesics provides a possible model for such understanding.

Secondly, one notes the increasing role being played by elements of complex analysis in the study of superstrings. Among others, Kahler manifolds, Riemann surfaces, and Dolbeault cohomology have been used in superstring theory. Twistor and ambitwistor theory could provide other useful tools from the complex analysis tool box.

In what form might ambitwistor-type techniques be used in superstring theory? While there has been speculation along various lines,[17] we shall discuss only one approach. Recalling the definition of $\mathbb{P}A_{(d,s)}$ and $\mathbb{P}\mathbb{C}_{(d,s)}$ in terms of null geodesic trajectories of massless particles in certain super spacetimes, we define

$$\mathbb{B}_{(d,s)} := \{\text{worldsheets of strings in } \mathbb{C}M_{(d,s)}\} \tag{14}$$

and

$$\mathfrak{B}_{(d,s)} := \{\text{worldsheets of strings in some}$$
$$\text{nonflat (complex)} \ \mathfrak{M}_{(d,s)}\} \tag{15}$$

These are infinite deminsional spaces whose properties are far from understood. By studying their holomorphic structure (and possibly examining holomorphic bundles over them) one hopes to obtain some insight regarding both a) the full (all-energy) superstring theory, and b) the low energy theory (manifest as gravitational and Yang-Mills fields). The former study might involve examining "string instantons." The latter could involve establishing relationships between $\mathfrak{B}_{(d,s)}$ and $\mathfrak{PB}_{(d,s)}$ (possibly for different spacetimes).

Perhaps $\mathfrak{B}_{(d,s)}$ and $\mathfrak{B}_{(d,s)}$ will not be useful in studying superstring theory. Perhaps there is no role for ambitwistor techniques in the analysis of superstrings. We believe, however, that there are enough signs to the contrary to make the relationship between ambitwistors and strings well worth pursuing.

Acknowledgements

This work was supported in part by the NSF under grant numbers DMS 83-03998 at the University of Oregon and DMS 85-04338 at Texas A & M University.

References

1] Comprehensive reviews of twistor theory appear in
   Penrose, R. and Ward, R.S., "Twistors for Flat and Curved Spacetime" in General Relativity and Gravitation (ed. A. Held), Plenum, 1980.
   Hughston, L.P. and Ward, R.S., Advances in Twistor Theory, Pitman, 1979.
   Wells, R.O., Complex Geometry in Mathematical Physics, SMS #78, Les Presses de l'Univ. de Montréal, 1982.

2] Ambitwistors are discussed in
   Isenberg, J., Yasskin, P.B., and Green, P., Phys. Lett. 78B, 462, 1978.
   Witten, E., Phys. Lett. 77B, 394, 1978.
   Isenberg, J. and Yasskin P.B., "Twistor Description of Non-Self-Dual Yang-Mills Fields, in Complex Manifold Techniques in Theoretical Physics (eds. D. Lerner and P. Sommers).
   Henkin, G.M. and Manin, Yu. I., Phys. Lett. 95B, 405, 1980.

3] Newman, E.T. and Hansen, R., Gen. Rel., Grav. 6, 361, 1975.

4] Ward, R., Phys. Lett. A61, 81, 1977.

5] Atiyah, M.F., Drinfeld, V.G., Hitchin, N.J., and Manin, Yu. I., Phys. Lett. A65, 185, 1978.
   Atiyah, M.F., Geometry of Yang-Mills Fields, Scuola Normale Superiore, 1979.

6] Penrose, R., Gen. Rel. Grav. 7, 31, 1976.
   Curtis, W.D., Lerner, P.F., and Miller, F.R., Gen. Rel. Grav. 10, 557, 1976.
   Hitchin, N.J., Math. Proc. Camb. Phil.Soc. 85, 465, 1979.

7] Penrose uses the term "googly" to refer to these attempts. See various issues of the Twistor Newsletter, compiled periodically at the Maths Institute, Oxford Univ.

8] Yasskin, P.B., and Isenberg, J., Gen. Rel. Grav. 14, 621, 1982.

9] LeBrun, C., Lett. Math Phys. 6, 345, 1982.
   LeBrun, C., Trans. Am. Math. Soc. 278, 209, 1983.
   LeBrun, C., Class Qtm. Grav. 2, 555, 1985.

10] Hurd, T., Private Communication.

11] Hughston, L., Private Communication.

12] Eastwood, M., "Supersymmetry, Twistors, and the Yang-Mills Field Equations," preprint, Oxford Univ., 1984.

13] Bryant, R., Duke Math. J. 51, 223, 1983.

14] See ref. 2 and also
   Witten, E., "Twistor-Like Transform in Ten Dimensions," preprint, Princeton Univ., 1985.

15] Manin, Yu. I., and Henkin, G.M., Sov. J. Nucl. Phys. 35, 941, 1982.
   Henkin, G.M., Sov. Math. Doklady 26, 224, 1982.

16] Eastwood, M., Penrose, R., and Wells, R. O., Comm. Math. Phys. 78, 305, 1981.

17] See Witten, ref. 14, and also
   Horowitz, G.T., "Introduction to String Theories," preprint, U. Cal. Santa Barbara, 1985.
   Shaw, W.T., "An Ambitwistor Description of Bosonic or Supersymmetric Minimal Surfaces and Strings in 4 Dimensions," preprint, MIT, 1985.

INFINITY CANCELLATION FOR O(32) OPEN STRINGS

P. H. Frampton

Institute of Field Physics, Department of Physics and Astronomy
University of North Carolina, Chapel Hill, NC 27514

This talk is in three subsections:

I.   General Introduction

II. Some Details[*]

III. Summary

1.   GENERAL INTRODUCTION

1.1 Strings for Strong Interactions

In 1968 Veneziano[1] guessed an explicit formula for the scattering amplitude with four external particles

$$A_4 = g^2 \frac{\Gamma(-\alpha_s)\Gamma(-\alpha_t)}{\Gamma(-\alpha_s - \alpha_t)} \qquad (1)$$

based on the postulates of (i) Regge behavior (ii) Resonances (iii) Analyticity and (iv) Duality and Crossing Symmetry. This A_4 was immediately generalized to A_M for an arbitrary number M of external spin-zero particles.

In 1969 an operator formalism was used to factorize these A_M and soon led to the interpretation as a theory of strings.[2] For this first, and still simplest, string theory there is an open string sector with leading Regge intercept $\alpha(0) = 1$, slope α', and closed

---

[*]Work done in collaboration with P. Moxhay and Y. J. Ng

strings with leading $\alpha(0) = 2$ and slope $\frac{1}{2}\alpha'$; the theory is unitary only for spacetime dimension d = 26. In 1971 a second string theory[3] was invented; it included also fermions and the critical spacetime dimension is d = 10.

Attempts to find new models led to such no-go theories as the "persistent photon" theorem for open strings and the "persistent graviton" theorem for closed strings. For strong interactions one needed phenomenologically leading intercepts $\alpha(0) \simeq \frac{1}{2}$ and $\alpha(0) \simeq 1$ respectively for the open and closed strings and this, together with the (then) embarrassing extra spacetime dimensions led to the demise of strings for strong interactions by 1975. This demise was obviously strongly encouraged by the success of QCD as a theory of strong interactions. Nevertheless, we should say that the strings are still relevant for strong interactions at large distances and/or large N, the number of colors.

1.2 Non-String Kaluza-Klein Theory

In order to obtain chiral fermions on M_4 after compactification of all but four dimensions we require that (i) the initial dimension = even = 6, 8, 10, $\cdots$ = 2n (ii) there are explicit gauge fields since

$$\gamma_{2n+1} \sim \gamma_5 \otimes \gamma_{2n-3}$$

$$\begin{array}{cc} M_4 & C_{2n-4} \\ + & - \\ - { \quad \quad \quad} \\ - & + \end{array} \qquad (2)$$

and asymmetry between the + and − eigenvalues of γ_5 needs a similar asymmetry for γ_{2n-3} which implies nontrivial topology on C_{2n-4}.

Such a chiral gauge theory has, in general, anomalies which destroy quantizability. For example, in d = 10 the hexagon diagram leads to such an anomaly. For nonexceptional simple groups with chiral fermions in the adjoint representation only 0(32) has $A_6 = 0$ in the decomposition

$$\mathrm{Tr}(\Lambda\Lambda\Lambda\Lambda\Lambda\Lambda) = A_6 \; \mathrm{Str}(\lambda\lambda\lambda\lambda\lambda\lambda)$$
$$+ A_6^{4,2} \; S\big(\mathrm{tr}(\lambda\lambda\lambda\lambda) \; \mathrm{tr} \; (\lambda\lambda)\big)$$
$$+ A_6^{222} \; S\big(\mathrm{tr}(\lambda\lambda) \; \mathrm{tr}(\lambda\lambda) \; \mathrm{tr}(\lambda\lambda)\big)$$
$$+ A_6^{333} \; S\big(\mathrm{tr}(\lambda\lambda\lambda) \; \mathrm{tr}(\lambda\lambda\lambda)\big) \tag{3}$$

where Λ, λ are generators in the adjoint, defining representations respectively. And $A_6^{4,2} \neq 0$ even for $0(32)$.

In 1983 we showed that the hexagon anomaly was finite, unique and calculable.[4] It appeared that all open superstrings were anomalous, unless the infinite numbers of massive states of the string changed the situation, as we considered "very unlikely."[5] In fact, the non-leading pieces may be cancelled by Chern-Simons terms in the H_{MNP} field strength.[6] Still, the cancellation of all gauge and gravity anomalies for $0(32)$ requires not one but four numerical coincidences.

There are now five superstring candidates: (1) $0(32)$ open (2) $0(32)$ heterotic (3) $E(8) \times E(8)$ heterotic (4) chiral $N = 2$ (5) nonchiral $N = 2$. The last four are closed string theories. Are these finite models of quantum gravity? The $0(32)$ open superstring looked easiest to kill but it turns out that a superstring miracle will save it. The finiteness of the six point function implies no anomaly and explains the coincidences.

2. SOME DETAILS

We shall use light-cone gauge where the action is given by

$$S = \int d\sigma \; d\tau \; \Big(- \frac{1}{4\pi\alpha'} \; \partial_\alpha X^i \partial^\alpha X^i + \frac{i}{4\pi} \; \bar{S}\gamma^-\rho^\alpha \partial_\alpha S\Big) \tag{4}$$

where all unexplained notation is as in the Schwarz review article.[7] In particular, the fermion operators in the expansion

$$S^a = \sum_{n=-\infty}^{\infty} S_n^a e^{-in\tau} \quad (\sigma=0) \tag{5}$$

satisfy

$$\{S^a_m, \bar{S}^b_n\}_+ = (\gamma^+ h^-)^{ab} \delta_{m+n,o} \tag{6}$$

with $h^- = \frac{1}{2}(1 - \gamma_{11})$. These S^a_n commute with all the bosonic operators a^i_n, $a^{i\dagger}_n$ occurring in $(\sigma = 0)$

$$X^i = x^i + p^i \tau + i \sum_{n=1}^{\infty} \frac{1}{\sqrt{n}} (a^i_n e^{-in\tau} - a^{i\dagger}_n e^{in\tau}). \tag{7}$$

A central role is played by the fermion zero modes which satisfy

$$S^a_o \bar{S}^b_o = \frac{1}{2}(\gamma^+ h^-)^{ab} + \frac{1}{4}(\gamma^{ij+} h^-)^{ab} R^{ij}_o \tag{8}$$

where $R^{ij}_o = \frac{1}{8} \bar{S}_o \gamma^{ij-} S_o$ are generators of $O(8)$ in the transverse space. The superstring vertex for massless vector emission is well known to be[7]

$$V(k) = g \zeta^i (P^i + k^j R^{ij}) V_o(k) \tag{9}$$

with $R^{ij} = \frac{1}{8} \bar{S} \gamma^{ij-} S$. It is easy to show that for the S_o traces there must be at least eight S_o's to obtain a nonzero result. This implies the one-loop superstring non-renormalization that the one loop amplitude for $M = 2$ or 3 gauge bosons vanishes, and also simplifies greatly the $M = 4$ case.

For $M = 4$, both the annulus and the Moebius strip diverge.[8] The only nonzero term is proportional to

$$K_4 = 16\pi^3 g^4 \; Tr(\Lambda^{a_1} \Lambda^{a_2} \Lambda^{a_3} \Lambda^{a_4}) \zeta_1^{i_1} \zeta_2^{i_2} \zeta_3^{i_3} \zeta_4^{i_4}$$
$$\begin{matrix} j_1 \; j_2 \; j_3 \; j_4 \\ k_1 \; k_2 \; k_3 \; k_4 \end{matrix} \; t^{i_1 j_1 i_2 j_2 i_3 j_3 i_4 j_4} \tag{10}$$

where

$$t = Tr(R_o^{i_1 j_1} R_o^{i_2 j_2} R_o^{i_3 j_3} R_o^{i_4 j_4}). \tag{11}$$

For gauge group $O(N)$ one finds results[8]]

$$A_4^{(Ann)} = NK_4 \int_0^1 \frac{dq}{q} F(q^2) \tag{12}$$

$$A_4^{(Mob)} = - 8K_4 \int_0^1 \frac{dq}{q} F(-\sqrt{\bar{q}}). \qquad (13)$$

If one puts $\lambda = q^2$, $\lambda = \sqrt{\bar{q}}$ respectively, one can add these and take the principal part according to

$$A_4^{(Ann)} + A_4^{(Mob)} = 16K_4 \int_{-1}^{+1} \frac{d\lambda}{\lambda} F(\lambda) \quad (PP) \qquad (14)$$

only for N = 32. This prescription for regularization is the working hypothesis we use for M > 4. Note that there is an ambiguity (putting, say, $\lambda = q$ in $A_4^{(Ann)}$) leads to a divergent answer) and some work is needed to justify this regularization.

The case of M = 4 external gauge bosons is quite special and one cannot see from it how the infinity cancellation works for general M: the factor 8 in A_4^{Mob} becomes 2^{M-1} in A_M^{Mob} so extra factors of 2^{-1} are necessary to effect the cancellation. The general case M ≥ 6 will be clear from the following discussion of M = 5.

For the pentagon M = 5 the leading divergence comes from pieces of the five vertices V which we may denote as

$$(R_o)^3 (S_n S_o)(\bar{S}_{-n} S_o) \qquad (15)$$
$$(R_o)^4 P .$$

Terms like $(R_o)^5$ do not contribute to the soft dilaton pole arising when the hole in the center of the annulus strikes to zero. In terms of the standard disc integration variable q, the momentum integral gives a factor $(\ell nq)^5$ while the Jacobian for transforming to disc variables gives $(\ell nq)^{-6}$ so only those terms giving an additional (ℓnq) factor correspond to the dilaton pole at q = 0.

The end result is conveniently expressed by introducing certain notations as follows. We define the planar mode sum F(c, w) as

$$F(c, w) = \sum_{n=1}^{\infty} \frac{c^n - (w/c)^n}{(1 - w^n)} \qquad (16)$$

and the trace

$$t^{i_1 j_1 i_2 j_2 i j} = \text{tr}(\gamma^{i_1 j_1} \gamma^{i_2 j_2} \gamma^{ij})$$

$$= 32 \, (\delta^{i_1 j_2} \delta^{i_2 j} \delta^{ij_1} - \delta^{j_1 j_2} \delta^{i_2 j} \delta^{ii_1}$$

$$- \delta^{i_1 i_2} \delta^{jj_2} \delta^{ij_1} + \delta^{ii_1} \delta^{i_2 j_1} \delta^{jj_2}$$

$$- \delta^{ii_2} \delta^{i_1 j_2} \delta^{jj_1} + \delta^{ii_2} \delta^{i_1 j} \delta^{j_1 j_2}$$

$$+ \delta^{ij_2} \delta^{i_1 i_2} \delta^{jj_1} - \delta^{ij_2} \delta^{i_2 j_1} \delta^{i_1 j}). \qquad (17)$$

Then for the annulus $A_5^{(Ann)}$ we obtain

$$A_5^{(Ann)} = \int \prod_{I=1}^{5} dx_I \, w^{-1} \left(\frac{-2\pi}{\ell nw}\right)^5 \prod_{I<J} (\psi_{IJ})^{k_I k_J}$$

$$\times \left[K_4(2345) \, \zeta_1^{i_1} \left(k_2^{i_1} F(x_2, \, w) + k_3^{i_1} F(x_2 x_3, w) \right. \right.$$

$$\left. - k_4^{i_1} F(x_5 x_1, \, w) - k_5^{i_1} F(x_1, \, w)\right) + \text{cyclic perms.}$$

$$+ \frac{1}{64} \zeta_1^{i_1} \zeta_2^{i_2} \zeta_3^{i_3} \zeta_4^{i_4} \zeta_5^{i_5} \, k_1^{j_1} k_2^{j_2} k_3^{j_3} k_4^{j_4} k_5^{j_5}$$

$$(- F(x_2, \, w) \, t^{i_1 j_1 i_2 j_2 i j} \, \text{tr}(R_o^{ij} \, R_o^{i_3 j_3} \, R_o^{i_4 j_4} \, R_o^{i_5 j_5})$$

$$- F(x_2 x_3, \, w) \, t^{i_1 j_1 i_3 j_3 i j} \, \text{tr}(R_o^{ij} \, R_o^{i_2 j_2} \, R_o^{i_4 j_4} \, R_o^{i_5 j_5})$$

$$+ \text{cyclic perms}) \, \Big\rangle$$

$$+ \text{finite terms.} \qquad (18)$$

For the Moebius strip amplitude $A_5^{(Mob)}$ the changes are that $x_1 \to -x_1$, $\psi_{IJ} \to \psi_{N,IJ}$ and, in particular,

$$F_N (c, \, w) = \sum_{n=1}^{\infty} \frac{c^n - (-w/c)^n}{1 - (-w)^n} . \qquad (19)$$

In the limit $q \to 0$, $F(c, \, w)$ and $F_N(c, \, w)$ have respectively the asymptotic expansions

$$F(c, w) = - \frac{\ln q}{2\pi} \cot \pi \nu \left(1 + 0(\frac{1}{\ln q})\right) \tag{20}$$

$$F_N (c, w) = - \frac{1}{2} \frac{\ln q}{2\pi} \cot(\frac{\pi \nu}{2}) \left(1 + 0(\frac{1}{\ln q})\right) \tag{21}$$

where $\nu = (\ln c)(\ln w)^{-1}$. It is the $1/2$ in the expansion of $F_N(c, w)$ that enables us to rewrite

$$A_5^{(Ann)} + A_5^{(Mob)} = N \int_0^1 \frac{dq}{q} F_5(q^2) - (\frac{1}{2}) 16 \int_0^1 \frac{dq}{q} F_5(-\sqrt{q}) + \text{finite} \tag{22}$$

as a principal part integral

$$A_5^{(Ann)} + A_5^{(Mob)} = 16 \int_{-1}^{+1} \frac{d\lambda}{\lambda} F_5(\lambda) \text{ (PP)} + \text{finite} \tag{23}$$

when $N = 32$, using the same changes of variable as for $M = 4$.

The number of pieces contributing to the dilaton pole is $(M-3)$ for $4 \leqslant M \leqslant 7$ and 5 for $M \geqslant 8$. There are $(M-4)$ mode sums of the type $F(c, w)$ for $A_M^{(Ann)}$ and of the type $F_N(c, w)$ for $A_M^{(Mob)}$ with the result that

$$A_M^{(Ann)} + A_M^{(Mob)} = N \int_0^1 \frac{dq}{q} F_M(q^2) - (\frac{1}{2})^{M-4} 2^{M-1} \int_0^1 \frac{dq}{q} F_M(-\sqrt{q})$$

$$+ \text{finite}$$

$$= 16 \int_{-1}^{+1} \frac{d\lambda}{\lambda} F_M(\lambda) \text{ (PP)} + \text{finite} \tag{24}$$

giving a finite result for $0(32)$.

3. SUMMARY

For this nonorientable $0(32)$ model there are divergences also in various zero- and one-loop graphs[8] involving <u>closed</u> strings for which one would like to calculate explicitly the infinity cancellation. For example, for external closed strings the one loop amplitudes include not only the torus which is not divergent but also the divergent annulus, Moebius and Klein bottle amplitudes between which the infinites are expected to cancel with O(N) only for N = 32.

For the other four purely-closed superstring models, there is only one one-loop, the torus. This has been shown to be finite[7,10] for M = 4 external lines and we are currently studying the situation[11] for general M ≥ 5. This is actually simpler to compute than the open string case we have already done. Very probably, these models are also one-loop finite.

The situation for two or more loops is an open question.

This work was supported by the U.S. Department of Energy under Grant No. DE-AS05-79ER-10448.

References

1. Veneziano, G., Nuovo Cim. 57A, 190 (1968).

2. Nambu, Y., Wayne State University Conference (1969); Copenhagen lectures (1970).

3. Ramond, P., Phys. Rev. D3, 2415 (1971).

4. Frampton, P. H., Phys. Lett. 122B, 351 (1983); Frampton, P. H., and T. W. Kephart, Phys. Rev. Lett. 50, 1343, 1347 (1983); Phys. Rev. D28, 1010 (1983).

5. Frampton, P. H. and Kephart, T. W., Phys. Lett. 131B, 80 (1983).

6. Green, M. B. and Schwarz, J. H., Phys. Lett. 149B, 117 (1984).

7. Schwarz, J. H., Phys. Reports 89, 223 (1982).

8. Green, M. B. and Schwarz, J. H., Phys. Lett. 151B, 21 (1985).

9. Frampton, P. H., Moxhay P., and Ng, Y. J., UNC-Chapel Hill Report IFP-255-UNC (1985, to be published.

10. Gross, D. J., Harvey, J. A., Martinec, E. and Rohm, R., Princeton University preprint (June 1985).

11. Frampton, P. H., Moxhay, P., and Ng, Y. J., UNC-Chapel Hill Report IFP-256-UNC (in preparation).

THE THERMODYNAMICS OF SUPERSTRINGS

Mark J. Bowick

Department of Physics

Yale University

New Haven, CT   06511

ABSTRACT

I review some work on the behavior of superstrings at high energy density and temperature. The density of states is determined for various superstring theories and attention is focused on a novel phase of closed superstrings.

1.   INTRODUCTION

Suppose for the moment that the interactions of nature are fundamentally those of one-dimensional extended objects (strings) rather than point-like objects (particles). It is naturally hoped that a stringy description will not be special to just some of the interactions known but will encompass all of them. In particular it should be capable of describing gravity at the short distances where quantum fluctuations become significant. This sets the length scale of the string to be approximately the Planck length $\ell_p = \sqrt{G} \sim 10^{-33}$ cm. It is only when such small lengths are probed that we can expect to see the microscopic differences in the dynamics of string-like versus point-like objects. One is led then to think about string dynamics at high energy densities and high temperatures where small distance fluctuations can be excited. In terms of a mode description of the string the zero modes are seen at low energy and long distances and interact according to effective field theories of point-like particles. Although the way in which a string theory gives a low energy effective field theory seems highly non-trivial it is nevertheless difficult or impossible to know that we are dealing with a string if we look only at low energy. At high energy we can excite more and more of the infinite tower of massive modes of the string. Now we have something new. Where could we look for evidence that nature is stringy? The early universe has been a popular and prolific testing ground in the last decade. Cosmologies based on string theories will involve an infinite number of effective particle species[1] and should differ markedly from those usually considered which entertain only a finite (and rather small) number of particles. Although I will not discuss cosmology in any detail in this talk (the possibilities seem immense) I will point out that the crudest analysis of the behavior of strings at finite temperature indicates that many interesting things could and probably do happen if and when strings are heated.

2. THERMODYNAMIC DESCRIPTION

I will use the most naive thermodynamic description possible – that of an ideal gas of superstrings. Although this may seem hopelessly inadequate I will give two justifications for the ideal gas approximation. The first is that the spectrum of the free string already contains some information about its linearized interactions.[2] This is like the situation in the statistical bootstrap model[3] where the mass spectrum simulates attractive strong interactions. Thus interactions are not ignored entirely. The second justification follows from considering the width of the excited states. Green and Veneziano[4] showed that for bosonic strings the width of an excitation of mass M is bounded by a number of order 1 times the string tension divided by M. Thus the higher mass resonances become increasingly more stable. In the limit of zero width string states[5] tree level interactions are sufficient and the ideal gas approximation becomes exact.[6] I expect then that the gross features of string thermodynamics will be accurately found in this picture. The width of massive excitations for superstrings should certainly be evaluated. Gleiser and Taylor[7] have made a different assumption – namely that the widths of states of mass M rises rapidly with M. Their analysis is then very different although they come to many of the same conclusions.

3. DENSITY OF STATES

The basic quantity needed to set up the string thermodynamics is the density of states Ω. In this section I determine Ω for various string theories of interest.[8] The mass spectrum (M) is given generically by $\alpha'M^2=N$ where the number operator N is an infinite sum over bilinears in bosonic and fermionic oscillators[9] and α' is the Regge slope parameter. The degeneracy of the eigenvalues of N is determined by constructing the generating function[10,1] $P(x) = \sum_{N=1}^{\infty} P(N)x^N$. Below I simply give $P(x)$ for various string theories

1. Veneziano Model (26-dimensional open bosonic string)[11]

$$P_B(x) = \{\pi_{n=1}^{\infty}(1 - x^n)\}^{-24} \tag{1}$$

2. Ten-dimensional Type I (open and closed non-oriented) Superstrings[9]

$$P_S(x) = \left[\pi_{n=1}^{\infty} \left(\frac{1 + x^n}{1 - x^n}\right)\right]^8 = [\theta_4(0,x]^{-8} \tag{2}$$

where θ_4 is the Jacobi theta function.[12]

3. The Heterotic String[13,8]

$$P_H(x) = P_B(x)\left[1 + \sum_{k \varepsilon 2\mathbf{Z}} 480 \ \sigma_7(k)x^{2k}\right]P_S(x) \tag{3}$$

where $\sigma_7(k)$ is the sum of the seventh powers of the divisors of m. Using a saddle-point evaluation of a contour integral for $P(N)$[8,2] one finds

$$P_B(N) \sim N^{-27/4}\exp[4\pi \ \sqrt{N}] \ ,$$

$$P_S(N) \sim N^{-11/4}\exp[\sqrt{8} \ \pi \ \sqrt{N}]$$

and

$$P_H(N) \sim N^{-11/2} \exp[(4 + 2\sqrt{2})\, \pi \sqrt{N}].$$

The corresponding values for the closed bosonic and superstring (Type II) are easily found from those above since they are essentially bound states of the open string theories. The density of states in mass space is then given by the generic form

$$\rho(m) = c m^{-a} \exp(bm) \tag{4}$$

with the following values for a and b:

Veneziano Model: $a = 25/2$, $b = \pi 4\sqrt{\alpha'}$

Type I Superstrings: $a = 9/2$, $b = \pi\sqrt{8}\,\sqrt{\alpha'}$

Type II Superstrings: $a = 10$, $b = \pi\sqrt{8}\,\sqrt{\alpha'}$ (5)

Heterotic Superstrings: $a = 10$, $b = \pi(2 + \sqrt{2})\,\sqrt{\alpha'}$

4. ENSEMBLE

I discuss first the pitfalls of the canonical ensemble applied to the string excitations. The canonical partition function Z in ten space-time dimensions is[8,14]

$$\ln Z = \frac{V}{(2\pi)^9} \int \rho(m)dm \int d^9k \, \ln\left[\frac{1 + \exp[-\beta\sqrt{k^2 + m^2}]}{1 - \exp[-\beta\sqrt{k^2 + m^2}]}\right] \tag{6}$$

$$\sim V \sum_{n=0}^{\infty} \left[\frac{1}{2n+1}\right]^5 \int_{\zeta}^{\infty} dm \, m^{-a+5} K_5[(2n+1)\beta m]e^{bm} \tag{7}$$

where ζ is an infrared cutoff below which the asymptotic form of the density of states Eq. (4) is no longer valid and $K_n(x)$ is a modified Bessel function. As the temperature approaches $T_0 = 1/b$, we see that

$$\ln Z = \underset{T \to T_0}{\longrightarrow} \left[\frac{TT_0}{T_0 - T}\right]^{-a + 11/2} \Gamma\left[-a + 11/2, \, \zeta\left(\frac{T_0 - T}{TT_0}\right)\right] \tag{8}$$

where $\Gamma(a,x)$ is the incomplete gamma function. Z diverges for $T > T_0$. Does this mean that T_0 is the maximum temperature for which there can be thermal equilibrium — the Hagedorn limiting temperature?[3] To answer this I examine the thermodynamic observables pressure P,

specific heat C_V and mean energy $\langle E \rangle$.[15,16] For a $\leq \frac{13}{2}$ ($\frac{D+3}{2}$
in D space-time dimensions) P_1, $\langle E \rangle$ and C_V diverge as $T \to T_0$.
In this case T_0 really seems to be a maximum temperature as it takes
an infinite amount of energy to heat the system past T_0. For
$\frac{13}{2} < a < \frac{15}{2}$, only the specific heat diverges at T_o, and for a $> \frac{15}{2}$,
not even the specific heat diverges. Why then can't the system be
heated above T_0? Perhaps it is merely the canonical ensemble
description which is breaking down above T_0.

The fundamental ensemble of statistical mechanics is the
microcanonical ensemble which is based solely on the postulate of
equal a priori probability of all microstates which yield a given
macrostate. The canonical ensemble is a steepest-descent
approximation to the microcanonical ensemble partition function and is
known to yield different results when the saddle-point evaluation is
not justified.[17] There are a number of ways the invalidity of the
canonical ensemble may be seen. The first is through the mean-square
energy fluctuation. This exceeds 1 for energy density $\rho = E/V$ greater
than the critical value $\rho_o = (\frac{1}{a - \frac{13}{2}})(\frac{1}{2\pi b})^{\frac{9}{2}} \zeta^{\frac{13}{2} - a}$. The second is
via the specific heat to which I will return after discussing the
microcanonical ensemble.

In the microcanonical ensemble (for an interesting examination
of microcanonical quantum field theory I refer the reader to a paper
of Strominger[18]) the total energy E is fixed and one counts the
density of states

$$\Omega(E,V) = \sum_{n=1}^{\infty} [\frac{V}{(2\pi)^9}]^n \frac{1}{n!} \prod_{i=1}^{n} \int_\zeta^\infty \rho(m_i)dm_i \int d^9 p_i \delta(\Sigma E_i - E)\delta(\Sigma p_i) \quad (9)$$

For a $> \frac{13}{2}$ and $\rho > \rho_o$[15,16] the density of states as a function of
energy has the same functional form as the level density $\rho(m)$ (it is a
solution of the bootstrap condition). Thus $\Omega(E,V) \simeq VE^{-a} \exp(bE)$.
From Eq. (5) we see that this situation applies to the closed
superstring theories – the Type II and Heterotic strings – but not to

the Type I superstring. They have very different thermodynamic behavior.

An analysis of the density of states $\Omega(E,V)$ for $a > \frac{13}{2}$ and $\rho > \rho_0$ reveals that the most probable number of strings ($\langle n \rangle$) is such that the favored thermodynamic configuration is for $\langle n \rangle - 1$ strings to carry as little energy (ζ) as possible and for one string to carry the remaining energy. The condition $\rho > \rho_0$ is equivalent to the condition $E \gg n\zeta$. Thus one string carries the majority of the energy. This is a highly inhomogeneous configuration and the source of the large energy fluctuations in the canonical ensemble.

Given $\Omega(E,V)$ I can now compute thermodynamic observables. Formally $T = \left(\frac{\partial \ln\Omega}{\partial E}\right)^{-1} = \frac{E}{bE-a} \geq T_o$. The specific heat $C = -\frac{1}{T^2}\left(\frac{\partial^2 \ln\Omega}{\partial E^2}\right)^{-1} = -\frac{1}{T^2}\left(\frac{E^2}{a}\right) \leq 0$. The temperature can exceed the Hagedorn temperature but we have passed into a phase of negative specific heat. This is the second indication that the canonical ensemble is inapplicable. There the specific heat is proportional to the mean-square energy fluctuation and is positive semi-definite. Negative specific heats have arisen in physical situations in the past.[19,20] The most interesting example I know is for a black hole where $\Omega(E) \sim \exp(E^2)$ and $C \sim -E^2$ (the temperature of a black hole is inversely related to its mass). Such systems can never be in thermal equilibrium with an infinite heat reservoir. They can, in contrast, come to equilibrium with a finite heat bath. A black hole can be in equilibrium with a thermal bath of radiation provided the total energy of the radiation is less than one quarter the mass of the black hole.[21] The corresponding result[8] for the massive excitations of the heterotic string (for example) to be in equilibrium with its massless modes is that

$$T < T_{max} = \frac{20bE - 9a \pm \sqrt{81\,a^2 + 40\,abE}}{20b(bE - a)} \tag{10}$$

The energy in the massless modes must then be less than $E + \frac{aT_{max}}{1-bT_{max}}$.

The maximum temperature may equivalently be formulated as a maximum volume for the system (as in the black-hole case).

The new string phase discussed in this talk may have novel cosmological implications. Similar thermodynamic considerations have also been applied to show that the string may be relevant in the last stages of black hole evaporation[22] and may provide some solution thereby to the associated Hawking puzzle. Strings should provide solutions not only to the problems of renormalizability and unitarity of gravity but also to these other issues of gravity at short distances.[23] Finally I refer the reader to some other work for further viewpoints on superstring behavior around the Hagedorn temperature.[24,25]

I wish to thank my collaborator in this work L.C.R. Wijewardhana and L. Smolin for valuable discussions. This research was supported in part by the U.S. D.O.E. under Contract No. DE-AC02-76ERO 3075.

REFERENCES

1. Huang, K., and Weinberg, S., Phys. Rev. Lett. 25, 895 (1970).
2. Sundborg, B., Nucl. Phys. B254, 583 (1985).
3. Hagedorn, R., Nuovo Cimento Suppl. 3, 147 (1965) and Cargèse Lectures in Physics, Vol. 6, edited by E. Schatzman (Gordon and Breach, New York, 1973) p. 643.
4. Green, M.B., and Veneziano, G., Phys. Lett. 36B, 477 (1971).
5. Tye, S.H.H., Phys. Lett. 158B, 388 (1985).
6. Dashen, R., Ma, S.K., and Bernstein, H.J., Phys. Rev. 187, 345 (1969).
7. Gleiser, M., and Taylor, J.G., King's College Preprint 85-0389 (1985).
8. Bowick, M.J., and Wijewardhana, L.C.R., Phys. Rev. Lett. 54, 2485 (1985); and Yale Preprint YTP 85/10, to be published in the Journal of General Relativity and Gravitation.
9. Schwarz, J.H., Phys. Reports 89, 223 (1982); Green, M.B., Surveys in High Energy Physics 3, 127 (1983); Brink, L., "Superstrings", CERN-TH 4006/84, to be published in Proceedings from Nato Advanced Study Institute, "Supersymmetry", Bonn 1984.

10. Hardy, G.H., and Ramanujan, S, Proc. London Math. Soc. <u>17</u>, 75 (1918).

11. Frampton, P.H., Dual Resonance Models (Benjamin, 1974).

12. Whittaker, E.T., and Watson, G.N., Modern Analysis (Cambridge Univ. Press, Cambridge, England, 1973); Erdelyi, A., et al., Higher Transcendental Functions (McGraw-Hill, New York, 1953).

13. Gross, D., Harvey, J.A., Martinec, E., and Rohm, R., Phys. Rev. Lett. <u>54</u>, 502 (1985), Nucl. Phys. <u>B256</u>, 253 (1985).

14. Alvarez, E., Phys. Rev. <u>D31</u>, 418 (1985).

15. Frautschi, S., Phys. Rev. <u>D3</u>, 2821 (1971).

16. Carlitz, R., Phys. Rev. <u>D5</u>, 3231 (1972).

17. Lax, M., Phys. Rev. <u>97</u>, 1419 (1955).

18. Strominger, A., Ann. Phys. <u>146</u>, 419 (1983).

19. Lynden-Bell, D., and Wood, R., Roy. Mon. Not. Astron. Soc. <u>138</u>, 495 (1968).

20. Thirring, W., Z. Phys. <u>235</u>, 339 (1970).

21. Hawking, S.W., Phys. Rev. <u>D13</u>, 191 (1975). See also General Relativity: An Einstein Centenary Survey, edited by S.W. Hawking and W. Israel (Cambridge Univ. Press, Cambridge, England, 1979), p. 746.

22. Bowick, M.J., Smolin, L., and Wijewardhana, L.C.R., Yale preprint in preparation.

23. Horowitz, G.T., Lectures given at the school on "Topological Properties and Global Structure of Spacetime", Ettore Majorana Center for Scientific Culture, Erice (May, 1985).

24. Olesen, P., CERN-TH 4249/85, CERN-TH 4175/85.

25. Salomonson, P., and Skagerstam, B.S., Goteborg Preprint 85-32 (1985).

TOWARD A COVARIANT STRING FIELD THEORY

Stuart Raby[*], Richard Slansky, and Geoffrey West

Theoretical Division, T-8, MS B285
Los Alamos National Laboratory
Los Alamos, New Mexico 87545 USA

ABSTRACT

We identify the physical degrees of freedom in the string field using the equations of motion, gauge conditions and residual gauge invariance of the system. We also present a first step towards obtaining a covariant string field theory.

INTRODUCTION

We want to discuss string <u>field</u> <u>theory</u> which we shall distinguish from the <u>particle</u> <u>theory</u> of the string. Moreover, we want to describe a string field theory which is both Lorentz invariant and covariant under local gauge transformations which we shall soon identify.

The point particle has canonical coordinates $x^\mu(\tau)$ which describe its motion along a world line (see fig. 1) in space-time. The parameter τ labels the position along the world line. The action

$$S = -m \int d\tau \sqrt{-\left(\frac{dx}{d\tau}\right)^2}$$

is invariant under arbitrary τ reparametrizations $\tau \to \tau + \xi(\tau)$. As a

---

[*]   Talk presented at the APS Meeting of the Division of Particles and Fields, Eugene, Oregon, August 12-16, 1985.

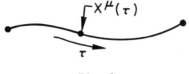

Fig. 1

result, we obtain the primary constraint

$$p^2 + m^2 = 0 \quad .$$

The field theory of the point particle satisfies the equations of motion

$$(\Box - m^2)\phi(x) = 0 \quad ,$$

where $\phi(x)$ is the real Klein-Gordon field which is a function of the space-time coordinate x^μ. The second quantized field ϕ creates and annihilates particles with the quantum numbers of the point particle. Note that the equations of motion are obtained from the primary constraints of the point particle theory, with the identification $p_\mu \to -i\partial/\partial x^\mu$. The reparametrization invariance of the point particle theory is not apparent in the corresponding field theory.

The particle theory of a one-dimensional string is described by the canonical coordinates

$$X^\mu(\tau,\sigma) \quad ,$$

where σ parametrizes the position along the string (at fixed τ) and τ parametrizes its motion in space-time (see fig. 2). The action for the string

$$S = \int d\tau d\sigma \sqrt{\left(\frac{\partial X^\mu}{\partial \sigma} \frac{\partial X_\mu}{\partial \tau}\right)^2 - \left(\frac{\partial X^\mu}{\partial \sigma} \frac{\partial X_\mu}{\partial \sigma}\right)\left(\frac{\partial X^\nu}{\partial \tau} \frac{\partial X_\nu}{\partial \tau}\right)}$$

is invariant under arbitrary σ,τ reparametrizations

$$\sigma \to \sigma + \xi_\sigma(\tau,\sigma) \quad ,$$

$$\tau \to \tau + \xi_\tau(\tau,\sigma) \quad .$$

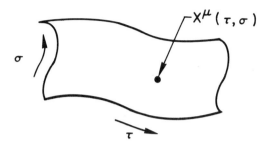

Fig. 2

As a result we obtain the primary constraints $L_n = 0$ where L_n are the generators of the Virasoro algebra. Free string field theory is then described by the equations

$$(L_0 - 1) \, \Phi \, [X(\sigma)] = 0 \quad ,$$

$$L_n \, \Phi \, [X(\sigma)] = 0 \quad , \quad n > 0 \quad ,$$

where the string field Φ is a hermitian functional of the coordinates $X^\mu(\sigma)$. Note we have used only half of the primary constraints. This is because our representation of the Virasoro algebra includes a central term which does not permit us to satisfy $L_n \Phi = 0$ for all n. Open and closed strings include gauge fields. One of the questions we shall address is, in what way are the local gauge symmetries of the string field theory related to the two-dimensional reparametriza- tion invariance of the string particle theory?

Why are we interested in string theories? The particle theory of the string provides an algorithm for calculating scattering ampli- tudes for the states of the string. It has been shown[1] that the ten dimensional superstring is 1) one-loop finite, 2) anomaly free, and 3) contains enough states for describing a truly "unified field theory" of nature. The particle theory of the string is quite compli- cated. The field theory of the string does not promise to be any simpler. Why then, do we want to construct a string field theory? a) Bosonic strings contain tachyons. In a scalar field theory, for example, a tachyonic state is usually a symptom that one is quantiz- ing the theory about the wrong ground state (see fig. 3). To quantize about the correct ground state, one minimizes the potential and then shifts the field by the appropriate amount. In the particle theory the background is fixed and thus finding a better ground state (should one exist) requires summing an infinite number of perturbation theory diagrams.

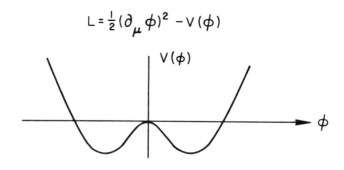

$$L = \frac{1}{2}(\partial_\mu \phi)^2 - V(\phi)$$

Fig. 3

b)   The closed string contains a massless symmetric tensor state - a "graviton".

i)   classical solutions of the closed string field theory may provide some insight as to the difference between string gravity and Einstein gravity at short distances.

ii)   A ten or 26 dimensional string must compactify to give us flat 4 dimensional Minkowski space. In Kaluza-Klein theories one looks for classical solutions to the field equations on a compact manifold. One would like to understand the vacuum geometry for the string.

iii) Finally and most importantly, it is clear that the standard general coordinate invariance of Einstein gravity is not a symmetry of the string. For example, there does not exist a local 4 point interaction for the graviton in string theories (unless one takes the zero slope limit). What is the "symmetry principle" which replaces general coordinate invariance as a symmetry of nature?

OPEN STRING

Coordinates:

$$X^{\mu}(\sigma) = x^{\mu} + 2 \sum_{\ell=1}^{\infty} \cos\ell\sigma \; x_{\ell}^{\mu} \quad ,$$

where $\mu = 1,\ldots,D$; $0 \leqq \sigma \leqq \pi$. The coordinates satisfy the boundary conditions

$$\left. \frac{dX^{\mu}}{d\sigma} \right|_{\sigma=0,\pi} \equiv \left. X'^{\mu}(\sigma) \right|_{\sigma=0,\pi} = 0 \quad .$$

The zero mode x^{μ} will be interpreted as space-time, while x_{ℓ}^{μ}, $\ell=1,\ldots,\infty$ describe an "internal" space.

String Field: $\Phi[X_{\mu}(\sigma)]$ with $\Phi^{+} = \Phi$. It describes an infinite component field when expanded in a Taylor series about the point x^{μ}.

Differential operators on functionals:

$$\frac{\delta}{\delta X^{\mu}(\sigma)} = \frac{1}{\pi} \sum_{\ell=0}^{\infty} \cos\ell\sigma \; \frac{\delta}{\delta x_{\ell}^{\mu}}$$

with ordinary space-time differentiation given by

$$\partial_{\mu} \equiv \frac{\delta}{\delta x_{0}^{\mu}} \quad .$$

Virasoro operators: We want to obtain a representation of L_n acting on string fields. In the first quantized string the L_n's are given by

$$L_n = \frac{1}{2} \; : \sum_{\ell=-\infty}^{+\infty} a_{\ell}^{\dagger} \cdot a_{\ell+n} \; :$$

where the creation and annihilation operators $a_n^{\mu\dagger}$, a_n^{μ} satisfy

i) $\qquad a_n^{\dagger\mu} \equiv a_{-n}^{\mu} \qquad , \quad n \geqq 1 \quad ,$

ii) $\qquad [a_n^{\mu}, a_m^{\dagger\nu}] = n\delta_{nm}\eta^{\mu\nu} \quad .$

The brackets : : denote normal ordering. The L_n's satisfy

i) $\quad L_n^{\dagger} = L_{-n} \quad ,$

ii) $\quad [L_n, L_m] = (n-m) \, L_{n+m} + \dfrac{D}{12} \, (n^3 - n) \delta_{n+m,0}$

A representation for the a's is obtained as follows. Define

$$p^\mu(\sigma) = i\frac{\delta}{\delta X_\mu(\sigma)} + X'^\mu(\sigma) \quad \left.\rule{0pt}{70pt}\right\} \quad 0 \leq \sigma \leq \pi$$

and

$$p^\mu(-\sigma) = i\frac{\delta}{\delta X_\mu(\sigma)} - X'^\mu(\sigma)$$

Then

$$a_n^\mu \equiv \frac{1}{\sqrt{4\pi}} \int_{-\pi}^{\pi} d\sigma \, e^{-in\sigma} \, p^\mu(\sigma) \quad .$$

We have

$$a_0^\mu \equiv \frac{i}{\sqrt{\pi}} \partial^\mu$$

$$a_n^\mu = \frac{i}{\sqrt{4\pi}}\left(\frac{\delta}{\delta x_{n\mu}} + 2\pi n \, x_n^\mu\right) \quad , \quad n \neq 0 \quad .$$

The Virasoro operators are then defined as operators on string fields using the above definition of the a's. For later use we write explicitly L_0, L_{-1}, L_{-2}.

$$L_0 = \frac{1}{2\pi} \, (2\pi N - \Box) \quad ; \quad N = \sum_{\ell=1}^{\infty} a_\ell^\dagger a_\ell \quad ,$$

$$L_{-1} = \frac{i}{\sqrt{\pi}} \, a_1^\dagger \cdot \partial + \sum_{\ell=1}^{\infty} a_{\ell+1}^\dagger \cdot a_\ell \quad ,$$

$$L_{-2} = \frac{i}{\sqrt{\pi}} \, a_2^\dagger \cdot \partial + \frac{1}{2} \, a_1^\dagger \cdot a_1^\dagger + \sum_{\ell=1}^{\infty} a_{\ell+2}^\dagger \cdot a_\ell \quad .$$

Physical degrees of freedom (PDOF):

We shall now identify the PDOF contained in Φ. We use, A) the field equations, B) the gauge conditions, and C) the residual local gauge invariance of A) and B).

A) $\Theta_0 \Phi = 0$; $\Theta_0 \equiv L_0 - 1$,

B) $L_n \Phi = 0$; $n \geq 1$.

A) and B) are invariant under the local transformation

C) $\delta\Phi = \sum_{m=1}^{\infty} L_{-m} \Lambda_m$

where

$\Lambda_m \equiv \Lambda_m[X(\sigma)]$ satisfy

$L_{+n} \Lambda_m = -(2n+m) \Lambda_{n+m}$, (a)

$(L_0 + m-1)\Lambda_m = 0$. (b)

This symmetry only exists for D = 26. As a result of (a) we see that

(1) Λ_n for $n \neq 0$ can be expressed in terms of Λ_1, i.e.,

$$\Lambda_n = \frac{2(-1)^{n-1}}{(n+1)!} L_1^{n-1} \Lambda_1 ,$$

and

(2) Λ_1 is constrained to satisfy

$L_0 \Lambda_1 = 0$

$L_{+2} \Lambda_1 = -\frac{5}{12} L_{+1}^2 \Lambda_1$, etc.

(3) Not all the independent parameters contained in Λ_1 are relevant, i.e., the above constraints on Λ_1 are invariant under the transformation

$\delta\Lambda_1 = \sum_{m=2}^{\infty} L_{-m} X_m$

where the $X_m \equiv X_m[X(\sigma)]$ satisfy

$L_{+n} X_m = -(2n+m) X_{n+m}$,

$(L_0 + m) X_m = 0$,

and

$$\delta^2\Phi = \sum_{m=1}^{\infty} L_{-m}\delta\Lambda_m \equiv 0 \quad .$$

The last equation says that not all the independent components of Λ_1 will change Φ, i.e., they are not all relevant. The above process continues. We have checked that using the results of A), B), and C), we find a one-to-one correspondence between the states of the first quantized string in the light cone gauge and the PDOF in Φ up to mass level 5. Let us discuss a few simple examples.

The solution to A) is given by the following construction. Define the ground state functional $\Phi_0[X(\sigma)]$ such that

$$a_n\Phi_0 = 0 \quad , \quad n \geq 0 \quad .$$

Φ_0 is a gaussian functional in the internal coordinates x_ℓ^μ , $\ell \neq 0$. Then a solution to A) is

$$\Phi[X(\sigma)] = \{\phi(x) + A_\mu^\ell(x) \, a_\ell^{\dagger\mu} + \frac{1}{2}h_{\mu\nu}^{\ell m}(x) \, a_\ell^{\dagger\mu}a_m^{\dagger\nu} + \ldots \} \, \Phi_0 \quad ,$$

$$\ell,m = 1, \ldots \infty \quad ,$$

where the coefficient functions satisfy the field equations

$$(\Box + 2\pi) \, \phi = 0 \quad ,$$

$$\Box A_\mu^1 = 0 \quad ,$$

$$(\Box - 2\pi) \, A_\mu^2 = 0 \quad ,$$

$$(\Box - 2\pi) \, h_{\mu\nu}^{11} = 0 \quad ,$$

etc.

ϕ is thus a tachyon with $m^2 = -2\pi$, A_μ^1 is a Maxwell field [the first mass level], and A_μ^2 and $h_{\mu\nu}^{11}$ have $m^2 = 2\pi$ [the second mass level]. The field equations A) enable us to identify an infinite set of local fields and mass levels. However, not all the fields are independent. We use B) + C) to identify the independent PDOF. Consider the first 2 mass levels.

$\underline{m^2 = 0}$

A) $\theta_0 \Phi = 0$ gives $\Box A_\mu^1 = 0$,

B) $L_1 \Phi = 0$ gives $\partial \cdot A^1 = 0$,

i.e., the Lorentz gauge condition. It is for this reason we have identified the equations B) as gauge conditions. Finally

C) $\delta \Phi = L_{-1} \Lambda_1 + \ldots$ gives $\delta A_\mu^1 = \partial_\mu \lambda_1$,

where

$$\Lambda_1 [x(\sigma)] = [\lambda_1(x) + \lambda_\mu^1(x) \, a_1^{\dagger \mu} + \ldots] \, \Phi_0 \quad ,$$

and the constraint $L_0 \Lambda_1 = 0$ implies $\Box \lambda_1(x) = 0$. Thus, the Maxwell field A_μ^1 has D components, of which only D-2 are physical since B) says one field is dependent and C) says that another can be gauged away. Note that to obtain a gauge invariant system we must remove the constraint on the gauge parameter, i.e., we would like

$\delta A_\mu^1 = \partial_\mu \lambda_1$ for arbitrary $\lambda_1(x)$.

At the second mass level $\underline{(M^2 = 2\pi)}$, we have

A) $\theta_0 \Phi = 0$ gives $(\Box - 2\pi) \, A_\mu^2 = 0$,

$$(\Box - 2\pi) h_{\mu\nu}^{11} = 0 \quad ,$$

or

$D + \dfrac{D(D+1)}{2}$ fields. They satisfy the gauge conditions

B) $L_1 \Phi = 0$ gives $2A_\mu^2 + \dfrac{i}{\sqrt{\pi}} \, \partial^\nu h_{\mu\nu}^{11} = 0$

$L_2 \Phi = 0$ gives $\dfrac{2i}{\sqrt{\pi}} \, \partial \cdot A^2 + \dfrac{1}{2} h^{11} = 0$

with $h^{11} \equiv h_\mu^{11\mu}$,   or

D + 1 gauge conditions. Finally we have the invariance

C) $\delta\Phi = (L_{-1} - \frac{1}{3} L_{-2}L_1)\wedge_1$ which gives

$$\delta A_\mu^2 = -\lambda_\mu^1 + \frac{1}{3\pi} \partial_\mu \partial \cdot \lambda^1 ,$$

$$\delta h_{\mu\nu}^{11} = \frac{-i}{\sqrt{\pi}} [\partial_\mu \lambda_\nu^1 + \partial_\nu \lambda_\mu^1 - \frac{1}{3}\eta_{\mu\nu}\partial \cdot \lambda^1]$$

with the constraint $L_0\wedge_1 = 0$ giving $(\Box-2\pi)\lambda_\mu^1 = 0$, or D independent gauge functions. We thus find

$$D + \frac{D(D+1)}{2} - (D + 1) - D = \frac{(D+1)(D-2)}{2}$$

independent PDOF. This is the correct count to describe a massive, traceless, symmetric tensor.

FIRST STEP TOWARD A COVARIANT FIELD THEORY

We shall obtain a solution to the following demands.

1) Remove the constraint on the gauge functional i.e., find a new equation of motion $\Theta\Phi = 0$ invariant under the local transformation

$$\delta\Phi = \sum_{m=1}^{\infty} L_{-m}\wedge_m$$

without the constraint

$$(L_0 + m - 1) \wedge_m = 0 .$$

2) The equations of motion are local in terms of the component fields.

3) No auxiliary fields are needed.

4) The gauge conditions $L_n\Phi = 0$ are obtained as a gauge choice. We find a solution for D = 26

$$\theta = L_0 - 1 - \sum_{n=1}^{\infty} \frac{L_{-n}L_n}{2n} .$$

As an example the equations of motion $\Theta\Phi = 0$ gives for the Maxwell field

$$\partial^\mu F_{\mu\nu} = 0 \text{ with } F_{\mu\nu} = \partial_\mu A_\nu^1 - \partial_\nu A_\mu^1 \quad .$$

This is invariant under the transformation $\delta\Phi = L_{-1}\Lambda_1 + \ldots$ which gives $\delta A_\mu^1 = \partial_\mu \lambda_1$ with arbitrary $\lambda_1(x)$. We note that the original gauge conditions $L_n\Phi = 0, n \geq 1$ may be obtained by a choice of Λ_1. It is however necessary to use the equations of motion $\Theta\Phi = 0$.

Closed String:

A closed string is defined by the coordinates

$$X^\mu(\sigma) = \sum_{\ell=-\infty}^{+\infty} e^{i\ell\sigma} x_\ell^\mu \quad , \quad -\pi \leq \sigma \leq \pi \quad ,$$

with

$$x_{-\ell}^\mu \equiv x_\ell^{\mu*} \quad , \quad \mu = 1, \ldots, D$$

(see fig. 4). $X^\mu(\sigma)$ describes a closed loop in space-time. Once again space-time is associated with the normal mode $x_0^\mu \equiv x^\mu$ and x_ℓ^μ, $\ell \neq 0$ are "internal spaces."

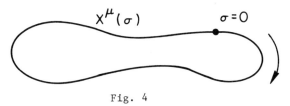

Fig. 4

The <u>string field</u> $\Phi[X_\mu(\sigma)]$ is a real functional of the coordinates; $\Phi = \Phi^\dagger$. Since the origin of the loop has no physical significance we shall require $\Phi[X]$ to be invariant under redefinitions of the origin, i.e. $\Phi[X(\sigma+c)] = \Phi[X(\sigma)]$ where c is an arbitrary constant.

Once again we shall define <u>differential</u> <u>operators</u> on closed string functionals and then use these operators to construct representations of the Virasoro algebra. We define the differential

$$\frac{\delta}{\delta X^\mu(\sigma)} = \frac{1}{2\pi} \sum_{\ell=-\infty}^{+\infty} e^{-i\ell\sigma} \frac{\delta}{\delta x_\ell^\mu}$$

with space-time differentiation given by

$$\partial_\mu \equiv \frac{\delta}{\delta x_0^\mu} \quad .$$

We then define two sets of creation and annihilation operators

$$P^\mu(\sigma) = i\frac{\delta}{\delta X_\mu}(\sigma) + X'^\mu(\sigma) = \frac{1}{\sqrt{\pi}}\sum_{\ell=-\infty}^{+\infty} e^{i\ell\sigma}\, a_\ell^\mu \quad,$$

$$\bar{P}^\mu(\sigma) = i\frac{\delta}{\delta X_\mu}(\sigma) - X'^\mu(\sigma) = \frac{1}{\sqrt{\pi}}\sum_{\ell=-\infty}^{+\infty} e^{-i\ell\sigma}\, \bar{a}_\ell^\mu \quad,$$

where in the particle theory a_ℓ^μ and $\bar{a}_\ell^\mu$ describe the states of right and left movers, respectively. Note that

$$a_0^\mu = \bar{a}_0^\mu = \frac{i}{\sqrt{4\pi}}\,\partial^\mu \quad .$$

We then obtain a Virasoro algebra for right (L_n) and left $(\bar{L}_n)$ movers. We have

$$L_0 = \frac{1}{8\pi}(-\Box + 8\pi N) \quad,$$

$$\bar{L}_0 = \frac{1}{8\pi}(-\Box + 8\pi N) \quad .$$

The generator of constant σ-translations is given by $L_0 - \bar{L}_0$. We thus require Φ to satisfy

$$(L_0 - \bar{L}_0)\Phi = (N - \bar{N})\Phi = 0 \quad,$$

where $N(\bar{N})$ are number operators for right (left) movers.

Let us now discuss the physical degrees of freedom contained in Φ. We use

A) the field equations $\theta_0\,\Phi = 0 \quad, \quad \theta_0 = L_0 - 1;$
B) the gauge conditions $L_n\,\Phi = \bar{L}_n\,\Phi = 0 \quad, \quad n \geq 1$, and
C) the residual gauge invariance of A) and B)

$$\delta\Phi = L_{-1}\bar{\Lambda}_1 + \bar{L}_{-1}\Lambda_1 + L_{-1}\bar{L}_{-1}\Lambda_{11}$$

where the gauge functionals satisfy the constraints

$$L_0\Lambda_{11} = L_0\bar{\Lambda}_1 = \bar{L}_0\Lambda_1 = 0 \quad .$$

They also satisfy

$$(N-\bar{N})\bar{\Lambda}_1 = \bar{\Lambda}_1 \quad , \quad (N-\bar{N})\Lambda_1 = -\Lambda_1 \quad ,$$

$$(N-\bar{N})\Lambda_{11} = 0 \quad ,$$

$$L_1\bar{\Lambda}_1 = \bar{L}_1\bar{\Lambda}_1 = L_1\Lambda_1 = \bar{L}_1\Lambda_1 = 0 \quad .$$

Note that this is not the complete residual invariance of the theory, but only that piece which affects the zero mass level which we shall now discuss. Consider the solution to A) the field equation. We define the ground state functional Φ_0, satisfying

$$a_n^\mu \Phi_0 = \bar{a}_n^\mu \Phi_0 = 0 \text{ for } n \geq 0 \quad .$$

Then

$$\Phi[X^\mu(\sigma)] = \{\phi(x) + h_{\mu\nu}^\ell \bar{a}_\ell^{\mu\dagger} a_\ell^{\nu\dagger} + \ldots\}\Phi_0 \quad ,$$

where the component fields satisfy $(\Box+8\pi)\phi(x) = 0$, a tachyon with mass^2 = -8π and $\Box h_{\mu\nu}^1 = 0$, a massless field with D^2 components; etc. Once again we can use the gauge conditions B) and residual gauge invariance C) to identify the PDOF. We find (at the zero mass level) that $h_{\mu\nu} \equiv 1/2(h_{\mu\nu}^1 + h_{\nu\mu}^1)$ describes a massless symmetric tensor field with $(D^2-3D)/2$ PDOF, i.e., a graviton and $A_{\mu\nu} \equiv 1/2(h_{\mu\nu}^1 - h_{\nu\mu}^1)$ describes a massless anti-symmetric tensor field with $(D^2-5D+6)/2$ PDOF, i.e., a Kalb-Ramond gauge field. Note that the massless dilaton which exists in the particle theory is missing.

We shall now obtain a new equation of motion given by A') $\theta_1\Phi = 0$ where Φ now satisfies the gauge conditions B') $L_n\Phi = \bar{L}_n\Phi = 0$ $n \geq 2$ and A') and B') are invariant under the local gauge transformation

$$\delta\Phi = L_{-1}\bar{\Lambda}_1 + \bar{L}_{-1}\Lambda_1 + L_{-1}\bar{L}_{-1}\Lambda_{11}$$

with the constraints removed i.e., $L_0\Lambda_{11} \neq 0$, $L_0\bar{\Lambda}_1 \neq 0$, and $\bar{L}_0\Lambda_1 \neq 0$. We find

$$\theta_1 = \frac{1}{2}[L_0 + \bar{L}_0 - 2 - L_{-1}L_1 - \bar{L}_{-1}\bar{L}_1 + L_{-1}\bar{L}_{-1}\left(\frac{1}{L_0+\bar{L}_0}\right)L_1\bar{L}_1] \quad .$$

The field equation in the massless sector is

$$\Box h^1_{\mu\nu} - \partial_\mu \partial^\lambda h^1_{\lambda\nu} - \partial_\nu \partial^\lambda h^1_{\mu\lambda} + \partial_\mu \partial_\nu \frac{1}{\Box} \partial^\lambda \partial^\zeta h^1_{\lambda\zeta} = 0 \quad .$$

Although it is non-local, by taking a trace, we obtain

$$\Box h^{1\mu}_{\mu} = \partial^\mu \partial^\nu h^1_{\mu\nu}$$

and thus it is equivalent to the local equation

$$\Box h^1_{\mu\nu} - \partial_\mu \partial^\lambda h^1_{\lambda\nu} - \partial_\nu \partial^\lambda h^1_{\mu\lambda} + \partial_\mu \partial_\nu h^{1\lambda}_{\lambda} = 0.$$

These are just the linearized equations of Einstein gravity for $h_{\mu\nu}$ and the gauge invariant equations for $A_{\mu\nu}$. Note this result is a necessary consequence of gauge symmetry $\delta\Phi$ (discussed above) which in component form implies

$$\delta h_{\mu\nu} = \partial_\mu \lambda_\nu - \partial_\nu \lambda_\mu \quad ,$$

$$\delta A_{\mu\nu} = \partial_\mu \lambda'_\nu - \partial_\nu \lambda'_\mu \quad ,$$

It is clear how to solve simultaneously the problem of the missing dilaton and the non-locality of the equations of motion. Consider the non-local linearized Einstein equations

$$\Box h_{\mu\nu} - \partial_\mu \partial^\lambda h_{\lambda\nu} - \partial_\nu \partial^\lambda h_{\mu\lambda} + \partial_\mu \partial_\nu \frac{1}{\Box} \partial^\lambda \partial^\zeta h_{\lambda\zeta} = 0 \quad .$$

This equation may be made local by introducing a new scalar field ϕ' satisfying

$$\Box h_{\mu\nu} - \partial_\mu \partial^\lambda h_{\lambda\nu} - \partial_\nu \partial^\lambda h_{\mu\lambda} + \partial_\mu \partial_\nu \phi' = 0 \quad ,$$

$$\Box \phi' = \partial^\lambda \partial^\zeta h_{\lambda\zeta} \quad .$$

Note that these equations are invariant under the transformations

$$\delta h_{\mu\nu} = \partial_\mu \lambda_\nu + \partial_\nu \lambda_\mu \quad ,$$

$$\delta h^\mu_\mu = 2\partial \cdot \lambda \quad ,$$

$$\delta\phi' = 2\partial \cdot \lambda \quad .$$

Thus, the linear combination $\phi' - h_\mu^{\;\mu}$ is gauge invariant and indeed one finds that it satisfies the equation of motion

$$\square(\phi' - h_\mu^{\;\mu}) = 0 \quad .$$

The non-locality of the string field equations can then be remedied by the same procedure. Introduce a new functional $\Phi'[X^\mu(\sigma)] = \phi'(x)\Phi_0 + \ldots$, with the coupled field equations

$$(L_0 + \bar{L}_0 - 2 - L_{-1}L_1 - \bar{L}_{-1}\bar{L}_1)\Phi - L_{-1}\bar{L}_{-1}\Phi' = 0 \quad ,$$

$$(L_0 + \bar{L}_0)\Phi' = -L_1\bar{L}_1\Phi \quad .$$

These equations are invariant under the gauge transformations

$$\delta\Phi = L_{-1}\bar{\Lambda}_1 + \bar{L}_{-1}\Lambda_1 \quad ,$$

$$\delta\Phi' = \bar{L}_1\bar{\Lambda}_1 + L_1\Lambda_1 \quad ,$$

Interactions:

Up until now we have discussed free string field theory. Interactions may be introduced into the action with terms of the form (see fig. 5)

$$S_I = \int \prod_{i=1}^{3} dX_i^\mu(\sigma)\Phi[X_1]\Phi[X_2]\Phi[X_3]V[X_1,X_2,X_3] \quad .$$

We define an energy-momentum tensor functional

$$T[X_1(\sigma)] = \frac{\delta S_I}{\delta\Phi[X_1(\sigma)]} \quad ,$$

$$= \int \prod_{i=2}^{3} dX_i^\mu(\sigma)\Phi[X_2]\Phi[X_3]V[X_1,X_2,X_3] \quad .$$

S_I is constrained to satisfy the linearized gauge transformations. As a result, we find the generalized conservation law

$$L_1 T[X_1(\sigma)] = \bar{L}_1 T[X_1(\sigma)] = 0 \quad , \qquad \text{(on shell)} \quad .$$

It will be through the study of the interacting string that we may learn the non-linear nature of the symmetry. It is at this point we will discover whether the symmetry is, for example, a space-time symmetry or rather an internal symmetry of the string.

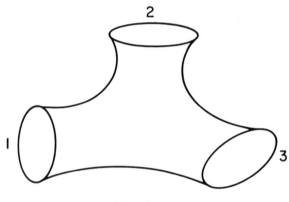

Fig. 5

FERMION FAMILIES IN SUPERSTRING THEORY*

Matt Visser
and
Itzhak Bars

Physics Department
University of Southern California
Los Angeles, CA 90089-0484

ABSTRACT

We perform a general analysis relating the number of fermion families in certain Kaluza-Klein theories to the index of an appropriate Dirac operator. In particular the analysis yields formulae applicable to hypothetical superstring compactifications that posess nonzero torsion and/or a spin connection that is not equal to the gauge connection.

1. INTRODUCTION

This year (1985) has witnessed an explosion of interest in Kaluza-Klein theories driven by the exciting recent results in superstring theory[1,2]. In addition to the increased interest in Kaluza-Klein theories this year has seen a major change of strategy. The "old" Kaluza-Klein theories (see e.g. ref. 3) used compact internal manifolds with large continuous symmetry groups, the continuous symmetries being related to the low energy gauge interactions. The "new" Kaluza-Klein theories use internal manifolds of as few continuous symmetries as possible (preferably none). Gauge interactions are now

---

* Invited talk delivered by M. Visser

assumed to occur ab initio in the higher dimensional spacetime (see e.g. ref. 2).

2. THE TREE LEVEL SPECTRUM

We exhibit a variant of a by now standard analysis, see (e.g.) ref. 3 and references therein. Consider a d-dimensional spacetime that compactifies according to the scheme $M_d \rightarrow M_4 \times M_{d-4}$, where M_4 is Minkowski space and M_{d-4} is compact. The d-dimensional spacetime is assumed to carry gauge fields belonging to some grand gauge group G, this grand gauge group being broken by the compactification to

$$G \rightarrow E \otimes B \qquad (1)$$

Here E is the effective low energy gauge group while B is the background gauge group. Specifically let x^μ denote co-ordinates on M_4 and let y^m denote co-ordinates on M_{d-4}. Then the vacuum expectation values of the G valued d-dimensional gauge fields define the background gauge group B by:

$$A^\mu = 0; \quad 0 \neq A^m(y) \; \epsilon \; B \qquad (2)$$

The effective gauge group E is defined as the maximal subgroup of G that commutes with B, the point being that among all fluctuations of the gauge fields around the vacuum, it is only the E valued fluctuations that do not interact with the background. A representation g of G will decompose under E⊗B as follows

$$(g)_G = \sum_{e,b} n_{e,b} \; (e,b)_{E \otimes B} \qquad (3)$$

The d-dimensional gamma matrices may be represented by

$$\Gamma^\mu = \gamma^\mu \otimes I; \quad \Gamma^m = \gamma_5 \otimes \gamma^m \qquad (4)$$

Finally, d dimensional spinorial fluctuations around the vacuum take the form

$$\eta_{d,G}(x,y) = \sum_i \eta_{4,E}^{(i)}(x) \otimes \eta_{d-4,B}^{(i)}(y) \tag{5}$$

Consequently the d-dimensional massless Dirac equation takes the form

$$0 = (\not{D}\eta)_{d,G} = \sum_{(i)} \left(\{ [\not{D}\eta^{(i)}]_{4,E} \otimes \eta_{d-4,B}^{(i)} \} \right.$$
$$\left. + \{ [\gamma_5 \eta^{(i)}]_{4,E} \otimes [\not{D}\eta^{(i)}]_{d-4,B} \} \right) \tag{6}$$

Then we see that any spinor field that is massless in the d-dimensional sense will give rise to an infinite tower of spinors in four dimensions whose tree level masses are

$$m_4^{(i)} = i\text{'th eigenvalue of } \not{D}_{d-4,B} \tag{7}$$

This formula will be valid at tree level in any of the "new" Kaluza-Klein theories. We expect the first nonzero eigenvalue of $\not{D}_{d-4,B}$ to be of order $M_{PLANCK} \sim 10^{19}$ GeV. Therefore the phenomenologically interesting case corresponds to the zero eigenvalues of this operator.

3. THE ZERO MODES

Each zero mode of $\not{D}_{d-4,B}$ gives rise (at tree level) to an E-multiplet of massless fermions in four dimensions. However if both left and right handed multiplets (in the four dimensional sense) are present, then they are not protected against radiative mass generation (presumably of order M_{PLANCK}). Accordingly we make the survival hypothesis[4] and identify the number of massless chiral generations N_e

$$N_e = n_{4,e,L} - n_{4,e,R}$$

$$= \left\{ \sum_b n_{e,b} \ n_{d-4,b,L} \right\} - \left\{ \sum_b n_{e,b} \ n_{d-4,b,R} \right\}$$

$$= \sum_b n_{e,b} \ \text{index}(\not{D}_{d-4,B})_b \tag{8}$$

Note that we have had to use the fact that a left-handed internal zero mode on M_{d-4} gives rise to a left-handed massless spinor in four dimensions. This is only true provided the original theory is chiral in d-dimensions and d=4n+2. See Witten[3]. If one attempts a similar analysis for non chiral theories, or for d ≠ 4n+2, then the survival hypothesis predicts that all modes become massive.

4.   THE DIRAC INDEX

The Atiyah-Singer index theorem[5], as applied to the Dirac operator, expresses the index of the Dirac operator on a compact manifold as an integral over the manifold of a certain polynomial constructed from the Riemann tensor and the gauge field strength.

$$\text{index}(\not{D})_b = \int \hat{A}(R) \ \text{ch}(F)_b = \int \hat{A}(R) \ \text{tr}_b(e^{F/2\pi}) \tag{9}$$

Here R and F are the curvature forms, matrix valued 2-forms defined by

$$R^a_{\ b} = R^a_{\ bmn} \ dy^m \wedge dy^n$$

$$F^i_{\ j} = F^i_{\ j \ mn} \ dy^m \wedge dy^n = (t_\alpha)^i_{\ j} \ F^\alpha_{\ mn} \ dy^m \wedge dy^n \tag{10}$$

Note that the trace is to be evaluated in the representation b of B. The object $\hat{A}(R)$ is the $\hat{A}$ genus, $\hat{A}$ may be expressed as a polynomial in R, or alternatively as a polynomial in the Chern classes.

5.   STRING CONSIDERATIONS

The first and simplest consequence of dealing with a superstring theory is that we have d=10. Accordingly on the six dimensional internal space we have[5]

$$\hat{A}(R) = 1 + \frac{c_2(R)}{24} \qquad (11)$$

The second important constraint coming from superstring theory is that

$$tr(R^2) - \frac{1}{30} Tr(F^2) = dH \qquad (12)$$

This equation arises from the Green-Schwarz modification of the Chapline-Manton low energy expansion of superstring theory and is essential for the anomaly cancellations. We note that for $tr(R^2)$ the trace is to be taken in the fundamental representation of $O(6)$, while for $Tr(F^2)$ the trace is to be taken in the adjoint representation of $G = SO(32)$ or $G = E_8 \otimes E_8$. Further observe that the first two Chern classes may be written

$$c_1(R) = \frac{1}{2\pi} \{tr(R)\}$$

$$c_2(R) = -\frac{1}{2} \left(\frac{1}{2\pi}\right)^2 \{tr(R^2) - (trR)^2\} \qquad (13)$$

Identical formulae may be written for $c_1(F)$, $c_2(F)$. The notation { } indicates that we should take equivalence classes modulo exact forms. With these conventions (12) implies

$$c_2(R) = \frac{1}{30} c_2(F) \qquad (14)$$

This means that the existence of a low energy expansion for the superstring thory places a topological constraint on the allowable compactifications. Going back to the index theorem we see

$$index(\not{D})_b = \frac{1}{6} \frac{1}{(2\pi)^3} \left[\int tr_b(F^3) - \frac{1}{8} \cdot \frac{1}{30} \int tr_b(F) \ Tr(F^2) \right] \qquad (15)$$

The number of massless multiplets in the representation e of E is now

$$N_e = \frac{1}{6} \frac{1}{(2\pi)^3} \sum_b n_{e,b} \left[\int tr_b(F^3) - \frac{1}{8} \cdot \frac{1}{30} \int tr_b F \; Tr(F^2) \right] \tag{16}$$

Hidden in the notation is the fact that this analysis is independent of whether or not the torsion [H] vanishes, and that this analysis is independent of whether or not the gauge connection "equals" the spin connection. Let us compare this formula with the analysis of Candelas et al.[2] Candelas et al. made three (independent?) assumptions

(a) the compactification preserves a N=1 supersymmetry

(b) the torsion (H) vanishes

(c) the spin connection "equals" the gauge connection.

With these assumptions equation (16) simplifies radically to yeild N = $\frac{1}{2}$|Euler characteristic|. Unfortunately it is not known at present whether assumptions (a), (b), and (c) are independent, or whether perhaps (b) and (c) can be derived from (a). Work in this area is continuing[6,7].

This work was supported by the Department of Energy under grant number DE-FG03-84ER40168.

6. REFERENCES

1. Green, M.B. and Schwarz, J.H., Phys. Lett. 136, 367 (1984), 149B, 117 (1984); Gross, D., Harvey, J., Martinec, E., and Rohm, R., Phys. Rev. Lett. 54, 502 (1985).

2. Candelas, P., Horowitz, G., Strominger, A. and Witten, E., Nucl. Phys. B258, 46 (1985).

3. Witten, E., in Proceedings of the Second Shelter Island Conference [June 1983],(MIT press,1985), eds. N. Khuri et al.

4. Georgi, H., Nucl. Phys. B156, 126 (1979).

5. Eguchi, T., Gilkey, P. and Hanson, A., Phys. Reports 66, 213 (1980) see esp. pp 331-334.

6. Bars, I. and Visser, M., Phys. Lett. B (1985).

7. Bars, I., Phys. Rev. D (1985).

E_6 SYMMETRY BREAKING IN SUPERSTRING MODELS

G. C. Segrè

Department of Physics, University of Pennsylvania
Philadelphia, Pennsylvania 19104

In this talk I would like to review some work done recently by John Breit, Burt Ovrut and myself on the problem of the breaking of the internal gauge group symmetry defined on superstrings.

As we know by the work of Green and Schwartz[1], the only allowed gauge symmetry that can be consistently introduced on the ten dimensional (d = 10) superstring is O(32) or $E_8 \times E_8$, i.e. only for these two examples can we have a well defined superstring theory[2] with anomaly cancellation. Since the lowest excitations of the superstring are the supergravity multiplet and a Yang-Mills supermultiplet, we can make contact with developments in the d = 10 field theory of these two supermultiplets[3]. Of course, all higher excitations of the superstrings are characterized by a mass scale $M_{Planck} \sim 10^{19}$ Gev and hence may be neglected.

To proceed further, we must compactify the d = 10 manifold

$$M_{10} \rightarrow M_4 \times K_o \tag{1}$$

where M_4 is ordinary Minkowski space and K_o is a compact six dimensional manifold. The three conditions placed on the compactification by Candelas, Horowitz, Strominger and Witten[4] are that (1) hold, that we have an unbroken N = 1 supersymmetry (abbreviated SUSY) in d = 4 and that the particle spectrum be realistic. The conditions for this to hold, as Candelas has reviewed in his talk here, are that K_o be Ricci flat, have SU(3) holonomy and be Kahler.

Naively, we may understand the SU(3) holonomy as follows: if a mode's field transforms non-trivially as we move around K_o, we typically expect it to have a mass of order of $(R_{compactification})^{-1} \sim M_{Planck}$. The gravitino in d = 10 leads to four gravitini in d = 4; the reason for this is that a spinor in 0(6), the compactified space, actually belongs to the 4 representation of SU_4 (isomorphic to 0(6)). If we look at the SU(3) subgroup of SU_4, the spinor transforms like 3 ⊕ 1. Holonomy is the group of transformations which parallel transport spinors on a manifold. If the holonomy is not the full SU_4, but an SU_3, subgroup we will have a spinor singlet under holonomy and hence a single massless gravitino when we compactify, i.e. we will be left with unbroken N = 1 SUSY.

As Candelas has pointed out in his talk, the integrability condition for the field strength H of the antisymmetric tensor B present in the supergravity multiplet forces us to identify the holonomy SU_3 with the background fields of a SU_3 subgroup of our gauge group. This clearly leads us to choose $E_8 \times E_8'$ over 0(32) since E_8 has a maximal subgroup decomposition of $SU_3 \times E_6$. E_6 is an interesting group for phenomenology[5] with chiral fermions placed in the fundamental 27 representation, whereas SO(32) fails in this respect.

For this lecture we regard the E_8' as a hidden sector and ignore it. Under $SU_3 \times E_6$, the 248 ten dimensional fields in the gauge supermultiplet transform as

$$(8,1) + (3,27) + (\bar{3},2\bar{7}) + (1,78) \qquad (2)$$

The (1,78) are the gauge superfields of E_6, the (8,1) are the gauge superfields of SU_3 and the remaining fields lead eventually to chiral matter superfields. As CHSW[4] show, the number of left handed minus right handed 27 superfields is fixed on topological grounds to be one-half the Euler characteristic $\chi(K_o)$ of the manifold

$$N_o = |n_L - n_R| = \frac{1}{2}|\chi(K_o)| \qquad (3)$$

Ricci flat, Kahler manifolds with SU(3) holonomy are called Calabi-Yau[6] manifolds. The simplest example is given by the solutions of the quintic polynomial

$$\sum_{i=1}^{5} z_i^{5} = 0 \tag{4}$$

In general Calabi-Yau manifolds have N_o values that are unacceptably large, e.g., N_o = 36, 64, 72, 100, since each multiplet contains one complete quark lepton family plus twelve other fields. The solution proposed by CHSW is to impose a discrete symmetry acting freely on the manifold. Examples of such discrete symmetries A and B are

$$A(z_1 z_2 z_3 z_4 z_5) \rightarrow (z_5, z_1, z_2, z_3, z_4)$$

$$+ \text{ permutations}$$

$$B(z_1, z_2, z_3, z_4, z_5) \rightarrow (\alpha z_1, \alpha^2 z_2, \alpha^3 z_3, \alpha^4 z_4, z_5)$$

$$\alpha^5 = 1 \quad . \tag{5}$$

Clearly A and B, isomorphic to $Z_5 \times Z_5$ leaves (4) unchanged. The twenty-five element discrete group changes the manifold K_o

$$K_o \rightarrow K = \frac{K_o}{G} \quad , \quad \chi(K_o) \rightarrow \chi(K) = \frac{\chi(K_o)}{N(G)} \tag{6}$$

i.e. when we apply the discrete group G to K_o, the Euler characteristic is reduced by dividing out the number of elements $N(G)$ in G, so e.g., for $N(G)$ = 25 we can have a four family model.

A second advantage of the discrete symmetry pointed out by CHSW and later elaborated by Witten[7] is that it provides a means of dynamically breaking the E_6 symmetry. A function acting on $K = K_o/G$ is equivalent to a function on K_o which satisfies

$$\psi(g(x)) = \psi(x) \ \forall \ g \in G \quad , \tag{7}$$

i.e. the points $g(x)$ and x are identified. Now we do not require (7) to hold because we have the freedom to do an E_6 rotation of the fields. What we do require is

$$\psi(g(x)) = U_g \psi(x) \tag{8}$$

where U_g is a particular E_6 rotation, Technically this is a homomorphism of G into E_6 since $U_g U_g'$ clearly equals $U_{gg'}$. Since G has $N(G)$ elements (e.g., twenty-five for $Z_5 \times Z_5$) we are selecting $N(G)$ elements of E_6. Under a constant gauge transformation in E_6 by a matrix V,

$\psi \to V\psi$; this is only compatible with (8) if

$$[V, U_g] = 0 \qquad (9)$$

so we conclude E_6 is broken to   , consisting of group of transformations which commute with U_g, V_g.

To see this somewhat more physically, consider a Wilson loop defined on a contour γ in K

$$W = \exp\ \{-i\int_\gamma\ T^a\ A^a_m dx^m\}$$

$$a = 1 \ldots\ldots 78$$

$$m = 5,6,7,\ldots 10 \qquad (10)$$

The W's are clearly elements of E_6; from the point of view of d = 4 space $I^a = \int_\gamma A^a_m\ dx^m$ is a spin zero field, i.e. a "Higgs boson" in the adjoint representation. If $I^a \neq 0$, the "Higgs boson" has a non zero v.e.v. In the vacuum the E_6 gauge boson field strength F_{mn} is zero but on a multiply connected manifold a loop may not necessarily be contractible so that we cannot use Stokes' theorem to relate $\int A \cdot d\ell$ to $\int F\ ds$. Since $A^a \neq 0$ necessarily, $I^a \neq 0$ and hence the symmetry is broken to the group   consisting of all transformations in E_6 which commute with all $W \neq 1$. For Z_5 one would expect to have five non-trivial Wilson loops and for $Z_5 \times Z_5$, the W's form a representation with twenty-five elements in general (some of them may be identical) of $G = Z_5 \times Z_5$.

A simple example of this procedure is the following. Consider gauge fields belonging to SU_3 defined on $K_o = S_2$ the two sphere. Let us now act with a discrete Z_2 on S_2 so $K = S_2/Z_2$, the real projective plane. Since antipodal points are identified, a contour which goes from the north to the south pole of K is closed, but clearly non-contractible. On the other the contour in which we cover this path twice is contractible

$$\pi_1(S_2/Z_2) = Z_2 \qquad . \qquad (11)$$

so W has in general two non-zero elements, $U_1 = 1$ and U_2 not necessarily equal to one. It depends on how we embed Z_2 in $SU(3)$. The possibilities are

$$
\text{(a)} \quad \begin{matrix} 1 \\ 1 \\ 1 \end{matrix} \quad \text{(b)} \quad \begin{matrix} 1 \\ -1 \\ -1 \end{matrix} \quad \text{(c)} \quad \begin{matrix} -1 \\ -1 \\ +1 \end{matrix} \quad \text{(d)} \quad \begin{matrix} -1 \\ 1 \\ -1 \end{matrix} \tag{12}
$$

For case (a) SU_3 is unbroken, while for (b), (c) and (d) it is broken to $SU_2 \times U_1$.

The analogous procedure of finding the embeddings of G in E_6 was examined by Witten[7], by Breit, Ovrut and Segre[8] and by Sen[9]. The unbroken subgroup must contain at least $SU_3 \times SU_2 \times U_1$ for phenomenological reasons. In fact, since the embedding leaves the rank of the group unchanged, must be at least as large as $SU_3 \times SU_2 \times U_1 \times U_1 \times U_1$.

If we classify E_6 by its $SU_3 \times SU_3 \times SU_3$ subgroups and let the first SU_3 be the color group, which must remain unbroken, an example of a possible embedding of $Z_5 \times Z_5$ in E_6 is

$$
U = \begin{matrix} 1 \\ 1 \\ 1 \end{matrix} \quad \times \quad \begin{matrix} \alpha^j \\ \alpha^j \\ \alpha^{-2j} \end{matrix} \quad \times \quad \begin{matrix} \beta^k \\ \beta^k \\ \beta^{-2k} \end{matrix}
$$

$$
\alpha^5 = 1 \quad \beta^5 = 1 \quad j,k = 1,2,3,4,5 \tag{13}
$$

It remains of course to be proven that this mechanism for E_6 dynamical breaking actually takes place. So far, all we know is that it is possible; analogous calculations for a far simpler case have been carried out by Hosotani[10].

In Ref. 8, we have actually studied this problem using the method of Weyl weights. The E_6 U's in which G is embedded are written generally as

$$
U = \exp \{iv_i H_i\}
$$

$$
i = 1,2\ldots6 \tag{14}
$$

where the H's are elements of the Cartan subalgebra of E_6 (maximal commuting set of generators) and v's are six dimensional real parameters $v = (a_1 \ldots a_6)$. The v's must have no projection along the direction of color SU_3 or weak SU_2; This fixes v as

$$v = (- c, c, a, b, c, 0) \tag{15}$$

[The mass of a gauge/boson coupled to an E_6 generator E_α, where $\vec{\alpha}$ is a root is given by[11]

$$M_{\alpha\beta}^2 \propto g^2 \, \delta_{\alpha\beta} \, (\alpha_i, v_i)^2 \quad] \; . \tag{16}$$

$G = Z_5 \times Z_5$ can be embedded in U by choosing e.g.

$$e^{ia} = \alpha^j \qquad e^{ib} = \alpha^j \beta^k \qquad e^{ic} = \beta^k \tag{17}$$

This allows us to solve the fine tuning problem, namely why are the SU_2 breaking Higgs doublets so light while the color triplets are so heavy. Let us return to the model with N_o, given in eq. (3), equal to four. A detailed analysis shows that we actually have one $2\overline{7}$ representation and five 27's, all left handed. Now the $2\overline{7}$ can in a sense couple to a linear combination of the five 27's via the effective adjoint Higgs boson present in I^a. If we call this linear combination of the five 27's, ψ, we see that all its components ψ_a for which

$$U \psi_a \neq \psi_a \tag{18}$$

will become superheavy and those for which $U\psi_a = \psi_a$ remain massless. We can now choose an embedding of g in U such that only color singlets satisfy $U\psi_a = \psi_a$. This is not fine tuning since U has at most twenty-five distinct elements.

The $2\overline{7}$ does not enter into the Yukawa couplings which are of the E_6 invariant form

$$yuk = h_{ijk} \, \phi_i \phi_j \phi_k \tag{19}$$

where h_{ijk} is a family matrix and ϕ_i's are chiral superfields belonging to the 27 representation of E_6: there is one invariant 27-27-27 coupling in E_6. In practice (19) is only   invariant, not E_6 invariant, as Witten[7] has pointed out. Naively one would think the couplings (19), which originate from an E_6 invariant Lagrangian would be E_6 invariant. The reason they are not is that the condition (8) tells us we have invariance under the direct sum of   + G. The number of massless modes on $K = K_o/G$ is fixed by the index theorem, but which components of the original supermultiplets survive is not, so we may e.g., find leptons and quarks originating in different multiplets.

Hence (19) will have   invariance, not E_6 invariance. The gauge couplings on the other hand have the full E_6 invariance since the gauge
bosons themselves all come from the (1,78) in the gauge boson supermultiplet, which is not acted on by G.

Clearly the desired solution, from the point of avoiding rapid
proton decay, would be to have the remaining full supermultiplets ψ_i
couple in the form $\psi_i \psi_j \psi$, i.e. no $\psi_i \psi_j \psi_k$ couplings since then we
would have no quark - anti-quark - color Higgs triplet couplings pre-
sent. Unfortunately, a way to implement this has not yet proved
possible.

In conclusion we have seen that imposing discrete symmetries on
the compact manifold has many interesting consequences:

 i) reduces numbers of families

 ii) breaks E_6 to   without breaking SUSY

 iii) can solve fine tuning problem

 iv) can solve phenomenological problems.

A new set of ideas and techniques is being developed[12] and we
may hope for exciting results in the future.

I would like to thank Dr. J. D. Breit for many helpful discussions.

References

1. Green, M.B. and Schwartz, J. H., Physics Lett. 149B, 117 (1983).

2. Gross, D. J., Harvey, J., Martinec, E. and Rohm, R., Phys. Rev.
   Lett. 52, 502 (1985).

3. Chapline, G. F. and Manton, N.W., Phys. Lett. 120B, 105 (1983)
   and earlier references.

4. Candelas, P., Horowitz, G., Strominger, A. and Witten, E., "Vacuum
   Configurations for Superstrings", Nucl. Phys. B (to be published)
   referred to in the text as CHSW.

5. Gursey, F. and Serdaroglu, M., Nuovo Cimento 65A, 337 (1981);
   Gursey, F., Ramond, P., and Sikivie, P., Phys. Lett. 60B, 177
   (1976).

6. Calabi, E., "Algebraic Geometry and Topology", p. 78 Princeton
   University Press (1957).

7. Witten, E., "Symmetry Breaking Patterns in Superstring Models',
   Nucl. Phys. B (to be published).

8. Breit, J.D., Ovrut, B.A., and Segrè, G., Phys. Lett. 158B, 33
   (1985).

9. Sen, A., Phys. Rev. Lett.

10. Hosotani, Y., Phys. Lett. 129B, 193 (1983).

11. Slansky, R., Phys. Rep. 79, 1 (1981).

12. As an example, see Strominger, A. and Witten, E., "New Manifolds for Superstring Compactification", Comm. Math. Phys. (to be published).

ON THE STATUS OF SUPERSTRING MODELS

Chiara R. Nappi[*]

Joseph Henry Laboratories
Princeton University
Princeton, New Jersey 08544

Recently, an enormous amount of progress has been made in superstring theory. Anomaly free d=10 superstring theories have been discovered[1] with gauge groups $SO(32)$ and $E_8 \times E'_8$. Moreover a starting point for phenomenology has been proposed[2] for the $E_8 \times E'_8$ theory.[3] If one requires unbroken supersymmetry in four dimensions, this theory must be compactified to $M_4 \times K$, where M_4 is four dimensional Minkowski space and K is a manifold of $SU(3)$ holonomy (a so called Calabi-Yau space). In this case the effective four-dimensional theory has $E_6 \times E'_8$ gauge symmetry. If one does not require unbroken supersymmetry, an alternative is a Ricci-flat six manifold with $O(6)$ holonomy, in which case E_8 would be broken down to $O(10)$, also a very nice group for phenomenology. Eventually the choice between these or other possible vacua might be made by checking which one actually satisfies the equations of motion of string theory with zero cosmological constant. All these choices so far look very promising since, aside from naturally reproducing the successes of grand unification, they solve the long-standing problem of higher dimensional theories, namely the problem of getting fermions in a chiral structure, and moreover seem to provide a

_____

[*]Research supported in part by NSF Grant PHY80-19754.

rationale for the family replication puzzle. Infact the fermions come automatically in a chiral structure since they belong to the $\underline{27}$ of E_6 or the $\underline{16}$ of $O(10)$ (both of them contain the the $\underline{10}+\underline{5}$ of $SU(5)$). Moreover the number of generations is provided in terms of the Euler characteristic of the compact manifold K.[2]

Due to the uniqueness of the underlying string theory, if one knew the true vacuum state $M_4 \times K$ and the proper pattern of E_6 breaking,[2,5] the theory would be fully determined, without any adjustable parameter left, including Yukawa couplings, which also might be determined topologically.[6] However, since currently it is not known how to deduce the right model from string theory, one can adopt the pragmatic approach of analyzing all possible models that one can get from string compactification. So far the analysis has been restricted to Calabi-Yau spaces only, since in this case it is easier to analyze what light particles are around after compactification, at least in specific cases. Infact, aside from the family content, one could be left with extra light particles due to the so-called incomplete multiplet mechanism.[7,8,9] This information of course is essential if one wants to carry on renormalization group analysis of the various superstring models, as done in [10] by M. Dine, V. Kaplunovsky, M. Mangano, N. Seiberg and myself. Therefore in [10] we restricted ourselves to E_6, in which case supersymmetry is unbroken at the compactification scale. We assumed that supersymmetry survived unbroken down to low energy (let us say down to 1 Tev). The scheme we had in mind was the hidden sector scenario. If the second E'_8 breaks down to some non-abelian subgroup, then gaugino condensation happens in that sector and supersymmetry is broken.[11] Although the scale of gaugino condensation in E'_8 might be fairly high ($>10^{13}$ Gev), since the breaking is communicated to the real world sector via gravitational and related couplings, it might well be that the effective scale of supersymmetry breaking in the real world is actually around the weak scale. Whether this mechanism works or not (if not, maybe a different one will do the job) I will still assume here that all superstring models are supersymmetric down to low energy (a desirable feature, if one hopes that supersymmetry will solve the gauge hierarchy problem).

The models we will analyze are those that can be gotten by break-ing E_6 via Wilson lines,[2,5] a breaking procedure that leaves supersym-metry unbroken. Since any phenomenological acceptable gauge group must contain $SU(3)_C \times SU(2)_L \times U(1)_Y$, out of the various subgroups of E_6 which can be gotten via Wilson lines we are interested only in those contain-ing the standard model. Moreover out of these we study only those which can be eventually broken to the standard model itself via the vacuum expectation values of the only $SU(3)_C \times SU(2)_L \times U(1)_Y$ invariant Higgses that we have at our disposal, namely the scalar partners of the singlets S_1 and S_2 available in the $\underline{27}$ representation of E_6. Infact, aside from standard quarks and leptons, in a $\underline{27}$ there are unobserved quarks (we call them g quarks) of charge -1/3, two doublets of leptons with the quantum numbers of Weinberg-Salam Higgses, and two electrical-ly neutral singlets S_1 and S_2, of which one has the quantum numbers of the antineutrino. Examples of groups G which satisfy the above requirement (there are 20 all together) are

$$SU(3)_C \times SU(2)_L \times U(1) \times (U(1) \times U(1)^2) \tag{1}$$

$$SU(3)_C \times SU(2)_L \times SU(2)_R \times U(1)^2 \tag{2}$$

$$SU(3)_C \times SU(3)_L \times U(1)^2 \tag{3}$$

$$SU(4)_C \times SU(2)_L \times U(1) \tag{4}$$

$$SU(5) \times U(1) \tag{5}$$

First of all, one notices[7] that one can never break E_6 down to the standard model itself. E_6 has rank six and "naturally" it tends to break down to a rank six group. However via "non-abelian Wilson lines" one can manage to break it down to rank five groups for instance $SU(3)_C \times SU(2)_L \times U(1)^2$, with the two U(1)'s here being predicted to be left and right hypercharge Y_L and Y_R.[7] These extra U(1)'s, unless they completely disappear at some intermediate energy scale, might be an important experimental test of superstring models. Neutral current data are within less than a few standard deviations from the $SU(3)_C \times SU(2)_L \times U(1)_Y$ predictions; however they might still be equally compatible with additional neutral gauge bosons of mass of a few hundred Gev.

The basic question about the various superstring models is at what scale they break down to the standard model, namely at what scale $S_{1,2}$ acquire their vacuum expectation values. Of course the answer to this question depends on the assumptions that we make on the superpotential. We know already that we do not need to assume it to be E_6 invariant, but only invariant under the subgroup G that E_6 breaks to.[7] In [10] we assumed that the superpotential contained renormalized as well as unrenormalized couplings

$$w(\phi) = \lambda\phi^3 + \bar{\lambda}\bar{\phi}^3 + O(M_{GUT}^{-1}) \, \phi\phi\bar{\phi}\bar{\phi} + \ldots$$

where ϕ are the superfields from the families or $\underset{\sim}{27}$ and $\bar{\phi}$ those from the antifamilies or $\overline{27}$. We also assumed that, after supersymmetry breaking, $S_{1,2}$ would get a negative square mass, i.e. in the potential for $S_{1,2}$ there was a term of the form $-M_W^2 \, S_{1,2}^2$. We concluded that $\langle S_{1,2}\rangle = \sqrt{M_W \, M_{GUT}}$ or M_W according to whether we could achieve D-flat directions for $S_{1,2}$ or not. In order to achieve D-flat directions we need the incomplete multiplet mechanism to provide in the model under investigation extra light $\underset{\sim}{\bar{S}}_{1,2}$ from the antifamilies $\overline{27}$. In case such particles are not available, then $\langle S_{1,2}\rangle = M_W$ and the outcome is the Grand Desert scenario, namely nothing happens between M_{GUT} and M_W. If particles from the antifamilies are available, we get instead $\langle S_{1,2}\rangle = \sqrt{M_W M_{GUT}}$, and this is the scale at which the various groups G will break down to the standard model (or to acceptable extensions of it). The intermediate scale scenario allows models with four generations to work (at the time of reference [10] only models with 1,2 and 4 generations were known, now additional examples of six manifold of SU(3) holonomy that lead to three generation models are available[12]). The problem with four generation models is that they have Landau poles, i.e. the gauge couplings diverge below unification. This is due to the fact that there are two many particles in the $\underset{\sim}{27}$ of E_6, as already discussed. There are no Landau pole problems if there are three generations only, or, in the case of four generations, if the family content is that of $\underline{10}+\underline{5}$ of SU(5) or $\underline{16}$ of O(10). Infact if one reduces the particle content of the family by making exotic fermions in the $\underline{27}$ heavy at the intermediate mass scale, one manages to avoid Landau poles even for four generation models.

Generic problems with those superstring models are proton decay and neutrino masses. The main cause of proton decay is the possible existence of dimension five operators due to the exchange of g quarks. Some of the symmetry groups G do force on us the Yukawa coupling which lead to unacceptable proton decay rate, and we need to exclude them. If the symmetry group does not force on us dangerous Yukawa couplings, we consider the model acceptable, although obviously these models might have the unwanted Yukawa couplings anyway. Infact it would be very desirable to have a more general mechanism to solve the proton decay problem, rather than hoping that the right superstring compactification scheme will set to zero the bad Yukawa couplings.

Similarly, with neutrino masses. Since there is a field with the quantum number of the right-handed neutrino, it is unnatural to set the neutrino Dirac mass to zero (this is actually impossible in some models, for instance those with left-right symmetry). To get rid of neutrino masses one could try to apply here a general mechanism like the Gell-Mann-Ramond-Slanski mechanism (or see-saw mechanism).[13] However this mechanism cannot be applied here in its simplest version, since no couplings are available to give the antineutrino a large Majorana mass. A generalization of this mechanism, which involves looking at the mass matrix of all neutral leptons around including heavy gauginos coming from the breaking of extra U(1)'s might however work.[14] In [10] we just selected the models where no Dirac neutrino masses or proton decay was forced on us from the symmetries of the gauge group. On the surviving models we performed the Georgi-Quinn-Weinberg one-loop renormalization-group analysis (this analysis can be performed here since gauge couplings, unlike Yukawa couplings, do obey the standard E_6 relations). In the Grand Desert scenario, only the simple extensions of the standard model in (1) give acceptable values of M_{GUT} and $\sin^2\theta_w$, namely

$$M_{GUT}=2.10^{17} \text{ Gev} \qquad \sin^2\theta_w=0.206 \qquad \alpha_{GUT}=0.11.$$

In the case of the intermediate scale scenario, aside from the models (1), a couple of other models, namely models (2) and (3), also give acceptable values of $\sin^2\theta_w$, M_{GUT} and α_{GUT}, both for three and four generations.

In conclusion, I would say that the current status of superstring models is fairly promising, surely comparable to the status of more standard unified models. Moreover, superstring models appear more interesting since potentially they have much more predictive power. However, it is fair to say that serious progress in the field cannot be made unless we reach a deeper understanding of the structure of the vacuum $M_4 \times K$, its discrete symmetries and its topological properties.

REFERENCES

[1]   M.B. Green and J.H. Schwarz, Phys. Lett. 149B, 117 (1984).

[2]   P. Candelas, G.T. Horowitz, A. Strominger and E. Witten, Nucl. Phys. B258, 46 (1985).

[3]   D.J. Gross, J.A. Harvey, E. Martinec and R. Rohm, Phys. Rev. Lett. 55, 502 (1985), and Princeton preprints (1985).

[4]   F. Gursey, P. Ramond, and P. Sikivie, Phys. Lett. 60B (1976); Y. Achiman and B. Stech, Phys. Lett. 77B, 389 (1978); Q. Shafi, ibid, 79B, 301 (1978).

[5]   Y. Hosotani, Phys. Lett. 129B, 193 (1984).

[6]   A. Strominger and E. Witten, Comm. in Math. Phys., in press; A. Strominger, NSF-ITP-85-105

[7]   E. Witten, Nucl. Phys. B258, 75 (1985).

[8]   J. Breit, B. Ovrut and G. Segré, Phys. Lett. B158, 33 (1985).

[9]   A. Sen, Phys. Rev. Lett. 55, 33 (1985).

[10] M. Dine, V. Kaplunovsky, M. Mangano, C. Nappi, N. Seiberg, Nucl. Phys. B, in press.

[11] M. Dine, R. Rohm, N. Seiberg and E. Witten, Phys. Lett. 156B, 55 (1985). See also J.P. Derendinger, L.E. Ibanez, and H.P. Nillen, CERN-TH 4123/85.

[12] S.-T. Yau and G. Tian, Proceedings on the Symposium on Anomalies, Geometry and Topology, ed. W.A. Bardeen and A.R. White (World Scientific Publishers, 1985).

[13] M. Gell-Mann, P. Ramond and R. Slansky, Supergravity, ed. D. Freedman (North-Holland, 1979); R.N. Mohapatra and G. Senjanovic, Phys. Rev. Lett. 44, 912 (1980).

[14] R.N. Mohapatra, University of Maryland Physics Publication No. 85-188.

UTTG-22-85

RADIATIVE CORRECTIONS IN STRING THEORIES

Steven Weinberg

Theory Group, Department of Physics, University of Texas
Austin, TX   78712

Supported in part by Robert A. Welch Foundation and NSF PHY 8304629.

In light of all the recent excitement about superstring theories, it is perhaps surprising that there has not yet appeared any complete prescription of how to calculate S-matrix elements in string or superstring theories beyond the tree approximation. The problem is partly one of infinities that arise from radiative corrections in external lines, but even where these infinities are absent, there are finite renormalization corrections that (as far as I know) have been universally ignored in published work on strings and superstrings.

The problem of infinities in such calculations was already noted [1] in the early 1970's. A closed string theory gives the total contribution of all L-loop graphs to any scattering amplitude as a single multiple integral over a compact Riemann surface with L handles. (For open strings, one has instead the sum of a few integrals). This is very nice, but unfortunately in addition to the diagrams we want, one also encounters the contribution of unwanted diagrams, in which radiative corrections are inserted into external lines. The trouble is that here we have an internal line, connecting the self-energy part to the rest of the diagram, whose momentum is forced by momentum conservation to lie on the particle's mass shell, so that the propagator $(k^2 + m^2)^{-1}$ is evaluated for $k^2 = -m^2$, and is therefore infinite.

Of course, in a quantum field theory we would not include diagrams of this sort. We can just throw them away, or if we are a little more careful, we can use renormalized fields and masses in the zeroth-order Lagrangian[2], so that counterterms cancel the effect of such diagrams. In a string theory, on the other hand, we do not have an opportunity of introducing mass and field-renormalization counterterms. Also, string theories do not give the scattering amplitude as a sum of separate terms, so even if we were willing just to throw away the unwanted diagrams, we would not know how to do it.

The infinities arising in this way are proportional to the two-point amplitude evaluated on the mass shell, i.e., to the mass shift of the incoming or outgoing particle. Hence, as is well-known, these infinites do not arise for processes involving only massless particles like gravitons and dilatons, for which gauge symmetries prevent mass shifts. For this reason, it is sometimes remarked that the whole problem of external-line divergences can be avoided by using the string theory only to calculate amplitudes for processes involving just massless particles, and then relying on unitarity to infer the scattering amplitudes for the massive particles that can appear in intermediate states of the massless particle reaction. This seems to me a pretty disappointing view of what string theories are good for. More important, even though the external-line divergences are absent for massless particles, we will see that they leave a trace, in the form of finite corrections.

Before turning to the detailed analysis of this problem, it may be useful to distinguish it from another that has attracted much attention recently. Momentum conservation fixes the squared momentum of an internal line not only when radiative corrections are inserted into external lines, but also when they generate tadpoles. Here the momentum has a fixed value of zero, so this generates a divergence if the dilaton, the massless scalar in these theories, has a non-vanishing tadpole. This infinity

appears to be absent for some superstring theories, where
cancellations between bosons and fermions cause the tadpole to
vanish[3]. In contrast, the infinity that concerns us here will
occur in superstring as well as string theories. ⊸me confusion
on this point may have arisen because studies of bosonic string
theories have generally emphasized processes involving tachyons,
which do have one-loop mass shifts, while for superstring theories
attention has mostly focused on amplitudes involving only massless
particles, which don't. In fact, for the problems that concern us
here there is not much difference between string and superstring
theories.

Now to details. For the sake of compactness and simplicity,
we shall restrict our analysis here to the simplest string theory:
the closed oriented bosonic string in 26 flat spacetime
dimensions. Also, we will limit our discussion to the one-loop
case, where the problems that concern us first arise. (This
theory has well known problems: not only is the flat
26-dimensional spacetime unrealistic, it is not even an
equilibrium configuration of the theory, as shown by the
non-vanishing dilaton tadpole. However, such problems can be kept
quite seperate from the divergence difficulties under study here.)

To make clear the theoretical framework adopted here, as well
as our notation, let us first recall the prescription provided by
the string theory for calculating the one-loop S-matrix element
for a reaction among a total of E incoming and outgoing particles
of arbitrary type. The particles are labelled with 26-momenta
$k_1^\mu \cdots k_E^\mu$ with $\mu = 0, 1, \cdots , 25$; other discrete labels for spin
and particle type are implicitly included in the k^μ's. We take
all k^μ as incoming; the particle is in the initial or final state
according as $k^0 > 0$ or $k^0 < 0$. We work with an invariant
amplitude Γ, defined by setting the S-matrix equal to

$$S = \frac{(2\pi)^{26} \delta^{26}(k_1 + \cdots + k_E) \; \Gamma \; (k_1, k_2, \cdots, k_E)}{(2\Pi)^{25E/2} \left(2|k_E^{\;o}|\right)^{1/2} \cdots \left(2 \; |k_E^{\;o}|\right)^{1/2}} \tag{1}$$

The string theory gives Γ in one-loop order as a multiple integral

$$\Gamma_1(k_1 \cdots k_E) = -i \int d\tau \; \mathscr{F}(\tau)$$

$$x \int d^2 z_1 \cdots d^2 z_E \; \langle V(z_1, k_1) \cdots V(z_E, \, k_E) \, \rangle_\tau \tag{2}$$

Here $V(z,k)$ is a "vertex function" of a string coordinate $x^\mu(z)$ and its derivatives at z, subject to certain symmetry and normalization conditions[4]; for instance, $V(z,k)$ for the tachyon is just proportional to $e^{ik \cdot x(z)}$. The average $\langle F [x] \rangle_\tau$ for any functional of $x(z)$ is defined as

$$\langle F [x] \rangle_\tau = \frac{\int\limits_{(\tau)} \prod\limits_{\mu, z} dx^\mu(z) \; e^{-I[x]} \; F [x]}{\int\limits_{(\tau)} \prod\limits_{\mu, z} dx^\mu(z) \; e^{-I[x]}} \tag{3}$$

where $I[x]$ is the string action

$$I[x] = T \int d^2 z \; \frac{\partial x^\mu}{\partial z} \; \frac{\partial x_\mu}{\partial z^*} \quad ; \tag{4}$$

T is the string tension; and the integrals are over all $x^\mu(z)$ satisfying the periodicity conditions for a torus:

$$x^\mu(z) = x^\mu(z + 1) = x^\mu(z + \tau) \tag{5}$$

The z-integrals may be taken over the parallelogram with corners at z values 0, 1, τ, and $1 + \tau$. We will not need to give the weight function[5] $\mathscr{F}(\tau)$ here.

In general, the pole in $\Gamma(k_1, \cdots, k_E)$, that occurs when the total 26-momentum of some N-particle subset of the k's approaches some mass shell, arises from the regions in z-space where either all N of the z's corresponding to these particles are close together, or all the E-N other z's are close together, or both. The divergences that concern us here occur because one-particle subsets of the E external lines automatically have 26-momenta on a mass shell, and since there is only one z corresponding to this subset, the divergence arises only from the region in which all the E-1 other z's are close together.

To deal with this divergence, we introduce a regulated one-loop amplitude[6] $\Gamma_1^{(\varepsilon)}$, by specifying that in Eq. (2) we include only z-configurations in which for each particle $i = 1, 2, \cdots, E$, we have $\sum |z_j - z_k|^2 > \varepsilon_i^2$, the sum running over all E-1 j and k not equal to i. It is not hard to show that this regulated one-loop amplitude has the ε-dependence, for $\varepsilon \to 0$

$$\Gamma_1^{(\varepsilon)}(k_1 \cdots k_E) = \frac{1}{8\pi T} \Gamma_o(k_1 \cdots k_E) \sum_{i=1}^{E} \delta m_i^2 \ln \varepsilon_i^2$$

$$+ \varepsilon - \text{independent terms} \qquad (6)$$

where Γ_o is the invariant amplitude in the tree (no handle) approximation, and δm_i^2 is the one-loop mass shift of the i-th particle. There is no reason to expect δm_i^2 to vanish except for a few massless particles like the graviton, so in general (6) blows up for $\varepsilon_i \to 0$.

This is the problem; now, what do we do about it? The ε-dependent term in the one-loop amplitude $\Gamma_1^{(\varepsilon)}$ is proportional to the tree amplitude Γ_o, so this ε-dependence could be cancelled if the vertex functions $V(z,k)$ that we used to calculate Γ_o had been multiplied by suitable ε-dependent factors. But why should we? We prefer finite theories, but that doesn't mean that we are entitled to modify a theory to get rid of infinities whenever they occur. Also, once we start modifying the vertex functions in this way, how do we know what finite corrections to include along with the infinite ones?

For the present, the one guide we have to the normalization of the vertex functions is the unitarity — in effect, the factorization — of the S-matrix. For instance, it is by applying this condition in the tree approximation that we learned[4] that the vertex function for the tachyon is g : exp (ik $\cdot$ x) : and that for the dilaton is $4\pi Tg(26)^{-1/2}$: $\dfrac{\partial x^{\mu}}{\partial z}\dfrac{\partial x_{\mu}}{\partial z*}$ exp (ik $\cdot$ x): where g is a constant related to the coefficient of the 2-curvature term in the string action, which for instance appears in Γ_0 through a universal factor $8\pi^2 Ti/g^2$. (The colons indicate that in the functional integral over x(z), we do not include contractions of x(z)'s in the same vertex function.) But these results for the vertex function were obtained in the tree approximation, and must be expected to suffer corrections in higher order.

For instance, let us calculate the one-loop corrections to the tachyon vertex function. Consider an E-tachyon amplitude, and suppose that the sum $k^{\mu} = k_1^{\mu} + \cdots + k_N^{\mu}$ of the first N momenta approaches the tachyon mass shell, $k^2 \to 8\pi T$. In accordance with the general rule cited above, the pole here arises from regions in which either (A) $z_1, \cdots, z_N$ are all close together, or (B) $z_{N+1} \cdots z_E$ are all close together, or (C) both. By a set $\mathcal{S}$ of z's being all close together, we may understand that

$$\sum_{j,k \in \mathcal{S}} | z_j - z_k |^2 < \varepsilon^2 \tag{7}$$

for some arbitrary small ε. It is not hard to work out the contributions of each of these three regions to the one-loop amplitude near the pole to leading order in ε:

$$\Gamma_{1A} (k_1 \cdots k_E) = \Gamma_0 (k_1 \cdots k_N, -k) D^{(\epsilon)}(k^2)$$
$$x \ \Gamma_1^{(\epsilon)}(k, k_{N+1} \cdots k_E) \tag{8}$$

$$\Gamma_{1B} (k_1 \cdots k_E) = \Gamma_1^{(\epsilon)}(k_1 \cdots k_N, -k) D^{(\epsilon)}(k^2)$$
$$x \ \Gamma_0(k, k_{N+1} \cdots k_E) \tag{9}$$

$$\Gamma_{1C}(k_1 \cdots k_E) = \Gamma_0(k_1 \cdots k_N, -k) D^{(\epsilon)}(k^2) \Pi(k^2) \tag{10}$$
$$x \ D^{(\epsilon)}(k^2) \Gamma_0(k, k_{N+1} \cdots k_E)$$

where

$$D^{(\epsilon)}(k^2) \equiv \frac{-i}{k^2 - 8\pi T} \ (\epsilon^2)^{(k^2 - 8\pi T)/8\pi T} \tag{11}$$

and Π is the 2-point function

$$\Pi(k^2) = - i\Gamma_1(k, -k)$$

It is important to note that the one-loop amplitudes $\Gamma_1^{(\epsilon)}$ in (8) and (9) are defined to <u>exclude</u> the regions (7) for the subsets $\mathscr{S} = \{N + 1, N + 2, \cdots E\}$ and $\mathscr{S} = \{1, 2 \cdots N\}$, respectively. This is just the same sort of regulator that was introduced earlier to avoid the external-line divergences, but the regulator appears here not because we want to suppress divergences, but to avoid double-counting; if Γ_1 had appeared in (8) or (9) without this regulator then the regions (A) and (B) would include the region (C), and the contribution (10) would be double or triple-counted.

Eq. (10) contains a double pole, corresponding to a one-loop shift in the position of the single pole:

$$\delta m^2 = - \Pi (-m^2) = \int d\tau \mathscr{F}(\tau)$$

$$x \int d^2z_1 d^2z_2 \left[< V(z_1,k) \ V(z_2, -k) >_\tau \right]_{k^2 = -m^2} \qquad (12)$$

In this form, this formula gives the mass shift for any sort of particle, not just for tachyons. For most particles δm^2 is complex, because one-loop corrections make these particles unstable.

Our concern here is with the simple pole in the amplitude $\Gamma_0(k_1 \cdots k_E) + \Gamma_1(k_1 \cdots k_E)$ at $k^2 = 8\pi T$. Unitarity requires this pole to take the form

$$-i(k^2 - 8\pi T)^{-1} \ \Gamma_R(k_1 \cdots k_N, -k) \ \Gamma_R(k, k_{N+1}, \cdots k_E) \qquad (13)$$

and tells us to interpret these Γ_R's as the renormalized N+1 and E − N + 1 tachyon amplitudes. Including the zero-th order term as well as (8) − (10), we easily see that the pole does have the form (13), with

$$\Gamma_R = \Gamma_0 + \Gamma_1^{(\varepsilon)} + \delta\Gamma_0 + \frac{1}{2} \ \Pi' \ (-m^2) \ \Gamma_0$$

$$- \frac{\delta m_T^2}{8\pi T} \ \Gamma_0 \ \ln \varepsilon^2 \qquad (14)$$

Here $\delta\Gamma_0$ is the change in the tree amplitude due to the change in squared mass of the exchanged tachyon, from $-8\pi T$ to $-8\pi T + \delta m_T^2$. Apart from this term, about which more presently, Eq. (14) can be interpreted to mean that in calculating Γ to one-loop order, the tree approximation term should be calculated using a vertex function

$$V_{TACHYON} = [1 + \frac{1}{2}\ \Pi'\ (-m^2) - \frac{\delta m_T^2}{8\pi T}\ \ln\ \varepsilon^2]\ g\ :\ e^{ik\cdot x}: \qquad (15)$$

where ε is the regulator introduced in calculating the one-loop term. The ε-dependent term in (14) or (15) (which arises from the ε-dependence of the $D^{(\varepsilon)}$ factors in (10)) cancels the ε-dependence (6) in the regulated one-loop amplitude $\Gamma_1^{(\varepsilon)}$, thus solving the infinity problem raised here.

For gauge particles like the graviton and dilaton there is no mass shift or external-line divergence problem, but there still is a vertex function correction like $\frac{1}{2}\ \Pi'$ in Eq. (15). Supersymmetry may cause some of these corrections to vanish in some superstring theories, but I know of no general theorems that would insure such cancellations.

Perhaps the most interesting term in (14) is $\delta\Gamma_0$, the change in the tree amplitude Γ_0 induced by evaluating it on the one-loop-corrected mass shell rather than the tree mass shell. This correction clearly must be included in the same order in g^2 as the ordinary one-loop corrections, but I do not know of any previous discussion of it, which is curious because this correction raises an important point of principle. The Polyakov path-integral formula[7] for Γ is conformal-invariant only on the tree mass shell (e.g., $k^2 = 8\pi T$ for tachyons). Hence if we used this formula to calculate Γ_0 on the corrected mass shell (e.g., $k^2 = 8\pi T - \delta m_T^2$) we would introduce a spurious dependence on the metric of the 2-sphere. But that is not what is meant here by

$\delta\Gamma_o$. In the derivation of Eqs. (8) – (10), the quantity which is here called Γ_o appears in the form

$$\Gamma_o(k_1 \cdots k_N, -k) = 8\pi^2 i T g^{N-1}$$

$$x \int d^2w_3 \cdots d^2w_N \exp \left[\frac{1}{4\pi T} \sum_{1 \leq i < j \leq N} k_i \cdot k_j \ln |w_i - w_j|^2 \right.$$

$$\left. - \frac{(k^2 - 8\pi T)}{8\pi T} \ln \sum_{1 \leq i < j \leq N} |w_i - w_j|^2 \right] \qquad (16)$$

with w_1 and w_2 taken at any fixed values $w_1 = a$, $w_2 = b$. For $k^2 \to 8\pi T$, this gives the usual $N+1$-tachyon amplitude in the tree approximation. (This can be seen by using a Mobius transformation in the usual formula for Γ_o to set $w_1 = a$, $w_2 = b$, and $w_{N+1} = \infty$)

However, for $k^2 \neq 8\pi T$, Eq. (16) provides an extension of the tree amplitude with one particle off the mass shell, that seems to have nothing to do with the Polyakov formula, but is required by unitarity. (The second term in the exponent of (16) does introduce a dependence on the measure used in Eq. (7) to define what we mean by a set of points being close together, but this dependence cancels in (14) with the dependence of $\Gamma_1^{(\epsilon)}$ on the measure used in regulating Γ_1.) It would be interesting to understand in a more geometric way why Eq. (16) is the correct extension of the tree amplitude with one particle off the mass shell.

I have not yet been able to extend this analysis to higher orders. Apparently, we must understand how the modular variables for a Riemann surface with L handles are converted near a pole into modular variables for Riemann surfaces with $L' < L$ and $L-L'$ handles. Until we learn how to deal with external line radiative corrections to all orders, it would seem premature to speak of superstring theories as being finite.

I am grateful for discussions on these problems with D. Gross, S. Mandelstam, G. Moore, J. Polchinski, J. Schwarz and E. Witten.

References

1. See e.g., S. Mandelstam, Phys. Reports 13, no. 6 (1974), 259
2. P.T. Matthews and A. Salam, Phys. Rev. 94, 185 (1954)
3. M. Green and J. Schwarz, Phys. Lett. 151B, 21 (1985)
4. S. Weinberg, Phys. Lett. 156B, (1985), 309
5. J. Polchinski, Texas preprint UTTG-13-85 (1985)
6. A similar regulator is used for open strings by P.H. Frampton, P. Moxhay, and Y.J. Ng, Harvard-North Carolina preprint HUTP-85-A059/IFP-255-UNC (1985)
7. A.M. Polyakov, Phys. Lett. 103B, (1981), 207

VI

DETECTORS AND ACCELERATORS

Session Organizers: M.A. Abolins and L. Teng

Review of Recent Progress in the Development of Čerenkov Ring Imaging Detectors*

David W.G.S. Leith

Stanford Linear Accelerator Center
Stanford University, Stanford, California 94305

1. INTRODUCTION

In the late 1970's Ypsilantis and Seguinot[1] proposed an extension of the classic binary "yes/no" Čerenkov counter to a new kind of particle identification system, with large dynamic range, robust performance and the potential to perform in the "heavy traffic" of the multi-particle jet environment. These new detectors were called Ring Imaging Čerenkov (RICH) counters[2] or Čerenkov Ring Imaging Detectors (CRIDs).[3]

The principle behind these devices involves focussing the Čerenkov light, emitted by a relativistic charged particle in passing through a radiator medium, onto a high efficiency photocathode which can in turn be read out with good spatial resolution, to localize the point of origin of the photoelectrons. This information permits the reconstruction of the circle of Čerenkov light for each particle above threshold, and hence the determination of the Čerenkov angle to an accuracy of a few percent. These principles are outlined in Figures 1(a)-(c).

However, there are a number of rather basic problem areas that have to be addressed before one can proceed with plans to implement a large ring imaging device.

---

* Work supported by the Department of Energy, contract DE-AC03-76SF00515.

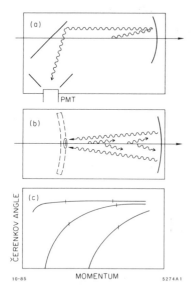

Fig. 1 Schematic representation of (a) a classic threshold counter, (b) a ring imaging detector, and (c) the separation of different species of particles through measurement of the Čerenkov angle.

- High Purity Fluid Systems:
  - At the wavelengths one chooses to work, a few parts per million of oxygen or water vapor in any of the gas or liquid systems would attenuate the Čerenkov light and the photoelectrons. Care has to be taken to use clean construction materials and to recirculate the fluids through filters.

- Mirror System:
  - The optical arrangement in most Čerenkov ring imaging devices requires large area mosaics of quartz or calcium fluoride window arrays to separate the photocathode gas from the radiator gas, and rather large area arrays of mirrors with good reflectance (i.e., > 75%) in the vacuum ultraviolet wavelength region.

- Photocathode:
  - The photocathode has to be very efficient, since for practical geometries the radiator dimensions are such that only a few dozen Čerenkov photons are produced.
  - The area of realistic photocathode surfaces tend to be large—of order of a few square meters to a few tens of square meters.
  - The photocathode should be capable of being read out electronically with pixell size of order 1 mm^3.

- Detectors:
  - The detectors have to be capable of efficiently detecting single photoelectrons and must have some protection from photon feedback.

    Photon feedback is the positive feedback phenomenon caused when a very efficient photoionizing agent is placed close to a MWPC which causes light emission from the electron avalanche near the anode wires.

Given acceptable solutions to these problems one should expect superior performance from a Čerenkov ring imaging particle identification system. An example, taken from the SLD Design Report[4] is shown in Fig. 2.

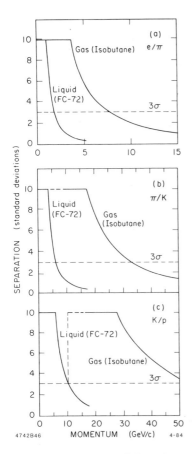

Fig. 2 The particle identification capability for a two radiator CRID, such as described in Ref. 4. The performance of the liquid (FC-72) and gas (isobutane) radiators is shown separately. The separation is shown for (a) e/π, (b) π/K and (c) K/p.

For the remainder of this talk we will look at which groups are currently working on these new detectors, examine the status of these projects and report in some detail on the recent R&D progress from the two 4π devices being prepared for physics at the Z^0.

2. WHO ARE IN THE GAME?

There are four different collaborations actively working on the implementation of large Čerenkov ring imaging devices. I will briefly describe each of the detectors and their status.

2.1 E605 At Fermilab

This is a forward spectrometer experiment in the fixed target program at Fermilab, designed to study very high energy hadron-hadron collisions. The best known result of this experiment is probably the spectacular dimuon mass spectrum produced in 800 GeV proton-nucleus collisions, but the spectrometer is equipped with a sophisticated ring imaging counter to provide π, K, p identification of secondary hadrons up to around 200 GeV/c.[5] A schematic of their RICH is shown in Fig. 3. The radiator is 15 meters of gaseous helium, and the Čerenkov light produced in this radiator is focused back onto the photon detectors by an array of 16 spherical mirrors. The detectors are isolated from the radiator volume by a mosaic of CaF_2 windows–each 4 mm thick and 10×10 cm^2 in area. The mirror reflectance is measured to be greater than 75% down to wavelengths of 150 nm, and over the same wavelength range the window transmission is measured to be greater than 70%.

Each detector measures (40×80) cm^2 in size, and is mounted one on each side of the radiator volume (see Fig. 3).

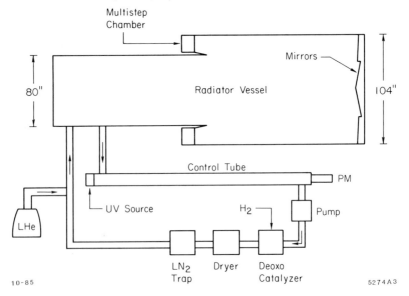

Fig. 3 Schematic of the Fermilab E-605 experiment ring imaging counter. See Ref. 5 for details.

The detector works by converting the Čerenkov light to free electrons via photoionization of an admixture of timethylamine (TEA) to the detector gas

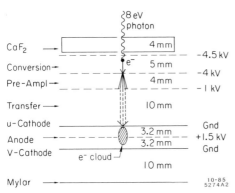

Fig. 4 A schematic of the grid structure in the photoelectron detector for experiment E-605 at Fermilab (Ref. 5).

mixture. The photoelectron then drifts under the action of electric field to a preamplification gap and then onto a multiwire proportional chamber. A schematic of the electro-static grid structure is given in Fig. 4. The detector gas mixture is 90 : 5 : 3 of helium : methane : triethylamine, (TEA). The absorption length for the vacuum U.V. Čerenkov light in this gas mixture is of order 1 mm. The total gain of the preamplification gap and the narrow gap PWC is about 10^7. The PWC anode is constructed of

20 μm wires wound on a 2 mm pitch, while the cathode planes are at ±45° with respect to the anode, and are wound with 50 μm wire on a 1 mm pitch. Both the cathode and the anode wires are read out to measure the electron coordinates.

On average, three photoelectrons are detected for each relativistic charged particle, and this enables a measurement of the radius of the Čerenkov circle (~ 70 mm) with an accuracy ~ 0.7 mm, which in turn, allows good particle identification up to 200 GeV/c. See Fig. 5.

This device has been running routinely in the Fermilab experiment for the past three years.

2.2 Rutherford's RICH at the Omega Spectrometer

The Omega spectrometer is yet another multi-particle forward spectrometer experiment, this time in the high energy secondary hadron beam at the SPS at CERN. A layout of the spectrometer is shown in Fig. 6. This experiment is equipped with a large ring imaging Čerenkov counter[6] to provide final state hadron identification. It is designed to identify pions from (5-80) GeV/c and providing K/p discrimination up to 160 GeV/c.

The radiator is 5 meters of nitrogen gas. Čerenkov light from the radiator volume is focussed back onto the photon detectors, which are mounted at the focal plane of a 28 square meter spherical mirror array. The reflectance in the (170-250) nm region was measured to be > 75%. The Čerenkov light is detected in 16 TPC devices, each (40 × 80) cm^2 in area, and shown schematically in Fig. 7. The Čerenkov photons photoionize an organic vapor (Tetrakis Dimethylamino Ethylene – TMAE) which is added to the chamber gas mixture of

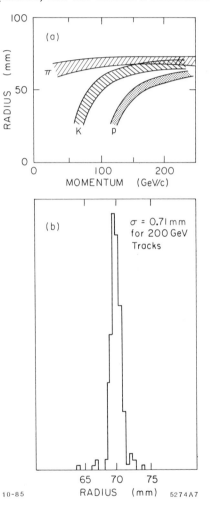

Fig. 5 (a) A schematic of the radius measurement for π, K, p as a function of momentum. (b) The measurement error for a sample of 200 GeV/c tracks.

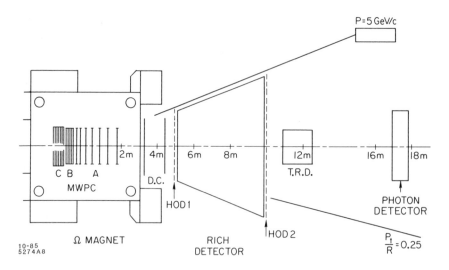

Fig. 6 A schematic of the Omega spectrometer experiment at CERN.

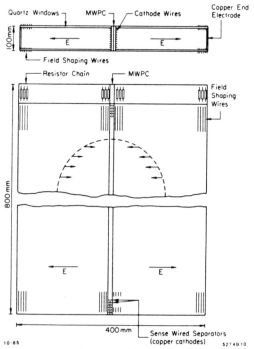

Fig. 7 A schematic of the photon detector for
the Omega experiment at CERN.

methane/isobutane 80/20. The chamber gas is isolated from the radiator gas by the quartz windows shown in Fig. 7.

Both windows and side walls of the TPCs are wound with an inner and outer field grid which establish a uniform electric drift field to transport photo-electrons from their creation point to the central MWPC detector. Measurement of the drift time and the wire address allows measurement of the Čerekov circle radius. (Note the intrinsic left-right ambiguity in this device).

The performance of this device is shown in Fig. 8. The accuracy of measurement of the Čerenkov ring radius is found to be $\sigma \sim 2.9$ mm, and allows good K/π identification up to nearly 80 GeV/c, and K/p identification up to about 160 GeV/c. The actual number of photoelectrons detected is about half that expected; they detect around 14 photoelectrons rather than the 28 calculated. The loss of photoelectrons is thought to be partly in the efficiency of detecting the single photoelectron and partly in the input electron optics to the MWPC shown in Fig. 7. This group is running again in the fall of this year with improved electronics and electron transport optics, and will also use a freon radiator gas. Improvements are expected from each of these changes. We wait, with great interest, their new results.

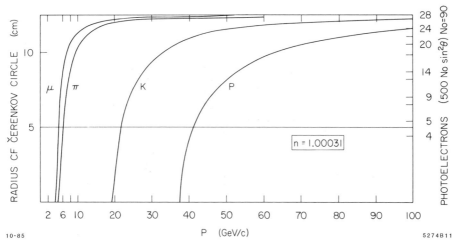

Fig. 8 The calculated performance of the Omega RICH.

2.3 The DELPHI Experiment at LEP and the SLD Experiment at SLAC

The DELPHI experiment is one of the four large experiments being built at CERN to study Z^0 production and decay at LEP. The usual quadrant view of the experiment is shown in Fig. 9.

870

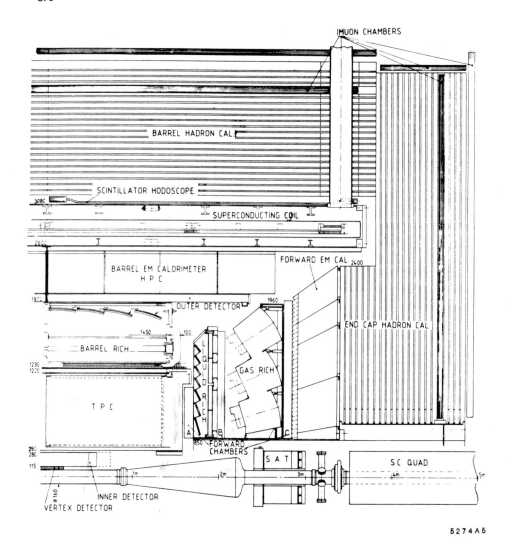

Fig. 9 General layout of the DELPHI experiment.

The SLD experiment is the large new detector being built at SLAC to fully exploit the physics opportunities of the SLC. The detail of the Čerenkov counter is shown in Fig. 10.

Both DELPHI and the SLD devices use a double radiator scheme to provide an extended momentum range over which particle identification may be achieved. See for example Fig. 2. The principle is described below.

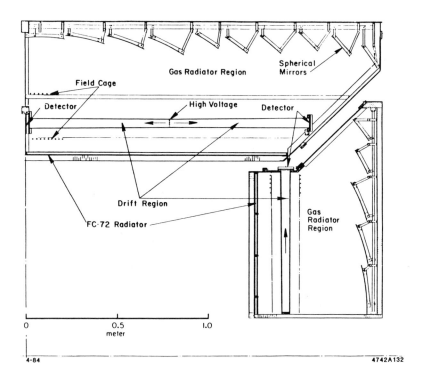

Fig. 10 Schematic of the CRID for the SLD experiment. For details see Ref. 4.

A relativistic charge particle passing through both radiators would generate patterns of light characterized in Fig. 11; the light from the liquid radiator, (C_5F_{12}), is proximity focussed on one side of the photon detector while the Čerenkov light from the gas radiator is focussed via spherical mirrors onto the other side of the photon detector. The image from the liquid is rather broad due to geometry and chromatic effects, and has a radius of about 17 cms. It is expected to have about 24 photoelectrons distributed around its circumference. The image from the gas radiator (isobutane or C_6F_{14}) should be about 3 cm radius, be rather sharp and have about 12 photoelectrons.

The photon detector is a double-sided quartz box (to transmit the Čerenkov light) with a double electric field cage wound on the inner and outer walls of the box to create an uniform electric drift field along the axis of the box. See Fig. 12. The photons are detected by photoionization of Tetrakis Dymethyl Amino

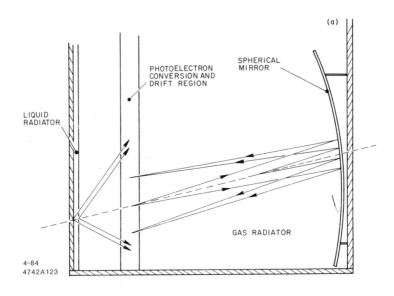

Fig. 11 A schematic representation of the two radiator geometry used in the Čerenkov ring imaging detectors in DELPHI and SLD.

Ethylene (TMAE), which is present at 0.1% level in a 70 : 30 methane : isobutane gas mixture. The TMAE has a very high photoionization cross section in the (1700-2400) Å region, and so acts as a very efficient photocathode. Photo-

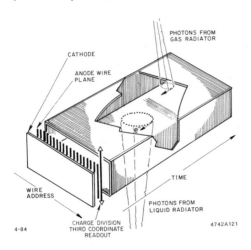

Fig. 12 Schematic of the photon detector boxes used in DELPHI and SLD.

electrons will drift under the action of the parallel electron and magnetic fields to a multi-wire proportional chamber detector. The drift time and wire address provide two of the coordinates of the creation point of the photoelectron. Information on the depth at which the photoionization took place is found from localization of the avalanche along the wire.

The Delphi and SLD devices have rather similar specifications. As an example of these systems, a summary of the parameters of the SLD device is given below in Table I.

Table I. Summary of Parameters of the SLD CRID

	Liquid	Gas
Solid Angle	98%	94%
Angular Acceptance: Endcap	$10.5° - 37°$	$8.5° - 33°$
Barrel	$37° - 90°$	$38° - 90°$
Radiator	FC-72	Isobutane
Index of Refraction (6.5 eV)	1.277	1.0017
Focusing	Proximity	Spherical mirror
Č Threshold (γ)	1.61	17.5
Č Ring Radius	17 cm	2.8 cm
N_{pe}	23	13
P_{thresh} (3 p.e.)		
e	~ 1 MeV/c	~ 10 MeV/c
π	0.23 GeV/c	2.6 GeV/c
k	0.80 GeV/c	9.5 GeV/c
p	1.50 GeV/c	17.8 GeV/c

A demonstration of the power of the full particle identification system in SLD, (*i.e.*, vertex detector plus CRID) is shown in Fig. 13. A Monte Carlo study of Z^0 decays within the SLD detector has been performed.[8] The charmed meson decay of the heavy boson was chosen as typical of the heavy quark decays and the $D \rightarrow K\pi\pi$ decay of the D as used to study the performance of the detector. The $(K\pi\pi)$ mass distribution is plotted in Fig. 13 for four separate selections of data; (a) displays the $K\pi\pi$ mass for all appropriately charged three-track combinations which belong to the same jet–very little sign of the D-meson is observed. (b) shows $K\pi\pi$ combinations that have been identified by the CRID; a clear D-meson peak is observed with a signal-to-noise of about 1 : 3. (c) displays the $K\pi\pi$ mass for unidentified tracks coming from a secondary vertex as measured by the high resolution CCD vertex detector in SLD; a very clear D peak is seen with a signal-to-noise of about 3 : 1. And finally in (d) the $K\pi\pi$ mass spectrum is plotted for identified π's and K's, that come from a separated, secondary vertex; an extremely clear D signal is observed with rather good tagging efficiency, ($\epsilon \sim 35\%$).

This ability to clearly and efficiently tag heavy quark decays (t, b, c), of the Z^0 should enable studies of very small branching ratio decays into heavy flavor mesons, allow detailed study of the cascade process $t \rightarrow b \rightarrow b \rightarrow c \rightarrow s$, and of $B - \bar{B}$ and $D - \bar{D}$ mixing.

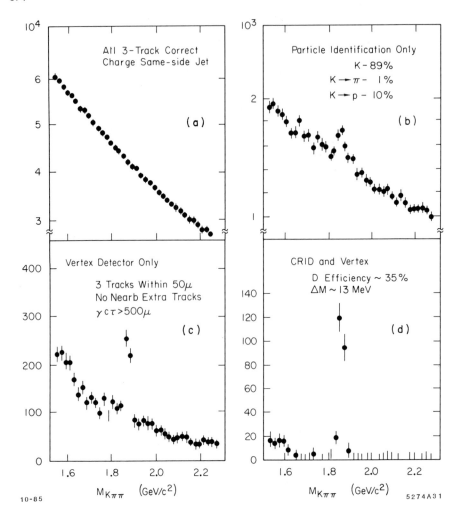

Fig. 13 Mass distribution of Kππ combination resulting from Z^0 decays. (a) All appropriately charged combinations. (b) Identified π and K candidates using the CRID. (c) Three appropriate charged particles coming from a secondary vertex. (d) Identified π's and K's which come from a separated secondary vertex.

3. R&D STATUS OF THE NEW 4π DETECTORS

There are two 4π ring imaging detectors currently under development. The SLD CRID being built at SLAC, and the DELPHI RICH detector being built at CERN. The R&D progress for both detectors is described below.

3.1 SLD R&D Progress

The SLD CRID group has been working on a 'proof-of-principle' prototype, designed to test most of the detector ideas in a full scale device, but not in the specific geometry or the actual construction techniques used in SLD. The device is shown schematically in Fig. 14.

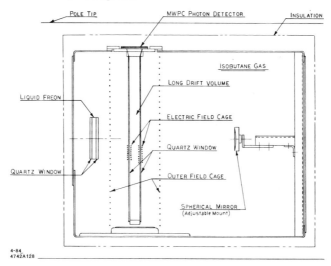

Fig. 14 Schematic of the "proof-of-principle" prototype of the CRID under test at SLAC.

The prototype elements are mounted within an aluminium box, 1 meter on each side, which can be inserted between the pole tips of a large electro magnet providing fields up to 18 kG. The test devices were:

- a liquid radiator cell, which is an all quartz cell 1.27 cms thick and connected to an external pump to circulate C_6F_{14}, perfluorohexane radiator liquid through external oxysorb filters to remove oxygen and water vapor.

- a secondary electrostatic field cage, wound with 1 mm diameter copper wire on a 2.5 cm pitch. This cage helped maintain electric field uniformity in the photon detector box, and isolate it from the nearby electrical grounds.

- the photon detector box, which is a quartz box with a double window of dimension 20 × 80 cms, 4 cms deep at one end, increasing to 6 cms depth at the detector end. A double field cage is wound on the inner and the outer surfaces of the detector box, with a 2.5 mm pitch, providing an adjustable electric field in the range (50-650) v/cm. The box is filled with a gas mixture

- $(CH_4$ (70%), isobutane (30%), TMAE ($\sim$ 0.1%) – which was circulated at flow rates of (5-40) ℓ/hr. The TMAE acts as the photocathode.

• a spherical mirror, with good ultraviolet reflectance, ($\gtrsim$ 86% in the region (1700 - 2400 Å), and adjustable so that it could image the Čerenkov light produced in the 43 cms of gas radiator back onto the photon detector box at any desired position along the length of the box.

• an electron detector, to efficiently detect the single photoelectrons released from the TMAE. The electrons are detected by a picket fence of multiwire proportional counters working at a gas gain of between 1 and 2 $\times 10^5$. In order to avoid a positive feedback situation one has to limit the illumination of the TMAE volume by the light emitted in the avalanche at the anodes of the MWPCs. This protection from photon feedback is achieved by mechanically limiting the solid angle and is discussed below under detector development.

The prototype was installed in a momentum analyzed secondary beam which could deliver either electrons or hadrons to the experiment. The response of the device to 11 GeV/c hadrons is shown in Fig. 15(a), a two-dimensional plot of time versus wire address integrated over 200 beam tracks. The plot shows the gas ring, the front and rear sectors of the liquid ring and the beam spot. Fig. 15(b) shows the gas ring by itself. The width of the gas ring is dominated by the parallax error from the lack of information on the third (depth) coordinate.

Figure 16 shows a single event display; (a) for the full event and (b) for the gas ring only. The solid lines are fiducial regions for the gas and liquid rings.

Figure 17 is the radius plot for the gas ring, and shows good signal-to-noise with very little background. The number of detected photoelectrons for each gas ring is shown in Fig. 18, and compared to a poisson distribution for $\bar{n} = 7$.

The number of photoelectrons detected from the liquid radiator is shown in Fig. 19, where the bottom scale is the actual number found for the leading sector of the ring, and the upper scale is the number of photoelectrons calculated for the full ring circumference. About $\sim$ 18 photoelectrons per particle are detected from the liquid radiator.

It is clear from the difference in populations of the leading and trailing sectors of the liquid ring in Fig. 15(a), that electrons are being lost as they drift down the photon detector box. We measured the attenuation lengths quantitatively by: i) varying the drift velocity (and therefore the drift time), of the photoelectron by changing the drift field, and ii) by changing the position of the focus of the gas ring by adjusting the mirror orientation. Both methods yield an electron attenuation length of approximately 35 cms (see Fig. 20). An average of 13 photoelectrons per incident particle for the gas radiator, and an average

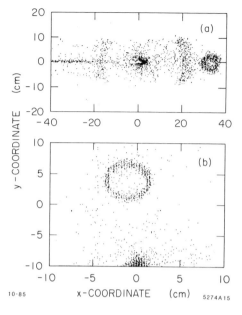

Fig. 15 (a) On-line residual plot of 200 beam tracks; y is the wire address and x is derived from the drift time. (b) A similar plot on an expanded scale showing the gas ring alone.

Fig. 16 Single event display of wire address against drift time.

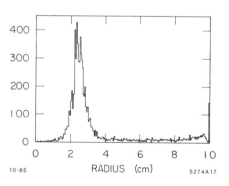

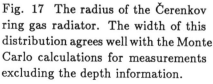

Fig. 17 The radius of the Čerenkov ring gas radiator. The width of this distribution agrees well with the Monte Carlo calculations for measurements excluding the depth information.

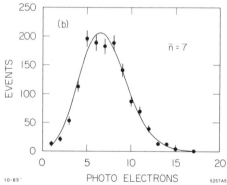

Fig. 18 The number of photoelectrons detected per event for the Čerenkov ring produced by the gas radiator.

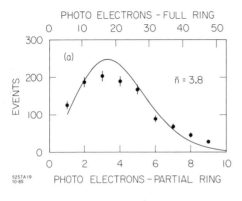

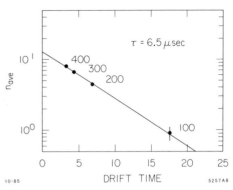

Fig. 19 The number of photoelectrons detected per event for the leading arc of the Čerenkov ring produced by the liquid radiator. The upper scale gives the scaled number that would be detected for the full ring.

Fig. 20 The average electron "lifetime", or attenuation length, derived from the number of electrons on a gas ring versus the electron drift time. The drift time is determined by the electric drift field which is shown in the figure in volts/cm.

of ~ 37 photoelectrons per incident particle for 1 cm of liquid radiator were observed after correcting for the electron attenuation length. These yields exceed our expectations.

The device was tested with incident pions from 4 GeV/c to 11 GeV/c. Figure 21 displays the Čerenkov angle measurements from the CRID prototype.

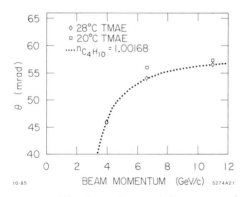

Fig. 21 The dependence of the measured Čerenkov angle for tagged pions as a function of momentum.

At 4 GeV/c, the beam was tuned to deliver a mixture of electrons and pions, the beam particle being tagged by a threshold Čerenkov counter. Figure 22 shows single event displays for a) an electron and b) a pion incident on the CRID prototype. The dotted circles indicate the expected Čerenkov circles for pions (inner) and electrons (outer), while the crosses are the detected photoelectron coordinates. The solid line is a fit to the data.

Figure 23 shows the distributions of the Čerenkov angles for the tagged pion and electron samples, as measured by the CRID prototype. A clear 3 σ separation is observed even without third coordinate (depth) measurement. It is interesting to note that a factor of four reductions of the width of these distributions is expected when the third coordinate information is available.

3.2 SLD Detector Development

The requirements for the CRID electron detector are:

— good single electron detection efficiency,

— efficient input optics from the drift region,

— protection against photon feedback,

— unaffected by $E \times B$ force.

The data discussed above was taken with the electrostatic structure shown in Fig. 24. Protection against photon feedback is achieved by the geometrical shielding of the TMAE drift volume by an array of thick wires; this reduces the feedback noise by a

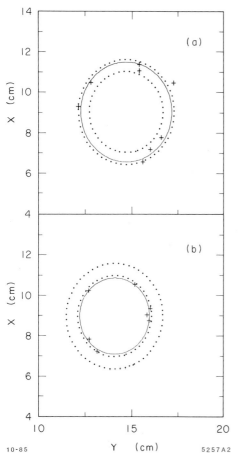

Fig. 22 The Čerenkov circles for single events with a) an electron, and b) a pion incident.

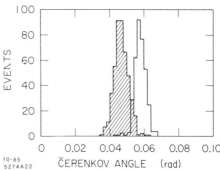

Fig. 23 The distribution of measured Čerenkov angles for tagged 4 GeV/c electron and pion incident on the SLD CRID prototype.

880

factor of four to five. The input optics efficiency is calculated to be essentially 100% and laser scanning measurements corroborate this value.

The main unanswered question in the detector design is how to obtain the third coordinate (depth) information.

This measurement dominates the accuracy of the angular resolution of the CRID. A variety of techniques can be employed to make this measurement (*e.g.*, cathode strips, wedge cathode, graded cathode, resistive cathode or resistive anode). The technique of choice should provide (i) resolution of order 1 mm, (ii) good solid angle coverage and (iii) good channel-to-channel signal isolation. Good progress has been made in implementing a resistive anode readout employing charge division on short ($\sim$ 5 cms) wires, to provide the third coordinate information.

Figure 25 shows the charge ratio measured in a single cell test chamber constructed with a 7 μm diameter carbon filament as the anode, and read out with two low noise HQV810 preamplifiers measuring the charge at opposite ends of the fiber. These results indicate a position measurement of better than 2% accuracy of the total length. This corresponds to a resolution $\sigma \sim 1$ mm for the depth measurement, and is found to be uniform along the length of the fiber.

We have constructed a chamber with 60 carbon fiber anodes, in the geometry shown in Fig. 24 above, and collected data with pions and electrons. The analysis of the Čerenkov rings data is currently in progress, but should be reported at the IEEE meeting in San Francisco in October 1985.

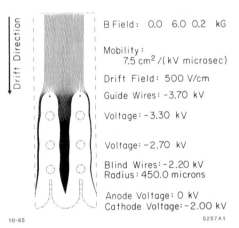

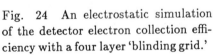

B Field: 0.0 6.0 0.2 kG

Mobility:
    7.5 cm^2/(kV microsec)

Drift Field: 500 V/cm

Guide Wires: -3.70 kV

Voltage: -3.30 kV

Voltage: -2.70 kV

Blind Wires: -2.20 kV
Radius: 450.0 microns

Anode Voltage: 0 kV
Cathode Voltage: -2.00 kV

10-85   5257A1

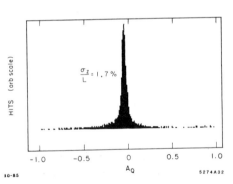

$\frac{\sigma_z}{L} = 1.7\%$

10-85 A_Q 5274A32

Fig. 24 An electrostatic simulation of the detector electron collection efficiency with a four layer 'blinding grid.'

Fig. 25 Charge division measurement.

3.3 DELPHI R & D Progress

The DELPHI experiment at LEP has an aggressive and comprehensive prototype and development program for their RICH work. They have successfully passed the CERN Program Committee milestones for demonstration of the RICH subsystem and are proceeding with the design and construction of the final RICH modules.

Figure 26 shows a perspective of the DELPHI RICH prototype. The liquid radiator is C_6F_{14}, 1 cm thick, and the gas radiator is 47 cm of isobutane. The drift tubes were filled with a gas mixture of 95/5 methane/isobutane, with about 0.1% of TMAE as the photoionizing agent. Čerenkov light from the single liquid radiator cell illuminates three drift boxes of 1 m 50 cm length, read out at one end by a MWPC. A uniform drift field is created by an inner and outer field cage wound on each drift box. A secondary field cage isolates the nearby electrical grounds in the prototype geometry. Light from the gas radiator volume is focused back on the top side of the drift tubes by three parabolic mirrors.

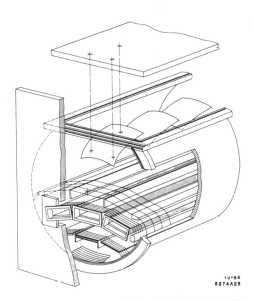

Fig. 26 Schematic of the DELPHI RICH prototype.

The response of this prototype to a 10 GeV/c hadron beam is shown in Figs. 27(a) and (b), two-dimensional plots of the drift time and wire address information. Figure 27(a) shows the liquid ring spread over three drift tubes, while Fig. 27(b) shows both the liquid and the gas ring. Figure 28 shows single event plots with the same information.

Figure 29 shows the radius plot for the liquid and the gas rings. Figure 30 indicates ~ 5.5 photoelectrons detected from the gas radiator and ~ 17 photoelectrons detected from the liquid radiator. Their best results from recent beam tests have achieved ~ 8.5 photoelectrons from gas and ~ 19 photoelectrons from the liquid radiators respectively. When corrected for electron attenuation and drift losses, these results imply ~ 12.7 photoelectrons produced in the gas ring, and ~ 22.7 photoelectrons produced in the liquid ring, in very good agreement with their expectations.

882

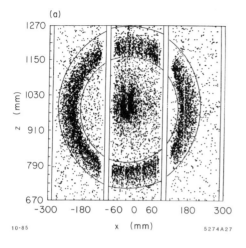

(a)

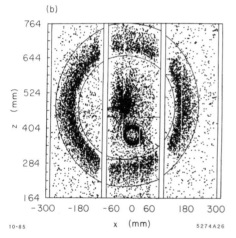

(b)

Fig. 27 Two dimensional plot of the photoelectron positions from the DEL-PHI prototype. The x–coordinate is the wire address and the z–coordinate is the drift time. The plot is an accumulation over several hundred beam tracks: (a) shows the liquid ring, (b) displays both the liquid and the gas rings.

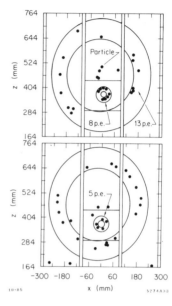

Fig. 28 Single event plots of wire address versus drift time.

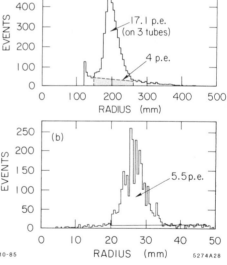

Fig. 29 The distribution of photoelectrons as a function of the radius from (a) the gas ring, and (b) the liquid ring.

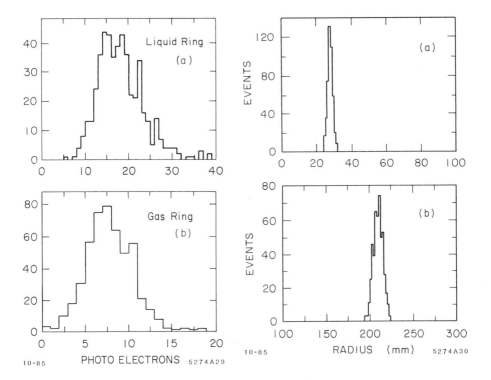

Fig. 30 The number of photoelectrons observed from the (a) liquid ring, and (b) the gas ring, in the DELPHI prototype.

Fig. 31 The distribution of corrected radius measurements for the gas ring and liquid ring, from the DELPHI prototype.

The accuracy of measuring the Čerenkov ring is shown in Fig. 31 for both gas and liquid radiators. The gas and liquid rings are measured to have a radius of 27.8 ± 1.5 mm and 209.6 ± 5.5 mm respectively. Both measurements are made without the third coordinate information and achieve the accuracy predicted from the Monte Carlo simulation.

The DELPHI group observed quite long attenuation lengths for drifting electrons in the methane/isobutane gas mixture ($\sim$ 2-5 m), in contrast to the experience at SLAC ($\sim$ 0.3 m). Additional experiments with ethane/methane mixtures instead of isobutane/methane results in even longer attenuation lengths, ($\gtrsim$ 15 meters). Further studies of electron attenuation lengths with different gas mixtures, and chemical studies of these gas mixtures are being pursued by both the CERN and SLAC groups.

4. SUMMARY AND CONCLUSIONS

4.1 Progress in the 'Problem Areas'

At the beginning of this review, I listed a number of problem areas, where progress had to be made before one could proceed to construct large ring imaging systems. Let us examine the progress over the past year.

4.1.1 Photocathodes

Two large fixed target experiments have demonstrated that both TEA and TMAE photocathodes are technologies that can be implemented. TMAE is rather more forgiving in terms of the allowable oxygen and water vapor contaminations, but seems to be chemically active and may cause poor electron drift lifetimes.

4.1.2 Optics

The optics issue branches into two subheadings - 'photon optics' and 'electron transport optics.' The former is no special problem, only demanding a careful choice of mirror fabrication process and of the optimum geometry for a given experiment. The latter still needs work; electrons can be efficiently and uniformly transported over the desired drift lengths under the influence of static electric and magnetic fields. (Both the DELPHI and SLAC experiments demonstrate transport over 70 cms with distortion of less than 1 mm, in circumstances where positive ion charge effects are negligible.) However, distortions due to charge build up on the quartz boxes and the positive ion cloud need further study.

4.1.3 Detectors

Several electrode geometries have been demonstrated to provide good single photoelectron detection efficiency, and have substantial protection against photon feedback. An additional factor of two rejection would be helpful, but is not absolutely required for an operational system. Studies have begun on the effect of $E \times B$ forces on detector efficiency, and will continue.

Third coordinate readout from both cathode coupled geometry and from charge division on resistive anodes have been demonstrated to give the required accuracy. Work will continue aggressively in this area.

4.1.4 Radiators

There were no special problems in the choice of radiators, beyond learning to keep them clean of oxygen and water vapor, both of which absorb strongly in the far U.V. region.

4.1.5 Engineering Considerations

Among the major problem areas in building new Čerenkov Ring Imaging Devices are: a) the choice of construction materials which will not react with TMAE or TEA, and which will not out-gas vapors which would absorb the Čerenkov photons or the photoelectrons, b) construction techniques which allow robust, economic assembly of devices without loss of coverage and c) imaginative and realistic geometric solutions which allow 4π coverage without compromising the possibility of servicing the device.

4.2 Summary of the Data

The measurements from the four groups working with ring imaging devices are summarized in Table II below.

Table II. Summary

Experiment	Radiator	Photocathode	Number of Photoelectrons			$\dfrac{N_0^{Achieved}}{N_0^{Expected}}$	Attenuation Length (cm)	$\Theta \pm \delta\Theta$ (mrad)	γ_{th}
			Detected	Zero Drift Length	Expected				
E-605	1500 cm [helium gas]	TEA	3	N.A.	6	25/50	N.A.	8.7 ± 0.09	116
RAL-Omega	500 cm [nitrogen gas]	TMAE	12	14	28	40/95	50	25.8 ± 0.5	41
DELPHI	1 cm [C_5F_{12} liquid]	TMAE	19	23	23	50/50	150-350 (> 500)	680 ± 80	1.5
	50 cm [isobutane gas]	TMAE	8	12	12	95/95	150-350 (> 500)	54.5 ± 3.0	17
SLD	1 cm [C_5F_{12} liquid]	TMAE	20	37	25	75/50	30-45	668 ± 90	1.5
	42 cm [isobutane gas]	TMAE	9	13	11	105/90	30-45	56.4 ± 3.2	17

4.3 Conclusions

- Čerenkov Ring Imaging Devices have been successfully used in two fixed target experiments at Fermilab and at the CERN SPS, demonstrating particle identification over a very large momentum range at high energy (*e.g.*, π/K separation at 3 σ level up to 200 GeV/c).

- Good progress has been achieved in the research and development efforts for the two 4π e^+e^- detectors at CERN and SLAC where the expected numbers of photoelectrons (12 from the gas, and 25 from the liquid radiators) have indeed been achieved (or even exceeded). The expected accuracy for the Čerenkov angle measurement has also been achieved.

It seems that the technique has come of age......but there is a long hard path ahead on engineering and systems optimization before such devices can be implemented as "standard equipment."

REFERENCES

1. Sequinot J., and Ypsilantis T., Nucl. Inst. and Meth. 142, 377 (1977).

2. Ypsilantis T., Urban M., Seguinot J., and Ekelof T., "Spherical RICH in Weak Magnetic Field", Isabelle Proc. 973 (1981);

   Barrelet E., Seguinot J., Urban M., Ypsilantis T., "A Rich Detector", Isabelle Proc. 1378 (1981);

   "Delphi Technical Proposal" CERN-LEPC-83-03.

3. Ekelof T., Ypsilantis T., Sequinot J., and Tocqueville J., "The CRID: Recent Progress and Future Development", Phys. Scripta 23, 371 (1980).

   Williams S., Leith D., Poppe M., Ypsilantis T., "An Evaluation of Detectors for a CRID", IEEE, NS-27, 91 (1980);

   "SLD Design Report", SLAC Report 274 (1984).

4. Ash, W. et al., "SLD Design Report", SLAC Report 273 (1984).

5. Glass, H., et al., IEEE, NS-32, 692 (1985); IEEE, NS-30, 30 (1983). Adams, M. et al., Nucl. Inst. and Methods 217, 237 (1983).

6. Apsimon, R.J., et al., IEEE, NS-32, 674 (1985);

   Apsimon, R.J., et al., RAL Report 85-014.

7. "DELPHI Proposal", CERN/LEPC 83-3; "DELPHI Progress Report", CERN/ LEPC 84-16.

8. Atwood, W., (private communication).

Uranium-Liquid Argon Calorimetry:
Preliminary Results from the D0 Tests

B. Cox *
Fermilab

Presented At the Meeting of the American Physical Society
University of Oregon, Eugene, Oregon

Abstract

The motivations for using uranium and liquid argon in sampling calorimetry are reviewed and the pros and cons of the technique are discussed. Preliminary results of the D0 uranium-liquid argon test program are presented.

---

* Members of the D0 Collaboration Participating in Calorimetry Tests:
S. Aronson[1], S. Augst[8], S. Catley[5], B. Cox[3], T. Ferbel[8], P. Franzini[2], E. Gardella[7], N. Giokaris[3], H. Gordon[1], P. Grannis[9], H. Greif[◊], W. Guryn[1], A. Ito[3], M. Johnson[3], A. Jonckheere[3], W. Kononenko[7], S. Linn[9] F. Lobkowicz[8], T. Marshall[4], M. Marx[9], P. O. Mazur[3], D. Owen[5], E. Prebys[8], B. G. Pope[5], R. Raja[3], J. Sculli[6], W. Selove[7], R. D. Schamberger[9], S. Stampke[5], G. Theodosiou[7], A. Zieminski[4], P. M. Tuts[2], H. Weerts[5]

Brookhaven National Laboratory[1], Columbia University[2],
Fermi National Accelerator Laboratory[3], Indiana University[4],
Michigan State University[5], New York University[6],
University of Pennsylvania[7], University of Rochester[8],
State University of New York at Stony Brook[9]

◊ Visiting scientist from Max Planck Institute

Introduction

Recently interest has been high in the possiblities for uranium-liquid argon calorimetry as the technique of choice for large detectors that are being planned for the Fermilab Collider, LEP, SLC and HERA. Much of this interest has been generated by the pioneering work of C. Fabian et ali whose prototype calorimeter achieved significantly better hadronic energy resolutions and more nearly equal electron and hadron response using uranium than had been achieved using Fe or other materials. These potential advantages coupled with the higher density (approaching 8 gm/cmℓ in real uranium plate calorimeters) has stimulated extensive R&D efforts intended to investigate further the properties of uranium-liquid argon calorimeters.

Uranium is under consideration for calorimeters not only because of its high density but also because of the hadron induced fission that takes place in the uranium during the shower process. Some of the fission products (neutrons, gammas and electrons) leave the uranium plates and produce signals in the liquid argon thereby increasing the visible hadronic signal and compensating for what otherwise would be the hidden energy lost in breaking up the nuclei. The better resolution of uranium calorimeters is thought to be partially due to these phenomena.

In addition to this effect which increases the size of the hadronic signal in a uranium-liquid argon calorimeter, there are several effects[2,3,4] which decrease the signals due to electron showers in high Z calorimeters. The first of these effects is due to the differences in critical energies of the uranium (6.0 MeV) and argon (30.5 MeV). The critical energy of a material is by definition the energy loss per radiation length due to collisions of electrons and positrons in that material. At the boundary between materials with very different critical energies such as uranium and argon, the collision losses will increase abruptly as the shower enters the argon, causing a loss of electrons from the shower. This will tend to lower the signal due to that shower. In an Fe-liquid argon calorimeter the difference in critical energies is not so great and the decrease of the electron shower signal will not be so great. A second effect that tends to decrease the electron signal in a uranium-liquid argon calorimeter relative to that in a Fe-liquid argon calorimeter is the increased track length in the uranium relative to the argon due to multiple scattering in the uranium. Finally a third effect which can decrease the electron pulse height in any calorimeter is the saturation of the active medium by low energy electrons in the later generations of an electromagnetic shower.

These effects, taken together, tend to raise the pulse height of hadronic showers and decrease the pulse height of electromagnetic showers in a uranium-liquid argon calorimeter relative to those observed in an Fe-liquid argon calorimeter. The net effect is in the direction of making the response of a uranium-liquid argon calorimeter to electromagnetic and hadronic showers more nearly equal than that of an Fe calorimeter. This equality of electromagnetic and hadronic signals has been advanced as the reason for the better hadronic resolution of uranium calorimeters. The degradation of the hadronic resolution due to variations in the relative size of the the em and hadronic components of the shower will be minimized if the calorimeter response to electromagnetic and hadronic showers is equal. Therefore the measurement of the ratio of the electromagnetic to hadronic response of a given calorimeter configuration is a prime objective of any calorimetry test program.

The suppression of the response of a high Z calorimeter to electromagnetic showers can be observed in the relative response of the calorimeter to electrons and muons. The observed electron signals in the calorimeter can be used to determine the apparent energy of the observed muon signal on the electron scale. From that apparent energy, the muon energy that is deposited in the liquid argon can be calculated using

$$S = \frac{(x \cdot dE/dx)_A}{(x \cdot dE/dx)_A + (x \cdot dE/dx)_U} = \text{fraction of shower energy visible in the liquid argon} \quad (1)$$

When the muon energy deposit calculated in this way is compared with the actual muon energy deposit measured directly using absolute amplifier gains or other measurements of dE/dx of muons in liquid argon, it is normally found to be considerably larger. The fact that this ratio (known as the 'μ/e' ratio) is greater than one is a manifestation of the 'transition effects'. Their suppression of the electron response of the calorimeter makes the apparent muon energy appear to be larger than it is in reality.

There are many contributions to the energy resolution of a uranium liquid argon calorimeter other than the ordinary sampling fluctuations due to hidden energy in the plate structure or the fluctuations in hadronic response due to an e/π ratio different from one. In general the energy resolution of such a device may be written as the sum in quadrature of three supposedly independent terms:

$$\sigma = \sqrt{A^2 E^2 + (B\sqrt{E})^2 + (C)^2} \quad (2)$$

There are several effects which contribute to each term. The A term which is proportional to the energy includes contributions from errors in the gain determination of each channel, contributions due to inductive or capacitative coupling of one section or channel of the calorimeter to another and variations in plate or gap thickness in a real calorimeter. Spread in beam momentum also contributes to this term. The B term which is proportional to the square root of the energy contains the sampling fluctuation contributions to the resolution. These include both the intrinsic shower fluctuations and fluctuations in energy deposit for a given shower between visible energy in the active medium and the hidden energy in the plates. Finally the C or constant term includes energy independent contributions to the resolution such as coherent and incoherent electronic noise, uranium radioactivity noise and drifts in electronic pedestals. Other effects such as the fluctuations between the electromagnetic and hadronic components of a hadronic shower in a calorimeter in which e/π is not equal to one or the leakage of shower energy out of a calorimeter will contribute to the resolution in a more complicated way that cannot be represented by a simple contribution to a single term of (2). Each of these contributions to the resolution function will occur in different amounts for a particular calorimeter and must be determined by suitable testing of that calorimeter.

Preliminary Results from the D0 Liquid Argon-Uranium Tests

Recently several institutions collaborating on the D0 experiment[5], E740, at Fermilab have conducted a series of tests to measure the response of a uranium-liquid argon test calorimeter to high energy hadrons and electrons. These tests were performed in the NW test beam at Fermilab at 10,15,25,50,100 and 150 GeV/c with electron and hadron data taken at each momentum. The data presented in this paper is that obtained using the thinnest uranium plate configuration that was tested in the course of this program. The preliminary results presented in this paper represent the state of the analysis of these data at the time of this talk (August,1985) and will be superseded by a more complete analysis to be published later. The liquid argon test cryostat and the uranium plate configuration that was used to obtain these data is shown in Fig. 1 and tabulated in Table I below:

Table I
DO "Thin" Plate Uranium-Liquid Argon
Test Calorimeter Configuration
(Argon Gap Thickness=1.6 mm)

Section	Plate Thickness	#Cells	#X_0	λ_a	L(cm)	Sampling Fraction
EM-1	2 mm U	4	2.6			
EM-2	2 mm U	4	2.6			
EM-3	2 mm U	8	5.3	0.8	20.4	.127
EM-4	2 mm U	14	9.2			
FH-1	4 mm U	24	-	1.1		
FH-2	4 mm U	64	-	2.8	99.0	.072
FH-3	4 mm U	24	-	1.1		
L-1	19 mm Cu	12	-	1.6		
L-2	19 mm Cu	12	-	1.6	57.0	.026
Totals		166	19.7	9.0	176.0	

The configuration of a cell (plate-argon gap-G10 readout board-argon gap) is shown in Fig. 2a. In Fig. 2b the 48 pad transverse readout structure of the G-10 readout boards is shown. The two pad sizes that were used were 2"x2" and 4"x4". The nine longitudinal readout sections listed in Table I were further subdivided by bringing out and digitizing separately the sum of every other gap in each of the nine sections. This allowed the major effects of doubling the plate thickness to be measured by observing the resolution achieved using signal from the sum of half the gaps. (Measurements of a test calorimeter in which the plate thickness was actually doubled were made to check that there were no subtle effects on the 'e/π' ratio or the hadronic resolution due to the presence of the extraneous argon gaps in the half the gap sum described above. No significant differences were detected between the results obtained summing every other gap and the double plate thickness test.)

For the transverse and longitudinal readout configuration described in Table I a total of 696 channels of electronics were required. The electronics is shown schematically in Fig. 3a. As indicated, it consisted of an charge sensitive preamplifier stages which used Toshiba 2SK147 FETs followed by a baseline subtractor stage. The output of the baseline subtractor stage was digitized by a LeCroy 2280 ADC system. The test calorimeter was operated with the uranium plates at negative high voltage in the configuration shown in Fig. 3b. Each uranium plate had a 100 nf buffering or blocking capacitor and a 50-108 ohm protection resistor in series to the high voltage. Both the blocking capacitor and the protection resistor were inside the test cryostat in the liquid argon.

The test calorimeter was positioned in a 'bathtub' inside the cryostat which was filled with liquid argon by condensation from a nitrogen cooling coil. The cryostat was maintained at 10 psi gauge during the cool down and fill process by regulating the liquid nitrogen flow to the cooling coil. This overpressure was maintained in order to minimize contamination of the 1000 liter liquid argon inventory by oxygen. The argon used in the test was obtained as bottles of liquid from commercial sources, allowed to boil and then recondense in the cryostat. Each bottle of liquid argon was checked using an alpha source in a small test cell to insure purity. Typical results of these measurements are shown in Fig. 4a in comparison with the published results of Willis and Radeka[7]. The purity of the liquid argon inventory in the cryostat was monitored by a similar alpha source mounted in the bathtub and by a oxygen monitoring device which sampled the boil off argon gas from the cryostat volume. The purity of the liquid argon inventory achieved with these precautions is indicated by the good quality of the high voltage plateau curves shown in Fig 4b. These were obtained with 50 GeV/c electrons and hadrons. For the majority of the data discussed in this paper an operating point of 1500 volts was used. The oxygen contamination monitor generally recorded 1-2 ppm of oxygen and the bathtub alpha source gave pulse heights quite similar to the test cell alpha source during the period of these tests.

The response of the test calorimeter to electrons, hadrons and muons is shown in Fig. 5a, b and c. Using the pulser calibration of the electronics and a 52.4 eV deposit per observed electron for argon, the deposited energy in the calorimeter corresponding to the average muon pulse height is calculated to be 89 MeV. This is to be compared to the expected energy deposit[8] of approximately 100 MeV for a minimum ionizing particle in liquid argon. However, if the observed pulse height of 150 GeV electrons is used along with the observed pulse height of muons in the electromagnetic section to determine the apparent energy deposit of the muons (using a sampling fraction of 12.6% in the electromagnetic section), then a μ/e ratio of 1.9 ±0.3 is obtained showing the suppression of the electron signal by the various effects mentioned in the introduction.

The linearity of the calorimeter for electrons and hadrons is shown in Fig. 6. The electron signal is defined for the linearity plot as the sum of the electromagnetic sections plus the first hadronic section of the calorimeter while the hadron signal is formed from the sum of all sections. In all cases when the electron and hadron pulse heights are formed from the signals observed in the different sections of the calorimeter, formula (1) is used to correct for the differing percentages of visible energy deposit in the different sections. As can be seen in Fig. 6, the electron pulse heights are larger than the hadronic pulse heights at all energies.

The resolutions for hadrons and electrons have been extracted from the raw data with the electron signals formed as indicated above from the sum of the em sections and the first fine hadronic section of the test calorimeter. The hadronic signals are formed from the sum of all sections of the calorimeter. These data have been corrected for amplifier gain differences, beam momentum on an event by event basis, and for the ringing of the signals due to inductive coupling between channels. In addition, all channels which were within 1.5 σ of their average pedestal values were deleted from the energy sum (the average noise per channel was approximately 6 MeV for the EM sections and approximately 12 MeV per channel for the hadronic sections). With these corrections and cuts the variations of the electron and hadron resolution as a function of $1/\sqrt{E}$ are shown in Figs. 7a and b. The approximate fits of $\sigma/E = \sqrt{A^2 + (B/\sqrt{E})^2 + (C/E)^2}$ are shown superimposed on the data. At this stage of analysis the coherent electronic noise dominates the resolution at low energies. At high energies the resolution is probably influenced by factors such as the uniformity of plate and gap thickness that we were able to achieve in the construction of the test calorimeter. We have also determined the resolution obtained by summing the signals from every other gap. The ratio of this resolution for electrons and hadrons to that determined from the sum of all gaps is shown in Fig. 7c. The dotted line at the $\sqrt{2}$ level is the ratio which would be expected if the resolutions were dominated by sampling fluctuations. While the electron resolution ratio is near that level, the hadronic resolution ratio is appreciably below that level indicating the presence of other contributions to the resolution.

The ratio of the electron to hadron pulse heights has been determined from these data with all of the above mentioned cuts and corrections except for the 1.5 σ noise cut. Formula (1) has been used to add signals from various parts of the calorimeter. The 1.5 σ cut was not used in this determination so that any effects on the absolute pulse heights of electrons and hadrons due to differences in the number of channels excluded for electrons and hadrons would be minimized. The values at the different energies for the most part lie between 1.0 and 1.1 with the definition of the electron pulse height used in the determination. There is little or no variation of this 'e/π' ratio with energy between 10 and 150 GeV/c.

Several other measurements were made with this test calorimeter. A determination of the uranium radioactivity noise was made with both negative and positive high voltage on the uranium plates by measuring the broadening of the pedestal as a function of high voltage. With negative high voltage on the uranium plates, we find that the uranium radioactive noise can be represented by

$$\sigma_U = 0.12 \sqrt{A\tau}/s \quad (MeV)$$

where s is the sampling fraction for the particular section of the detector in which the uranium noise is observed, A is the area in square meters of the readout pads which are being summed and τ is the sampling time of the digitizing system in nanoseconds. Reversing the high voltage polarity reduces the uranium noise by 25%.

Finally we have tried adding photosensitive dopants[9] to the liquid argon in attempt to modify the 'e/π' ratio. We first added isobutylene (C_4H_8) at the 27 ppm level. While the electron and hadron pulse heights and resolutions at 25 and 100 GeV/c did not change, the pulse height of the monitor alpha source increased by over 50% and the maximum voltage at which the plate array could be operated without sparking decreased from 2400 V to approximately 1300 V. The same general behavior was observed with allene(C_3H_4). The increase of the alpha particle pulse height and simultaneous lack of change of the hadronic pulse height is taken as evidence that the component of hadronic showers due to heavily ionizing particles is negligible.

Conclusions

Uranium-liquid argon calorimetry has many attractive features for hadronic calorimetry. These include the high density relative to other types of calorimetry and the approximate equality (e/π≈1.0->1.1) of electron and hadron signals as measured by the D0 collaboration. The resolutions for hadrons in the 10 to 150 GeV/c range are somewhat better than those obtained by comparable Fe-liquid argon calorimetry. This is due at least partially to the near equality of the response to electrons and hadrons. This equality minimizes the contribution to the resolution due to fluctuations between the electromagnetic and hadronic components of the hadronic showers. Noise due to uranium radioactivity has been found to contribute $\sigma = 0.12\sqrt{A\tau}/s$ (MeV) to the energy resolution where A is the area of the plate in m^2, τ is the sampling time of the digitizing electronics in nanoseconds and s is the sampling fraction of the calorimeter. If the calorimeter is operated with positive high voltage on the uranium plates this noise is found to be 25% less. Finally, the addition of photosensitive dopants to the liquid argon is found to have little effect on the electron or hadron showers but causes major changes in the pulse height of an alpha source monitor. This observation leads to the conclusion that heavily ionizing particles do not constitute a large component of hadronic showers in uranium-liquid argon calorimeters.

Acknowledgments

We would like to acknowledge the substantial help of the Fermilab Research Division, the Fermilab Physics Department and the Experimental Support Department of the Fermilab Accelerator Division in the execution of these tests. The support of the Department of Energy and the National Science Foundation for the various universities involved in the D0 collaboration is gratefully acknowledged. The interest and help of institutions outside this collaboration, in particular the Max Planck Institute of Munich and the University of Wisconsin, is acknowledged. Finally, the assistance of W. Willis and C. Fabian of CERN both in helpful conversations and the loan of equipment is acknowledged.

References

1. C.W. Fabian et al, Nuclear Instruments and Methods, 141(1977)61.

2. J. E. Brau et al, Nuclear Instruments and Methods, A238(1985)489.

3. C.W. Fabian, 'Calorimetry in High Energy Physics', CERN-EP/85-54.

4. S. Iwata, "Calorimeters (Total Absorption Detectors) for High Energy Experiments at Accelerators", unpublished, Nagoya University Preprint, (1979)

5. D0 Design Report, Fermilab, Nov 1984.

6. P. Franzini, D0 Internal Note 222.

7. W. J. Willis and V.Radeka, Nuclear Instruments and Methods, 120 (1974) 221.

8. Atomic and Nuclear Properties of Materials, Reviews of Modern Physics, Vol 56, 2,II, (1984)S53

9. Following a suggestion by D. Anderson.

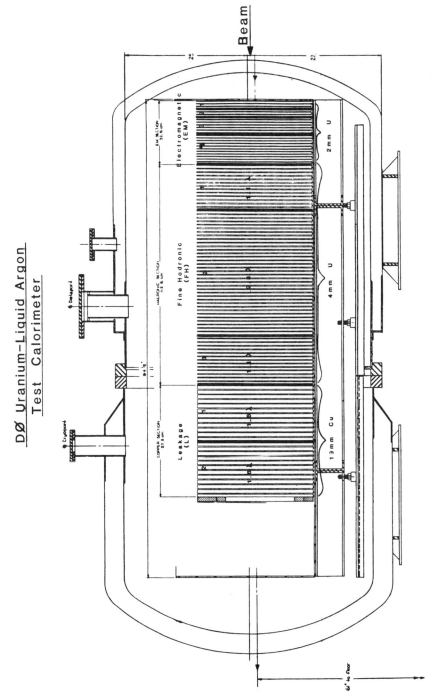

Fig. 1: DØ test calorimeter and cryostat

**CELL STRUCTURE
DØ URANIUM—LIQUID ARGON
TEST CALORIMETER**

2a)

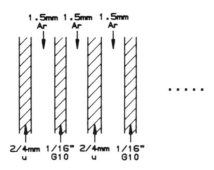

Fig. 2a: Cell structure for "thin" uranium
plate assembly used in DØ tests.

**G10 READOUT BOARD
PAD CONFIGURATION**

2b)

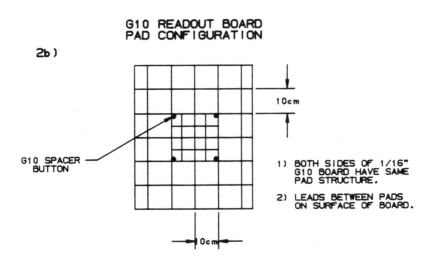

Fig. 2b: G-10 readout board pad configuration used
in DØ tests. Both sides of the 1/16" G-10
readout boards have identical pad structures.

3a) LIQUID ARGON TEST AMPLIFIER ELECTRONICS

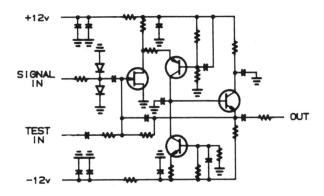

STAGES OF ELECTRONICS

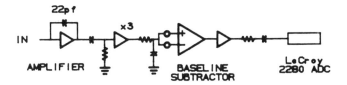

3b) HV HOOKUP FOR TEST CALORIMETER

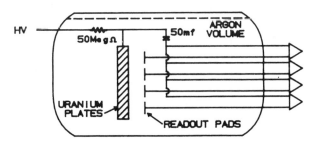

Fig. 3: a) Schematic of the DØ liquid argon amplifier and the general configuration of the integrating amplifier-baseline subtractor-ADC system, b) configuration of the high voltage hookup of the uranium plates for the DØ calorimetry tests.

900

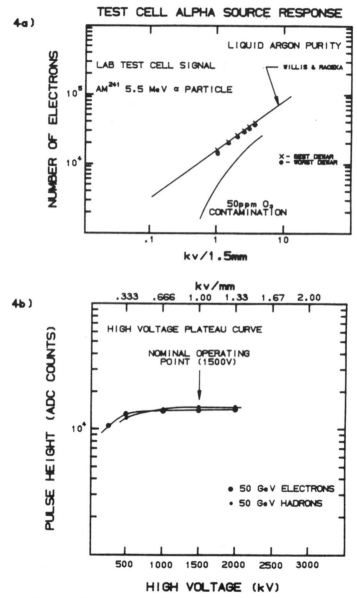

Fig. 4: a) Variation of the test cell alpha source
signal as a function of voltage for several
bottles of commercial grade liquid argon
used in the DØ tests, b) high voltage
plateau curves for 50 GeV hadrons and
electrons.

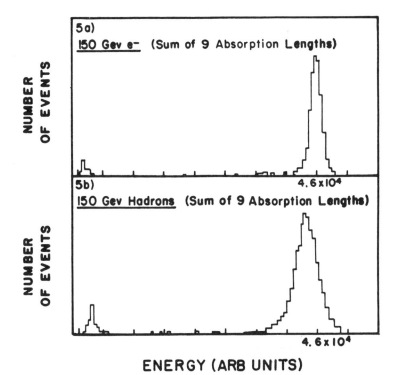

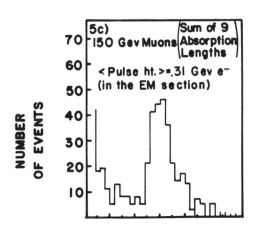

Fig. 5: a) 150 GeV electron pulse height distribution,
b) 150 GeV hadron pulse height distribution,
c) 150 GeV muon pulse height distribution

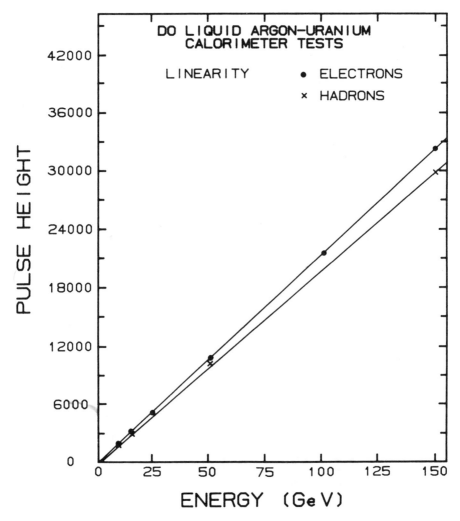

Figure 6: Linearity of the DØ test calorimeter for electrons and hadrons.

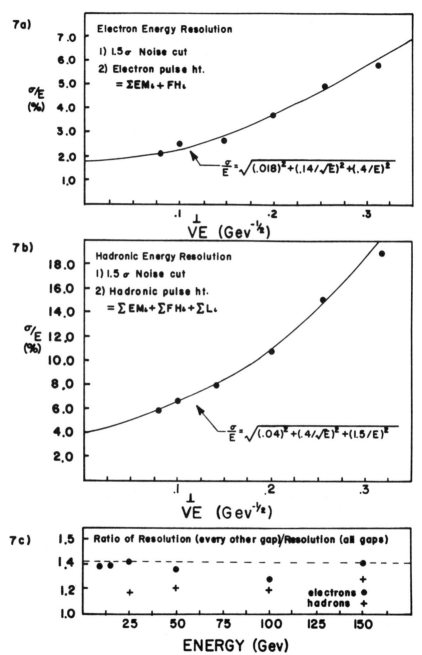

Fig. 7: a) Variation of electron resolution as a factor of $1/\sqrt{E}$, b) variation of the hadron resolution as a function of $1/\sqrt{E}$, c) ratio of the resolution obtained from summing every other gap to that obtained from summing all gaps.

Some New Development in Instrumentation

David F. Anderson
Fermilab
Batavia, Il 60510
U.S.A.

Abstract

The latest results of BaF_2 coupled to photosensitive wire chambers are presented. Also a new work on the addition of photosensitive dopants to liquid argon is discussed.

1. INTRODUCTION

The subject of new developments in instrumentation is much too broad of a topic to discuss with any completeness here. We will limit ourselves to just two topics 1) the scintillator BaF_2 coupled to a low pressure, photosensitive wire chamber, and 2) photosensitive dopants for liquid argon.

These two subjects are at very different points in their development. BaF_2 with a wire chamber readout has been under development for 4 or 5 years, and thus is not new. But, with some recent developments, and the tentative approval for its use in an experiment at Brookhaven National Laboratory, it has now become a mature technique.

The subject of photosensitive dopants for liquid argon is completely new. There is still a great deal of work to be done to determine the usefulness of the technique, and its shortcomings. But, at the moment it offers the prospects of some very interesting developments for the future.

For an interesting discussion on these subjects, as well as other instrumentatal ideas, the contribution by George Charpak at the 1985 International Symposium on Lepton and Photon Interactions at High Energies in Kyoto Japan is recommended.

2. BaF_2 WITH WIRE CHAMBER READOUT

BaF_2 has the distinction of being the only known solid scintillator whose emission spectrum is energetic enough to be detected by the photosensitive gas TMAE in a wire chamber. This scintillator has some other desirable characteristics. As a comparison, Table I lists some properties of BaF_2 as well as BGO and NaI(Tl).[1] Unlike the other two, BaF_2 has two emission spectra with the "fast" component, containing about 20% of the light, being about 500 times faster than for BGO or NaI(Tl). Also, TMAE is only sensitive to this fast component. As will be discussed below, BaF_2 also appears to be the most radiation resistant scintillator.

Table 1
Properties of three scintillators[*]

	BaF_2	BGO	NaI(Tl)
Density(g/cm³)	4.9	7.1	3.7
Radiation length(cm)	2.1	1.1	2.6
dE/dx(M.I.P.)(MeV/cm)	~ 6	8	4.8
Peak emission(nm)	225 310	480	410
Decay constant(ns)	0.6 620	300	250
Index of refraction	1.56	2.15	1.85
Light yield(photons/MeV)	2x10³ 6.5x10³	2.8x10³	4x10⁴
Hygroscopic	No	No	Yes

2.1 Earlier Results

The first successful joining of a solid scintillator to a wire chamber readout was a small BaF_2 crystal coupled to a single wire proportional counter filled with argon(90%), methane(10%) and a small amount of TMAE.[2] Although, the results were marginal, it started a long development that included a large variety of readout techniques.

One such technique was the "liquid photocathode" which consisted of a thin condensed layer of TMAE below a low pressure wire chamber. Condensed TMAE, forming a cathode, has a thresholds somewhat lower than its 5.36 eV thresholds in the gas phase, and thus is much more sensitive to the BaF_2 light. With such a counter, joined to a 2.5cm thick BaF crystal, filled with TMAE and 3 to 9 Torr of isobutane, signals of only 10ns rise time were measured.[3] A timing resolution for 540ps FWHM was also achieved with 350 MeV alpha particles. It had been known for sometime that the emission spectrum of BaF_2 overlaps with the absorption spectrum of TMAE in the gas phase. It was also found that at room temperature TMAE tends to adhere to some surface making them slightly photosensitive, though this effect has proven to be very non-reproducable. A BaF_2 calorimeter tower was constructed using TMAE gas and low pressure wire chambers.[4] The instrument consisted of 14 crystals with a total BaF_2 thickness of 40.5 cm, making 19.3 radiation lengths. Each crystal was prreceded by its own wire chamber.

The resolutions for 108 MeV and 200 MeV electrons were 30% and 20% FWHM, respectively. But, when corrected for the energy leakage out the side of the tower, a resolution of $\sigma/E = 2.5\%$ E^- (GeV) was determined. This correction was made using the EGS Monte Carlo simulation program.

2.2. New Results

There have recently been tests made at BNL (C. Woody, C. Petridon and G. Smith) and at Fermilab (D. Anderson and S. Majewski) using low pressure were chambers.

DETECTOR CONFIGURATIONS

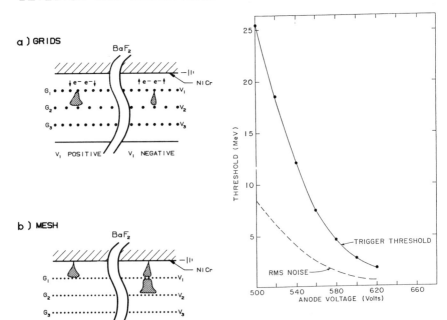

Figure 1. Counter configurations
a) BNL design with grids, and
b) Fermilab design with meshes.

Figure 2. Threshold as a function of anode voltage.

2.2.1 BNL results. The BNL counter configuration is shown in Figure 1a.[5] The BaF$_2$ crystal was a hexagon 46mm flat-to-flat and 10 cm long, wrapped with teflon tape to improve light collection. The lower surface had a NiCr coating maintained at ground potential. The 2 grids, G1 and G3, were made of 50 μm wires with a 0.5mm pitch, held at the same potential. Grid G2 was made of 12 μm wires with a 1mm pitch. This grid was operated at a potential V2 and served as the annode plane. An additional grounded cathode was included for symmetry of the field. All electrod spacings were 3mm. In this test detector, the grids were smaller than the crystal, giving, a reduction in the sensitivity to the scintillation photons of a factor of 3. The counter gas was 10 Torr of isobutane and TMAE vapor at room temperature. The counter was tested in two operating modes. In the first mode V1 and V3 were positive. Thus photoelectrons were collected from the entire detector volume. This gives the largest signal but with a rise time of over 50 ns.

In the second operating mode, V1 and V3 were operated negative. Thus photoelectrons liberated between G1 and BaF$_2$ crystal, and G3 and the grounded cathode, drifted away from the amplification region and were not detected. Only electrons liberted between G1 and G3 are amplified. At low pressures, much of the amplification between the grids is parallel plate amplificaion. Thus in this mode the number of detected photoelectrons is smaller, but the signal rise time is very fast. The signal for minimum ionizing particles was only about 30ns at the base and reflects the response of the amplifier.

Although the measured energy resolution was a little worse than measured earlier, partly due to poor photon conversion in the design, the signal was very good. Figure 2 shows the threshold for partial detection in MeV as a function of anode voltage, V2. For these measurements V1 = V2 = -100V and the threshold was taken to be three times the amplifier noise. Figure 2 reflects the increase in gain with increase in voltage. For this configuration a threshold, of 1.8 MeV was achieved.

Using simple leading edge discrimination they measured a timing resolution of 1.8ns FWHM. With a detector design that allows for better light collection, and operation at slightly higher temperatures to increase the TMAE pressure, a considerable improvement in energy resolution, threshold, and timing resolution should be achieved.

2.2.2 Fermilab results. The Fermilab detector, Figure 1b, used 3 meshes rather than wire grids as its electrodes. The advantages of meshes are what they make construction much easier, and there is no problem with breaking wires. The BaF$_2$ consisted of two stacked crystals making 6 cm in length. Like the BNL detector, the surface of the BaF$_2$ was coated with NiCr. The meshes were made of 50 μm wires with a 500 μm spacing. The electrode spacing was 3mm. The gas filling was 3 Torr of methylal and 1.4 Torr of TMAE (40 °C). The detector was placed in a small oven at 50 °C to prevent condensation of the TMAE.

This detector was also operated in two modes. The first is the single step mode, where only the photoelectrons from the space above G1 is amplified. Because of the parallel plate amplication, only the photoelectrons liberated near the BaF_2 receive full amplication. The number of photoelectrons is small in this mode. The signal is taken from G1.

In the double step mode, the electrons from two gaps are amplified, with almost full amplication of the electrons liberated in the first gap. The advantage is a much larger number of photoelectrons detected. The signal is also taken from G2 but is of the opposite sign than in the single step mode.

Using the single step mode, the signal has a rise time of 10ns but with a gain of only 10^4. Also using constant fraction discrimintion, a timing resolution of 1.1ns FWHM was measured using 35 GeV pions. The voltages for this measurement were V1 = 480V, V2 = 800V, and V3 = 600V. Using the double step mode, the signal has a rise time of over 20ns but a gain of 10^5. A gain of 10^6 is possible at higher gas pressures.

2.3 BaF_2 Radiation Hardness

In an early work we exposed a 6mm thick BaF_2 to 1.3×10^7 Rad, using 800 GeV protons.[6] The scintillation was unchanged and there was little effect to the transmission of the crystal. Unfortunately, the evaluation was after some months and thus not sensitive to damage with self recovery.

Fermilab is now working with Idaho National Engineering Laboratory (C.R. Heath and P. Ritter) to measure the radiation hardness of BaF_2. The test crystals are 2.5cm in diameter and 2.5 cm long. After exposure of 10^6 Rad of gamma rays, a reduction of 20% was seen in the scintillation and 5% in transmission, for the fast component. At a radiation of 10^8 Rad the scintillation and transmission had not changed from the previous measurement. Thus it appears that the color centers are limited in number and probably related to the purity of the crystal.

These results are consistant with our earlier work but inconsistent with the results of Murashita et.al.[7] In their work there is much poorer performance for BaF_2. But, their unradiated crystals showed strong absorption in the 200-220 nm range indicating a high level of impurities in the crystals.

Our work is continuing, but the early results imply that BaF_2 is the most radiaion hard of all the solid scintillators.

2.4 Experiment E787

At the moment, BaF_2 is being tentatively approved for use in experiment E787 at BNL. This experiment, is a search for rare K+ decay modes. The BaF_2 will be used as a photon veto in the 20-200 MeV range. The first "wall" will consist of about 260 hexaginal cyrstals, 5cm face-to-face and 15cm long. These will be followed by a low pressure wire chamber readout.

BaF_2 is proposed for E787 because of its performance at low proton energies and its good high rate characteristic. It is presently proposed to build one "wall" followed by a second, after evaluation of the first's performance. This will be the first use of BaF_2 coupled to a photosensitive wire chamber in a high energy physics experiment.

2.5 BaF_2 - Conclusion

The work on BaF_2 has now come to fruition, with its inclusion in E787. BaF_2 is the fastest of the high density scintillators. When coupled to a low pressure, photosensitive wire chamber, it makes an instrument that is fast, compact , and which can work in a magnetic field.

BaF_2 also appears to be the most radiation hard of the scintillators. One must ask the question if the wire chamber is also so resistant. This has not been demonstrated. But, the Fermilab detector was operated with methylal, which is believed to be non-polymerizing. Following the work of Sipila and Jarvinen,[8] one can hope that the addition of H_2 will also greatly extend the life of these chambers.

In high energy physics, there is, an increasing need for very fast, radiation hard detectors for particle detection. This will be especially true if the SSC is approved. BaF_2 coupled to a photosentitive wire chamber is such an instrument.

3. DOPED LIQUID ARGON

In the past, liquid argon, LAr, has been doped with hydrocarbons to increase the electron drift velocity. Xenon has aslo been added to increase the ionization yield by the conversion of excitons. This is similar to the Penning effect in gases. (See the paper by Doke[9] for a review). Now, photosensitive dopants have been successfully added to LAr to convert the scintillation photons into detected charge.

The first dopants were triethylamine (TEA) and trimethylamine (TMA).[10] We now know of thirteen such dopants. As an example, Figure 3 shows the charge collected as a function of electric field for alpha particles in pure LAr, and in LAr with several different dopants. A dopant with a response similar to TMGe is Allene (C_3H_4) not shown here. Thus for alpha-particles, which give high scintillation yields because of recombination, the charge collected is over a factor of two higher at 1kV mm^{-1} for some dopants.

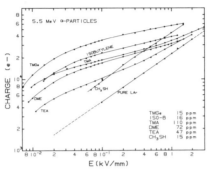

Figure 3. Charge collected as a function of electric field for alpha-particles in pure LAr with various dopants.

For minimum ionizing particles, the charge increase with the doped LAr is only a few per cent at 1kVmm^{-1}. This is because there is little or no recombination, and thus few scintillation photons.

The difference between the response of doped LAr for minimum ionizing events and heavily ionizing events suggested a probe for compensation in uranium-LAr calorimetry. If compensation were due to heavily ionizing events, such as from the neutrons, a dopant should change the e/π ratio of the calorimeter. We doped the D-0 test calorimeter with isobutylene and allene in two separate tests. In both tests the e/π ratio remained unchanged while the absolute pulse heights of both electrons and pions increased by about 8%. can say that heavily ionizing processes do not play a substantial role in uranium-LAr compensation.

We have also seen a small amount of charge gain around a 10μm wire in LAr doped with TMA. If this gain can be made stable, and increased in magnitude, it may be possible to build LAr drift chambers with position resolutions much greater than for gaseous detectors.

Photosensitive dopants for LAr are still in their infancy. There is much work to be done to understand how they work and what uses can be made of them. But, the potential is there for some very interesting work.

References

1. M. Laval et al., Nucl. Instrum. Methods 208 (1983) 169; R. Allemand et. al., "New Developments in Fast Timing with BaF$_2$ Scintillator", communication LETI/MCTE/82-245, Grenoble, France (1982); BGO-NaI(Tl) Comparison, paper distributed at the International Workshop on Bismuth Germanate, Princeton University, 1982; M.R. Farukhi and C.F. Swinehart, IEEE Trans. Sci. NS-18 (1971) 200.

2. D.F. Anderson IEEE Trans. Nucl. Sci. NS-28 (1981) 842.

3. D.F. Anderson et. al., Nucl. Instrum. Methods 217 (1983) 217.

4. D.F. Anderson et. al., Nucl. Instrum. Methods 228 (1984) 33.

5. C.L. Woody, C.I. Petridou and G.C. Smith, BNL Experiment 787 Technical Note No. 84, September 18, 1985.

6. S. Majewski and D.F. Anderson, "Radiation Damage Test of Barium Fluoride Scintillator", Fermilab - Pub - 85/67, submitted to Nucl. Instrum. and Methods in Phys. Rev. A.

7. M. Murashita et. al., "Performance of a BaF$_2$ Scintillator as an Electromagnetic Shower Calorimeter for 0.5-5 GeV Electrons and its Radiation Resistivity. KEK Report 84-24, Appendix A, February 1985.

8. H. Sipila and M.L. Jarvinen, Nucl. Instrum. and Methods, 217 (1983) 301.

9. T. Doke, Portugal Phys. 12 (1981) 9.

10. D.F. Anderson, "Photosensitive Dopants for Liquid Argon", submitted to Nucl. Instrum. and Methods.

IMPROVEMENT PROGRAMME OF THE UA2 DETECTOR

The UA2 Collaboration

Bern - Cambridge - CERN - Milan - Orsay (LAL)

Pavia - Pisa - Saclay

Presented by J.-M. GAILLARD

(LAL Orsay, France)

1. INTRODUCTION

The UA2 experiment took its first data in November 1981 and has been operating smoothly since then. The very successful performance of the Spp̄S collider has given us the opportunity to collect a large amount of very fruitful data[1-7].

After a long shut-down in 1986, the Spp̄S collider will resume operation with a substantially higher luminosity than presently available. The recently approved Antiproton Collector (ACOL) should allow for a total integrated luminosity of nearly 10 pb^{-1} by the end of 1989. In order to make the best use of the increased luminosity in terms of physics results the performance of the present UA2 detector has to be simultaneously improved.

Detailed descriptions of the modifications of UA2 detector have been given elsewhere[8]. The main components are new calorimeter end caps and an upgraded vertex detector.

We will first describe the components of the new vertex detector. The design of its most novel part, a scintillating fibre detector will be discussed and recent results obtained with a prototype of that detector will be described.

2. UPGRADED VERTEX DETECTOR

Moving outwards from the interaction region the main components of the new design include (fig. 1) :

i) a small diameter beryllium vacuum chamber to be installed in 1985,

ii) a jet chamber vertex detector (JVD) providing an accurate measurement of the longitudinal and transverse positions of the event vertex,

iii) a matrix of silicon counters with the double purpose of measuring ionisation and helping in pattern recognition,

iv) a pair of transition radiation detectors (TRD) providing an additional rejection factor of at least an order of magnitude against fake electrons,

v) a multilayer scintillating fibre detector (SFD) for tracking and for measuring early electromagnetic showers developing after a 1.5 radiation length thick converter.

The new design is expected to substantially reduce the two most significant sources of backgrounds faking electrons: converted photons from π^0 decays and narrow π^0-charged hadron pairs (overlaps).

The insertion of a transition radiation detector, a key feature of the new design, is at the price of dedicating to it a major fraction of the space available in the central region of UA2 (about half of the radial range). In turn this implies very efficient and compact tracking devices fitting in the remaining available space: the JVD and silicon array in the inner region, the SFD in the outer region. Both have the ability to measure track segments. In addition the outer cathodes of the TRD chambers are equipped with strips to help match the two regions.

While inner tracking provides direct localisation in space (charge division on the JVD wires and silicon pads), the SFD measures three stereoscopic projections. The stereo angle has been chosen small enough to reduce ghost tracks to a satisfactory level while retaining sufficient localisation accuracy. The outer layers of the SFD act as a preshower counter, thus providing a compact and unified design in the angular range covered by the central UA2 calorimeter. Modifications to the design of the extreme electromagnetic calorimeter cells have been necessary to enlarge the space available for the UA2 central detector.

914

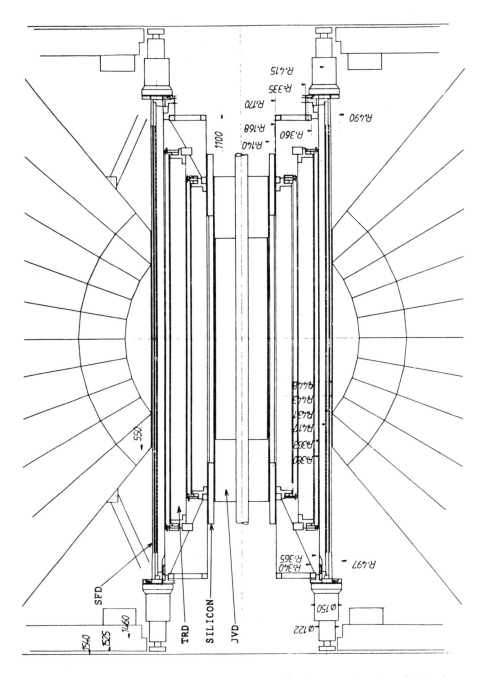

Figure 1 : The new central detector : schematic layout, longitudinal view.

3.   SCINTILLATING FIBRE DETECTOR (SFD)

Recent developments in scintillating plastic fibres made at Saclay
offer an attractive possibility to construct a compact position detector
with good track reconstruction efficiency and adequate spatial accuracy.
As emphasized above compactness is absolutely necessary to provide space
within the UA2 apparatus for the TRD. The present design of the SFD is
based upon:

   i) extensive tests that we have made with fibres image intensifiers
      and CCD since the middle of 1984.

  ii) Monte Carlo studies of the track reconstruction properties using
      simulated p$\bar{\text{p}}$ events.

The SFD whose general layout is shown in fig. 1, has a radial thickness
of 60 mm at an average radius of 410 mm. It consists of about 60000
fibres, with a total length of 150 km, arranged in eight groups of three
layers (triplets). In each triplet, the angles of the fibres with
respect to the beam axis are -α, 0, and α for the three layers
respectively. This provides space coordinates from three stereoscopic
projections. Within the angular range covered by the central UA2
calorimeter, a converter (1.5 radiation lengths) is inserted in front of
the two outer triplets, which are thus used as a preshower counter.

The choices for the design of the detector are based on the standard
fibres which we have tested . They are made of a core of polystyrene,
doped with POPOP and butyl-PBD, with a thin cladding ($\cong$10 μm) and have a
numerical aperture of 0.62. These fibres are currently produced by
Saclay (STIPE) in lengths of 600 m with a 1 mm diameter.

We plan to view the fibres from one end only and to place a
reflector at the other end. We expect on average 91% efficiency per
fibre for particles at θ = 90°. For other values of θ the average
efficiency becomes closer to 100%. A schematic representation of the
read-out system is shown in fig. 2. It is composed of three parts. The
light output from the fibres is amplified with a photon gain of 10^4 by
the image intensifiers (II). The light is then converted by a charge
coupled device (CCD) into a signal which is read by the digitizer and
stored in a memory. The photon gain of the II system is chosen such that
the average signal per CCD element (pixel), within the area
corresponding to a fibre hit by a charged particle, is about ten times

larger than the noise level of the CCD. The principle of operation is also indicated in fig. 2. The gate of the II system is normally open and information accumulates in the CCD. About 1 μs after each crossing an external trigger decision (not using the fibre information) leads to one of the two following sequences:

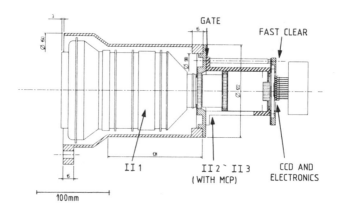

GATE

FAST CLEAR

II 1   II 2 ` II 3   CCD AND
          (WITH MCP)   ELECTRONICS

100mm

Figure 2 : Principle of operation of the readout and digitization of the SFD system.

  i) no-trigger case: a fast-clear pulse is applied to the CCD, draining away the accumulated charge in ≤ 1.5 μs. The II gate stays open and new information is collected in the CCD.

 ii) trigger case: the gate on the II system is closed to prevent illumination from subsequent p$\bar{\text{p}}$ interactions while reading the CCD information related to the event associated with the trigger (≅4 ms). When the CCD reading is completed the II gate is reopened.

From our investigations, it has become clear that three amplifying stages are necessary to meet our gain specifications : a moderate gain unit, a high gain II tube with microchannel plate, and an other moderate gain unit. The first and the third units are demagnifying the fibres image. For the complete detector about 30 II chains will be used.

In commercial II's, the light output is provided by a phosphor3 screen with a decay time of several milliseconds. These have to be replaced by fast phosphor screens (τ ≤ 0.5 μs) which are less efficient by factor ~ 0.5.

The fibre optic window at the exit of the II chain is directly coupled to a CCD where the light is converted to electrical charges which are stored. It is essential that unwanted information can be cleared off from the device during the time available between two $p\bar{p}$ crossings. This can be achieved by using the antiblooming system with which some commercial CCD's are equipped. In collaboration with the manufacturer[9], we have shown that CCD TH7852 which we plan to use can be fast-cleared by applying a pulse of ≤ 1.5 µs duration to the antiblooming input.

The CCD TH7852 has an optical area of 5.8×4.3 mm^2 with 144×208 sensitive elements (pixels). Each pixel has an active area of 19×30 µm^2 and a geometrical area of 28×30 µm^2. With the demagnification of the II chain , a 1 mm diameter fibre may illuminate up to about 15 pixels.

For events satisfying the trigger requirement, the content of each CCD is read out. We consider that pulse height information is useful in pattern recognition, and necessary for the preshower part of the SFD. In addition, since in a typical event particle tracks cross $\cong 2\%$ of the fibres, the compaction of data is essential for an efficient and economical utilization of memory.

4.   TEST OF A SFD PROTOTYPE

A bundle made of 16 layers of 40 scintillating plastic fibres ($\phi = 1$ mm, $L = 2$ m) has been exposed to hadrons and to electrons of 40 GeV/c with the beam direction perpendicular to the layers. Each layer, made of 40 contiguous fibres on a scotch tape support, has a total thickness of about 1.2 mm. The bundle is viewed from one end by the chain of image intensifiers shown in fig. 2, the other end of the fibres is equipped with reflecting tape. All the II's have fibre optic (FO) windows (input and output), and fast phosphor anode screens ($\tau \simeq 0.2$ µs). The measured photon gain of the full II system is 1.5×10^4.

A fast gate is applied to the second II tube in the chain. The FO output of the II system is coupled to the CCD which is operated with fast clear. The principle of operation with the long spill of the test beam is somewhat different from that with the p$\bar{\text{p}}$ collider. The logic of the electronic pulses is shown in fig. 3. As in the collider operation, fast phosphor anode screens for the II's, fast clear of the CCD and, to a lesser extent, fast gating of the II chain are essential ingredients for a good performance of the detector. Under those conditions the noise of the II chain is negligible.

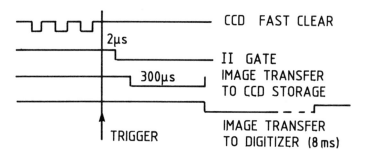

Figure 3 : Logic of the electronic pulses for the test beam operation. The fast clear pulse has a duration of 1 μs every 8 μs.

The image of an hadron track in the CCD plane (140 × 208 sensing elements or pixels) is shown in fig. 4a. The CCD is positionned in a way such that the tracks are parallel to the column direction. The image of a fibre covers about (3.5 × 3.5) pixels. The cumulative distribution of the charge per CCD line, obtained from ~ 100 tracks, is shown in fig. 4b. The 16 individual layers of fibres are clearly separated, an indication of the good quality of the II chain resolution.

In order to test the preshower performance of the SFD, slabs of lead had been inserted in two regions along the bundle. In each position the lead covered the full width of the layers over a length of 10 cm. The corresponding configurations along the beam direction were :
  i) 10 layers of fibres, 1 X_0 of lead, 3 layers of fibres, 0.5 X_0 of lead, 3 layers of fibres.
  ii) 10 layers of fibres, 1.5 X_0 of lead, 6 layers of fibres.

With the bundle position such that the 40 GeV electron beam traverses region i), fig. 5a shows the image of an electron track. The cumulative distribution of ~ 100 electron tracks is shown in fig. 5b.

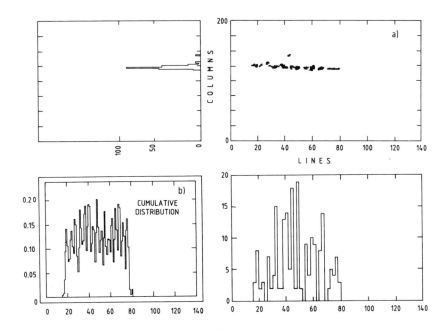

Figure 4 : CCD images of hadron tracks. a) CCD lines vs columns display of an hadron track. b) Cumulative distribution of the charge per CCD line for ~ 100 hadrons.

The charge distributions derived from the pulse height outputs of the CCD are shown for hadrons in the layers in front of the lead (fig. 6a) and for electrons after 1.5 X_0 of lead (fig. 6b).

From fig. 4b where individual fibres are clearly separated, it is already apparent that the resolution of the II-CCD system is rather good. In a separate experiment a single fibre ($\phi = 1$ mm) has been used to measure the resolution of the system, with the result that at the CCD output 80% of the signal is within a square corresponding to (1.3 × 1.3) mm^2 at the system input.

920

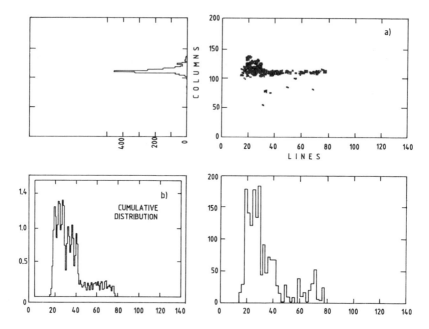

Figure 5 : CCD images of electron tracks with the two converters configuration.   a) CCD lines VS columns display of an electron track.   b) Cumulative distribution of the charge per CCD line for ~ 100 electrons.

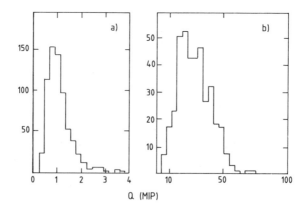

Figure 6 : a) Measured charge distribution for hadrons.
b) Measured charge distribution for electrons after 1.5 X_0 of lead.

5. CONCLUSION

We are in the process of modifying the present UA2 detector with the aim to match its performance to that of the improved Spp̄S collider in 1987.

The upgrade consists mainly of :
a) new calorimeter end caps to improve the quality of the missing transverse energy measurement.
b) a new vertex detector which aims at ensuring that multivertex events can be reconstructed and at improving the detector performance in relation with electron identification.

A component of this new vertex detector is a compact tracking device made of scintillating fibres coupled to image intensifiers and CCD read out. Recent prototype tests have shown that this novel technique is likely to provide a very good detector.

REFERENCES

1. Banner, M. et al., Phys. Lett. 118B, 203 (1982).
2. Bagnaia, P. et al., CERN-EP/84-12.
3. Bagnaia, P. et al., Z. Phys. C 20, 117 (1983).
4. UA2 Collaboration, CERN 83-04 (1983), p. 190.
5. Banner, M. et al., Phys. Lett. 122B, 476 (1983).
6. Bagnaia, P. et al., CERN-EP/84-39.
7. Bagnaia, P. ot al., Phys. Lett. 129B, 130 (1983).
8. Merkel, B., Proceedings of the Fifth Topical Workshop on pp̄ Collider Physics, St Vincent, Italy.
   Gaillard, J.-M., Proceedings of the International Symposium on Physics of pp̄ Collision, Tsukuba, Japan.
9. Thomson-CSF, CCD TH7852.

SILICON MICROSTRIP DETECTORS†

Michal Turala

Institute for Particle Physics

University of California at Santa Cruz*

ABSTRACT

A short overview of the history of Silicon Microstrip Detectors is given as well as a brief description of the basics of their operation. Examples of their application in fixed target experiments(NA11/NA32 at CERN SPS) and future collider spectrometers (MARK II at SLC) are presented.

INTRODUCTION

Historical overview

For many years, silicon detectors have been used in high energy physics for energy measurements[1,2] and to give crude information about the position or angle of scattering[2]. They have also been considered as possible precission position measuring devices[3].

The recent interest in silicon microstrip detectors was triggered by the discovery of short-lived particles($\tau = 10^{-12} - 10^{-13}s$). The first high spatial resolution devices were produced and tested in 1980[4,5] and since then they have received increasing attention by experimentalists. Silicon microstrip detectors of pitch $50\mu m$ or even $25\mu m$ have become "standard" and can be obtained from several sources(Technische Universität Munich, Micron Semiconductor, Hamamatsu, Enertec-Schlumberger, Hughes). In the meantime, it was also shown that commercially available CCDs can also be used as efficient detectors of minimum ionizing particles[6]. In addition, new kinds of devices, such as a silicon drift chamber, have been built[7]. The readout electronics evolved from simple preamplifiers[5] to VLSI custom designed chips[8], which open new possibilities of application. The knowledge of silicon devices among experimental physicists has increased significantly - technological secrets, radiation damage problems, signal processing have become better understood. Most of these topics are extensively discussed in the recent report of C.Damerell[9].

A number of experiments have used, are currently using, or are planning to use silicon microstrip detectors(or CCD's) for precise determination of particle trajectories. Among these one can list:

- at CERN: NA11/NA32, WA71, WA76,...ALEPH, DELPHI,
- at Fermilab: E653, 687, 691, 706,...CDF,
- at SLAC: MARK II, SLD.

---

† Work supported by the US Department of Energy

* Visitor from the Institute of Nuclear Physics, Krakow, Poland

In the next sections the application of silicon microstrip detectors in the ACCMOR (NA11/NA32) spectrometer at the SPS, and a proposed application in the MARK II detector at the SLC will be discussed.

Basic principles of detection of short lived particles

In Fig.1 a simplified diagram of the decay of a particle with a lifetime τ is drawn.

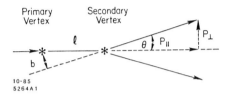

Fig.1. Schematic diagram of particle decay.

The mean decay path l is(for velocity $\beta \cdot c$ and *energy/mass* $= \gamma$):

$$l = c \cdot \tau \cdot \beta \cdot \gamma$$

Given the opening angle Θ between the primary and secondary particles, the impact parameter b(the distance by which the decay trajectory misses the primary vertex) is given by:

$$b = l \cdot \Theta$$

which for a simplified scheme of decay(symetric decay into 2 light particles),

$$< \Theta > \sim \frac{m}{p}$$

can be written as:

$$< b > \approx c \cdot \tau \cdot \frac{p}{m} \cdot \frac{m}{p} = c \cdot \tau$$

This shows that the measurement of the impact parameter is equivalent to the measurement of decay length, and that to detect a particle with a short lifetime it is necessary to identify tracks with b non-zero. In the table below a few particles of interest and their $c \cdot \tau$ values are listed.

particle	τ	$D^{\pm}$	D^o	F	B	B_s
$c \cdot \tau (\mu m)$	90	240	120	90	300	?

A good vertex detector must be able to separate the primary and secondary verticies. A rule of thumb is that such detector should have an impact parameter resolution given by $\Delta b \leq c.\tau/4$.

924

The basics of silicon detectors

The operation of silicon detectors for the detection of charged particles is based on the principle of direct collection of free charges left in the bulk of the material by the traversing ionizing particle. Even in a thin layer of silicon a significant charge is created - for minimum ionising particles its most probable value is about 24 thousand electron-hole pairs per 300μm.

In this respect a silicon detector resembles an ionisation chamber, but with one essential difference: at normal conditions silicon is a semiconductor(ideally pure it has the resistivity of about 200kohm.cm) and it will show too much leakage current to allow for detection of small signals(the ratio of signal to background current would be about 10^{-4}!). To stop this leakage current, a barrier of a p-n(or n-p) junction has to be created - in practice the leakage current is reduced by a factor $10^4 - 10^5$. By choosing the right distribution and the density of dopants one can obtain a diode structure in which under the application of a reverse bias voltage, the depletion region will extend through almost the entire thickness of the detector, thus allowing for the efficient collection of created charges - see Fig.2.

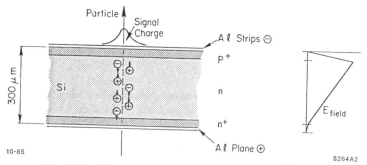

Fig.2. The distribution of dopants and electrical field in
a diode junction of a silicon detector.

The next most critical parameter of a silicon detector is its thickness. It should be as thin as possible to avoid unnecessary material but on the other hand it should be thick enough to provide a sufficient amount of charge for efficient registration. Because the noise level of standard charge amplifiers is of the order of 10^3 electrons, one needs about a 15-20 times larger signal if further processing is to be done(eg., charge division) - this corresponds to a detector with a thickness of 300μm(0.3% of radiation length). Such a detector is fast - the charge collection time is below 10ns(even for holes, whose mobility is three times lower than that for electrons) and its intrinsic precision is very high - 90% of the time a track shift due to δ-rays doesn't exceed 5μm.

To obtain information about the track position, a detector has to be produced as a set of narrow strips, each of which is an individual diode. For the fabrication of a silicon microstrip detector, the full power of the planar technology of modern electronics can be used. This gives extremly high placement accuracy for the strips(below 1μm). As some authors have described /10/ the technology looks "simple". In fact, the process is quite complicated and it is still an art to make good detectors(eg. with low, uniform leakage current at full depletion voltage).

SILICON STRIP DETECTORS IN FIXED TARGET EXPERIMENTS

A good example of the application of silicon strip detectors is given by the ACCMOR (Amsterdam-Bristol-CERN-Cracow-Munich-Rutherford) experiments on charmed particle production at the CERN SPS[11]. The scheme of the silicon strip vertex detector of the ACCMOR spectrometer is presented in Fig.3.

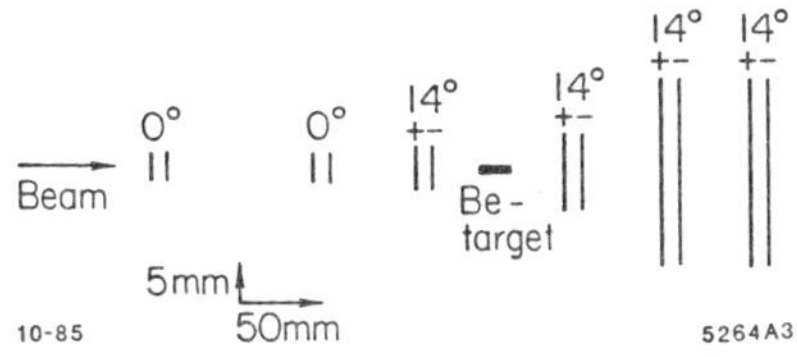

Fig.3. Detector arrangement for the NA11 vertex telescope.

The vertically narrow beam is defined by 6 microstrip detectors(BMSDs) with a pitch of 50μm and 20μm. The beam hits a Be target and the products of the interactions are measured by 6 vertex microstrip detectors(VMSDs), all with a pitch of 20μm. The BMSDs have an active area of 36 x 2.4mm^2 and 36 x 8mm^2; the size of the VMSDs is 36x24mm^2. Two of the BMSDs which are in front of the target and all of the VMSDs are combined in stereo pairs with an orientation of $\pm14°$ with respect to the horizontal plane. The detectors with 50μm pitch are connected to digital electronics; the detectors with 20μm pitch use charge division readout. The latter scheme allows one to obtain high spatial resolution by charge interpolation, and at the same time reduces significantly the amount of electronics needed(only every third or every sixth strip is connected). Altogether in the NA11 experiment about 200 channels of digital electronics and about 800 channels of analog electronics were used. Most of the detectors and electronics were constructed at the Technische Universität Munich and the Werner Heisenberg Institute of MPI Munich - the details of their construction can be found elsewhere[12].

Spatial resolutions of 4.5μm and 7.8μm(rms) were measured for the VMSD detectors with readout at every 3th and every 6th strip, respectively. The resolution for reconstructed vertices projected on the plane perpendicular to the beam direction varied from 21μm for particles with momenta below 5 GeV/c and to 8μm for tracks above 15 GeV/c.

The prime goal of the NA11 experiment was to look for charm particles using prompt electrons for triggering[11]:

$$\pi^- Be \to D\overline{D} + \text{other particles}$$
$$\qquad \quad \Big|\ \hookrightarrow e^-\nu + \text{other particles}$$
$$\qquad \quad \hookrightarrow K^-\pi^+\pi^+ \text{ (eg.)}$$

An example of such event as recorded by the silicon vertex detector of the ACCMOR spectrometer is shown in Fig.4 - the tracks reconstructed off-line are superimposed on the data. In Fig.4a the whole target region is presented: the vertical lines represent Si detectors and the short horizontal segments correspond to the signal positions and amplitudes recorded from these detectors. Fig. 4b shows the interaction region in an expanded scale. The separation of primary and secondary vertices is clear.

926

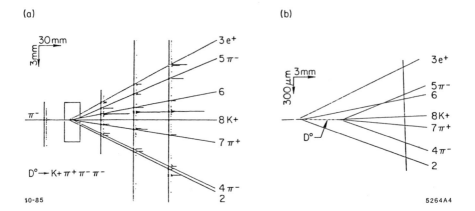

(a)   (b)

Fig.4 A charm event as measured by the silicon vertex detector(a)
and details of the reconstructed vertex(b).

With the silicon strip vertex detectors, the ACCMOR group has measured the
lifetimes of D's and F in several decay channels. The analysis of these data is presented in another talk at this meeting(M.Bosman). The reconstructed lifetimes are
given below[12,13]:

$$\tau_{D\pm} = \left(10.6^{+3.6}_{-2.4}\right) \cdot 10^{-13}s, \qquad \tau_{D^o} = \left(3.7^{+1.0}_{-0.7}\right) \cdot 10^{-13}s, \qquad \tau_F = \left(3.1^{+1.2}_{-0.8}\right) \cdot 10^{-13}s$$

The ACCMOR collaboration has explored the power of this new technology in
greater depth in the new experiment(NA32) where the Be target was replaced with
an active silicon target(AT). The scheme for this device is presented in Fig.5[11].

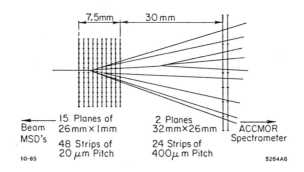

Fig.5. The schematic layout of the silicon active target
of the ACCMOR spectrometer.

The target module is assembled out of 15 silicon strip detectors of transverse size 26x1mm², thickness of 280μm, and pitch of 20μm. Downstream from it two larger silicon microstrip detectors were placed (active area 32x9.6mm, pitch 400μm) to look for the interaction products. Each strip of all detectors was connected to analog electronics and an ADC - in total 768 channels. The active target provides direct information about the interaction point along the beam, measures precisely the coordinates of tracks close to the vertex(a resolution of 2.8μm has been found), and can be used for triggering because it recognises decays as a step in multiplicity of secondaries. This last task, although very attractive, has proven to be also very difficult - to reduce the danger of a trigger bias, frequent and complicated callibrations were required. But the active target has proven to be very useful for reconstruction of events and the understanding of the background.

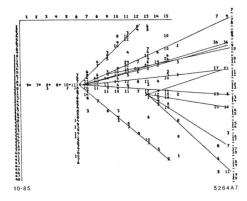

Fig.6. A D⁰ event as seen by the active target of NA32.

In Fig.6 an example of a D⁰ decay is shown as recorded by the active target. The reconstruction of tracks and the identification of particles is done using information from the forward spectrometer. The picture corresponds to the scheme of Fig.5 with a distortion of the linear dimensions. The vertical column on the left shows the strip number of 15 active target detectors(1-48), and the column on the right gives the strip number of the two "multiplicity" counters. The numbers in the center correspond to the pulse heights of signals recorded from the detectors - the average pulse height of a minimum- ionizing particle is equal to 10 counts. For this particular event the first five detectors show a beam track - an interaction took place in the sixth one, which shows a signal of large amplitude plus a recoil. The D⁰ decays in plane 13 - the following counters show increased multiplicity.

SILICON MICROSTRIP DETECTORS FOR COLLIDER EXPERIMENTS

The experimental enviroment at colliders is very different from the one faced in external beams. The biggest difference comes from the fact that a vertex detector has to fit into a small confined space near the intersection point. The acceptance of such a detector has to be large which requires many detectors with the number of channels in the range of 10-100 thousand. Since it is impossible to bring out so many signal lines from a small detector using conventional technology, it is necessary to use special electronics. At e^+e^- machines the whole detector and electronics must withstand a high level of synchrotron radiation.

The development of special integrated circuits for silicon microstrip detector[14] made possible thoughts about silicon vertex detectors for colliders. The "microplex" readout chip has a size of 5x7mm^2 and contains analog electronics for 128 channels including:
- low noise preamplifiers(rms noise of about 1500 electrons and gain of 300),
- analog storage(double correlated sampling),
- multiplexing(all channels on to one output line)
The density of "microplex" electronics allows for one-to-one connections for detectors with 50μm pitch or every second strip connection for detectors with 25μm pitch(the intermediate strips are attached to a second chip at the other end of the detector).

The "microplex" chips have been connected to silicon detectors and tested with a β source[16] and with high energy beams at CERN and SLAC. One such test assembly is presented in Fig.7. It shows a silicon detector with 256 strips and two "microplex" chips at its ends mounted on a large frame for tests. The extra circuitry includes an amplifier-driver, power lines and clock signals.

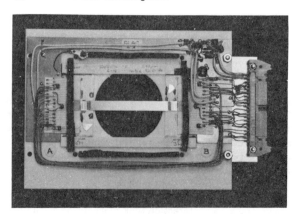

Fig.7. A silicon strip detector with the "microplex" readout
mounted on its test frame.

The results of tests performed at SLAC in a beam of 3 GeV/c positrons are shown in the next few figures. Examples of single and double track events are demonstrated in Fig.8 - for a clearer presentation only a limited number of channels is selected. The signals from the "microplex" chips were recorded by a BADC[17] read out by an LSI-11

microcomputer. The BADC performed pedestal subtraction so only the noise level and
the signals are seen.

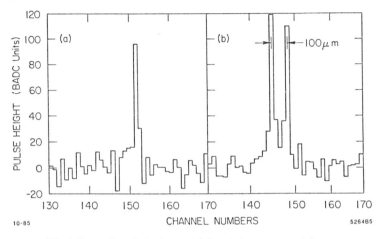

Fig.8 Examples of single track(a) and double track(b) events
as recorded by "microplex" electronics.

Fig.9a shows the pulse height spectrum of the recorded signals and a fit to the data
points(the fit is a convolution of Landau and gaussian distributions). The noise is not
visible because of a software cut at a pulseheight level of 42 counts. More quantative
information about the performance of the detector with VLSI readout is given Fig.9b, in
which the efficiency and noise level are presented as a function of pulseheight threshold.
The plateau efficiency of 98.5% is due to one malfunctioning channel in the beam region.
For threshold above 42 counts the noise rate is below 10^{-5} per strip.

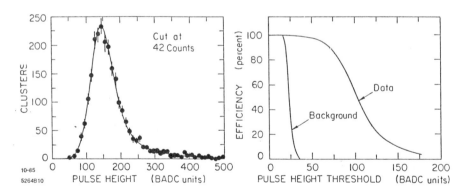

Fig.9. Pulse height distribution of the signals from the silicon strip
detector generated by 3GeV/c positrons(a.) and efficiency and
noise as a function of the pulseheight threshold.

The SLAC Linear Collider(SLC) offers unique opportunities to use the silicon strip detectors for construction of vertex detectors : a very small beam spot, small radius of the beam pipe and low repetition rate. These properties reduce many constraints and allow the design to be significantly simplified. Groups from the University of Hawaii and the University of California at Santa Cruz propose to build such a vertex detector for the Mark II spectrometer[18].

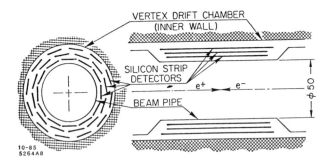

Fig.10. The geometry of silicon strip vertex detector for Mark II.

The geometry of the silicon strip vertex detector(SSVD) for the MARK II is shown in Fig.10. The detector is designed to fit within the small available space around the intersection point and it will be remotely moved in and out. It will consist of three layers of 12 silicon strip detectors each with length 85-93mm, width of 12.8-16.9mm and strip pitch of 25-33μm. To allow for installation and removal it will be subdivided into two semi-cylinders. Although the total number of strips is equal to 18 thousand, the number of cables is very much reduced by using the "microplex" electronics - in total a few tens of thin flat cables will be needed.

ACKNOWLEDGEMENTS

The results presented in this paper are due in general to the two groups that I have worked with: the ACCMOR at CERN and the Silicon Strip Vertex Detector group at SLAC. I acknowledge all of my collegues for providing me with information, in particular: M.Bosman, C.Damerell, R.Horisberger, L.Hubbeling, B.Hyams, R.Klanner, G.Lutz, P.Weilhammer, F.Wickens from the ACCMOR group and C.Adolphsen, A.Breakstone, A.Litke, S.Parker and A.Schwarz from the SSVD group. I also thank S.Amendolia, T.Ferbel and P.Shepard for their private communications.

I am very much obliged to UC Santa Cruz and particulary to A.Litke for organizing my stay in the US.

REFERENCES

1. G.Bellini et al., Nucl.Instr. and Meth. 107(1973),85
2. V.Bartenev et al., Sov.J.Nucl.Phys. 23(1976),400
3. A.Kanofsky, Nucl.Instr. and Meth. 140(1977),429
4. S.Amendolia et al., Nucl. Instr. and Meth. 176(1980), 449
5. E.H.M.Heijne et al., Nucl. Instr. and Meth. 178(1980), 331
6. R.Bailey et al.,Nucl.Instr. and Meth. 213(1983), 201
7. E.Gatti et al., Nucl.Instr. and Meth. 226(1984), 129
8. J.T.Walker et al., Nucl. Instr. and Meth. 226(1984), 200
9. C.J.S.Damerell, Report RAL-84-123, Dec.1984
10. J.Kemmer, Nucl.Instr. and Meth. 226(1984), 89
11. CERN/SPS/78-14, CERN/SPS/82-57
12. B.Hyams et al., Nucl.Instr. and Meth. 205(1983), 99
    R.Bailey et al., Nucl. Instr. and Meth., 226(1984), 56
13. E.Belau, thesis, MPI-PAE/Exp E1-151, June 1985
14. R.Bailey et al., Phys.Lett., 139B(1984), 320
    E.Belau et al., NIKHEF-H/85-5, June 1985
15. R.Hofmann et al., Nucl. Instr. and Meth. 226(1984), 196
16. G.Anzivino et al., Nucl. Instr. and Meth. to be published
17. M.Breidenbach et al., Trans. IEEE, NS-25(Feb.1978),706
18. A.Breakstone et al., Proposal for the addition of a silicon microstrip detector to the Mark II at the SLC, SLAC, July 1985

SCINTILLATING FIBER DETECTORS

Randal C. Ruchti
Department of Physics, University of Notre Dame
Notre Dame, Indiana 46556

INTRODUCTION

High resolution tracking and micro-vertex detection have proved to be essential for the study of heavy particle production and decay in high energy reactions, as well as for the pattern recognition of events in both fixed-target and colliding-beam geometries. Because of the proximity of such devices to the interaction region (for active targets the beam interacts within the fiducial volume of the material), the requirements for the performance of these devices are very exacting and include good signal to noise, fast response, fine segmentation for excellent single-track and multi-track resolution, radiation resistance, simple multiplexing, and modest (or controlled) power dissipation. Conventional solutions to vertex detection, such as silicon microstrips, CCD's, high-pressure drift chambers, bubble chambers, streamer chambers, and emulsions, satisfy some, but not all of the above requirements. This has lead a number of groups to pursue an alternative approach to tracking based on scintillating optical fibers.[1]

The concept of scintillation imaging is an old one,[2] but advances over the last decade in fiber-optic technology, image-intensification, and image-processing systems have revitalized the field to a point where several of these detectors will be used in experiments.

The fundamental structure of a scintillating-fiber detector is the fiber-optic waveguide. A schematic diagram of a generic guide is shown in Fig. 1 and it consists of a scintillating core material of index n_1, a non-scintillating cladding of index $n_2 < n_1$, and an optional coating called extra-mural absorber (EMA) which is typically an opaque glass layer or aluminum reflective layer and which resides on the external surface of the cladding. Desirable properties for this waveguide system are: the quantum efficiency of the core material should be as high as possible; the index difference $\Delta n = n_1 - n_2$ should be as large as possible to optimize light collection by total internal reflection; the cladding should be thick enough to contain the evanescent wave (i.e. keep the wave propagating in the guide), yet sufficiently thin that most of the volume of the fiber is active (scintillation) material; and finally the EMA should absorb away the untrapped light.

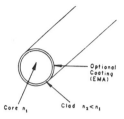

Fig. 1 Schematic of a fiber
       optic waveguide.

Among the choices for the scintillating (core) material are: glass, plastic, liquid, and crystal structures. Several groups are currently active in developmental work in the first three areas:

Glass Scintillator Development:[3]

Group (1) ND/LBL/SDD/CHI/SES (Ruchti and Bross) development of an efficient Cerium glass of low density for tracking, and the fabrication of fibers and coherent plates from this material.

Group (2) LBL/SES (Bross) development of dense, high-index Cerium glasses.

Group (3) ND/FNAL/SDD/CHI/LH (Ruchti) development of Cerium active targets for Tevatron experiment E687.

Group (4) ND/SDD/CHI/LH/SES (Ruchti and Rogers) development of new glass compositions for high efficiency and long attenuation length utilizing (Ce, Pr, Bi) scintillation.

Group (5) CERN/MPI/RAL/LH (Kirkby) development of low density Cerium glasses for tracking and long attenuation length.

Plastic Scintillator Development:[4]

Group (6) BNL (Strand and Borenstein) fibers for tracking.

Group (7) SACLAY/CERN (Bourdinaud and Fabre) development of fibers for DELPHI/LEP calorimetry and tracking for the UA2 upgrade.

Group (8) WUSL/FODS (Binns) fibers for detection of nuclear fragments.

Group (9) FNAL/WUSL (Bross and Binns) large Stoke's shift plastics.

Liquid Scintillator Development:

Group (10) Carnegie-Mellon University (Potter) Scintillation liquids in capillary arrays.

I will restrict my remarks in this paper to the current efforts in glass and plastic fiber detectors.

Tracking with Glass Fibers

Of the possible glass scintillators available, the material which provides fastest response and good efficiency is Cerium (3+) in a silicate host. The scintillation emission is associated with allowed 4f–5d electric dipole transitions of the Cerium (3+) ion, yielding a fluorescent decay of $\tau \sim 50$ nsec, and light emission which is local to the trajectory of a passing charged particle. The particle track may therefore be reconstructed from a lattice of fibers of very small cross section (15μm). Additionally, Cerium oxides have traditionally been added to glasses as radiation "hardeners" and considerable literature exists on the subject.[5] Thus Cerium (3+) glass, in principle, simultaneously satisfies the requirements of localized tracking and radiation resistance, attributes which make it attractive as a microvertex detector.

During the last 12 months a new family of scintillation glasses has been developed and studied by group (1) with base composition designated GS1/SC20. This material has substantially better quantum efficiency than a previously known composition known as NRL glass (factor of 4 improvement). The new material has also been drawn with a commercial cladding glass into individual optical fibers of diameter 1 mm and into multifibers and coherent glass plates containing 25μm, 30μm, and 40μm fibers. The fluorescence properties of several of the GS1/SC20 family of glasses are shown in Fig. 2 and indicate some overlap of emission and absorption spectra, a fluorescence decay which increases monotonically with wavelength (with $\tau \sim 50$ ns at peak fluorescence), and quantum efficiencies as measured in ^{90}Sr source tests which are substantially above NRL glass. In Table I are presented properties of the Cerium glasses of the GS1/SC20 type (low density and long radiation length) as measured by Group (1), and for a number of high density variants produced by Group (2) in an attempt to increase refractive index and reduce radiation length. Of these materials, the base GS1/SC20 composition is the most efficient.

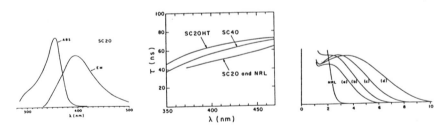

Fig. 2 Flourescence Spectra for the GS1/SC20 glass family.
Left: emission and absorption spectra.
Center: exponential decay times.
Right: raw light yield relative to NRL glass for
(a)SC20HT,(b)SC40,(c)SC20D,(d)SC20 glass.

TABLE 1

PROPERTIES OF SCINTILLATING GLASSES CONTAINING Ce(3+)

PROPERTY	NRL	GS1/SC20	SC20HT	SC40	SC40HT	SC56	SC61
Refractive index	1.58	1.56	1.51	1.58	1.55	--	--
Density (g/cc)	2.69	2.62	2.40	3.00	2.70	3.34	4.34
Efficiency(NRL=1)	1	4-5	2-3	3-4	2-3	2	0.3
Radiation Length(cm)	9	9.7	10.4	6.5	7.2	4.55	2.28
Emission maximum(nm)	395	395	403	404	--	403	458
Fluorescence decay at emission maximum(ns)	48	48	64	56	--	--	--

Fiber-optic plates of this material have been produced initially for groups (1) and (3) and subsequently for group (5). Tracks of minimum ionizing particles, nuclear fragments, as well as particle interactions have been recorded on film and by electronic means (CCD cameras) by groups (3) and (5) using multistage image-intensification systems illustrated schematically in Fig. 3. For the tracking tests scintillation glass fiber optic plates of dimension 1 cubic inch are optically greased onto the input fiber-optic faceplate of an image intensifier, the first two stages of which are electrostatically focussed diodes (group (3)) or proximity photodiodes (group (5)). For group (3), the side of fiber-optic target away from the photocathode was aluminized to reflect light back into the system. Typically 6% of the scintillation light is trapped within a given fiber. The remaining 94% is absorbed by EMA, and black epoxy (or wax) which holds (fuses) the multifibers together (see Fig. 4). The third stage of intensification is provided by a microchannel plate wafer tube which is quiescently reverse-biased) and activated with a trigger pulse derived from external fast logic. The typical gain in these systems is 10^5.

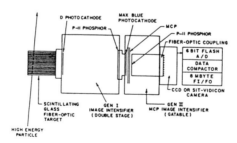

Fig. 3 Schematic of the scintillation imaging system. (Group(3))

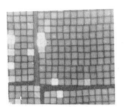 

Fig. 4 Optical microscope images of fiber lattices of GS1
glass: left, 40μm with EMA; right, 25μm with EMA.
Border between multifibers is dark epoxy fill.

Data has been recorded photographically using KODAK 2475 film and
electronically using CCD/SID cameras. Figs. 5 and 6 are sample
photographs which indicate that tracks and interactions are definitely
seen in the GS1 glass. Typical tracking resolution is shown in Fig.
7. Group (3) has recorded data with a custom built Video Data
Acquisiton System (VDAS) which is capable of operation at rates up to
100 MHz. The system includes a 6-bit flash A/D, data compactor, and 8
MByte FI/FO memory. Examples of cosmic rays digitized with this system
are shown in Fig. 8. The number of detected photo electrons is $\gtrsim 4/mm$
with good single track resolution $\sigma \lesssim 28$ μm and two track resolution
typically 80 μm (the diameter of the dots in the film and CCD
recordings). Attenuation length for the GS1/SC20 glass is quite short
$(\lambda \lesssim 6$ cm).[6] This suggests that this particular material, in its
current form, is unsuitable for meter-long tracking systems envisioned
for colliding-beam detectors. For such devices, a different or refined
composition must be found. Groups (4) and (5) are currently pursuing
this.

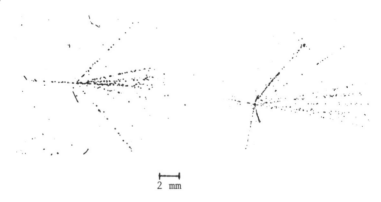

├──────┤
2 mm

Fig. 5 Interactions of 50 GeV/c pions recorded in GS1 glass
using the Fermilab NH beam: left, 40μm fibers with
EMA; right, 25μm fibers with EMA. (Group(3))

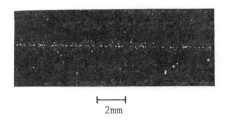

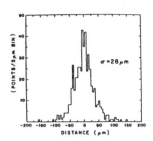

Fig. 6 Minimum ionizing track recorded
in a GS1 target with EMA, at
CERN/PS test beam (Group(5)).

Fig. 7 Residual distribution for track
fitting using digitized film
data (Group(3)).

Fig. 8 Examples of electronically digitized images of cosmic rays
using a CCD3000F camera (Group (3)). Dot size is 80-100μm.

Tracking with Plastic Fibers

Plastic scintillation detectors have been standard devices for
high energy physics because of their fast response and high efficiency.
Initial efforts to utilize plastic scintillator in fiber form were
directed toward high-rate colliding beam experiments by group (6).
Current efforts to produce plastic fibers include single-strand
(Saclay) fibers of diameter 1-2 mm by group (7) and multidraw fibers of
100 μm diameter by group (8). The single-strand fibers are being
utilized for tracking by both UA2 at CERN and by E787 at BNL, the
latter group using a bundle of fibers as an active target in the search
for the rare decay $K^+ \rightarrow \pi^+ \nu\bar{\nu}$. A detailed description of the UA2 plans
are presented elsewhere in these proceedings.[7] Their group intends to
use 1.8 x 10^3 fibers of 1 mm cross section and of 3 m length as a
part of the central tracking system and for shower preconversion. An
example of a track observed during beam tests (Fig. 9)
shows $\gtrsim$ 6 photoelectrons detected per millimeter diameter fiber, when
the beam crosses the fibers at a distance of 50 cm from the
photocathode of a three stage image intensification system. Unlike the
reversed-biased/triggerable systems used by fixed target groups (3,5),
here the image intensification system was operated forward-biased, but
with a 1 μsec fast clear capability in the CCD imager (by pulsing the
antiblooming drain).[8] This method is particularly well suited for
collider operation, where beam crossing occurs at a well-defined time
interval.

Fig. 9 Track recorded in a bundle of 1mm diameter Saclay fibers
of 1m length. (Each clump is approx. 1mm across.) Tests
by Group (7).

Development of plastic multifibers has been pursued by group (8)
and bundles containing working fibers of 100 μm diameter (no EMA) have
been produced for studies of the isotopic iron abundance in cosmic
rays. Fig. 10 shows the trajectory of a stopping iron fragment in a
bundle of 7 x 10⁴ polysytrene PBD/POPOP fibers, with the image
recorded electronically using a CHEVRON microchannel plate image
intensifier coupled to a CID camera. The figure indicates that a
considerable "cross talk" exists between fibers close to the particle
path, which is associable to the mean-free path for wave shifting in
the plastic (typically 1 mm). For the cosmic ray work, in which single
track nuclear fragments are being detected, this distributed pulse
height actually improves track centroiding. However, for high energy
interactions involving multiparticles and jets, this would be
disastrous in terms of two-track resolution. An important, and as yet
unanswered question, is whether or not plastic fibers of small cross
section ($\lesssim$25μm with EMA) can be used for particle tracking because of
this wave shift property. A promising solution may be the use of large
Stoke's shift materials such as 3-HF in polystyrene.[9] Group (9) is
pursuing this possibility.

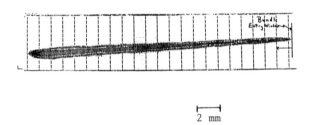

2 mm

Fig. 10 Stopping Fe fragment in a bundle of polystyrene multifibers of 100μm diameter. Recorded at LBL by Group 8.

SUMMARY

Table II summarizes and compares the properties of currently used plastic and glass fiber-optic detectors. The important remark to be made is: <u>tracks</u> <u>have</u> <u>been</u> <u>seen</u> <u>in</u> <u>these</u> <u>materials.</u> Currently the best spatial resolution has been achieved with glass fibers ($\sigma \lesssim 28\mu m$), and the best attenuation length with plastic fibers ($\lambda \gtrsim 1$ m). Radiation resistance is good for cerium glass, and is observed to also be excellent for polystyrene fibers of the Saclay type.[10] These materials represent a potentially elegant solution to vertex detection and tracking for both colliding beam and fixed target geometries. The high energy community should be taking fibers seriously now. It's no longer speculation – fiber-optic tracking works.

The author would like to thank A. Bross, W. Binns, J. Kirkby, R. Strand, M. Bourdinaud, and J. M. Gaillard for contributing their most recent results for this summary.

TABLE 2

COMPARISON OF PLASTIC AND GLASS SCINTILLATING FIBER DETECTORS USED FOR TRACKING

PROPERTY	PLASTIC (Saclay)	GLASS (GS1/ SC20)
Fiber size	1mm	35μm
Photoelectrons/mm (at detector)	13/mm(1mm dia)	4.1/mm(25μm dia)
Resolution per measurement	0.3mm	15μm
Two track resolution	1mm	70μm
Fluorescence decay time	few ns	50ns
Fluorescence emission maximum	420nm	395nm
Attenuation length	1m(1mm dia)	6cm(25μm dia)
Crosstalk	yes(need EMA)	yes(need EMA)
Tracks seen	yes	yes
Radiation resistance	$\leq 10^6$ Rads	$\leq 10^7$ Rads
Density	1g/cc	2.6g/cc
dE/dx	2MeV/cm	4.1MeV/cm
Radiation length	40cm	9.7cm

REFERENCES

1. R. Ruchti, et. al. IEEE Transactions on Nuclear Science, Vol. NS-32, No. 1 (1985) 590-594, R. Ruchti, et. al. IEEE Transactions on Nuclear Science, Vol. NS-31 (1984) 69-73, and R. Ruchti, et. al. IEEE Transactions on Nuclear Science, Vol. NS-30 (1983) 40-43.

2. G. T. Reynolds, IRE Transactions on Nuclear Science, Vol. 7 (1960) 115, and the papers which follow in that journal.

3. Abbreviations used:
   ND   University of Notre Dame
   LBL Lawrence Berkeley Laboratory
   SDD Synergistic Detector Designs
   CHI Collimated Holes, Inc.
   SES SES Technology Consultants
   LH   Levy-Hill Laboratories
   MPI Max Planck Institute
   RAL Rutherford Appleton Laboratory

4. Abbreviations used:
   WUSL Washington University of St. Louis
   FODS Fiber Optic Development Systems

5. B. McGrath, et. al. "Effects of Nuclear Radiation on the optical properties of Cerium-Doped Glass", CERN 75-16 (1975) and R. Ruchti, et. al. IEEE Transactions on Nuclear Science, Vol. NS-32, No. 1 (1985) 590-594, and G. H. Sigel, Jr. and B. D. Evans, "Prospects for Radiation Resistant Fiber Optics", First European Conference on Optical Fibre Communication, 16-18 September (1975) 48-50, and B. D. Evans and G. H. Sigel, "Radiation Resistant Fiber Optic Materials and Waveguides", IEEE Transactions on Nuclear Science, Vol. NS-22, No. 6 (1975) 2462-2467.

6. For GS1/SC20 glass, our measurement in small fiber-optic targets indicates that the attenuation length is greater than 1 cm. A. Bross estimates $\lambda \sim 6$ cm in long 30 μm multifibers of SC20 glass. J. Kirkby measures $\lambda \lesssim 18$ cm in bulk GS1 glass. (A. Bross and J. Kirkby, private communication).

7. See J. M. Galliard, these proceedings.

8. This beautiful idea is due to J. P. Fabre, CERN.

9. C. L. Renschler and L. A. Harrah, Nucl. Instr. and Meth. A235 (1985) 41-45.

10. M. Bourdinaud, private communication.

TRANSITION RADIATION

A REVIEW

J. Richard HUBBARD

D.Ph.P.E., CEN-Saclay

ABSTRACT

Transition radiation as a means of particle identification is reviewed. Emphasis is placed on certain practical problems encountered in building these devices.

1 - THEORETICAL REVIEW

In 1946, Ginzburg and Frank predicted that soft X-rays were produced when charged particles traversed a boundary between two materials with different dielectric constants.[1] For values of $\gamma = E/M$ greater than about 500, X-rays are emitted in a small cone with $\theta \simeq 1/\gamma$. The radiation yield for a single surface is small, roughly equal to $\alpha = 1/137$. The total energy radiated is proportional to γ.

For a **single foil** of thickness ℓ, transition radiation is produced at both surfaces. Interference between the radiation produced at the two surfaces leads to an interference pattern in the energy spectrum of the X-rays produced. The energy spectrum of the single-foil radiation yield, shown in Fig. 1, can be expressed as[2]

$$\frac{dW}{d\omega} = \frac{2\alpha}{\pi} G(\nu, \Gamma)$$

where ω is the photon energy and ν and Γ are scaling variables defined by $\nu = \omega/\omega_1$ and $\Gamma = \gamma/\gamma_1$, with $\gamma_1 = \ell\omega_p/2$ and $\omega_1 = \gamma_1\omega_p$; ω_p is the plasma frequency of the medium. Peaks are present for $\nu = 1/n\pi$. The largest yield is obtained by choosing the foil thickness such that the lowest order peak, at $\nu = 1/\pi$, coincides with the maximum detection efficiency of the experimental apparatus.

The **multi-foil** yield is obtained by multiplying the single-foil yield by the effective number of foils. Multiple-foil interference effects are usually small. Saturation effects become important, however, if the gap between foils is comparable to the distance required for the readjustment of the electromagnetic fields (the formation zone). The energy spectrum is now a function of three scaling variables[2]

$$\frac{dW}{d\omega} = \frac{2\alpha}{\pi} N_{eff} G_{many}(\nu,\Gamma,\tau)$$

where $\tau = \ell_2/\ell_1$, and $N_{eff} = (1-e^{-N\sigma})/(1-e^{-\sigma})$ is the effective number of foils (σ is the absorption probability for one foil and one gap). Formation zone saturation is visible in Fig. 2. Saturation sets in above $\Gamma_{sat} \simeq 1.2\sqrt{\tau}$.

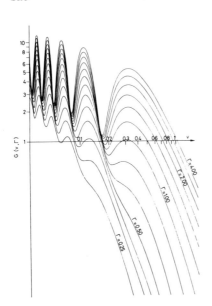

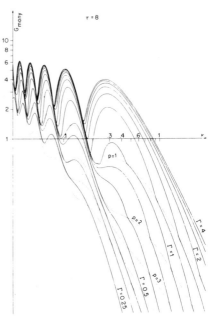

Figure 1. Single-foil yield.
Single foil yield $G(\nu,\Gamma)$
Vs. $\nu = \omega/w_1$ from Artru,
et. al.[2]

Figure 2. Multi-foil yield.
Multi-foil yield $G(\nu,\Gamma,\tau=8)$
Vs. ν for $\tau = \ell_2/\ell_1 = 8$ from
Artru, et. al.[2]

2 - DESIGN CONSIDERATIONS

The **radiator material** is normally chosen to maximize the number of transition-radiation photons detected in a particular experimental apparatus. The effective number of foils is $N_{eff} \simeq 1/\sigma \sim \omega^2/Z^3$ for a full stack ($N\sigma \gg 1$), or $N_{eff} \simeq N$ for a transparent stack ($N\sigma \ll 1$). The number of photons detected varies as $1/Z$ for a full stack, but is independent of Z (but proportional to ρ) for a transparent stack. Materials from deuterium to carbon have been considered as radiators for particle identification, but most groups have concentrated their efforts on Lithium (ω_p=14 eV) or polypropylene (CH_2, ω_p=19 eV) foils.

The **noble gas**, which acts as the photo-detector, is chosen to maximize the photon detection efficiency while minimizing the number of delta-rays produced. We can describe a quality factor for the noble gas as

$$Q.F. = \frac{(dN/dx)_\gamma}{(dN/dx)_\delta} \sim \frac{Z^3}{\omega^2}$$

This form is only valid, however, if ω, the photon energy, is greater than 35 keV, the Xenon K-edge, and if the absorption probability is small. For practical detectors, with transition radiation photons between 3 and 15 keV, the quality factors of the noble gases are roughly Xenon : Krypton : Argon : Neon = 10 : 3 : 5 : 1.

3 - EXPERIMENTAL VERIFICATION

Interference effects in the integrated energy spectrum were observed by Cherry[3] using polypropylene foils of varying thickness as radiators, a magnet to separate the charged particles from the X-ray emission, a diffraction crystal to analyze the X-ray energy, and a Xenon/CO_2 proportional chamber to detect the X-rays. Results for three foil thicknesses are shown in Fig. 3. In all cases, the gap thickness was large (l_2 = 1.4 mm), so the interferences observed were those due to the two surfaces of each foil.

Formation zone saturation effects have been measured by the E715 collaboration at Fermilab.[4] Again the radiators were polypropylene foils. This time the X-ray detector was a Xenon/methane multi-wire proportional chamber. Pions were used to investigate low values of γ, and electrons for large values. Their results are shown in Fig. 4. At low energy, the number of photons detected increases as γ^3, in agreement with their

Monte-Carlo calculations. At high energy, the number of photons detected is constant.

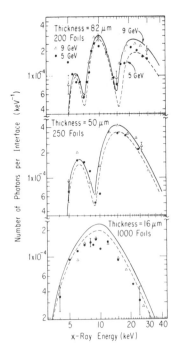

Figure 3. Interference.
Measurements by Cherry.[3]

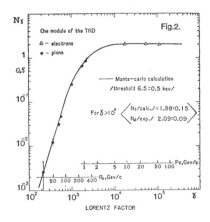

Figure 4. Saturation effects.
Formation zone saturation reported by Kulikov.[4] The radiator consisted of 210 layers of 17 μm poly-propylene with 1 mm gaps. The detector was a MWPC with 16 mm of Xenon/CH$_4$.

4 - ANALYSIS TECHNIQUES

A **magnetic field** can be used to separate the transition radiation photons from the ionization produced in the photon detector by the charged track. The problem is that the magnet requires space which is often not available. The **angular distribution** of the transition radiation itself separates the photons from the ionization,[5] but the opening angle of the transition radiation cone is so small, $\theta \simeq 1/\gamma \simeq 1$ mrad, that an appreciable lever arm is still required to separate the photons from the track.

The **total charge** can be used to analyze tracks with overlapping ionization and transition radiation. A disadvantage of this technique is that the ionization has a long tail toward high energies, the Landau tail.

The **cluster-counting** technique, in which the number of energy clusters above some threshold is counted, gives a value which is Poissonian-distributed for δ-rays as well as photons. The separation power of a TRD can be further improved by using **maximum-likelihood** techniques.

A **comparison of different techniques** has been published by Ludlam, et. al.[6]. Figure 5 shows the separation efficiency obtained by total energy (Q), cluster counting (N), and maximum likelihood (W). They found that cluster counting gave an appreciable improvement over total energy, but that maximum likelihood, using the distance of each photon from the entrance window, gave little further improvement.

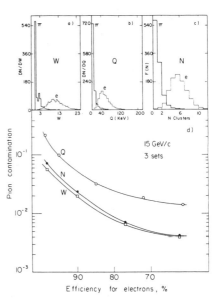

Figure 5. Analysis techniques.
Comparison of total charge (Q), cluster counting (N), and maximum likelihood (W) by Ludlam, et. al.[6] The radiator contained 1500 Lithium foils (30 μm foils, 200 μm gaps). The detectors were Xenon/CO_2 longitudinal drift chambers. Three events were combined to approximate three radiator-detector sets with a total length of 110 cm. Electrons and pions were identified at 15 GeV/c.

Improved cluster counting has been obtained by the NA34 group, with Dolgoshein, Fabjan, Willis, et. al. Their transition radiation detector for NA34 is currently under test at CERN.[7] The radiators are made of polypropylene foils, and the detectors are Xenon/isobutane MWPC's. The TRD contains 8 modules in a total of 70 cm. They obtain a trigger rejection factor of better than 1000 for pions above 3 GeV/c. They obtain this excellent performance with specific cluster-counting electronics, instead of using flash ADC's. Some of their techniques include: an intelligent, or variable, threshold on the cluster energy such that low-energy clusters far from the entrance window are disallowed, analog pairing of the 2-mm spaced anode wires to avoid splitting clusters into smaller pieces, and elimination of high-energy δ-rays by excluding energy deposited on more than two wires. The NA34 group has measured the longitudinal diffusion in different gas mixtures, and chosen 5% isobutane as a low-diffusion quencher.

Some **typical TRD projects** are shown in Table I. The first four experiments in the table have already taken data and produced physics results. The next two are currently being tested in their final locations, and the last two are in the planning stage.

Table I. Typical TRD Projects.

Group	Part A/B	γ min	L cm	N set	Radiator	Anal	Rjt (B)	Eff (A)
Chicago	e/p	10000	100	6	CH_2	Q		
ISR	e/π	1000	55	2	Li foils	Q		
EHS	K/π	200	350	20	Mylar foil	Q	15	90%
E715	e/π	10000	360	12	CH_2 foils	N	1500	99%
NA34	e/π	5000	70	8	CH_2 foils	N	1000	90%
Ω	K/π	1000	200	12	CH_2 foils	N+Q	20	85%
UA2	e/π	10000	21	2	Li foils	FADC	20	80%
D0	e/π	10000	30	3	CH_2 foils	FADC	20	90%

5 - PROBLEMS ENCOUNTERED IN TRD

Cluster separation requires that the longitudinal diffusion be small compared to the distance between photons. There is a thermodynamic relationship between the longitudinal and the transverse diffusion in a gas in which the electrons suffer only elastic collisions[8]:

$$\frac{D_L}{D_T} = \frac{\partial(\ln v)}{\partial(\ln E)}, \qquad \text{for elastic scatters only,}$$

where v is the electron drift velocity in the electric field E. Thus, for "saturated" drift velocities, the longitudinal diffusion is minimum. This happy coincidence also holds in certain complex gases.

Electron attachment is usually held responsible for a frequently-observed loss of gain for electron clusters with long drift times. The main attachment reaction at low electron energies is the three-body sequence

$$e^- + O_2 \longrightarrow O_2^{-*}$$

$$O_2^{-*} + M \longrightarrow O_2^- + M^*$$

The initial electron capture is unstable, with a relaxation time of about 100 psec. The electron is only effectively captured if the excitation energy (0.4 eV) is removed by interaction with another molecule.

Gas purity requirements can be deduced from this three-body attachment process. If the detector were filled entirely with humid air, the electron lifetime would be about 10 nsec. To obtain electron lifetimes greater than 100 μsec, as required for TRD, we require less than 160 ppM of Oxygen for any Xenon filled detector, less than 40 ppM if the detector contains 10% CO_2, and less than 30 ppM if the detector gas is saturated with water vapor (relative humidity 100%). These are not excessively stringent requirements, except that Xenon is very expensive, so that most users choose recirculating gas systems, rather than once-through flow.

Two-body dissociative attachment is important at higher electron energies. As an example, the dissociative process

$$e^- + CO_2 \longrightarrow O^- + CO$$

has resonant peaks at 4.4 and 8.2 eV. Using a Monte-Carlo technique, Binnie has shown that electron loss in Argon + 5% CO_2 in a drift field of 2000 V/cm amounts to 33% per cm.[9]

Space charge, either local or bulk, can also influence the uniformity of chamber gain. In the case of local space charge, the charge reaching the anode from the nearest segment of a given track reduces the effective gain for later segments of the same track. In the case of bulk space charge, it is the integrated flux which influences the gain of the chamber. In our tests at D0, we were bothered by local space charge effects, which depend on the angle of the track with respect to the anode wires.

Recombination of electrons with the positive ions is usually considered unimportant for chambers operating in the proportional mode. It should be noted, nonetheless, that the recombination rate for Xenon ions is three times larger than that for the more common Argon ions. Both space charge and recombination should be helped by choosing a quencher gas with an ionization potential lower than the Xenon ionization potential at 12.1 eV. In this case the positive charge is quickly transfered to the quencher molecules, which can be expected to have higher mobilities (i.e., higher ion drift velocities) and lower recombination rates. From this point of view, ethane and isobutane should be better candidates than CO_2 and methane.

Uniform spacing of radiator foils is required for maximum transition-radiation yield. For Lithium, this is obtained by marking the foils so they becoming self-supporting.[10] for polypropylene foils, spacers are used. Note that the cylindrical geometry of UA2 and D0 renders this problem more difficult.

Hydrostatic pressure due to the high density of the Xenon gas (four times heavier than air) causes thin windows to sag, and chamber uniformity suffers if the windows form cathodes for the detectors. This problem becomes important for window dimensions larger than about 50 cm. Various solutions to this problem have been found. NA34 uses a double window to rectify their exit cathode. EHS diluted their Xenon with four parts of Helium to obtain a chamber gas with the same density as air.

REFERENCES

1. Ginzburg, V.L. and Frank, I.M., JETP 16 (1946) 15.

2. Artru, X., Yodh, G.B., and Mennessier, G., Phys. Rev. D12 (1975) 1289. See also Cherry, M.L., et. al., Phys. Rev. D10 (1974) 3594.

3. Cherry, M.L., Phys. Rev. D17 (1978) 2245.

4. Kulikov, A.V., preprint (1984).

5. Deutschmann, M., et. al., NIM 180 (1981) 409.

6. Ludlam, T., et. al., NIM 180 (1981) 413.

7. Dolgoshein, B., Fabjan, C.W., Willis, W., et. al., private communications (1985).

8. Robson, R.E., Aust. J. Phys. 25 (1972) 685. See Huxley, L.G.H. and Crompton, R.W., The Diffusion and Drift of Electrons in Gases, Wiley Interscience Publications (1974), p. 119.

9. Binnie, D.M., NIM A234 (1985) 54.

10. Cobb, J., et. al., NIM 140 (1977) 413.

INITIAL OPERATION OF THE FERMILAB ANTIPROTON SOURCE

Stephen D. Holmes

Fermi National Accelerator Laboratory*
P.O. Box 500
Batavia, IL   60510

1. INTRODUCTION

The Fermilab Tevatron I (TeV I) project has as its prime objective initiation of proton-antiproton collisions at 2.0 TeV in the center-of-mass, with a luminosity of 10^{30} cm^{-2} sec^{-1} by September 1, 1986. The two components needed to produce such collisions are the Fermilab 1 TeV (Tevatron) accelerator and the Antiproton Source. The Fermilab Tevatron has been in operation for more than a year providing high energy protons for fixed target experiments while the Antiproton Source is currently being commissioned. Progress made so far in the commissioning of the source has encouraged us to form an intermediate goal of observing proton-antiproton collisions in the Tevatron before the scheduled shutdown in October, 1985.

2. OPERATION OF THE ANTIPROTON SOURCE

The Antiproton Source is made up of three main components--the target station, the Debuncher ring, and the Accumulator ring. The location of the source relative to the Fermilab Main Ring and Tevatron, as well as the interconnecting beamlines are shown in Figure 1. The Debuncher and Accumulator rings reside within a common tunnel. Antiprotons are produced through the interaction of 120 GeV protons with a tungsten target. Negatively charged secondaries are delivered to the Debuncher ring where a longitudinal phase space rotation reduces the momentum spread as a DC beam is produced. Antiprotons accumulation takes place in the Accumulator. The accumulator has the same energy (8.9 GeV) and circumference (474 m) as the Fermilab Booster. Once a sufficient number of antiprotons have been accumulated they are extracted and sent back to the Main Ring. The total flux of antiprotons through the system is designed to be 10^{11}/hour.

The sequence by which antiproton accumulation takes place is as follows:

---

*Operated by the Universities Research Association under contract with the U.S. Department of Energy.

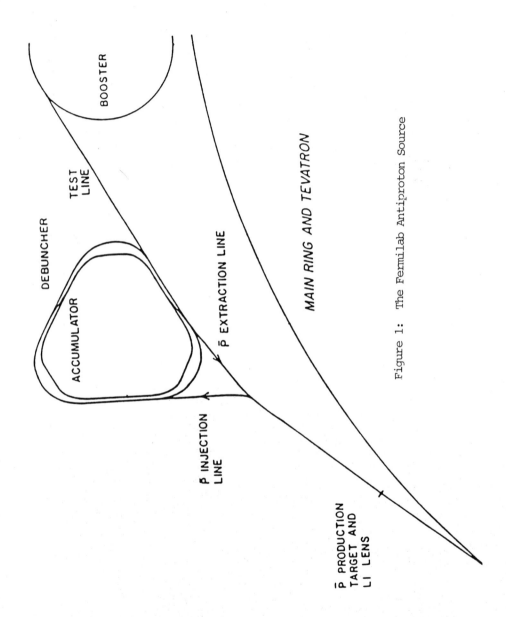

Figure 1: The Femilab Antiproton Source

1) 2×10^{12} protons are accelerated in the Main Ring to 120 GeV. These protons are contained in 80 bunches separated by 20 nsec. RF bunch rotation in the Main Ring produces a bunch length of less than 1 nsec prior to extraction. These protons are then extracted and sent to the target station.

2) The 120 GeV protons are focussed to a spot of .06 mm^2 on a target composed of either tungsten, a tungsten alloy, or copper. About 7×10^7 antiprotons are produced into an acceptance of 20π x 20π (horizontal x vertical) mm-mrad and a momentum spread of 3% with 8.9 GeV of longitudinal momentum. The antiprotons of course retain the 1 nsec bunch length of the incoming protons. The antiprotons are focussed through a lithium lens (focal length 20 cm) into the beamline which carries them to the Debuncher. They are accompanied on their journey by several hundred times as many pions, muons, and electrons.

3) The antiprotons enter the Debuncher with an energy of 8.9 GeV and an energy spread of 3%. The bunches are rotated by an RF system to trade the short bunch length and large momentum spread for a small (0.2%) momentum spread and a long bunch length (DC). Transverse cooling systems in the Debuncher also reduce the beam size to 7π x 7π prior to sending the antiprotons on to the Accumulator. Antiprotons are extracted from the Debuncher and injected into the Accumulator on a cycle which repeats steps 1)-3) every two seconds. No accumulation takes place in the Debuncher.

4) Antiproton accumulation takes place by stacking in momentum space in the Accumulator. A core with density 10^5/MeV containing 5×10^{11} antiprotons is built up over a period of five hours. Antiprotons in the core are also cooled transversely to an emittance of 2π x 2π. A total of 10^{11} antiprotons with a longitudinal emittance of 2 eV-sec are required to achieve the design luminosity of the collider. These antiprotons are extracted from the core and sent back to the Main Ring in thirteen 20 nsec bunches. The remaining 4×10^{11} antiprotons are preserved in the core. The removed antiprotons can be replenished in an hour to allow extraction from the Accumulator every hour once the initial core has been created.

5) The antiprotons delivered from the Accumulator are coalesced into a single bunch and then accelerated to 150 GeV in the Main Ring. They are then transfered to the Tevatron where they are accelerated, along with a single proton bunch to 900 GeV.

3. COMMISSIONING THE ANTIPROTON SOURCE

Several modes of operation have been used in the commissioning of the antiproton source. Initial operations were carried out with 8 GeV primary protons delivered directly to the Debuncher from the Main Ring via the AP1 and AP2 beamlines.   These protons were subsequently sent on to the Accumulator via the Debuncher to Accumulator (D to A) line. In this mode of operation both rings and all beamlines were at the opposite polarity to that required for circulating antiprotons but with the correct sense of circulation. This was used for setting up the rings and for doing high intensity measurements on the stochastic cooling systems.

Three other modes of operation have also been used which include (in historical order): 1) Secondary protons produced from 120 GeV primary protons; 2) 8 GeV primary protons run backwards through the entire system starting with injection into the Accumulator via the nominal extraction beamline (AP-3) and ending up at the production target; and 3) Secondary antiprotons.

A list of commissioning milestones is given in Table 1. Commissioning began on January 29, 1985 with the first extraction of beam from the Main Ring and transport to the target.   Installation of the Debuncher was completed in mid-April and circulating beam was achieved soon thereafter. During the period of April and May time was divided between Debuncher commissioning and Accumulator installation. Following the completion of Accumulator installation in early June and the removal of a 'pizza pan' on June 19 beam circulated in the Accumulator. Other notable milestones chronicled in the table include the first stacking of antiprotons on September 8, the first antiprotons accelerated in the Main Ring and transferred to the Tevatron on October 8, and the first proton-antiproton collisions at 1.6 TeV in the center-of-mass on October 13.

Table 1 -- Antiproton Source Commissioning Milestones

January 29, 1985	Beam transported from Main Ring to target.
April 12	Beam transported to end of AP2.
April 21	Circulating beam in the Debuncher.
May 15	Antiproton yield measured at end of AP2.
May 25	Secondary proton beam debunched.
June 6	30 turns observed in Accumulator (via AP3).
June 15	Beam cooled in Debuncher.
June 19	Remove pizza pan from Accumulator.

June 20	Circulating beam in the Accumulator.
June 23	Accumulator beam decelerated to stacking orbit.
July 20	Beam transferred from Debuncher to Accumulator.
August 23	Replace six collapsed bellows in the Accumulator.
August 30	Core cooling system commissioned.
September 3	First stochastic stacking of protons.
September 7	Reverse polarities. First antiprotons seen in Debuncher.
September 8	10^8 antiprotons accumulated.
September 9-14	CDF detector installed.
September 25	10^9 antiprotons accumulated.
October 3	First antiprotons extracted.
October 5	First antiprotons observed in Main Ring.
October 8	9×10^9 antiprotons accumulated. 1.3×10^9 antiprotons accelerated in the Main Ring and observed in the Tevatron.
October 13	First proton-antiproton collisions at 1.6 TeV.

3.1 The Debuncher

The Debuncher has a circumference of 505 meters, giving it a revolution frequency at 8 GeV which is exactly the 90th subharmonic of the Main Ring RF frequency at 120 GeV. Both the Debuncher and the Accumulator are triangular in shape with lattices which are sixfold symmetric. The Debuncher lattice consists of three arcs and three long straight sections. One straight section contains stochastic cooling pickups and extraction, a second contains the stochastic cooling kicker⁻, and the third contains RF and injection.

The Debuncher is operating at its design tune of $(Q_x, Q_y) = (9.77, 9.78)$ with the chromaticity zeroed by two families of sextupoles. The measured aperture is 4.5% in momentum, and $13\pi \times 13\pi$ transversely as compared to the design values of 4.0% longitudinally and $20\pi \times 20\pi$ transversely. The closed orbit distortion in the ring is less than 3 mm (rms). The vertical dispersion in the ring is everywhere less than 5 cm, and the horizontal dispersion is measured to be less than 7 cm everywhere within the nominally dispersion-free straight sections. The beam lifetime in the machine is greater than one hour.

Both the debunching and the stochastic cooling systems are operating close to design specifications. The debunching RF system is producing a reduction in the momentum spread of a factor of twenty for a momentum bite of 3% while the stochastic cooling system is reducing the beam emittance by about a factor of six in 4 seconds.

3.2 The Accumulator

The Accumulator is located in the same tunnel as the Debuncher but has a slightly smaller circumference of 474 meters. The circumference and extraction energy are exactly the same as the Fermilab 8 GeV booster. The ring contains three high dispersion and three zero dispersion straight sections. Two sets of stochastic cooling systems are used in the Accumulator. One, the stack tail system, is used to move particles quickly away from the point in momentum space where they are deposited after injection from the Debuncher toward the high density core. The core cooling system collects antiprotons into the high density core. Both systems provide both longitudinal and transverse cooling. When the flux of antiprotons into the system is held constant the system builds up a stack of antiprotons whose density varies exponentially with momentum, culminating in a core with a density of 10^5 particles/eV. Such an accumulated antiproton stack is shown in Figure 2. The transverse cooling systems in the Accumulator reduce the beam emittance to 2π x 2π. Stochastic cooling pickups and kickers occupy four of the straight sections. Injection and extraction both occur in the same (high dispersion) straight section, while the RF is located in a fifth straight section.

The Accumulator is not operating at its design tune of (6.611,8.611) at the moment because of a large quadratic dependence of the tune on momentum which we are unable to remove with the presently installed octupoles. This tune dependence causes a tune excursion of .01 both horizontally and vertically as the antiprotons travel from the injection orbit into the stack. As a result we are currently operating at the tune point (6.641,8.641). The aperture of the machine is measured to be >2% longitudinally, and 3π x 3π transversely. The design values are 2% and 10π x 10π respectively. The vacuum in the ring is below 10^{-9} Torr everywhere and the beam lifetime ranges between a few hours with the stochastic cooling systems turned of to a few hundred hours with the core systems turned on.

The RF and stochastic cooling systems are all working at a level of greater than 50% of their design specifications with the exception of the stack-tail betatron cooling system which has yet to be commissioned. The stack-tail system has been observed to move particles away from the stacking orbit at a rate of 4 MeV/sec. The stacking rf system has operated at close to 100% efficiency at times.

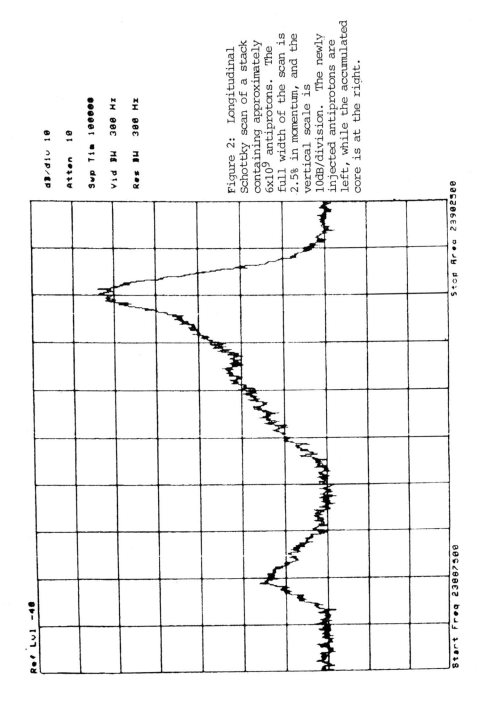

Figure 2: Longitudinal Schottky scan of a stack containing approximately 6x10⁹ antiprotons. The full width of the scan is 2.5% in momentum, and the vertical scale is 10dB/division. The newly injected antiprotons are left, while the accumulated core is at the right.

The injection efficiency into the Accumulator from the Debuncher is very dependent on the size of the beam being transferred since the Debuncher aperture is so much larger than the Accumulator's. Transfer efficiencies of >75% are observed for beams having an emittance of $2\pi \times 2\pi$.

4. OPERATIONAL EXPERIENCE WITH THE ANTIPROTON SOURCE

Approximately one month has been spent operating the source as an antiproton accumulator. During this period many antiproton stacks were accumulated with the maximum number of antiprotons ever accumulated 10×10^9. Figure 2 shows the momentum distribution of antiprotons in a stack containing about 5×10^9 particles. By the end of the month the accumulation rate was approximately 8×10^8/hour. The accumulation rate is a factor of 100 below the design value of the source. The various components of this factor are reasonably well understood and are summarized in Table 2.

Table 2 -- Current Performance of the Antiproton Source

	Design	Actual	Missing Factor
Flux of 120 GeV protons from Main Ring	1×10^{12}/sec	2.5×10^{11}/sec	4
Lithium lens strength	500 KAmps	250 KAmps	2.0
Debuncher Aperture	$400\pi^2$	$170\pi^2$	2.5
Debuncher RF Acceptance	4%	3%	1.3
Deb. to Accumulator transfer	100%	75%	1.3
Accumulator stacking RF	100%	50%	2
Stacking efficiency	100%	67%	1.5
Accumulator acceptance	$100\pi^2$	$15\pi^2$	(6.7)
Accumulation Rate	10^{11}/hour	10^9/hour	

As can be seen from the table the missing factor of 100 is made up of lots of factors of two. Some of the factors given in the table are somewhat redundant. For example the bad accumulator aperture has no affect on the accumulation rate at present since the lowered flux of primary protons and low Debuncher aperture produce a smaller emittance out of the Debuncher. In contrast the weaker lithium lens being used at the moment would be less crucial if the Debuncher aperture were larger.

For commissioning of the collider a mode of operation was developed in which two (eight hour) shifts were spent accumulating somewhere between 6-10x10^9 antiprotons, followed by a shift in which these were extracted and sent back to the Main Ring and Tevatron in batches of 1x10^9. This allowed somewhere between four and seven shots to the Main Ring per day (at least on days on which the stack was not lost as happened twice). Approximately twenty shot were required before antiprotons successfully accelerated in the Main Ring and injected into the Tevatron.

5. BRINGING THE ANTIPROTON SOURCE UP TO DESIGN SPECIFICATIONS

The Fermilab accelerator complex began a nine month shutdown on October 14 for installation of the BO overpass and construction of the DO experimental hall. The Main Ring and Tevatron will be brought up again in the summer of 1986 with proton-antiproton collisions at the design luminosity scheduled to begin on September 1, 1986. By mid-November of this year a beamline will be installed to transport 8 GeV protons directly from the Booster into the Debuncher (see Figure 1). This beamline will be used for machine studies to bring the Debuncher and Accumulator up to specification while the Main Ring is off. Modifications to the Main Ring should allow upgrading of the primary proton flux by the required factor of four. A 500 KAmp lithium lens has been built already, however it failed during the earliest antiproton accumulation run. The failure is understood and the lens will be rebuilt.

Our early experience with the Antiproton Source has provided us with invaluable information for the upcoming down period. There is every expectation that the source will be working at a level sufficient to provide the design luminosity in the collider one year from now.

THE STATUS OF SLC[*]

STANLEY D. ECKLUND

Stanford Linear Accelerator Center
Stanford University, Stanford, California, 94305

ABSTRACT

The current construction status of the Stanford Linear Collider (SLC) is described along with a brief overview of the project. Tests of completed parts of the machine are summarized.

1. INTRODUCTION

The Stanford Linear Collider Project is intended to serve primarily two goals. First it is to make available electron positron collisions at higher energies in order to explore the next area of interest in elementary particle physics. This sets the energy goal of 100 GeV in the center of momentum system, in order to reach the Z^0. The second goal is to explore the feasibility of linear colliders for producing high energy collisions of electrons and positrons. This second goal is important in that this type of machine is different in many aspects from existing colliders and the technical difficulties encountered are often new to the accelerator physics community. If these technical problems are solvable at reasonable cost the linear collider would provide an attractive alternative to electron positron storage rings which suffer increasingly from synchrotron radiation losses as the energy increases.

The SLC project[1] is an adaptation of SLAC with additions to provide for bringing small intense beams into collision. Rather than aiming two linacs at each other, SLC utilizes the one existing linac to accelerate both the positron and electron beams. This saves building a second 2–mile long linac or alternately doubling the accelerator gradient. It does of course require transport lines to bring the beams into head–on collisions. A schematic view of SLC is shown in Fig. 1. The various subsystems of the SLC are: an Electron Source to provide two high intensity short pulses; a sector (1 of 30) of acceleration to

---

[*] Work supported by the Department of Energy, contract $DE - AC03 - 76SF00515$.

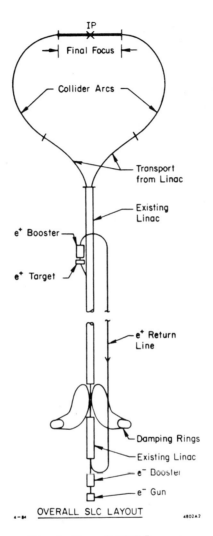

IP

Final Focus

Collider Arcs

Transport from Linac

Existing Linac

e⁺ Booster

e⁺ Target

e⁺ Return Line

Damping Rings

Existing Linac

e⁻ Booster

e⁻ Gun

OVERALL SLC LAYOUT

Fig. 1. Overall SLC Layout.

bring the energy up to 1.21 GeV; two Damping Rings to reduce the phase space occupied by the beams; the existing Linac modified to provide stronger focusing and beam guidance capability; higher power Klystrons for higher energy acceleration in the Linac; a new Positron Source system; an Arc transport for each beam; and a Final Focus system to produce small beams and bring them into collision. In addition there is a new control system with much improved capabilities needed for this kind of machine. Also a considerable amount of conventional construction is required for the ring housings, the Arc tunnels and the Collider Experimental Hall (CEH).

The machine cycle of SLC sends the beams around the above mentioned systems in the following manner. Each of the Damping Rings has stored in it two bunches of their respective positrons and electrons injected in previous cycles and suitably damped. Each cycle one of the positron bunches is kicked out of the Positron Ring and both of the electron bunches from the Electron Ring. Each of the Ring–To–Linac transports contains a bunch compressor to match the longitudinal phase space from the rings to the Linac. The timing of the ring extraction systems put the positron bunch first in the Linac followed by the two electron bunches. The three bunches are approximately half the ring circumference apart, 17 m or 59 nanoseconds, in the linac. The first two are accelerated to the end of the linac and transported around their respective Arc and Final Focus transports to collide in the interaction point. After collision the bunches are kicked into a dump and discarded. The third bunch, designated the scavenger bunch as it uses up some of the remaining r.f. in the linac, is kicked out at the two-thirds point and targeted to produce positrons. The positrons are collected and accelerated to 200 MeV in an high gradient strongly focused linac. They are then sent down a 2 Km return transport to the beginning of the linac for subsequent acceleration to 1.21 GeV. Just prior to the arrival of the positrons the electron gun is fired producing two bunches to be accelerated and supplied to the electron ring. The three bunches are sent to their respective rings completing the machine cycle. On alternate cycles alternate positron bunches are kicked out of the positron ring so that each of the positron bunches stays in that ring for two machine cycles. This gives the required amount of damping for their larger injected emittance.

2. STATUS

The Stanford Linear Collider Project is presently 60% through its planned 3 year construction cycle. Some areas of the project are complete and others are well along, anticipating completion in 9 to 14 months. In the following sections we will look briefly at each subsystem of SLC and note its present status.

3. THE ELECTRON SOURCE[2-5]

The Electron Source has been tested and performs at full SLC specification producing two single S–Band bunches 59 nanoseconds apart with intensity over 5×10^{10} electrons per bunch. The system consists of a gun, two subharmonic velocity modulator cavities at 178.5 MHz, and a S–Band buncher and accelerator. This source provides beams for non–SLC use as well as for injection into the electron damping ring. Single bunch beams of 7×10^{10} have been produced and accelerated to 1.2 GeV. Addition of 80 quadrupoles between the source and the damping ring provided the strong focusing needed to overcome the transverse wake fields.

4. THE DAMPING RINGS[6–13]

The electron damping ring was built as an R&D machine to check its performance as a fast damper. To date it has stored single beams of 4×10^{10} electrons and two beams of 2×10^{10} each. No unexpected storage ring phenomena have been encountered and it is expected that the necessary damping will be possible. Study of this ring has been valuable in indicating improvements to be made to the ring and in improving the design of the positron ring. Indicated improvements are: increasing the sextupole family to better control chromaticity, adding more beam position monitors to better control orbits, and improvements to the injection and extraction septum magnet cooling. The original designs planned for both electron and positron rings to be housed in the same vault, one ring mounted above the other. The actual complexity of these compact rings however made it evident that both installation of the second ring and serviceability of both rings would be improved sufficiently to justify the cost of separate housings. This lead to the present design with two symmetric rings on each side of the linac. The second positron ring housing is complete and being prepared for installation of the ring next October. Newly designed transports for Linac–To–Ring and Ring–To–Linac are installed and working in the south electron ring and will be installed in the north positron ring before October. Testing of the north ring will first be done with electrons in January of 1986.

5. THE LINAC UPGRADE[14–22]

In order to accelerate the intense single bunches for the SLC the quadrupole focusing has been increased in strength and a system of beam position monitors has been installed in the first third of the linac. The last two–thirds of the linac will likewise upgraded in the next year. The increased focusing and beam monitoring is to reduce effects of transverse wake fields produced by the beam being off center in the wave–guide. These fields deflect the tail of bunch and cause a effective increase in transverse emittance. The present upgraded configuration has a quadrupole and position monitor between each forty foot linac section. Beams of intensity 1.5×10^{10} are now obtained at the one–third point when injected from the damping ring.

Because of the desire to increase the bunch length at the interaction point to effect a larger beam-beam pinch, the bunch length out of the RTL transport will be increased from a σ of 1 mm to 1.5 mm. This will cause an undesired increase in the transverse wake force on the bunch tail. To counteract this, the linac focusing will be increased to one quadrupole and monitor every ten feet in sector 2, just downbeam of the damping rings and one every 20 feet in sectors 3 and 4. This requires shortening the ten foot sections to 9.5 feet in the locations of the added quadrupole magnets. This work is scheduled to completed in the next year.

6. KLYSTRON UPGRADE[23-27]

The existing S–Band r.f. drive of the linac uses 34 MW klystrons to produce beams of up to 33 GeV. In order to explore the particle physics of Z^0 production and decay, beams of about 51.5 GeV are needed at the end of the linac. This increase in energy will be achieved by replacing the 230 klystrons down beam of the damping rings by new higher power ones. The peak power will be increased to 50 MW and the pulse length to 5 microsec. Lengthening the pulse length increases the SLED gain factor from 1.4 to 1.78. The new system will initially run at a maximum rate of 120 Hz in order to keep average power within existing modulator capability. The thyratron tubes and the output transformer are being changed to provide the higher pulse power of 315 kV and 354 amperes. Future modifications to the modulator will allow operation at a 180 Hz rate.

The klystrons, designated as model 5045 for 50 MW peak and 45 KW average power,are being manufactured at SLAC. As of November 1985, 120 tubes have meet specifications. The tube factory produces four klystrons per week and has an accepted tube yield of 73%. Extensive quality control measures are responsible for this improved yield which started out at around 40%. The production schedule is planned to provide sufficient linac energy by January 1987. Failures of acceptance tests are typically due to window breakage from overheating and insufficient stable operating region.

Twelve of the thirty sectors or 94 stations have been outfitted with the new tubes for in–field testing. The beam energy as measured in a spectrometer indicates the energy gain per klystron is correct and full 51.5 GeV will be obtained. By the end this year, modulators in 16 sectors will be converted for operation with the new klystron tubes leaving 13 to be done in the following year.

7. THE POSITRON SOURCE[28]

The positrons are produced by pair production in a tantalum – 10% tungsten target. Down beam of the target a 5 Tesla pulsed magnet and DC focusing solenoids is used to collect a large transverse emittance. A high gradient accelerator immediately follows the pulsed magnet to achieve a good longitudinal capture. After acceleration to 200 MeV the positrons are transported from the source location back to the sector one. This return transport consists of two 180 degree turn–arounds, one at each end and a FODO array 2 Km long with quadrupoles, position monitors, and steering every 12.7 meters. Present status has about 70% of the return transport installed and the remainder scheduled to be complete by October 1985. The target area is scheduled for completion by May 1986 at which time testing with beam will begin.

8. THE ARC SYSTEM[29-32]

The purpose of the Arc transport is to bend the high energy electron and positron beams from the linac around to allow head–on collisions with a minimum of phase space dilution and energy loss. Quantum effects of synchrotron radiation will cause growth proportional to L^3/r^4 where L is the length of a single focusing or defocusing magnet and r is the bending radius. This suggests a dipole guidance field of modest strength for large bending radius and many short but high gradient magnets. The design utilizes alternating gradient magnets with guide fields of 5.98 KG at 50 GeV and a corresponding gradient of 7.02 KG/cm. A betatron phase advance of 108 degrees per cell is used with 10 cells grouped together to form an achromat. A total 940 magnets are used in the Arc system. Each achromat lies in a plane but at a slope to allow the tunnel to follow the earths terrain. This terrain following simplifies the civil construction.

Magnets for the Arc transport are now in production at a rate of six per day with 540 completed. Extensive measurements are being made to insure correct gap dimensions and correct field components. Only about 5% of the magnets fail specifications and are unusable. Installation of the magnets in the tunnels is scheduled to begin July 1985 and be completed by July 1986.

9. THE FINAL FOCUS[33-34]

The Final Focus system has the job of producing micron sized beams and bringing two such beams into head on collision. The transport optics design utilizes telescopic modules with simultaneous point–to–point and parallel–to–parallel focusing. This feature tends to minimize the magnitude of higher–order optical distortions. Because the beams will have a finite momentum spread of 0.2 to 0.5% the line requires a chromatic correction section. The final set of quadrupoles near the interaction point need to be as strong as feasible. Optical designs exist for both superconducting and conventional quadrupoles in this location. Initial operation will be with conventional magnets to alleviate worries of beam induced quenches. Later operation at the highest luminosity will require the higher gradient of the superconducting magnets.

10. THE CONTROL SYSTEM[35−42]

The SLC project requires good monitoring and control of a large number of systems. A new control system was developed and now runs all the new SLC installations. It consists of a Vax 780 with a Broadband cable network connecting multibus microcomputers which in turn control CAMAC crates. One of the new features is the control and monitoring of the r.f. systems on the linac. At present one–third of linac is equipped with control and readback of the amplitude and phase of each klystron in relation to a reference line. This has improved the ease with which beams are setup and maintained by the operators. Full control of the klystrons will be installed this summer.

11. PROJECTED PERFORMANCE

Checkout running of all systems upbeam of the arcs will begin in May 1986 and continue throughout the summer. At the completion of the construction project, Oct. 1, 1986, the arcs and final focus systems will be also tested with beam. The initial luminosity is expected to be around 5×10^{28}. The final design value is 6×10^{30}. To reach this it will be necessary to: increase the intensity to 7.2×10^{10}; increase the repetition rate to 180 Hz; and improve the final focus quadrupole triplet using superconducting magnets.

12. ACKNOWLEDGEMENTS

This report is a summary of the work of many people. The references noted from the recent Particle Accelerator Conference contain more specific and detailed reports on the various aspects of the SLC systems.

13. REFERENCES

1. SLC Design Handbook, Stanford Linear Accelerator Center, December 1984.

2. M.C. Ross, M.J. Browne, J.E. Clendenin, R.K. Jobe, J.T. Seeman, J.C. Sheppard, and R.F. Stiening, "Generation and Acceleration of High Intensity Beams in the SLC Injector," IEEE Trans. Nucl. Sci. NS-32, 3160 (1985).

3. M.J. Browne, J.E. Clendenin, P.L. Corredoura, R.K. Jobe, R.F. Koontz, and J. Sodja, "A Multi-Channel Pulser for the SLC Thermionic Electron Source," IEEE Trans. Nucl. Sci. NS-32, 1829 (1985).

4. J. Judkins, J.E. Clendenin, and H.D. Schwarz, "A Solid-State High-Power Amplifier for Driving the SLC Injector Klystron," IEEE Trans. Nucl. Sci. NS-32, 2909 (1985).

5. H. Hanerfeld, W.B. Herrmannsfeldt, M.B. James, and R.H. Miller, "SLC Injector Modeling," IEEE Trans. Nucl. Sci. NS-32, 2510 (1985).

6. R.A. Early and J.K. Cobb, "TOSCA Calculations and Measurements for the SLAC SLC Damping Ring Dipole Magnet," IEEE Trans. Nucl. Sci. NS-32, 3654 (1985).

7. J.P. Delahaye and, L. Rivkin "SLC Positron Damping Ring Optics Design," IEEE Trans. Nucl. Sci. NS-32, 1695 (1985).

8. J.E. Spencer, "Some Uses of REPMM's in Storage Rings and Colliders," IEEE Trans. Nucl. Sci. NS-32, 3666 (1985).

9. I. Almog, J. Jager, M. Lee, and M. Woodley, "On Line Model-Driven Control of the SLC Electron Damping Ring," IEEE Trans. Nucl. Sci. NS-32, 2098 (1985).

10. P.L. Morton, J.-L. Pellegrin, T. Raubenheimer, L. Rivkin, M. Ross, R.D. Ruth and W.L. Spence, "A Diagnostic for Dynamic Aperture," IEEE Trans. Nucl. Sci. NS-32, 2291 (1985).

11. L. Rivkin, J.P. Delahaye, K. Wille, M. Allen, K. Bane, T. Fieguth, A. Hofmann, A. Hutton, M. Lee, W. Linebarger, P. Morton, M. Ross, R. Ruth, H. Schwarz, J. Seeman, J. Sheppard, R.F. Stiening, P. Wilson, and M. Woodley, "Accelerator Physics Measurements at the Damping Ring," IEEE Trans. Nucl. Sci. NS-32, 2626 (1985).

12. A.M. Hutton, W.A. Davies–White, J.–P. Delahaye, T.H. Fieguth, A. Hofmann, J. Jager, P.K. Kloeppel, M.J. Lee, W.A. Linebarger, L. Rivkin, M. Ross, R. Ruth, H. Shoaee, and M.D. Woodley, "Status of the SLC Damping Rings," IEEE Trans. Nucl. Sci. NS-32, 1659 (1985).

13. J–C. Denard, K.L. Bane, J. Bijleveld, A.M. Hutton, J–L. Pellegrin, L. Rivkin, P. Wang, and J.N. Weaver, "Parasitic Mode Losses Verses Signal Sensitivity in Beam Position Monitors," IEEE Trans. Nucl. Sci. NS-32, 2000 (1985).

14. G.A. Loew and J.W. Wang, "Minimizing the Energy Spread within a Single Bunch by Shaping Its Charge Distribution," IEEE Trans. Nucl. Sci. NS-32, 3228 (1985).

15. J.R. Bogart, N. Phinney, M. Ross, and D. Yaffe, "Beam Position Monitor Readout and Control in the SLC Linac," IEEE Trans. Nucl. Sci. NS-32, 2101 (1985).

16. J.C. Sheppard, M.J. Lee, M.C. Ross, J.T. Seeman, R.F. Stiening, and M.D. Woodley, "Beam Steering in the SLC Linac," IEEE Trans. Nucl. Sci. NS-32, 2180 (1985).

17. K.L.F. Bane, "Landau Damping in the SLAC Linac," IEEE Trans. Nucl. Sci. NS-32, 2389 (1985).

18. J.W. Wang and G.A. Loew, "Measurements of Ultimate Accelerating Gradients in the SLAC Disk-Loaded Structure," IEEE Trans. Nucl. Sci. NS-32, 2915 (1985).

19. J.C. Sheppard, J.E. Clendenin, M.B. James, R.H. Miller, and M.C. Ross, "Real Time Bunch Length Measurements in the SLC Linac," IEEE Trans. Nucl. Sci. NS-32, 2006 (1985).

20. J.T. Seeman, M.C. Ross, J.C. Sheppard, and R.F. Stiening, "RF Beam Deflection Measurements and Corrections in the SLC Linac," IEEE Trans. Nucl. Sci. NS-32, 2629 (1985).

21. J.T. Seeman, M.C. Ross, J.C. Sheppard, and R.F. Stiening, "Observations of Accelerated High Current Low Emittance Beams in the SLC Linac," IEEE Trans. Nucl. Sci. NS-32, 1662 (1985).

22. M.C. Ross, J.T. Seeman, R.K. Jobe, J.C. Sheppard, and R.F. Stiening, "High Resolution Beam Profile Monitors in the SLC," IEEE Trans. Nucl. Sci. NS-32, 2003 (1985).

23. G.A. Loew, M.A. Allen, R.L. Cassel, N.R. Dean, G.T. Konrad, R.F. Koontz, and J.V. Lebacqz, "The SLC Energy Upgrade Program at SLAC," IEEE Trans. Nucl. Sci. NS-32, 2748 (1985).

24. R.K. Jobe, M.J. Browne, and K.P. Slattery, "Hardware Upgrade for Klystrons in the SLC," IEEE Trans. Nucl. Sci. NS-32, 2107 (1985).

25. R.K. Jobe, K.A. Thompson and N. Phinney, "Klystron Control Software in the SLC," IEEE Trans. Nucl. Sci. NS-32, 2110 (1985).

26. K.Eppley, S. Yu, A. Drobot, W.B. Herrmannsfeldt, H. Hanerfeld, D. Nielson, S. Brandon, and R. Melendez, "Results of Simulations of High Power Klystrons," IEEE Trans. Nucl. Sci. NS-32, 2903 (1985).

27. S.S. Yu, P. Wilson, and A. Drobot, "2-1/2-D Particle-In-Cell Simulation of High-Power Klystrons," IEEE Trans. Nucl. Sci. NS-32, 2918 (1985).

28. F. Bulos, H. DeStaebler, S. Ecklund, R. Helm, H. Hoag, H. LeBoutet, H. Lynch, R. Miller, and K. Moffeit, "Design of a High Yield Positron Source," IEEE Trans. Nucl. Sci. NS-32, 1832 (1985).

29. G.E. Fischer, M. Anderson, R. Byers, and K. Halbach, "SLC Arc Transport System – Magnet Design and Construction," IEEE Trans. Nucl. Sci. NS-32, 3657 (1985).

30. W.T. Weng, M. Anderson, R. Byers, J. Cobb, G. Fischer, and V. Hamilton, "SLC Arc Transport System – AG-Magnet Measurement and Performance," IEEE Trans. Nucl. Sci. NS-32, 3660 (1985).

31. W.T. Weng and A. Chao, "Vibrational Modes of the Pedestal Support System for the SLC Arc Magnets," IEEE Trans. Nucl. Sci. NS-32, 3663 (1985).

32. K.L. Brown and R. Servrancks, "Applications of the Second–Order Achromat Concept to the Design of Particle Accelerators," IEEE Trans. Nucl. Sci. NS-32, 2288 (1985).

33. D.R. Walz and W.O. Brunk, "High and Ultra-High Gradient Quadrupole Magnets," IEEE Trans. Nucl. Sci. NS-32, 3651 (1985).

34. S. Yencho and D.R. Walz, "A High-Resolution Phosphor Screen Beam Profile Monitor," IEEE Trans. Nucl. Sci. NS-32, 2009 (1985).

35. H.D. Schwarz, "Computer Control of RF at SLAC," IEEE Trans. Nucl. Sci. NS-32, 1847 (1985).

36. M. Crowley-Milling, "Control Problems in Very Large Accelerators," IEEE Trans. Nucl. Sci. NS-32, 1874 (1985).

37. J. Jäger, M. Lee, R. Servranckx, and H. Shoaee, "GIANTS: a Computer Code for General Interactive Analysis of Transport Systems," IEEE Trans. Nucl. Sci. NS-32, 1877 (1985).

38. J.E. Linstadt, "A Programmable Delay Unit Incorporating a Semi-Custom Integrated Circuit," IEEE Trans. Nucl. Sci. NS-32, 2112 (1985).

39. J.E. Linstadt, "Dissecting the COW," IEEE Trans. Nucl. Sci. NS-32, 2115 (1985).

40. N. Phinney, "Report on the SLC Control System," IEEE Trans. Nucl. Sci. NS-32, 2117 (1985).

41. N. Spencer, J. Bogart, N. Phinney and K. Thompson, "Error Message Recording and Reporting in the SLC Control System," IEEE Trans. Nucl. Sci. NS-32, 2120 (1985).

42. K.A.Thompson and N. Phinney, "Timing System Control Software in the SLC," IEEE Trans. Nucl. Sci. NS-32, 2123 (1985).

PROGRESS WITH HERA

Peter Schmüser

II. Institut für Experimentalphysik der Universität Hamburg
Notkestrasse 85, 2000 Hamburg 52
W.-Germany

ABSTRACT

The electron-proton colliding beam facility HERA
is described and the status of the superconducting
magnet programme is reviewed.

1. GENERAL DESCRIPTION OF THE HERA MACHINE

The electron-proton colliding beam facility HERA[1] which is now
under construction at DESY consists of a 30 GeV electron storage ring
and an 820 GeV proton ring. Both machines will be mounted on top of
each other in an underground tunnel of 6336 m circumference and 5.2 m
diameter which has four 90° arcs joined by four 360 m long straight
sections. The experiments will be installed in large underground
halls which contain also machine equipment and control rooms.
Presently, electron-proton collisions are foreseen in three of the
halls leaving the interaction region on the DESY site free for
injection and a proton beam abort system.

The electron ring is equipped with conventional iron magnets.
The dipole magnets are excited by a single aluminum current bar. The
vacuum chamber is made from a copper alloy and reinforced with lead
to contain the synchrotron radiation. Ion getter pumps are integrated
in the beam pipe of the dipole and quadrupole magnets. The electron
RF system will be taken over from PETRA. For energies above 26 GeV a
system with 500 MHz superconducting cavities is envisaged.

The proton storage ring consists of superconducting dipole and
quadrupole magnets connected in series. The nominal proton energy of
820 GeV requires a dipole field of 4.64 T and a quadrupole gradient
of 90 T/m. The beam pipe is at liquid helium temperature and will be
copper-plated on the inside to reduce the ohmic heating due to image
currents accompanying the proton bunches. The quadrupole and
sextupole correction magnets as well as the correction dipoles are

also superconducting. The superconducting coils are cooled by single phase helium which traverses one octant, is then expanded through a Joule-Thomson valve and is returned to the refrigerator as a two-phase mixture. In the dipole magnets there is heat exchange between the single and two-phase system. The radiation shields in the cryostats are cooled by helium gas of 40-80 K.

Two refrigerator plants [2] each providing 6.5 kW at 4.6 K and 20 kW at 40-80 K are used for steady state operation. A third plant is needed for cool down.

The RF in the HERA proton ring is 52 MHz while the 210 bunches are injected from PETRA. After injection the voltage of a 208 MHz system is raised to reduce the bunch length and the protons are then accelerated. The bunch length at 820 GeV is 35 cm (Ref.3).

The injection scheme is shown in Fig. 1. Several stages of acceleration are needed before the electrons or protons can be injected into HERA: separate linear accelerators and synchrotrons and the modified PETRA storage ring to accelerate electrons to 14 GeV and protons to 40 GeV. The proton injection chain starts with negative hydrogen ions which are produced in an 18 keV ion source. A radio frequency quadrupole accelerates the H⁻ beam to 750 keV and focuses and bunches it at the same time. The next step is a 50 MeV linear accelerator. From here the H⁻ ions are injected into the rebuilt synchrotron (DESY III), stripped, and the protons are then accelerated to 7 GeV/c. Due to the statistical nature of the stripping process, Liouville's theorem does not apply and the proton intensity in the synchrotron is not limited by the phase space of the preaccelerators. The bunch spacing in DESY III and PETRA is 28.8 m like the final spacing in HERA. The estimated filling time is 20 min.

2.   INTERACTION REGIONS AND POLARIZATION

In the present design the electron and proton beams collide head-on because particle tracking calculations [4] have shown that with a crossing angle of more than a few milliradians synchro-betatron oscillations would be excited when the electron bunches penetrate the much longer proton bunches. The zero crossing angle geometry requires that the electron and proton machines have a number of dipole and quadrupole magnets in common. The electrons are bent away from the protons by an angle of 10 mrad in the horizontal plane by means of displaced quadrupoles and a dipole magnet. The proton quadrupoles and dipoles in this region have to be normal iron magnets because the synchrotron radiation generated in the 10 mrad bends would put an intolerable heat load on the cryogenic system of superconducting magnets. This radiation may also cause a severe background in the experiments. It seems possible, however, to arrange collimators and shielding in such a way that the synchrotron   radiation can enter the central tracking chambers of the detectors only after double scattering. The resulting flux of photons with energies above 50 keV has been estimated [5] to be below 10^7 per second which is a factor of 10 lower than the flux observed in the TASSO drift chamber at PETRA.

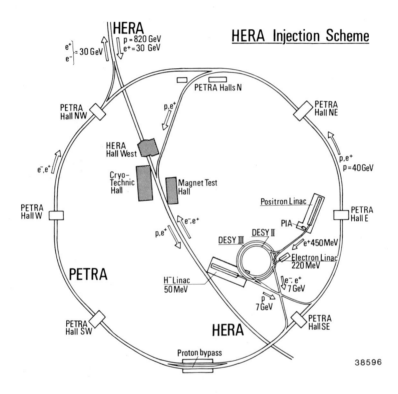

<u>Fig. 1</u>: The preaccelerators needed for the injection of electrons and protons into the HERA ring.

Considerable efforts are needed to achieve longitudinal electron polarization in the interaction regions. The emission of synchrotron radiation leads to a natural transverse polarization which is built up with a time constant of 25 min at 30 GeV and may reach a theoretical maximum of 92.4 %. The machine has to be very well aligned for this purpose and depolarizing resonances have to be avoided. The spin can be rotated into the longitudinal direction by a magnetic deflection. For ultrarelativistic electrons, the spin precession frequency can be much higher than the cyclotron frequency, the difference frequency being given by

$$\Delta\omega = E/mc^2 \cdot (g-2) \cdot \omega_{cycl} \text{ , g Landé factor of the electron.}$$

At 30 GeV a deflection by 1.32° is sufficient to rotate the spin by 90°. A "spin rotator" has been developed [6] which by successive vertical and horizontal deflections allows to obtain electrons with either positive or negative helicity in all interaction regions. The principle is explained in Fig. 2. Particle tracking calculations [7] with a linear optics program indicate that a degree of longitudinal polarization of 80 % seems feasible. With only slight mechanical adjustments of the magnets the spin rotator can cover the energy range from 27.5 to 35 GeV. The 10 mrad bend near the interaction point is part of the spin rotator.

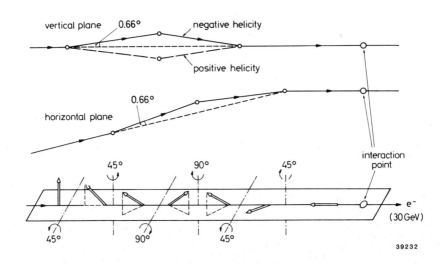

Fig. 2: Schematic view of the spin rotator in front of the interaction point. The graphs at the top and in the middle indicate the deflections of the electron beam in the vertical and horizontal plane. At the bottom the orientation of the electron spin relative to the horizontal plane is shown. A similar arrangement behind the interaction point restores the transverse orientation in the arcs.

3. SUPERCONDUCTING MAGNETS

When the superconducting magnet program was initiated at DESY several years ago it was decided to follow closely the successful design of the FNAL Tevatron magnets. A dipole magnet with warm iron yoke has been developed at DESY and ten 6 m long units have been built with tooling suitable for series production. All magnets tested so far have exceeded the nominal field of 4.53 T and have shown good field homogeneity [8]. In collaboration with DESY an industrial company (BBC Mannheim) has designed and constructed three 6 m long dipoles with a cold iron yoke which directly clamps the superconducting coils[9]. The first magnet has been tested and has reached a field of 5.75 T. Both magnet types have their relative virtues and drawbacks. The warm iron dipole offers good field quality without any iron saturation but has a fairly large heat flux into the cryogenic system because the coil has to be centered inside the yoke by many supports. The static heat load in the cold iron dipole can be made much smaller because only few supports are needed and in addition about one third of the superconductor can be saved but this magnets suffers from large field distortions due to iron saturation above 4.5 Tesla.

Early 1984 it was proposed [10] to consider a magnet which combines the positive features of both concepts and avoids most of the drawbacks. In this new type of dipole the coil of the DESY magnet, collared with aluminum clamps, is surrounded with a cold iron yoke and a cryostat similar to that of the BBC magnet. In the meantime, the design of this "HERA dipole" has been completed, two 1 m long units have been built and tested and four full size magnets of 9 m magnetic length are close to completion. A cross section of the magnet is shown in Fig. 3. The coil consists of two current shells with longitudinal wedges for improved field homogeneity. It is wound from a Rutherford type cable with 24 strands and copper to superconductor ratio of 1.8. The diameter of the NbTi filaments is 14 μm and the critical current at 5.5 Tesla and 4.6 K exceeds 8000 A. The coil is clamped by precision-stamped aluminum collars which define the exact geometry and take up the huge Lorentz forces. The collars have four noses which slide into notches in the iron yoke and thereby provide an accurate alignment. The yoke laminations are surrounded by a stainless steel tube which serves as the liquid helium container and at the same time defines the curvature of the magnet with a radius of 580 m. The cold part has to be supported only at three locations over the length of 9 m. The computed static heat flux is 2.4 W into the liquid helium vessel and 21 W into the radiation shield.

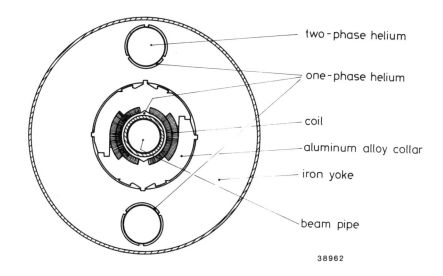

two-phase helium

one-phase helium

coil

aluminum alloy collar

iron yoke

beam pipe

38962

Fig. 3: Cross section of the cryogenic part of the superconducting
HERA dipole magnet. The coil is clamped by an aluminum collar
and then surrounded by a cold iron yoke. A detailed
description is given in Ref. 11.

The new magnet concept has several obvious advantages:

- compared to the warm iron dipole one gains 12 % in the central
field but saturation effects are very small even at 6 T

- the static head load is low

- a passive quench protection by means of parallel diodes is
possible.[12]

The test results [11] obtained with the 1 m long prototype
magnets are extremely encouring. Both magnets exceeded the nominal
current of 4990 A at the first excitation and required just one
training step to arrive at the critical current of the
superconductor. The higher harmonics measured at the nominal field
are in general less than 1×10^{-4} and thus well within the allowed
limits. Fig. 4 shows the current dependence of the normal sextupole
coefficient b_3. Besides the well-known hysteresis at low current
which is caused by persistent eddy currents in the NbTi filaments a
definite curvature is observed at large excitation. This indicates a
slight saturation of the iron yoke. These systematic sextupole fields
will be minimized in future magnets by some additional shimming of
the coils and the remaining contributions can be compensated using
the sextupole correction coils.

974

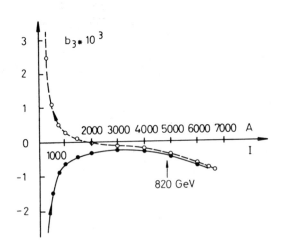

Fig. 4:

The sextupole coefficient of a 1 m long prototype dipole as a function of the current in the coil. The continous curve is obtained when the current in the magnet is increased, the dashed curve when the current is decreased. The hysteresis is caused by eddy currents in the superconducting NbTi filaments.

One disadvantage which is present for any type of cold iron magnet is of course the large cold mass. Detailed calculations [13] have shown, however, that the cool down or warm up to a whole HERA octant be done in about 40 hours.

Superconducting quadrupole magnets have been designed and built at Saclay[14]. Two warm iron quadrupoles have reached a gradient of 120 T/m, well above the design value of 90 T/m. Recently, quadrupole magnets with stainless steel collars and a cold iron yoke have been constructed and have been excited to 120 T/m without quench. These quadrupoles are well matched to the HERA cold iron dipoles.

The quadrupole and sextupole correction coils are 6 m long and are mounted on the cold beam pipe inside the main dipoles. Several prototypes have been built by Dutch industry. The measured quench currents in an external field of 5T were a factor of 2.5 above the operating current. The dipole for orbit correction is a 60 cm long "superferric" magnet with laminated iron yoke and two saddle-shaped superconducting coils. A prototype magnet reached a field of 3.2 T compared to the required value of 1.4 T.

A more detailed description of the HERA project, in particular of its present status and time schedule can be found in Ref. 15.

The work reported here is being carried out by many people, both at DESY and other institutions. I want to thank my colleagues for many stimulating discussions.

References

1) Study on the Proton-Electron Storage ring Project HERA
   ECFA report 80/42 (1980), HERA Proposal DESY HERA 81/10 (1981)

2) Horlitz, G., DESY HERA 84-02 (1984)

3) Maidment, J.R., Planner, C.W. and Rees, G.H., HERA report in preparation

4) Piwinski, A., Contribution to the 11th Particle Accelerator Conference, Vancouver, 1985

5) Bartel, W., et al., DESY-HERA 85-15 (1985)
   Foster, B., private communication

6) Buon, J. und Steffen, K., DESY HERA 85-09 (1985)

7) Barber, D., private communication

8) Horlitz, G. et al., Journ. de Physique C1-255 and C1-259 (1984)

9) Dustmann, C.H. and Kaiser, H., private communication

10) Balewski, K., Kaiser, H. and Schmüser, P., DESY internal note (1984)

11) Wolff, S., Invited Talk at the 9th International Conference on Magnet Technology, Zürich 1985, DESY HERA 85-21 (1985)

12) Mess, K.H. and Otterpohl, U., private communication

13) Horlitz, G., Lierl, H. and Schmüser, P., DESY HERA 85-20(1985)

14) Auzolle, R. Patoux, A., Perot, J. and Rifflet, J.M., Journ. de Physique C1-263 (1984)

15) Wiik, B.H., Invited Talk at the 11th Particle Accelerator Conference Vancouver (1985), DESY HERA 85-16 (1985)

EXPERIMENTATION AT p̄p COLLIDERS*

David B. Cline

Physics Department
University of Wisconsin-Madison
Madison, WI   53706

ABSTRACT

We describe some of the issues encountered in experiments at the Sp̄pS collider using the UA1 detector. We indicate how such experiments rely on new types of measurements in detectors (lepton isolation for example). We indicate how the CDF detector will trigger on forward muons and the possible upgrade of this system to include a $\cos\theta$ dipole-TRD spectrometer. Finally we discuss the possibility of searching for rare B decays and of precise (M_Z-M_W) mass determination in future experiments at the p̄p colliders.

CONTENTS

1. Introduction

2. Detectors at p̄p Colliders

3. Operating Experience with the UA1 μ Detector

4. Isolation of Tracks and Top Quark Physics

5. Examples of New Physics: Search for Heavy Leptons and B_s^0 Mixing With Dimuon Events

6. Separated Function Detectors: CDF with Forward Detector Upgrade

7. Future: Search for Rare B Decays and Precise M_Z-M_W Mass Determination

---

*Invited talk at the Oregon DPF Meeting August 12-13, 1985.

1. Introduction

The initial goal of the CERN and FNAL $\bar{p}p$ colliders was the detection of W and Z particles.[1] With the tremendous success of the W, Z discovery, the goals have been greatly enlarged to include the search for supersymmetry heavy quarks, heavy leptons, Higgs Boson, etc. We will add to this list in this paper the search for rare B decays and the precise measurements of the W and Z mass difference.

Along with the enlarged physics goals of the $\bar{p}p$ colliders has come the realization of the need to increase the luminosity and to improve the detectors. Therefore an ambitious program of detector improvement is underway at CERN and a high resolution detector (at 0°) has now been approved for the Tevatron I at FNAL.

Operation of the detectors at the higher luminosity ($\sim 10^{30}$ -10^{31} cm^{-2} sec^{-1}) will be problematic if the present performance in the low luminosity at the $S\bar{p}pS$ is any guide. In this note we will address some of these issues.

2. Detectors at $\bar{p}p$ Colliders

We will briefly describe the two detectors for which our group has first hand experience, namely the UA1 detector at CERN and the CDF detector soon to be operating at Fermilab. These detectors also demonstrate the possibility of separated function detectors which we turn to later. The aim of the experimental physicsts working at a collider is to describe individual collisions in as much detail as possible.

Fig. 1 shows a schematic of the UA1 detector at CERN. The UA1 detector at the CERN $Sp\bar{p}S$ Collider uses several signatures. An artist's view, as shown in Fig. 1, allows identification of the various elements. Figure 2 gives a view along the beam, where the structure can be recognized. An attempt to capture as much energy as possible leads to a complex structure in the forward directions as exemplified in Fig. 2.

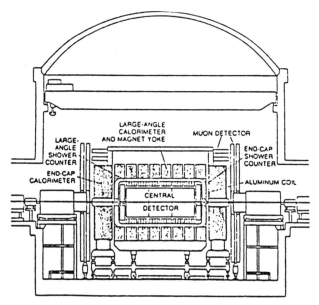

Fig. 1

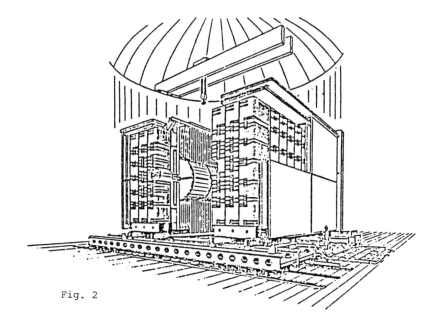

Fig. 2

Figure 3 shows a schematic of the CDF detector at FNAL. For Tevatron I at Fermilab, a general facility (the Collider Detector Facility, CDF) is under design and construction. Figure 3 gives a general overview. The basic structure is the same as the UA1 detector, although the magnetic field is parallel to the beam and some iron toroids allow measurement of the muons in the foward directions. Figure 4 shows the elevation view of half of the detector. The design of the central calorimeter is similar to that of UA2 with a tower structure of comparable segmentation.

3. Operating Experience with the UA1 Detector

The possibility of using very high luminosity pp collisions at the SSC or LHC depends crucially on the possibility of multilevel-real time triggers. One of the first such triggers is operating in the UA1 experiment for the μ system. This trigger allows a calculation of the angle of a track to within $\pm$ 150 mv in ~2 μS. A second level trigger that uses the drift time information gives a resolution that is approximately a factor of ten improved.

The challenges of the use of luminosity are due to the superposition of events, pile-up in individual detecting elements, distortions due to positive ions, d.c. shift, etc. In general, one can get around these limitations by using a large enough number of cells. Inversely a given detector is limited: for instance the large drift gaps used in the UA1 central detector exclude operation above its design of a few 10^{30} cm^2/s, a value which may eventually be reached by the Sp$\bar{p}$S Collider.

Another important consequence of luminosity is the need for multilevel triggers. Already at 2 x 10^{29} cm^2/s, there is a necessity of going from "local" triggers (e.g. local high-E$_T$ deposition or muon track) to "global" triggers (such as e + jet or μ + jet). This requires rather flexible and complex systems based on powerful on-line computers (68000, 168E, etc.).

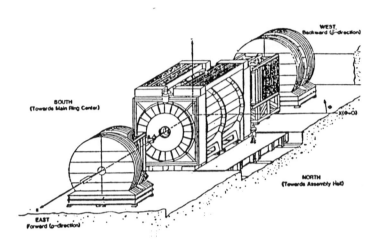

Fig. 3

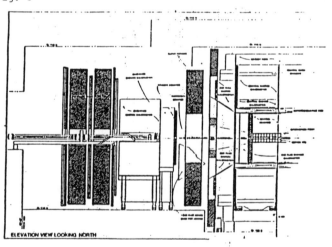

Fig. 4

4. Isolation of Tracks and Top Quark Physics

One concept that has influenced the most recent physics from the SppS collider is the possibility that some leptons from heavy quark decays will be "isolated" from nearby hadronic particles. This occurs for the decay of a high mass quark such as the top quark due to the large energy release in the decay process

$$
\left. \begin{array}{l}
W \rightarrow t + \bar{b} \\
\quad \lfloor \\
\quad \lfloor\rightarrow \mu + \nu + b
\end{array} \right\rbrack \mu\nu \text{ jet jet}
$$

The major background for this signal comes from lower mass quarks giving prompt leptons via

$$
\left. \begin{array}{l}
\bar{p}p \rightarrow b + \bar{b} + gluon \\
\quad \lvert \qquad \lvert \quad \lfloor\rightarrow \text{jet } \mu\nu \text{ jet jet} \\
\quad \lvert \qquad \lfloor\rightarrow \mu\nu \text{ ..} \\
\quad \lfloor\rightarrow \text{jet}
\end{array} \right\rbrack
$$

In the electron case the isolation requirements are very difficult to impose since any stray electromagnetic energy (γ) that are nearby the electron gets "sucked up" into the electron EM cascade in the shower counter.

In contrast for the μ case the energy deposited in the detector is due to ionization and can be effectively separated from any nearby EM energy.

Thus the definition of isolation of lepton tracks would appear to be depend on the particle (e or μ). This effect has been observed in the UA1 experiment and provides a concrete example of the need to improve the detector for the next round of experiments.

The UA1 detector will be improved with the addition of a Uranium plate--TMP (warm liquid) electron drift system. This new calorimeter will vastly improve the resolution and will clarify the question of the isolation of lepton tracks.

The UA1 detector has made use of calorimetry which was unprecedented at the time of its conception. The detector has been the first device in which one has applied a "hermetic" calorimeter geometry in order to identify neutrino emission, crucial for the discovery of the $W^{\pm}$ particle. However, a fundamental limitation in calorimetry is still present and represents the main uncertainty in the jet energy and missing energy determinations, namely the differences of response of ordinary calorimeters for hadronic and electromagnetic showers, which amounts to a 40% enhancement factor for a purely electromagnetic deposition. A very ingenious method has been proposed and tested by Willis, which makes use of the fission amplification of depleted Uranium in order to "boost" the hadronic response to the level of the EM response. A considerable improvement of the hadronic energy resolution, σ_E/E, is also observed which drops from $0.7\sqrt{E}$ to $0.4/\sqrt{E}$ because of internal compensation of the EM/hadronic fluctuations of the cascade.

Uranium is not only the most appropriate material for a hadronic calorimeter; it has also the shortest radiation length (3.2 mm) and the lowest critical energy (6.5 MeV). This suggests a completely integrated structure in which both hadronic and EM calorimetery are performed in the same device. In any case, the distinction between EM and hadronic calorimetry is somewhat arbitrary, since on average over 80% of the transverse energy of a miniumum bias event is absorbed in the 26 radiation lengths of the "gondolas" and "bouchons" of our detector.

An ideal detector would then consist of a uniform, "hermetic" stack of Uranium sheets, sufficiently deep to absorb all hadronic debris of the collision. Unfortunately this is not possible at present, for at least two reasons: namely (1) finanacial, in view of the cost of the Uranium and (2) geometrical, since the space available between the present central detector and magnet is limited. However, making use of the extraordinarily high density of Uranium, (18.95 g/cm^3) it is possible to pack enough collision lengths within the available space in order to collect the bulk of energy in the

Uranium, and to use the present hadron calorimeter to correct for the leakage. The high containment fraction and improved resolution are essential requirements for triggering on events with relatively low missing transverse energy which could include new physics.

The new ionization will use a room temperature liquid such as Tetramethylsilane (TMS) or TMP. Studies at CERN (still in progress) indicate that this system will have superior properties to liquid Argon or scintillation detectors.

5. Examples of New Physics: Search for Heavy Leptons and the Low Mass Dimuon Events

We will now describe two searches that are made possible at the $\bar{p}p$ colliders: (1) The Search For High Mass Heavy Leptons and (2) The Search for B_s^o Mixing Using Low Mass Same Sign Dimuons.

The production and decay of Intermediate Vector Bosons produced in proton-antiproton collisions is now accomplished at CERN and will be studied at FNAL. The recent observation of events that are attributed to the Intermediate Vector Boson brings closer the possibility of using the $W^{\pm}$ particles to search for new leptons and heavy quarks. Owing to the expected asymmetry of the masses of the different species of heavy-flavour quarks and leptons, the $W^{\pm}$ decays are sensitive to heavy-flavour particle masses near the actual mass of the $W^{\pm}$. For example, the decay modes

$$W^{\pm} \to t + \bar{b} \text{ (where t is a Q = + 2/3 heavy quark}$$
$$\text{and } \bar{b} \text{ is the bottom quark)}$$

and

$$W^{\pm} \to L + \nu_L \text{ (where L is a new sequential lepton)}$$

have an adequate phase-space factor for t-quark masses up to $\sim(50-60)$ GeV/c^2 and for L leptons to masses up to ~ 60 GeV/c^2 if the associated neutrino (ν_L) is very low-mass.

In order to identify these new leptons we now consider the consequences of the maximal $W^{\pm}$ helicity in production, and the effects on the leptons in the decay. As is well known, this is expected to yield a charge asymmetry in the charged leptons at high transverse momentum. The asymmetry is a strong function of the transverse momentum and laboratory angle of the lepton. The asymmetry has been studied in several theoretical works. In this report we show that the introduction of heavy leptons in the $W^{\pm}$ decay final states modifies this asymmetry in a characteristic manner, and how the observation of the decay lepton P and angular distribution provides a technique for searching for the decay chain

$$W \rightarrow L + \nu_L \ (L \rightarrow \ell + \nu_\ell + \bar{\nu}_L)$$

up to L mass vaues of ~60 GeV/c^2.

It may be possible to search for the L lepton through hadronic decays of the lepton, i.e.

$$L \rightarrow \nu_L + (\text{hadronic jet}).$$

In this case events with a sngle high-p_T jet and large missing energy would be observed. However, there are potential backgrounds from heavy-flavour production and calorimetry fluctuation that must be addressed. Of course, the observation of the correct number of "wrong asymmetry" leptons in the intermediate p_T, as well as the corresponding number of jets plus missing energy events, would provide additional evidence for $L^{\pm}$ hypothesis. M. Mohammadi (U. Wisconsin) is studying this topic for his thesis.

The search for B_s^0 mixing is more straightforward. We must first describe the detection of B production at the collider.

The physics of the search for rare B decays is independent of the search for rare K decays due to the different kinematics and the fact that the b quark is in the third generation of quarks whereas the K (strange) quark is in the second. Since there is little understanding of the role of the kinematic or dynamic factors (i.e., the nature of the exchanged particles that give the rare decay process) it is logically consistent that large effects could occur in B decays whereas only very small effects exist for K decays. Thus it is important to search for new B decay modes at <u>any</u> level below the present search.[3]

5a. <u>BB̄</u> <u>Production</u> <u>at</u> <u>the</u> <u>Sp̄pS</u> <u>Collider</u> <u>and</u> <u>Cross</u> <u>Section</u> <u>Estimates</u>
    <u>at</u> <u>√s̄</u> <u>=</u> <u>2</u> <u>TeV</u>

The production of BB̄ pairs can be observed through the decay of the B's into final states with two leptons of the same electric charge. Such events have been detected in the UA1 experiment and lead to a determination of the cross section for BB̄ production in the central region.

The events that are most characteristic of BB̄ production are those with muons that that have nearby jet activity. One such event is shown in Fig. 5. Table 1 gives the most recent breakdown of the dimuon events into isolated and non-isolated, same charge or opposite charge.[5,6] From these events using the non-isolated sample the UA1 group obtains the following cross section[1,2]

$$\sigma_{BB} \simeq 2.4 \pm 0.2 \ \mu b \qquad \text{(Using the ISAJET program for detection efficiency)}$$

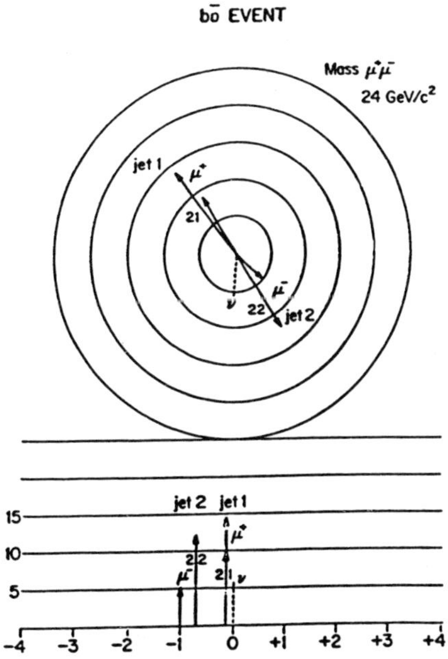

bō EVENT

Fig. 5

Table 1

Breakdown of UA1 Dimuon Events Into Same Charge and Opposite
Charge Categories

	Isolated	Non Isolated
$\mu^+\mu^-$	44	106
$\mu^{\mp}\mu^{\mp}$	7	55

These estimates are limited to $|\mu| < 2.5$ and $P_{\perp_B} > 5$ GeV/c due to the detector costs.

Therefore the UA1 group concludes that the cross section for $B\bar{B}$ production is of the order of several μb at the $\sqrt{s} > 630$ GeV, which is almost three orders of magnitude larger than the limits for $B\bar{B}$ production at the FNAL/SPS energies.

The large $B\bar{B}$ production cross section is in good agreement with expectations from theory and phenomenology. The angular distributions of the jets and the dilepton events and the transverse momentum characteristics are also in good agreement with the $B\bar{B}$ production hypothesis.

The UA1 group has also reported preliminary evidence for a ratio of some charge to opposite charge dimuon events that is consistent with large B^0 mixing. These results are shown in Fig. 6 and are compared to the expectations of large B_s^0 mixing (assuming negligible B_d^0 mixing).[8] Of course there is no direct evidence for mixing or even that the mixing is due to B_s^0 and not B_d^0. This exciting result indicates the potential of $\bar{p}p$ collisions to produce large quantities of $B\bar{B}$ pairs and enables the study of rare decay processes in a new generation of dedicated experiments.

Due to the excellent agreement between the UA1 results and the ISAJET Monte Carlo, we henceforth will use this program to simulate the various rare B decay modes discussed in this report.

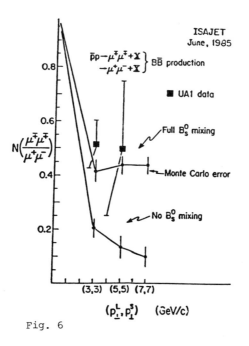

Fig. 6

6. Separated Function Detectors: CDF With a Forward Detector Upgrade

The CDF detectors allows for additions to construct a sophisticated forward and backward spectrometer. This separated function option is likely to be important for a LHC or SSC detector as well.

One possible version of a forward spectrometer utilizes a TRD electron identifier and a $\cos\theta$ dipole along with additional muon coverage (supertoroids).

7. Future: Search for Rare B Decays and a Precise M_Z-M_W Mass Determination

The study and search for rare decay modes of K mesons played an important role in the development of the standard model of weak interactions. New experiments on K rare decay modes are being planned now. We believe that the study of rare B decay modes could play an equally important role because of the larger B mass and the larger number of kinematically allowed decay channels. Of course we are assuming that very sensitive searches can be carried out to low enough levels to search for new physics.

In Table 2 we give a partial list of B decay modes that would be very interesting to detect. We also give some of the reasons for studying these channels.

7a. $B^+ \to (K^{\pm}_{\pi^{\pm}}) \mu^+\mu^-$ Decay Detection

This decay mode is an example of a GIM breaking decay. The expected branching ratio for this decay has been estimated by Kane to be $\sim 10^{-5}$ and in other works to be less. The decay also represents an efffective B flavor changing weak netural current process. There could be new physics contributions to this decay in the form of horizontal symmetry breaking. A branching ratio in the range of 10^{-5} – 10^{-7} due to horizontal symmetry breaking could represent new physics in the multi TeV mass range. In this sense the detection of such decay processes extends the physical range of new physics beyond that available to ultrahigh energy colliding beam machines.

Table 2

Rare Decays of B Mesons

Process	Estimated B ratio	Remarks
B → u + ...	< 10^{-2}	Measures$\|U_{bu}\|^2$ important for CP
B → τ + ν_τ	~1/3 x $10^{-4}(fB/200)^2$ MeV	CP Violation Tests
$B_s^0 \not\supseteq \bar{B}_s^0$	~1	K-M Matrix –Possible CP Violation if 4th Generation Exists
$B_s^0 \not\supseteq \bar{B}_d^0$	0.3 to ≪ 0.1	Possible Test of CP Violation $(\gamma\gamma)$ = $\frac{N^{++}-N^{--}}{N^{+-}}$ ≠ 0 if 4th Generation Exists
$B^\pm \to K^\pm \Psi + ...$	~1.5%	Possible Large CP Violation $B(B^- \to L^-\Psi) \neq B(B^+ \to K^+\Psi)$
$B_{s/d}^0 \to \mu^+\mu^-$	~10^{-8}	Flavor Changing Neutral Currents
$B^\pm \to K^\pm\mu^+\mu^-$	~(10^{-5}–10^{-8})	GIM Breaking Flavor Chainging Neutral Currents

The experimental signature for this decay mode is the observation of a dimuon with mass beyond the J/ψ mass region and a K or μ^+ or μ^- with a impact parameter indicating a B decay lifetime. A further test would be the tagging of the $K^\pm$ particles from the decay. If the $K^\pm$ momentum can be measured the invariant mass of the $K_{\mu\mu}$ system should have a sharp cutoff at the $B^\pm$ meson mass.

We have carried out preliminary calculations of the decay kinematics of the $K^\pm$ and $\mu^\pm$ from the $B^\pm$ produced in $\bar{p}p$ collisions at $\sqrt{s}$ = 2 TeV. The ISAJET Monte Carlo simulation suitably modified to include the rare $B \rightarrow K_{\mu\mu}$ decay mode was used in these simulations. In Fig. 7 we show the results of these calculations.

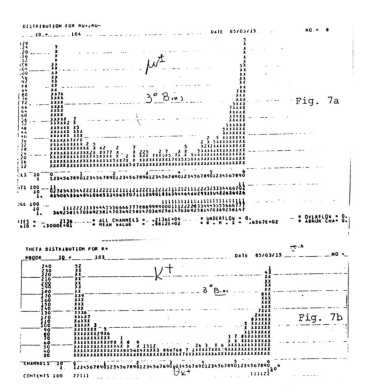

Fig. 7a

Fig. 7b

7b. An Experimental Detector – M_Z-M_W Mass

We have studied the possibility of carrying out a dedicated experiment to detect one or more of the previously mentioned rare B decays. The detector described here is specifically designed to observe these decays to operate in a large background rate situation.

We consider the three elements needed to detect a rare B decay to be

  (a) Tagging of the B vertex
  (b) Detection of the final state products
  (c) Measurement of the invariant mass of the decay products to reconstruct the B mass!

To estimate the possible rate for the detection of a B rare decay we assume that the TeV I $\bar{p}p$ collider has reached an average luminosity of $\sim 10^{31}$ cm^{-2} sec^{-1}. The B$\bar{B}$ production cross section is assumed to be 10 μb. For three months of data taking this gives a total number of B$\bar{B}$ events of

$$N_{B\bar{B}} = 10^{31} \times 8 \times 10^6 \text{ sec} \times 10^{-29} = 8 \times 10^8 \text{ events.}$$

The following processes would be interesting to study with such a detector

$$B_s^0 \rightleftarrows \bar{B}_s^0$$
$$B_d^0 \rightleftarrows \bar{B}_d^0$$
$$B^{\pm} \rightarrow K^{\pm} \mu^+ \mu^-$$
$$B^{\pm} \rightarrow \psi \, K^+$$
$$\rightarrow \psi + X$$
$$B_{s,d}^0 \rightarrow \mu^{\pm} e^{\mp}$$

The measurement of the mass difference M_Z-M_W will be of great importance to the future study of the electroweak model and the SU(7) x U(1) description of the world. The measurement of the M_Z-M_W mass difference to high precision will play a similar role in the test of the electroweak theories as the g-2 of the e and μ did for the QED theory.

In order to measure the M_Z-M_W mass difference the UA1 experiment will use an ionization calorimeter. The determination of the energy deposition is reduced to a measurement of the absolute charge deposited in the detector capacitance. At present it is conceivable to measure and to maintain calibration of this quantity to a precision better than 1/1000, corresponding to a precise and absolute energy calibration that is related to the ability of measuring the masses of the W and Z particles to an accuracy primarily determined by the statistics of the sample, or about 100-200 MeV. This is a factor of 20-30 better that the presently quoted systematic measurement error. The simultaneous measurement of the two masses has great relevance to the verification of the validity of the Standard Model and of the electro-weak radiative corrections.

The accurate calibration of individual calorimeter cells, associated with the excellent energy resolution of our spectrometer for high energy electrons, namely $0.09/\sqrt{E(GeV)}$, should make possible to detect for the first time the broadening of the line due to the natural Z^0 width, which is predicted to be $\Gamma = 2.8$ GeV. Any "new" physics in the decay amplitude of the Z^0 must have a corresponding effect on the natural width.

References

1. C. Rubbia, P. McIntyre and D. Cline, Proc. Int. Neutrino Conference, Aachen, 1976 (Vieweg, Braunschweig), 1977, p. 683.
2. G. Arnison et al., Phys. Lett. 122B (1983) 103.
3. M. Banner et al., Phys. Lett. 122B (1983) 476.
5. G. Bauer, Thesis, University of Wisconsin, 1985 (unpublished).
6. These results have been presented at various meetings, for example:
   C. Rubbia, proceedings of the Kyoto International Conference on High Energy Physics;
   K. Eggert, proceedings of New Particles, 1985;
   J. Rohlf, proceedings of the Oregon DPF meeting, 1985;
   D. Cline, proceedings of the 1985 Aspen Winter Conference series (published by the New York Academy of Sciences Press, 1986).
8. These results were obtained by modifying the ISAJET Monte Carlo program to include B_s^0. D. Cline and J. Rhoades, UA1 TN 84-65.

VERY HIGH ENERGY POLARIZED BEAMS FOR THE SSC[*]

Y. I. Makdisi
Alternating Gradient Synchrotron Department
Brookhaven National Laboratory
Associated Universities, Inc.
Upton, New York 11973 United States of America

This talk represents a summary of the Workshop on Polarized Protons at the SSC that was held from June 10-15, 1985, in Ann Arbor, Michigan. My remarks are quite general; for additional details the reader is referred to the proceedings of the workshop which will be published soon.

The workshop dealt with a number of topics divided along functional hardware lines in addition to the theory group. Periodically all the groups met together to discuss common issues that transcended their divisions.

The main topics are:

1. Spin-Physics and Motivation
2. Booster I (1-70) GeV
3. Booster II (70 GeV - 1 TeV)
4. The Main Ring (1-20) TeV
5. Polarimeters
6. Demonstration Snake

I shall expand on these items separately with some remarks on the depolarizing resonances and techniques to cope with them, and, at the end, present the conclusions of the workshop.

---

[*] Work performed under the auspices of the U.S. Department of Energy.

PHYSICS MOTIVATION

It is fair to say that spin, which is a fundamental parameter in Physics, is still not fully understood. It has always been a hypothesis that spin effects should become negligible as we go higher in energy, and in particular, as the momentum transfer gets larger. However, there are certain experimental results (surprises) that seem to deviate from such a prediction:

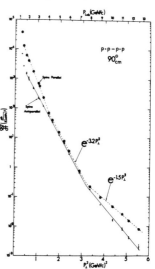

a) Two spin measurements in pp elastic scattering[1] at the Argonne ZGS show the cross section when two protons spins are parallel to be consistently higher than when the spins are antiparallel (Fig. 1). The ratio between these cross sections tends to grow with $P_\perp$ reaching about a factor of 4 at the highest $P_\perp$ attainable then. These measurements will be extended at the AGS with the new polarized proton beam capability.

Fig.1 $d\sigma/dt$ for two-spin orientations in pp elastic scattering

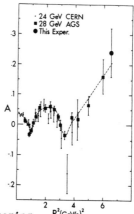

b) Single spin measurements in pp elastic scattering at the AGS[2] using an unpolarized 28 GeV/c beam impinging on a polarized target, show a striking asymmetry, Fig. 2, that continues to increase with larger momentum transfer. Perturbative QCD assumes that helicity is conserved in these reactions. This implies that the asymmetry $A_n = 0$ at

Fig. 2. The asymmetry A_n in pp elastic scattering.

high $P_\perp$ when the quark mass is small compared to the available energy in the center-of-mass (Mq/√s ≪ 1). Similarly, measurements on the pn elastic scattering[3] show large asymmetries that have not yet been taken to such high energies.

c) Inclusive Lambda polarization measurements utilize the self-analyzing power of the lambda decays. This polarization exhibits the following characteristics: independence of target nucleus and beam energies (it has been observed at KEK, PS, AGS, FNAL, ISR [not necessarily in this order]), it is still persistent at relatively large $P_\perp$[4] (Fig. 3). Other hyperons exhibit large polarization and we still do not understand the underlying mechanism. Strong spin dependence was observed in inclusive $\pi^\pm$, $K^\pm$ and π° production.

One expects that the new AGS polarized proton beam and the proposed polarized proton beam at Fermilab would enrich our data samples and bring us closer to the high energy domain.

From here on, one has to extrapolate to the usefulness of polarized proton beams at SSC energies. Due to fast falling cross sections, spin measurements in elastic scattering will be limited to relatively low t. Figure 4 shows some predictions[5] for pp elastic scattering.

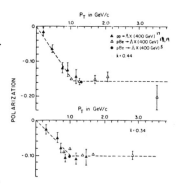

Fig. 3. Polarization in inclusive lambda production vs $P_\perp$.

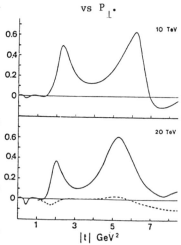

Fig. 4. Predicted asymmetries A_n in pp elastic scattering at SSC energies.

In addition, longitudinally polarized beams would allow us to distinguish between left and right handed charged currents. Most interesting will be to test for right-handed W production (W_R); a lower bound on its mass has been established[6] to be around 400 GeV. A_{LL} asymmetry measurements could help establish the onset of SUSY thresholds. This asymmetry is expected to change sign as we cross the SUSY threshold. Further calculations are needed to study these effects in more detail.

BEAM DEPOLARIZATION AND SOME CURES

In circular accelerators, the spin vector can be made to precess about its preferred orientation due to the presence of perturbing magnetic field components perpendicular to the spin vector. Depolarization occurs when the spin precession frequency matches the frequency of the driving magnetic fields. Two basic types of resonances are seen:

Intrinsic resonances result from horizontal fields experienced as the particles execute betatron oscillations. They occur when $G\gamma = kp \pm \nu_y$ is satisfied, where $G = (g-2/2)$, γ the Lorentz factor, p the periodicity of the machine, ν_y the harmonic tune, and k is an integer. In lower energy accelerators such as the AGS, a fast shift in ν at the appropriate γ does the trick. This can be seen in Fig. 5.

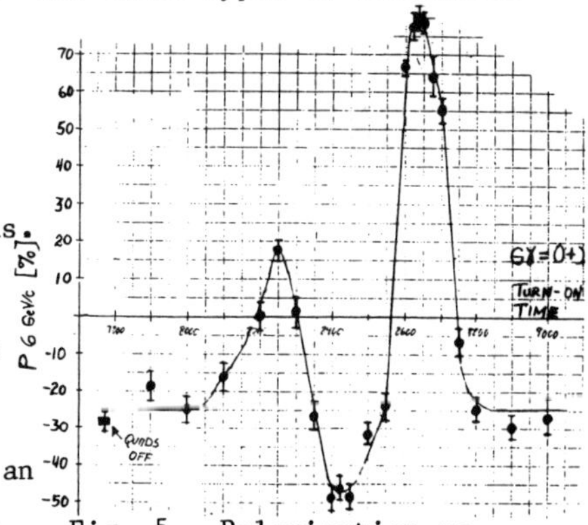

Fig. 5. Polarization vs. quadrupole pulse timing to jump $G\gamma = 0 + \nu_y$.

Another type of resonance is due to magnet misalignment, thus the name imperfection resonances. The condition $G\gamma = n$ has to be met. Dipoles are used to correct the offending nth harmonic at the appropriate γ. An example of these is shown in Fig. 6.

Both of the above corrections scale with beam energy. Experience has it that these techniques become difficult at energies above 25 GeV. For example, at the AGS a Δν shift of .25 units requires 2300 Amps of current in each of 10 pulsed quadrupoles at 23 GeV.

A cure-all solution (dubbed the Siberian snake) proposed by Derbenev and Kondratenko[7] utilized localized magnetic fields in a ring to precess the components of the spin vector appropriately during one or two revolutions, so that the system closes upon itself. Its action is shown in Fig. 7a,b,c, where the spin vector components execute a multiple of π rotations. The single snake shown requires two revolutions for closure. A ring equipped with two snakes requires one revolution.

This precession can be accommplished either by a longitudinal solenoidal field or transverse fields. In a solenoid, a precession angle of π for protons requires a $B\ell = 35.2$ $\beta\gamma$ kGauss- meters, an excitation that depends on γ. In transverse fields a factor of γ is gained. Thus the excitation needed for a π/2 rotation is $B\ell = \frac{\pi}{2}\frac{B\rho}{\gamma G} = 27.5$ kGauss-m per magnet independent of energy.

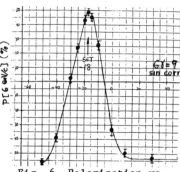

Fig. 6. Polarization vs dipole current to correct the $G\gamma = 9$ resonance.

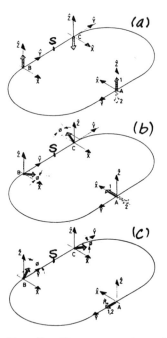

Fig. 7. The precession of the spin vector components over one revolution starting at A. The snake is between B & C.

Having prescribed various depolarization cures, let us examine the proposed SSC complex with its three ring lattices as defined by the Central Design Group. No depolarization effects should occur in the LINAC. This is born out by our experience at the AGS where we measure relatively high beam polarization at 200 MeV approaching that predicted from the source.

1-70 GeV Booster I

Figure 8a shows the predicted resonance strengths as calculated by the DEPOL[8] program.

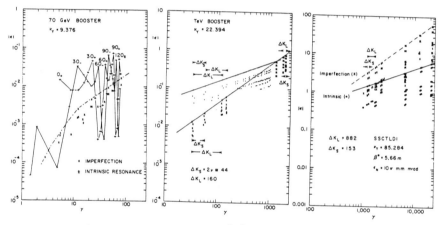

Fig. 8. Resonance strength $|\varepsilon|$ for the three rings.

Two resonance jumping techniques will be employed here. The conventional tune-shifting quadrupoles for γ from 1-20 followed by two transverse Siberian snakes to preserve the beam polarization up to 70 GeV. The snakes will be adiabatically energized during the early part of the acceleration cycle. This is not expected to appreciably dilute the beam polarization. The two snakes allow the proton beam to be injected vertically at a spin tune of $G\gamma = n + 1/2$. The strings of 8 magnets per snake fit easily in the allotted 16 meter straight sections.

The Booster II (1 TeV)

Figure 8b shows the expected resonance strengths to be well behaved with $|\varepsilon| < 1$. Here again two or four transverse snakes will be adequate to preserve the beam polarization up to 1 TeV. This lattice with two straight sections seems ideal for low resonance strengths.

The Main Ring (20 TeV)

Estimates of the resonance strengths appear in Fig. 8c. The DEPOL calculations used the following criteria for magnet misalignment: $\Delta_x = \Delta_y = .1$ mm in the arcs, and $\Delta_x = \Delta_y \simeq .05$ mm for the insertion quads. It should be noted that:

1. $|\varepsilon|$ scales with γ for imperfection resonances and with $\gamma^{1/2}$ for intrinsic resonances.

2. The resonances with strengths $|\varepsilon| > 10$ may be problematic.

3. It appears that the number of snakes needed is proportional to $|\varepsilon|^2$; however, studies are underway to determine the proper functional dependence or the acceptable minimum number.

4. Assuming that 100 snakes will be needed, about one snake per per mile of circumference, a reasonable estimate for the snake length impacts the circumference at the (1-2)% level.

5. A Terrain following machine is acceptable. It is expected that 1/2 snakes will be used to compensate for each vertical bend.

6. It follows from 5 that an over/under ring configuration is acceptable as well. However, it suffices to say that the preferred design for polarized proton beams is a flat accelerator.

INTENSITY

We aim for a polarized proton luminosity of $10^{33} \text{cm}^{-2}\text{sec}^{-1}$ at 20 TeV. This requires 10^{12} polarized protons per pulse. Currently the present source at the AGS provides $(1-2) \times 10^{10}$ protons/pulse. The proposed accumulator booster will give a factor of 20 and a modest expectation of source improvement should provide another factor of 5.

POLARIMETERS

These are an absolute necessity for the proper operation of a polarized beam facility. At 20 TeV the most promising technique seems to be the measurement of the asymmetry in the Coulomb-nuclear interference region in pp elastic scattering. A 4% asymmetry is expected at $t < .003$ $(GeV/c)^2$ and $P_\perp$ of .05 GeV/c. The scattering angles are small $< 2 \times 10^{-6}$ radians, requiring long experiments to get away from the beam. At $10^5 - 10^6$ events per second, one expects an error in the measurement of 8%/sec. A polarized gas jet will be needed for calibration. Other potential schemes include inclusive hyperon production, pe↑ scattering, as well as Coulomb diffractive production of N^* as proposed for the FNAL polarized beam.

A DEMONSTRATION SNAKE

The snake concept is heavily relied upon in this venture; however, while theoretically feasible, it has never been tested. The demonstration group is proposing to build a solenoid snake to be installed and tested at the AGS. The experiment will attempt to correct two imperfection resonances ($G\gamma = 8$ and 9) and one instrinsic $G\gamma = 0 + \nu_y$ up to 5 GeV. The solenoid would be energized up to 185 kGauss-meter. The following tests will be carried out: A longitudinally polarized beam should preserve its polarization up to 5 GeV/c; with vertical spin alignment the snake is expected to precess the spin to a longitudinal direction; finally the snake will test the adiabatic turn-on concept which is vital for the operation of the (1-70) GeV booster.

SUMMARY

The prospect of accelerating polarized protons in the SSC complex looks quite promising. The physics is exciting and studies are underway to further pinpoint the areas of interest, as well as estimate the expected asymmetries. Technically, the first two rings look quite good. The resonance strengths in the main ring that are above 10 look

1000

potentially troublesome and additional
studies are needed to get a better
grip on the situation. We think an
investment in providing the proper
lattices and space for the snakes
in the machines is well worth it.
Finally, the idea of strong fixed
target spin physics using polarized
jets and polarized beams is worth
pursuing. In the adjacent figure
we show the SSC complex as seen from
the polarized beam perspective.

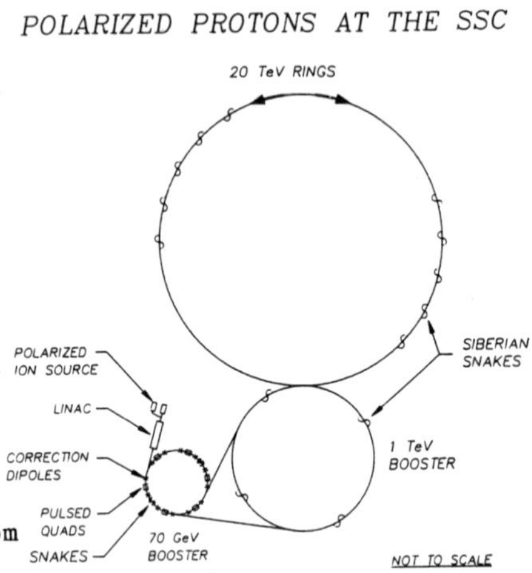

POLARIZED PROTONS AT THE SSC

20 TeV RINGS

POLARIZED
ION SOURCE

LINAC

CORRECTION
DIPOLES

PULSED
QUADS

SNAKES

70 GeV
BOOSTER

SIBERIAN
SNAKES

1 TeV
BOOSTER

NOT TO SCALE

Fig. 9. SSC Complex

REFERENCES

1.   E.A. Crosbie et al., Phys. Rev. D23, 600 (1981).

2.   P.R. Cameron et al., Measurement of the Analyzing Power for
     $p + p\uparrow \rightarrow p + p\uparrow$ at $P_\perp^2 = 6.5$ (GeV/c)^2 UM-HE 85-17.

3.   Y. Makdisi et al, Phys. Rev. Lett. 45, 1529 (1980).

4.   K. Heller. Inclusive Hyperon Polarization: A Review. Proceed-
     ings of the 6th International Symposium on High Energy Spin
     Physics, Marseille, France, September 12-19, 1984. J. Soffer,
     Editor.

5.   C. Bourrely. Elastic Proton Proton Polarization in the TeV
     Energy Domain, ibid.

6.   D.P. Stoker et al. Phys. Rev. Lett. 54, 1887 (1985).

7.   For a good description see B.W. Montague, Polarized Beams in High
     Energy Storage Rings. Physics Reports 113, (1984).

8.   E.D. Courant and R.D. Ruth, BNL-51270, (1980).

ACCELERATORS FOR THE STUDY OF MANY PARTICLE SYSTEMS*

Jose R. Alonso

Lawrence Berkeley Laboratory
Berkeley, CA 94720

ABSTRACT

Higher energy accelerators continue to play an important role in nuclear physics, probing ever more deeply into the properties and behavior of the constituents of nuclear matter. Three main projectile-types currently used are electrons, light hadrons (protons, mesons) and heavy ions; each addresses different aspects of the reaction process. Current and planned accelerators for each of these probes are discussed.

1. INTRODUCTION

Accelerators have played a key role in nuclear physics from the earliest days of the field. Each technological advance in accelerator design has opened up new discoveries, and opportunities for new views of nuclear phenomena. For example, the exquisite beam quality and fine energy resolution of modern Van de Graaffs and cyclotrons have permitted very high precision studies in nuclear structure and nuclear reaction dynamics not at all possible with earlier generations of machines addressing the same energy range[1]. In recent years the trend has been towards using higher energy beams, with the aim of probing smaller details of nuclear constituents, or of achieving higher levels of excitation of nuclear matter. Three types of beams have emerged as being most useful for these studies; electrons, light hadrons (protons and mesons), and heavy ions. Each of these exhibits a different interaction with nuclear matter, providing complementary information about fundamental nuclear processes. In this paper we shall discuss the production of each of these beams, touching on particular characteristics of accelerators needed for each. We shall also briefly describe existing facilities, and current plans for the next generation of accelerators.

* Work supported by the Director, Office of Energy Research, Office of High Energy and Nuclear Physics, Nuclear Science Division, U.S. Department of Energy under contract number DE-AC03-76SF00098.

2. ELECTRONS

Interacting only through the electromagnetic force, electrons probe the charge
carriers inside nucleons. As indicated in
Figure 1, this implies interactions only with
the quarks. The energy range currently of
interest, up to 4 GeV, covers de Broglie
wavelengths down to 0.1 fm, and hence spans
the interesting range from where electrons
scatter off clusters of quarks (nucleons) to
where they interact with individual quarks.

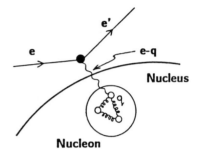

Fig. 1. Electrons couple only to
charge-carriers in the nucleus.

Present-day electron accelerators in this energy range are mostly linacs, and
have a very poor duty cycle; very short, intense beam bursts strike the target. This
low duty cycle is mandated by the large amount of RF power needed for the
acceleration process; availability of the power and the need to dissipate it in the
accelerating structure have been serious technical problems. Because of the poor
duty cycle, most experiments to date have been "inclusive", that is only one reaction
product is detected, with the loss of detailed information about the remainder of
the participants in the interaction.

There is general agreement[2] that coincidence experiments must be performed
next, requiring, ideally, a 100% duty factor. In addition, the interaction cross
sections of interest are quite low, requiring high beam currents to produce
acceptable data rates; currents of 100 to 200 µa are requested. In summary, the
desired parameters for a new electron accelerator are: variable energies, up to 4
GeV, long duty cycle, and high intensity.

Several technologies exist today for achieving these goals. None however is
without risks. Superconducting linacs, with lower power needs and very low losses,

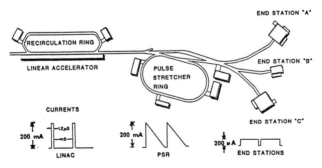

Fig. 2. Schematic layout of the CEBAF accelerator complex.

are attractive, but nothing of the required size has ever been built before. Microtrons, usually in the form of a modest linac with large 180° magnets at each end to provide multiple recirculating paths through the linac are an attractive source of CW electrons; a lower energy (60 MeV) facility at the University of Illinois stands out as an excellent demonstration of their practicability. Extending the technique to higher energies presents questions of transverse and longitudinal beam stability which are difficult to answer. A more attractive approach is to use a conventional high-current linac, followed by a pulse stretcher ring to provide a longer duty cycle. Even in this case, control of slow extraction from the stretcher ring, plus coherent and incoherent instabilities in the ring, present difficult problems. At the 1983 Particle Accelerator Conference, a full session was dedicated to the various options for CW electron accelerators[3]. It provides interesting background for those desiring further information on the subject.

After an intensive competition, the design proposed by the Southeastern Universities Research Association (SURA) was selected by the NSAC subpanel empowered to review the various proposals. This design[4], now called CEBAF (Continuous Electron Beam Accelerator Facility) is to be located at Newport News VA. Shown in Figure 2, it follows the third approach discussed above, namely of a linac (with one recirculating loop to double the intensity), followed by a pulse stretcher ring. The beam current profiles at various stages of the accelerator are shown in the traces below the schematic. The key to a continuous, uniform beam on target is the ability to carefully control the rate of spill of the beam from the stretcher ring.

3. LIGHT HADRONS

The interaction of hadronic probes with nuclear matter, shown schematically in Figure 3, is more complex, involving contributions from quark-quark, quark-gluon and gluon-gluon processes. Although more complicated, such strongly interacting probes have successfully yielded much experimental data on basic nuclear and nucleon properties. Different projectiles can be employed, either protons, available directly from the accelerator, or secondary particles such as mesons produced by directing the proton beam into a production target. To maximize the yield

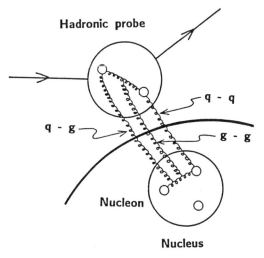

Fig. 3. Hadron projectiles interact with with both quarks and gluons in nucleons.

of such projectiles one requires energy sufficiently above the production threshold to ensure adequate phase space, and very high beam currents to overcome low production cross sections and the wide spread in production momentum and divergence. Three facilities specifically designed for high intensity meson production, LAMPF (Los Alamos), SIN (Zurich) and TRIUMF (Vancouver) have energies of 600-800 MeV, well suited for pion production, and beam currents approaching 0.5 to 1 milliamp, two to three orders of magnitude higher than previous accelerators in this energy range.

The next logical step is to provide beams of the same intensity but at an energy of around 30 GeV, to produce good fluxes of kaons and antiprotons in addition to the lighter mesons. It is not surprising that the three above-mentioned facilities all have proposals for adding boosters to their existing machines to reach the desired energy[5]. These plans make economic sense, as a significant portion of the very high cost of such a high energy facility lies in providing the high intensity beam suitable for injection into the final accelerator, a beam already available at all three of these laboratories.

Preserving the beam intensity presents some formidable problems in accelerator design. The present accelerators achieve their large currents by being continuous-beam machines, two cyclotrons (SIN and TRIUMF) and one linac (LAMPF), but continuation of these technologies to the desired energies is not practical. All are proposing synchrotrons as boosters, but must go to great lengths to achieve high intensities. A synchrotron typically injects beam over a short period, (short compared to the acceleration and spill cycle), thus losing much of the intensity available from a CW injector. To overcome this, a series of beam accumulation and bunching rings, and final beam stretchers must be employed. An example of such a design is shown in Figure 4, from the proposal for TRIUMF II[6]. A total of five new rings are proposed, first an accumulator to store 10^5 turns from the cyclotron, then a booster to raise the energy to 3 GeV. Several booster bunches are transferred to the

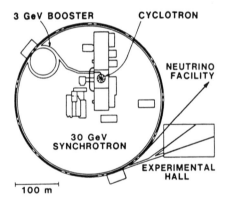

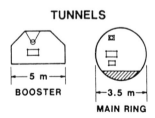

Fig. 4. Layout of TRIUMF Kaon Factory. Two new rings are located in the Booster Hall, three in the Main Ring Tunnel.

large tunnel into the collector ring, are then accelerated to 30 GeV in the driver ring and are put into the stretcher ring which peels beam out to a production target while the acceleration cycle is repeated in the other rings. This multi-stage acceleration will produce 100 μamps at 100% duty factor. (The total beam power is 3 MW!)

Although all aspects of the design are within the state of the art, there are many technical problems which must be addressed; space charge limits at injection of the rings, proper matching between rings, and most importantly, minimizing beam loss at all stages. In dealing with beam currents of this magnitude, the problems of activation of accelerator components and environs is particularly acute, both for component lifetime and for required maintenance access and handling.

4.   HEAVY IONS

The interaction of a high energy heavy ion with a target nucleus is of a different nature from the reactions described above. In central (head-on) collisions, the principal aspect is the transfer of projectile kinetic energy into excitation of the overlapping nuclei. This excitation energy can be high enough to drive nuclear matter into a region of temperature and density far removed from its normal state, potentially creating environments thought to exist only in the primordial universe,

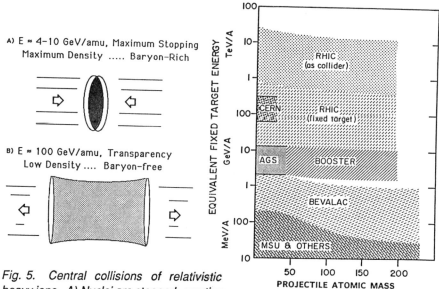

Fig. 5.   Central collisions of relativistic heavy ions. A) Nuclei are stopped, creating a region of maximum density. B) Nuclei pass through each other, leaving a region of high temperature vacuum.

Fig. 6.   Characteristics of present and planned high energy heavy ion facilities.

or in the interior of neutron stars or supernovae[7]. Energy densities achievable are also believed to be high enough to cause quark deconfinement, producing a quark-gluon plasma in the interaction region. Study of this new form of matter, and of these astrophysical environments provides a strong justification for building relativistic heavy ion accelerators.

Two distinct reaction types are envisioned, depending on the total kinetic energy available (see Figure 5). A maximum-density region is produced when the projectile has sufficient energy to just come to rest in the target nucleus. Since all the nucleons of the two nuclei remain inside the interaction region, the plasma produced is referred to as the "baryon-rich plasma". At the other extreme, a low density, high temperature region results when the kinetic energy of the incident nuclei is so high that the nuclei pass through each other, leaving behind a superheated, "baryon-free" region of space.

At the LBL Bevalac, the only facility where beams of uranium ions to lab-energies of 1 GeV/nucleon are presently available[8], experiments have demonstrated that nuclear transparency sets in for light nuclei ($A \approx 40$) at energies of a few hundred MeV/nucleon, but that top-energy mass-200 beams are stopped[9]. Best estimates of the optimum energy for production of the baryon-rich plasma are between 1 and 2 GeV/nucleon center-of-mass energy. Although Bevalac energies are generally felt to be too low to produce the plasma, nuclear densities about five-times normal have been inferred from observations[9], supporting the belief that at the right energy the plasma can in fact be created. Energies at least one to two orders of magnitude higher are deemed most desirable for studying the baryon-free plasma region.

Facilities existing and planned for study of these phenomena are shown in Figure 6. At energies above the Bevalac, the AGS at Brookhaven will be injecting light ions from their Tandem accelerators in mid 1986, and will be able to accelerate gold ions when a booster ring between the Tandem and the AGS is built, in about three to four years. To reach the energy needed to study the quark-gluon plasma the RHIC facility must be built[10]. This superconducting collider, to be located in the ISABELLE tunnel, will access the baryon-rich region when operated in a fixed-target mode, and the baryon-free region as a collider with 100-on-100 GeV/nucleon gold beams injected from the AGS. On a much earlier time scale, experiments on the CERN SPS with oxygen and possibly sulphur beams will take place in late 1986, utilizing a heavy-ion injector built by a GSI-LBL-CERN collaboration[11].

As is seen in the Brookhaven layout (Figure 7), achieving beams of these energies is a multi-stage process. (Note that the total energy for a mass 200 ion is 20 TeV!) Significant differences between heavy-ion and proton facilities come primarily in the size of the injector needed. Hydrogen is easily ionized, making an ion of q/A = 1. Producing highly ionized heavy ions of suitable intensity from an ion source is difficult; the normal route is to take lower charge states from the source and go through various stages of acceleration and stripping to eventually end up with fully stripped ions (with q/A still << 1) in the final accelerator or storage ring. Handling the stiffer ions requires larger accelerating structures, higher intensities in the earlier acceleration stages to overcome losses in stripping, and a very high synchrotron vacuum to prevent beam losses due to pickup or loss of electrons with residual gas. None of these problems is serious; there is a large body of expertise in the production and acceleration of heavy ion beams around the world.

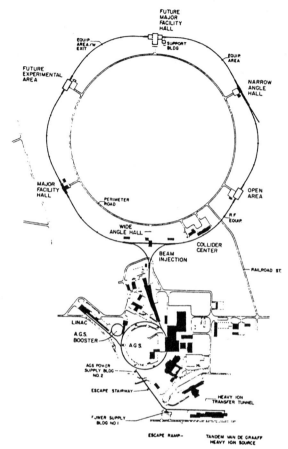

Fig. 7. Layout of the RHIC project. Ions will go from the Tandem to the Booster to the AGS and finally to the Collider.

5. SUMMARY: STATUS AND PROSPECTS FOR MAJOR NUCLEAR PHYSICS FACILITIES

The large construction projects discussed above are all in the $100M to $200M cost range; it is clear that all cannot be built at the same time in one country. In the United States, the Nuclear Science Advisory Committee (NSAC) has established priorities for the various initiatives; first continuous electron beams, then relativistic heavy ions. Although CEBAF has been a well-specified project for about two years now, it has received no construction funding to date. It is expected that the Booster for the AGS will be funded in the next few years, but prospects for construction money for RHIC are not good in the near term.

The brightest possibilities for a Kaon factory now are at TRIUMF, where enthusiastic reception by the community and funding sources seem to indicate a construction start within about two years.

6. ACKNOWLEDGEMENTS

The author would like to acknowledge the assistance of Hermann Grunder and Bev Hartline in the preparation of this paper.

7. REFERENCES

1) See for example, "Proceedings of the International Conference on Reactions Between Complex Nuclei", Robinson, R.L., McGowan, F.K., Ball, J.B., and Hamilton, J.H., eds, North Holland, (1974).

2) "The Role of Electromagnetic Interactions in Nuclear Science", report of the DOE/NSF - NSAC Subcommittee on Electromagnetic Interactions, Barnes, P.A., (Chairman), (1982), unpublished.

3) IEEE Trans. Nucl. Sci. NS-30, pp. 3250-3298, (1983).

4) "Conceptual Design Report: Continuous Electron Beam Accelerator Facility", Southeastern University Research Association, (1984), unpublished.

5) Thiessen, H.A., "A Review of Kaon Factory Proposals", IEEE Trans. Nucl. Sci. NS-32, 1601, (1985).

6) Craddock, M.K., et al, "The TRIUMF Kaon Factory", IEEE Trans. Nucl. Sci. NS-32, 1707, (1985).

7) "The Tevalac, A National Facility for Relativistic Heavy-Ion Research to 10 GeV per Nucleon with Uranium", LBL Accelerator & Fusion, and Nuclear Science Divisions, LBL-PUB-5081, (1982), unpublished.

8) Alonso, J.R., "Relativistic Uranium Beams--The Bevalac Experience", IEEE Trans. Nucl. Sci. NS-30, 1988, (1983).

9) Renfordt, R.E., Schall, D., Bock, R., Brockmann, R., Harris, J.W., Sandoval, A., Stock, R., Strobele, H., Bangert, D., Rauch, W., Odyniec, G., Pugh, H.G., Schroeder, A., Phys. Rev. Lett. 53, 763, (1984).

10) Samios, N.P., "Physics Opportunities with Relativistic Heavy Ion Accelerators", IEEE Trans. Nucl. Sci. NS-32, 3824, (1985).

11) Angert, N., et al, "A Heavy Ion Injector for the CERN Linac I", Proc. 1984 Linac Conference, Seeheim, Germany, p.374, (1984).

A BISMUTH GERMANATE ELECTROMAGNETIC CALORIMETER : CUSB-II

Paolo Franzini
Columbia University, New York, N.Y. 10027, USA

Juliet Lee-Franzini
SUNY at Stony Brook, Stony Brook, N.Y. 11794, USA

ABSTRACT

The properties and design philosophy of the first
bismuth germanate electromagnetic calorimeter used in
high energy physics are discussed. Presented also
are the test and physics results obtained with a
partial calorimeter.

1. INTRODUCTION

We describe in the following a bismuth germanate (BGO) detector,
CUSB-II[1], which was installed at the North Area of CESR in Sept. 1985.
The original CUSB detector[2], an electromagnetic (e.m.) calorimeter
built out of NaI crystals in layers for a total thickness of eight
radiation lengths (λ_o) of sodium iodide crystals backed by seven λ_o's
of lead glass, has the distinction of having discovered over ten T
states in the past five years[3]. Designed as the first calorimeter to
explore the T energy region at CESR, it has limitations due to
compromises between good photon energy resolution and the capability
for other physics, as well as to time and financial constraints. The
NaI-Pb glass crystal array is shown at the top of figure 1. The
detector has poor projective geometry and many gaps and and cracks, due
to its square geometry and the necessity of hermetically sealing the
highly hygroscopic NaI crystals which also resulted in ≥ 0.15 λ_o of
inactive material between layers of NaI. Finally, the total thickness

of NaI is too small for full shower containment. The longitudinal segmentation provides however excellent particle identification (hadrons vs photons and electrons) because of the five energy deposit measurements. Radioactive sources between crystal layers allow real time calibration during data taking, making it possible to achieve the theoretical resolution of the detector.

For the second generation of T physics experiments, which include the precision measurements of fine and hyperfine splittings, a spectrometer with improved photon resolution and perfect projective geometry is necessary. To accomplish these purposes we need more λ_o's, no inactive materials, while retaining the longitudinal segmentation and source imbedding features, occupying the radial space 10cm < r < 25cm. These are possible using bismuth germanate scintillating crystals. Table 1 lists some properties of NaI and BGO[4].

Table 1: Comparison of NaI and BGO Properties

ρ gr/cm^3	λ_o cm	Relative photo-electron yield	Resolution, r.m.s. @ 0.66 MeV	1.1 MeV
NaI	3.7	2.59	"1"	6.5%	
BGO	7.1	1.12	≈0.1	12.6%	8.8%

2. CUSB-II GEOMETRY

From examining the above one might be deterred from using BGO because of its lower light output hence poorer resolution for low energy γ's. However, its better material properties, including that of being nonhygroscopic, allow us to fit 12 λ_o's of BGO with ideal geometry into the available space. Each element has trapezoidal cross section and subtends 10° in ϕ and 45° to 90° (or 90° to 135°) in θ. Five such elements, of increasing size, form one ϕ-θ sector; the details of their mechanical specifications are listed in figure 2. Each crystal is viewed at one end by a miniature photomultiplier tube (p.m.), permanently glued on. All crystals are uniform, in light response, to better than 2% across their length (some after compensation, most as delivered from the Shanghai Institute of

Ceramics). Teflon and aluminized mylar wrapping (less than three mils thick) maximize light output and remove optical cross talk. Figure 3 shows one polar half of the BGO assembly (180 elements) viewed from the free crystal end. One notes 5 free standing concentric rings, constructed in a time honored (Roman Arch) method. Figure 4 shows the crystal assembly viewed from the phototube end.

CUSB-II COLUMBIA STONY BROOK

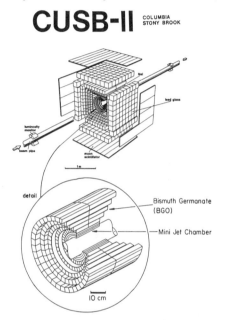

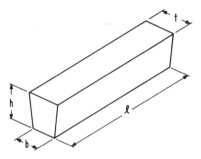

MATERIAL: BGO
DIMENSIONS IN MILLIMETERS
TOLERANCES: ±.025mm (±.005 INCH)

DETAIL		b	t	h	ℓ
1-PLANE	0	14.42	18.27	22	86
2-PLANE	1	18.97	22.82	22	113
3-PLANE	2	23.52	27.37	22	140
4-PLANE	3	27.72	33.15	31	165
5-PLANE	4	33.50	38.92	31	199

Fig. 1. Schematic View of CUSB-II. Fig. 2. Specifications of BGO.

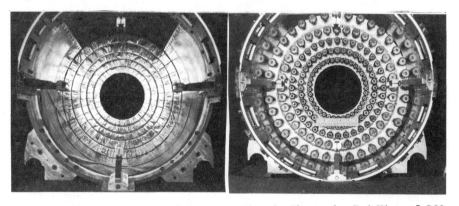

Fig. 3. Crystal End View of BGO.   Fig. 4. Phototube End View of BGO.

3. CUSB-II ELECTRONICS

Noise considerations require the readout electronics to be placed in close proximity to the detector elements. This requires a miniature "voltage divider, preamp, differential line driver" assembly which runs at very low power levels to reduce local heating. The physical size of this package relative to the phototubes and crystals is shown in figure 5. A module composed of 11 ϕ sectors, complete with electronic readout, is shown in figure 6.

Fig. 5. 1 ϕ Sector, BGO unwrapped.   Fig. 6. Eleven Assembled Sectors.

Figure 7 gives a schematic diagram of the signal flow and typical values, from phototube and amplifier, through differential sending and receiving. P.m. voltages are typically 600 V and divider currents ≈100 μA. Peak and average anode full scale signal (2-4 GeV) currents are respectively 100 and 30 μA. Long term anode average current is ≈10-100 pA. The equivalent amplifier noise input charge is ≈0.5 fC, or 3100 electrons, equivalent to 50 keV energy deposit (i.e. <1/20th of the calibration source signal). At the line driver output the full scale differential signal is ≈1 V.

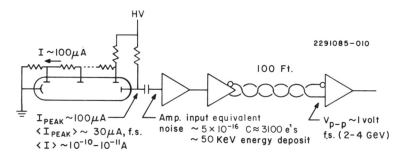

Figure 7. Schematic Diagram of Signal Flow.

4. CUSB-II CALIBRATION

All CUSB calibration is done in real time during data taking with sources which are imbedded between crystal layers. In figure 8 we show ^{137}Cs and ^{65}Zn spectra in one of our BGO crystal. The online calibration is accomplished by having a dual signal path, as shown in figure 9, such that two diffferent sensitivity channels measure collision events and source signals. The complete calibration cycle for all crystals is ≤ 30min. We use for this purpose a (286/310) stand alone Intel micro computer. All programs are written in FORTRAN, but the histograms are accumulated in memory by an 8089 Intel I/O processor. We have achieved ≤0.1% channel to channel calibration during a test run at CESR (see next section) and can achieve ≈ 1-2% absolute calibration as checked with Bhabha events and shower leakage calculations.

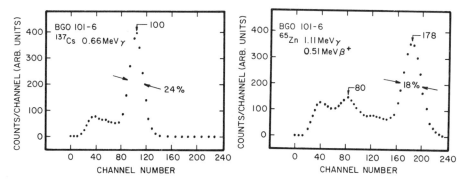

Figure 8. ^{137}Cs Source Signal (left) and ^{65}Zn Source Signal (right).

Figure 9. (right) Schematic
Diagram of Dual Signal
Path for Real Time
Calibration.

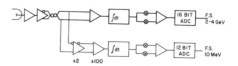

5. TEST RESULTS

During last fall we ran on
the T(1S) with a sector of
reduced thickness (9 radiation
lengths instead of 12) of the BGO
cylinder, covering $110°$ in ϕ
while the rest was covered by
sodium iodide. Approximately a
total of 420,000 T's were
produced during the 22 pb^{-1} run.
The BGO array provided the
expected improvement (a factor of
2) in resolution over the equal
radiation length NaI array for
≈ 4.7 GeV electrons from Bhabha
scattering, figure 10.

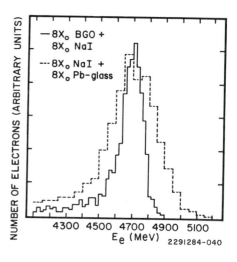

Fig. 10. Resolution for parts
of the Test Module.

The BGO and NaI data provided independent measurements of the
inclusive photon spectra from $T \to \gamma + X$. These were examined in detail, no
monochromatic signal consistent with our known resolutions was found in
any of the distributions. Therefore we calculate the 90% c.l. upper
limit BR ($T \to \gamma + X$) for the BGO and NAI data separately by fitting the
structureless spectra to a smooth function ($\chi^2 \approx 1/$d.o.f.) and use the
square root of the number of events within a 3.7σ interval as the
r.m.s. uncertainty on the presence of a possible signal, a more
conservative procedure than usual. The upper limit is given for the
two samples combined. In figures 11-13 we show our limits for $T \to \gamma X$,
where X stands for anything, for three photon energy regions of
interest: 30-500 MeV, 150-1325 MeV, and 740-5000 MeV.

LOWER BOUND ON THE ETA(B) MASS

The η_b's, the singlet bound ($b\bar{b}$)'s are expected to lie quite close to their triplet S partners. See, for example, the hyperfine splitting for this system predicted by several methods in the literature[5] which range from 30 to 100 MeV. The branching ratio for the M1 transition $T \to \gamma \eta_b$ is given by:

$$BR(T \to \gamma \eta_b) \approx E_\gamma^3 (4 \times 10^{-8}/\Gamma_{tot}(T)),$$

E_γ in MeV, $\Gamma_{tot}(T)$ in keV. The dashed line in figure 11 give the expected value of BR vs E_γ. From the experimental upper limit we obtain $E_\gamma < 168$ MeV at 90% c.l. and $M(\eta_b) > 9292$ MeV.

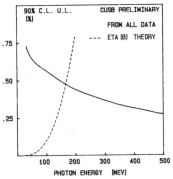

Fig. 11. η_b Limit.

LIMIT ON SQUARKONIUM

Tye and Rosenfeld[6] have proposed that there may exist scalar quark bound states for which the 3P state overlaps with the T(9.46). Under these conditions they conclude that there would be an "apparent" branching ratio for radiative decays of events with the T mass. These decays are from E1 transitions of the 3P state to the 3S state ($E_\gamma = 0.18$ GeV, $BR \approx 0.5\%$), to the 2S state ($E_\gamma = 0.52$ GeV, $BR \approx 0.3\%$), to the 1S state ($E_\gamma = 1.08$ GeV, $BR \approx 0.3\%$). These BR's are indicated as the dotted lines in figure 12. Our 90% upper limit rules out this proposal.

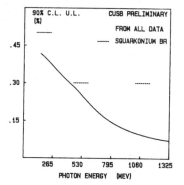

Fig. 12. Squarkonium Limit.

LIMIT ON HIGGS BOSONS PRODUCTION

The Standard model of the electroweak interaction requires the existence of a scalar neutral particle, the Higgs meson

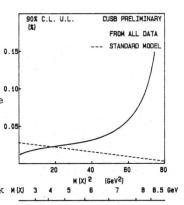

Fig. 13. Higgs Boson Limit.

(H). Theory can not uniquely predict the mass of the Higgs and searches for these elusive states remain unsuccessful and to a large extent inconclusive. Should the Higgs have a mass of the order of or less than M(T), we can search for them in the latter's radiative decays. The branching ratio for T→γ+H via the Wilczek[7] mechanism (without QCD corrections) is given by:

$BR(T→γH) = [BR(T→\mu\mu)M_q^2 G_F/(\alpha\pi\sqrt{2})](1-M_H^2/M_T^2)$. The dashed line in fig. 13 depicts this branching ratio. We find $m_H > 4.2$ GeV at 90% c.l.. Our limit is preliminary because intensive Monte Carlo studies are underway to determine our efficiency as a function of assumed H decay modes.

6. ACKNOWLEDGEMENTS

The authors wish to emphasize that CUSB-II is a group enterprise, albeit a small group by today's high energy physics standards, and thank all collaboration members for the successful realization of CUSB-II. In particular, Drs. R. D. Schamberger and P. M. Tuts especially deserve credit for their leadership during its construction and installation, and Mr. T. Zhao for much work on this project. CUSB-II is supported by U. S. National Science Foundation Grants Phy-8310432 and Phy-8315800.

7. REFERENCES

1. CUSB-II members include: P. Franzini, P. M. Tuts, S. Youssef and T. Zhao of Columbia University; J. Lee-Franzini, T. M. Kaarsberg, D. M. J. Lovelock, M. Narain, S. Sontz, R. D. Schamberger, J. Willins and C. Yanagisawa of SUNY at Stony Brook.
2. P. Franzini and J. Lee-Franzini, Physics Reports 81, 239 (1982).
3. See P. Franzini and J. Lee-Franzini, Ann. Rev. of Nuc. and Part. Sci. 33, 1 (1983) for earlier references, and J. Lee-Franzini, Physics in Collision V, Autun, France, eds. B. Aubert and L. Montanet (Edition Frontieres, France, in press) for recent references.
4. P. M. Tuts and P. Franzini, International Workshop on Bismuth Germanate, ed. C. Newman Holmes, 596 (Princeton Univ. 1982).
5. A. Martin, Phys. Lett. 100B, 511 (1981); L. J. Reinders et al., Phys. Lett. 104B, 305 (1981); W. Buchmuller et al., Phys. Rev. D24, 3003 (1981); R. M. McClary and N. Byers, Phys. Rev. 28, 1692 (1983); and J. Lee-Franzini, see previous reference.
6. S.-H. H. Tye and C. Rosenfeld, Phys. Rev. Lett. 53, 2215 (1984).
7. F. Wilczek, Phys. Rev. Lett. 40, 220 (1978).

LIST OF OTHER PAPERS PRESENTED

M. Adams — *$\mu^+ \mu^-$ Pair Production in the Υ Region.*

S. Ahn — *Forward Charged Particle Distributions in pp Collisions with High E_T at $\sqrt{s} = 27.4$ GeV.*

D. Akers — *Magnetic Monopole Interactions: Shell Structure of Meson and Baryon States.*

J. Amjorn — *Monte Carlo Simulations of Configurations with Stastic Charges.*

T. Andrews — *A Wave System Theory of Elementary Particles.*

ARGUS Collab. — *An Upper Limit on the Mass of the Tau Neutrino.*

ARGUS Collab. — *Direct Evidence for W Exchange in Charmed Meson Decay.*

ARGUS Collab. — *Inclusive Electrons from the Υ (4S) Resonance.*

R. Arnowitt — *N=1 Supergravity and Monojets.*

S. Ayub — *SU (3) Unification Ramifications.*

K. S. Babu — *Constraints on the Higgs Boson Masses.*

C. Baltay — *Report on the 1985 HEPAP Study* (Plenary talk)

G. Baranko — *Charm Production at TASSO.*

G. Baranko — *Review of PETRA experiments on b Lifetime.*

P. Baringer — *K^o and Λ Production in Jets at PEP*

R. Bland — *A Cryogenic System for a Superconducting Indium Solar Neutrino Detector.*

R. W. Bland — *A Prototype Supercond. Indium Detector for Solar Neutrinos.*

D. Blockus — *Comparison of D^o and D^+ Lifetime Measurements.*

R. D. Brooks — *Propagation of a Laser Beam Through a Plasma and Electron Plasma Wave Generation and Acceleration.*

C. N. Brown — *e+e- Pair Production in the Υ Region.*

R. Bryan — *Internal Quantum Numbers of Quarks and Leptons.*

P. R. Burchat   *Measurement of the Branching Fraction for $\tau \to 5\pi + \nu$ and an Upper Limit on the ν_τ Mass.*

W. B. Campbell   *Vacuum and False-Vacuum States from a Vector Bag Model.*

G. Carboni   *A Measurement of the Decay of Beauty Particles into Charm Particles.*

L. Carson   *Deuterium in the Skyrme Model.*

S. Chiangir   *The Super Solenoid.*

M. D. Corcoran   *Triggering on High P_T Jets.*

J. Crittenden   *The Angular Dependence of Single Hadron Production in p-Nucleus Collisions at High x_T.*

M. Derrick   *KNO Scaling Distributions in e^+e^- Annihilation at $\sqrt{s} = 29$ GeV.*

J. L. Diaz-Cruz   *Decays of Heavy Charged Higgs Bosons.*

R. Driscoll   *The Chromostatic Potential: Correspondence between Quantum and Neo-Ritzian Theories.*

J. Dubisch   *Infrared Topology on a Zero Rest Mass Field.*

G. Eilam   *Skyrmion Solutions to the W. S. Model.*

A. R. Erwin   *An Experimental Constraint on the Gluon Structure Function of the Pion.*

A. R. Erwin   *Invariant Cross Section at $\sqrt{s} = 28$ GeV for Coplanar High P_T Clusters Selected by a Harware Trigger.*

G. Fitzpatrick   *Quarks as Microscopic Unruh Detectors.*

J. Fleischman   *A Search for a Correlation between Small Angle Charged Particle Multiplicity and the Jettiness of 400 GeV pp Collisions.*

F. Flensburg   *Strings and SU (3) Lattice Gauge Theory.*

P. Franzini   *B^* Production in e^+e^- Interactions from $\sqrt{s} = 10.6$ to 11.2 GeV and the B^* Mass.*

P. Franzini   *Search for Higgs with the CUSB I.5 Detector.*

D. M. Frazer   *CP Violating Phases in a Minimal $SU(2)_L \times SU(2)_R \times U(1)$ Electroweak Model.*

R. Frey	*D* ± Production in Jets at the CERN Collider.*
K.K. Gan	*Review of New τ Physics.*
A. Garcia	*Radiative Corrections with Vertices Scalar, Psuedoscalar, and Tensorial (S,T, and P) to the Neutron β Decay.*
H. Glass	*High P_T Particle Ratios from Hydrogen and Deuterium.*
S. Godfrey	*Some Puzzles in Meson Spectroscopy.*
A. T. Goshaw	*Charm Production in 360 GeV/e $π^-$ p and 400 GeV/c pp Interaction.*
F. Graziani	*Fluctuation Modified Dynamics of Relaxation.*
J. Gunion	*Four Jet Processes: Gluon-Gluon Scattering to Non-Identical Quark-Antiquark Pairs.*
T. J. Haines	*Recent Results form the IMB Nucleon Decay Detector.*
X. G. He	*CP Non-Conservation with Four Generations.*
R. Herczeg	*On Muon Decay in Left-Right Symmetric Electroweak Models.*
W. Hiscock	*Constraints on Quantum Instabilities of the Cosmological Constant.*
B. Holdom	*17 KeV and 30 eV Dirac Neutrinos and a Techniphoton*
Y. Hseung	*A Dependence of the Inclusive Production of Large P_T Hadrons.*
S. Hsueh	*Determination of the Form Factors in $Σ^-$ Beta Decay.*
R. Huerta	*Test of Right Handed Currents in Baryon Semileptonic Decays.*
G. Intemann	*An Analysis of the Hadronic Decays of $η_c$ in a Pseudoscalar Mixing Model.*
J. Isenberg	*Ambitwistors and Strings.*
D. Issler	*Leptons and Horizontal Scalars from Topological Particle Theory.*
J. Iwai	*Anomalous Features of Cosmic Ray Interactions in Heavy Targets from JACEE.*
J. Izen	*The ALEPH Detector.*
J. Izen	*The ALEPH TPC.*

K.A. Johns	*Study of Gluon Bremsstrahlung Effects in Hadron-Hadron Collisions at $\sqrt{s} = 27 GeV$.*
C. Jung	*Lifetime Measurement of F + Meson.*
M. Kalelkar	*Rapidities and Multiplicities from 200 GeV/c Hadron Interactions on Au, Ag and Mg.*
P. Kalyniak	*Nonstandard Contributions to the Two Photon Production of a Higgs Boson.*
P. Karchin	*First Results on the Performance of a Silicon Microstrip Vertex Detector Used in a High Energy Photo-Production Experiment.*
M. Karliner	*The Baryon Spectrum of the Skyrme Model.*
R. Kass	*Recent Results from the CLEO Vertex Detector.*
R. Kennet	*A Comparison of Diffractive Λ and Ξ Photoproduction.*
P. Kesten	*Measurement of the Ratio of $\alpha_s(q^2)$ for Light and Charm Quarks.*
M. Kwan	*Leptonic Signals for Compositeness at Hadron Colliders.*
J. Lee-Franzini	*Search for Higgs and Squarkonium States with the CUSB I.5.*
J. Lee-Franzini	*Search for the n_b in Radiative Υ Decay with CUSB I.5.*
R. R. Lewis	*On the Macroscopic Effects of Low Energy Neutrinos.*
J. T. Linnemann	*Triple Jet Production in High Transverse Energy Events at the CERN ISR.*
W. C. Louis	*Upper Limits on D^0-$\bar{D}^0$ Mixing and on the Decay $D^0 \rightarrow \mu^+\mu^-$*
R. J. Loveless	*Search for Neutrino Oscillations in BEBC.*
S. Mani	*A Silicon MicrostripVertex Detector for Direct Photon Physics.*
P. Mannheim	*Classical Spin and its Quantization.*
H. Matis	*Threshold Behavior of Direct Lepton Production by Protons at 2.9 and 5.8 GeV/c.*
D. Meyer	*Comparison of Light Quark and Charm Quark Fragmentation.*
P. McIntyre	*TheTexas Accelerator Center andthe Accelerators of the Future.*
H. E. Miettinen	*Study of High Transverse Energy Proton-Nucleus Interactions at Fermilab.*

D. Millers	Q^2-Dependence of Exclusive Meson Production in Two-Photon Reactions.
S. L. Mintz	Weak Semileptonic Interactions in Proton-Nucleus Collisions.
R. Mir	New High Gain Thin Gap Detectors for the OPAL Hadron Calorimeter.
M. Mugge	A Neutrino Mass Experiment Using Frozen Tritium.
C. R. Ng	Measurement of b-quark Production in e^+e^- Annihilations at $\sqrt{s} = 29$ GeV.
B. S. Nielsen	Di-Jet Production Cross Section and the Structure of Jets in pp Collisions at $\sqrt{s} = 63$ GeV.
C. Pan	T>0 Thermal Properties of the Three-State Potts Model.
H. Peng	Gravito-Magnetohydrodynamics.
A. Petersen	Inclusive Charged Particle Distribution of Large Angle 3-Jet Events.
C. Peterson	Direct Observation of String Vibrations in Three-Dimensional Compact QED.
L. E. Piilonen	Branching Ratios for Rare Muon Decays.
J. P. Ralston	Coherent Enhancement of the High Energy Neutrino Total Cross Section.
B. Ram	Calculations of Bound State Relativistic Eigenvalues for Non-Confining Potentials.
L. Rangan	A Test of the LUND Fragmentation Model.
R. E. Ray	Search for Narrow Lines in the Gamma Spectra Resulting from $\bar{p}p$ Annihilation.
A. Read	Leptonic Decay of Tau.
R. W. Robinett	Searching for Gluinos in Cosmic Ray Interactions.
D. Rust	Lifetime of D^0 from Low z D^* Events.
M. Samuels	Test of QCD in pp and $p\bar{p}$ Collisions-The Radiation Amplitude Zero and the Magnetic Moment of the W Boson.
M. Sawicki	Relativistic Bound State Wave Functions for Scalar Field Model Quantized on the Light-Cone.

T. Schaad — *Upper Limit on B^0-$\bar{B}^0$ Mixing in e^+e^- Annihilation at 29 GeV.*

V. Schegelsky — *High Statistics Measurement of Polarized Σ^- Beta Decay.*

W.B. Schmidke — *Study of the Decay $\tau^\pm \to \pi^\pm \pi^+ \pi^- \nu_\tau$*

K. Schwitkis — *Measurement of the Two-Photon Production of the η'.*

M. Shepko — *Preliminary Results from the Texas A & M GUT Monopole Search.*

A. Sill — *Measurement of Elastic Electron-Proton Scattering at High Momentum Transfer.*

P. Simic — *QCD at Large N_c-Skyrme or the Bag?*

P. Sinervo — *A Study of Light Quark Systems Containing Strange Quarks in 11 GeV/c K^-p Interactions.*

M. Slaughter — *The Vector Glueball and Algebraic Approach.*

J. C. Slonczewski — *Effects of the Axion Halo on Bound Electrons.*

D. Snyder — *Observation.*

R. Stabler — *Helical vs. Chiral Projection Operators for the Weak Interaction.*

A. B. Steiner — *Search for Quarks Produced by 800 GeV Protons.*

A. Stern — *General Acton Principle for Supersymmetric Particles and Strings.*

D. Strom — *A Measurement of the D^0 Lifetime.*

D. Strom — *The TASSO Vertex Detector.*

K. Sugano — *Measurement of the Mean Charge Multiplicity in the 3-Jet Events in the e^+e^- Annihilations at $\sqrt{s} = 29$ GeV.*

F. R. Tangherlini — *Isospin Multiplet Mass Splitting Algarithm.*

M. J. Tannenbaum — *Do E_T Spectra Obey KNO Scaling in High Energy pp and Nucleus-Nucleus Collisions?*

L. C. Teng — *New Concepts in Particle Acceleration. (Plenary talk)*

R. S. Van Dyck — *Single Proton in a Penning Trap.*

E. Vella   *Event Structure and Jet-Jet Cross Section in pp Collisions at $\sqrt{s}$ =63 GeV.*

H. von Gersdorff   *Correlation between Transverse Momentum and Multiplicity for Spherically Exploding Quark-Gluon Plasmas.*

W. Walker   *Hadron-Nuclear Interactions at Energies of 100 and 320 GeV.*

Z. Wang   *Symmetry Orbits and the Relationship between Various Anomalies in Supersymmetric Yang-Mills Theories.*

J. Weinstein   *Scalar Mesons and the Iota: A Paradox Explained.*

W. Wilcox   *The Lattice Electric Form Factor.*

J. F. Wilkerson   *Status of the LANL Free Atomic Tritium Beta Decay Experiment.*

P. Yasskin   *Introduction to Ambitwistors.*

P. C. Yin   *EMC Effect and Nucleon Size in Nucleus.*

G. Zapalac   *Measurement of the Σ^- Magnetic Moment from E715 Using both the $n\pi$ and $ne\bar{\nu}$ Decay Modes.*

LIST OF PARTICIPANTS

ABOLINS, Maris
Michigan State University

ADAMS, Mark R.
SUNY

ADKINS, Gregory
Franklin & Marshall College

ADLER, Stephen L.
Institute for Advanced Study

AHN, Seung-Chan
Rutgers University

AKERS, David
Los Gatos

ALBERS, James R.
Western Washington University

ALONSO, Jose
Lawrence Berkeley Laboratory

AMMAR, Ray
University of Kansas

ANDERSON, E. Walter
Iowa State University

ANDERSON, Roger H.
Seattle Pacific University

ANDERSON, David
Fermilab

ANDREWS, Thomas
NYNEX

ARNOWITT, Richard
Northeastern University

AUVIL, Paul R.
Northwestern University

BABU, Kaladi S.
University of Hawaii at Manoa

BAGGETT, Neil
Brookhaven National Laboratory

BAGGETT, Millicent
Brookhaven National Lab.

BALTAY, Charles
Columbia University

BARANKO, Greg
University of Wisconsin

BARDON, Marcel
National Science Foundation

BARGER, Vernon
University of Wisconsin

BARINGER, Philip S.
Indiana University

BARNETT, Michael
Lawrence Berkeley Lab.

BARNHILL, Maurice V. III
University of Delaware

BARTLET, John E.
SLAC

BATES, Ross
Univ. of British Columbia

BAUR, Ulrich
Max-Planck Institute for Physics

BENNETT, Bradley F.
URA

BENSON, Richard S.
University of Minnesota

BERKELMAN, Karl
Cornell University

BERNARD, Claude
UCLA

BLAND, Roger
San Francisco State Univ.

BLOCKUS, David L.
Indiana University

BODWIN, Geoffrey T.
Argonne National Laboratory

BOSETTI, Peter
RWTH Aachen

BOSMAN, Martine J.
Max-Planck Institute-Munich

BOWICK, Mark
Yale University

BOWLES, Thomas J.
Los Alamos National Laboratory

BRAATEN, Eric
Northwestern University

BRAU, James E.
University of Tennessee

BROOKS, Bob
University of Washington

BROWN, Stanley G.
Physical Review

BROWN, Chuck
Fermilab

BRYAN, Ronald
Texas A & M University

BURCHAT, Pat
SLAC

CALDWELL, Allen
University of Wisconsin

CALDWELL, David V.
University of California

CAMPBELL, William B.
University of Nebraska-Lincoln

CANDELAS, P.
University of Texas, Austin

CARBONI, Giovani
CERN

CARLITZ, Robert
University of Pittsburgh

CARLSON, Per
. University of Stockholm

CARR, John
University of Colorado

CARSON, Larry J.
Northwestern University

CASELLA, Russell C.
National Bureau of Standards

CASSEL, David G.
Cornell University

CHANDLEE, Clark D.
University of Rochester

CHAO, A.
Lawrence Berkeley Lab.

CHAPMAN, J. W.
University of Michigan

CHEN, Wei
University of Oregon

CHENG, Ta-Pei
Univ. of Missouri-St. Louis

CHENG, Hai-Yang
Brandeis University

CHRIST, Norman H.
Columbia University

CHUNG, S. U.
Brookhaven National Lab.

CIHANGIR, Selcuk
Texas A & M University

CLAUS, Richard
University of Michigan

CLIFFORD, H. James
University of Puget Sound

CLINE, David
Univ. of Wisconsin, Madison

COLLINS, John
Illinois Institute of Technology

COOPER, Richard
Los Alamos National Lab.

CORCORAN, M. D.
Rice University

CORK, Bruce
Lawrence Berkeley Laboratory

COUTURE, Gillis
TRIUMF

COWARD, David H.
SLAC

COX, Bradley
Fermilab

CRICHTON, James H.
Seattle Pacific University

CRITTENDEN, James A.
Columbia University

CRONIN, J. W.
Enrico Fermi Institute

CUDELL, Jean-Rene
University of Wisconsin

DARLING, Byron T.
Universite Laval

DAVIS, Robin E. P.
University of Kansas

DEERY, Robert F.
Lewis and Clark College

DERRICK, Malcolm
Argonne National Laboratory

DESHPANDE, N. G.
University of Oregon

DETAR, Carleton
University of Utah

DIEBOLD, Robert
DOE

DONOGHUE, John
University of Massachusetts

DORFAN, Jonathan
SLAC

DRAPER, Terrance A. J.
University of California, Irvine

DUBISCH, Russell
Siena College

DUCK, Ian
Rice University

ECKLUND, Stan
SLAC

EILAM, Gad
University of Oregon

EINHORN, Martin B.
University of Michigan

EKELIN, Svante
Royal Inst. of Technology

ERWIN, Albert
University of Wisconsin

FACKLER, Orrin
Lawrence Livermore Lab.

FELDMAN, Gary
Brown University

FERBEL, T.
University of Rochester

FISHER, John C.
Carpinteria

FITZPATRICK, Gerald L.
Sigma Research Inc.

FLEISCHMAN, Jack G.
University of Pennsylvania

FOLEY, K. J.
Brookhaven National Laboratory

FRAMPTON, Paul
University of North Carolina

FRANZINI, Paula
SLAC

FRANZINI, Paolo
Columbia University

FRASER, Dan
Los Alamos National Laboratory

FREUND, Peter
University of Chicago

FREY, Raymond
Univ. of California, Riverside

FRISKEN, W. R.
York University

GAILLARD, Jean-Marc
LAL, Orsay France

GAISSER, Thomas
University of Delaware

GAN, K. K.
Purdue University

GARWIN, Charles A.
Santa Monica

GERSDORFF, Henrique von
Fermilab

GILCHRIESE, M.
Newman Lab.

GODFREY, Stephen
TRIUMF

GOLLIN, George
Princeton University

GOSHAW, Alfred T.
Duke University

GOTTSCHALK, Thomas D.
Caltech

GRAZIANI, Frank
University of Colorado

GROSS, David
Princeton University

GUNION, Jack
University of California, Davis

GURYN, Wlodzimierz
Brookhaven National Lab.

HAINES, Todd
University of Pennsylvania

HALZEN, Francis
University of Wisconsin

HANSL-KOZANECKI, Traudl
SLAC

HARVEY, Jeff
Princeton University

HE, Xiao-Gang
University of Hawaii

HEDGES, Christopher
San Francisco State University

HEMPSTEAD, Martin
Cornell University

HEPBURN, Charles Darwin
UC Los Angeles

HERCZEG, Peter
Los Alamos National Laboratory

HISCOCK, William Allen
Montana State University

HITLIN, David
Caltech

HOLDOM, Bob
University of Toronto

HOLMES, Stephen D.
Fermi National Accelerator Lab.

HSIUNG, Yee Bob
Columbia University

HSUEH, Shao Yuan
University of Chicago

HUBBARD, Richard
Saclay

HWA, Rudolph C.
University of Oregon

INNES, Walter
SLAC

INTEMANN, Gerald W.
University of Northern Iowa

ISENBERG, Jim
University of Oregon

ISSLER, Dieter
UC Berkeley

IWAI, Junsuke
University of Washington

IZEN, Joseph
University of Wisconsin

JACKIW, Roman
MIT

JACOB, Richard J.
Arizona State University

JAFFE, David
SUNY

JAZWAHERY, Abolhassan
Syracuse University

JOHNS, Kenneth A.
Rice University

JOHNSON, Robert
Syracuse University

JUNG, Chang Kee
Indiana University

KAJINO, Fumiyoshi
Virginia Tech.

KALELKAR, Mohan S.
Argonne National Laboratory

KALINOWSKI, Jan
Warsaw University

KALLIANPUR, Kalpana
University of Texas, Austin

KALYNIAK, Patricia
Carleton University

KANE, Gordon
University of Michigan

KARCHIN, Paul E.
Univ. of Calif. at Santa Barbara

KARLINER, Marek
SLAC

KARSCH, Frithjob
Univ. of Illinois at Chicago

KAYSER, Boris
National Science Foundation

KENNETT, Rosemasry Gilliard
Calif. State Univ. Northrop

KESTEN, Philip
University of Michigan

KEUNG, Wai-Yee
Univ. of Illinois at Chicago

KIRK, Thomas B. W.
Fermilab

KISTIAKOWSKY, Vera
Mass. Institute of Technology

KITAZUWA, Yoshihisn
Enrico Fermi Insitute

KLIWER, James
University of Nevada, Reno

KLOPFENSTEIN, Chris
Lawrence Berkeley Laboratory

KNAPP, David
Los Alamos National Laboratory

KOGUT, John B.
University of Illinois

KONIUK, Roman
York University

KOVACS, Eve
Rutgers University

LACH, Joseph
Fermilab

LANG, Karl
University of Rochester

LAVINE, Theodore
University of Wisconsin

LAZARUS, Donald M.
Brookhaven National Lab.

LEE, T. D.
Columbia University

LEE-FRANZINI, Juliet
SUNY

LEITH, David W. C. S.
SLAC

LEUNG, Chung Ngoc
Fermilab

LIND, V. Gordon
Utah State University

LINNEMANN, James T.
Michigan State Univesity

LO, Peter
University of Oregon

LOBKOWICZ, Frerick
University of Rochester

LOUIS, William C.
Princeton University

LOVELESS, Richard
University of Wisconsin

LUBATTI, Henry J.
University of Washington

MAKDISI, Yousef
Brookhaven National Laboratory

MANI, Sudindra
University of Pittsburgh

MANNHEIM, Philip
University of Connecticut

MARCIN, Martin R.
Rice University

MARGOLIS, Bernard
McGill University

MARX, Jay
Lawrence Berkeley Laboratory

MATIS, Howard
Lawrence Berkeley Laboratory

MCCARTHY, Robert L.
SUNY

MCINTYRE, Peter M.
Texas A & M University

MCLERRAN, Larry
Fermilab

MENG, Rubin
University of Oregon

MERRITT, Frank S.
University of Chicago

MIETTINEN, Hannu E.
Rice University

MIGNERON, Roger
University of Western Ontario

MINTZ, Stephen L.
Florida International University

MIR, Ronen
Weizmann Institute

MONTH, Melvin
Brookhaven National Laboratory

MOORE, Fred L.
University of Washington

MORFIN, Jorge G.
Fermilab

MOROMISATO, Jorge
Northeastern University

MORRIS, T. W.
Brookhaven National Lab.

MUELLER, Alfred
Columbia University

MUGGE, Marshall
Lawrence Livermore National Lab.

MUNCZEK, Herman
University of Kansas

MUNDY, George
San Francsico State University

MURTAGH, Michael J.
Brookhaven National Laboratory

MYERS, Eric
Brookhaven National Laboratory

NAPPI, Chiara R.
Princeton University

NARAYANAN, K. Sowmya
University of Western Ontario

NELSON, Donald
retired-Boeing Aerospace Co.

NELSON, Ken
University of Wisconsin

NG, Chokuen
University of Illinois

NG, C. Ray
Purdue University

NIELSEN, Borge S.
SLAC & CERN

NORDSTROM, Dennis
Physical Review

OLSSON, Martin
Univ. of Wisconsin-Madison

OLTMAN, Ed
Nevis

OVRUT, Burt
Rockefeller University

PAIGE, Frank
Brookhaven National Laboratory

PALMER, Robert B.
Utah State University

PAN, Ching-Yan
Utah State University

PECCEI, R.
DESY

PENG, Huei
Utah State University

PERKINS, D. H.
University of Oxford

PERRIER, Jacques
UC Santa Cruz

PETERSEN, Alfred
SLAC

PETERSON, Carl
Univ. of Lund

PHILLIPS, G. C.
T. W. Booner Nuclear Lab.

PI, So-Young
Boston University

PIANO-MORTARI, Giovanni
University "La Sapienza" ROMA

PIETRZYH, Z. Adam
University of Washignton

PIILONEN, Leo
Los Alamos National Laboratory

PLANTS, Donald
University of Pittsburg

POLLAK, Greg
Lawrence Livermore Nat'l. Lab.

PONDROM, Lee
University of Wisconsin

PRICE, Lawrence
Argonne National Laboratory

PROTOPOPESCU, Serban D.
Brookhaven National Laboratory

PUGH, Robert
University of Toronto

PULONEN, Leo
Los Alamos National Laboratory

PUROHIT, Milind V.
Fermilab

QUIGG, Chris
Fermilab

RABY, Stuart
Los Alamos National Laboratory

RADFORD, Stanley F.
Wayne State University

RALSTON, John P.
University of Kansas

RAM, B.
New Mexico State Univ.

RATRA, Bharat
SLAC

RAY, Ron
UC Irvine

READ, Alex
University of Colorado

REARDON, Paul J.
Brookhaven National Laboratory

REEDER, Don D.
University of Wisconsin

REES, C. David
University of Washignton

REID, J. H.
Oklahoma State University

REITER, Albrecht
SLAC

ROBINETT, Richard
Univ. of Massachusetts-Amherst

ROE, Natalie
SLAC

ROHLF, J.
Harvard University

ROLNICK, William B.
Wayne State University

ROSEN, S. Peter
Los Alamos National Laboratory

ROSEN, Jerome
Northwestern University

ROSNER, Jonathan L.
Enrico Fermi Insitutute

ROSZKOWSKI, Leszek
University of Washington

RUCHTI, Randal
University of Notre Dame

SACHS, Robert G.
University of Chicago

SAMUEL, Mark
Oklahoma State University

SAVAGE, Maureen
San Francisco State University

SCHAAD, Theo P.
Harvard University

SCHARFSTEIN, H.
Trumbull, CT

SCHELLMAN, Heidi
Enrico Fermi Institute

SCHINDLER, Rafe H.
Caltech

SCHMIDKE, Bill
Lawrence Berkeley Laboratory

SCHMUESER, Peter
DESY

SCHWITKIS, Kent
UC Santa Barbara

SCKEGELSKY, Valery
Fermilab

SCOTT, W.
CERN

SEGRE, Gino
University of Pennsylvania

SEIDEL, Sally
University of Michigan

SENNHAUSER, Urs
Swiss Inst. for Nuclear Research

SHAEVITZ, Michael H.
Columbia University

SHAPERO, DON
National Academy of Sciences

SHEPKO, Michael J.
Texas A & M University

SILL, Alan
The American University

SIMIC, Peter
Rockefeller University

SIMPSON, Ken
San Francisco State University

SINERVO, Pekka K.
SLAC

SIVOS, Dennis
Argonne National Laboratory

SLAUGHTER, Milton D.
Los Alamos National Laboratory

SLONCZEWSKI, John C.
IBM Watson Research Center

SNYDER, Arthur E.
Indiana University

SOLEMAN, Sandra
San Francisco State University

SONI, A.
UCLA

SOPER, Davison E.
University of Oregon

STABLER, Robert C.
Princeton, N. J.

STALLARD, Bryan
Litton Applied Technology

STANISLAUS, Shirvel
University of British Columbia

STEINER, Al
San Francisco State Univ.

STERN, Allen
University of Oregon

STORK, Donald H.
Univ. of Calif. at Los Angeles

STROM, David
University of Wisconsin

SUGANO, Katsu
Argonne National Laboratory

SUSSKIND, Leonard
Stanford University

TANGHERLINI, Frank R.
College of the Holy Cross

TANNENBAUM, M.J.
Brookhaven Naitonal Laboratory

TATA, Xerxes
University of Oregon

TAYLOR, Frank E.
MIT (at Fermilab)

TENG, Lee C.
Fermilab

THALER, Jon J.
University of Illinois

TRUEMAN, T. L.
Brookhaven National Laboratory

TRYON, Edward P.
Hunter College of CUNY

TURALA, Michal
University of Santa Cruz

UY, Zenaida E. S.
Millersville University

VELLA, Eric
University of Washington

VENUTI, John
University of Houston

VISSER, Matt
Univ. of Southern California

WAGONER, David E.
Fermilab

WAI, Chung Fai
SUNY at Buffalo

WALKER, W. D.
Duke University

WALLENMEYER, William
Department of Energy

WANG, P.
University of Oregon

WATTS, T. L.
Rutgers University

WEINERG, Erick
Columbia University

WEINBERG, Steven
Univ. of Texas at Austin

WEINSTEIN, John
University of Toronto

WEISBERGER, William I.
SUNY at Stony Brook

WENZEL, William A.
Lawrence Berkeley Laboratory

WHITMORE, Jim
Penn State University

WIJEWARDHANA, L. C. R.
Yale University

WILCOX, Walter
TRIUMF

WILKERSON, John F.
Los Alamos National Lab.

WILLEY, Raymond
University of Pittsburg

WILSON, Robert
SLAC

WINSTON, Roland
University of Chicago

WISE, Mike
Caltech

WISNIEWSKI, William
University of Illinois

WITHERELL, Michael
UC Santa Barbara

WOODY, Craig L.
Brookhaven National Laboratory

WU, San Lan
University of Wisconsin

YAMIN, Peter
Department of Energy

YANG, Ming-Jen
University of Chicago

YIN, Pong-Chen
Utah State University

YOSHIDA, Takuo
Kyoto University

ZACHOS, Cosmas
Argonne National Laboratory

ZAPALAC, Geordie
University of Chicago

ZELLER, M. E.
Yale University

ZHU, Renyuan
Caltech

ZI, Wang
Utah State University

ZIEMINSKI, Andrzej
Indiana University

ZIEMINSKI, Daria
Indiana University

ZMUIDZINAS, Jonas S.
Caltech

AUTHOR INDEX

Alonso, J. R.	1001	Keung, W. Y.	207
Anderson, D. F.	904	Kogut, J. B.	584
Baranko, G. J.	258	Lang, K.	370
Barger, V.	95	Lee T. D.	135
Barnett, R. M.	353	Lee-Franzini, J.	1009
Berkelman, K.	150	Leith, D.	863
Bosman, M.	232	Leung, C. N.	319
Bowick, M. J.	805	Makdisi, Y. I.	992
Bowles, T. J.	418	Margolis, B.	648
Braaten, E.	707	McLerran, L.	689
Caldwell, D. O.	542	Merritt, F. S.	335
Candelas, P.	737	Moromisato, J. H.	327
Carlson, P.	680	Mueller, A. H.	634
Carr, J.	512	Murtagh, M. J.	378
Chao, A. W.	170	Nappi, C. R.	843
Cheng, H. Y.	240	Ovrut, B. A.	760
Christ, N. H.	593	Paige, F. E.	566
Cline, D. B.	976	Peccei, R. D.	1
Collins, J. C.	505	Perkins, D. H.	113
Coward, D. H.	219	Piano-Mortari, G.	615
Cox, B.	887	Price, L. E.	389
Davis, R.	602	Protopopescu, S.	671
Ecklund, S.	958	Raby, S.	813
Feldman, G.	411	Reeder, D. D.	715
Frampton, P. H.	797	Rohlf, J.	66
Franzini, P.	310	Ruchti, R. C.	932
Franzini, P. J.	1009	Schmuser, P.	968
Freund, P. G. O.	727	Scott, W.	520
Gaillard, J. M.	25, 912	Segre, G. G.	835
Gaisser, T. K.	643	Shaevitz, M. H.	426
Gan, K. K.	248	Soni, A.	287
Gollin, G. D.	299	Taylor, F. E.	343
Gross, D. J.	49	Turala, M.	922
Gunion, J. F.	445	Visser, M.	829
Halzen, F.	529	Weinberg, S.	186, 850
Harvey, J. A.	753	Wilson, R.	362
Holmes, S. D.	949	Winston, R.	215
Hubbard, J. R.	941	Wise, M. B.	279
Hwa, R. C.	699	Wisniewski, W. J.	657
Isenberg, J.	787	Witherell, M. S.	437
Jackiw, R.	772	Wu, S. L.	471
Jawahery, A. H.	267	Zachos, C.	779
Karsch, F.	575	Zeller, M. E.	495
Kayser, B.	397	Zhu, R. Y.	552